NanoScience and Technology

For further volumes:
http://www.springer.com/series/3705

The series NanoScience and Technology is focused on the fascinating nano-world, mesoscopic physics, analysis with atomic resolution, nano and quantum-effect devices, nanomechanics and atomic-scale processes. All the basic aspects and technology-oriented developments in this emerging discipline are covered by comprehensive and timely books. The series constitutes a survey of the relevant special topics, which are presented by leading experts in the field. These books will appeal to researchers, engineers, and advanced students.

Vladimir G. Dubrovskii

Nucleation Theory and Growth of Nanostructures

Springer

Vladimir G. Dubrovskii
Saint Petersburg Academic University
Saint Petersburg
Russia

ISSN 1434-4904 ISSN 2197-7127 (electronic)
ISBN 978-3-642-39659-5 ISBN 978-3-642-39660-1 (eBook)
DOI 10.1007/978-3-642-39660-1
Springer Heidelberg New York Dordrecht London

Library of Congress Control Number: 2013945850

Printed on acid-free paper

Springer is part of Springer Science+Business Media (www.springer.com)

Dedicated to the memory of my father,
Professor German V. Dubrovskii

Preface

Nucleation theory provides a powerful tool for growth modeling of a variety of objects: from liquid droplets to thin solid films and biological structures. Theoretical approaches based on nucleation theory apply whenever the particles, or "nuclei," of a new phase emerge spontaneously in a metastable surrounding of an old phase. Typical examples include water condensation from supersaturated vapors and growth of vapor-deposited surface islands. In most cases, nuclei of interest consist of at least several tens of atoms and are terminated by distinct boundaries with a metastable phase. Such systems can be well described within the frame of the classical approach involving macroscopic approximations for the formation energy and a kinetic equation for time-dependent distribution of nuclei over sizes, coupled with the material balance. With the known thermodynamics of a particular system and at the given growth conditions, classical nucleation theory allows for a detailed modeling in terms of size distribution. Perhaps the most important result of this approach in the Zeldovich nucleation rate, showing how many stable nuclei would emerge per unit time in a given system at the known supersaturation.

Nucleation theory has been proved very practical, capable of answering (qualitatively as well as quantitatively) important technological questions such as: what happens to the size distribution if the growth temperature is raised or the influx is terminated? Consequently, many excellent reviews and books have been published devoted to nucleation theory and applications. So, the question arises: why this new book on nucleation theory might be useful? Of course, it is the readers who will decide. The main intention in this work, which makes it rather different from others, was to link the two subjects that I have been studying for many years: analytical research in general nucleation theory and growth modeling of semiconductor nanostructures obtained by modern epitaxy techniques.

The general topic was my "first love" in the 1980s, when I was a student at the Department of Statistical Physics of St. Petersburg State University under the supervision of Professors F. M. Kuni and A. P. Grinin. Later on, I started my own research in the growth kinetics of thin films, largely inspired by the excellent works of S. A. Kukushkin and A. V. Osipov, who applied nucleation theory for surface islands. The subject of semiconductor nanostructures attracted my attention much later, at the beginning of the 2000s, after I entered Zh. I. Alferov Laboratory of Physics of Semiconductor Heterostructures at Ioffe Institute and then have been working at St. Petersburg Academic University. It turned out that nucleation theory in open systems helps to substantially extend our knowledge about the growth behavior and properties of sophisticated semiconductor nanostructures such as quantum dots and nanowires. In this research, I was supported by my Ioffe teachers and friends, R. A. Suris, V. M. Ustinov, and G. E. Cirlin. Many studies have been performed together with my former pupil N. V. Sibirev. I would like to express my sincere thanks to all these people for their help in different aspects of my work. Over time, my own group has grown into a theoretical laboratory dealing mainly with nucleation theory and its applications in physics of nanostructures. With other groups in our city also continuing their research, St. Petersburg can be truly called "the nucleation city," a nice place to work in this field.

However, our research would not be possible without external collaboration both in nanostructure growth theory and modeling experimental data from other groups. Here, I would like to thank my main collaborators, namely, F. Glas and J. C. Harmand from LPN CNRS (France) for a long and fruitful joint work on semiconductor nanowires, and C. J. Chang-Hasnain from UC Berkeley (USA) for collaboration in the field of III–V nanowires and nanoneedles. Other colleagues include P. Pareige (University of Rouen, France) and B. Grandidier (IEMN Lille, France) who invited me to work on Si and Ge nanowires, E. Gil (University of Clermont-Ferrand, France) on hydride vapor phase epitaxy of nanostructures, H. Riechert (PDI Berlin, Germany) on self-induced GaN nanowires, D. Bimberg (TU Berlin, Germany) on InAs quantum dots, D. Zeze (Durham University, UK) on applications, and X. Ren (BUPT, China) on III–V nanowires. Especially, it was Frank Glas who not only made an outstanding contribution to the entire field of nanostructure modeling (which I use extensively throughout the book) but has also stimulated my own research by fruitful discussions over many years. Of course, I was also influenced by other theorists with whom I have not had a chance of working together, particularly J. Tersoff, D. Kashchiev and J. Johansson. I am grateful to many colleagues and friends for inviting me for short stays with their groups, providing me with new data and a little more "free time" to work.

The content of the book, defined by the general subject and the topics of my own research over the last 20 years, is the following. Chapter 1 presents a systematic introduction to classical nucleation theory. It is written in such a way that the readers who are looking for only a brief preview of the subject can skip after this chapter. Nevertheless, the introduction contains original material: theoretical considerations of the droplet stability in unusual configurations and some exact solutions to the discrete rate equations. The main goal of the chapter is to formulate and justify the macroscopic approach, the basic kinetic equations, the stationary state, and the Zeldovich nucleation rate. Chapter 2 is the most mathematized part of the book. Here, I consider in detail different stages of the entire nucleation–condensation process (usually in open systems), separated from each other due to a timescale hierarchy. A special emphasis is given to the double-exponential distribution shape at the nucleation stage. After that, the regular growth stage, the Ostwald ripening and the direct coalescence stages are studied, based on the continuum theory. At the end of the chapter, I present some exactly solvable cases of discrete rate equations for irreversible growth, with interesting mathematics and applications. In Chap. 2, the main ideas of this work are formulated in the general case, showing how the morphology of emerging nanoparticles can be tuned by technologically controlled conditions such as temperature and flux.

Chapter 3 is devoted to self-induced nanoislands of three types: the Stranski–Krastanow semiconductor quantum dots, III–V nanoneedles on lattice mismatched substrates and metal islands on insulators. Here, the exposition becomes friendlier for non-theorists. I start the analysis with a general description of the Stranski–Krastanow growth and consider the driving force: elastic relaxation on free side facets. General theory is then applied for growth modeling of different nanostructures and formulating the kinetically controlled knobs to tune the resulting morphology. Theoretical predictions are compared with many experimental data (which is also the case in the foregoing chapters), making the whole composition of easier access for a more general audience.

Almost half of the book is devoted to modeling of semiconductor nanowires. In Chap. 4, the "vapor–liquid–solid" growth method of nanowire synthesis is considered catalyzed by metal particles. The chapter starts with a physical introduction explaining the essentials of this method. After that, I present the advanced growth modeling of "vapor–liquid–solid" nanowires. This includes the size-dependent limitations on the growth rate, the mononuclear growth, the diffusion-induced contributions, and the two-dimensional nucleation at the liquid–solid interface or at the triple phase line. In many cases, general theory can be reduced to simple models for complex growth phenomena such as nonlinear growth effects and self-consistency between the material transport and the nucleation-mediated nanowire elongation. I consider in detail the length-time, length-radius and other important theoretical dependences in comparison with relevant experimental data.

Chapter 5 concerns more specific aspects of nanowire modeling: elastic relaxation and plastic deformation in nanowires on lattice mismatched substrates, nanowire shapes, doping, self-catalyzed growth, self-induced GaN nanowires on silicon substrates and cooperative growth effects. Here, the conventional nucleation theory is considerably modified in many respects. In particular, one of the most interesting features of nucleation in confined volumes of catalyst droplets is a self-regulatory oscillatory behavior of supersaturation. This feature originates from a special type of timescale hierarchy where nucleation and growth becomes much faster than the refill stage, and has an interesting impact on nucleation statistics and the morphology of nanowires. In Chap. 6, I consider polytypism of III–V nanowires, a surprising effect which appears to be closely related to peculiarities of nucleation during the vapor–liquid–solid growth. Crystal structures of III–V materials are carefully described, along with relevant data on the bulk energy differences and surface energies of relevant nanowire sidewalls. This part goes far beyond the nucleation topic, but is absolutely necessary for modeling. After that, I describe theoretical approaches that allow us to understand and control crystal structures in different nanowires. At the end, while considering the Ga-catalyzed formation of GaAs nanowires and their zincblende structure, I return to the very beginning of the book and further develop a model of the wetting growth mode of nanowire growth.

Each chapter begins with a brief introduction describing its subject and structure. I use the numbering of mathematical expressions and figures of the (x,y) type, where x is the chapter number and y is the number of this formula within the chapter, for example, (4.35) is formula No. 35 of Chap. 4. I have tried to make the book self-consistent so that the reader could find all the necessary material in the text, which should be convenient for students. However, some parts are explained only briefly, with the reference list containing almost 600 citations. Not many acronyms are used in the text (so that nanowire remains nanowire but not NW) due to my personal dislike of acronyms.

I sincerely hope that this work and my modest contribution to growth modeling of nanostructures will attract more researches, especially young scientists, to this rapidly growing field in the future.

Finally, I wish to thank Springer Edition and in particular Executive Editor in Physics, Claus Ascheron, for inviting me to write this book.

Saint Petersburg Vladimir G. Dubrovskii

Contents

Chapter 1
Fundamentals of Nucleation Theory

Theoretical analysis of nucleation rates and distributions of nuclei over sizes is of great importance in connection with clustering and condensation in vapors, crystallization of liquid alloys, phase separation in solid solutions, kinetics of colloidal and biological systems and many other growth-related phenomena such as thin film condensation, epitaxy of semiconductor quantum dots and freestanding nanowires. This chapter gives a systematic introduction to the classical nucleation theory treating the kinetics of phase transition as a spontaneous birth and follow-up growth of embryos in a metastable environment. Below in this book, we will consider both three-dimensional (3D) and two-dimensional (2D) systems or even their combinations such as 3D surface islands growing from a 2D environment. To account for different system configurations, we formulate all necessary expressions in arbitrary dimension $d = 2$ or 3.

The core assumptions of classical approach are: (i) low enough temperature below T_c and (ii) low enough supersaturation of a metastable phase. The entire system can then be subdivided into an ensemble of dense nuclei and a dilute, almost perfect vapor. The nuclei of interest contain many monomers and are separated from vapor by distinct interfacial boundaries that can be described by macroscopic surface energies. However, the applicability of classical nucleation theory is not always guaranteed, and non-classical scenarios are often observed. To give a feeling of how different these scenarios might be, we start with a detailed analysis of phase transition thermodynamics, presented in Sect. 1.1 in the frame of the lattice gas model. The model is relevant for 3D vapor–liquid phase transitions and is especially well justified for island nucleation on solid surfaces. The equation of state is obtained directly from the statistical sum within the frame of mean field approximation and used for the description of phase equilibrium. Different forms of the equation of state such as chemical potential—density, pressure-volume or magnetization—external field isotherms are discussed. Although not within the scope of our study, we mention the critical phenomena and critical scaling which the mean field theory fails to describe correctly. Section 1.2 presents a classification of the phase transition pathways and relevant theoretical methods of their description, either in terms of size and number of nuclei

V. G. Dubrovskii, *Nucleation Theory and Growth of Nanostructures*, NanoScience and Technology, DOI: 10.1007/978-3-642-39660-1_1,

or by means of field theories for a fluctuational order parameter. We also show how the central notions of classical nucleation theory such as supersaturation are derived from the mean field equation of state when the temperature is well below the critical value.

In Sect. 1.3, we consider the Laplacian pressure and the corresponding Gibbs–Thomson modification of chemical potential in the nanoparticles, caused by the curvature of their surface. It is this effect which is responsible for the barrier character of nucleation leading to irreversible first order phase transition. On the other hand, the Gibbs–Thomson effect has a tremendous importance in the "vapor–liquid–solid" growth of nanowires, one of the major applications considered later. Section 1.4 might look trivial at first glance, as it studies the well-known Young's condition for the contact angle of a droplet resting on a solid surface. We show, however, that the usual mechanical balance of the surfaces forces acting upon the triple phase line is not sufficient for stable equilibrium. A remarkable example is given by a spherical droplet seated on top of a cylindrical nanowire. When the droplet slides down along the nanowire sidewalls, the Young's angle yields the balance of forces but corresponds to the maximum rather than the minimum surface energy!

The most important thermodynamic characteristic of any condensing system, the macroscopic formation energy of a nucleus containing i monomers at a given supersaturation, is considered in Sect. 1.5. We discuss different cases such as droplets and crystal nuclei in gas-vapor environments, 2D and 3D surface islands growing on similar and dissimilar substrates, as well as heterogeneous nucleation on condensation seeds. The nucleation barrier and the number of monomers in the critical nucleus are then obtained in arbitrary space dimension. We only briefly analyze the known renormalizations of quasi-equilibrium distribution since, in open systems, the normalization can be obtained in a self-consistent manner. Section 1.6 contains an overview of useful results regarding the adsorption isotherms and thin films on solid surfaces. In particular, we discuss the elementary processes such as adsorption, desorption and surface diffusion, as well as thin film growth either by step flow or islanding. Section 1.7 is very important, as it presents systematization of the known growth scenarios for super-critical nuclei in different systems: surface layers, droplets and islands in different metastable environments. We show that, in most cases, the growth rate is a power law function of the number of monomers with the corresponding index ranging from zero to one. This feature is of fundamental importance for constructing the analytical approximations for the nucleation probabilities and the size spectra.

Finite difference rate equations of nucleation theory in the monomolecular reaction schemes are given in Sect. 1.8. From the very beginning, we consider the open systems where the monomers are delivered by a material influx from a surrounding and monomer sinks are also possible. In the open systems studied below in this book, supersaturation changes in time due to the monomer consumption by the growing nuclei and the material influx, the so-called nucleation under dynamical conditions. This is the major difference of our approach (which, for example, is essential in studying the epitaxy of nanostructures) from those considering the nucleation which originates from a fixed supersaturation established at zero moment of time. We also discuss the principle of detailed balance and the curvature effect on the growth rates. Section 1.9 presents the unique cases of i-linear rate constants where the discrete set of rate equations can be integrated exactly. It is shown that the exact solution is given by the time-invariant Polya size distribution, reduced to the Poisson and geometrical distributions in the particular cases. Although irrelevant for the general analysis of nucleation, the obtained solutions are very important for checking the validity of different approximations and for the foregoing analysis of fluctuation-induced broadening of size spectra. In Sect. 1.10, we derive the continuum approximation reducing the infinite set of discrete rate equations to the Fokker-Plank type kinetic equation in partial derivatives for the continuum distribution.

Section 1.11 presents well-known results describing the stationary state in the near-critical region of sizes. It is shown that, at high enough nucleation barrier, the entire size axis can be divided into the subcritical region where the nuclei are at the quasi-equilibrium state, the near-critical region in the stationary state with a size-independent nucleus flux and the super-critical region where the nuclei can grow irreversibly. We take special care to describe in detail the major assumptions and the small parameters justifying the famous Zeldovich expression for the nucleation rate. Although it might seem a little annoying, the nucleation rate and the corresponding stationary distribution are obtained twice based on the discrete and continuum theories to show how and why their results match. We also present the explicit representations for the nucleation rates of 3D and 2D islands in different cases, and discuss the origins of differences in the expressions used previously. It is shown that the stationary distribution reaches the size-independent drift form at the boundary of the near-critical and super-critical regions, if expressed in terms of a certain invariant size. This important feature provides the boundary condition for studying the follow-up growth stages. Section 1.12 starts with a discussion of the mononuclear growth regime in nano-volumes, where the Zeldovich formula for the nucleation rate alone provides sufficient basis for modeling. We then give a preview of consecutive nucleation-condensation stages that are separated in time due to the timescale hierarchy. In brief, the stationary state in the near-critical region of sizes is established almost instantaneously and is then adjusted to the time dependent supersaturation. The nucleation stage under the material influx is short-scale, followed by a much longer stage of regular growth. After that, the nuclei can either coalesce directly or be affected by a special type of indirect interaction leading to the Ostwald ripening, depending on the character of material influx and the island growth law. We discuss the evolution of the mean size of vapor-deposited surface islands at the regular growth stage under the assumption of a narrow size distribution. Finally, we analyze the large time asymptotes of the mean size and supersaturation at different growth indices.

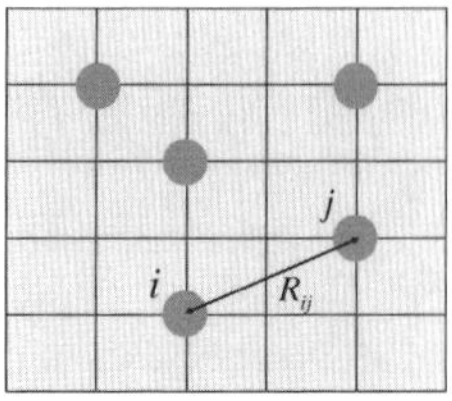

Fig. 1.1 2D lattice gas on a square lattice with the average occupation of $5/36 = 0.14$

1.1 Thermodynamics of Phase Transition

Nucleation theory treats the kinetics of first order phase transitions. The hallmark of such a transition is a discontinuous change of density, the condensation of supersaturated vapor into the liquid droplets in 3D space being a typical example. Aggregation of metastable 2D system of adsorbed atoms (adatoms) into the surface islands is the corresponding analogue in 2D space. Thermodynamics of phase transition provides important information about the equilibrium chemical potential, pressure, and densities of the two phases at a given temperature. Thermodynamic properties of any system undergoing a transition are entirely determined by the equilibrium equation of state. The latter gives a relationship between any three independent thermodynamic parameters, for example: $f(P, V, T) = 0$ or $f(\mu, n, T) = 0$, where P is the pressure, V is the volume, T is the temperature, n is the atomic density (or the concentration), and μ is the chemical potential. A particular form of the equation of state should be obtained from statistical integral with a given Hamiltonian.

One of the most known Hamiltonians which is widely used for the description of different phase transitions in 3D or 2D space is the lattice gas Hamiltonian [1, 2]. In this model, the atoms are placed in the discrete sites $i = 1, 2 \ldots N_0$ of 3D or 2D lattice (for example, cubic or rectangular lattice, as shown in Fig. 1.1). The sites are characterized by the occupation numbers $\alpha_i = 0$ or 1 depending on whether the site is empty or occupied. Repulsive interactions are taken into account automatically, since the sites cannot accommodate more than one atom. Attractive interatomic potential $U_{ij} = U(R_{ij}) < 0$ depends on the distance between sites i and j. In the grand canonical ensemble with the fixed chemical potential μ, total number of sites N_0 and temperature T, the configuration part of the statistical sum is given by

$$\Xi(\mu, N_0, T) = \sum_{\alpha_i=0,1} \exp\left(\frac{\mu}{k_B T}\sum_{i=1}^{N_0}\alpha_i - \frac{1}{k_B T}\sum_{(i,j)} U_{ij}\alpha_i\alpha_j\right), \tag{1.1}$$

where k_B is the Boltzmann constant. The summation in the non-diagonal interaction term includes all non-identical atom pairs (i, j). The statistical averaging accounts for all possible states of N_0 sites with different occupation numbers. The average occupation of sites $<\alpha>$ is readily obtained from the statistical sum by

$$\theta \equiv <\alpha> = \frac{N}{N_0} = \frac{k_B T}{N_0}\left(\frac{\partial \ln \Xi}{\partial \mu}\right)_T. \tag{1.2}$$

Here, N denotes the total number of atoms or molecules in the system. While the occupation number takes only one of the two discrete values, the average occupation can take any value between zero and one. Obviously, the atomic density is related to θ as $n = n_0\theta$, where $n_0 = 1/\Omega$ corresponds to the dense lattice packing and Ω is the elementary volume. We will therefore call θ the normalized density, or just density for brevity.

The average volume per one atom in the system with occupation θ equals $1/(n_0\theta)$. Integrating the known thermodynamic relationship $d\mu = dP/n_0\theta$ at $T = const$ along the isotherm, we arrive at

$$P/n_0 = \theta\mu(\theta) - \int_0^\theta d\theta' \mu(\theta'). \tag{1.3}$$

In integration, we use the known asymptotic behaviors of any perfect system without interactions: $\theta\mu(\theta) \to 0$ and $P(\theta) \to 0$ at $\theta \to 0$. This relationship allows one to obtain the equation of state in the form of $P(V, T)$ dependence at the known $\mu(\theta, T)$.

The mean field approximation [1–3] for calculation of the statistical sum given by (1.1) assumes the effective absence, in the temperature range of interest, of local density fluctuations. Applicability of this assumption will be discussed later. With neglect of fluctuations, the local potential field acting upon the atom in a given site i can be approximated by the effective mean field with the average occupation θ:

$$\sum_{(i,j)} U_{ij}\alpha_i\alpha_j \cong \theta\alpha_i \sum_j U_{ij} = -U_0\theta\alpha_i. \tag{1.4}$$

Here, $U_0 = -\sum_j U_{ij} = -\sum_j U_{0j} > 0$ is the effective energy of attractive interactions, obtained upon the summation of pair potentials over all the interatomic distances. Of course, the potential $U(R_{ij})$ should be sufficiently short-range to enable such a summation. As in the ideal case with no attractive interactions, the insertion of (1.4) into (1.1) leads to the diagonalization of the statistical sum:

$$\Xi = \left[1 + \exp\left(\frac{\mu}{k_B T} + \varphi\theta\right)\right]^{N_0}, \tag{1.5}$$

in which $\varphi = V_0/(k_B T)$ is the dimensionless interaction constant. Using (1.2) and (1.5), we obtain the mean field equation of state that can be put in two equivalent forms

$$\frac{\mu}{k_B T} = -\varphi\theta + \ln\left(\frac{\theta}{1-\theta}\right); \tag{1.6}$$

$$\frac{\mu}{k_B T_c} \equiv \overline{\mu} = -4\theta + \frac{T}{T_c}\ln\left(\frac{\theta}{1-\theta}\right). \tag{1.7}$$

Here, by definition, $T_c = (\varphi T)/4 = V_0/(4k_B)$. The second terms in the right hand sides describe the perfect lattice gas with repulsion, while the first terms arise due to the attractive interactions. In dilute systems at $\theta \to 0$, the chemical potential tends to $\mu = k_B T \ln \theta$, which is the known perfect gas limit.

Differentiation of (1.6) with respect to θ at a constant temperature yields: $(\partial \mu/\partial \theta)_T \propto 1 - \varphi\theta + \varphi\theta^2$, showing that the mean field chemical potential monotonously increases when $\varphi < 4\,(T > T_c)$, and reaches its local maximum and minimum at the points

$$\theta_{1s} = \frac{1}{2}\left(1 - \sqrt{1 - \frac{4}{\varphi}}\right); \quad \theta_{2s} = \frac{1}{2}\left(1 + \sqrt{1 - \frac{4}{\varphi}}\right) \tag{1.8}$$

when $\varphi > 4\,(T < T_c)$. The value $\varphi_c = 4$ relates to the inflection of the $\mu(\theta)$ curve at the critical density $\theta_c = 1/2$. Therefore, T_c is the critical temperature. The critical chemical potential is given by: $\mu_c = -V_0/2 = -2k_B T_c$ (yielding $\overline{\mu}_c = -2$) at the inflection point. The two densities defined by (1.8) are the boundaries of the so-called spinodal region. Obviously, they are symmetric relative to the critical occupation of 1/2.

Graphs of chemical potential given by (1.7) at different temperatures are presented in Fig. 1.2. The wavy regions in the $\mu(\theta)$ dependences below T_c (the so-called van der Waals loops) have no physical sense in equilibrium in view of the known thermodynamic inequality $(\partial \mu/\partial \theta)_T \geq 0$. The loops must therefore be substituted by the horizontal lines relating to the discontinuous change from equilibrium vapor to equilibrium liquid at a constant chemical potential [1–3]. The correct choice of the equilibrium chemical potential cannot be made in an absolute fashion by considering the $\mu(\theta)$ isotherms alone, but requires additional considerations. As is known [1], the two phases with densities θ_{1e} and θ_{2e} at a fixed temperature T are in equilibrium with each other provided that $\mu(\theta_{1e}) = \mu(\theta_{2e})$ and $P(\theta_{1e}) = P(\theta_{2e})$. The equality of chemical potentials implies the absence of diffusion fluxes. The equality of

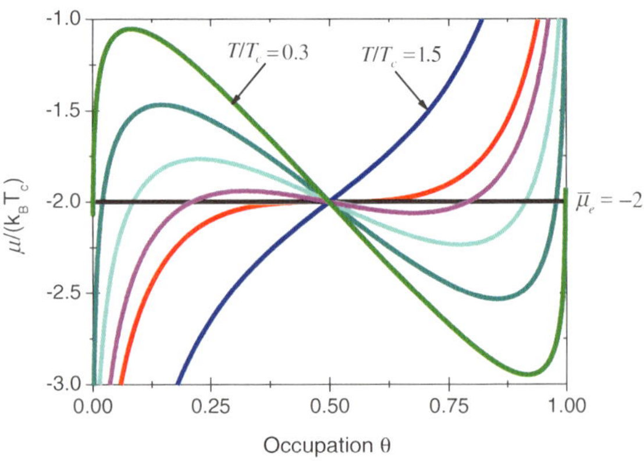

Fig. 1.2 Normalized chemical potential at different T/T_c decreasing from 1.5 to 0.3. The *horizontal line* is the equilibrium chemical potential corresponding to the equality of pressures in both phases

pressures yields the compensation of forces acting between the two phases. These two equations enable the unique determination of the phase equilibrium parameters at a given T. Whatever is the form of $\mu(\theta)$, we can write down (1.3) for θ_{1e} and θ_{2e}:

$$p(\theta_{1e}) = \theta_{1e}\overline{\mu}(\theta_{1e}) - \int\limits_0^{\theta_{1e}} d\theta\mu(\theta); \quad p(\theta_{2e}) = \theta_{2e}\overline{\mu}(\theta_{2e}) - \int\limits_0^{\theta_{2e}} d\theta\overline{\mu}(\theta), \tag{1.9}$$

where $p \equiv P/(n_0 k_B T_c)$ is the normalized pressure. Subtraction of the first equation from the second one, along with the condition $\overline{\mu}(\theta_{1e}) = \overline{\mu}(\theta_{2e}) \equiv \overline{\mu}_e$, readily gives

$$p(\theta_{2e}) - p(\theta_{1e}) = (\theta_{2e} - \theta_{1e})\overline{\mu}_e - \int\limits_{\theta_{1e}}^{\theta_{2e}} d\theta\overline{\mu}(\theta). \tag{1.10}$$

From the equality of pressures, the equilibrium chemical potential in both phases thus amounts to the averaged value of $\overline{\mu}(\theta)$ within the interval between θ_{1e} and θ_{1e}:

$$\overline{\mu}_e = \frac{1}{(\theta_{2e} - \theta_{1e})} \int\limits_{\theta_{1e}}^{\theta_{2e}} d\theta\overline{\mu}(\theta). \tag{1.11}$$

Therefore, the equilibrium chemical potential satisfies the Maxwell's rule [1] stating that the areas of the van der Waals loops above and below μ_e must equal each other.

For the specific form of $\mu(\theta)$ given by (1.7), the Maxwell's rule yields

$$\overline{\mu}_e = \overline{\mu}_c = -2, \tag{1.12}$$

the value corresponding to the horizontal line shown in Fig. 1.2. Using this in (1.7), the equilibrium densities are obtained as

$$\theta_{1e} = 1/2 - \Delta\theta_e; \quad \theta_{2e} = 1/2 + \Delta\theta_e, \tag{1.13}$$

where $\Delta\theta_e$ is the solution to the transcendent equation

$$\frac{1}{\Delta\theta_e} \ln\left(\frac{1/2 + \Delta\theta_e}{1/2 - \Delta\theta_e}\right) = \varphi = \frac{4T_c}{T}. \tag{1.14}$$

Obviously, the left hand side contains only the even powers of $\Delta\theta_e$, so that (1.14) has two anti-symmetric solutions at $T < T_c$. We therefore take the value of $\Delta\theta_e > 0$ in (1.13). The resulting equilibrium densities of the dilute and dense phases are symmetric relative to 1/2. The density discontinuity in the transition, also called the equilibrium order parameter, equals $s_e = 2\Delta\theta_e$. The variable $s = 2\Delta\theta = 2\theta - 1$

ranges between -1 and 1 at $0 \le \theta \le 1$. The s value can thus be associated with the magnetization (or the averaged spin variable) in the Ising model of magnetism, showing the known analogy between a magnetic system and a lattice gas in grand canonical ensemble [1, 4] .

In terms of the spin variable s and the chemical potential counted off its equilibrium value, the above equations of state take the anti-symmetrical form. For example, (1.6) and (1.7) can be re-arranged as

$$h = \frac{\mu - \mu_c}{k_B T} = -\frac{\varphi}{2}s + \ln\left(\frac{1+s}{1-s}\right), \quad \overline{\mu} - \overline{\mu}_c = -2s + \frac{T}{T_c}\ln\left(\frac{1+s}{1-s}\right), \tag{1.15}$$

or in a more conventional inverted form

$$s = \tanh\left(\frac{h}{2} + \frac{T_c}{T}s\right). \tag{1.16}$$

Here, h stands for the appropriately normalized external magnetic field. Graphical representation of phase transition at $h = 0$ is shown in Fig. 1.3, where the functions s and $\tanh[(T_c/T)s]$are plotted against the magnetization s. The curves meet only at the trivial point $s_e = 0$ above T_c, corresponding to a stable disordered state. At $T < T_c$, (1.16) has two stable anti-symmetric solutions $\pm s_e$ describing two anti-symmetric ordered states with non-zero magnetization that rapidly approach ± 1 as the temperature decreases.

Let us now see how the mean field equation of state looks like in terms of the more usual pressure-volume isotherms. Inserting (1.7) into (1.3) and integrating, we arrive at

$$\frac{P}{n_0 k_B T_c} \equiv p = -2\theta^2 - \frac{T}{T_c}\ln(1-\theta). \tag{1.17}$$

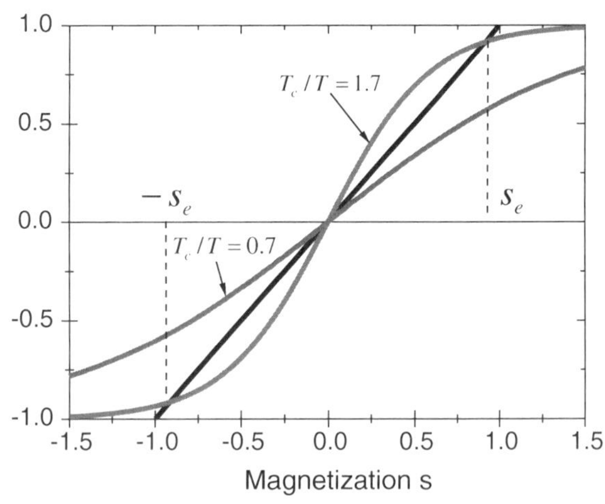

Fig. 1.3 Illustration of magnetic phase transition at zero external field

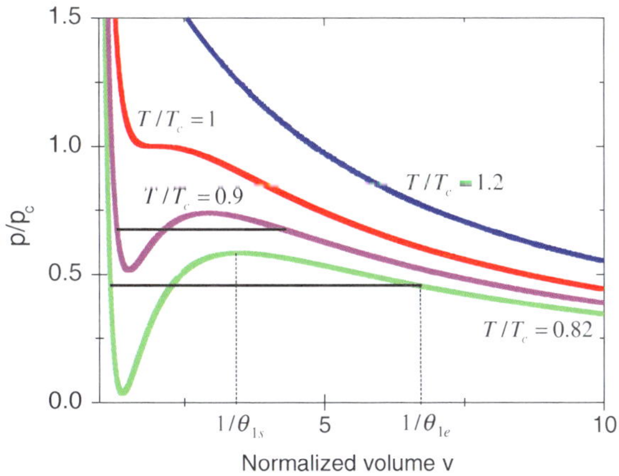

Fig. 1.4 Pressure-volume diagrams at different T/T_c. The *horizontal lines* correspond to the Maxwell's rule. The values of θ_{1e} and θ_{1s} are the equilibrium occupation and the spinodal boundary, respectively

This equation gives correctly the perfect gas limit $P = nk_BT$ at $\theta \to 0$ and yields an infinite pressure at $\theta \to 1$. The critical pressure at $T = T_c$ and $\theta_c = 1/2$ is readily obtained as $p_c = \ln 2 - 1/2$. Introducing the normalized volume $v = V/\Omega = 1/\theta$, (1.17) can be put in the form [3]

$$\frac{p}{p_c} = -\frac{2b_0}{v^2} - \frac{b_0 T}{T_c} \ln\left(1 - \frac{1}{v}\right) \tag{1.18}$$

with $b_0 = 1/(\ln 2 - 1/2) \cong 5.177$. The pressure-volume isotherms obtained from (1.18) at different T/T_c are shown in Fig. 1.4. The curve at the critical temperature has the inflection point at $\theta_c = 1/2$ (corresponding to $v_c = 2$) and $p = p_c$. The spinodal points are obtained by putting the derivative of (1.15) to zero and are the same as given by (1.8). The horizontal lines in Fig. 1.2 at $T < T_c$ correspond to the Maxwell's rule and yield the same equilibrium densities as those given by (1.13) and (1.14). Finally, we underline universality of these equations of state. Indeed, as follows from (1.7) and (1.18), all systems having the same T/T_c would obey exactly identical thermodynamic laws in terms of the appropriately normalized variables.

As mentioned above, neglect of local fluctuations of the order parameter, which is the core assumption in any mean field theory (the van der Waals gas, the Weiss theory of magnetism [1], the Landau theory of phase transitions and its spatially inhomogeneous generalization [4]), is not always valid. Thermodynamic fluctuations are known to increase drastically as temperature approaches T_c, so that the correlation radius tends to infinity at $T \to T_c$. This gives rise to a special type of fluctuational behavior, called the critical behavior, where the fluctuation-induced interactions become more important than those defined by the deterministic interaction potential [4]. As a consequence, thermodynamic properties near T_c only weakly depend on the details of interatomic potential. Rather, they are affected by the type of symmetry in the system, the number of components of the order parameter, the

Table 1.1 Scaling exponents for gas–liquid transitions in 3D space

Physical value	Order parameter, s_e	Compressibility, K_T	Heat capacity, C_P	Pressure, $\Delta P = P_c - P$	Correlation radius, r_c	Correlation function, $G(r)$
Notation of exponent	β	γ	α	δ	ν	η
Definition	$s_e \propto \tau^{\beta}$	$K_T \propto \tau^{-\gamma}$	$C_P \propto \tau^{-\alpha}$	$\Delta P \propto s^{\delta}$	$r_c \propto \lvert\tau\rvert^{-\nu}$	$G \propto r^{-d+2-\eta}$
Experiment	0.34 ± 0.01	1.22 ± 0.02	0.12 ± 0.04	4.6 ± 0.2	0.63 ± 0.01	0.05 ± 0.01
Mean field values	0.5	1	0	3	0.5	0
ε-expansion	0.34	1.244	0.077	4.46	0.626	0.037

cutoff radius of interactions (short-range or long-range forces) and, very importantly, by the space dimension. Such a behavior naturally leads to the universality, where the asymptotic power law dependences of different thermodynamic values at $T \to T_c$ are well described by the critical exponents that are independent of interaction potential.

For a spatially homogeneous system, the four critical exponents are introduced by definitions summarized in Table 1.1. In terms of gas–liquid transition, they reveal the leading power law dependences of the density discontinuity $s_e = 2\Delta\theta_e$, isothermal compressibility $K_T = \theta^{-1}(\partial\theta/\partial P)_T$ heat capacity at a constant pressure $C_P = T(\partial S/\partial T)_P = -T(\partial^2\mu/\partial T^2)_P$ on $\tau = (T_c - T)/T_c$, and pressure-density dependence $P(s_e)$ along the critical isotherm. The other two exponents describe the properties of density–density correlation function [4] and will be discussed in the next paragraph.

Table 1.1 summarizes the known experimental values of the critical exponents for 3D vapor–liquid transitions according to the data of [4]. Let us now consider the critical behavior in the mean field theory. First, the Taylor expansion of (1.16) in s at $T \to T_c$ leads to

$$\overline{\mu}(s) - \overline{\mu}_c \cong -2\tau s + \frac{2s^3}{3}. \tag{1.19}$$

Putting the left hand side to zero at $\tau > 0$, we obtain $s_e = (3\tau)^{1/2}$ at $T < T_c$, yielding $\beta = 1/2$. Above T_c, the only possible solution is $s_e = 0$. Similar expansion of p defined by (1.17) gives

$$p(s) - p_c \cong -\tau s - \frac{\tau s^2}{2} + \frac{s^3}{3}. \tag{1.20}$$

At $\tau = 0$ relating to the critical isotherm, this is reduced to $p - p_c \propto s^3$, corresponding to $\delta = 3$. The leading linear term in the right hand side shows that $K_T \propto \tau^{-1}$, yielding $\gamma = 1$. Further, from (1.19) and (1.20) it follows that

$$\overline{\mu} - \overline{\mu}_c \cong 2(p - p_c) + \tau s^2. \tag{1.21}$$

At $\tau < 0$ where $s_e = 0$, the heat capacity $C_P = 0$. At $\tau > 0$, $s_e^2 = 3\tau$, resulting in $-T_c(\partial^2\overline{\mu}/\partial T^2)_p = 6/T_c$. Returning to the dimensional chemical potential $\mu = k_B T_c \overline{\mu}$, we get the value of $C_P = 6k_B$ below T_c, reflecting a discontinuous change of the heat capacity while passing the critical temperature. Such a discontinuity relates, however, to $\alpha = 0$ by its definition given in Table 1.1. It is noteworthy that the mean field exponents do not depend on the space dimension, in contrast to their experimental values and those obtained by computer simulations.

The mean field exponents summarized in Table 1.1 are obviously far beyond the error bars in experimental data. As expected, neglect of fluctuations is incorrect near the critical point, leading to the wrong values of critical exponents. This discrepancy has led to an impressive development of a principally new approach in studies of critical phenomena based on the scaling hypothesis for the fluctuational field of the order parameter [5, 6]. The turning point in these studies has been the so-called ε-expansion for the critical indices that demonstrates a surprisingly good quantitative correlation with the experiment (see Table 1.1). Summarizing, the mean field theory fails to quantify the near-critical behavior, although predicting correctly all thermodynamic values in the critical point itself [1], as well as giving a reasonable qualitative description of phase separation process in general.

In practice, nucleation theory always deals with systems at low enough temperatures, where the fluctuations are week and the mean field thermodynamics is exactly applicable. Indeed, the very notion of a "nucleus" can only be used if the entire system can be subdivided into dilute and dense phases separated by a distinct boundary, a liquid droplet surrounded by an almost perfect vapor being the typical case. Such a separation requires a large density discontinuity between the equilibrium phases, which occurs only at temperatures well below T_c. However, non-classical kinetic scenarios of phase transition can be observed even at low Tin certain cases. The most common scenarios are discussed in the next section.

1.2 Scenarios of Phase Transition

As is seen from Figs. 1.2 and 1.4, as temperature decreases, the equilibrium densities θ_{1e} and θ_{2e} rapidly approach zero and one, respectively. The boundaries of spinodal region also do so, but their relaxation is slower. This feature is illustrated in Fig. 1.5, where the$\Delta\theta_e$ and the halfwidth of the spinodal region $\Delta\theta_s = \sqrt{1 - 4/\varphi}/2$ are plotted against the interaction constant $\varphi = 4T_c/T$. At low enough temperatures, $T < 0.5T_c$ (corresponding to $\varphi > 8$), the two solutions to (1.14) simplify to the following expressions

$$\theta_{1e} = \exp\left(-\frac{2T_c}{T}\right); \quad \theta_{2e} = 1 - \exp\left(-\frac{2T_c}{T}\right). \tag{1.22}$$

In this low-temperature domain, the occupation of the dilute phase is smaller than 0.02, while the condensed phase is denser than 0.98. The first of (1.22) is the analogue

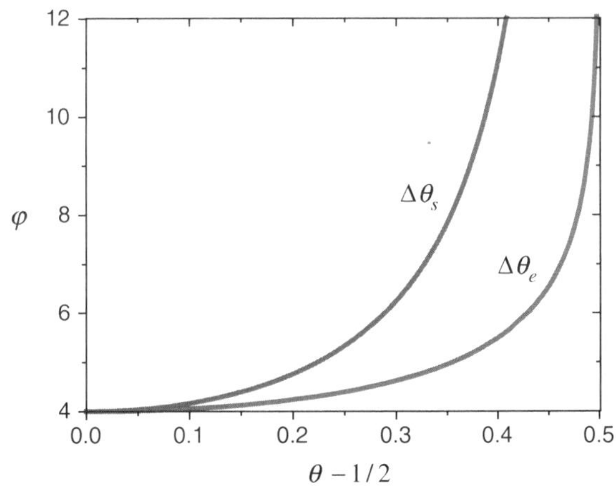

Fig. 1.5 Halfwidth of spinodal region and density discontinuity versus φ

of the Clapeyron-Clausius equation for the equilibrium density in vapor. The quantity $2k_BT_c$ is called the hidden condensation heat.

In thermodynamics, the van der Waals loops on the $\mu(\theta)$ isotherms at $T < T_c$ are substituted by the horizontal lines corresponding to the phase equilibrium. However, the loops are useful in the kinetic theory of phase transitions [3], as demonstrated in Fig. 1.6. Considering a condensation type of transition, the region between θ_{1e} and θ_{1s} relates to the metastable states where the vapor chemical potential is larger than the equilibrium. The driving force for an isothermal phase transition from a metastable state to a liquid (which is at the equilibrium chemical potential) is determined by

$$\Delta\mu = \mu(\theta) - \mu_e. \tag{1.23}$$

In a dilute metastable system, the driving force writes down as $\Delta\mu = k_BT\ln(\zeta + 1)$, where

$$\zeta = n_1/n_{1e} - 1 \tag{1.24}$$

is the supersaturation, n_1 is the concentration in a metastable phase and $n_{1e} = \theta_{1e}/\Omega$ is its equilibrium concentration. The maximum possible supersaturation at a given temperature is determined by $\zeta_{\max} = \theta_{1s}/\theta_{1e} - 1$. The spinodal region between θ_{1s} and θ_{2s} corresponds to unstable states with a negative derivative of chemical potential with respect to density.

Let us know consider possible scenarios of phase transition, schematized in Fig. 1.6. In order to better illuminate our qualitative analysis, we adopt the following model for the free energy (measured in the units of k_BT) of forming a 3D nucleus consisting of i monomers (atoms, molecules, adatoms etc.): $F(i) = ai^{2/3} - \ln(\zeta + 1)i$. The first term stands for the surface energy of 2D interphase boundary, whose area scales as $i^{2/3}$. It is always positive because the formation of nucleus surface in energetically unfavorable. The second term gives the change in the volume energy of

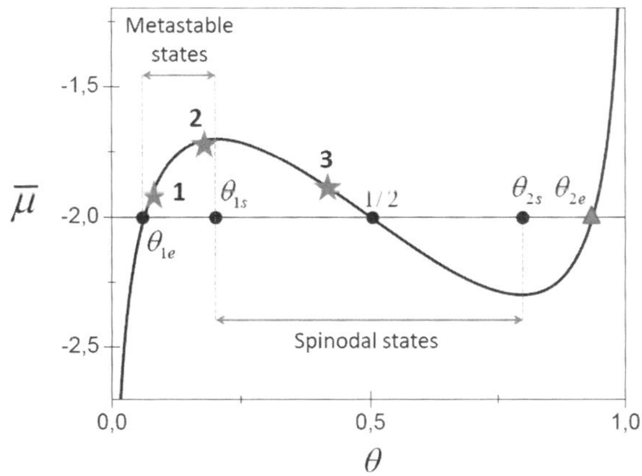

Fig. 1.6 *1* low supersaturation, classical nucleation theory; *2* high supersaturation, field theory of scalar order parameter; *3* spinodal decomposition via increasing density oscillations

i monomers transferred from a metastable vapor into a liquid. This term is negative at $\zeta > 0$. Therefore, condensation process is kinetically suppressed due to an energetically unfavorable process of building the nucleus surface. A more detailed analysis of different types of the formation energies will be given later on. Since the two terms scale differently with i, the formation energy reaches its maximum $F \equiv F(i_c) = (4a^3)/[27 \ln^2(\zeta + 1)]$ at the critical number of monomers $i_c = (8a^3)/[27 \ln^3(\zeta + 1)]$. The maximum value of the formation energy F is called the activation energy for nucleation, or the nucleation barrier. This barrier should be surpassed to form an irreversibly growing nucleus. The particle consisting of i_c monomers is called the critical nucleus, which is in unstable equilibrium with a supersaturated environment. The nucleation barrier and the critical number of monomers tend to infinity at $\zeta \to 0$ (no transition) and to zero at $\zeta \to \infty$ (neither barrier to surpass nor unstable critical nucleus). As shown in Fig. 1.6, a metastable vapor can in principle be "prepared" in any state between θ_{1e} and θ_{2s}. When the initial supersaturation is low enough to ensure that the critical size i_c and the nucleation barrier F are much larger than one, and the density discontinuity $2\Delta\theta$ is close to one, the classical nucleation theory [3, 7–30] is directly applicable. Its major assumptions, such as macroscopic nuclei, the distinct boundaries that can be characterized by their surface energies, and the smallness of thermodynamic fluctuations in the subcritical region below i_c, can be well justified. This case, corresponding to point 1 in Fig. 1.6, is studied throughout this book.

Increasing the initial concentration of vapor toward θ_{1s} (point 2 in Fig. 1.6) and ultimately entering the spinodal region (point 3 in Fig. 1.6) leads to a rather different behavior [3]. First, the correlation radius can get larger than the characteristic size of nuclei. Second, thermodynamic fluctuations perturb the subcritical region so that the size distribution there cannot remain quasi-equilibrium. Third, internal structure of nuclei might be so sophisticated that the usual notions of the interfacial boundary and surface energy would no longer apply. Whenever the initial spatially homogeneous state falls within the spinodal region (for example, at $\theta_0 = 1/2$), the system is thermodynamically unstable from the beginning. In view of $d\mu/d\theta < 0$, the diffusion flux in the spinodal region is directed against rather than along the density gradient, leading to the so-called "uphill" diffusion [31]. When the diffusion goes uphill, the amplitude of fluctuation-induced density oscillations will increase with time. Such system cannot be described in terms of size and number of nuclei, at least at the beginning of phase separation. The phase transition scenario via the increasing density oscillations is usually called the spinodal decomposition [32].

Theoretical analysis of highly metastable systems requires quite different approaches from those used in the case of low supersaturations. One of the non-classical approaches treats the phase transition as the relaxation of spatially inhomogeneous scalar field of the order parameter $s(\vec{r}, t) = 2\Delta\theta(\vec{r}, t)$, where $\vec{r}$ is the spatial coordinate and t is time [3, 33–35]. The general form of the field equation can be written down as

$$t_* \frac{\partial s}{\partial t} = -\frac{\delta F}{\delta s} + f_*. \tag{1.25}$$

Here, t_* is the kinetic coefficient, $F[s]$ is the free energy functional [4] in the units of $k_B T$ per site, and f_* is the external field. The change of spatially inhomogeneous part of F at the fixed T and the total number of sites N_0 writes: $dF = h(s)ds$, where $h(s)$ is defined in (1.15). Therefore, the free energy functional can be presented in the form

$$F[s] = \int d\vec{r} \left[\frac{R_s^2}{2} (\nabla s)^2 + E(s) \right]; E(s) = \int ds h(s), \tag{1.26}$$

in which R_s is the scale factor determined by the cutoff radius of interaction potential. The first contribution under the integral of (1.26) describes the leading term of the series expansion of the field Hamiltonian in ∇s according to the Ginsburg-Landau theory [4]. Taking the functional derivative of (1.26), (1.25) is reduced to

$$t_* \frac{\partial s}{\partial t} = R_s^2 \Delta s - h(s) + f_*. \tag{1.27}$$

Realistically, (1.27) makes sense only for small enough $s_1 = s - s_0$, where s_0 is the initial spatially homogeneous state. We first consider the case where point 2 in Fig. 1.5 corresponds to the left boundary of the spinodal region: $s_0 = -\sqrt{1 - 4/\varphi}$. The function h here can be presented as $h(s_1) \cong a - bs_1^2$ with positive constants a and b. Upon appropriate rescaling, (1.27) can be put as

$$\frac{\partial s_1}{\partial t} = \Delta s_1 + 2(s_1^2 - 1) + f_*. \tag{1.28}$$

This equation describes the behavior of highly metastable systems near the spinodal boundary. As shown in [36] in the case of 2D adatom systems, (1.28) is capable of describing the stationary configuration corresponding to the critical nucleus, the relaxation of the order parameter in the linear approximation, and, most importantly, the nucleation rate which is given by $I(\zeta) = I(\zeta_{\max}) \exp\left[-const(\zeta_{\max} - \zeta)^2\right]$.

When point 3 in Fig. 1.6 is fixed at $\theta_0 = 1/2$ and the temperature is close to T_c, the behavior of $h(s)$ is given by the right hand side of (1.19). Using this in (1.27) and rescaling, the equation for the order parameter takes the form

$$\frac{\partial s}{\partial t} = \Delta s + s(s_e^2 - s^2) + f_*. \tag{1.29}$$

Here, s_e corresponds to the equilibrium spatially homogeneous states. As expected, (1.29) has periodic solutions already in the linear approximation in u. Separation of the initially space-uniform state at $s_0 = 0$ into the phases with $s_e = \pm\sqrt{3\tau}$ evolves via the increasing density fluctuations, starting from linear but then acquiring an essentially non-linear character. More details on the kinetics of spinodal decomposition in the systems described by (1.29) can be found, for example, in [37].

Stationary (1.29) in the linear approximation above T_c takes the form

$$-\Delta s + s/r_c^2 = f_*, \tag{1.30}$$

where $r_c = 1/\sqrt{3\,|\tau|}$. The Green function of this equation in 3D space, corresponding to the point source $f_*(r) = \delta(r)$ in the origin, is given by

$$G(r) = \frac{\exp(-r/r_c)}{4\pi r} \quad . \tag{1.31}$$

Clearly, the r_c parameter plays a role of the correlation radius. The obtained Green function is closely related to the correlation function of the order parameter $< s(\vec{r})s(0) >$. As discussed above, r_c discontinuously increases near T_c. The mean field theory thus predicts the critical exponent $\nu = 1/2$ for its temperature behavior. At $r << r_c$, $G(r)$ scales as r^{-1}, yielding $\eta = 0$. This completes the set of mean field critical exponents given in Table 1.1. We note that the correlation function given by (1.31) is of the Ornstein–Zernike type, as in the standard Landau theory of phase transitions [4]. This is not surprising, since the Taylor expansion of chemical potential near the critical point reduces the lattice gas model to the Landau-type Hamiltonian.

1.3 Laplacian Pressure and Gibbs–Thomson Effect

Thermodynamic parameters of nanoparticles such as pressure and chemical potential are essentially size-dependent. One of the most important manifestations of size-dependent effects is a tremendous increase of the equilibrium pressure inside a nanoparticle, and the corresponding elevation of its chemical potential. Both effects are driven by the surface curvature [38]. Let us briefly discuss how the surface energy is introduced in thermodynamics. Whenever the system consists of two or more phases, for example, liquid droplets or solid particles surrounded by vapor, all thermodynamic potentials are modified by the surface energy term $\sum_k \gamma_k S_k$. The index k accounts for different surfaces of the area S_k having different surface energies γ_k (J/m^2). Typical values of γ in liquids and solids are of the order of 1 J/m^2. The surface energy must be positive, because the minimum of thermodynamic potential is reached with the minimum surface area at a given volume. Otherwise, the two phases could not be separated at all. In the case of a solid–solid heterogeneous interface, the corresponding γ is usually called the interface energy. The microscopic origin of the surface energy is the dangling bonds created, for example, by the dissection of crystal with a certain crystallographic plane. Some dangling bonds can be passivated by the species adsorbed from vapor environment. That is why the surface energy generally depends on the chemical composition of both phases. The interface energy originates from microscopic interactions between dissimilar atoms at the interface. In isotropic liquids, the surface energy does not vary along the surface, while the surface energy of crystal solids depends on the crystallographic orientation of the corresponding planes terminating the particle.

We now consider a particular case of liquid or solid particle (phase 2) of surface area S surrounded by vapor (phase 1). At a constant temperature in both phases and under the constrain of the fixed volume $V = V_1 + V_2$, the particle of a given size is in equilibrium with vapor when the grand potential of the entire system

$$\Psi = -P_1 V_1 - P_2 V_2 + \gamma S \tag{1.32}$$

reaches the extremum. It is noteworthy that this extremum relates not to the minimum, but to the maximum of Ψ, showing that the equilibrium state is unstable. Taking the differential of Ψ at $V = const$,

$$d\Psi = (P_1 - P_2) dV_2 + \gamma dS, \tag{1.33}$$

and putting it to zero, the pressure difference is obtained in the form

$$P_2 - P_1 = \gamma \frac{dS}{dV_2}. \tag{1.34}$$

Obviously, the dS/dV_2 scales inversely proportional to the linear size of particle R, yielding the equality of pressures in the bulk limit $R \to \infty$. Otherwise, P_2 is larger than P_1, with the difference rapidly increasing as the particle size shrinks. The additional pressure in the condensed phase $\Delta P = P_2 - P_1$ given by (1.34) is called the Laplacian pressure.

We now show how the Laplace formula allows one to find the pressures in the two phases separately, as well as the chemical potential corresponding to the unstable equilibrium. We write down the conditions of equilibrium between the vapor and condensed phases with finite and infinite R as

$$\mu_1(P_1) = \mu_2(P_2) \tag{1.35}$$

and

$$\mu_{1\infty}(P_e) = \mu_{2\infty}(P_e) = \mu_e. \tag{1.36}$$

Here, ∞ denotes the $R \to \infty$ limit, where the two phases are at the same equilibrium pressure P_e and chemical potential μ_e. Subtracting the second equation from the first one, we note that the vapor chemical potential depends on its pressure as $k_B T \ln P_1$, yielding $\mu_1(P_1) - \mu_1(P_e) = k_B T \ln(P_1/P_e)$. The chemical potential in the condensed phase at a constant temperature is given by $d\mu_2 = \Omega dP_2$, where Ω is the elementary volume. Assuming that the particle is incompressible, we get: $\mu_2(P_2) - \mu_2(P_e) = \Omega(P_2 - P_e)$. For a small enough particle, P_2 should be much larger than P_e. This allows us to write approximately $P_2 - P_e \cong P_2 - P_1$ and to use (1.34) for $P_2 - P_e$. In this approximation, the Gibbs–Thomson modification of the equilibrium chemical potential in both phases is obtained in the form

$$\mu_1(P_1) = \mu_2(P_2) = \mu_e + \Omega\gamma \frac{dS}{dV_2}. \tag{1.37}$$

As for the vapor pressure, it increases exponentially with decreasing the particle size due to a high compressibility of vapor:

$$P_1 = P_e \exp\left(\frac{\Omega\gamma}{k_B T}\frac{dS}{dV_2}\right). \tag{1.38}$$

The derivative dS/dV_2, entering the expressions for the Laplacian pressure and the Gibbs–Thomson effect on the chemical potential, depends strongly on the system geometry. Some typical geometries are shown in Fig. 1.7. In the case of a spherical liquid droplet of the radius R surrounded by vapor, $S = 4\pi R^2$ and $V_2 = (4\pi/3)R^3$, yielding $dS/dV_2 = 2/R$. If the "particle" is a solid cylinder of the radius R and length L, its sidewall surface area and volume are $S = 2\pi RL$ and $V_2 = \pi R^2 L$, respectively. If this cylinder can extend only radially ($L = const$), we get $dS/dV_2 = 1/R$. When a liquid droplet is seated on top of a cylindrical solid rod of the radius R, it has the shape of a spherical cap with the contact angle β. The surface area and the volume of a spherical cap are given by

$$S = \frac{2\pi R^2}{1 + \cos\beta} \tag{1.39}$$

and

$$V_2 = \frac{\pi R^3}{3} f(\beta); \quad f(\beta) = \frac{(1 - \cos\beta)(2 + \cos\beta)}{(1 + \cos\beta)\sin\beta}. \tag{1.40}$$

At $R = const$, the droplet size can change only by varying the contact angle β. In view of $df(\beta)/d\beta = 3/(1 + \cos\beta)^2$, we obtain $dS/dV_2 = (2\sin\beta)/R = 2/R_L$, where R_L is the droplet radius (see Fig. 1.7c). In all the cases considered, the size-dependent terms are inversely proportional to the radius of the surface curvature.

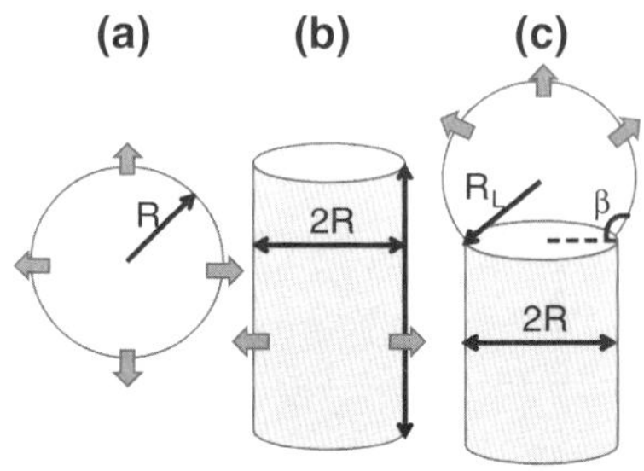

Fig. 1.7 Model geometries of particles surrounded by vapor, the arrows show the moving surfaces

Using the obtained relationships in (1.37), the Laplacian pressure ΔP and the corresponding Gibbs–Thomson modification of chemical potential $\Delta\mu = \mu - \mu_e$ are given by

$$\text{Spherical droplet of the radius } R: \Delta P = \frac{2\gamma}{R};\ \Delta\mu = \frac{2\Omega\gamma}{R}; \tag{1.41}$$

$$\text{Cylinder of the radius } R: \Delta P = \frac{\gamma}{R};\ \Delta\mu = \frac{\Omega\gamma}{R}; \tag{1.42}$$

$$\text{Spherical cap on a cylinder of the radius } R: \Delta P = \frac{2\gamma \sin\beta}{R};\ \Delta\mu = \frac{2\Omega\gamma \sin\beta}{R}. \tag{1.43}$$

To give a numerical estimate of the curvature effect, consider a 10 nm radius GaAs nanowire growing from a liquid Au-Ga droplet (with neglect of As due to its low solubility in Au), at $\beta \cong \pi/2$. The droplet surface energy in this case must be between that of pure liquid Ga and Au (0.72 and 1.14 J/m^2, respectively [39]), yielding a huge Laplacian pressure between 0.14 and 0.23 GPa. As discussed later on, this has a major impact of nanowire growth: at a given vapor pressure, smaller droplets will desorb more atoms than larger ones, which will suppress the nanowire growth from droplets smaller than a certain critical radius.

1.4 Contact Angle

Consider a liquid droplet seated on a planar solid surface and surrounded by vapor, as illustrated in Fig. 1.8. The surface energies of such a three-phase system are denoted as γ_{LV}, γ_{SV} and γ_{SL} at the liquid–vapor, solid–vapor and solid–liquid phase boundaries, respectively. The surface energy of the system writes down as

$$G = \gamma_{LV}\frac{2\pi R^2}{1+\cos\beta} + \gamma_{SL}\pi R^2 + \gamma_{SV}(S_0 - \pi R^2), \tag{1.44}$$

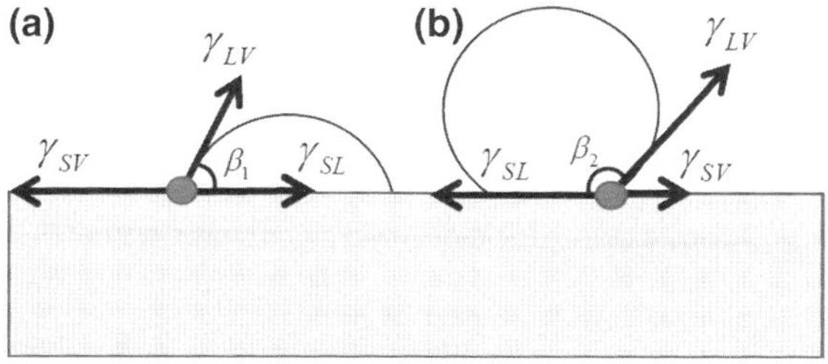

Fig. 1.8 Liquid droplet on solid surface with equilibrium contact angle $\beta_1 < \pi/2$ (**a**) and $\beta_2 > \pi/2$ (**b**). The *dot* denotes the triple phase line

where β is the contact angle, R is the radius of the base and $S_0 = const$ is the surface area of the solid. We aim at finding the equilibrium contact angle relating to the minimum of surface energy at the fixed droplet volume V given by (1.40).

The surface energy G is generally a function of two variables, R and β, which are related to each other by the condition $V = const$. Writing the volume differential in the form

$$dV = \pi R^2 \left[f(\beta) dR + \frac{R}{3} \frac{df(\beta)}{d\beta} d\beta \right],$$

we derive:

$$R d\beta = -(2 + \cos\beta)\sin\beta dR. \tag{1.45}$$

After simple calculations, the differential dG amounts to

$$dG|_{V=const} = (\gamma_{SL} - \gamma_{SV} + \gamma_{LV}\cos\beta) 2\pi R dR. \tag{1.46}$$

Therefore, the system is in equilibrium when the contact angle satisfies the Young's equation [38]:

$$\gamma_{SV} = \gamma_{SL} + \gamma_{LV}\cos\beta_*. \tag{1.47}$$

From (1.46), it can be easily seen that the second differential of G is always positive at $\beta = \beta_*$. The Young's contact angle therefore corresponds to the minimum surface energy. The mechanical interpretation of the Young's equation is quite obvious: it requires the compensation of horizontal surface forces acting upon the triple phase line, as shown in Fig. 1.8. At the same time, the compensation of vertical surface force directed upward should lead to a certain reconstruction of the solid structure under the droplet.

With arbitrary surface energies, (1.47) can be satisfied with a positive contact angle only if $\gamma_{SV} < \gamma_{SL} + \gamma_{LV}$. In this case, the droplet does not wet the solid surface. The case of $\gamma_{SV} > \gamma_{SL}$ yields $\beta_* < \pi/2$. The condition $\gamma_{SV} < \gamma_{SL}$ corresponds to $\beta_* > \pi/2$, with the extreme case of $\beta_* = \pi$ being reached at $\gamma_{SV} = \gamma_{SL} - \gamma_{LV}$. From these considerations, the three surface energies must satisfy the inequality $\gamma_{SV} \le \gamma_{SL} - \gamma_{LV}$, which is not at all guaranteed in the general case. However, the solid–vapor surface energy could be modified by monolayer adsorption, and such an adsorption will necessarily take place if the last inequality is not met without it [38].

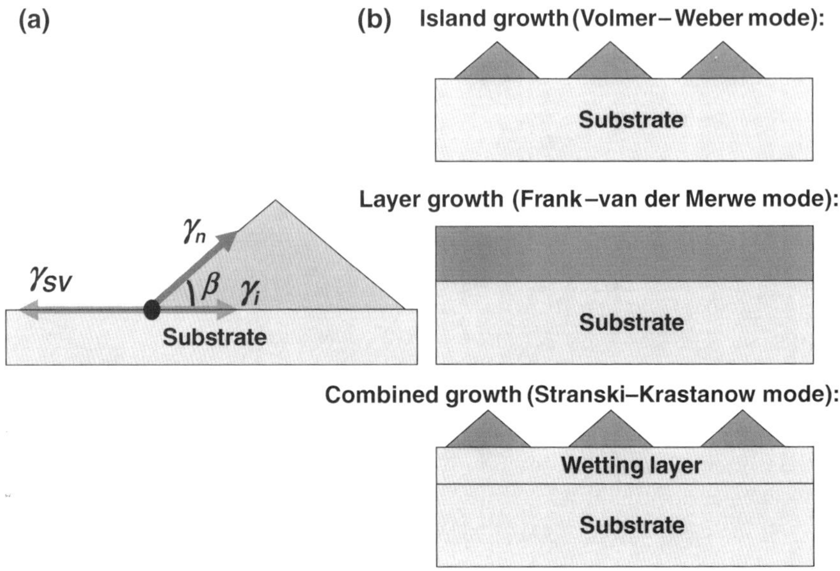

Fig. 1.9 Surface energetics of the crystal island (**a**) and the surface growth modes (**b**)

When $\gamma_{SV} > \gamma_{SL} + \gamma_{LV}$, the Young's equation cannot be satisfied, and the liquid wets the solid surface. The wetting occurs, for example, when the solid–vapor surface is much more energetic than the liquid–vapor one. Formation of a 2D liquid film is then energetically preferred to the liquid agglomeration into 3D droplets where some part of the solid surface remains exposed to vapor.

The Young's equation is sometimes applied to the case of solid surface islands, for example, in the elementary classification of the growth modes of thin solid films [40–42], as shown in Fig. 1.9a. Surface film can either grow layer-by-layer in the Frank–van der Merwe mode, or form 3D islands on a bare surface (the Volmer-Weber mode). In some specific cases, the combined Stranski-Krastanow growth is observed, where the film forms the so-called 2D wetting layer at the beginning, but then the growth is transformed into a 3D island mode. These modes are schematized in Fig. 1.9b. The Young's equation can now be re-written in the form

$$\gamma_{SV} = \gamma_i + \gamma_n \cos\beta. \tag{1.48}$$

Here, γ_i is the interface energy at the planar solid–solid boundary under the island and γ_n is the surface energy of the island facet plane (Fig. 1.9a). The facets should relate to the lowest among the local minima of the crystal surface energy, while the global minimum is usually reached in the in-plane direction. In the case of homoepitaxy, $\gamma_i = 0$ and $\gamma_n = \gamma_{SV}$ for the substrate plane, showing that homoepitaxial layers should be stable against faceting and grow in 2D form (however, nucleation of upper layers before the completion of the lower ones is not always guaranteed). The Volmer-Weber islands form during heteroepitaxy if $\gamma_{SV} < \gamma_i + \gamma_n$, when the deposit-substrate interface is highly energetic, and the surface energy of the deposit is also sufficiently large compared to the substrate. Heteroepitaxial layer will grow in 2D form in the opposite case $\gamma_{SV} > \gamma_i + \gamma_n$, where the substrate is more energetic than the "sandwich" composed of the deposit-vapor surface and the deposit–substrate interface.

The Stranski-Krastanow growth on a dissimilar surface cannot be fully explained on purely surface energetic grounds. Indeed, while 2D growth of the first few monolayers can be energetically preferred, it is difficult to see why 3D islands prefer to form on the similar substrate, which is effectively the homoepitaxy case. One can thus suggest that 3D islands gain some decrease in the volume energy compared to 2D thin film. This is exactly what happens in lattice mismatched material systems, where the driving force for 2D-3D growth transformation is the relaxation of elastic stress (induced by the lattice mismatch) on free side facets. Very importantly, sufficiently small 3D islands remain dislocation-free. The Stranski-Krastanow growth of coherent strained islands will be studied in detail in Chap. 3.

We now consider the freestanding "nanowire" geometry with the liquid droplet seated on the nanowire top. The standard system configuration, where the droplet wets the top facet but not the sidewalls (the non-wetting case) [43, 44] is shown schematically in Fig. 1.10a. We assume the wire as being a straight cylinder of a constant radius and the droplet in the form of a spherical cap with the contact angle $\beta_0 > \pi/2$. The choice of β_0 will be discussed later on. We consider five surface energies shown in Fig. 1.10a–c: those of the horizontal solid–liquid surface γ_{SL}, the liquid–vapor surface γ_{LV}, the top facet in contact with the vapor γ_{SV}, the nanowire sidewalls in contact with the vapor γ_{WV}, and the vertical solid–liquid surface γ_{LV}^l. Let us first study the stability of the triple phase line position at the outer periphery of the nanowire top. When the triple phase line is shifted horizontally toward the wire center (Fig. 1.10b), the back surface force $F_h = \gamma_{SV} - \gamma_{LV}\cos\beta_0 - \gamma_{SL}$ is positive in view of $\gamma_{SV} > \gamma_{SL}$ (this inequality should pertain in most cases) and thus returns the triple phase line to its original position. When the triple phase line is randomly shifted downward (Fig. 1.10c), the vertical force $F_0 = \gamma_{LV}\sin\beta_0 + \gamma_{SL}^l - \gamma_{WV}$ is directed upward provided that

$$\gamma_{WV} < \gamma_{LV}\sin\beta_0 + \gamma_{SL}^l. \tag{1.49}$$

This inequality, obtained here as the condition for stability of the triple phase line at the top periphery, is exactly identical to the Nebol'sin-Shchetinin [43] and Glas [45] conditions that are very important in studying the so-called vapor–liquid–solid growth of nanowires [46, 47]. This will be discussed further in Chap. 4.

Let us now see what happens if the inequality (1.49) is not satisfied, that is, when

$$\gamma_{WV} > \gamma_{LV}\sin\beta_0 + \gamma_{SL}^l. \tag{1.50}$$

Opposite to the previous case, whenever the triple phase line moves slightly downward, the surface force is also directed downward and tends to increase the part of nanowire height y wetted by the liquid (the wetting case), as shown in Fig. 1.11a.

Under the constraint of a fixed liquid volume V and radius R, the penetration of wire cylinder into the droplet leads to the increase of contact angle β. As β increases, the vertical force $F = \gamma_{WV} - \gamma_{LV}\sin\beta - \gamma^{l}_{SL} > 0$ can only increase. Therefore, any configuration with y between zero and $y_{\max} = -2R\cot an\beta_{\max}$ is unstable, and the droplet will continue sliding down until the cylinder hits the droplet surface (Fig. 1.11b).

Configuration shown in Fig. 1.11b relates to the minimum surface energy with a spherical droplet under the condition given by (1.50). This is the consequence of the following theorem [48]:

The surface energy of the system incorporating the droplet and the wire of length $L = const$,

$$G = \frac{2\pi R^2}{1+\cos\beta}\gamma_{LV} + \pi R^2\gamma_{SL} + 2\pi Ry\gamma^{l}_{SL} + 2\pi R(L-y)\gamma_{WV}, \tag{1.51}$$

under the condition of a fixed liquid volume

$$V = \frac{\pi R^3}{3}\frac{(1-\cos\beta)^2(2+\cos\beta)}{\sin^3\beta} - \pi R^2 y = const \tag{1.52}$$

and at a constant R, has the minimum either at $\beta = \beta_0$ (relating to $y = 0$) or at $\beta = \beta_{\max}$ (relating to $y = y_{\max}$) when the inequality (1.49) is satisfied and only at $\beta = \beta_{\max}$ otherwise. To prove the theorem, we first express the height y in (1.51) through β by (1.52) and then differentiate the resulting $G(\beta)$ with respect to β. The derivative is given by

$$\frac{dG}{d\beta} = \frac{2\pi R^2}{(1+\cos\beta)^2}(\gamma_{LV}\sin\beta + \gamma^{l}_{SL} - \gamma_{WV}), \tag{1.53}$$

showing that the surface energy increases in the non-wetting case (1.49) and decreases otherwise. When G decreases at $\beta = \beta_0$, it will continue decreasing at larger β, with the minimum being reached at $y = y_{\max}$. When G increases at $\beta = \beta_0$, the minimum

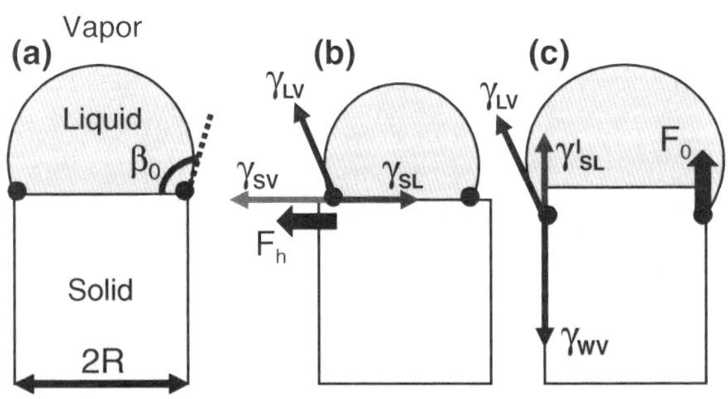

Fig. 1.10 Standard vapor–liquid–solid configuration involving surface energies described in the text: the triple phase line (*black dots*) at the outer periphery of top facet (**a**) and its stability under random horizontal (**b**) and vertical (**c**) shifts

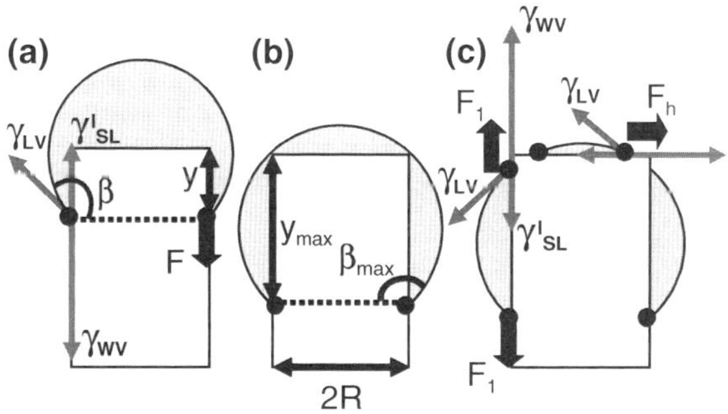

Fig. 1.11 Vapor–liquid–solid wire configuration in the wetting case: unstable configuration with intermediate y (**a**); system geometry at $y = y_{max}$ (**b**) and the directions of surface forces after the droplet is torn apart (**c**). From geometrical considerations, the maximum of y in spherical geometry is given by $y_{max} = -2R \cot an\beta_{max}$

energy is reached either in the standard configuration with $y = 0$ or at $y = y_{max}$, depending on whether the value of $G(\beta_0) - G(\beta_{max})$ is negative or positive.

Surprisingly, the conventional Young's equation on the sidewalls, $\gamma_{WV} = \gamma_{LV} \sin\beta_* + \gamma^l_{SL}$, in the geometry considered relates not to the minimum but to the maximum surface energy. This property follows directly from

$$\left(\frac{d^2G}{d\beta^2}\right)_{\beta=\beta_*} = \frac{2\pi R^2 \gamma_{LV} \cos\beta_*}{(1+\cos\beta_*)^2} < 0 \tag{1.54}$$

for any $\beta_* > \pi/2$. The instability of Young's configuration between $y = 0$ and $y = y_{max}$ is physically explained by the constraint of a fixed liquid volume, with any deviation of contact angle from β_* creating an uncompensated surface force. At $\beta_{max} < \beta_*$, the standard vapor–liquid–solid configuration with $y = 0$ relates to the global energy minimum. Whenever, $\beta_{max} > \beta_*$, the two minima of surface energy are separated by the barrier $G(\beta_*)$. The stability of non-wetting configuration is therefore only local.

The next step of system evolution could be the droplet torn apart, as shown in Fig. 1.11c. Let us consider the directions of surface forces in this case. The vertical force acting at the top triple phase line on the wire sidewall, $F_1 = \gamma_{WV} - \gamma_{LV} \sin\beta_{max} - \gamma^l_{SL}$, is always directed upward in view of (1.50) and $\beta_{max} > \beta_0$ and. The horizontal force acting upon the triple phase line resting on the top facet, $F_h = \gamma_{SV} + \gamma_{LV}\cos\beta_{max} - \gamma_{SL}$, returns it back to the outer periphery provided that $\gamma_{SV} > \gamma_{SL} - \gamma_{LV}\cos\beta_{max}$. This inequality gives the second condition for the stability of droplet at the wire top.

However, as seen from Fig. 1.11c, the force F_1 acting upon the bottom triple phase line on the wire sidewall remains uncompensated and tends to decrease the contact angle β_{max}. Such force gives rise to further evolution of system morphology, with the droplet losing its spherical geometry. The new, stationary droplet shape should be obtained from the minimization of surface energy functional, a complex problem studied in Chap. 6.

We now turn to the discussion of droplet contact angle β_o in the non-wetting vapor–liquid–solid configuration of Fig. 1.10a. In some works, for example, in [43, 44, 49], it has been assumed that the contact angle could be obtained from the balance of horizontal forces at the triple phase line:

$$\gamma_{SL} = -\gamma_{LV} \cos \beta_0, \tag{1.55}$$

yielding that the droplet is always more than a hemisphere. However, this cannot be proved by the minimization of the total surface energy of the wire-droplet system given by (1.51) at $y = 0$. Indeed, varying the two variables, β and R at fixed V would require a shift of sidewalls. The shift would change the sidewall surface energy $2\pi RL\gamma_{WV}$, which is of course dominant at $L >> R$. Equation (1.55) can only be obtained from the minimization of the reduced surface energy associated with the droplet:

$$G_0 = \frac{2\pi R^2}{1+\cos\beta}\gamma_{LV} + \pi R^2 \gamma_{SL}, \tag{1.56}$$

at a constant V given by (1.52) at $y = 0$ [43], where the sidewall energy term is omitted. This should be considered as a hypothesis rather than the rigorous proof of (1.55).

1.5 Formation Energy

Central to nucleation theory is the expression for the free energy of nucleus formation $F_i = \Delta G_i/(k_B T)$ [3, 7–30, 50–53]. Determination of F_i is the core step for construction of any nucleation-condensation model. As discussed already in Sect. 1.2, the macroscopic ($i >> 1$) change of free energy associated with the formation of a nucleus consisting of i monomers (the formation energy for brevity) always contains the volume term $-(\Delta\mu/k_B T)i$ and the surface energy term. In homogeneous nucleation theory, the $F(i)$ can be thus written down as [3, 7, 9, 50–53]

$$F_i = F_s(i) - (\Delta\mu/k_B T)i = F_s(i) - \ln(\zeta + 1)i, \tag{1.57}$$

where the last expression corresponds to the perfect gas approximation. The entire system is considered to be isothermal, with all nuclei having the same temperature T which is equal to the vapor temperature. This is usually ensured by the presence of a passive gas (in the case of vapor condensation) or the thermostatic effect of the substrate (in the case of surface islands), which is kept constant during the entire growth process. In heterogeneous condensation on the nucleation seeds such as charged ions in gases, (1.57) should be modified to account for the interactions of the liquid with the seed. Particular form of $F_s(i)$ depends on the system geometry, surface energies involved in the transition and, most importantly, the nucleus dimension.

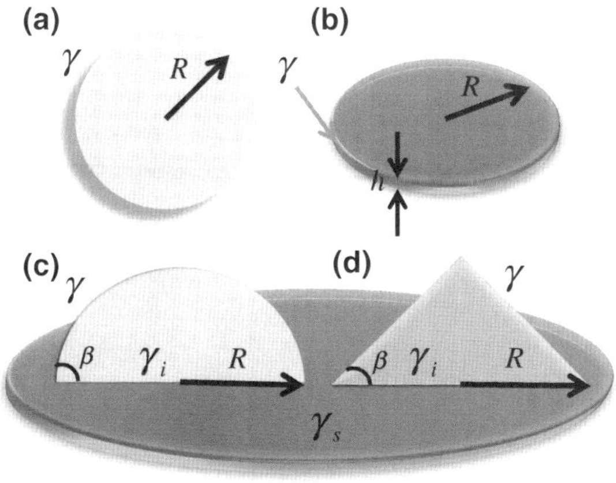

Fig. 1.12 Different shapes of nuclei: 3D liquid droplet (**a**), 2D surface island on similar substrate (**b**), spherical cap island (**c**) and pyramidal island (**d**) with contact angle β on dissimilar substrate

Let us now consider some typical nucleation scenarios schematized in Fig. 1.12. In the case of 3D spherical droplet of the radius R growing from a supersaturated vapor (Fig. 1.12a), the surface energy contribution is given by $F_s = (\gamma/k_BT)4\pi R^2$, where γ is the droplet surface energy. Introducing the elementary volume in the liquid phase such that $\Omega i = (4/3)\pi R^3$, we obtain

$$F_s(i) = ai^{2/3}; \quad a = \frac{4\pi\gamma}{k_BT}\left(\frac{3\Omega}{4\pi}\right)^{2/3} = \frac{4\pi\gamma}{k_BT}r_0^2, \tag{1.58}$$

where $r_0 = (3\Omega/4\pi)^{1/3}$ is the interatomic distance in the liquid. For a crystal 3D nucleus growing from vapor, the surface energy term should be re-written as $F_s = (1/k_BT)\sum_k \gamma_k S_k$. This yields the same equation as (1.58), where the constant a sums up the contributions from different crystallographic planes with the corresponding shape constants.

When 2D island of the monolayer height h is formed from an adatom sea or a liquid alloy on the similar substrate (Fig. 1.12b), the surface energy depends on the island shape. The latter is determined by the crystallographic orientation of the substrate and might depend on the growth conditions used. Introducing the linear size R, the island perimeter and surface area can be generally put as c_1R and c_2R^2, respectively, with the corresponding shape constants c_1 and c_2. In particular, $c_1 = 2\pi$, $c_2 = \pi$ for circles and $c_1 = 8$, $c_2 = 4$ for squares with side $2R$. The island surface area equals σi, with σ being the elementary surface area. Since homoepitaxial growth should not change the in-plane surface energy, the F_s term corresponds to the formation of lateral surface of monolayer height of the area c_1Rh. Therefore,

$$F_s(i) = ai^{1/2}; \quad a = \frac{c_1\gamma h}{k_BT}\left(\frac{\sigma}{c_2}\right)^{1/2} = \frac{c_1\gamma h}{k_BT}r_0, \tag{1.59}$$

where γ is the surface energy of the lateral interface in contact with a given environment and $r_0 = (\sigma/c_2)^{1/2}$ is the lattice spacing.

In the case of heteroepitaxy from vapors or liquid alloys, islands can emerge either in 2D or 3D shape. For a 2D island, the lateral surface energy is exactly the same as given by the previous expression, unless it is modified by short-range edge interactions that are not considered here. Besides, the total surface energy contains the in-plane contribution. This is given by $(\gamma_d + \gamma_i - \gamma_s)\sigma i$, where γ_d is the surface energy of the deposited material (created by the nucleation), γ_i is the energy of the deposit-substrate interface (created by the nucleation) and γ_s is the surface energy of the bare substrate (eliminated by the nucleation). Overall, the surface energy of 2D island on the dissimilar substrate is given by

$$F_s(i) = ai^{1/2} + \frac{(\gamma_d + \gamma_i - \gamma_s)\sigma i}{k_B T}. \tag{1.60}$$

Using this in (1.57), it can be re-written in the form

$$F_i = ai^{1/2} - \ln(\zeta + 1)i \tag{1.61}$$

upon the re-normalization of the equilibrium adatom concentration

$$n_{1e}^{eff} \equiv n_{1e} \exp\left[\frac{(\gamma_d + \gamma_i - \gamma_s)\sigma}{k_B T}\right]. \tag{1.62}$$

The latter is modified by the change of in-plane surface energy. Otherwise, the formation energies of 2D islands in homo- and heteroepitaxy are identical.

When a 3D island (liquid or solid) with the base dimension R and contact angle β is formed on a dissimilar substrate (Fig. 1.12c and d), the surface energy change equals

$$F_s(i) = \frac{\gamma}{k_B T} c_1 R^2 + \frac{(\gamma_i - \gamma_s)}{k_B T} c_0 R^2. \tag{1.63}$$

Here, γ is the surface energy of the island, assumed as being constant also for crystal. This property holds for the equivalent lateral facets. The c_1 and c_0 are the shape constants such that $c_1 R^2$ is the lateral surface area and $c_0 R^2$ is the base area of the island. The last term in (1.63) accounts for the change in the in-plane surface energy after nucleation. The island volume amounts to $c_2 R^3$, with c_2 as the corresponding shape constant. For example, $c_0 = \pi$, $c_1 = 2\pi/(1 + \cos\beta)$, and $c_2 = \pi f(\beta)$ for a spherical cap island, with $f(\beta)$ given by (1.40). For a crystal island in the form of pyramid with the rectangular base of side $2R$, $c_0 = 4$, $c_1 = 4/\cos\beta$, and $c_2 = (3/4)\tan\beta$. Using the relationship $\Omega i = c_2 R^3$, (1.63) takes the form

$$F_s(i) = ai^{2/3}; \quad a = \frac{[c_1\gamma + c_0(\gamma_i - \gamma_s)]}{k_B T}\left(\frac{\Omega}{c_2}\right)^{2/3}. \tag{1.64}$$

A more complex case of a lattice mismatched substrate, where the mismatch-induced strain modifies the volume energy depending on the island shape and the aspect ratio, will be considered in Chap. 3.

Summarizing, the surface energy contribution can be put as $ai^{2/3}$ for a 3D nucleus and $ai^{1/2}$ for a 2D island in all the cases considered. We can therefore present the macroscopic formation energy F_i in the form [52]

$$F_i = ai^{(d-1)/d} - \ln(\zeta + 1)i, \tag{1.65}$$

where $d = 2, 3$ is the space dimension of the nucleus. The first, positive contribution describes the energetically unfavorable process of the formation of new surfaces, with coefficient a being proportional to the characteristic surface energy in thermal units. This coefficient is always much larger than one, showing that the surface is stable against thermal fluctuations. For example, $a = 9.6$ in the case of water condensation at room temperature. In the case of heterogeneous condensation of vapor on the charged ions ($d = 3$), (1.65) is modified as [9]

$$F_i = ai^{2/3} - \ln(\zeta + 1)i + ci^{-1/3}, \tag{1.66}$$

where $c = q^2(\varepsilon_v^{-1} - \varepsilon_l^{-1})/(2k_B Tr_0)$, q is the electrical charge of the ion, and ε_v, ε_l are the dielectric constants in the gas-vapor system and in the liquid, respectively.

The function $F(i)$ defined by (1.65) reaches its maximum

$$F(\zeta) \equiv F[i_c(\zeta)] = \frac{(d-1)^{d-1}}{d^d} \frac{a^d}{\ln^{d-1}(\zeta + 1)} = \frac{a}{d} i_c^{\frac{d-1}{d}} \tag{1.67}$$

at the critical size

$$i_c(\zeta) = \left[\frac{(d-1)a}{d\ln(\zeta + 1)}\right]^d. \tag{1.68}$$

The quantity $F(\zeta)$ is the nucleation barrier at a given supersaturation. The critical nucleus is in unstable equilibrium with a metastable phase; the nuclei with $i > i_c$ tend to grow, while smaller nuclei with $i < i_c$ tend to evaporate. Of course, there is an almost zero probability that i_c is an integer. The half-width of the near-critical region Δi_c, defined by $\left(\left|F''(i_c)\right|/2\right)\Delta i_c^2 = 1$, is readily obtained as

$$\Delta i_c = d\left(\frac{2}{(d-1)a}\right)^{1/2} i_c^{\frac{d+1}{2d}}. \tag{1.69}$$

If the critical size is large enough, (1.68), (1.69) can be used for the charged droplets as well. However, in heterogeneous nucleation, the formation energy has a minimum at small i, corresponding to the energetically preferred stable state. In this case, the nucleation barrier is modified to $\Delta F(\zeta) = F(\zeta) - F_{\min}(\zeta)$.

Let us now discuss a very important question of quasi-equilibrium distribution normalization and the corresponding size-independent corrections to the macroscopic formation energy [3, 7, 54]. Such modifications do not change (1.68) and (1.69) but do influence the nucleation barrier and related values. The quasi-equilibrium concentration of nuclei with i atoms n_i^{eq} is given by the Boltzmann exponent of the formation energy [38]. When vapor is at the concentration n_1, the n_i^{eq} is given by [7, 9]

$$n_i^{eq} = n_1 \exp(-F_i). \tag{1.70}$$

It should be noted here that the quasi-equilibrium distribution n_i^{eq} relates to a metastable vapor, while the global equilibrium is reached at $n_1 = n_1^{eq} \equiv n_{1e}$, that is, at zero supersaturation. It is easily seen that this equation is self-inconsistent when F_i is given by (1.65): since $F_1 \neq 0$, it does not give correctly the monomer concentration at $i = 1$. Kashchiev [7] proposed to use $F_i - F_1$ instead of F_i for the formation energy, which modifies (1.70) to

$$n_i^{eq} = n_1 \exp[-(F_i - F_1)]. \tag{1.71}$$

This equation is self-consistent at $i = 1$. The corresponding corrections to the macroscopic (1.65), (1.67) are given by

$$F_i = a[i^{(d-1)/d} - 1] - \ln(\zeta + 1)(i - 1); \tag{1.72}$$

$$F(\zeta) \equiv F[i_c(\zeta)] = \frac{(d-1)^{d-1}}{d^d} \frac{a^d}{\ln^{d-1}(\varsigma + 1)} - a + \ln(\zeta + 1). \tag{1.73}$$

In view of $a >> 1$, such a correction will lead to increasing the size distribution normalization by the orders of magnitude.

Long ago, Lothe and Pound [55] proposed another correction to the formation energy originating from the entropy factor in distributing n_1 monomers over n_0 sites in the lattice (where $n_0 = 1/\Omega$ for 3D and $n_0 = 1/\sigma$ for 2D lattice):

$$F_i = ai^{(d-1)/d} - \ln(\zeta + 1)i + \ln(n_1/n_0). \tag{1.74}$$

The logarithmic term change the pre-exponential factor of the equilibrium distribution from n_1 to n_0. In view of $n_1/n_0 << 1$, the normalization increases again by the orders of magnitude. Although we do not see any reasonable justification for the Lothe-Pound factor in the general case, such a correction is often used in modeling adatom nucleation and thin film growth [3]. Together with the F_1 correction as given by (1.70), (1.74) gives the result which is obviously incorrect at $i = 1$. However, without the F_1 term, the large factors $\exp(a)/(\zeta + 1)$ and $n_0/n_1 = 1/[\theta_{1e}(\zeta + 1)]$ might be of the same order. Overall, the problem of distribution normalization in the macroscopic theory requires special attention. Truly microscopic approaches are needed to describe correctly the formation energy of small clusters consisting of only

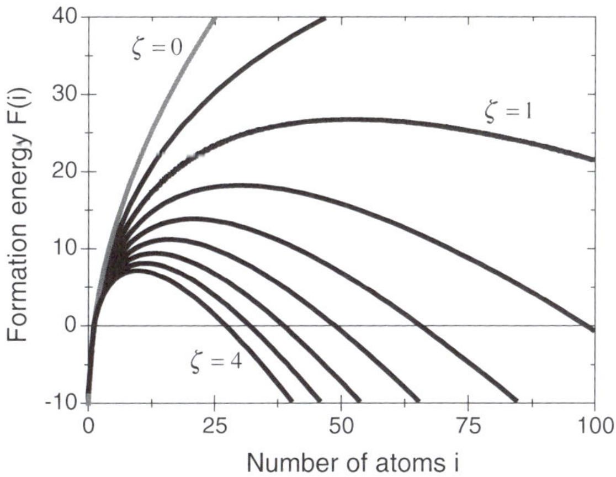

Fig. 1.13 Formation energy of 2D island at $a = 10$ and different supersaturations from 0 to 4

a few monomers. Surprisingly, the problem simplifies in the case of nucleation in open systems due to a special type of self-consistency between the external material influx and the material sink by nucleation, as described in detail in Chap. 2.

Figure 1.13 shows the formation energy of 2D island at the fixed $a = 10$ and different supersaturations, obtained from (1.72). The equilibrium case at $\zeta = 0$ relates to infinite nucleation barrier where the transition is thermodynamically forbidden. Increasing supersaturation from 0.5 to 4 leads to decrease of the critical number of monomers from 55 to 11, while the nucleation barrier decreases from 27 to 7. Further increase of supersaturation will break the possibility of using the macroscopic approach. We note, however, that if supersaturation increases gradually due to a material influx into the system or, more generally, by smoothly changing the macroscopic parameters, supersaturation would not increase to these high val-

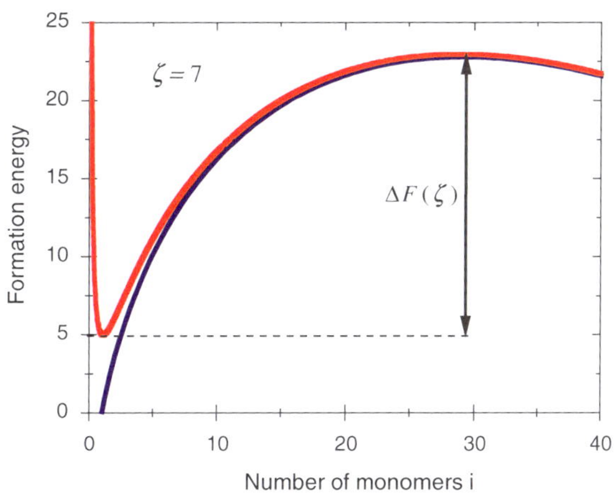

Fig. 1.14 Formation energies of 3D droplets in homogeneous and heterogeneous cases at $\zeta = 7$, $\Delta F(\zeta)$ showing the decreased barrier for heterogeneous nucleation

ues. Rather, it would reach a maximum at a certain stage due to a dynamic balance between the material influx and the monomer sink caused by nucleation and growth of embryos. This maximum supersaturation would correspond to a high enough nucleation rate which can outweigh the material influx, but is still low enough to be reasonably described within the macroscopic theory.

In Fig. 1.14, we show how the condensation seeds decrease the nucleation barrier at a given supersaturation value. As discussed later on, the nucleation rate, the value giving the number of irreversibly growing nuclei per unit volume per unit time, is extremely sensitive to supersaturation. The decrease of ζ by only a few percent results in decreasing the nucleation rate by the order of magnitude. This naturally leads to a certain threshold value of supersaturation in a given material system, below which the nucleation process is too slow to be observed on a realistic time scale. The existence of nucleation seeds such as charged ions in 3D gas–vapor systems or surface steps, kinks or chemical impurities on a 2D surface can therefore significantly decrease this threshold value.

1.6 Adsorption and Thin Films

In many practical applications of nucleation theory such as modeling the growth properties of nanostructures synthesized by different epitaxy techniques, we deal with metastable adatom systems, surface islands and layers. An "ideal" substrate surface without defects, steps or chemical impurities can be treated as a 2D lattice of a given geometry that can accommodate atoms in the adsorption sites. The 2D lattice gas is a good model to describe the monolayer film on a crystal surface in the case of strongly localized adsorption. The film is usually grown from vapor or, more generally, from a gaseous system characterized by the deposition rate (or the arrival rate) I (1/cm^2 s) onto the substrate [3, 56, 57].

In the case of vapor phase with pressure P and temperature T_v above the surface, the arrival rate can be calculated from the gas kinetic theory [38]. As shown in Fig. 1.15, the number of vapor molecules arriving onto the unit surface area per unit

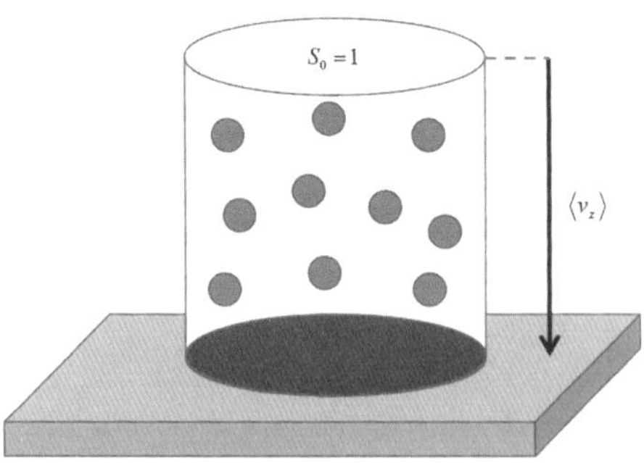

Fig. 1.15 To the calculation of vapor flux impinging the elementary area $S_0 = 1$

time is given by $I = n_v \langle v_z \rangle$. Here, $\langle v_z \rangle$ is the averaged component of molecule speed in the direction normal to the substrate. The latter can be easily found from the Maxwell distribution at temperature T_v:

$$\langle v_z \rangle = \left(\frac{m}{2\pi k_B T_v} \right)^{1/2} \int_0^{\infty} dv v \exp\left(-\frac{mv^2}{2k_B T_v} \right) = \sqrt{\frac{k_B T_v}{2\pi m}}. \tag{1.75}$$

with m as the molecule mass. The integration over speeds is performed from zero to infinity, because the molecules with $v_z < 0$ fly away from the surface. The vapor flux can be therefore presented in two equivalent forms

$$I = \sqrt{\frac{k_B T_v}{2\pi m}} n_v = \frac{P}{\sqrt{2\pi m k_B T_v}} \tag{1.76}$$

by using the perfect gas equation of state $P = n_v k_B T_v$.

When the vapor and the substrate temperatures equal each other, the gaseous phase is in equilibrium with the film having the occupation θ (called the coverage in the case of adsorbed surface layers) provided that $\mu_g = \mu(\theta)$, where $\mu(\theta)$ is the lattice gas chemical potential defined by (1.6). The chemical potential of gaseous phase is given by $\mu_g/k_B T = \ln(P/P_0) = \ln(I/I_0)$, with certain constants P_0 and I_0. We thus arrive at the Fowler-Guggenheim adsorption isotherm [58]

$$\ln\left(\frac{I}{I_0}\right) = -\varphi\theta + \ln\left(\frac{\theta}{1-\theta}\right); \quad \frac{I}{I_0} = \frac{\theta}{1-\theta} \exp(-\varphi\theta). \tag{1.77}$$

For very small $\theta << 1$, (1.77) is reduced to the Henry's law stating that the coverage is proportional to the vapor flux:

$$\theta = I/I_0. \tag{1.78}$$

For an ideal adsorbate with no attractive interactions ($\varphi = 0$), (1.77) gives the Langmuir isotherm

$$\theta = \frac{I/I_0}{1 + I/I_0} \tag{1.79}$$

for any θ between zero and one.

The adsorption isotherm of non-ideal monolayer adsorbate given by (1.77), describes three principally different equilibrium states in the adsorbate-vapor system at $T < T_c$ [38, 59, 60], plotted in Fig. 1.16. When $\ln(I/I_0) < -\varphi/2 = \mu_e/k_B T$, the equilibrium state relates to a sub-monolayer coverage, with $\theta_0 << 1$ at temperatures well below T_c. The vapor at $\ln(I/I_0) > -\varphi/2$ is in equilibrium with a dense adsorbate, whose coverage rapidly tends to one at low temperatures. The unique situation with $\ln(I_e/I_0) = -\varphi/2$ corresponds to the coexistence of the three phases: vapor with the equilibrium flux I_e, and two adsorbate phases with the equilibrium coverages θ_{1e} and θ_{2e}. When θ_{1e} and θ_{2e} are close to zero and one, respectively, we

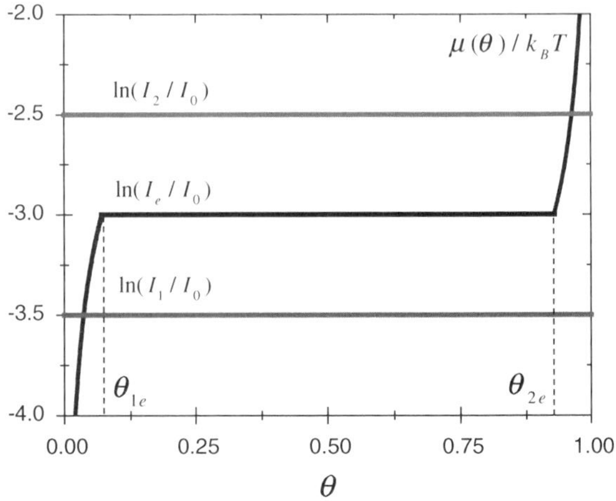

Fig. 1.16 Three possible equilibrium states in the adsorbate-vapor system: stable adatom system with $\theta_0 << 1$ at $I/I_0 < \mu_e$, stable monolayer at $I/I_0 > \mu_e$ and coexistence of vapor with two adsorbate phases having the equilibrium densities θ_{1e} and θ_{2e} at $I_e/I_0 = \mu_e$. The $\mu(\theta)$ curve corresponds to $\varphi^{\cdot} = 6$, the horizontal part corresponds to the density discontinuity

deal with a dilute adatom sea which is in equilibrium with vapor (due to a dynamic balance of adsorption and desorption processes) and, at the same time, coexists with the monoatomic surface step. The condition for the film growth clearly yields $I > I_e$, the inequality which should hold in any epitaxy process. In this case, supersaturation of the adatom sea will increase gradually with time due to deposition until it hits the threshold value necessary to initiate the film growth. The value of critical supersaturation depends on whether the substrate is planar or vicinal. In the former case, the surface steps should be created by nucleation, while in the latter case they are present initially.

Let us now briefly consider the most important elementary processes on a solid surface [3], illustrated in Figs. 1.17 and 1.18. The first process shown in Fig. 1.17 is the physical adsorption with probability χI per unit time per unit area. The χ is the adsorption coefficient showing which percentage of atoms from the impinging vapor flux actually becomes adatoms. In chemical deposition methods using precursors to deliver the required growth species onto the surface, χ is defined by the pyrolysis

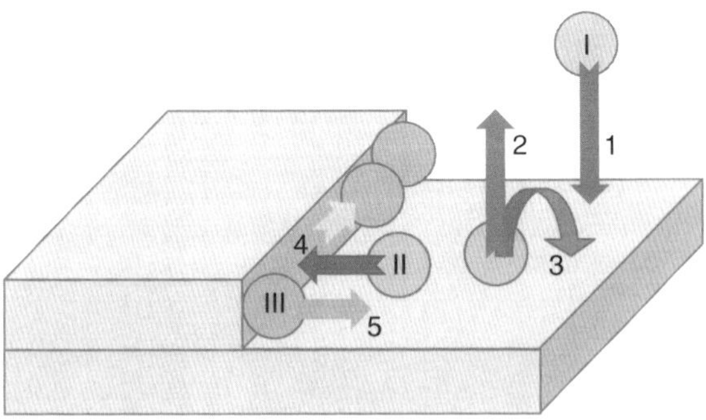

Fig. 1.17 *1* adsorption; *2* desorption; *3* surface diffusion; *4* incorporation to monoatomic step (2D condensation); *5* 2D evaporation from the step. *I* denotes the vapor atom, *II* is the adatom on a planar terrace and *III* is the incorporated adatom

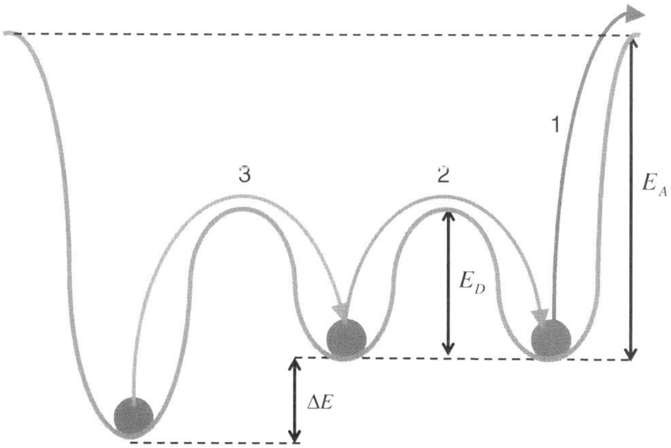

Fig. 1.18 Energy profile of a solid surface and the activation energies of different processes: E_A—desorption (*1*); E_D—surface diffusion (2); ΔE—escape from a kink (*3*)

efficiency at a given substrate temperature [61]. In molecular beam epitaxy [62, 63], it is primarily determined by the source temperature and the incident angle of the beam to the substrate. Usually, χ is very close to one, unless the beam approaches the glazing incidence. The second process shown in Fig. 1.17 is the desorption whose probability is characterized by the mean adatom lifetime on the surface before re-evaporation

$$\tau_A = \nu_A^{-1} \exp\left(\frac{E_A}{k_B T}\right). \tag{1.80}$$

Here, E_A is the activation energy for desorption (see Fig. 1.18), and ν_A is the pre-exponential factor which is of the order of thermal oscillation frequency of the lattice (10^{11}–10^{13} s^{-1}).

Whenever $I < I_e$ and, therefore, $n_1 < n_{1e}$, the adatom concentration obeys the simple kinetic equation

$$\frac{dn_1}{dt} = \chi I - \frac{n_1}{\tau_A}. \tag{1.81}$$

Obviously, n_1 tends to its maximum value $\chi I \tau_A$ with the relaxation time τ_A:

$$n_1(t) = \chi I \tau_A \left[1 - \exp\left(-\frac{t}{\tau_A}\right)\right] \tag{1.82}$$

Comparing (1.81) and (1.78) we obtain $I_0 = 1/(\chi\sigma\tau_A)$, where $\sigma$is the area of adsorption site.

The third elementary process in Fig. 1.17 is the surface diffusion of adatoms. In the case of strongly localized adsorbates, adatoms oscillate near the points of minima of the surface potential and jump to the neighboring sites. Probability of the jump per unit time is given by

$$t_D = \nu_D^{-1} \exp\left(\frac{E_D}{k_B T}\right). \tag{1.83}$$

Here, E_D is the activation energy for diffusion (see Fig. 1.18), and ν_D is the corresponding pre-exponential factor. The diffusion coefficient can be defined as

$$D = \frac{l_D^2}{\xi t_D}, \tag{1.84}$$

where l_D is the length of diffusion jump ($l_D^2 \cong \sigma$) and ξ is the lattice coordination number. The adatom diffusion length is the average square distance passed by the adatom on a bare substrate before it desorbs:

$$\lambda = \sqrt{D\tau_A}. \tag{1.85}$$

Usually, E_A is at least two times larger than E_D. Since both energies are much larger than $k_B T$, the adatom can make many diffusion jumps before its desorption. Equation (1.85) is sometimes used with the effective lifetime limited by incorporation into the surface steps or the boundaries of growing islands rather than just by desorption. In this case, the values of τ_A and λ depend on the surrounding and can no longer be attributed to a single adatom.

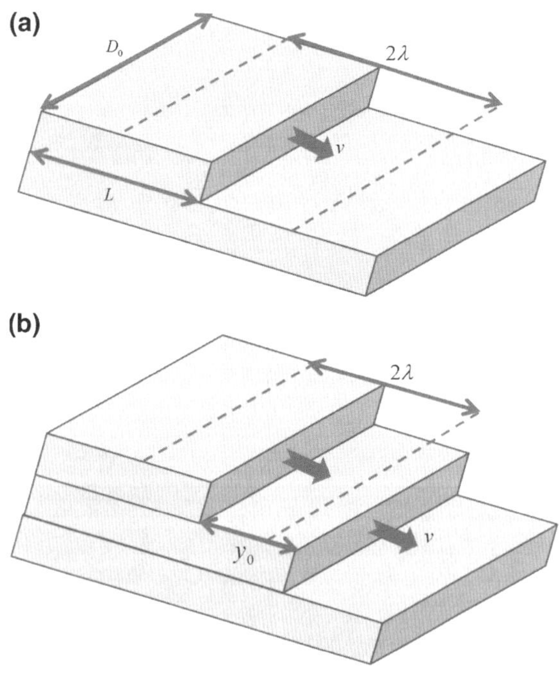

Fig. 1.19 Growth rate of a single step (**a**) and an echelon of steps (**b**)

Let us now see how a straight step grows from a supersaturated adatom "sea". As shown in Fig. 1.19a, a single step of the width $D_0 = const$ is fed from the terrace area $2D_0\lambda$. Assuming that all $2\chi ID_0\lambda$ atoms arriving from vapor to this collection area per unit time are subsequently attached to the step (process 4 in Fig. 1.17), and taking the reverse flux from the step to the terrace (process 5 in Fig. 1.17) in the form $2\chi I_e D_0\lambda$, the growth rate $v_s = dL/dt$ of the step is obtained as

$$v_s = 2\sigma\lambda\chi I_e\Phi. \tag{1.86}$$

Here, $\Phi = I/I_e - 1$ is the vapor supersaturation with respect to the surface. When rather than a single step, the echelon of steps is growing separated by the spacing y_0 (Fig. 1.19b), the steps will compete for the diffusion flux from the terraces. Now, a function of $y_0/(2\lambda)$ must enter the growth equation. This function, first derived by Burton, Cabrera and Frank [63], is given by $\tanh(y_0/2\lambda)$, modifying the previous equation to

$$v_s = 2\sigma\lambda\chi I_e \tanh\left(\frac{y_0}{2\lambda}\right)\Phi. \tag{1.87}$$

Obviously, the growth rate decreases toward smaller y_0 and is reduced to (1.86) at $y_0/(2\lambda) >> 1$.

Whenever the actual adatom concentration on the terrace is smaller than $\chi I\tau_A$, which usually happens due to nucleation of 2D islands on the terrace, the Φ in the above equations should be changed to the adatom supersaturation $\zeta = n_1/n_{1e} - 1$. In particular, in view of $\chi I_e = n_{1e}/\tau_A$, (1.86) should be re-written as

$$v_s = \frac{2\sigma\lambda n_{1e}}{\tau_A}\zeta. \tag{1.88}$$

Equation (1.88) for a straight step can be used to obtain a simplified formula for the growth rate of 2D island whose size is much larger than critical, where the curvature effects can be neglected. Indeed, the growth equation can be re-arranged as $\sigma^{-1}dS/dt = di/dt = P_{isl}(\lambda n_{1e}\zeta)/\tau_A$, where S is the island area, i is the number of atoms in the island and P_{isl} is the perimeter. Adatoms of the second layer are not taken into account here. For a circular island with $P_{isl} = 2\pi R$, the growth rate writes down as

$$\frac{di}{dt} = \frac{2\pi R D n_{1e}}{\lambda}\zeta. \tag{1.89}$$

The island growth rate in this approximation is proportional to R and, therefore, to $i^{1/2}$.

If we formally put $\lambda = l_D$, thus accounting only for such adatoms that can reach the island boundary in one diffusion jump, and use (1.84) for D together with the relationship $\pi R^2 = \sigma i$, (1.89) takes the well-known form [3, 65, 66]

$$\frac{di}{dt} \cong \frac{2\sqrt{\pi\sigma}l_D}{\xi t_D}n_{eq}i^{1/2}\zeta = \frac{2i^{1/2}}{\tau}\zeta. \tag{1.90}$$

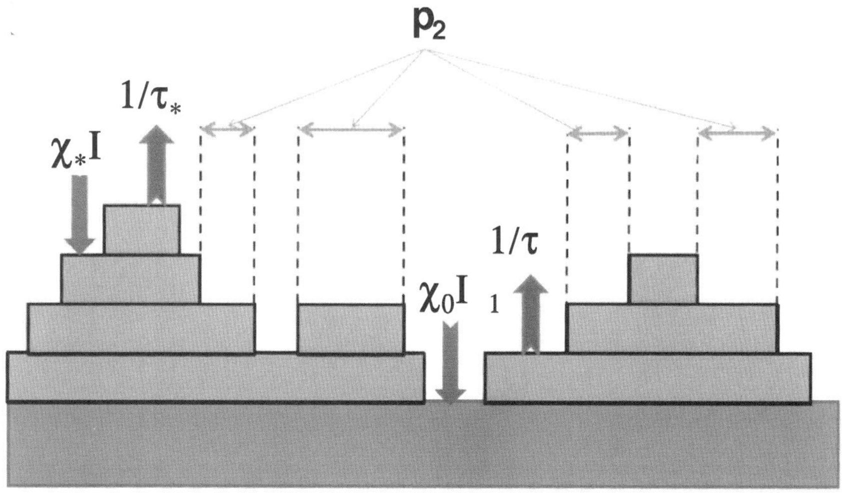

Fig. 1.20 BET model of multilayer adsorption with the parameters described in the text

Here, by definition, τ is the characteristic time of island growth which is of the order of t_D/θ_{1e}. Equation (1.90) pertains only if the diffusion length is much smaller than the island linear size. More general expressions for the growth rate of 2D islands will be given in the next section.

The approach presented above is principally limited to the monolayer adsorption. To access the case of thicker films, we consider the multilayer surface structure shown in Fig. 1.20. Different layers are labeled by indices $i = 1, 2, 3 \ldots$; the index $i = 0$ relates to the surface. Each layer is characterized by the coverage θ_i giving the percentage of the surface area covered by this layer (yielding $\theta_0 \equiv 1$). In absence of vacancies and overhangs in the film, the quantities

$$p_0 = 1 - \theta_1; \quad p_1 = \theta_1 - \theta_2; \quad \ldots \quad p_i = \theta_{i-1} - \theta_i \ldots \tag{1.91}$$

are the probabilities that the layer i is exposed to the vapor. Kinetic balance equations for p_i can be written down as

$$\frac{dp_0}{dt} = -\chi_0 \sigma I p_0 + \frac{p_1}{\tau_1}, \quad i = 0;$$

$$\frac{dp_i}{dt} = \chi_{i-1} \sigma I p_{i-1} - \frac{p_i}{\tau_i} - \chi_i \sigma I p_i + \frac{p_{i+1}}{\tau_{i+1}}, \quad i = 1, 2, 3 \ldots \tag{1.92}$$

The quantities χ_i and τ_i are the adsorption coefficients on the surface of layer i and the characteristic desorption times from layer i, respectively. The adsorption on the surface of layer $i - 1$ must be compensated by the desorption from layer i in equilibrium:

$$p_i = \chi_{i-1}\sigma I \tau_i p_{i-1} \quad , \quad i = 1, 2, 3, \ldots, \tag{1.93}$$

where we write p_i instead of p_i^{eq} for brevity.

Following the idea of Brunauer, Emmett, Teller [67], we now assume that the characteristics of all the layers beginning from the second one are identical. However, the first layer is different due to the influence of a dissimilar substrate surface: $\chi_0 \neq \chi_1 = \chi_2 = \chi_3 = \cdots \equiv \chi_*$; $\tau_1 \neq \tau_2 = \tau_3 = \tau_4 = \cdots \equiv \tau_*$. Introducing the normalized vapor flux b and the parameter c accounting for the difference of the first and the upper layers by definitions

$$b = \chi_* \sigma I \tau_*; \quad c = \frac{\chi_0 \tau_1}{\chi_* \tau_*}, \tag{1.94}$$

(1.93) simplifies to

$$p_i = cb^i p_0. \tag{1.95}$$

The probabilities p_i are normalized to one from geometrical considerations. The mean thickness of multilayer adsorbate h (measured in the units of a monolayer height) is the statistical average of i with probabilities p_i. Therefore,

$$\sum_{i=0}^{\infty} p_i = 1; \quad h = \sum_{i=0}^{\infty} i p_i. \tag{1.96}$$

Inserting (1.95) into (1.96) and performing the respective summation of geometrical progressions, one obtains the known two-parametric Brunauer-Emmett-Teller (BET) isotherm [67]:

$$h = \frac{cb}{(1-b)(1-b+cb)}. \tag{1.97}$$

Graphs of the film thickness h versus b at different c are plotted in Fig. 1.21. The case of $c = 1$ corresponds to the identical characteristics in the first and the upper layers, that is, the adsorption on the similar substrate. When $c << 1$, the film does not wet the surface due to a small adsorption coefficient or an enhanced desorption. The thickness of a non-wetting film is the lowest at given vapor pressure. The case of $c >> 1$ corresponds to a good wetting, where the first layer forms easily and the total thickness is limited by formation of the upper layers. The film thickness at a given pressure is much higher than in the previous cases. Since the thickness increases monotonously with the flux, the BET isotherm does not give any phase transition, i.e., describes a perfect multilayer adsorbate. Some generalizations of the BET model have been studied, for example, in [31], where the attractive interactions are introduced within the frame of the mean field theory in each layer and between the layers, while the entire adsorption process is studied kinetically. More details on the kinetic lattice gas models for adsorption phenomena can be found, for example, in [68–70].

1.7 Growth Rates

Growth rate of a nucleus depends strongly on its dimension (2D or 3D) as well as on the dimension and characteristics of supersaturated environment. We start the analysis with a simplified picture shown in Fig. 1.22. The nucleus considered is assumed to be a disc or a sphere of radius R fed from a supersaturated 2D or 3D environment, respectively. The radius is assumed as being sufficiently larger than the critical radius. This allows us to neglect the curvature effects. We thus consider the growth rates of essentially super-critical nuclei so that the derivative of formation energy defined by (1.65) with respect to i tends to $-\ln(\zeta+1)$. The vapor concentration equals n_1 at the distance Λ from the interface, and decreases towards the interface due to a diffusion flux onto the nucleus. The equilibrium concentration n_{1e} must be reached at the interface. The quantity Λ therefore plays a role of the characteristic length of spatial inhomogeneity in supersaturated environment.

Stationary diffusion equation with the corresponding boundary conditions

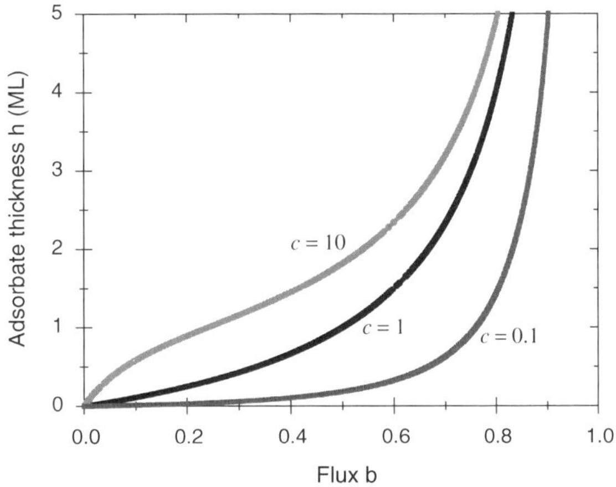

Fig. 1.21 BET isotherms at different c

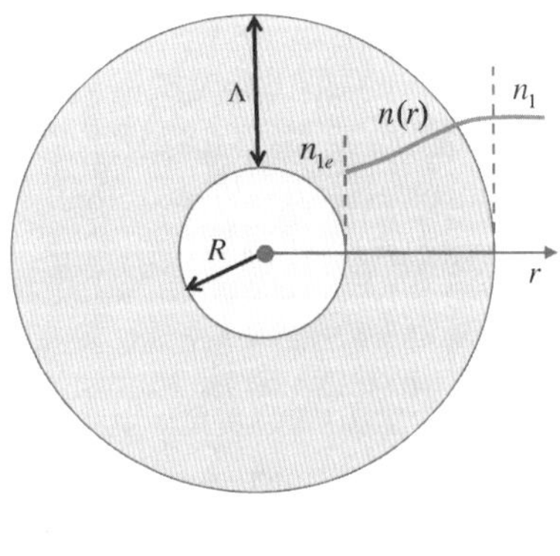

Fig. 1.22 Diffusion-induced growth of nucleus with the vapor concentration decreasing from n_1 at distance Λ to n_{1e} at the interface

$$\Delta n(r) = 0; \quad n(r = R) = n_{1e}; \quad n(r = R + \Lambda) = n_1 \tag{1.98}$$

describes the concentration profiles in absence of external fluxes; Δ denotes the Laplace operators in 2D ($d = 2$) or 3D ($d = 3$) space. Solutions to this equation are given by

$$n(r) = n_{1e} + \frac{(n_1 - n_{1e})}{\ln(1 + \Lambda/R)} \ln\left(\frac{r}{R}\right), \quad d = 2; \tag{1.99}$$

$$n(r) = n_{1e} + (n_1 - n_{1e})\left(1 + \frac{R}{\Lambda}\right)\left(1 - \frac{R}{r}\right), \quad d = 3. \tag{1.100}$$

The corresponding material fluxes onto the surface

$$\frac{di}{dt} = D2\pi R \left.\frac{dn}{dr}\right|_{r=R}, \quad d = 2; \tag{1.101}$$

$$\frac{di}{dt} = D4\pi R^2 \left.\frac{dn}{dr}\right|_{r=R}, \quad d = 3, \tag{1.102}$$

give the diffusion-induced growth rates. Straightforward calculation leads to the following results

$$\frac{di}{dt} = \frac{2\pi D n_{eq}\zeta}{\ln(1 + \Lambda/R)}; \quad d = 2; \tag{1.103}$$

$$\frac{di}{dt} = 4\pi D n_{eq}\zeta R\left(1 + \frac{R}{\Lambda}\right); \quad d = 3. \tag{1.104}$$

Despite its simplicity, our model illuminates well two major features. First, the growth rates of super-critical nuclei are proportional to the supersaturation, yielding the phase coexistence at $\zeta = 0$, growth at $\zeta > 0$ and evaporation at $\zeta < 0$. Second and less obvious, it is seen that the growth rates are proportional to the perimeter in 2D case and surface area in 3D case at $\Lambda << R$, where the nucleus does not disturb significantly the spatial homogeneity of the gas–vapor environment. This mode, corresponding to $di/dt \propto i^{1/2}$ in 2D and $di/dt \propto i^{2/3}$ in 3D space, is called the ballistic growth regime [3, 10, 15, 17, 50–53]. The opposite case of $\Lambda >> R$, where the inhomogeneity length is much larger than the particle linear dimension, relates to $di/dt \propto 1/\ln(\Lambda/R)$ in a 2D system. This weak logarithmic dependence at $\Lambda/R >> 1$ can often be neglected, yielding an almost size-independent growth rate of small enough islands. The growth rate of 3D island in the same limit is proportional to $i^{1/3}$. Such a growth is said to proceed in the diffusion regime [50–53, 69].

When the impingement and desorption processes are included, the diffusion equation for the adatom concentration around a 2D island writes down as

$$\lambda^2 \Delta n + n_{1\infty} - n = 0. \tag{1.105}$$

Here, λ is the adatom diffusion length and $n_{1\infty}$ is the concentration far away from the island. General solution to this equation is given by

$$n(r) = n_{1\infty} + c_1 I_0(r/\lambda) + c_2 K_0(r/\lambda), \tag{1.106}$$

with $I_0(z)$ and $K_0(z)$ as the modified Bessel functions of zero order, and the two constants determined by the boundary conditions. Since $I_0(z)$ increases at large z, we must put c_1 to zero to ensure the correct behavior far away from the island. Boundary condition at the interface remains the same as in (1.98), yielding

$$n(r) = n_{1\infty} - (n_{1\infty} - n_{1e}) \left[\frac{K_0\,(r/\lambda)}{K_0\,(R/\lambda)} \right]. \tag{1.107}$$

Calculation of the diffusion flux onto the island by means of (1.101) readily gives

$$\frac{di}{dt} = \frac{2\pi R D n_{1e} \zeta}{\lambda} \frac{K_1\,(R/\lambda)}{K_0(R/\lambda)}, \tag{1.108}$$

where $\zeta = n_{1\infty}/n_{1e} - 1$ and $K_1(z)$ is the modified Bessel function of the first order. Compared to (1.89), this expression contains the ratio of the two Bessel functions. The latter tends to one at $R/\lambda >> 1$ and to $\lambda/[R \ln(\lambda/R)]$ at $R/\lambda << 1$, reducing (1.108) to (1.89) and

$$\frac{di}{dt} = \frac{2\pi D n_{1e}}{\ln\,(\lambda/r)} \zeta, \tag{1.109}$$

respectively. The last expression is identical to (1.103) at $R/\Lambda << 1$ and $\Lambda = \lambda$. The characteristic scale of spatial inhomogeneity thus appears to be exactly equivalent to the adatom diffusion length, which is indeed a natural and well anticipated result.

Figure 1.23 shows the growth rates obtained from (1.108) with different diffusion lengths at otherwise identical parameters. Black lines are the approximations given by (1.109) and (1.89). It is clearly seen that the logarithmic approximation works very well everywhere at large $\lambda = 800$ nm, even if the radius is smaller than λ. The radius dependence of the growth rate is indeed weak. The linear approximation at small $\lambda = 50$ nm predicts correctly the slope angle, but is shifted downward compared to the exact expression. Therefore, (1.108) at $R/\lambda >> 1$ can be well fitted by a linear function with a certain additive correction.

For a spherical droplet in the ballistic growth regime in 3D space, the condensation rate equals the vapor flux impinging the entire droplet surface $S = 4\pi R^2$ per unit time, multiplied by the corresponding adsorption coefficient at the surface: $W_i^+ = \chi I S$. In isothermal conditions, the I is given by (1.76) at $T_v = T$ and $n_v = n_1$, because the vapor molecules now play the role of condensing monomers. This yields

$$W_i^+ = 4\pi R^2 \chi n_1 \sqrt{\frac{k_B T}{2\pi m}} = \frac{3 i^{2/3}(\zeta + 1)}{\tau}. \tag{1.110}$$

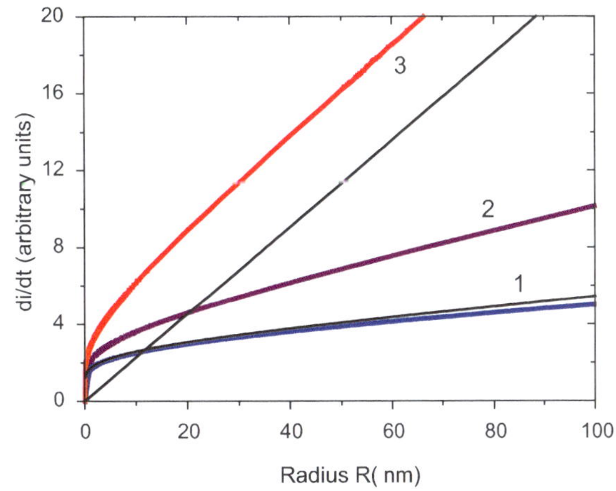

Fig. 1.23 Dependences of the growth rate on R at different $\lambda = 800$ nm (*curve 1*), 200 nm (*curve 2*), and 50 nm (*curve 3*). *Black lines* show approximations described in the text

The kinetic time constant τ, defined as $\tau^{-1} = \left(4\pi\Omega^2/3\right)^{1/3} \chi n_{1e} \left(k_B T/2\pi m\right)^{1/2}$, gives the characteristic time of droplet growth. The evaporation rate of super-critical droplets W_i^- must equal the condensation rate at $\zeta = 0$. The growth rate of super-critical droplets is therefore given by

$$\frac{di}{dt} = \frac{3i^{2/3}\zeta}{\tau}. \tag{1.111}$$

This expression is also valid for a crystal nucleus growing from a supersaturated vapor, where only the constant τ is changed.

The ballistic growth regime in 3D space holds as long as the nucleus does not disturb the gas–vapor environment. This is ensured by the following inequality:

$$\frac{\chi R}{l} << 1, \tag{1.112}$$

where l is the free path length of the vapor molecules. For larger nuclei, the growth proceeds in the diffusion regime. Generalized expression accounting for both mechanisms as well as the transition between them can be obtained in the form [69]

$$\frac{di}{dt} = \frac{3i^{2/3}\zeta}{\tau} \frac{1}{1 + (i/i_l)^{1/3}}. \tag{1.113}$$

Here, i_l denotes the number of molecules in the droplet having the radius l/χ. Obviously, this expression has the same asymptotic behaviors as those given by (1.104).

As follows from our analysis, the growth rates of 2D islands and 3D droplets obey the scaling law $di/dt \propto W_i^+ \propto i^\alpha$ in most cases, where the exponent α depends on the space dimensions of the metastable system ($d' = 2, 3$) and the nucleus ($d = 2, 3$), as well as on the mechanisms of mass transport into the nucleus. Let us now briefly

Table 1.2 Schematization of growth scenarios and growth indices for different systems

No	System description	d'	d	α	m	Invariant size ρ
1	Thin film growth at low temperatures	3	2	0	1	i
2	Growth of 2D islands from 2D adatom sea, ballistic regime	2	2	1/2	2	$i^{1/2}$
3	Growth of 2D islands from 2D adatom sea, diffusion regime	2	2	0	1	i
4	Condensation of 2D island from 3D environment	3	2	1	∞	$\ln i$
5	Growth of 3D islands from 2D system, ballistic regime	2	3	1/3	3/2	$i^{2/3}$
6	Growth of 3D islands from 2D system, diffusion regime	2	3	0	1	i
7	Condensation of 3D droplets from 3D vapor, ballistic regime	3	3	2/3	3	$i^{1/3}$
8	Condensation of 3D droplets from 3D vapor or liquid, diffusion regime	3	3	1/3	3/2	$i^{2/3}$

discuss other relevant growth scenarios [50]. As follows directly from (1.92) at i-independent χ_i and $\tau_i \to \infty$, the low temperature growth of multilayer thin film without surface diffusion and desorption proceeds at $\alpha = 0$. Growth of 3D surface islands from a 2D metastable system ($d' = 2, d = 3$) relates to $\alpha = 1/3$ in the ballistic regime (the case of the Stranski-Krastanow quantum dots [25]) and $\alpha = 0$ in the diffusion regime. A more exotic case of surface island fed from a vapor or liquid phase though its surface but growing in a 2D form ($d' = 3, d = 2$), corresponds to $\alpha = 1$.

Different growth regimes are summarized in Table 1.2, where the growth index m is introduced by definition

$$m = \frac{1}{1-\alpha}. \tag{1.114}$$

As discussed, the exponent α usually ranges from zero to one, yielding $1 \le m \le \infty$. The α values between 0 and 2/3 relate to $1 \le m \le 3$. At $\alpha < 1$, we write down the condensation rate in two equivalent forms:

$$W_i^+ = \frac{(\zeta+1)}{(1-\alpha)\tau} i^\alpha = \frac{m(\zeta+1)}{\tau} i^{(m-1)/m}. \tag{1.115}$$

At $\alpha = 1$, we do not introduce the factor $1-\alpha$ in the denominator. In view of $W_i^- = W_i^+(\zeta = 0)$, the growth rate of essentially super-critical particles, $di/dt =$

$W_i^+ - W_i^-$, equals to

$$\frac{di}{dt} = \frac{\zeta}{(1-\alpha)\tau} i^\alpha = \frac{m\zeta}{\tau} i^{(m-1)/m}. \tag{1.116}$$

The invariant size introduced in Table 1.2 is defined as follows: ρ is an i-dependent variable such that the regular growth rate of super-critical nuclei in terms of ρ is ρ-independent. As discussed in detail in Chap. 2, this variable is of paramount importance for the analytical description of different growth stages [3, 10, 17, 50–52]. For now, we just note that, due to $d\rho/dt = (d\rho/di)(di/dt)$, the invariant size is given by

$$\rho = i^{1-\alpha} = i^{1/m}, \alpha < 1; \quad \rho = \ln i, \alpha = 1. \tag{1.117}$$

These relationships with different growth indices are also presented in Table 1.2. It is seen that the invariant size equals the particle radius (measured in the units of interatomic distance) in systems no. 2 and 7. In terms of growth indices, 3D surface islands growing in the ballistic regime from a 2D system appear to be equivalent to 3D droplets consuming a 3D vapor in the diffusion regime. Using (1.116), the growth rate in terms of the invariant size takes the form

$$\frac{d\rho}{dt} = \frac{\zeta}{\tau}, \tag{1.118}$$

thus being proportional to the supersaturation with a certain kinetic time constant τ.

1.8 Rate Equations of Nucleation Theory

Classical nucleation theory is based on the discrete set of rate equations for the time-dependent concentrations n_i of nuclei containing i monomers [7–9, 27, 50–52, 72–74]. In the case of homogeneous nucleation, we consider the monomolecular condensation and evaporation processes

$$A_i + A_1 \Leftrightarrow A_{i+1}, \quad i = 1, 2, 3 \ldots \tag{1.119}$$

where A_i denotes the nucleus with i monomers and A_1 is the free monomer of a metastable phase. It is assumed that the probability of collision between two nuclei is negligible, as well as many-body interactions. This essential property is normally ensured by a low concentration of nuclei (in vapors) or their disability to move (in surface systems). As above, the rate constants for the direct and reversed process (1.119) are denoted as W_i^+ and W_{i+1}^-, respectively. The entire reaction chain is illustrated in Fig. 1.24a. For a given "level" $i \geq 2$, the concentration of nuclei containing i monomers increases due to the condensation on the nuclei with $i-1$ monomers and by the evaporation from the nuclei containing $i+1$ monomers, and decreases due to the condensation–evaporation processes on the nuclei with i monomers. The concentration of vapor changes in any of these condensation–evaporation processes. Production of dimer requires two monomers, while the decay of dimer creates two monomers. In most cases, the nucleation takes place under the dynamical conditions, where the total concentration of monomers can change in time, for example, due to a material influx into the system or desorption [3, 17, 50–52, 56]. To describe these processes, we introduce positive ("pumping") and negative ("sink") sources of monomers, denoted as P and $-Qn_1$, respectively. From balance considerations, the "quasi-chemical" set of rate equations for the concentrations $n_i(t) = [A_i]_t$ can be written down as

$$\frac{dn_1}{dt} = -2J_2 - \sum_{i=3}^{\infty} J_i + P - Qn_1 \tag{1.120}$$

$$\frac{dn_i}{dt} = J_i - J_{i+1}, \quad i = 2, 3, 4 \ldots; \tag{1.121}$$

$$J_i = W_{i-1}^{+} n_{i-1} - W_i^{-} n_i, \quad i = 2, 3, 4 \ldots \tag{1.122}$$

Obviously, J_i is the flux of monomers from the level $i-1$ to i. The condensation rate constants W_i^{+} are always proportional to the monomer concentration n_1: $W_i^{+} = n_1 k_i^{+}$, with k_i^{+} independent of n_1.

From (1.120) and (1.121), it is easily seen that the total number of monomers in vapor and nuclei is changed in time only by pumping:

$$\frac{dn_1}{dt} + \frac{d}{dt}\sum_{i=2}^{\infty} in_i = P - Qn_1, \tag{1.123}$$

and remains constant at zero sources. Integration leads to the integral equation of material balance

$$n_{tot}(t) - n_{1e} = \int_0^t dt' \left[P(t') - Q(t')n_1(t')\right] = n_1(t) - n_{1e} + \sum_{i=2}^{\infty} in_i(t). \tag{1.124}$$

Here, the moment $t = 0$ corresponds to the equilibrium monomer concentration and, therefore, to zero n_i for all $i \geq 2$. Dividing (1.124) to n_{1e}, the material balance takes the non-dimensional form

$$\Phi(t) = \zeta(t) + G(t). \tag{1.125}$$

The left hand side is the so-called "ideal supersaturation" [10, 17, 50–52] that would be reached in the system without nucleation: $\Phi(t) = n_{tot}(t)/n_{1e} - 1$. Its time dependence is entirely determined by the external fluxes. The right hand side is the sum of the actual vapor supersaturation at time t and the total number of monomers in the

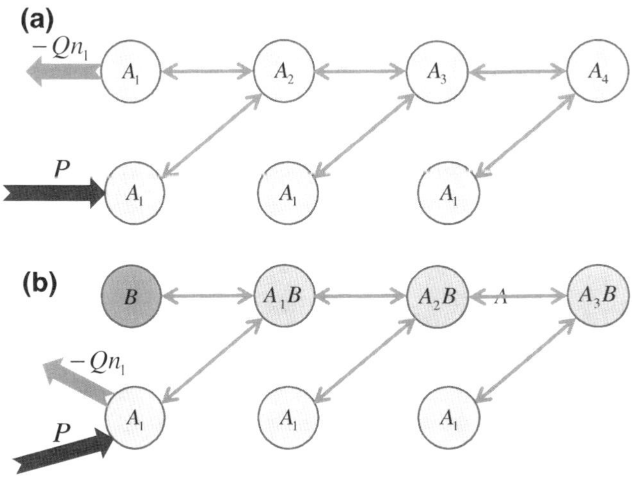

Fig. 1.24 Monomolecular reaction schemes for homogeneous (**a**) and heterogeneous (**b**) nucleation

condensed phase per unit volume, normalized to the equilibrium vapor concentration

$$G(t) = \frac{1}{n_{1e}} \sum_{i=2}^{\infty} i n_i(t). \tag{1.126}$$

The most common monomolecular reaction scheme describing heterogeneous nucleation [10, 73, 74] writes down as

$$A_i B + A_1 \Leftrightarrow A_{i+1} B, \quad i = 0, 1, 2, 3 \ldots \quad (A_0 B \equiv B). \tag{1.127}$$

Here, B denotes the condensation center or seed, for example, a charged ion in the gas-vapor system or a catalyst particle on the substrate surface; A_1 is the monomer and $A_i B$ with $i = 1, 2, 3 \ldots$ are the heterogeneous clusters consisting of i monomers and one seed. Similarly to the previous case, the reaction chain is schematized in Fig. 1.24b. The set of kinetic equations for the concentrations of monomers $n_A(t) = [A_1]_t$ and clusters $n_i(t) = [A_i B]_t$ (at $i = 0$, $n_0(t) = [B]_t$ is the concentration of free seeds) is given by

$$\frac{dn_A}{dt} = -\sum_{i=1}^{\infty} J_i + P - Q n_A; \tag{1.128}$$

$$\frac{dn_i}{dt} = J_i(1 - \delta_{io}) - J_{i+1}, \quad i = 0, 1, 2, 3 \ldots; \tag{1.129}$$

$$J_i = W_{i-1}^{+} n_{i-1} - W_i^{-} n_i, \quad i = 1, 2, 3 \ldots \tag{1.130}$$

As above, P and Q describe the positive and negative sources of monomers. Kroneker's symbol δ_{io} in (1.29) disables the decay of condensation seed B. The flux given by (1.130) has the same form as in (1.122), however, the heterogeneous rate constants might be different, particularly for smaller clusters. The condensation rates W_i^+ are proportional to the monomer concentration n_A: $W_i^+ = n_A k_i^+$, with k_i^+ independent of n_A.

Summing up (1.129) with different i, it is seen that the total concentration of heterogeneous clusters (including the free nucleation centers) is preserved in time, because it equals the total number of available seeds per unit volume at any time. This is the main difference from the homogeneous case, where the total number of nuclei is not conserved. Similarly to (1.123), the total number of the monomers in vapor and nuclei is changed in time only by external sources. Therefore, (1.128) and (1.129) yield two conservation laws

$$\sum_{i=0}^{\infty} n_i = n_B^{tot}; \quad \frac{dn_A}{dt} + \frac{d}{dt}\sum_{i=1}^{\infty} i n_i = P - Q n_A, \tag{1.131}$$

where n_B^{tot} is the total concentration of nucleation centers.

All fluxes must cancel in the quasi-equilibrium state at a given vapor concentration n_1, leading to the detailed balance

$$W_{i-1}^+ n_{i-1}^{eq} = W_i^- n_i^{eq}. \tag{1.132}$$

Here, the index i runs from 3 to infinity and from 2 to infinity in the homogeneous and heterogeneous cases, respectively. At $i = 2$, the principle of detailed balance yields $W_1^+ n_1 = W_2^- n_2^{eq}$ in the homogeneous case and $W_0^+ n_B = W_1^- n_1^{eq}$ in the heterogeneous case. Clearly, (1.132) also apply in the global equilibrium, where $n_1 = n_1^{eq}$ or $n_B = n_B^{eq}$ and i runs from 2 or 1. The quasi-equilibrium distribution (in the homogeneous case) can be presented in the form

$$n_i^{eq} = \prod_{k=2}^{i} (W_{k-1}^+ / W_k^-) n_1. \tag{1.133}$$

Comparing it to (1.70), we obtain the relationship between the formation energy and the ratios of evaporation–condensation rate constants:

$$F_i = \sum_{k=2}^{i} \ln\left(W_k^- / W_{k-1}^+\right). \tag{1.134}$$

Since the evaporation rates depend on many complex processes inside the nuclei and their independent determination is rather obscure, (1.132) is normally used the other way round. Indeed, while the W_i^- are generally unknown, the W_i^+ can be obtained by considering the mass transport into the particle, as demonstrated in

Sect. 1.8. The macroscopic formation energy is also well known, as discussed in Sect. 1.6. Hence, the principle of detailed balance enables the determination of the unknown evaporation rates by

$$W_i^- = W_{i-1}^+ \exp(F_i - F_{i-1}), \tag{1.135}$$

where F_i is given by (1.65) at $i >> 1$.

In practice, the macroscopic description of nucleation process applies beginning from $i = i_c - \Delta i_c$, where i_c is the number of monomers in the critical nucleus and Δi_c is the half-width of the near-critical region, the quantities defined in Sect. 1.6. As shown in Sect. 1.8, the W_i^+ is a power law function of i, so that the W_{i-1}^+ in (1.135) can be changed to W_i^+ with a negligible relative error of $1/i$. As for the exponential term, the small parameter of the Taylor expansion

$$F_i - F_{i-1} = \frac{dF(i)}{di} - \frac{1}{2}\frac{d^2F(i)}{di^2} + \cdots , \tag{1.136}$$

actually equals $1/\Delta i_c$ in the near-critical region $i_c - \Delta i_c < i < i_c + \Delta i_c$, where the relative error of neglecting the higher order terms is the largest. It is also clear that the relative error of substituting the integer i to real i in $\exp(-F_i)$ and, therefore, in quasi-equilibrium distribution, is smaller than $1/\Delta i_c$. Hence, (1.135) can be reduced to

$$W^-(i) = W^+(i) \exp[dF/di] \tag{1.137}$$

with small relative error in the range of interest at $1/\Delta i_c << 1$. The parameter $1/\Delta i_c$ is one of the two small parameters of classical nucleation theory, as discussed later on.

Further, using (1.132) in (1.130), we obtain a very useful representation for the flux:

$$J_i = W_{i-1}^+ n_{i-1}^{eq} (\tilde{n}_{i-1} - \tilde{n}_i). \tag{1.138}$$

Here, $\tilde{n}_i = n_i/n_i^{eq}$ is the concentration of nuclei of a given size, normalized to the corresponding quasi-equilibrium concentration.

1.9 Exact Solutions

Let us now consider the unique cases where the discrete set of rate equations can be solved exactly [50]. Although restricted to the specific choice of rate constants, these solutions are relevant for the foregoing analysis of real size spectra. With neglect of non-linearity caused by the time dependence of monomer concentration and in absence of evaporation, (1.129) can be simplified to

$$dp_0/dt = -p_0;$$

$$dp_i/dt = i^{\alpha} p_{i-1} - (i+1)^{\alpha} p_i, \quad i = 1, 2, 3 \ldots \tag{1.139}$$

Here, $p_i(t)$ are the normalized $(\sum_{i=0}^{\infty} p_i(t) = 1)$ probabilities $p_i(t)$ to find a particle with i monomers at time t. The normalized time t is now measured in the units of the characteristic time of particle growth. The growth rates di/dt of differently sized particles equal i^{α} with the exponent α ranging between 0 and 1, as discussed in Sect. 1.7. The model given by (1.139) can be called the linear irreversible growth at a constant supersaturation. The initial condition

$$p_i(t = 0) = \delta_{io} \tag{1.140}$$

means that all particles at $t = 0$ are concentrated at state $i = 0$ with zero size and probability $p_0(t = 0) = 1$. In other words, we are searching for the solution for the Green function describing the time evolution of the initial delta-like package. Clearly, the time-dependent mean number of atoms in the nuclei (the mean size for brevity), $i_*(t)$, and the dispersion, $\psi(t)$, are given by

$$i_*(t) = \sum_{i=0}^{\infty} i p_i(t); \quad \psi(t) = \sum_{i=0}^{\infty} [i - i_*(t)]^2 p_i(t). \tag{1.141}$$

We now note that the discrete system of (1.139) has the unique exact solutions at integer $\alpha = 0$ and $\alpha = 1$, obtained in [74] (in a more general case) and containing some of earlier results by Montroll and Shuler [75], Rankin and Light [76] and Glauber [77] as the particular cases. The solutions to (1.139) with the initial condition (1.140) at $\alpha = 0$ are given by the Poisson distribution:

$$p_i(t) = e^{-t} \frac{t^i}{i!}. \tag{1.142}$$

This can be easily proved by the direct insertion of (1.142) into (1.139) at $\alpha = 0$, while meeting the initial condition is evident. For this distribution, $i_* = \psi = t$. At large t (and i), the Poisson spectrum asymptotically tends to the Gaussian

$$p(i, t) = \frac{1}{\sqrt{2\pi t}} \exp\left[-\frac{(i-t)^2}{2t}\right], \tag{1.143}$$

the property seen directly by applying the Stirling formula $i! \cong \sqrt{2\pi i}(i/e)^i$ to (1.142) at large i.

The second order continuum approximation of (1.139) at $\alpha = 0$ is obtained simply by the Taylor expansion of their right hand side at large i:

$$p_{i-1} - p_i = \left[\exp\left(-\frac{\partial}{\partial i}\right) - 1\right] p_i \cong -\frac{\partial p(i, t)}{\partial i} + \frac{1}{2} \frac{\partial^2 p(i, t)}{\partial i^2}. \tag{1.144}$$

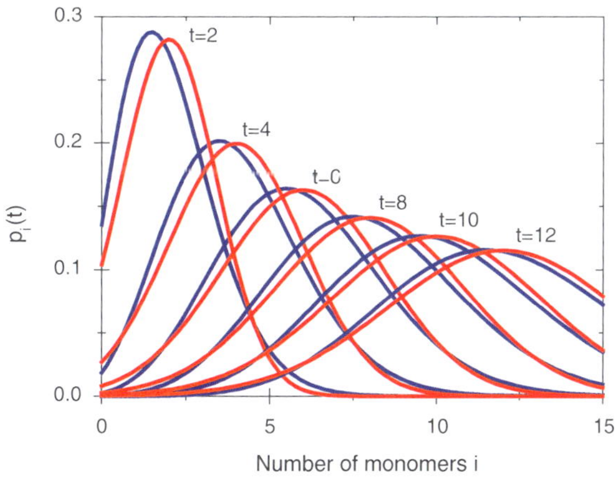

Fig. 1.25 Comparison between the Poisson (*blue curves*) and Gaussian (*red curves*) distributions at different t. Continuum Poisson spectra are guide for the eye only, since the Poisson distribution is defined only for integer i

The resulting Fokker-Plank type kinetic equation for the continuum size spectrum has the form

$$\frac{\partial p(i,t)}{\partial t} = -\frac{\partial p(i,t)}{\partial i} + \frac{1}{2}\frac{\partial^2 p(i,t)}{\partial^2 i};$$

$$p(i, t = 0) = \delta(i). \tag{1.145}$$

It is noteworthy that the diffusion-like spreading Gaussian given by (1.143) is the exact solution to the continuum (1.145). This is indeed the natural and well-known result for the diffusion-induced broadening of a delta-like package. Therefore, the discrepancy between the Poisson and Gaussian distributions originates as the difference in the exact solutions to the discrete set of kinetic equations and its continuum form. This difference rapidly becomes negligible toward larger i, as demonstrated in Fig. 1.25.

Very importantly, the system of (1.139) at $\alpha = 0$ describes exactly the low temperature growth of vapor-deposited thin film shown in Fig. 1.20, where the desorption and surface diffusion are negligible. The growth on the similar surface in homoepitaxy proceeds at the rate $V = \chi\sigma I$ (in monolayers per second (ML/s)). All layers $i = 0, 1, 2\ldots$ are identical. Such a growth leads to the Poissonian surface with the probabilities to find a random point of surface relief at the height $h = i$ MLs given by

$$p_i(t) = \exp[-\overline{h}(t)]\frac{\overline{h}^i(t)}{i!} \cong \frac{1}{\sqrt{2\pi\overline{h}(t)}}\exp\left[-\frac{(i-\overline{h}(t))^2}{2\overline{h}(t)}\right]. \tag{1.146}$$

The mean film thickness $\overline{h}(t) = \psi(t) = Vt$ equals the squared surface roughness. The last (1.146) applies for thick enough film with $\overline{h} >> 1$.

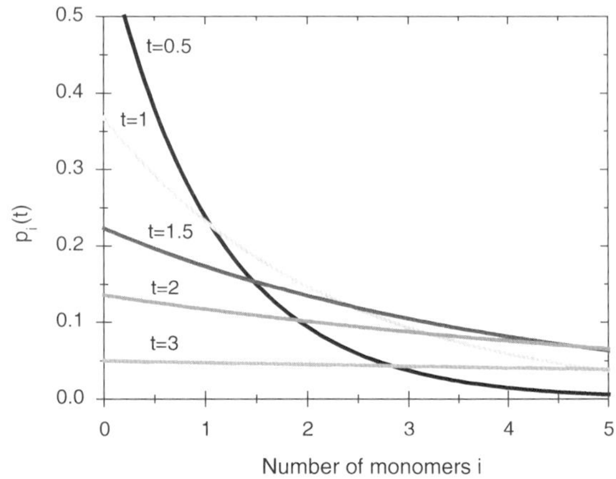

Fig. 1.26 Time evolution of geometrical distribution

Exact solutions to the discrete set of (1.139) at $\alpha = 1$ with the initial condition (1.140) are given by the geometrical distribution

$$p_i(t) = \frac{1}{[i_*(t)+1]}\left[\frac{i_*(t)}{i_*(t)+1}\right]^i, \tag{1.147}$$

with the time-dependent mean size $i_*(t) = \exp(t) - 1$. The dispersion of the geometrical distribution is given by $\psi = i_*(1 + i_*)$. This distribution is much wider than the Poisson spectrum or the spreading Gaussian with the same mean size. Time evolution of this distribution, presented in Fig. 1.26, is distinctly different from the previous case at $\alpha = 0$. The Poissonian spectra are almost symmetrical and are well centered near the most representative size, whereas the geometrical distribution reaches its maximum at zero size at any time. In terms of variable i, the spectra shown in Fig. 1.26 do not feature any propagation of the most representative size toward larger i, but rather rapidly become flat already at $t = 3$.

Neglect of evaporation and the time dependence of monomer concentration is not essential for obtaining the exact solutions to the growth equations. The critical assumption regards the form of kinetic rate constants. The exact solution can be obtained in a simple analytical form when the rate constants are linear in i [74]. Let us show this in the case of heterogeneous nucleation described by (1.128–1.130). In the following, it is convenient to consider the normalized size distribution $f_i(t) = n_i(t)/n_B^{tot}$, and the normalized monomer concentration $f_A(t) = n_A(t)/n_B^{tot}$. As discussed, the condensation rate constants are proportional to the monomer concentration and can be put in the form $W_i^+ = f_A(t)k_i^+$, where k_i^+ is time-independent. Changing the notation of W_i^- to k_i^-, (1.128–1.130) take the form

$$\frac{df_A}{dt} = -\sum_{i=1}^{\infty} J_i + P - Qf_A; \tag{1.148}$$

$$\frac{df_i}{dt} = J_i(1 - \delta_{io}) - J_{i+1}, \quad i = 0, 1, 2, 3 \ldots; \tag{1.149}$$

$$J_i = k_{i-1}^+ f_A f_{i-1} - k_i^- f_i; \tag{1.150}$$

$$f_i(t = 0) = \delta_{io}; \quad f_A(t = 0) = M(0), \tag{1.151}$$

with the obvious re-normalization of influx P. The quantity $M(0) = n_A^{tot}(0)/n_B^{tot}$ is the ratio of the total number of monomers to the number of available condensation seeds at $t = 0$. The two conservation laws for this system are given by

$$\sum_{i=0}^{\infty} f_i(t) = 1; \quad \frac{di_*}{dt} + \frac{df_A}{dt} = P - Qf_A(t), \tag{1.152}$$

where $i_*(t) = \sum_{i=0}^{\infty} if_i(t)$ is the mean size of clusters.

We now consider the case of i-linear rate constants

$$k_i^+ = k^+(1 + \delta i); \quad k_i^- = k^- i, \tag{1.153}$$

where δ is an arbitrary positive constant. Since the direct integration of kinetic equations is much less obvious than in the previous cases, we introduce the generating function for probabilities $f_i(t)$ by definition

$$f(z, t) = \sum_{i=0}^{\infty} f_i(t) z^i. \tag{1.154}$$

Obviously, $f_i(t)$ are obtained as the coefficients of Taylor expansion for $f(z, t)$ at $z = 0$. From its definition, the generating function features the following important properties:

$$f(1, t) = 1; \quad \left.\frac{\partial f(z, t)}{\partial z}\right|_{z=1} = i_*(t). \tag{1.155}$$

Differentiating $f(z, t)$ with respect to time and using (1.149) and (1.150) leads to the following equation in the general case

$$\frac{\partial f(z, t)}{\partial t} = (z - 1) f_A(t) \sum_{i=0}^{\infty} k_i^+ f_i(t) z^i + \left(\frac{1}{z} - 1\right) \sum_{i=1}^{\infty} k_i^- f_i(t) z^i. \tag{1.156}$$

This is useful only if the right hand side can be expressed through $f(z, t)$ and its derivatives with respect to z, resulting in the closed equation in partial derivatives for $f(z, t)$. This is exactly what happens when the rate constants are given by (1.153). In this case, (1.156) is reduced to the first order partial differential equation of the form

$$\frac{\partial f(z,t)}{\partial t} + (z-1)\left[k^- - \delta k^+ f_A(t) z\right] \frac{\partial f(z,t)}{\partial z} = (z-1)k^+ f_A(t) f(z,t). \quad (1.157)$$

Differentiating this with respect to z and putting $z = 1$, we get the differential equation for the mean size. Together with the second (1.152), it forms the closed system for the determination of $f_A(t)$ and $i_*(t)$ at given monomer sources:

$$\begin{aligned} &\frac{di_*(t)}{dt} + [k^- - \delta k^+ f_A(t)] i_*(t) = k^+ f_A(t); \\ &\frac{di_*(t)}{dt} + \frac{df_A(t)}{dt} = P(t) - Q(t) f_A(t); \\ &f_A(t=0) = M(0); \quad i_*(t=0) = 0. \end{aligned} \quad (1.158)$$

Solution to (1.157) can be found in analytical form, because one of its first integrals is obtained from the special type of integrable Ricatti equation having the partial solution $z = 1$ [74]. In the case of mono-dispersive initial condition $f(z, t = 0) = 1$, the solution is given by

$$f(z,t) = [1 + \delta i_*(t)(z-1)]^{-1/\delta}. \quad (1.159)$$

More generally, this solution holds whenever the initial condition to (1.157) has the form (1.159) with arbitrary mean size $i_*(0)$ at $t = 0$.

The Taylor expansion of the generating function given by (1.159) near $z = 0$ produces the Polya distribution of probabilities $f_i(t)$:

$$f_i(t) = [1 + \delta i_*(t)]^{-1/\delta} \frac{(1+\delta)(1+2\delta)\ldots(1+(i-1)\delta)}{i!} \left[\frac{i_*(t)}{1+\delta i_*(t)}\right]^i. \quad (1.160)$$

We thus have proved the following theorem:

If the initial conditions to the discrete set of kinetic (1.148–1.150) with the size-linear rate constants (1.153) are given by the Polya distribution with arbitrary mean size $i_*(0)$, its form remains time-invariant during growth with only the mean size changing in time, the mono-dispersive initial distribution with $i_*(0) = 0$ being the particular case.

This interesting property is called the canonical invariance of the Polya distribution of probabilities. At $\delta = 1$, (1.160) is reduced to the geometrical distribution given by (1.147). At $\delta \to 0$, it is transformed to the Poisson distribution (1.142), where t is changed to $i_*(t)$. The dispersion of the Polya distribution is related to the mean size as

$$\psi(t) = i_*(t)[1 + \delta i_*(t)]. \quad (1.161)$$

The previous results therefore follow from (1.160) and (1.161) in the limit cases.

The time evolution of the mean size is primarily determined by the monomer fluxes. In particular, as follows from the second (1.158), the equilibrium monomer concentration at $t \to \infty$ is given by

$$f_A^{eq} = f_A(\infty) = P(\infty)/Q(\infty), \tag{1.162}$$

provided that the quantity in the right hand side remains finite at large t. From the first (1.158), we get the large time asymptote of the mean size

$$i_*(\infty) = \frac{k^+ f_A(\infty)}{k^- - \delta k^+ f_A(\infty)}. \tag{1.163}$$

This value is finite at $P(\infty)/Q(\infty) < k^-/\delta k^+$, the inequality giving the condition for the existence of the asymptotic equilibrium state with a finite mean size. Otherwise, the mean size increases infinitely at $t \to \infty$ and the global equilibrium is never reached. In the former case at a finite $i_*(\infty)$, the equilibrium Polya size distribution obeys the principle of detailed balance

$$k^+[1 + \delta(i-1)]f_A(\infty)f_{i-1}(\infty) = k^- i f_i(\infty). \tag{1.164}$$

Interestingly, this equation shows that the equilibrium Polya distribution will be established at $t \to \infty$ for a more general system with the rate constants $k_i^+ = k^+(1 + \delta i)Y(i)$, $k_i^- = k^- i Y(i-1)$. Here, $Y(i)$ is an arbitrary function of i that cancels in (1.164).

The system of (1.158) can be integrated exactly in some particular cases. For example, in the isolated system with $P = Q = 0$, the total number of monomers is conserved in time: $f_A(t) + i_*(t) = M = const$. The time evolution of the mean size is then determined by the non-linear equation

$$\frac{di_*(t)}{dt} = -\delta k^+ i_*^2(t) + (\delta k^+ M - k^+ - k^-)i_*(t) + k^+ M. \tag{1.165}$$

Solution to (1.165) with an arbitrary initial $i_*(0)$ has the form

$$i_*(t) = \frac{a_1 + |a_2| \exp[-\delta k^+(a_1 + |a_2|)t]}{1 - C\exp[-\delta k^+(a_1 + |a_2|)t]} \tag{1.166}$$

with

$$a_{1,2} = \frac{1}{2}\left(M - \frac{k^+ + k^-}{\delta k^+}\right) \pm \frac{1}{2}\left[\left(M - \frac{k^+ + k^-}{\delta k^+}\right)^2 + \frac{4M}{\delta}\right]^{1/2};$$

$$C = 1 - \frac{a_1 + |a_2|}{i_*(0)}. \tag{1.167}$$

It is seen that the mean size tends to its equilibrium value a_1 at $t \to \infty$, while the monomer concentration tends to $M - a_1$. At $\delta = 0$, (1.165) becomes linear in $i_*(t)$, with the corresponding solutions given by

$$f_A(t) = c^+M + [f_A(0) - c^-M]\exp[-(k^+ + k^-)t)];$$

$$i_*(t) = c^-M - [f_A(0) - c^-M]\exp[-(k^+ + k^-)t)]. \tag{1.168}$$

Here, $c^{\pm} = k^{\pm}/(k^+ + k^-)$ are the relative values of the condensation and evaporation rate constants. Obviously, the mean size and the monomer concentration tend to c^+M and c^-M, respectively, regardless of the initial conditions.

Finally, the solutions to the linear system of (1.158) at $\delta = 0$ with the time-independent monomer sources $P = const$, $Q = const$ and arbitrary initial conditions can be obtained in the form

$$f_A(t) = \frac{P}{Q} + C_1\exp(-\omega_1 t) + C_2\exp(-\omega_2 t);$$

$$i_*(t) = \frac{k^+P}{k^-Q} + \left(\frac{\omega_2}{k^-} - 1\right)C_1\exp(-\omega_1 t) + \left(\frac{\omega_1}{k^-} - 1\right)C_2\exp(-\omega_2 t). \tag{1.169}$$

The constants are determined by

$$\omega_{1,2} = \frac{1}{2}\left(k^+ + k^- + Q\right) \mp \frac{1}{2}\left[\left(k^+ + k^- + Q\right)^2 - 4k^-Q\right]^{1/2};$$

$$C_1 = M(0) - \frac{\omega_2[f_A(0) + P/Q] - P - k^-M(0)}{\omega_2 - \omega_1};$$

$$C_2 = \frac{\omega_2 f_A(0) + \omega_1 P/Q - P - k^-M(0)}{\omega_2 - \omega_1}. \tag{1.170}$$

Obviously, $\omega_2 > \omega_1 > 0$, yielding the asymptotic equilibrium values $f_A(\infty) = P/Q$ and $i_*(\infty) = (k^+P)/(k^-Q)$, which is consistent with (1.162) and (1.163) at $\delta = 0$.

To summarize the results of this section, the exact solutions to the discrete kinetic equations of nucleation theory can be obtained in the case of i-linear condensation-evaporation constants. The size spectrum in terms of i variable features a remarkable time invariance of the Polya distribution. The form of this distribution is preserved in time, and depends on time only through the mean size of particles. At $\delta = 0$, the Polya spectrum takes the Poissonian form, which is rapidly converted to the Gaussian as the particles grow. At $\delta = 1$, it is reduced to the geometrical distribution. The dispersion of size spectrum increases with δ, featuring the linear dependence on the mean size at $\delta = 0$ and the quadratic dependence $i_*(1 + \delta i_*)$ at $\delta > 0$. The distribution shape also changes drastically with increasing δ. While the Poissonian and Gaussian spectra are centered near the most representative size (which is very close to the mean size for the Poisson distribution and equals the mean size for the symmetrical Gaussian), the geometrical distribution is essentially asymmetric. Its most representative size equals zero at any time, so that the spectrum rapidly flattens in terms of i variable as the particles grow.

The exact solutions presented here can be very useful in analyzing the model systems such as the low temperature thin films, vapor clusters in supersonic free jet expansions, population dynamics and many others, as well as for checking the validity of various approximate analytical methods, as discussed later on. They can describe quite well the irreversible growth at high enough supersaturations. However, the model (1.153) have one major difference from most real systems, whose typical feature is the maximum of formation energy at the critical size, the situation shown in Fig. 1.13. Indeed, since the condensation constant is positive and the evaporation constant starts from zero at $i = 0$, the curves k_i^+ and k_i^- can either never meet or intercept at a certain point i_0 such that $k_i^+ > k_i^-$ at $i < i_0$ and $k_i^+ < k_i^-$ at $i > i_0$. The condition $k_i^+ > k_i^-$ at any i corresponds to the irreversible growth, as described above. The interception case relates to the stable nucleus with i_0 monomers, since smaller particles tend to grow and larger particles tend to evaporate. This is exactly opposite to the critical size of nucleation theory i_c, which is in unstable equilibrium with a supersaturated environment. This yields the thermodynamically preferred evaporation at $i < i_c$ and growth at $i > i_c$. Therefore, the linear model for the rate constants cannot adequately describe the nucleation instability where the particles surpass the nucleation barrier due to thermodynamic fluctuations and start growing irreversibly only after entering the super-critical region.

1.10 Continuum Approximation

In the vast majority of applications, we deal with large enough nucleus consisting of many monomers. The continuum approximation at $i >> 1$ [3, 7–19, 22–25, 50–52] converts the infinite set of ordinary differential equations to just one second order partial differential equation, which is clearly of great help for theoretical analysis. Let us write down the discrete kinetic (1.121) and (1.122) in the form

$$\frac{dn_i}{dt} = W_{i-1}^+ n_{i-1} + W_{i+1}^- n_{i+1} - (W_i^+ + W_i^-) n_i. \tag{1.171}$$

Continuum approximation is obtained simply by the Taylor expansion of the non-diagonal terms in the right hand side around i and leaving only the first and the second order derivatives with respect to i:

$$\begin{aligned} W_{i-1}^+ n_{i-1} &\cong W_i^+ n_i - \frac{\partial}{\partial i}(W_i^+ n_i) + \frac{1}{2}\frac{\partial^2}{\partial i^2}(W_i^+ n_i); \\ W_{i+1}^- n_{i+1} &\cong W_i^- n_i + \frac{\partial}{\partial i}(W_i^- n_i) + \frac{1}{2}\frac{\partial^2}{\partial i^2}(W_i^- n_i). \end{aligned} \tag{1.172}$$

As discussed, the condensation rate is usually a power law function of i, while the evaporation rate is related to W_i^+ by (1.137). Hence, the differentiation of the rate constants gives the multiplying factor of the order of $1/i$ in the corresponding terms.

This factor is of the order of $1/i_c$ in the near-critical region. If n_i was the quasi-equilibrium distribution, its derivatives would decrease as $1/\Delta i_c$. The same factor appears upon differentiating the Gaussian with the half-width Δi_c. Therefore, the small parameter of the expansions given by (1.172) is $1/\Delta i_c$. The Taylor expansion of kinetic constants is $i_c/\Delta i_c$ times more accurate than that of n_i in typical cases.

Inserting (1.172) into (1.171) and introducing the continuum distribution function $n(i, t)$ that equals $n_i(t)$ in the discrete integer points, we arrive at the Fokker-Plank type kinetic equation of the form

$$\frac{\partial n(i,t)}{\partial t} = -\frac{\partial}{\partial i}\left[A(i)n(i,t) - B(i)\frac{\partial n(i,t)}{\partial i}\right]. \tag{1.173}$$

Kinetic coefficients at the first and second derivative with respect to size

$$A(i) \equiv \frac{di}{dt} = W^+(i)[1 - \exp(dF(i)/di)];$$
$$B(i) = \frac{1}{2}W^+(i)[1 + \exp(dF(i)/di)] \tag{1.174}$$

describe the regular growth rate and the kinetic fluctuations, respectively. Equation (1.173) with coefficients (1.174) is also called the Becker–Doering–Zeldovich equation. In the essentially super-critical region $i > (5-6)i_c$, where $dF(i)/di \cong -\ln(\zeta + 1)$, the kinetic coefficients are simplified to

$$A(i) = \frac{W^+(i)\zeta}{\zeta + 1} = \frac{m\zeta}{\tau}i^{(m-1)/m};$$
$$B(i) = \frac{W^+(i)(\zeta + 2)}{2(\zeta + 1)} = \frac{m(\zeta + 2)}{2\tau}i^{(m-1)/m}. \tag{1.175}$$

The last expressions in the right hand side apply when the condensation rate is given by the scaling (1.115). Equations (1.175) do not account for the interactions of the critical size with the spectrum and are not applicable either in the near-critical region or at the late Ostwald ripening stage. At $\Delta i_c/i_c << 1$ (this inequality will be discussed in more detail later on), the kinetic coefficients $A(i)$, $B(i)$ in (1.173) can be taken out of derivatives, at least at the regular growth stage.

Continuum kinetic equation of nucleation theory (1.173) contains the unknown supersaturation entering its coefficients as given by (1.174). Therefore, it should be coupled with the integral equation of material balance. In the case of homogeneous nucleation, the corresponding (1.125) and (1.126) are re-written as

$$\Phi(t) = \zeta(t) + G(t); \quad G(t)\frac{1}{n_{1e}}\int_0^\infty diin(i,t). \tag{1.176}$$

Here, as above, the function $G(t)$ represents the total number of monomers distributed in all nuclei. Equations (1.173) and (1.176) give the closed system of equations for finding $n(i, t)$ and $\zeta(t)$ at the known monomer sources.

1.11 Stationary State

It has already been shown that classical nucleation theory applies when the nucleation barrier is high enough. In this case, we can divide the entire size axis into three regions with distinctly different behaviors, as shown in Fig. 1.27. Due to the smallness of fluctuations, the quasi-equilibrium state at a given supersaturation should be observed in the subcritical region $i \le i_c - \Delta i_c$ [12]:

$$n_i = n_i^{eq}, \quad i \le i_c - \Delta i_c. \tag{1.177}$$

The near-critical region $i_c - \Delta i_c < i < i_c + \Delta i_c$ is expected to quickly evolve to the stationary state n_i^s such that the particle flux through it is size-independent: $J_i = J^s \equiv J$. In fact, the stationary flux propagates even further toward $i_+ \sim (5-6)i_c$, the region where the curvature effects become negligible and the derivative of the formation energy with respect to i can be approximated by a constant (see Fig. 1.27). It is this stationary state that will be studied in detail in this section. In the essentially super-critical region $i > i_+$, the nuclei grow irreversibly at the rate given by the first (1.175). We underline that the stationary state relates to a particular supersaturation value and changes almost instantaneously with ζ. This important feature holds because the time needed to establish the stationary size distribution in near-critical region at a given ζ is much shorter than the macroscopic time required to change the supersaturation itself.

The stationary flux through the near-critical region gives the nucleation rate [3, 7–10, 17, 27], also called the Zeldovich nucleation rate [12], probably the most important characteristic of any nucleation process. At a given supersaturation, J equals the number of irreversibly growing nuclei emerging in the unit volume (or on the unit surface area) per unit time. These super-critical nuclei emerge due to thermodynamic fluctuations in the subcritical region, surpassing the nucleation barrier at the rate J.

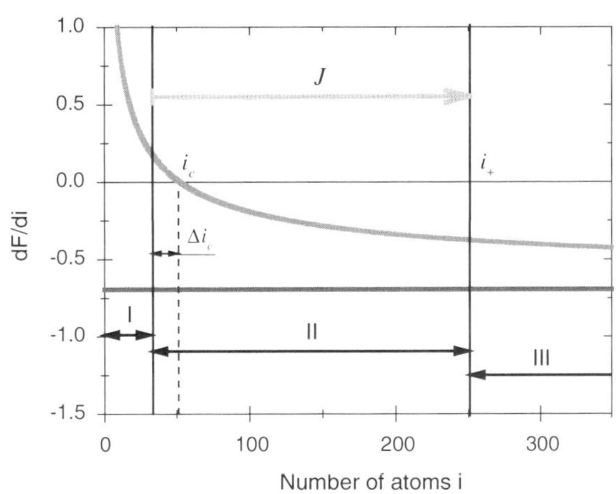

Fig. 1.27 Schematic representation of the subcritical (the quasi-equilibrium) region *I*, the extended near-critical (the fluctuational) region *II* and the essentially super-critical (the growth) region *III*. The *red line* shows dF/di for 2D islands at $a = 10$ and $\zeta = 1$, the *blue line* is its super-critical asymptote. Stationary flux J through the near-critical region is the nucleation rate at a given supersaturation

Let us now see how the stationary distribution and the corresponding nucleation rate can be obtained in the frame of discrete theory. We consider (1.138) for the flux assuming that the stationary spectrum n_i^s tends to zero at large enough i, the property that should hold for any reasonable physical distribution. Summing up (1.138) with $J = const$ at $k = i, i+1, i+2\ldots$ beginning from $i \sim i_c - \Delta i_c$, we obtain

$$n_i^s = J n_i^{eq} \sum_{k=i}^{\infty} \frac{1}{W_k^+ n_k^{eq}}. \tag{1.178}$$

From (1.177), the condition $n^s_{i_c-\Delta i_c} = n^{eq}_{i_c-\Delta i_c}$ should be met at the boundary of near-critical and subcritical region. This can be ensured only by the specific choice of constant J in (1.178), given by [27]

$$J = \left(\sum_{k=i_c-\Delta i_c}^{\infty} \frac{1}{W_k^+ n_k^{eq}} \right)^{-1}. \tag{1.179}$$

As discussed already in Sect. 1.9, the equilibrium distribution at integer i can be substituted to the continuum function $n^{eq}(i)$ with the relative error $1/\Delta i_c << 1$. The sum in (1.179) can be then changed to the integral with the same accuracy, yielding

$$J = \left(\int_{i_c-\Delta i_c}^{\infty} \frac{di}{W^+(i) n^{eq}(i)} \right)^{-1}. \tag{1.180}$$

The continuum analogue of (1.178) for the stationary distribution is given by

$$n^s(i) = J n^{eq}(i) \int_i^{\infty} \frac{di'}{W^+(i') n^{eq}(i')}, \quad i \geq i_c - \Delta i_c. \tag{1.181}$$

Let us consider the integral in the right hand side of (1.180) for the nucleation rate. Due to a very sharp dependence of quasi-equilibrium distribution on i, the formation energy can be approximated by the parabolic dependence near the critical size i_c as

$$F(i) \cong F - \frac{(i-i_c)^2}{\Delta i_c^2}. \tag{1.182}$$

Inserting this into the exponent under the integral, transforming it to the variable $x = (i - i_c)/i_c$ and using the Laplace method for asymptotic calculation of integrals containing a large parameter in the exponent, we obtain

$$\int_{i_c-\Delta i_c}^{\infty} \frac{di}{W^+(i) n^{eq}(i)} = \frac{\exp(F)}{n_1} i_c \int_{-\Delta i_c/i_c}^{\infty} \frac{dx}{W^+(x)} \exp\left[-\left(\frac{i_c}{\Delta i_c}\right)^2 x^2 \right] \cong \frac{\exp(F)}{n_1} \frac{\Delta i_c}{W^+(i_c)} \sqrt{\pi}. \tag{1.183}$$

It is seen that the parameter $\Delta i_c/i_c$ should be much smaller than one to justify the final result.

Using this in (1.180), we arrive at the famous Zeldovich formula for the nucleation rate [3, 7–10,12]

$$J = \frac{1}{\sqrt{\pi}\,\Delta i_c} n_1 W^+(i_c)\exp(-F). \tag{1.184}$$

The nucleation rate is thus proportional to the inverse exponent of the nucleation barrier at a given supersaturation, the condensation rate on the critical nucleus and the monomer concentration. The factor $1/\sqrt{\pi}\Delta i_c$ originating from the asymptotic integration is called the non-equilibrium Zeldovich factor. In view of $1/\Delta i_c << 1$, it is usually much smaller than one. Using the definition for the half-width of nucleation energy in the near-critical region, $\Delta i_c = \sqrt{2/|F''(i_c)|}$, (1.184) can be put in the equivalent form [3]

$$J = \sqrt{\frac{|F''(i_c)|}{2\pi}} n_1 W^+(i_c)\exp(-F). \tag{1.185}$$

As follows from the analysis of stationary state in the near-critical region, the two small parameters of macroscopic nucleation theory are given by [78]

$$1/\Delta i_c << 1; \quad \Delta i_c/i_c << 1. \tag{1.186}$$

The first strong inequality is required to justify the transformation from discrete to continuum distribution, while the second one is needed to use the Laplace method leading to the Zeldovich expression for the nucleation rate. Let us check the validity of these inequalities in some typical cases. For a 3D droplet or crystal nucleus in a 3D or 2D environment, Δi_c is given by (1.69) at $d = 3$, yielding

$$\Delta i_c = \frac{3}{\sqrt{a}} i_c^{2/3}; \quad \frac{\Delta i_c}{i_c} = \frac{3}{\sqrt{a} i_c^{1/3}}, \quad d = 3. \tag{1.187}$$

It is clear that the inequalities (1.186) are indeed valid at typical values of a of the order of ten and i_c of the order of several tens of monomers. For a 2D island, (1.69) gives

$$\Delta i_c = 2\sqrt{\frac{2}{a}} i_c^{3/4}; \quad \frac{\Delta i_c}{i_c} = 2\sqrt{\frac{2}{a}} \frac{1}{i_c^{1/4}}, \quad d = 2, \tag{1.188}$$

showing that the inequalities defined by (1.186) are again relevant at large enough a and i_c of the order of several tens. We note, however, that the first inequality (1.186) is always satisfied much better than the second one.

In the continuum theory, the size distribution obeys the Fokker-Plank-type (1.173), where the nucleus flux is defined as

$$J(i) = A(i)n(i) - B(i)\frac{\partial n(i)}{\partial i}. \tag{1.189}$$

In the near-critical region $i_c - \Delta i_c < i < i_c + \Delta i_c$ where dF/di is small, (1.174) are simplified to

$$A(i) \cong -W^+(i)dF/di; \quad B(i) \cong W^+(i) \tag{1.190}$$

with the accuracy $1/\Delta i_c$. Substitution of these expressions into (1.189) gives the stationary flux in the form

$$J = -W^+(i)n^{eq}(i)\frac{\partial \tilde{n}^s(i)}{\partial i}, \tag{1.191}$$

where $\tilde{n}^s$ is the stationary distribution normalized to the equilibrium one. Obviously, the discrete (1.138) yields the same result for J at $\tilde{n}_{i-1} - \tilde{n}_i \cong -\partial\tilde{n}(i)/\partial i$, the approximation valid in the near-critical region with the adopted accuracy. Straightforward integration of (1.191) with the boundary conditions $n^s(i_c - \Delta i_c) = n^{eq}(i_c - \Delta i_c)$ and $n^s(i) \to 0$ at $i \to \infty$ yields exactly the same results for the nucleation rate and stationary distribution as those given by (1.180) and (1.181), respectively.

As for the stationary distribution, (1.178) can be considerably simplified when the particle size approaches the boundary of the near-critical and super-critical regions i_+. At large enough $i \geq i_+$, the formation energy can be approximated by $F_i \cong -\ln(\zeta + 1)i$, therefore

$$n_i^s = J\sum_{k=i}^{\infty}\frac{1}{W_k^+}e^{F_k - F_i} \cong \frac{J}{W_i^+}\sum_{k=i}^{\infty}e^{-\ln(\zeta+1)(k-i)} = \frac{J}{W_i^+}\frac{(\zeta+1)}{\zeta}. \tag{1.192}$$

Taking the rate constant out of summation is justified by its power law size dependence with the adopted accuracy. The end result is obtained simply by summation of the geometrical progression. Comparing the obtained expression to (1.115) and (1.116), it is seen that the stationary distribution in the essentially super-critical region has the "drift" form

$$n_i^s = \frac{J}{di/dt}. \tag{1.193}$$

This formula shows simply that, in the stationary state, the nuclei emerge at the rate J and surpass the boundary of the near-critical and super-critical regions with the speed di/dt. Of course, the latter should be treated as the nucleus growth rate without the curvature effects, as given by (1.116). For the power law dependence of the condensation rate constant as defined in (1.115), the stationary distribution takes the form

$$n_i^s = \frac{J\tau}{m\zeta i^{(m-1)/m}}, \tag{1.194}$$

where m is the growth index and τ is the characteristic growth time of super-critical nuclei. This expression is of major importance in the foregoing analysis, as it provides the boundary condition for the size distribution in the growth region. The same representation for $n^s(i)$ at $i \geq i_+$ can also be derived from the continuum (1.181).

Several works [10, 29, 79] have been studied how the stationary state is established in the near-critical region and what is the characteristic time t_s of this process in particular. The average estimate for t_s is given by

$$t_s \sim \frac{1}{|F''(i_c)|\, W^+(i_c)}. \tag{1.195}$$

This time is extremely short for most systems of interest. For example, it is not longer than 10^{-4} s in the case of 2D adatom condensation, which is evidently much shorter than the typical nucleation times. As mentioned already, the smallness of t_s on the time scale of interest justifies well the major assumption of a time-independent supersaturation while considering the stationary state. Since the latter is adjusted almost instantaneously to the current value of supersaturation, (1.184) and (1.194) can be treated as being parametrically time-dependent through $\zeta(t)$. The supersaturation is a "slow" variable at the stage considered. In the timescale hierarchy of the entire nucleation-condensation process, the initial step of establishing the stationary state in the near-critical region is the shortest.

Explicit representation for the nucleation rate as a function of supersaturation is obtained from the general (1.184) using (1.69) for Δi_c and (1.67) for $F(\zeta)$. We allow for the Kashchiev and Lothe-Pound modifications of nucleation barrier as defined by (1.71–1.73) and (1.74), respectively by writing the nucleation rate as $F(\zeta) + \Delta$. Here, $\Delta = -F_1 = -a + \ln(\zeta + 1)$ in the Kashchiev's case and $\Delta = \ln(\sigma n_1) = \ln\theta_{1e} + \ln(\zeta + 1)$ in the Lothe-Pound case, with σ being the elementary volume in 3D case or surface area in 2D case, as discussed in Sect. 1.5. Obviously, the Δ-shift changes only the pre-exponent of the Zeldovich formula for the nucleation rate and does not affect either the ζ-dependence of the nucleation barrier or the coefficients of kinetic equation for the size distribution. After simple calculations, one obtains

$$J(\zeta) = \frac{n_{1e}}{\tau}\sqrt{\frac{a}{2\pi}}\frac{m\sqrt{d-1}}{d}(\zeta+1)^2\left[\frac{d\ln(\zeta+1)}{(d-1)a}\right]^{(1-d)/2+d/m}\left(\frac{e^a}{\zeta+1}\right)^{\kappa}\frac{1}{[\sigma n_{1e}(\zeta+1)]^{\vartheta}}$$
$$\times\exp\left[-\frac{(d-1)^{d-1}a^d}{d^d\ln^{d-1}(\zeta+1)}\right]. \tag{1.196}$$

Here, κ and ϑ equal one if the Kashchiev's or the Lothe-Pound corrections are included and zero otherwise.

Let us consider some particular cases where (1.196) is reduced to the commonly used expressions. In their studies of 3D droplet condensation, Kuni *et al.* [10, 17] simply put $\Delta = 0$, the assumption used also in [14, 19, 53]. In the case of 3D droplets or islands growing in the ballistic regime ($d = m = 3$) with $\kappa = 0$ and $\vartheta = 0$, this yields the nucleation rate in the form

$$J(\zeta) = \frac{n_{1e}}{\tau}\sqrt{\frac{a}{\pi}}(\zeta + 1)^2 \exp\left[-\frac{4a^3}{27\ln^2(\zeta + 1)}\right]. \qquad (1.197)$$

The corresponding stationary distribution written in terms of invariant size $\rho = i^{1/3}$ is given by

$$f_s(\zeta) = \frac{I(\zeta)}{d\rho/dt} = \frac{I(\zeta)\tau}{\zeta} = n_{1e}\sqrt{\frac{a}{\pi}}\frac{(\zeta + 1)^2}{\zeta} \exp\left[-\frac{4a^3}{27\ln^2(\zeta + 1)}\right]. \qquad (1.198)$$

When modified by the Kashchiev's re-normalization, the nucleation rate of 3D particles takes the form

$$J(\zeta) = \frac{n_{1e}}{\tau}\sqrt{\frac{a}{\pi}}e^a(\zeta + 1) \exp\left[-\frac{4a^3}{27\ln^2(\zeta + 1)}\right]. \qquad (1.199)$$

Similar expressions apply for liquid or solid 3D islands on a solid surface, with the corresponding modification of the surface energy constant a depending on the island shape.

In application to nucleation of vapor-deposited thin films, Kukushlin and Osipov [3] used the Lothe-Pound entropy correction. In the case of 2D surface islands growing in the ballistic regime ($d = m = 2$), $\kappa = 0$ and $\vartheta = 1$ (1.196) is reduced to the result of [3, 66]:

$$J(\zeta) = \frac{1}{\sqrt{\pi\sigma\tau}}(\zeta + 1)\ln^{1/2}(\zeta + 1) \exp\left[-\frac{a^2}{4\ln(\zeta + 1)}\right]. \qquad (1.200)$$

When the renormalization is of the Kashchiev's type, (1.200) is changed to

$$J(\zeta) = \frac{n_{1e}}{\sqrt{\pi\tau}}e^a(\zeta + 1)\ln^{1/2}(\zeta + 1) \exp\left[-\frac{a^2}{4\ln(\zeta + 1)}\right]. \qquad (1.201)$$

Very importantly, the Zeldovich expressions for the nucleation rate and stationary size distribution feature an extremely steep exponential dependence on supersaturation through $F(\zeta)$. This property follows directly from (1.73) for the nucleation barrier containing a large factor $a >> 1$. The nucleation rate is so sensitive to supersaturation that a relatively small variation of the latter changes $J(\zeta)$ by the order of magnitude. The effect is demonstrated in Fig. 1.28, where the nucleation rate of 2D surface islands given by (1.200) is plotted against supersaturation at fixed $\sigma = 0.16\,\mathrm{nm}^2$, $\tau = 3 \times 10^{-3}\,\mathrm{s}$ and different surface energy coefficients a. The threshold character of nucleation is clearly seen, with the threshold values of supersaturation increasing approximately from 1.4 to 3 with a changing from 7.8 to 10. When supersaturation exceeds the threshold value, the $J(\zeta)$ curve rapidly becomes

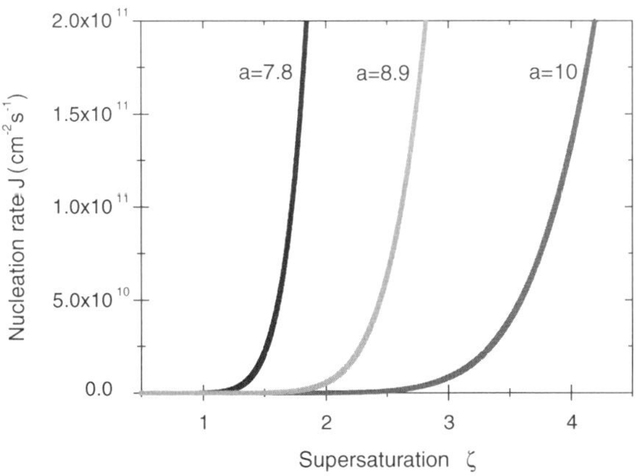

Fig. 1.28 Nucleation rate $J(\zeta)$ at three different a

almost vertical. At $a = 7.8$, the increase of supersaturation only from 1.7 to 1.8 results in $\sim$4 times increase in the nucleation rate.

The Zeldovich formula for the nucleation rate has been verified experimentally in many material systems by measuring the number of emerging nuclei at different conditions such as the growth time, temperature, or material influx. At a known supersaturation, such measurements may be helpful in deducing important information about the surface energy for example. Indeed, writing (1.197) in the case of 3D surface islands in the simplified form

$$J = J_0 \exp\left[-\frac{b}{(k_BT)^3 \ln^2(\zeta + 1)}\right] \tag{1.202}$$

with $b = (4/27)a^3(k_BT)^3$ and assuming that the initial adatom concentration equals It_A, we obtain $\zeta + 1 = n_1/n_{1e} = I\tau_A/n_{1e}$. Using the temperature dependences $n_{1e} = n_0 \exp(-2T_c/T)$ and $\tau_A = \tau_A^0 \exp(E_A/k_BT)$, we arrive at

$$\ln\left(\frac{J}{J_0}\right) = -\frac{T_I}{T}; \quad T_I = \frac{b}{k_B[E_A + 2k_BT_c + k_B\tilde{T}\ln(I\tau_A^0/n_0]^2}, \tag{1.203}$$

where $\tilde{T}$ is the average temperature within the analyzed temperature range. Linear dependence of $\ln J$ on $1/T$ has been observed in many experiments with surface islands, the corresponding data can be found, for example, in [3]. Measuring the slope angle of the experimental curve enables finding the constant b and therefore the surface energy.

When plotted against supersaturation, (1.202) takes the form

$$\ln\left(\frac{J}{J_0}\right) = -\frac{b'}{\ln^2(\zeta + 1)} \cong -\frac{b'}{\zeta^2}, \tag{1.204}$$

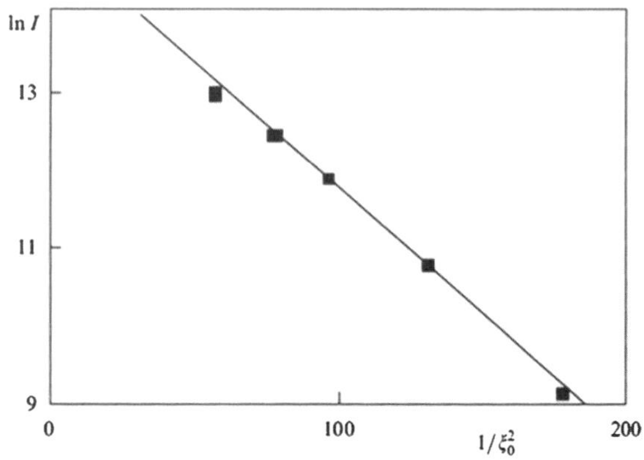

Fig. 1.29 Experimental (*symbols*) and theoretical (*line*) dependences of the nucleation rate ($m^2\,s^{-1}$) on the relative super-heating [3]

with $b' = (4/27)a^3$. The last expression applies when the supersaturation is much smaller than one. In this case, the nucleation process is slow and the emerging islands are large, which makes them easier to measure experimentally. Figure 1.29 shows the experimental dependence of $\ln J$ on $1/\zeta_0^2$ for the system of the lead zirconate–titanate ferroelectric islands, where ζ_0 is the relative super-heating (an analogue of initial supersaturation) [85]. It is seen that the experimental points are well fitted with the linear dependence following from the Zeldovich equation.

To summarize, at high enough nucleation barriers, the quasi-equilibrium state is thoroughly maintained in the subcritical region of sizes. The stationary state with the size-independent nucleus flux through the near-critical region is established very rapidly and pertains during the follow-up nucleation and growth stages by adjusting itself to the time-dependent supersaturation. The nucleation rate depends on the supersaturation primarily via the inverse exponent of nucleation barrier. This leads to an extremely high sensitivity of the entire nucleation process to the value of supersaturation. In the imperfect systems with chemical potential other than $\Delta\mu = k_B T \ln(\zeta + 1)$, all the results presented above remain valid upon substitution of to $\Delta\mu / k_B T$ in the corresponding expressions. In particular, this applies to (1.184), (1.185), (1.196), (1.197), (1.199) to (1.201) for the nucleation rate and (1.193), (1.194), (1.198) for the stationary size distribution. However, the equation of state $\Delta\mu(\zeta)$ should be known for performing concrete calculations, while the unknown ζ must be obtained from the material balance condition in any case.

1.12 Stages of Nucleation-Condensation Process

This section presents a semi-quantitative preview of different stages of nucleation-condensation process, which will be considered in more detail in Chap. 2. As discussed, the shortest nucleation step is the establishment of stationary state in the near-critical region. The nucleation rate J depends on supersaturation, material constants and temperature. In some scenarios of nanostructure formation, the Zeldovich formula for J alone provides a sufficient basis for modeling the entire growth process. Let us consider a very important case of nucleation in restricted nano-volumes or on nano-surfaces [45, 80–83]. Typical examples of such systems include a droplet of supersaturated liquid alloy (Fig. 1.30a), a vertical nanowire growing directly from vapor (Fig. 1.30b) or from a liquid alloy in the nanodroplet seated atop the nanowire (Fig. 1.30c). Provided that the available volume of a metastable phase or the surface area of the nanowire top facet is small enough, the nucleation will proceed in the so-called mononuclear mode [7]. In this mode, the characteristic time between two consecutive nucleation events is much longer that the time required for a nucleus to spread over the entire available volume or surface area. In the nanowire case, the vertical growth rate in the mononuclear mode is determined by the waiting time between the two nucleation events rather than by the lateral growth time of the island.

By its physical sense, the Zeldovich nucleation rate in the mononuclear mode gives the nucleation probability per unit time per unit volume. The total nucleation probability per unit time is therefore proportional to the available volume or surface area

$$V_{nucl}(\zeta, R) = I(\zeta, R)cR^{d} \tag{1.205}$$

with c as the corresponding shape constant, R as the linear size and $d = 3$ or 2 depending on the system dimension. The supersaturation value should be generally obtained by balancing the mass transport into the nanoparticle and the material sink caused by the nucleation. In some cases such as the steady state "vapor–liquid–solid" growth of nanowires [82], the supersaturation remains almost constant during the entire growth process due to a repeated material refill from vapor after each nucle-

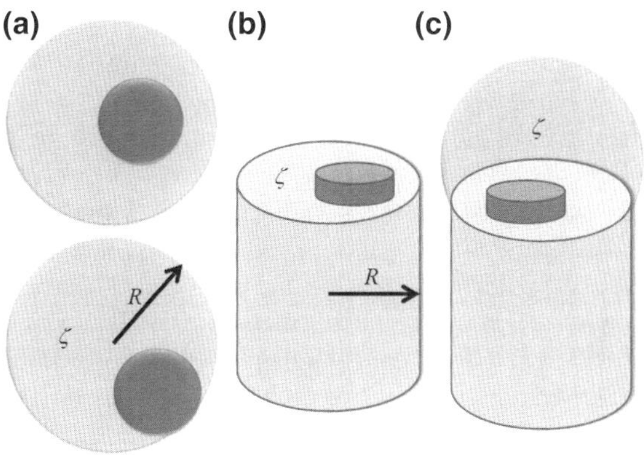

Fig. 1.30 Nucleation of islands in the volume of nanodroplet (**a**), on the top facet of self-induced nanowire (**b**) and of the vapor–liquid–solid nanowire (**c**) in the mononuclear mode. The nucleation rate is generally a function of supersaturation ζ in the droplet (**a** and **c**) or in the adatom system (**b**), and the radius R. **a** shows two different positions of the nucleus either entirely surrounded by the liquid or emerging partly at the liquid–vapor interface

ation event. Such a growth can be considered as the re-iterative, nucleation-mediated building of nanowire monolayers under the droplet, which proceeds under exactly identical conditions.

Very importantly, the nucleation rate in (1.205) may become radius-dependent for the following reason. While considering the formation energy in Sect. 1.5, it has always been assumed that an infinitely large metastable phase is not affected by nucleation of a very small particle. This natural assumption becomes irrelevant for nanoparticles. Indeed, when a crystal island nucleates inside or under the nanodroplet, it consumes the growth material from the droplet almost instantaneously, while the refill stage lasts for a much longer time. Depending on the position of the island and the ratio of elementary volumes in the liquid and solid phases, the nucleation may result in changing the droplet volume and consequently its surface area. When such a change occurs, the free energy of island formation ΔG must include the modification of the Laplacian pressure caused by nucleation [82]:

$$\Delta G(i) = -\Delta \mu i + \gamma_{eff} S_{isl} + \gamma \Delta S_L. \tag{1.206}$$

Here, the first two terms are the same as for an infinitely large metastable phase, with $\Delta\mu$ being the difference in chemical potentials between the bulk phases, γ_{eff} the effective surface energy of the island and S_{isl} the island surface area. It should be noticed that the surface energy now depends on the island position. In particular, it is distinctly different for the island entirely surrounded by liquid or replacing some part of the pre-existing liquid–vapor interface by the solid–vapor interface [45]. The position dependence of the formation energy is another interesting feature of nucleation in nano-volumes. The last term in (1.206) thus stands for the modification of a metastable phase itself, where γ is the liquid surface energy and ΔS_L is the change of droplet surface area caused by nucleation. Clearly, this contribution is proportional to $1/R$ and therefore modifies the surface energy term. Our simple considerations show that, while a complex story of modeling the size distribution is absent for just one island emerging in the mononuclear regime, relevant modifications of the formation energy are required for the correct determination of the nucleation probability. This will be further developed in Chaps. 4–6 in connection with the growth mechanisms of semiconductor nanowires.

When the volume of metastable phase is large enough, the nucleation proceeds in the more usual polynuclear mode. An ensemble of nuclei emerging at different times and growing at different rates should be described by a time-dependent size distribution function. Let us briefly consider the growth stages in this case. To specify the analysis, we adopt the common scenario where the supersaturation is achieved due to a material influx into the system. The influx gradually increases the total number of available monomers. The ideal supersaturation $\Phi(t)$ defined in (1.125) is an increasing function of t, for example, a linear function at a time independent P and zero Q. Since the nucleation process requires reaching a certain threshold (or critical) supersaturation ζ_*, the actual supersaturation $\zeta(t)$ follows $\Phi(t)$ below the threshold, as shown in Fig. 1.31. The time period required to establish the critical supersaturation can be called the incubation time t_*. The critical supersaturation

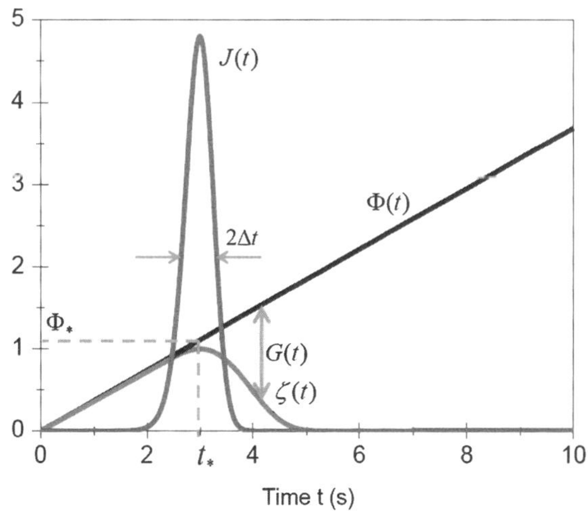

Fig. 1.31 Qualitative representation of nucleation process under a material influx. Φ_* denotes the ideal supersaturation at time t_* corresponding to the maximum supersaturation ζ_*

is also the maximum supersaturation in a given growth process. The maximum is reached due to the dynamic balance between the monomer influx and their sink caused by nucleation, which starts just before the threshold. The maximum supersaturation corresponds to the maximum nucleation rate and the most representative size in the size distribution (in terms of the invariant size ρ) at the follow-up growth stages.

Since the dependence of the nucleation rate on supersaturation is extremely steep, the nucleation process is enabled only in a very narrow region near the supersaturation maximum and exponentially decreases outside this region (see Fig. 1.31). It might be expected that, in the case of nucleation under the material influx, the nucleation stage is short scale compared to the macroscopic time t_*. This feature, explored in more detail in Chap. 2, is of paramount importance for obtaining the narrow and symmetric size distributions of different self-induced nanoobjects such as nanodroplets, colloidal particles and coherent strained islands in heteroepitaxial systems. It is also clear that the characteristics of this short scale nucleation stage in largest measure determine the observed size distributions. During the entire nucleation stage, the mean size of nuclei remains very small and the normalized total number of monomers in the nuclei $G(t)$ is much smaller than supersaturation.

At $t > t_* + \Delta t$, where $2\Delta t$ is the duration of the nucleation stage, the nucleation rate decreases drastically and the nucleation process is effectively turned off. The nuclei having emerged within the time interval $2\Delta t$ continue growing by consuming monomers which are supplied from the surrounding. The nuclei growth leads to the decrease of supersaturation (usually to zero in the large time limit), as shown in Fig. 1.31. The stage of regular, independent growth of nuclei lasts for time t_R, which is usually much longer than Δt. The maximum of size distribution propagates toward larger sizes, while the size dispersion may change as the nuclei grow. The number of monomers in the nucleus $G(t)$ almost matches $\Phi(t)$ at the asymptotic growth stage. The total number of nuclei remains constant at the regular growth stage.

As supersaturation tends to zero, the critical size defined by (1.68) begins to increase. The next stage of system evolution depends on the character of material influx and the growth rate of nuclei. If the material influx is turned off at a certain time or is slow enough, the critical size reaches the pre-existing size distribution and larger nuclei start growing at the expense of smaller ones. This process, called the Ostwald ripening [3, 86–92], has a major impact on the size spectrum: the number of nuclei is no longer conserved and the distribution dispersion increases much faster than at the regular growth stage. The late and slow Ostwald ripening stage leads to the establishment of a universal asymptotic size distribution [86], as discussed in Chap. 2. When the mean size increases faster than the critical one, the Ostwald ripening process is disabled.

In many dense systems such as vapor-deposited thin films or crystallizing liquid alloys, the independent growth of islands is followed by their direct coalescence. The coalescence process can be liquid-like, where two droplets form one bigger droplet of the same shape, or solid-like, leading to the formation of a complex "mosaic" crystal structure. The latter can no longer be described in terms of the number and size of nuclei. Special geometrical and probabilistic approaches should be used at the coalescence stage, treating the crystallizing system in terms of the mean filling factor, surface area or perimeter of crystalline boundaries and so on. In the case of solid-like coalescence, a very useful approach is based on the so-called Kolmogorov–Johnson–Mehl–Avrami crystallization model [93–96]. This model, discussed in detail in Chap. 2, is of great importance in modeling the growth of thin solid films, semiconductor nanowires and many other crystals [65, 80, 97–100].

Main stages of nucleation-condensation process in the case of 2D thin film are shown schematically in Fig. 1.32. As discussed in more detail in Chap. 2, the Ostwald ripening stage is disabled, for example, in the case of 2D islands growing under a constant deposition rate, regardless of their growth mechanism. In this case, the time scale hierarchy of different growth stages is given by

$$t_s << 2\Delta t << t_R << t_c. \tag{1.207}$$

The shortest stage is establishing the stationary state in the near-critical region of sizes, followed by the nucleation stage with the duration $2\Delta t$, then the regular growth with the characteristic time t_R, which is much shorter than the coalescence time t_c at which the islands start touching each other. For example, at the deposition rate of 0.1 ML/s and a small effective desorption rate, the total time required to form a complete monolayer is 10 s, while the coalescence process starts roughly at 5 s, the moment when the film starts reaching continuity.

The size distribution established at the nucleation stage will be studied in detail in Chap. 2. While the particular spectrum shape is not known by this end, some useful information regarding the growth law can be deduced under the assumption that the distribution width is much smaller than the mean size. In open systems, this important property is usually ensured by the shortness of nucleation stage under the material influx, as shown in Fig. 1.31. Taking the example of 2D islands growing in the ballistic regime, the invariant size defined by (1.117) and (1.118) is the island

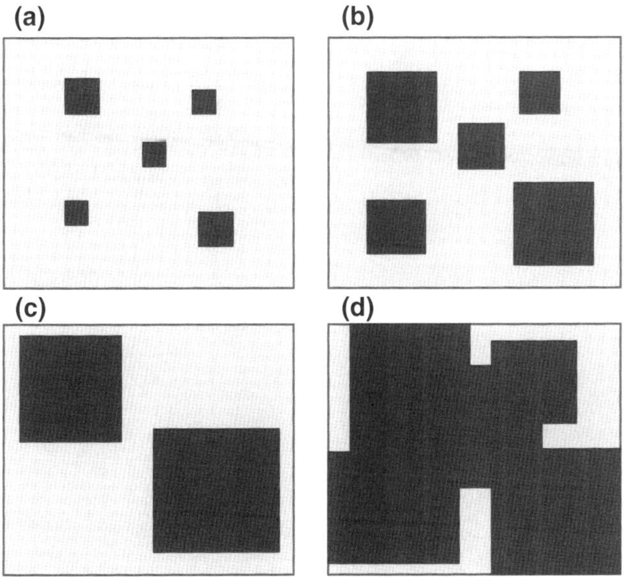

Fig. 1.32 A part of substrate surface containing 5 square islands upon the completion of nucleation stage (**a**). The islands then grow independently at the regular growth stage (**b**). If the deposition flux is turned off at a certain time, the Ostwald ripening is observed, where larger islands grow at the expense of smaller ones (**c**). If the flux remains on, the islands coalesce into a continuous film (**d**)

radius in the units of lattice spacing, $\rho = i^{1/2}$. We assume that the size distribution in terms of ρ variable is narrow. For example, it can be approximated by the Gaussian

$$f(\rho, z) = \frac{N}{\sqrt{2\pi\psi(z)}} \exp\left[-\frac{(\rho - z)^2}{2\psi(z)}\right], \tag{1.208}$$

in which $z(t) \equiv \rho_*(t)$ is the time-dependent mean size, N is the surface density of islands upon the completion of nucleation stage and $\sqrt{2\psi(z)}/z << 1$ during the entire growth stage. In this case, we can use the δ-like approximation for the distribution function in the equation for the total number of adatoms in the islands:

$$G(t) = \frac{1}{n_{1e}} \int_0^\infty d\rho \rho^2 f(z, \rho) \cong \frac{N}{n_{1e}} z^2(t). \tag{1.209}$$

On the other hand, the growth rate of the mean size is given by (1.118):

$$\frac{dz}{dt} = \frac{\zeta}{\tau}. \tag{1.210}$$

The ideal supersaturation in the case of vapor deposition with the time-independent arrival rate χI and the desorption time τ_A is determined by

$$\Phi(t) = \frac{1}{n_{1e}} \int_{t_e}^{t} dt' \left[\chi I - \frac{n_1(t')}{\tau_A} \right] \cong \Phi_* + \Phi_{max} \frac{(t - t_*)}{\tau_A} - \frac{\tau}{\tau_A} z(t), \quad t \geq t_* \tag{1.211}$$

Here, $\Phi_{max} = (\chi I \tau_A)/n_{1e} - 1$ is the maximum ideal supersaturation that would be reached on the surface in absence of nucleation. In the last expression, we assume that ζ equals Φ_* at $t = t_*$, where the time moment t_* corresponds to the supersaturation maximum as shown in Fig. 1.31. This happens when the deposition flux is large enough, that is, at $\Phi_{max} >> 1$. Using (1.209–1.211) in the material balance condition $\Phi(t) = \zeta(t) + G(t)$, and assuming that the mean size equals zero at $t = t_*$, we arrive at the closed equation for $z(t)$:

$$\tau \frac{dz}{dt} + \frac{N}{n_{1e}} z^2 + \frac{\tau}{\tau_A} z = \Phi_* + \frac{\Phi_{max}}{\tau_A}(t - t_*);$$
$$z(t = t_*) = 0. \tag{1.212}$$

Upon several straightforward substitutions of variables, this non-linear growth equation can be reduced to the linear second order equation for the Airy functions, yielding the exact solution in the form

$$z(t) = [U(x(t)) - \delta]\, z_0. \tag{1.213}$$

The parameters are defined as follows

$$z_0 \equiv \left(\frac{n_{1e}}{N}\right)^{2/3} \left(\frac{\tau}{\tau_A} \Phi_{max}\right)^{1/3}; \tag{1.214}$$

$$\delta \equiv \frac{1}{2} \frac{\Phi_*^2}{\Phi_{max}^{5/3}} \left(\frac{\tau}{\tau_A}\right)^{2/3} \left(\frac{n_{1e}}{N}\right)^{1/3}. \tag{1.215}$$

The function $U(x)$ is given by the ratio of linear combinations of Airy functions and their derivatives

$$U(x) = \frac{Bi'(x) - k\left|Ai'(x)\right|}{Bi(x) + kAi(x)}; \quad k \equiv \frac{Bi'(x_0) - \delta Bi(x_0)}{|Ai'(x_0)| + \delta Ai(x_0)}, \tag{1.216}$$

where $Ai(x)$ is the decreasing and $Bi(x)$ is the increasing Airy function in standard notations [101]. The variable x is the linear function of time

$$x(t) = \beta \left(1 + \nu + \frac{\Phi_{max}}{\Phi_*} \frac{(t - t_*)}{\tau_A}\right) \tag{1.217}$$

with coefficients

$$\beta = \frac{\Phi_*}{\Phi_{\max}^{2/3}} \left(\frac{\tau_A}{\tau}\right)^{2/3} \left(\frac{N}{n_{1e}}\right)^{1/3}; \tag{1.218}$$

$$\nu = \frac{1}{4} \frac{\Phi_*^5}{\Phi_{\max}^4} \frac{n_{1e}}{N} \left(\frac{\tau}{\tau_A}\right)^2. \tag{1.219}$$

The quantity x_0 in the second (1.216) is defined as $x_0 = x(t = t_*) = \beta(1 + \nu)$.

This solution describes the time evolution of the mean size of surface islands at the known N, Φ_* and t_*, which values are defined at the nucleation stage. The typical dependence of $r_*(t) = \sqrt{\sigma}z(t)/4$ for square islands with side $2r_*$ is shown in Fig. 1.33. The curve is obtained for the parameters approximately corresponding to the Ga-limited growth of GaAs islands on the GaAs(100) substrate at the temperature $T = 580\,°\mathrm{C}$ with the Ga deposition rate $V = \chi I\sigma \cong I\sigma$ of 0.1 ML/s [102]: $N = 6\times 10^9\,\mathrm{cm}^{-2}, n_{1e} = 3.8\times 10^{12}\,\mathrm{cm}^{-2}, \sigma = 0.16\,\mathrm{nm}^2, \theta_{1e} = 6\times 10^{-3}, \tau = 3\times 10^{-3}\,\mathrm{s}$, $\tau_A = 2.6\,\mathrm{s}$, $\Phi_{\max} = 42$, $\Phi_* = 1.9$, and t_* =0.17 s. It is seen that the mean size increases much faster at the beginning and then reaches the asymptotic regime where z is roughly proportional to $t^{1/2}$.

These features can be understood from the general solution. Indeed, when $(\Phi_{\max} + 1)/(\Phi_* + 1) = n_{1*}/(\chi I\tau_A) >> 1$, only a tiny fraction of adatoms desorb from the surface (the so-called regime of complete condensation of thin films). The latter inequality is also consistent with $\tau/\tau_A << 1$, showing that the islands grow much faster than re-evaporate. Under such conditions, the parameters δ and ν in the above general expressions can be put to zero, while the constant β is larger than 1.5 (it equals 1.62 for the parameters considered). Since $x > \beta$, we can use the known asymptotes of the two Airy functions in (1.216) at large x: $Ai(x) \cong [1/(2\pi^{1/2}x^{1/4})]\exp[-(2x^{3/2})/3]$; $Bi(x) \cong [1/(\pi^{1/2}x^{1/4})]\exp[(2x^{3/2})/3]$. This reduces (1.213) to

$$z(t) = \sqrt{\frac{\Phi_*\theta_{1e} + (1-\alpha_{des})V(t-t_*)}{\sigma N}} \tanh\left[C\left(\left(\frac{\Phi_*\theta_{1e} + (1-\alpha_{des})V(t-t_*)}{\Phi_*\theta_{1e}}\right)^{3/2} - 1\right)\right] \tag{1.220}$$

with $C \equiv (2/3)\beta^{3/2}$. Here, V is the deposition rate in ML/s, θ_{1e} is the equilibrium adatom coverage at a given temperature and $\alpha_{des} = n_{1e}/(\chi I\tau_A) = \theta_{1e}/(V\tau_A)$ is the small desorption term. The latter equals to only 0.023 in our numerical example. It is seen that the time dependence of mean size is actually expressed via the normalized time $t/t_{ML} = (1 - \alpha_{des})V(t - t_*)$ taking account for both a short waiting time t_* required to reach the maximum adatom supersaturation and a small percentage of desorbed adatoms α_{des}.

Since the constant C is large, the increase of $z(t)$ is faster for shorter times close to t_*. After the relaxation time $\sim 4t_g$, where

$$t_g = \frac{\Phi_*\theta_{1e}}{V} \cong \frac{\Phi_*}{\Phi_{\max}}\tau_A, \tag{1.221}$$

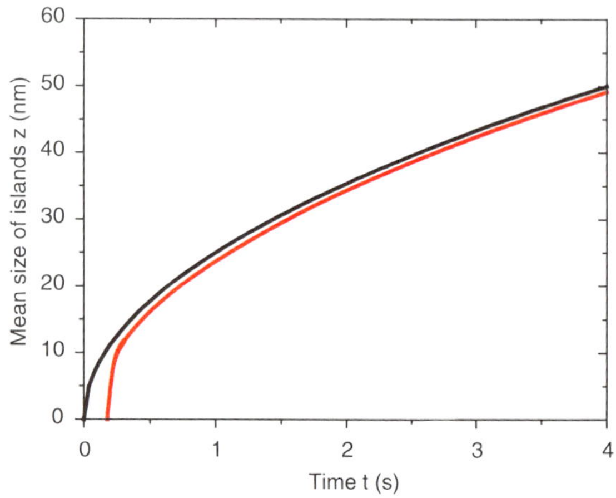

Fig. 1.33 Mean size of 2D square islands obtained with the parameters described in the text from the general (1.213) to (1.219) (*red line*) and the simplified formula (1.222) (*black line*)

the mean size given by (1.220) reaches the asymptotic regime

$$z(t) = \sqrt{\frac{\Phi_* \theta_{1e} + (1 - \alpha_{des}) V (t - t_*)}{\sigma N}} \cong \sqrt{\frac{(1 - \alpha_{des}) V t}{\sigma N}}. \tag{1.222}$$

This dependence, also shown in Fig. 1.33, contains only the most important characteristic of nucleation stage: the island density N. In our numerical example, the time t_g equals 0.11 s, so that the asymptotic regime is reached already after $\sim$0.5 s after the beginning of deposition. The islands therefore obey the simple growth law given by (1.222) most of the time before they coalesce (we remind that the deposition time of 1 ML equals 10 s in our example).

Finally, using the "mono-disperse" approximation given by (1.209) at an arbitrary growth index m:

$$G(t) = \frac{1}{n_{1e}} \int_0^\infty d\rho \rho^m f(z, \rho) \cong \frac{N}{n_{1e}} z^m(t), \tag{1.223}$$

we are able to deduce important information about the asymptotic behaviors of the mean size and supersaturation. Let us consider the case with $Q = 0$, and P being a power law function of time, where the ideal supersaturation is given by

$$\Phi(t) = \frac{1}{n_{1e}} \int_0^t dt' P(t') = \left(\frac{t}{t_\infty} \right)^q. \tag{1.224}$$

Here, q is the flux index and t_∞ is the characteristic macroscopic time required to reach the equilibrium vapor concentration n_{1e}. For example, at $P(t) = I = const$, $t_\infty = n_{1e}/I = \theta_{1e}/V$ is the time of establishing the equilibrium adatom coverage at a given deposition rate V with neglect of desorption. Writing the material balance given by (1.176) in the form

$$\left(\frac{t}{t_\infty}\right)^q = \zeta(t) + \frac{N}{n_{1e}} z^m(t) \tag{1.225}$$

and assuming that $\zeta \rightarrow 0$ at the asymptotic growth stage, the leading power law asymptote of the mean invariant size is obtained as

$$z(t) = \left(\frac{n_{1e}}{N}\right)^{1/m} \left(\frac{t}{t_\infty}\right)^{q/m}. \tag{1.226}$$

Equation (1.222) is the particular case of (1.226) at $m = 2$ and $q = 1$. Obviously, the supersaturation $\zeta(t) = \tau(dz/dt)$ asymptotically decreases to zero only if $q/m < 1$. Under a constant material influx corresponding to $q = 1$, this happens for all $m > 1$, while the growth at $m = 1$ leads to a constant value of supersaturation in the large time limit. Obviously, such a behavior disables the Ostwald ripening process at $m = 1$ under a material influx.

Chapter 2
Theoretical Description of Condensation Stages

This most mathematized chapter of the book presents a systematic account of relevant theoretical models that are capable of describing different stages of nucleation-condensation process in an analytical, physically transparent way. We are not at all trying to give a complete overview of the entire body of knowledge about nucleation theory, but rather restrict the analysis to the approaches that will be useful in the growth modeling of epitaxial nanostructures, the main subject of this work. In particular, we usually do not consider small nuclei (except for Sects. 2.13, 2.14 and 2.15), because the nanoislands of interest usually consist of at least a few tens of atoms. Heterogeneous nucleation on the condensation seeds is not discussed since it is not used in what follows. The surface nanostructures are solid, which is why we do not consider the liquid-like coalescence of droplets, an interesting process where the Kolmogorov approach is not directly applicable. Our main concern will be identifying the kinetics turning knobs that can be used for tuning the morphology of self-induced nanostructures to the desired properties, particularly by tailoring the size distribution. Of course, the lines of our own research have played a role in the selection of material. We do not stick to a concrete material system in theoretical analysis, so that the numerical examples should be thought of only as illustrations of the effects considered.

The chapter is organized as follows. In Sect. 2.1, we describe the transformation of the continuum kinetic equation of classical nucleation theory to the so-called invariant size. Such an invariant representation eliminates the trivial deformation of size spectrum in the deterministic limit, originating from different growth rates of differently sized islands. The time invariance is essential for obtaining the universal double exponential spectrum at the nucleation stage, an outstanding contribution by Kuni and co-authors considered in Sect. 2.2. The unique possibility to find the solution in a simple analytical form originates from an extreme non-linearity of the problem due to the exponential dependence of the Zeldovich nucleation rate on supersaturation. In particular, we obtain the so-called critical supersaturation corresponding to a dynamic balance of the monomer influx from a material source and their sink due to nucleation. This critical value will correspond to the maximum

V. G. Dubrovskii, *Nucleation Theory and Growth of Nanostructures*,
NanoScience and Technology, DOI: 10.1007/978-3-642-39660-1_2,

of size spectrum in the future. The double exponential distribution is characteristic of nucleation under the material influx. In Sect. 2.3, we show how the spectrum shape changes when the influx is terminated below the critical supersaturation, a triangle-like shape in the essentially subcritical nucleation mode being a limit case. It is shown how the characteristics at the short scale nucleation stage determine the nucleus density and the initial spectrum width, probably the most important parameters of self-induced nanostructures.

In the deterministic approximation without the second derivative with respect to size in the kinetic equation, the nucleation spectrum would maintain its time-invariant shape at the follow-up growth stage, with only the mean size changing in time. We start the analysis of Sect. 2.4 with considering whether the exact solutions presented in Chap. 1 feature such a behavior, and show that the time invariance at the growth stage is an exception. When the second derivative describing kinetic fluctuations is included, the Green function at the growth stage can be obtained under rather general assumptions in the form of spreading Gaussian with a time dependent dispersion. As a function of the most representative size, the dispersion obeys a self-consistent differential equation whose form depends on the characteristics of monomer influx via the material balance. Section 2.5 justifies the neglect of the second derivative at the nucleation stage, where the fluctuation-induced spectrum broadening is much smaller than the deterministic nucleation width. Section 2.6 presents useful analytical and numerical results regarding the time evolution of size spectra at the growth stage. Two systems of 2D surface islands growing in the ballistic and the diffusion regime under a constant vapor flux are considered as the examples. The flux can be either always on or turned off at a certain moment of time, leading to distinctly different behaviors. In any case, the tendency for the fluctuation-induced broadening increases at a smaller growth index, as suggested by the analytical findings. That is why the system with $m = 1$ features a Poissonian-like behavior where the size dispersion scales proportionally to the mean size with no small pre-factors, while the spectra at $m = 2$ remain much narrower.

In Sect. 2.7, the asymptotic growth stage is considered, where the supersaturation tends to zero. This allows one to obtain the large time asymptotes for the mean size, supersaturation and dispersion from the material balance. By analyzing different growth scenarios, we map out the combinations of the growth and material flux parameters where the size distributions are time invariant or changing at different rates. We also discuss the Kukushkin–Osipov condition for the Ostwald ripening under a material influx. It shows, in particular, that the only exclusive system undergoing the Ostwald ripening under a constant flux is 3D droplets growing in the ballistic regime. Section 2.8 is devoted to the Ostwald ripening upon the termination of material influx. We calculate numerically the time evolutions of size spectra by means of different kinetic equations, and compare the results with the famous Lifshitz–Slezov shapes.

Section 2.9 is the most important in this chapter since, based on the analysis of nucleation mechanisms, we identify the kinetic control tools to tune the size distribution of nanostructures to the desired properties. We analyze quite generally the temperature, flux and deposition thickness dependences of the mean size, density and size distribution width of self-induced nanostructures. Ii is demonstrated that these dependences have a rather different character below and above the critical deposition thickness. The morphology of essentially super-critical nanostructures appears to be controlled by the growth kinetics rather than by thermodynamics. This important feature brings about additional kinetic tools to tailor the size distribution, an approach that will be further explored in Chap. 3 while considering the strain-induced 3D islands in lattice mismatched material systems.

Section 2.10 presents the Kolmogorov-Johnson-Mehl-Avrami model crystallization model that enables finding the fraction of crystallized volume or surface at the known nucleation and growth rates of nuclei at an early stage. Several important particular cases are considered, leading to different power indices of the Kolmogorov exponent. Typical time dependences of the crystallized fraction and the boundary of crystallization front are discussed. In Sect. 2.11, the hierarchical Kolmogorov approach is applied to studying the growth kinetics of 3D thin films. A particular emphasis is given to the time dependences of the mean thickness and surface roughness in the large time limit. Section 2.12 is devoted to studying the vertical growth rate of a crystal facet of arbitrary size, mediated by the 2D nucleation at the interface. Two approaches are considered: the Kashchiev interpolation formula and the Kolmogorov expression generalized to include the boundary effect. Both models show that the growth rate of a smaller facet is always slower than that of an infinitely large one. The size-dependent limitation on the growth rate is very important in the formation of nanowires by the vapor-liquid-solid mechanisms, as will be further explored in Chap. 4.

Section 2.13 concerns an isolated irreversible growth which occurs if the evaporation is neglected. In such models, the entire growth process stops when all the monomers are consumed by growing clusters. We consider some exactly solvable discrete models with size-linear and size-independent growth rates and show how the results are influenced by the non-linear monomer consumption. We also study in detail the spectrum shapes and their continuum approximations. In Sect. 2.14, we discuss an interesting application of the irreversible growth model to understanding the formation of linear peptide chains. Finally, Sect. 2.15 treats open irreversible systems with a material influx and sink, a model which is usefull in growth modeling of surface islands. We discuss the universal soolutions for the adatom concentration, surface density and size spectra which depend on only one scaling parameter.

2.1 Transformation to Invariant Size

Continuum nucleation theory is based on the Fokker-Plank type kinetic equation (1.173) with coefficients (1.174), coupled with the integral equation of material balance (1.176) at a given time dependence of $\Phi(t)$. Below in this chapter, we assume that the ideal supersaturation has a scaling dependence on t with the corresponding "flux index" $q > 0$ and is stabilized at time t_0 when the material influx is switched off:

$$\Phi(t) = \begin{cases} (t/t_\infty)^q, \ 0 \le t \le t_0 \\ (t_0/t_\infty)^q \equiv \Phi_0, \ t > t_0 \end{cases}. \tag{2.1}$$

As above, the macroscopic time t_∞ determines the characteristic rate of material influx, i.e., the time needed to reach the equilibrium concentration n_{1e} at a given monomer source. The most usual case of $q = 1$ corresponds to a constant influx into the system producing a linearly increasing total concentration of monomers. Obviously, (2.1) assumes the effective absence of the monomer sink such as desorption from the substrate surface in the case of thin films.

The kinetic equation (1.173) is obtained as a continuum approximation of finite-difference rate equations in the monomolecular limit, where the condensation-evaporation rates are related via the detailed balance, as described in Sect. 1.10. The formation energy of differently sized nuclei is given by (1.65). The steep exponential dependence of nucleation rate on supersaturation makes the entire system strongly non-linear. The non-linearity of kinetic equations, the size dependence of growth-evaporation rate constants, the variation of growth thermodynamics and kinetics in different time-size domains, the presence of transient growth steps, the fluctuation-induced effects and many other factors make the nucleation theory rather complex for the analysis. Such an analysis is considerably simplified upon the transformation of nucleation equations to a certain invariant size for which the regular growth rate of supercritical nuclei becomes size-independent and equals ζ/τ [3, 10, 17, 50–52]. As discussed in Sect. 1.7, when di/dt is given by (1.116), $di/dt \propto i^{(m-1)/m}$ with a certain growth index m, the invariant size is determined by $\rho = i^{1/m}$. For 3D or 2D islands growing from 3D or 2D environment in the ballistic regime ($m = 3$ or 2, respectively), the invariant size simply equals the radius of nucleus expressed in the units of interatomic distance in the condensed phase. More generally, if

$$\frac{di}{dt} = \phi(i)\frac{\zeta}{\tau} \tag{2.2}$$

with an arbitrary function $\phi(i)$, the invariant size is obtained as

$$\rho = \int_0^i \frac{di'}{\phi(i')}. \tag{2.3}$$

We now introduce the new distribution in terms of the invariant size, $f(\rho, t)$ such that

$$f(\rho, t)d\rho = n(i, t)di. \tag{2.4}$$

This important property preserves the number of nuclei in the intervals from i to $i + di$ and from ρ and $\rho + d\rho$, respectively. From (2.4), we obtain

$$n(i, t) = \frac{d\rho}{di} f(\rho, t) = \frac{1}{m\rho^{m-1}} f(\rho, t). \tag{2.5}$$

The last formula applies when di/dt is given by (1.116). Differentiation with respect to i should be then substituted to

$$\frac{\partial}{\partial i} = \frac{d\rho}{di}\frac{\partial}{\partial \rho} = \frac{1}{m\rho^{m-1}}\frac{\partial}{\partial \rho}. \tag{2.6}$$

In view of (1.65), (1.68) and (1.117), the derivative of the formation energy with respect to i [entering the kinetic coefficients given by (1.174)] can be put as

$$dF(i)/di = -(1-U)\ln(\zeta+1). \tag{2.7}$$

Here, the function U is defined by

$$U \equiv \left(\frac{\rho_c}{\rho}\right)^{m/d} = \frac{(d-1)a}{d\ln(\zeta+1)}\frac{1}{\rho^{m/d}}, \tag{2.8}$$

with

$$\rho_c(\zeta) = \left[\frac{(d-1)a}{d\ln(\zeta+1)}\right]^{d/m} \tag{2.9}$$

as the critical invariant size in space dimension d.

By using the rules (2.5) and (2.6) together with (2.7), the kinetic equation (1.173) can be re-arranged in terms of $f(\rho,t)$:

$$\frac{\partial f(\rho,t)}{\partial t} = -\frac{\partial}{\partial \rho}\left[\frac{(\zeta+1)-(\zeta+1)^U}{\tau}f(\rho,t) - \frac{\zeta+1+(\zeta+1)^U}{2\tau}\frac{\partial}{\partial \rho}\left(\frac{f(\rho,t)}{m\rho^{m-1}}\right)\right]. \tag{2.10}$$

Differentiation of $1/\rho^{m-1}$ in the second order fluctuational term results in a minor correction of the order $1/\rho^m$ at $m>1$ and becomes ineffective at $m=1$, so that the $1/\rho^{m-1}$ factor can be put before the derivative at large enough ρ. More importantly, $(\zeta+1)^U \to 1$ at $U\to 0$, i.e., where the size ρ is much larger than the critical size ρ_c. This reduces the regular growth rate to ζ/τ in the essentially super-critical region of sizes. At $\zeta\to 0$ and finite U, the regular growth rate becomes $[\zeta(1-U)]/\tau$. Linearization of (2.10) in U results in

$$\frac{\partial f(\rho,t)}{\partial t} = -\frac{\partial}{\partial \rho}\left[\frac{\zeta(1-U)}{\tau}f(\rho,t) - \frac{[2+\zeta(1+U)]}{2\tau}\frac{\partial}{\partial \rho}\left(\frac{f(\rho,t)}{m\rho^{m-1}}\right)\right]. \tag{2.11}$$

The boundary conditions to (2.10) or (2.11) are given by

$$f(\rho=\rho_+,t) = f_s[\zeta(t)] = \frac{J[\zeta(t)]\tau}{\zeta(t)} \tag{2.12}$$

at the boundary of the near-critical and super-critical regions ρ_+ and $f(\rho, t) \to 0$ at $\rho \to \infty$ (usually, an exponential decay is considered). The stationary distribution $f_s(\zeta)$ over the invariant sizes has the ρ-independent, drift form following from (1.194). Equation (2.12) shows simply that the nuclei emerge at the rate $J(\zeta)$ given by the Zeldovich formula (1.185) and surpass the boundary of the near-critical and super-critical region with the speed ζ/τ. As for the material balance, (1.176) in view of (2.4) takes the form

$$\Phi = \zeta + G; \tag{2.13}$$

$$G(t) = \frac{1}{n_{1e}} \int_0^\infty d\rho \rho^m f(\rho, t). \tag{2.14}$$

Equation (2.10) or its U-linear approximation (2.11) together with (2.13) and (2.14) constitute the closed system for the determination of the unknown $f(\rho, t)$ and $\zeta(t)$.

The factor $1 - U$ in the regular growth rate of (2.11) accounts for the variation of the critical size with time and is responsible for its interaction with the size spectrum, while the second derivative with respect to ρ describes the influence of fluctuations. Let us now consider the case with $U \to 0$, where the critical size is negligibly small compared to the nuclei of interest and therefore does not affect considerably the size spectrum. In the deterministic limit without the second derivative, (2.11) takes the conventional form [3, 10, 17, 25]

$$\frac{\partial f(\rho, t)}{\partial t} = -\frac{\zeta}{\tau} \frac{\partial f(\rho, t)}{\partial \rho}. \tag{2.15}$$

To further simplify the analysis, we introduce the most representative invariant size $z(t)$ corresponding to the maximum of size distribution at a given moment of time [17]. The most representative size equals the mean size for symmetrical distributions such as the Gaussian spectrum. Obviously, the time evolution of z obeys the equation $dz/dt = \zeta/\tau$; the most representative nuclei are born at the time moment t_* where the supersaturation reaches its maximum: $\zeta(t_*) = \zeta_{\max}$. Introducing the new variable z instead of t in (2.15), we get

$$\frac{\partial f(\rho, z)}{\partial z} = -\frac{\partial f(\rho, z)}{\partial \rho}. \tag{2.16}$$

This equation has the first integral $z - \rho$, so that any particular solution must be a function of $x(\rho, t) = z(t) - \rho$. Physically, since the growth rate of differently sized nuclei is ρ- independent, the entire size spectrum will evolve in time preserving its time-invariant shape, with only the mean size changing. Once formed, the time-invariant spectrum is maintained throughout the regular growth stage, until the Ostwald ripening or direct coalescence starts. This important feature is demonstrated

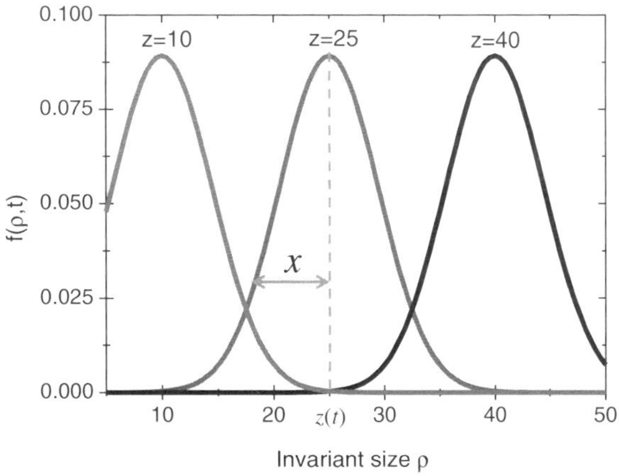

Fig. 2.1 Time-invariant Gaussian spectra $f(\rho, t)$ with dispersion $\psi = 20$ at $z = 10$, 25 and 40. The spectrum shape at any time depends only on the distance $x = z - \rho$ from the mean size z

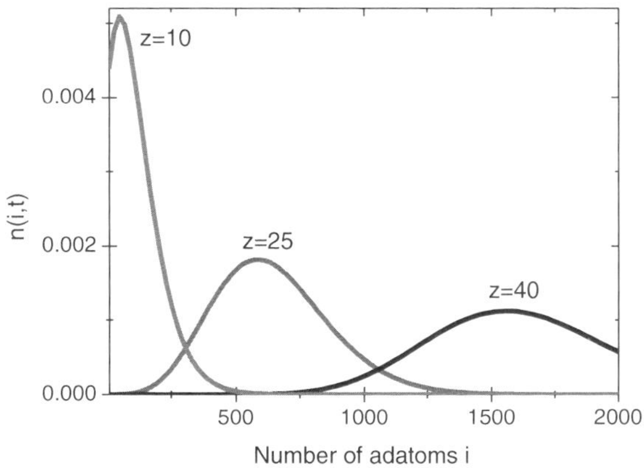

Fig. 2.2 Same distribution as in Fig. 2.1 plotted in terms of the number of atoms i

in Figs. 2.1 and 2.2 for a hypothetical system of 2D islands growing in the ballistic regime ($m = 2$, $\rho = i^{1/2}$). Figure 2.1 shows the time-invariant Gaussian spectra in terms of the invariant size ρ that equals the island radius in the units of lattice spacing. When plotted in terms of the number of adatoms in the islands i, the spectra loose the time invariant character: the distribution in Fig. 2.2 rapidly spreads and becomes asymmetric as the islands grow. Of course, the time invariance holds only in the deterministic approximation neglecting the fluctuation-induced broadening.

2.2 Nucleation Stage

In this section, we consider nucleation under the material influx, in other words, the case where the monomer source is turned on until the nucleation is effectively completed. Such a situation, schematized in Fig. 1.31, is rather typical for the epitaxy of nanostructures, the strain-induced Stranski–Krastanow islands [40, 41] being a typical example. In this case, the moment of time t_0 in (2.1), corresponding to the stabilization of the ideal supersaturation, is sufficiently larger than the time $t_* + \Delta t$ where the nucleation is disabled. We can therefore write down simply

$$\Phi(t) = \left(\frac{t}{t_\infty}\right)^q \tag{2.17}$$

on the time scale of interest. The nucleation stage is treated in the deterministic limit, which will be justified below.

As mentioned already, the solution to the deterministic kinetic equation (2.16) is a function of $x = z - \rho$. The particular form of this function is determined by the boundary condition. Assuming that the critical size remains small compared to the essentially super-critical nuclei during the entire nucleation stage, and that the subcritical and near-critical nuclei do not contribute significantly to the material balance owing to their small size, we put the boundary condition (2.12) at $\rho_+ = 0$. In this approximation, the super-critical nuclei are assumed to emerge with zero size. The solution to the kinetic equation

$$\frac{\partial f(\rho, z)}{\partial z} = -\frac{\partial f(\rho, z)}{\partial \rho};$$

$$f(\rho = 0, z) = f_s(z) \tag{2.18}$$

is given by

$$f(\rho, z) = f_s(x)\Theta(z - x), \tag{2.19}$$

where $\Theta(z - x)$ is the step function. The most representative size is now defined as

$$\frac{dz}{dt} = \frac{\zeta}{\tau}; \quad z(t = t_c) = 0, \tag{2.20}$$

where $t_c = t_*$ corresponds to the supersaturation maximum under the material influx. We note that this definition yields $z < 0$ at $t < t_c$ and $z > 0$ at $t > t_c$, showing that the spectrum maximum is reached at zero size before the moment of time t_c and becomes larger than zero after t_c.

The asymptotic Γ-method first developed by Kuni [17] and then widely used in different nucleation models [3, 10, 25, 51, 52, 66] relies upon an extremely high sensitivity of the Zeldovich nucleation rate and the stationary distribution to supersaturation, as discussed in Sect. 1.11. As follows from (1.185), the leading dependence of J and the corresponding $f_s = (J\tau)/\zeta$ on ζ is given by the exponential term $\exp[-F(\zeta)]$, while the pre-exponential factors only weakly depend on ζ. We can therefore introduce the following approximations for the nucleation barrier, the nucleation rate and the stationary distribution within the essential part of size spectrum:

$$F(\zeta) \cong F(\Phi_c) - \Gamma \frac{(\Phi_c - \zeta)}{\Phi_c}; \tag{2.21}$$

$$J(\zeta) \cong J(\Phi_c) \exp\left[-\Gamma \frac{(\Phi_c - \zeta)}{\Phi_c}\right]; \tag{2.22}$$

$$f_s(\zeta) \cong f_s(\Phi_c) \exp\left[-\Gamma \frac{(\Phi_c - \zeta)}{\Phi_c}\right]. \tag{2.23}$$

Here, $\Phi_c = (t_c/t_\infty)^q$ is the ideal supersaturation at the maximum supersaturation point $t = t_c$. In the above expressions, we assume that the difference between the ideal and actual supersaturation at the point of maximum is relatively small: $(\Phi_c - \zeta_c)/\Phi_c << 1$.

The quantity

$$\Gamma \equiv \Gamma_c = -\Phi_c \left(\frac{dF}{d\zeta}\right)_{\zeta=\Phi_c} >> 1 \tag{2.24}$$

is considered as the large parameter of the theory. From (1.67) and (1.68) it follows that

$$\Gamma = \left(\frac{\Phi_c}{\Phi_c + 1}\right) i_c(\Phi_c). \tag{2.25}$$

Therefore, the numerical value of Γ is of the order of the monomer number in the critical nucleus at $\zeta = \Phi_c$ and should be indeed of the order of ten or more. This justifies (2.22) and (2.23) at the nucleation stage (continuing as long as $(\Phi_c - \zeta)/\Phi_c \leq 1/\Gamma$) with only the leading exponential dependence on ζ being taken into account, since at a smaller supersaturation the unknown nucleation rate is very small anyway.

In (2.22), the supersaturation ζ may be considered as a function of time or the most representative size z. Very importantly, z increases linearly with time during the short range nucleation stage. This property follows directly from (2.20) at $\zeta \cong \Phi_c$:

$$z = \frac{1}{\tau} \int_{t_c}^{t} dt' \zeta(t') \cong \frac{\Phi_c}{\tau}(t - t_c). \tag{2.26}$$

This relationship enables presenting different characteristics at the nucleation stage either in terms of z or $t - t_c$ by the linear substitution of variables. Furthermore, the time dependence of ideal supersaturation (2.17) can be approximated by the linear function near the point t_c:

$$\Phi \cong \Phi_c + \frac{\Phi_c q(t - t_c)}{t_c} = \Phi_c + \frac{q\tau}{t_c} z, \tag{2.27}$$

where the last expression is obtained from (2.26). Using this in the material balance equation (2.13), we obtain the following representation for the indicator of exponents in (2.22) and (2.23)

$$\Gamma\frac{(\Phi_c - \zeta)}{\Phi_c} = \frac{\Gamma}{\Phi_c}G - cz. \tag{2.28}$$

The constant c is defined as

$$c = \frac{\Gamma q}{\Phi_c}\frac{\tau}{t_c} = \frac{\Gamma q}{\Phi_c^{(q+1)/q}}\frac{\tau}{t_\infty}. \tag{2.29}$$

Since the ratio of the microscopic time of island growth τ to the macroscopic pumping time t_∞ is very small, the constant c is much smaller than one in spite of strong inequality $\Gamma >> 1$.

In view of $f(\rho, t) = f(x) = f_s(x)$ and $x = z - \rho$, the substitution of variables in (2.14) leads to

$$G(z) = \frac{1}{n_{1e}}\int_{-\infty}^{z} dx(z - x)^m f(x). \tag{2.30}$$

From (2.23) and (2.28), $f(x) = f_s(\Phi_c)\exp[cx - (\Gamma/\Phi_c)G(x)]$. Inserting this expression into the right hand side of (2.30) we arrive at the closed integral equation for the total concentration of monomers in the nuclei $G(z)$:

$$G(z) = \frac{f_s(\Phi_c)}{n_{1e}}\int_{-\infty}^{z} dx(z - x)^m \exp\left[cx - \frac{\Gamma}{\Phi_c}G(x)\right]. \tag{2.31}$$

This equation is solved by iterations. The iterations rapidly approach the exact solution so that the first iteration $G^{(1)}(z)$ (corresponding to $G^{(0)}(x) = 0$ under the integral in the right hand side) gives already a good approximation to the exact solution with only slightly overestimated $G(z)$. In this approximation, $G(z)$ increases exponentially with z:

$$G(z) = \frac{f_s(\Phi_c)}{n_{1e}}\frac{m!}{c^{m+1}}\exp(cz), \tag{2.32}$$

where we use the notation $m!$ for the gamma-function $\Gamma(m+1)$ at arbitrary fractional m.

Differentiating (2.32) with respect to z and putting $z = 0$, we obtain $G'(0) = [f_s(\Phi_c)m!]/(c^m n_{1e})$. On the other hand, differentiating (2.28) and putting $z = 0$ in view of $\zeta'(z = 0) = 0$ gives $G'(0) = (\Phi_c/\Gamma)c$. Comparing the two results for $G'(0)$, we arrive at the non-trivial relationship between the thermodynamics of phase transition and the growth kinetics in the form

$$f_s(\Phi_c) = \frac{J(\Phi_c)\tau}{\Phi_c} = n_{1e}\frac{\Phi_c}{m!\Gamma}c^{m+1}. \tag{2.33}$$

This gives the transcendent equation for the self-consistent determination of the maximum supersaturation Φ_c, the time moment t_c and, consequently, the size distribution normalization [17, 51]. In some cases, (2.33) can be considerably simplified. Indeed, from (1.196) it follows that the quantity $[J(\Phi_c)\tau]/n_{1e}$ or $J(\Phi_c)\tau\sigma$ (depending on the normalization used) contains a very small value $\exp[-F(\Phi_c)]$, while the pre-exponential factor in many important systems is of the order of one. Dividing (2.33) by n_{1e} or $n_0 = 1/\sigma$, the corresponding small factor in the right hand side is $(\tau/t_\infty)^{m+1}$ or $\theta_{1e}(\tau/t_\infty)^{m+1}$, because both the characteristic time ratio τ/t_∞ and the equilibrium filling factor θ_{1e} are always much smaller than one. Therefore, the corresponding approximate solutions to (2.33) with the logarithmic accuracy are given by [16, 25]

$$F(\Phi_c) \cong (m+1)\ln\left(\frac{t_\infty}{\tau}\right), \quad F(\Phi_c) \cong (m+1)\ln\left(\frac{t_\infty}{\theta_{1e}^{1/(m+1)}\tau}\right) \tag{2.34}$$

where the both sides are much greater than one. Together with (1.67), this allows one to find the unknown maximum (also called as critical) supersaturation at the given kinetic parameters m and t_∞/τ.

Substitution of (2.33) into (2.32) yields the following expression:

$$G(z) = \frac{\Phi_c}{\Gamma}\exp(cz) \tag{2.35}$$

Inserting this into (2.22) and (2.23) gives the Kuni double exponential solutions for the stationary flux and the size distribution [10, 15, 17, 25, 51]

$$J(z) = J(\Phi_c)\exp(cz - e^{cz}) \tag{2.36}$$

$$f(x) = f_s(\Phi_c)\exp\left(cx - e^{cx}\right)\Theta(z-x) \tag{2.37}$$

According to (2.26) and (2.29),

$$cz = \frac{c\Phi_c}{\tau}(t - t_c) = q\Gamma\frac{(t-t_c)}{t_c} \tag{2.38}$$

during the entire nucleation stage. The time dependence of the nucleation rate is therefore given by

$$J(t) = J(\Phi_c)\exp(\alpha_c\bar{t} - e^{\alpha_c\bar{t}}); \quad \alpha_c = \frac{q\Gamma}{t_c} \tag{2.39}$$

in terms of the shifted time $\bar{t} = t - t_c$. The total density of nuclei is obtained as

$$N(z) = \int_{-\infty}^{z} dx f(x) \tag{2.40}$$

Integration of the size distribution given by (2.37) readily yields

$$N(z) = N_c \left[1 - \exp(-e^{cz})\right], \quad N_c = \frac{f_s(\Phi_c)}{c} = \frac{J(\Phi_c)}{\alpha_c}. \tag{2.41}$$

Using (2.29) and (2.32), the maximum density can be put in the form

$$N_c = n_{1e} \frac{q^m}{m!} \frac{\Gamma^{m-1}}{\Phi_c^{m(q+1)/q-1}} \left(\frac{\tau}{t_\infty}\right)^m, \tag{2.42}$$

the expression completing the analytical description of nucleation under the material influx.

Let us now analyze the obtained expressions. First, (2.37) and (2.39) show that the x- dependence of the time-invariant size distribution follows exactly the time dependence of the nucleation rate:

$$f(x) = \frac{\tau}{\Phi_c} J\left[\alpha_c \bar{t} \to cx\right]. \tag{2.43}$$

Thus, the size spectrum is obtained from the time dependence of J upon the above substitution of variables. This important property is a consequence of the deterministic kinetic equation with the stationary boundary condition at zero size. Second, the double exponential distribution is asymmetric, with a broader right tail. It might be approximated by the symmetric Gaussian

$$f(x) \cong [f_s(\Phi_c)/e] \exp[-(cx)^2/2] \tag{2.44}$$

with the dispersion $\psi_{nucl} = 1/c^2$ and the spectrum half-width $\Delta x \cong \sqrt{2}/c$. The characteristic duration of the nucleation stage equals

$$2\Delta t = \frac{2\sqrt{2}}{q\Gamma} t_c << t_c. \tag{2.45}$$

This important inequality demonstrates the timescale hierarchy: the nucleation stage appears much shorter than the time needed to reach the maximum supersaturation at a given monomer source. The nucleation stage is completed already at $z = z_{nucl} \cong 2/c$, where the total density of nuclei reaches N_c. A continuing monomer flux does not produce any new nuclei after this time. Third, (2.42) for N_c contains the kinetic factors such as the characteristic time ratio τ/t_∞ and the growth index m. This feature is of paramount importance for the controlled fabrication of nanostructures by different epitaxy techniques, showing that the nanostructure density can be tuned kinetically. Fourth, (2.36) at $z = 0$ demonstrates that $(\Phi_c - \zeta_c)/\Phi_c = 1/\Gamma$ at the maximum supersaturation point, which justifies the assumption made in writing (2.21)–(2.23).

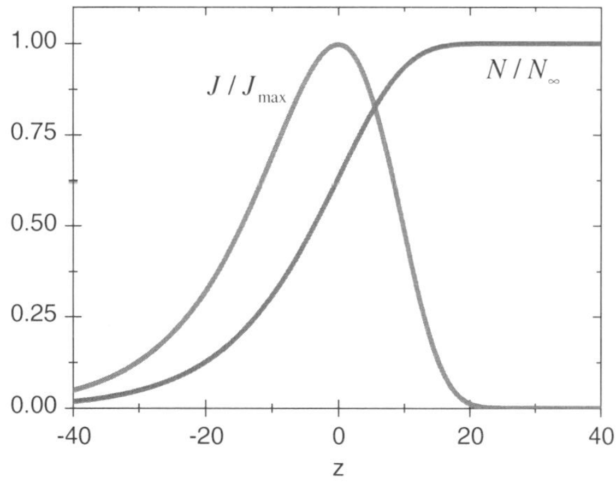

Fig. 2.3 Normalized nucleation rate and total density of nuclei plotted against the most representative invariant size z

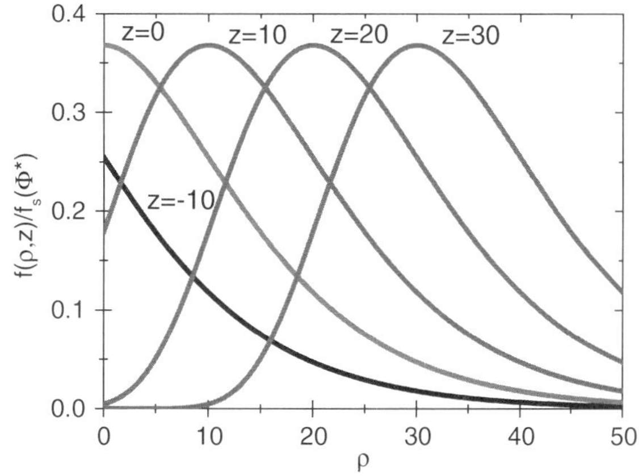

Fig. 2.4 Double exponential size spectra before ($z = -10$), at ($z = 0$) and after ($z = 10, 20$ and 30) the "nucleation catastrophe" at $t = t_c$. The spectrum is fully formed already at $z = 20$

Finally, the z dependences of the nucleation rate, the nuclei density and the corresponding size distributions at different z for a model system with $c = 0.1$ are presented in Figs. 2.3 and 2.4. Figure 2.3 clearly shows the density saturation after $z = 20$. Time evolution of the size spectrum in Fig. 2.4 demonstrates that the time moment t_* at which $z = 0$ corresponds to the so-called "nucleation catastrophe", where the maximum of size distribution develops and starts moving along the size axis. As discussed above, the double exponential shape of the size spectrum is preserved in time: the spectrum is transferred as a whole along the size axis as the nuclei grow. However, the time invariance can be broken by the fluctuation-induced spreading described by the second derivative with respect to size in the kinetic equation, as discussed later on.

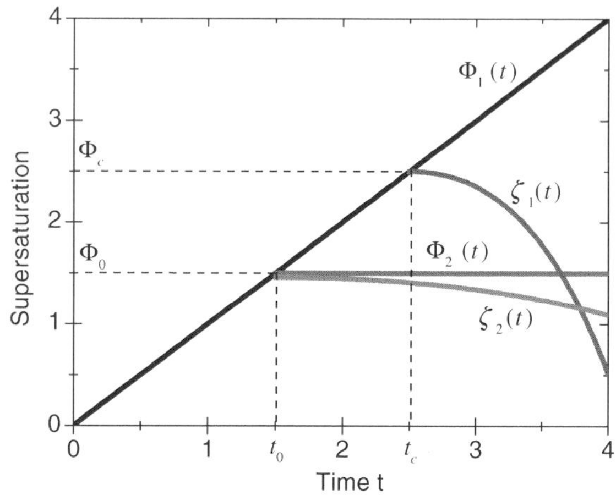

Fig. 2.5 Graphs schematizing the super-critical (1) and subcritical (2) modes of material influx into the system. When the monomer source is switched off at time t_0 before t_c, the maximum supersaturation equals $\Phi_0 < \Phi_c$. The system metastability is lower than in the super-critical case, resulting in a much slower decrease of supersaturation, a longer nucleation stage and a lower density of nuclei

2.3 Nucleation Modes

In this section, we consider the dependence of size distribution shapes on the moment of time t_0 (or the total amount of material Φ_0 introduced into the system by the moment of time t_0) where the ideal supersaturation given by (2.1) is stabilized. We saw earlier that the size distribution acquires a rather narrow double exponential shape when t_0 is noticeably larger than the critical time t_c corresponding to the supersaturation maximum under the material influx. Intuitively, when the monomer source is turned off well before the moment of time t_c, as shown in Fig. 2.5, one can anticipate a much longer nucleation stage, a lower density of nuclei and a rather asymmetric nucleation spectrum following a weakly decreasing nucleation rate at $t > t_0$. We will now construct a general analytical solution that contains the double exponential spectrum in the essentially super-critical range $\Phi_0 > \Phi_c$, the triangle-like spectrum in the essentially subcritical range $\Phi_0 < \Phi_c$ as well as the transition between them at $\Phi_0 \cong \Phi_c$ [51].

In the foregoing analysis, we use the Γ-method in terms of time variable instead of z, which is the equivalent representation in view of (2.43). Since $d\rho/dt = \zeta/\tau$ for any ρ and $f(\rho, z) = f_s[\zeta(z-\rho)]$, the total concentration of monomers in the condensed phase given by (2.14) can be presented in the equivalent Kolmogorov-Johnson-Mehl-Avrami [93–96] form:

$$G(t) = \frac{1}{n_{1e}} \int\limits_0^t dt' J(t') \rho^m(t', t). \qquad (2.46)$$

Here,

$$\rho(t', t) = \frac{1}{\tau} \int\limits_{t'}^t dt'' \zeta(t'') \qquad (2.47)$$

is the invariant size of a nucleus having emerged with zero size at time t'.

Analytical solutions for the nucleation rate will be obtained below in the case of a time-independent rate of material influx [$q = 1$ in (2.1)], which is typical, e.g., for most epitaxy techniques. Generalization for arbitrary flux index is straightforward. The equation of material balance given by (2.13) in view of (2.1) at $q = 1$ and $\Phi_* = t_*/t_\infty$ can be written down in the form

$$\Phi_* - \zeta(t) = \begin{cases} G(t) - \frac{(t-t_*)}{t_\infty}, & t \le t_0 \\ G(t) - \frac{(t_0-t_*)}{t_\infty}, & t > t_0 \end{cases}. \tag{2.48}$$

Here, t_* and Φ_* denote the moment of time and the value of ideal supersaturation at the point where the supersaturation ζ reaches its maximum. These values could be smaller or equal t_c and Φ_c, respectively, as shown in Fig. 2.5. The Γ-representation of the nucleation rate now writes down as

$$J(\zeta) \cong J(\Phi_*) \exp\left[-\Gamma\frac{(\Phi_* - \zeta)}{\Phi_*}\right], \tag{2.49}$$

where the large parameter Γ is given by (2.25) at $\Phi_c = \Phi_*$. Substitution of (2.48) into (2.49) readily yields

$$J(t) = J(\Phi_*) \begin{cases} \exp\left[\frac{\Gamma}{\Phi_*}\left(\frac{t-t_*}{t_\infty} - G(t)\right)\right], & t \le t_0 \\ \exp\left[\frac{\Gamma}{\Phi_*}\left(\frac{t_0-t_*}{t_\infty} - G(t)\right)\right], & t > t_0 \end{cases}. \tag{2.50}$$

Linearllization of (2.47) near the point of supersaturation maximum gives

$$\rho(t', t) \cong \frac{\Phi_*}{\tau}(t - t'). \tag{2.51}$$

Using (2.50) and (2.51) in (2.46), we arrive at the closed integral equation for $G(t)$:

$$G(t) = \frac{I(\Phi_*)}{n_{1e}}\left(\frac{\Phi_*}{\tau}\right)^m \times \begin{cases} \int\limits_0^t dt'(t-t')^m \exp\left[\frac{\Gamma}{\Phi_*}\left(\frac{t'-t_*}{t_\infty} - G(t')\right)\right], \quad t \le t_0 \\ \int\limits_0^{t_0} dt'(t_0-t')^m \exp\left[\frac{\Gamma}{\Phi_*}\left(\frac{t'-t_*}{t_\infty} - G(t')\right)\right] \\ + \exp\left(\frac{\Gamma}{\Phi_*}\frac{(t_0-t_*)}{t_\infty}\right) \int\limits_{t_0}^t dt'(t-t')^m \exp\left[-\frac{\Gamma}{\Phi_*}G(t')\right], \\ t > t_0 \end{cases} \tag{2.52}$$

This equation is solved by iterations, as in the super-critical case. The first iteration to the exact solution is again obtained by putting $G(t') = 0$ in all integral terms in the right hand side. In this approximation, the result for $G(t)$ in terms of the shifted time $\bar{t} = t - t_*$ has the form

$$G(\bar{t}) = \frac{\Phi_*}{\Gamma}\beta \begin{cases} \exp(\alpha\bar{t}), \ \bar{t} < \bar{t}_0 \\ \exp(\alpha\bar{t}_0)\left[1 + \dfrac{(\alpha(\bar{t} - \bar{t}_0))^{m+1}}{(m+1)!}\right], \ \bar{t} > \bar{t}_0 \end{cases} . \tag{2.53}$$

Here, $\alpha\bar{t}_0 = \Gamma(\Phi_0/\Phi_* - 1)$, $p! \equiv \int_0^\infty d\xi \xi^p \exp(-\xi)$ is the gamma-function $[p! \equiv \Gamma(p+1)]$ and the coefficients α and β are given by

$$\alpha = \frac{\Gamma}{\Phi_* t_\infty}; \quad \beta = m!\frac{\tau J(\Phi_*)}{n_{1e}}\left(\frac{\Phi_*^2}{\Gamma}\right)^m \left(\frac{t_\infty}{\tau}\right)^{m+1} . \tag{2.54}$$

Numerical studies of (2.52) verify a reasonable accuracy of the solution given by (2.53) for the calculations of the size distribution and the nucleus density. Equation (2.53) pertains only at the nucleation stage where the function $G(t)$ remains relatively small.

As shown in Fig. 2.5 and discussed in Sect. 2.2, at large enough $\Phi_0 > \Phi_c$, the maximum supersaturation $\Phi_* = \Phi_c$ is reached at the time moment $t_* = t_c = \Phi_c t_\infty < t_0$ due to the dynamic balance of the material influx and the monomer consumption by the emerging nuclei. The value of Φ_c is the critical supersaturation (e.g., the critical deposition thickness of the Stranski–Krastanow quantum dots at which the 2D to 3D growth transformation occurs [24, 25, 41, 42]). The case where $\Phi_0 > \Phi_c$ (and $\bar{t}_0 = t_0 - t_c > 0$) is usually referred to as the super-critical nucleation mode. All indexes "*" are replaced to "c" at the critical supersaturation. Since $\bar{t}_0 > 0$, we can use the first of (2.53) to find the derivative of $G(\bar{t})$ with respect to $\bar{t}$ at $\bar{t} = 0$: $G'(\bar{t} = 0) = (\Phi_c/\Gamma_c)\beta_c\alpha_c = \beta_c/t_\infty$. On the other hand, $\zeta'(\bar{t} = 0) = 0$ by definition. Upon differentiation of (2.13) and putting $\bar{t} = 0$, in view of (2.1) at $q = 1$ one obtains: $G'(\bar{t} = 0) = 1/t_\infty$. Comparing the two results for $G'(\bar{t} = 0)$ and using (2.54) for $\beta_{c,}$, we arrive at the self-consistent expression for the normalization factor of the nucleation rate (and the size distribution) in the super-critical nucleation mode

$$\beta_c = 1; \quad \tau J(\Phi_c) = \frac{n_{1e}}{m!}\left(\frac{\Gamma_c}{\Phi_c^2}\right)^m \left(\frac{\tau}{t_\infty}\right)^{m+1} . \tag{2.55}$$

This equation is exactly equivalent to (2.33) in view of (2.29) at $q = 1$.

When $\Phi_0 < \Phi_c$, the maximum supersaturation $\Phi_* = \Phi_0$ is reached simply due to the termination of material flux at time $t_* = t_0$, relating to $\bar{t}_0 = 0$. We can therefore write down quite generally

$$\Phi_* = \begin{cases} \Phi_0, \ \Phi_0 < \Phi_c \\ \Phi_c, \ \Phi_0 \geq \Phi_c \end{cases}. \tag{2.56}$$

The case with $\Phi_* = \Phi_0 < \Phi_c$ is referred to as the subcritical nucleation mode and denoted below by subscript "0" instead of "c". Equation (2.55) is no longer valid, which is why we should use the general Zeldovich expression for the nucleation rate and (2.54) for β_0 at $\Phi_* = \Phi_0$. Since $J(\Phi)$ is a very steep function, the value of β_0 exponentially decreases with decreasing Φ_0:

$$\beta_0 = m! \frac{\tau}{n_{1e}} J(\Phi_0) \left(\frac{\Phi_0^2}{\Gamma_0} \right)^m \left(\frac{t_\infty}{\tau} \right)^{m+1} = \left(\frac{\Gamma_c \Phi_0^2}{\Gamma_0 \Phi_c^2} \right)^m \frac{J(\Phi_0)}{J(\Phi_c)}, \tag{2.57}$$

where $\Gamma_0 = \Gamma(\Phi_0)$.

Analytical solutions for the nucleation rate are obtained by substitution of (2.53), (2.55) (in the super-critical nucleation mode) or (2.57) (in the subcritical nucleation mode) into (2.59). The distribution function is then found from (2.43): $f(x) = (\tau/\Phi_*) J \left[\alpha \bar{t} \to cx \right]$. The density of nuclei is obtained by integration of the nucleation rate over time:

$$N(t) = \int_0^t dt' J(t'). \tag{2.58}$$

After some straightforward calculations, we get the following results:

(1) In the super-critical nucleation mode $\Phi_0 > \Phi_c$:

$$J(\bar{t}) = J(\Phi_c) \begin{cases} \exp\left(\alpha_c \bar{t} - e^{\alpha_c \bar{t}}\right), \ \bar{t} \leq \bar{t}_0 \\ \exp\left[\alpha_c \bar{t}_0 - e^{\alpha_c \bar{t}_0} - \left(\delta_c (\bar{t} - \bar{t}_0)\right)^{m+1}\right], \ \bar{t} > \bar{t}_0 \end{cases}; \tag{2.59}$$

$$f(x) = f_s(\Phi_c) \begin{cases} \exp\left(cx - e^{cx}\right), \ x \leq z_0 \\ \exp\left[cz_0 - e^{cz_0} - (d_c(x - z_0))^{m+1}\right], \ z_0 < x \leq z \end{cases}; \tag{2.60}$$

$$N(\bar{t}) = N_c \begin{cases} 1 - \exp\left(-e^{\alpha_c \bar{t}}\right), \ \bar{t} \leq \bar{t}_0 \\ 1 - \exp\left(-e^{\alpha_c \bar{t}_0}\right) + \dfrac{[(m+1)!]^{1/(m+1)}}{(m+1)} \\ \times \exp\left[\dfrac{m}{(m+1)} \alpha_c \bar{t}_0 - e^{\alpha_c \bar{t}_0}\right] \gamma \left(\dfrac{1}{m+1}, \delta_c (\bar{t} - \bar{t}_0) \right), \\ \bar{t} > \bar{t}_0 \end{cases} \tag{2.61}$$

$$N_c = \frac{J(\Phi_c)}{\alpha_c} = \frac{n_{1e}}{m!} \frac{\Gamma_c^{m-1}}{\Phi_c^{2m-1}} \left(\frac{\tau}{t_\infty} \right)^m. \tag{2.62}$$

The shifted time is now given by $\bar{t} = t - t_c$. The coefficients in (2.59) to (2.61) are defined as follows:

$$\alpha_c = \frac{\Gamma_c}{\Phi_c t_\infty}; \quad \delta_c = \left[\frac{\exp(\alpha_c \bar{t}_0)}{(m+1)!}\right]^{1/(m+1)} \alpha_c;$$
$$c = \frac{\Gamma_c}{\Phi_c^2}\frac{\tau}{t_\infty}; \quad d_c = \left[\frac{\exp(\alpha_c z_0)}{(m+1)!}\right]^{1/(m+1)} c; \quad cz_0 = \alpha_c \bar{t}_0 = \Gamma_c\left(\frac{\Phi_0}{\Phi_c} - 1\right). \tag{2.63}$$

The function $\gamma(p, y) \equiv \int_0^y d\xi \xi^{p-1} \exp(-\xi)$ in (2.61) and below denotes the incomplete gamma-function. As above, in (2.60) and in all further expressions for the size distribution we always put $f(x) \equiv 0$ at $x > z$.

(2) In the subcritical nucleation mode $\Phi_0 < \Phi_c$

$$J(\bar{t}) = J(\Phi_0)\begin{cases} \exp\left(\alpha_0 \bar{t} - \beta_0 e^{\alpha_0 \bar{t}}\right), \ \bar{t} \le 0 \\ \exp\left[-\beta_0 - \left(\delta_0 \bar{t}\right)^{m+1}\right], \ \bar{t} > 0 \end{cases}; \tag{2.64}$$

$$f(x) = f_s(\Phi_0)\begin{cases} \exp\left(c_0 x - \beta_0 e^{c_0 x}\right), \ x \le 0 \\ \exp\left[-\beta_0 - (d_0 x)^{m+1}\right], \ 0 < x \le z \end{cases}; \tag{2.65}$$

$$N(\bar{t}) = \frac{I(\Phi_0)}{\alpha_0 \beta_0}\begin{cases} 1 - \exp\left(-\beta_0 e^{\alpha_0 \bar{t}}\right), \ \bar{t} \le 0 \\ 1 - \exp(-\beta_0) + \dfrac{[(m+1)!]^{1/(m+1)}}{m+1} \exp(-\beta_0)\beta_0^{m/(m+1)} \\ \times \gamma\left(\dfrac{1}{m+1}, \delta_0 \bar{t}\right), \ \bar{t} > 0 \end{cases} \tag{2.66}$$

The shifted time now equals $\bar{t} = t - t_0$. The coefficients in these equations are given by

$$\alpha_0 = \frac{\Gamma_0}{\Phi_0 t_\infty}; \quad \delta_0 = \left[\frac{\beta_0}{(m+1)!}\right]^{1/(m+1)} \alpha_0 = \left[\frac{\Gamma_0 \tau J(\Phi_0)}{(m+1)n_{1e}}\right]^{1/(m+1)} \Phi_0^{(m-1)/(m+1)} \frac{1}{\tau};$$

$$c_0 = \frac{\Gamma_0}{\Phi_0^2}\frac{\tau}{t_\infty}; \quad d_0 = \left[\frac{\beta_0}{(m+1)!}\right]^{1/(m+1)} c_0 = \left[\frac{\Gamma_0 \tau J(\Phi_0)}{(m+1)n_{1e}\Phi_0^2}\right]^{1/(m+1)}. \tag{2.67}$$

These results allow one to describe analytically the time dependences of the nucleation rate and nucleus density as well as the size distribution at the nucleation stage at the given material constants and in different regimes of material influx.

We now show how the obtained results enable to distinguish the two major nucleation modes leading to the triangular and the double exponential size distributions depending on the regime of material influx, and to find a non-trivial crossover between them. We first note that (2.59) and (2.64) for the nucleation rate, (2.60) and (2.65) for the size distribution and (2.61) and (2.66) for the nucleus density are consistent at $\Phi_0 = \Phi_c$, which follows from the relationships $\alpha_0 = \alpha_c$, $\beta_0 = \beta_c = 1$ and $\alpha_c \bar{t}_0 = cz_0 = 0$ at $\Phi_0 = \Phi_c$. If the material influx is switched off exactly at $\Phi_0 = \Phi_c$, the general solutions for the nucleation rate, the size distribution and the island density are reduced to (2.64)–(2.66) with $\beta_0 = 1; \alpha_0 = \alpha_c; c_0 = c; \delta_0 = \alpha_c/[(m+1)!]^{1/(m+1)}$ and $d_0 = c/[(m+1)!]^{1/(m+1)}$. The right tail of the size distribution at $x \leq 0$ is approximately Gaussian near the maximum: $f_+(x) \propto \exp\left[-(x/\Delta_+ x)^2\right]$. The left part of size spectrum decreases exponentially with the power index *m+1*: $f_-(x) \propto \exp\left[-(x/\Delta_- x)^{m+1}\right]$ The corresponding spectrum half-widths are given by $\Delta_+ x = \sqrt{2}/c: \ \Delta_- x = [(m+1)!]^{1/(m+1)}/c; \ \Delta_+ x \sim \Delta_- x \propto (\Phi_c^2/\Gamma_c)(t_\infty/\tau)$. The characteristic nucleation times are obtained from these half-widths upon the replacement of c to α_c:$\Delta_- t = \sqrt{2}/\alpha_c; \ \Delta_+ t = [(m+1)!]^{1/(m+1)}/\alpha_c; \ \Delta_- t \sim \Delta_+ t \propto (\Phi_c/\Gamma_c)t_\infty << t_\infty$. The distribution function and the nucleation rate are asymmetric; however, the left and the right half-widths are of the same order $\sim 1/c$. The characteristic duration of nucleation stage $\Delta t = \Delta_- t + \Delta_+ t$ is much smaller than t_∞.

In the essentially super-critical mode of material influx where $cz_0 = \alpha_c \bar{t}_0 = \Gamma_c (\Phi_0/\Phi_c - 1) >> 1$, the general solutions given by (2.59)–(2.62) take the Kuni double exponential form defined by (2.37), (2.39) and (2.41), (2.42) at $q = 1$. As discussed in Sect. 2.2, the nucleation rate and the size spectrum are approximately Gaussian near the point of maximum: $I(\bar{t}) \propto \exp\left[-(\bar{t}/\Delta t)^2\right]$; $f(x) \propto \exp\left[-(x/\Delta x)^2\right]$ with almost symmetric half-widths. The characteristic duration of the nucleation stage equals $\Delta t = 2\sqrt{2}/\alpha_c$ and the width of size distribution is given by$\Delta x = 2\sqrt{2}/c$. The total density of nuclei reached upon the completion of the nucleation stage tends to N_c. The value of N_c does not depend on the total amount of material in the system Φ_0.

In the essentially subcritical nucleation mode, the value of parameter β_0 defined by (2.57) is much smaller than one. The asymptotes of (2.64)–(2.67) at $\beta_0 \to 0$ are given by

$$J(\bar{t}) = J(\Phi_0) \begin{cases} 0, \bar{t} \leq 0 \\ \exp\left[-\left(\delta_0 \bar{t}\right)^{m+1}\right], \quad \bar{t} > 0 \end{cases}; \tag{2.68}$$

$$f(x) = f_s(\Phi_0) \begin{cases} 0, x \leq 0 \\ \exp\left[-\left(d_0 x\right)^{m+1}\right], \quad 0 < x \leq z \end{cases}; \tag{2.69}$$

$$N(\bar{t}) = \frac{1}{[(m+1)!]^{1/(m+1)}} \frac{J(\Phi_0)}{\delta_0} \begin{cases} 0, \bar{t} \leq 0 \\ \gamma\left(\frac{1}{m+1}, \delta_0 \bar{t}\right), \quad \bar{t} > 0 \end{cases}. \tag{2.70}$$

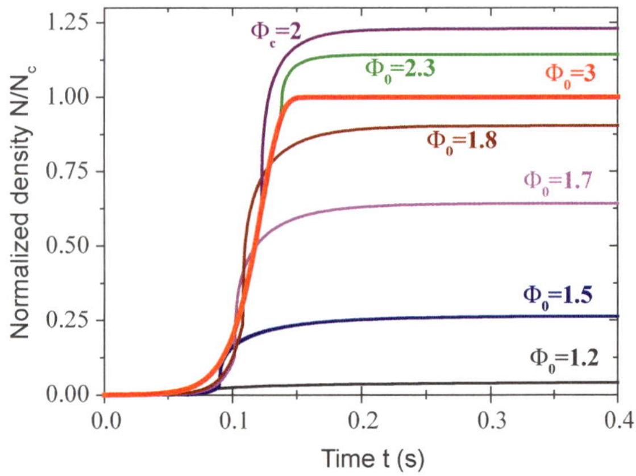

Fig. 2.6 Time dependences of normalized island density $N(t)/N_c$ at different values of the total amount of material Φ_0 changing from 1.2 to 3; the critical ideal supersaturation $\Phi_c = 2$

The duration of the nucleation stage in the essentially subcritical regime is much longer than the time t_0. The solutions given by (2.68)–(2.70) describe the nucleation at an instantaneous set of nucleation pulse [15, 53, 103]. All important characteristics of the nucleation process are driven by thermodynamics (the maximum nucleation rate $I(\Phi_0) \propto \exp[-F(\Phi_0)]$) and also by the power index m of the nucleus growth rate. Neither of these characteristics depends on the material flux. Of course, the maximum nucleus density is much lower than in the super-critical case.

Typical time dependences of the nucleus density $N(\bar{t})$ at different Φ_0, obtained from (2.61). (2.62) and (2.66), are presented in Fig. 2.6. In calculations, we used the typical parameters of 2D surface islands ($d = 2$) growing from the adatom sea in the ballistic regime (with $m = 2$) [51]: $a = 7.8$, $\tau = 3 \times 10^{-3}$ s, $\theta_{1e} = 6 \times 10^{-3}$, $\sigma = l_0^2$, $l_0 = 0.37$nm, $t_\infty/\tau = 20$. The critical supersaturation and the surface density in the essentialy super-critical nucleation mode are found to be: $\Phi_c = 2.0$ and $N_c = 5 \times 10^9 \text{cm}^{-2}$. Figure 2.6 demonstrates that the maximum density increases by the factor of 40 when Φ_0 increases from 1.2 to its critical value of 2. After that, the maximum density slightly decreases and stabilizes at N_c at $\Phi_0 = 3$. The bold curve at $\Phi_0 = 3$ does not change when Φ_0 is further increased, as it should be in the overcritical case. The characteristic nucleation time increases at lower Φ_0, however, the nucleation process is always a short scale compared to the time of deposition of 1 ML of 10 s.

Very importantly, the decrease of Φ_0 from the super-critical to subcritical range not only leads to the exponential decrease of the island density, but also develops the asymmetry and a considerable broadening of size distribution. As follows from (2.37) in the essentially super-critical and (2.69) in the essentially subcritical regimes of material influx, the distribution function transforms from only slightly symmetrical double-exponential to the triangle–like shape with an abrupt right tail at $x = z$ and a very broad distribution tail at $x < z$ This shape modification is well illustrated by the curves in Fig. 2.7, obtained from the general (2.60) and (2.65) with the same values of parameters as those used in Fig. 2.6, with Φ_0 decreasing from 3 in the

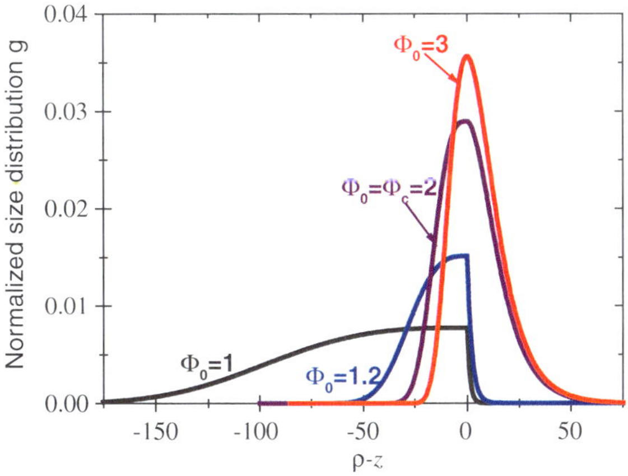

Fig. 2.7 Transition from the double exponential to the triangular size distribution with decreasing Φ_0

super-critical regime to 1 in the subcritical regime. The distributions are presented in the normalized form $g = f/N(\infty)$ as the functions of $\rho - z$, where the most probable size $z(t)$ would be of course different at different Φ_0.

2.4 Green Function at the Growth Stage

After the completion of the nucleation stage, the nuclei continue to grow by consuming the remaining monomers from the vapor phase. Central to modeling the size distribution at this regular growth stage is the question of whether the distribution over the invariant sizes preserves its time invariant shape that follows from the deterministic kinetic equation without the second derivatives. Let us first study the influence of kinetic fluctuations on the size spectra for a model system given by (1.139), where the evaporation and the material balance are neglected [50]. As discussed in Sect. 1.9, the exact solution to the finite difference kinetic equations at $\alpha = 0$ (or $m = 1$) is given by the Poisson distribution (1.142). The latter tends to the Gaussian (1.143) with the spreading dispersion $\psi(t) = t$, which is the exact solution for the Green function of the continuum kinetic equation (1.145). In the deterministic limit, the corresponding Green function would be given by $\delta(i - t)$. We note that the invariant size $\rho = i$ whenever $\alpha = 0$ (see Table 2.2), so that the spectrum shape is expected to persist in time in terms of i, which is in qualitative contradiction with (1.142). The exact solution therefore breaks the major property of (2.16): the width of the size spectrum now increases with t as $\sqrt{t}$, converting the initially narrow Gaussian package to a rather broad distribution as the particles grow. Therefore, the deterministic approximation is qualitatively incorrect at $\alpha = 0$.

The exact solution to the discrete (1.139) at $\alpha = 1$ is given by the geometrical distribution (1.147), while the invariant size is determined by the second (1.117): $\rho = \ln i$. Using the known definition of exponent, $[1 + i\exp(-t)/i]^i \to \exp[i\exp(-t)]$ at finite $i\exp(-t)$ and $i \to \infty$, introducing the invariant size ρ and the new distribution $g(\rho,t) = p[i(\rho),t]\exp(\rho)$, we arrive at the following asymptotical behavior of the size spectrum at large ρ:

$$g(\rho,t) = \exp\left[\rho - t - e^{(\rho-t)}\right]. \tag{2.71}$$

Interestingly, we have obtained the double exponential distribution as the large time asymptote of the exact solution to the discrete kinetic equations at $\alpha = 1$. This solution is a function of $\rho - t$ and preserves all the properties of time-invariant growth. In particular, (2.71) can be further simplified to the Gaussian $g(\rho,t) \propto \exp[-(\rho-t)^2/2]$ with the time-independent dispersion $\psi = 1$. Therefore, the exact solution and the Green function $\delta(\rho - t)$without fluctuations at $\alpha = 1$ lead to very similar results. No distribution spreading takes place and the fluctuation term in the continuum kinetic equation can be safely neglected.

Since the exact solutions exhibit spreading at $\alpha = 0$ and no spreading at $\alpha = 1$, one may suggest that the tendency for spreading somehow decreases toward larger α. To investigate this point, we now proceed to the analysis of the continuum kinetic equation in the case of arbitrary fractional α in the interval $0 < \alpha < 1$. By using the general rules given by (2.5) and (2.6), the continuum second order equation for the normalized distribution $g(\rho,t) = [di/d\rho]p(i,t)$ in terms of the invariant size $\rho = i^{1/m}$ is obtained from (1.139) in the form

$$\frac{\partial g(\rho,t)}{\partial t} = -\frac{\partial g(\rho,t)}{\partial \rho} + \frac{\partial}{\partial \rho}\left[\frac{1}{2m\rho^{m-1}}\frac{\partial g(\rho,t)}{\partial \rho}\right], \tag{2.72}$$

where, as above, $m = 1/(1-\alpha)$. The second order derivative with the multiplying factor $1/\rho^{m-1}$ corrects the deterministic equation for fluctuations. This equation cannot be integrated analytically with arbitrary $m > 1$. We note, however, that the tendency to suppress spreading at larger α follows directly from (2.13): the fluctuation term is of the same order as the first derivative at $m = 1$ $(\alpha = 0)$ but disappears at $m \to \infty (\alpha \to 1)$.

To analyze the spreading at arbitrary index α, we search for the asymptotical solutions to (2.72) at large t in the form of the diffusionlike Gaussian

$$g(\rho,t) = \frac{1}{\sqrt{2\pi C t^{\beta}}}\exp\left[-\frac{(\rho-t)^2}{2Ct^{\beta}}\right] \tag{2.73}$$

with the mean size t and the scaling spectrum width $\sqrt{Ct^{\beta}}$. Obviously, (2.73) contains the earlier results (1.143) at $C = \beta = 1$ and (2.71) at $\beta = 0$. The unknown index $\beta \geq 0$ describes the power-law time dependence of the dispersion. Of course, (2.73) is reasonable only if the spectrum width increases with time slowly than t, i.e., when $\beta < 2$. For such well-centered distributions, (2.72) can be simplified by the substitution of $1/\rho^{m-1}$ to $1/t^{m-1}$ within the essential part of the spectrum:

$$\frac{\partial g(\rho,t)}{\partial t} = -\frac{\partial g(\rho,t)}{\partial \rho} + \frac{1}{2mt^{m-1}}\frac{\partial^2 g(\rho,t)}{\partial \rho^2}. \tag{2.74}$$

Upon substitution of (2.73) into (2.74), it can be easily seen that the cancellation of all terms is exactly equivalent to the asymptotic condition at large t

$$\beta = \frac{1}{mCt^{m-2+\beta}}. \tag{2.75}$$

At $m < 2$ (or $\alpha < 1/2$), this requires two equalities for β and C of the form

$$\beta = 2 - m, \quad C = \frac{1}{m(2-m)}, \quad m < 2. \tag{2.76}$$

At $m > 2$ (or $\alpha > 1/2$), (2.75) cannot be satisfied at any positive β, because the right hand side vanishes at $t \to \infty$. The only possibility therefore yields

$$\beta = 0, \quad m \geq 2, \tag{2.77}$$

with an arbitrary C defined by the initial dispersion. The width of the size spectrum increases with time as $t^{\beta/2}$ at $\alpha < 1/2$ and remains constant at $\alpha > 1/2$. The transition between the two regimes occurs at $\alpha = 1/2$, where β becomes zero. This corresponds to the logarithmic spreading. The limit cases at $\alpha = 0$ and $\alpha = 1$ directly follow from (2.76) and (2.77). Looking at the growth indices summarized in Table 1.2, we conclude that only two of the eight systems considered, the exotic case 4 of a 2D island growing directly from vapor, and droplet condensation in 3D space in the ballistic regime (case 7), will withstand spreading and remain stable under kinetic fluctuations at a time-independent supersaturation. The growth of 2D islands in the ballistic regime corresponds to the crossover at $\alpha = 1/2$ with a week logarithmic spreading. Other five systems will definitely undergo spreading, with the maximum broadening corresponding to the Poissonian case at $\alpha = 0$.

The above analysis applies to the model systems at a fixed and very high supersaturation. In real condensing systems, the tendency for the fluctuation-induced broadening will be even stronger, because ζ usually tends to zero at the asymptotic growth stage, the effect leading to the increase of the fluctuational term in the kinetic equation [50–52]. Let us now consider the general kinetic equation (2.11) where the second derivative with respect to ρ is included, under the assumption of a small critical size ($U \to 0$). As in Sect. 2.2, we introduce the most representative invariant size z by definition

$$\frac{dz}{dt} = \frac{\zeta}{\tau}; \quad z(t = t_*) = 0, \tag{2.78}$$

where the moment of time t_* corresponds to the supersaturation maximum. Assuming that the nucleation stage is already completed and the nucleus density has reached its maximum value $N \equiv N(\infty)$, we introduce the normalized size distribution $g(\rho, z) = f(\rho, z)/N$. The kinetic equation for the distribution $g(\rho, z)$ now writes down as

$$\frac{\partial g(\rho, z)}{\partial z} = -\frac{\partial g(\rho, z)}{\partial \rho} + \frac{\partial}{\partial \rho}\left[\frac{\varphi(z)}{m\rho^{m-1}}\frac{\partial g(\rho, z)}{\partial \rho}\right]. \tag{2.79}$$

The function $\varphi(z)$ is determined by

$$\varphi(z) = \frac{1}{2} + \frac{1}{\zeta(z)}, \tag{2.80}$$

where the supersaturation ζ must be expressed in terms of z. This equations shows an infinite increase of φ at $\zeta \to 0$. The material balance given by (2.13) and (2.14) becomes

$$\Phi = \zeta + G; \tag{2.81}$$

$$G(z) = \frac{N}{n_{1e}}\int_0^\infty d\rho\rho^m g(\rho, z), \tag{2.82}$$

Following [51], we search for the asymptotic solution to (2.79) at large z (and ρ) in the form of the Gaussian with a z-dependent dispersion $\psi(z)$:

$$g_0(\rho, z) = \frac{1}{\sqrt{2\pi\psi(z)}}\exp\left[-\frac{(\rho - z)^2}{2\psi(z)}\right]. \tag{2.83}$$

It can be easily seen that this Gaussian is the exact solution to the equation

$$\frac{\partial g(\rho, z)}{\partial z} = -\frac{\partial g(\rho, z)}{\partial \rho} + \frac{\varphi(z)}{mz^{m-1}}\frac{\partial^2 g(\rho, z)}{\partial \rho^2}, \tag{2.84}$$

if the dispersion obeys the differential equation

$$\frac{d\psi}{dz} = \frac{2\varphi(z)}{mz^{m-1}}. \tag{2.85}$$

For the Green function with the initial condition $g_0(\rho, z_{nucl}) = \delta(\rho - z_{nucl})$, this equation should be solved with the boundary condition $\psi(z = z_{nucl}) = 0$, where z_{nucl} is the most representative size upon the completion of nucleation.

It is well understood intuitively that, at large enough z and for sufficiently narrow distributions, the function $1/\rho^m$ in the right hand side of (2.79) can be changed to $1/z^m$, which reduces (2.79) to (2.84). More precisely, (2.79) in the essential part of the Gaussian spectrum $|\rho - z| \le \sqrt{2\psi}$ can be written as

$$\frac{\partial g(\rho, z)}{\partial z} = -[1 + O(z^{-m})]\frac{\partial g(\rho, z)}{\partial \rho} + \frac{\varphi(z)}{mz^{m-1}}[1 + O(\sqrt{2\psi}/z)]\frac{\partial^2 g(\rho, z)}{\partial \rho^2}. \tag{2.86}$$

Here, the term $O(\sqrt{2\psi}/z)$ is the correction for the substitution $\rho \to z$ in the fluctuational term with second derivative with respect to size and $O(z^{-m})$ is a negligible fluctuation-induced correction for the regular growth rate. Transformation of (2.79) to the simplified (2.84) and the validity of the asymptotic solution given by (2.83) and (2.85) are therefore justified whenever

$$\sqrt{2\psi}/z << 1 \tag{2.87}$$

at large z, i.e., when the spectrum width increases slower than the mean size. In particular, for the power law dependence $\psi \sim z^{\nu}$ with the index $\nu < 2$, the symbolic form of (2.86) is given by

$$z^{\nu/2} + Az^{\nu-1} = z^{\nu/2} + Cz^{\nu/2-m} + Bz^{\nu-1} + Dz^{3\nu/2-2} \tag{2.88}$$

with certain coefficients A, B, C and D. Trivial identity at $A = B = C = D = 0$ is equivalent to the deterministic kinetic equation. The equality $A = B$ cancels the terms $z^{\nu-1}$, which corresponds to the Gaussian solution given by (2.83) and (2.85). The next power law correction $z^{3\nu/2-2}$ (which is always weaker than $z^{\nu-1}$ as long as $\nu < 2$) increases at $4/3 < \nu < 2$ and decreases absolutely at $\nu < 4/3$, while the correction for the regular growth rate $z^{\nu/2-m}$ always decreases absolutely at $m \geq 1$.

The solution to the kinetic equation (2.84) with arbitrary initial condition at $z = z_{nucl}$

$$g(y, z = z_{nucl}) = \varphi(y), \tag{2.89}$$

where $y = \rho - z_{nucl}$, is now obtained by the convolution of the Green function $g_0(\rho, z)$ with the initial condition:

$$g(y, z) = \frac{1}{\sqrt{2\pi\psi(z)}} \int_{-\infty}^{\infty} dy' \varphi(y') \exp\left[-\frac{(z - z_{nucl} - y + y')^2}{2\psi(z)}\right]. \tag{2.90}$$

This equation gives the general solution to the second order kinetic equation over the entire growth stage at the known size distribution after the end of the nucleation stage. In particular, if nucleation proceeds under the material influx, the initial nucleation spectrum is given by the double exponential distribution (2.37) at $y = -x = \rho - z_{nucl}$ and $z_{nucl} \cong 2/c$. In this case, the general solution is obtained in the form

$$g(y, z) = \frac{c}{\sqrt{2\pi\psi(z)}} \int_{-\infty}^{\infty} dy' \exp\left[-cy' - e^{-cy'}\right] \exp\left[-\frac{(z - z_{nucl} - y + y')^2}{2\psi(z)}\right]. \tag{2.91}$$

In the Gaussian approximation for the nucleation distribution $\varphi(y)$, (2.91) is reduced to

$$g(\rho, z) = \frac{1}{\sqrt{2\pi\psi(z)}} \exp\left[-\frac{(\rho - z)^2}{2\psi(z)}\right], \tag{2.92}$$

where the dispersion $\psi(z)$ obeys (2.85) with the initial condition $\psi(z = z_{nucl}) = \psi_{nucl}$. This formula holds because the result of the convolution of two Gaussians is also the Gaussian where the mean sizes and dispersions of the initial distributions are summed up.

Substitution of the Gaussian distribution given by (2.92) into (2.82) in view of strong inequality (2.87) leads to the mono-disperse approximation for $G(z)$ [17, 25, 50–52]

$$G(z) = \frac{N}{n_{1e}} \frac{(2\psi)^{m/2}}{\sqrt{\pi}} \int\limits_{-z/\sqrt{2\psi}}^{\infty} dx (x + z/\sqrt{2\psi})^m \exp(-x^2) \cong \frac{N}{n_{1e}} z^m. \tag{2.93}$$

This formula shows simply that the total number of monomers in the nuclei approximately equals the nucleus density multiplied by the mean number of monomers in the nuclei z^m. Substitution of this result into (2.81) with the known time dependence of ideal supersaturation and $\zeta(t) = \tau dz/dt$ allows one to find the time evolution of mean size $z(t)$ on the growth stage. This also determines the dependences of supersaturation as a function of t and z. After that, one can obtain the function $\varphi(z)$, which is required for the self-consistent calculation of the dispersion $\psi(z)$ by integration of (2.85).

2.5 Absence of Fluctuation-Induced Spreading at the Nucleation Stage

While accounting for kinetic fluctuations is absolutely necessary for the adequate description of the near-critical nucleus (for which the regular growth rate is small by definition), there has been relatively few attempts to consider the second order kinetic equation in the super-critical region of sizes [14, 19, 89]. We will now estimate the fluctuational effect at the nucleation stage using the results obtained in Sects. 2.3 and 2.4. Among the nucleation modes discussed in Sect. 2.3 and presented in Fig. 2.7 (obtained in the deterministic approximation), the minimum dispersion at the end of the nucleation stage evidently relates to the double exponential distribution given by (2.37):

$$\psi_{nucl} \cong 1/c^2. \tag{2.94}$$

From (2.29), the coefficient c contains a very small ratio of microscopic and macroscopic times τ/t_∞ and therefore must be much smaller than one. This result looks quite natural, because the quantity $1/c$ gives the characteristic size scale of the distribution function. In order to compare the minimum nucleation dispersion ψ_{nucl} to the fluctuational broadening during the nucleation stage, we assume that all embryos emerge simultaneously at $t = t_c$ with zero initial dispersion. This allows us to use the Green function defined by (2.83) and (2.85) at the nucleation stage. As follows from the results of Sect. 2.2, the change of supersaturation during the entire nucleation stage remains relatively small. Equation (2.85) for the z-dependent dispersion originating from the fluctuation-induced spreading can be now integrated at a constant $\zeta \cong \Phi_c$:

$$\psi_{fluct}(z) = \frac{(1 + 2/\Phi_c)}{m(2-m)} z^{2-m}. \tag{2.95}$$

The nucleation stage is fully completed at $z_{nucl} \cong 2/c$, which follows directly from (2.36), (2.37) and (2.41).

Comparison of (2.94) and (2.95) at $z = z_{nucl} = 2/c$ gives the ratio of the fluctuation-induced to the nucleation dispersion of the size spectrum in the form

$$\psi_{fluct}(z_{nucl})/\psi_{nucl} = \frac{(1 + 2/\Phi_c)}{m(2-m)} 2^{2-m} c^m \sim c^m << 1. \tag{2.96}$$

For all interesting particular condensing systems, the growth index m obeys the inequality $m \geq 1$, so that the fluctuation effects can be safely neglected at the nucleation stage. This important conclusion validates the results obtained in Sects. 2.2 and 2.3 in the deterministic approximation, as well as the earlier results of [10, 15, 17, 25, 103]. The strong inequality following from (2.96) does not depend on the flux index q, because the quantity $c = [\Gamma q/\Phi_c^{(q+1)/q}](\tau/t_\infty)$ in the general case remains much smaller than one regardless of q. Obviously, the fluctuation-induced effects would be even less noticeable for a broader triangle-like distribution formed in the subcritical nucleation mode. The fluctuation spreading is switched on only at the growth stage, for which the normalized Green function given by (2.83) and (2.85) is directly applicable. To complete the analytical description of the growth stage, we need to consider the time dependences of the most representative size z, the supersaturation ζ and the size dispersion ψ, which is the aim of the next section.

2.6 Time Evolution of Size Spectrum

The dependences $z(t)$, $\zeta(t)$ and $\zeta(z)$ should be found from the equation of material balance given by (2.81). In view of (2.93) for $G(z)$ at $\sqrt{2\psi}/z << 1$ and $\zeta = \tau dz/dt$, this equation can be put in the form

$$\tau \frac{dz}{dt} = \Phi(t) - \frac{N}{n_{1e}} z^m. \tag{2.97}$$

Substitution of the known time dependence of the ideal supersaturation given by (2.1) into (2.97) reduces the latter to the closed non-linear differential equation for the most representative size $z(t)$.

Let us first consider the case of a constant rate of material influx, relating to $q = 1$ in (2.1). In the case of nuclei growth under the material influx (i.e., at $t_0 \to \infty$ and $\Phi(t) = t/t_\infty$ for all t of interest), (2.97) can be re-written as

$$\frac{du}{d\eta} = 1 - u^m + \varepsilon\eta; \quad u(\eta = 0) = 0. \tag{2.98}$$

Here, we introduced the normalized invariant size $u \equiv z/z_r$ and the normalized time $\eta \equiv \bar{t}/t_r$, with $z_r \equiv (\Phi_c n_{1e}/N)^{1/m}$ as the characteristic relaxed size and $t_r \equiv (\tau/\Phi_c)z_r$ as the relaxation time. The parameter $\varepsilon \equiv t_r/t_c = t_r/(\Phi_c t_\infty)$ must be much smaller than one, because the growth stage (at which the supersaturation decreases to zero) is much shorter than the macroscopic time t_∞. Equation (2.98) does not have analytical solutions at arbitrary growth index m. Besides the trivial case of $m = 1$, its exact solution is known at $m = 2$ (2D islands growing from the adatom sea in the ballistic regime), where the mean size can be obtained in the form of combination of Airy functions and their derivatives. This solution is given in Sect. 1.12. In certain cases, the time scale hierarchy ($\varepsilon << 1$) allows one to use approximate solution of the form [52]

$$u = u^{(0)}[1 + \varepsilon\eta]^{1/m}, \tag{2.99}$$

where $u^{(0)}$ is the solution to (2.98) at $\varepsilon = 0$

$$\frac{du}{d\eta} = 1 - u^m. \tag{2.100}$$

Analytical or numerical solution of (2.98) and expressing the mean size as a function of time allows us to find the time dependence of supersaturation, $\zeta(t) = \tau dz/dt$. Rescaling it in terms of z we obtain the function $\varphi(z)$ in (2.80) and finally the dispersion $\psi(z)$ by integrating (2.85).

In particular, at the asymptotic growth stage where the supersaturation is much smaller than $G(t)$, the leading power law asymptote of the most representative size is readily obtained from (2.97) in the form

$$z(t) = \left(\frac{n_{1e}}{N}\frac{t}{t_\infty}\right)^{\frac{1}{m}}. \tag{2.101}$$

The supersaturation $\zeta = \tau dz/dt$ therefore scales as

$$\zeta = \frac{1}{m}\frac{\tau}{t_\infty}\left(\frac{n_{1e}}{N_c}\right)^{\frac{1}{m}}\left(\frac{t}{t_\infty}\right)^{\frac{1}{m}-1} = \frac{1}{m}\frac{\tau}{t_\infty}\left(\frac{n_{1e}}{N_c}\right)z^{1-m}. \tag{2.102}$$

Using the last expression in (2.80), inserting the result for $\varphi(z)$ into (2.85) and integrating it with zero initial condition at $z = z_{nucl}$, we arrive at

$$\psi(z) = \frac{1}{m(2-m)}\left(z^{2-m} - z_{nucl}^{2-m}\right) + \frac{2\omega}{q}\left(z - z_{nucl}\right), \qquad (2.103)$$

which is the asymptotic result of [50] in the particular case of $q = 1$. The asymptotic expressions at arbitrary q will be presented in section 2.8, along with the analysis of the conditions for the Ostwald ripening. The parameter ω in (2.103) is given by

$$\omega = \frac{N}{n_{1e}}\frac{t_\infty}{\tau} = \frac{1}{m!}\frac{\Gamma^{m-1}}{\Phi_c^{2m-1}}\left(\frac{\tau}{t_\infty}\right)^{m-1}. \qquad (2.104)$$

The second equality in the right hand side is obtained by using (2.42) for the density of nuclei $N = N_c$ in the super-critical regime of material influx. Equation (2.103) demonstrates that the dispersion generally contains two contributions. The first term grows with z at $m \le 2$ [the case of $m = 2$ corresponds to weak logarithmic spectrum spreading as $\ln^{1/2}(z/z_{nucl})$], and disappears at $m > 2$. For example, the corresponding broadening of the spectrum width of 3D droplets growing in the diffusion regime at $m = 3/2$ scales with z as $z^{1/4}$, while the spectrum of 2D surface islands growing in the diffusion regime at $m = 1$ spreads as $z^{1/2}$.

At $m > 1$, the second term in (2.103) increases faster than the first one, giving the spectrum width $\sqrt{\psi} \propto z^{1/2}$. Actual spreading depends, however, on the numeric value of ω. Smallness of the pre-factor can lead to a negligible fluctuation-induced effect, particularly at $m \ge 2$. Indeed, (2.104) for ω contains the small factor τ/t_∞ (as the ratio of microscopic to macroscopic times) multiplied by the large factor Γ (the quantity of the order of the critical size at supersaturation Φ_c), both to the power $m - 1$. The value of ω is always of the order of one at $m = 1$, the case where the spectrum spreads as $z^{1/2}$ with no small pre-factors in the dispersion. The effect of fluctuation-induced broadening therefore strongly depends on the material constants, the growth mode and the rate of material influx, because ω can be of the order or much smaller than one depending on these parameters, as demonstrated by the numerical examples given below.

We first consider the semi-analytical solutions given by (2.91) for the size distribution, (2.97) for the most representative size, and (2.80) and (2.85) for the dispersion. Results of numerical calculations of their time evolutions are presented in Figs. 2.8, 2.9, 2.10 and 2.11. In calculations, we used the typical parameters of 2D surface islands ($d = 2$) growing from the adatom sea [51, 52, 102]: $\tau = 3 \times 10^{-3}$ s, $t_\infty/\tau = 20$, $a = 7.8$, $n_{1e} = 4.4 \times 10^{12}\,\mathrm{cm}^{-2}$ in two different growth modes: the ballistic regime ($m = 2$) and the diffusion regime ($m = 1$) with $\Phi_c = 2.0$, $c = 0.1$, $\Gamma = 8.1$ at $m = 2$ and $\Phi_c = 2.8$, $c = 0.041$, $\Gamma = 6.3$ at $m = 1$. All calculations are performed under a constant material influx at $q = 1$. The island density $N = N_c$, obtained by (2.62), equals $5 \times 10^9\,\mathrm{cm}^{-2}$ at $m = 2$ and to $7.7 \times 10^{10}\,\mathrm{cm}^{-2}$ at $m = 1$. Figure 2.8 shows that the dispersion at $m = 2$ is rather

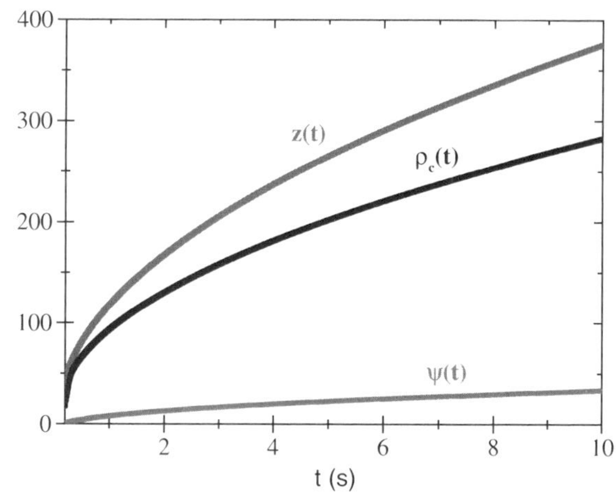

Fig. 2.8 The most representative size $z(t)$, the critical size $\rho_c(t)$ and the dispersion of the Green function $\psi(t)$ of 2D islands at $m = 2$, obtained from the semi-analytical expressions

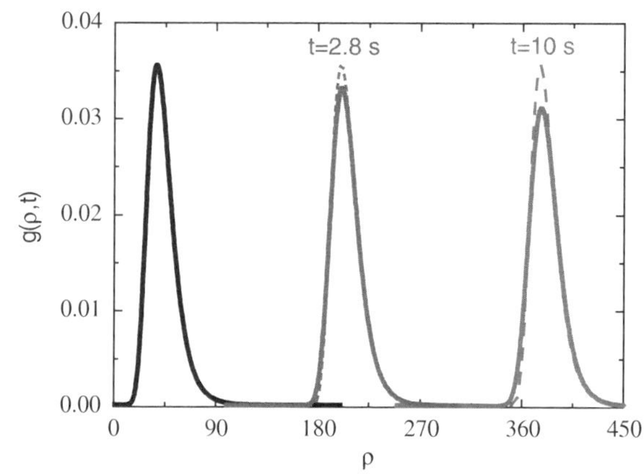

Fig. 2.9 Time evolution of size spectrum at $m = 2$, the *dashed lines* show the time-invariant double exponential spectrum in the deterministic approximation

small, which is also demonstrated by an almost negligible broadening of the corresponding size distributions in Fig. 2.9. This is because the first term in (2.103) for the dispersion is only logarithmical, and the pre-factor ω in the second term is much smaller than one ($\omega = 0.025$ for the parameters considered). The critical invariant size $\rho_c = i_c^{1/m}$, also shown in Fig. 2.8, is always smaller than the left boundary of the spectrum $z - \sqrt{2\psi}$. However, its increase with time can be clearly seen.

At $m = 1$, the critical size is so small that it cannot be seen in Fig. 2.10. The Ostwald ripening stage is therefore absent for the both cases considered, as will be discussed in more detail in Sect. 2.7. However, the decrease of growth index m from 2 to 1 has the major effect on the size spectra. Indeed, Figs. 2.10 and 2.11 clearly demonstrate a huge fluctuation-induced spreading at $m = 1$. The dispersion $\psi(t)$ in Fig. 2.10 is always larger than the mean size, because the corresponding ω in (2.103) equals 0.36. Both z and ψ increase linearly with time (the Poissonian case), in agreement with (2.101) and (2.103) at $m = 1$. As is seen from Fig. 2.11, the spectrum

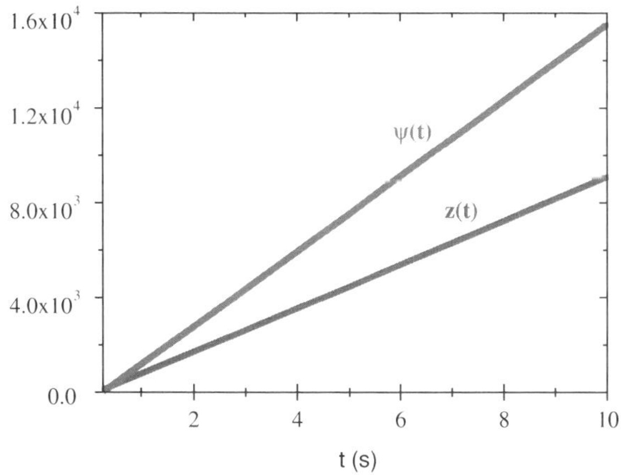

Fig. 2.10 Same as in Fig. 2.8 at $m = 1$

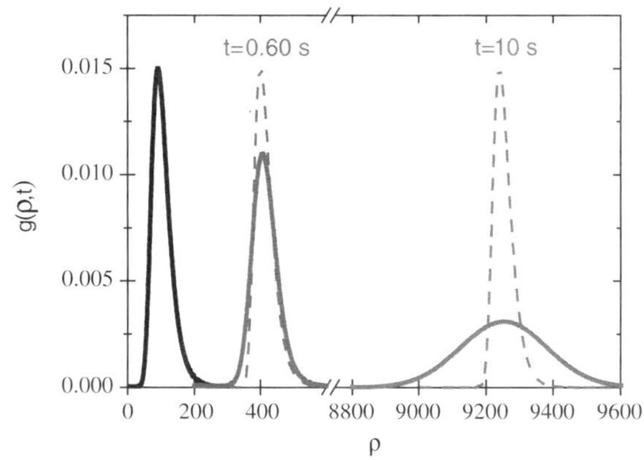

Fig. 2.11 Same as in Fig. 2.9 at $m = 1$

starts broadening immediately upon the completion nucleation stage. The difference between the fluctuational and deterministic distributions becomes significant already at $t = 0.6$ s. At the asymptotic growth stage, the spectrum converts to a rather broad Gaussian which forgets completely about the initial condition.

It is noteworthy that the tendency for no spreading at $m = 2$ and spreading at $m = 1$ does not strongly depend on the external parameter t_∞/τ, which equals 20 in our calculations but can be tuned by changing the surface temperature and the deposition rate. This is illustrated by the curves in Fig. 2.12, showing the dependences of ω on t_∞/τ at different m within a plausible range from 10 to 400. The curves are obtained from (2.104), (2.25), (2.29) and (2.33) in the super-critical nucleation mode. The value of ω is always smaller than 0.03 at $m = 2$ but larger than 0.3 at $m = 1$, so that the described behavior must remain qualitatively correct within a wide range of deposition conditions.

If the material influx is switched at time t_0 soon after the supersaturation reaches its maximum (the difference between t_0 and t_* should be sufficient, however, to form

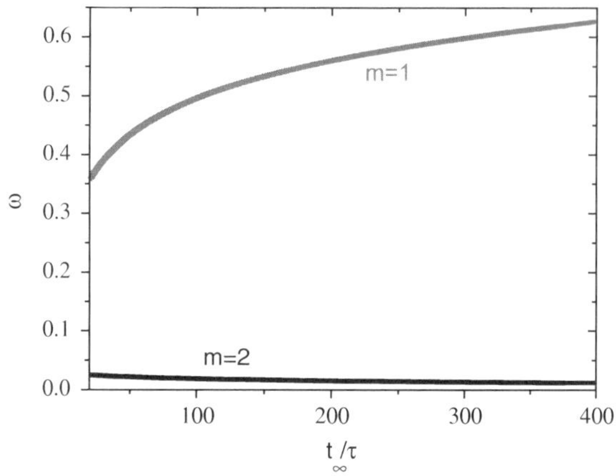

Fig. 2.12 Dependences of parameter ω defined by (2.104) on t_∞/τ at $m = 1$ and 2

the double exponential spectrum at the nucleation stage), the slow term $\varepsilon\eta$ in (2.98) can be neglected with a reasonable accuracy. The time evolution of the mean size is now well described by (2.100) at $u \cong u^{(0)}$, studied in [17, 19, 25] at different m. The solution to (2.100) is obtained in the form of reverse dependence

$$\frac{t - t_c}{t_r} = \frac{1}{m} P_m(u); \tag{2.105}$$

$$P_m(u) \equiv m \int_0^u \frac{dx}{1 - x^m} = \begin{cases} -\ln(1-u), & m = 1 \\ \ln\left(\dfrac{1+u}{1-u}\right), & m = 2 \\ \ln\left[\dfrac{(1+u+u^2)^{1/2}}{1-u}\right] + \sqrt{3}\arctan\left(\dfrac{\sqrt{3}u}{2+u}\right), & m = 3 \\ \ln\left[\dfrac{(1+\sqrt{u}+u)^{1/2}}{1-\sqrt{u}}\right] - \sqrt{3}\arctan\left(\dfrac{\sqrt{3u}}{2+\sqrt{u}}\right), & m = 3/2 \end{cases} . \tag{2.106}$$

These expressions summarize the known analytical results for the integral $P_m(y)$ at different growth indices. The size dependence of supersaturation is given by

$$\zeta(u) \cong \Phi_c(1 - u^m) \tag{2.107}$$

Obviously, the function $u = z/z_r$ exponentially tends to one and the size z tends to the maximum size z_r at $t \sim 3t_r$. Therefore, the time t_r determines the characteristic time scale of the size relaxation stage. From definitions for t_r and z_r, in view of (2.62) for N/n_{1e} and $t_c = \Phi_c t_\infty$, we obtain

$$t_r = \frac{(m!)^{1/m}}{\Gamma_c^{(m-1)/m}} t_c. \tag{2.108}$$

Comparing it to (2.45) for the characteristic duration of the nucleation stage, we arrive at the following systems of strong inequalities

$$2\Delta t << t_r << t_c. \tag{2.109}$$

This confirms the timescale hierarchy discussed in Sect. 1.12: the nucleation stage is much faster than the size relaxation stage, while the latter is much shorter than the macroscopic time required for establishing the critical supersaturation (the last inequality (2.109) is not satisfied only at $m = 1$). However, it is not at all guaranteed that the size relaxation stage will be fully completed before the onset of the Ostwald ripening, as discussed below.

Using (2.107) in (2.80) for $\varphi(z)$ and inserting the result into (2.85) leads to

$$z_r^{m-2}\frac{d\psi}{du} = \frac{1}{mu^{m-1}} + \frac{2}{\Phi_c mu^{m-1}(1-u^m)}. \tag{2.110}$$

Integration of this equation with zero initial condition at $z = z_{nucl}$ gives the z-dependent dispersion in the form

$$\psi(z) = \frac{z^{2-m} - z_{nucl}^{2-m}}{m(2-m)} + \frac{2z_r^{2-m}}{m\Phi_c}Q_m(u, u_{nucl}) \tag{2.111}$$

$$Q_m(u, u_{nucl}) = \int\limits_{u_{nucl}}^{u} \frac{dx}{x^{m-1}(1-x^m)} = \begin{cases} -\ln\left(\frac{1-u}{1-u_{nucl}}\right), m = 1 \\ -\frac{1}{2}\ln\left(\frac{1-u^2}{1-u_{nucl}^2}\right) + \ln\left(\frac{u}{u_{nucl}}\right), m = 2 \end{cases}, \tag{2.112}$$

where $u_{nucl} \equiv z_{nucl}/z_r$ is much smaller than one. Expressions (2.112) summarize the known analytical representations of the integral in the right hand side of (2.111). The first term in the right hand side of (2.111) increases only at $m \leq 2$, as we saw earlier in the case of a constant rate of material influx. However, the decrease of supersaturation due to the monomer consumption by the growing nuclei after the monomer source is turned off always leads to the asymptotic growth stage where $\zeta \to 0$ and $z \to z_r$. At this stage, the second term in the right hand side of (2.111) continues to increase logarithmically regardless of m. This process, however, is usually much slower than the increase of critical invariant size $\rho_c(\zeta)$. The Ostwald ripening stage, which always occurs in absence of material influx, will start when ρ_c becomes comparable with the typical nucleus sizes in the spectrum.

To investigate this point, we consider a typical time evolution of condensing system in the regime with termination of the material influx, illustrated in Figs. 2.13, 2.14, 2.15 and 2.16. The parameters of the initial distribution are the same as before at $m = 2$ with $c = 0.14$, but the influx is now switched off at $\Phi_0 = 2.8$. The calculations were performed for the ballistic ($m = 2$) and diffusion ($m = 1$) growth regimes, for the same initial distribution. The mean size was obtained from the corresponding (2.106), the dispersion was calculated by (2.112), and the supersaturation

Fig. 2.13 Time dependences of the most representative size, dispersion, left boundary of the spectrum and the critical size in the regime with termination of the material influx at $m = 2$; crossing of the curves $\rho_-(t)$ and $\rho_c(t)$ corresponds to the onset of the Ostwald ripening

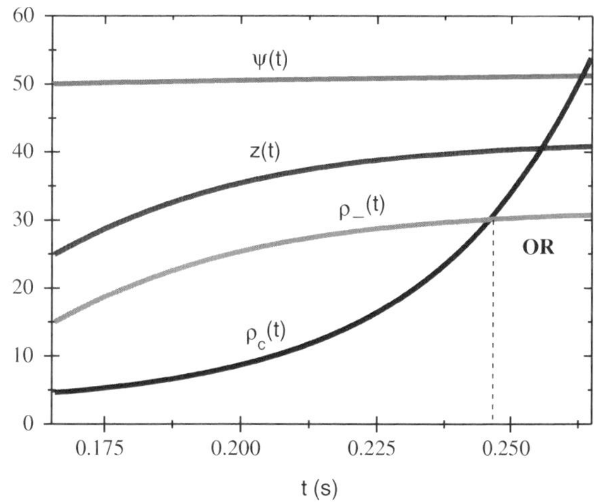

Fig. 2.14 Size spectra at the end of nucleation stage and at the beginning of the Ostwald ripening at $m = 2$, *horizontal arrows* show the corresponding critical size

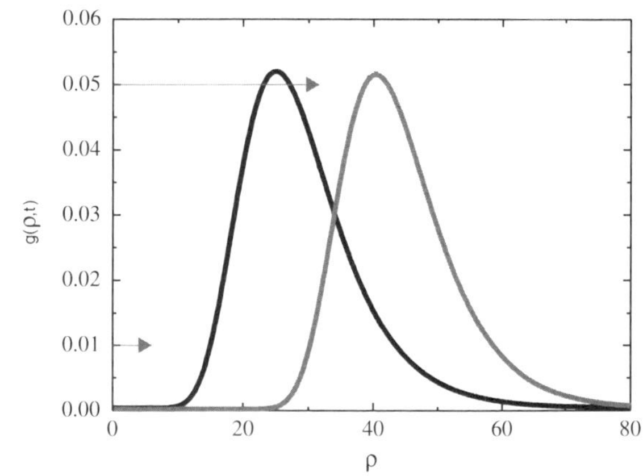

was then found by means of (2.107). The critical invariant size was obtained from (2.9). Figure 2.13 at $m = 2$ shows that the initial dispersion (in our example $\psi_{nucl} = 1/c^2 \cong 50$), acquired upon the completion of nucleation stage at $t = 0.16$ s, is not changed before the onset of Ostwald ripening. The latter starts when the critical size ρ_c becomes equal to the left boundary of the size spectrum, $\rho_- = z - \sqrt{2\psi}$. This happens already at $t = 0.25$ s, i.e., soon after the nucleation stage is completed. The difference between the deterministic and fluctuational spectra cannot be seen in Fig. 2.14, which is in agreement with the results of [3, 17, 19].

As in the previous case at a constant material influx, the spectrum behavior is rather different at $m = 1$, which is clearly seen from Figs. 2.15 and 2.16. The nucleation time now equals 0.20 s, and the Ostwald ripening onsets at 0.46 s (Fig. 2.15). By this time, the double exponential initial spectrum shown in Fig. 2.16 changes noticeably,

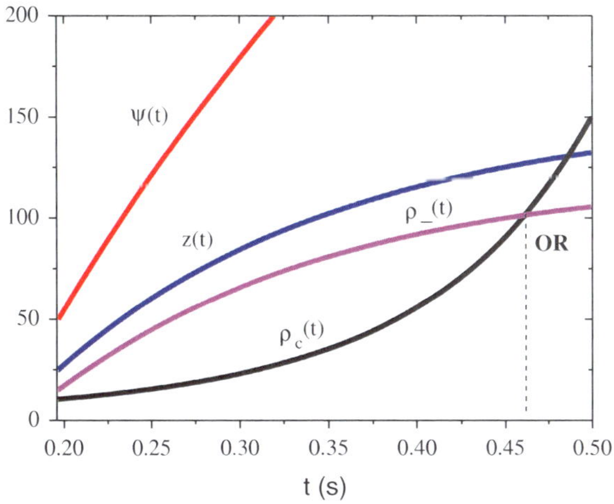

Fig. 2.15 Same as in Fig. 2.13 at $m = 1$

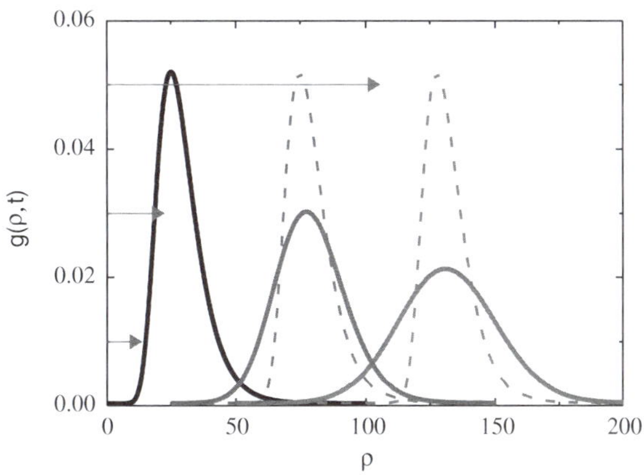

Fig. 2.16 Time evolution of size spectrum at $m = 1$ with the corresponding critical sizes indicated by *arrows*; *dotted lines* show the deterministic spectra

so the fluctuation-induced spreading cannot be neglected even in the regime where the growth stage is rather short.

The next set of numerical calculations is performed directly with the general kinetic equation (2.10), the linearized kinetic equation (2.11), the kinetic equation at $U = 0$ and the deterministic kinetic equation without second derivative with respect toρ, at a constant monomer source with $q = 1$ [104]. The initial conditions are given by the normalized double exponential distribution (2.37). The modified parameters of the initial distributions of 2D islands at $m = 2$ and 1 are summarized in Table 2.1. They are obtained from the corresponding expressions of Sect. 2.2, with the Kashchiev's re-normalization of the nucleation rate given by (1.196) at $\kappa = 1$ and $\vartheta = 0$. The initial supersaturation and the ideal supersaturation upon the completion of the short range nucleation stage are selected as $\zeta_{nucl} \cong \Phi_c(1 - 3/\Gamma)$, $\Phi_{nucl} \cong \Phi_c + (3\tau)/(t_\infty \Phi_c c)$. Numerically solving the second order kinetic equation (2.10) or (2.11) has been performed by applying the explicit scheme of the Crank–Nicolson

Table 2.1 Parameters of the initial size distributions of 2D islands in different growth regimes

m	a	τ (s)	t_∞ (s)	Γ	c	N_c/n_{1e}	z_{nucl}	ζ_{nucl}	Φ_{nucl}
2	7.8	3×10^{-3}	0.06	9.8	0.18	2.6×10^{-3}	20	1.23	2.30
1	7.8	3×10^{-3}	0.06	9.5	0.16	2.9×10^{-2}	22	1.71	2.35

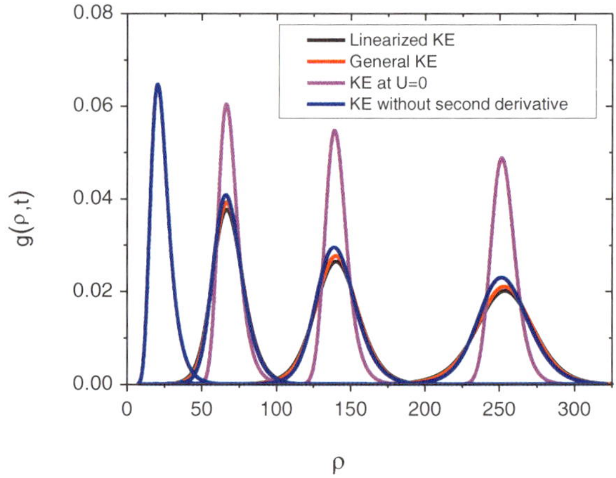

Fig. 2.17 Size distributions at four different times beginning from the initial nucleation spectrum, calculated from different kinetic equations (KEs) for the system with $m = 2$ and the parameters summarized in Table 2.1

finite difference method [105], on a regular grid $\rho_k = \rho_c + k\delta\rho$. In this algorithm, the vector $g(t+\delta t)$ is obtained from $g(t)$ as

$$g(t+\delta t) = -g(t) + 2[1-(1/2)T\delta t]^{-1}g(t). \tag{2.113}$$

The matrix T is obtained by the discretization of kinetic coefficients on the grid. Inversion of the propagator given by (2.113) requires the solution of a system of equations with three-diagonal matrix, mastered by the standard sweep method. Initial number of points k in the ρ axis is fixed within the range from 3000 to 6000. At the fast growth stage, the array pitch $\delta\rho$ is constant. Since the spectrum width is always much less than the mean size, the fluctuational broadening leads to only several times increase in the number of points in the grid. After the calculation of time-dependent size spectra, the mean size $Z(t)$ and the dispersion $\psi(t)$ are obtained as

$$Z(t) = \int_0^\infty d\rho\rho g(\rho,t); \quad \psi(t) = \int_0^\infty d\rho[\rho - Z(t)]^2 g(\rho,t). \tag{2.114}$$

Figures 2.17, 2.18 and 2.19 show the solutions obtained with different kinetic equations in the case of $m = 2$. The effect of kinetic fluctuations of supercritical nuclei at $U = 0$, although noticeable, leads to only a minor spectrum spreading in the ballistic regime, as we saw earlier in Figs. 2.8 and 2.9. Therefore, the spectra

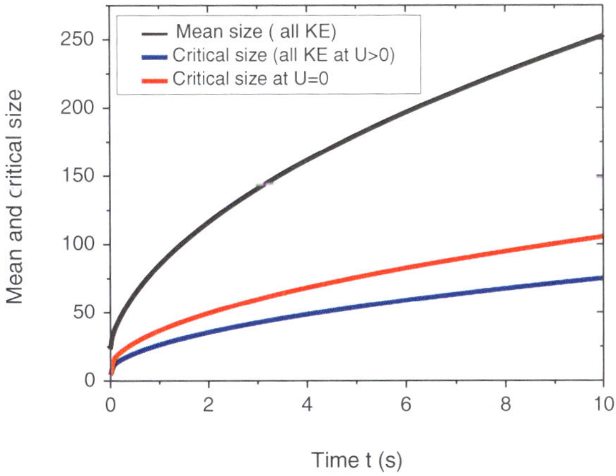

Fig. 2.18 Mean and critical sizes at $m = 2$ obtained from different kinetic equations

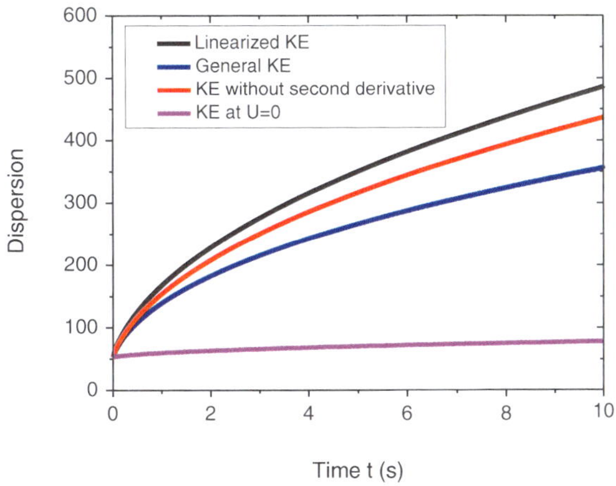

Fig. 2.19 Spectrum dispersions at $m = 2$ obtained from different kinetic equations

obtained with the deterministic kinetic equation without second derivative (Fig. 2.17) are rather close to that calculated from the non-linear and linearized second order equations at $U \neq 0$. In contrast, a purely deterministic effect of the increasing critical size ρ_c on the size spectrum (i.e., keeping the U term in the kinetic equations) is very large. This is demonstrated by the three spectra at $U \neq 0$ in Fig 2.17 and also by the dispersion curves in Fig. 2.19. Such an additional spreading is well understood: the closer is a given nucleus to the critical size the smaller is its growth rate. This leads to rather different growth rates of differently sized nuclei and broadens the spectrum. Figure 2.18 shows that all four kinetic equations considered provide almost identical results for the mean size (the four curves in the figure cannot be resolved). The critical size predicted by the kinetic equation at $U = 0$ is overestimated. We note that the Ostwald ripening stage is absent in all the cases. The total number of nuclei is therefore conserved at the entire growth stage until the direct coalescence into a

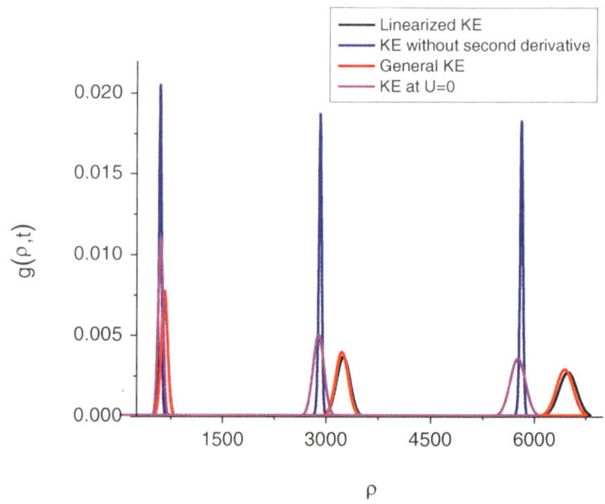

Fig. 2.20 Same as in Fig. 2.17 at $m = 1$

continuous 2D layer, as demonstrated by Fig. 2.17 with the areas below the curves being always equal one.

Figures 2.20 and 2.21 show the same kinetic characteristics for the system of 2D islands growing in the diffusion regime ($d = 2, m = 1$). As discussed already, the case of $m = 1$ is the worst concerning the size homogeneity. The solutions to the second order kinetic equations are Poissonian with the dispersion being proportional to the mean size. The spectra and the time dependences of the dispersion obtained from the deterministic kinetic equation are qualitatively incorrect, as is seen from the corresponding curves in Figs. 2.20 and 2.21. Unlike in the system with $m = 2$, the fluctuation broadening is much larger than that due to $U > 0$. However, the approximation $U = 0$ gives a slower increase of average size compared to the real one. Surprisingly, kinetic equations at $U > 0$ predict a larger regular growth rate, which must be due to the increase of supersaturation caused by the decay of some near-critical nuclei. Further, the linearized kinetic equation is found to be a good approximation in all our numerical runs, as demonstrated by the corresponding curves in Figs. 2.17 and 2.20 for the parameters considered.

2.7 Asymptotic Growth Stage

In this section, we consider the asymptotic growth stage under a material influx with arbitrary flux index q, with a particular emphasis to the large time asymptotes of the most representative size, the critical size and the spectrum width. This will allow us to access the spectrum broadening and also to establish the general criterion for the Ostwald ripening under a material influx. From (2.13) at $\zeta << G$, (2.17) and (2.93), the leading power-law asymptote of the mean size at large t is given by

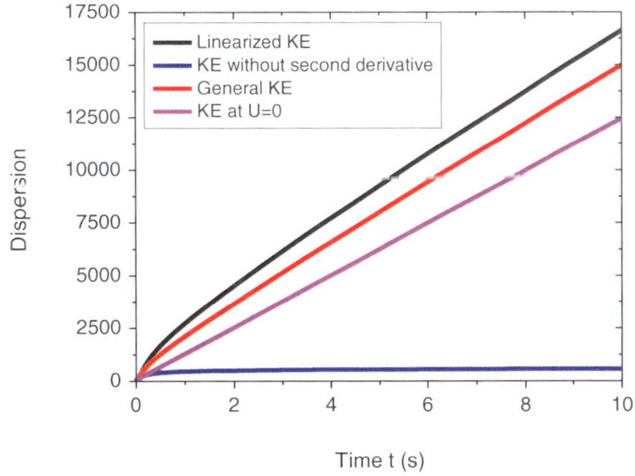

Fig. 2.21 Same as in Fig. 2.19 at $m = 1$

$$z(t) = \left(\frac{n_{1e}}{N}\right)^{\frac{1}{m}} \left(\frac{t}{t_\infty}\right)^{\frac{q}{m}}. \tag{2.115}$$

This result generalizes (2.101) to the case of arbitrary q. Integration of the equation $\zeta = \tau dz/dt$ at large t with $z(t)$ given by (2.115) yields the following dependencies of supersaturation on t and z:

$$\zeta = \frac{q}{m}\frac{\tau}{t_\infty}\left(\frac{n_{1e}}{N}\right)^{\frac{1}{m}} \left(\frac{t}{t_\infty}\right)^{\frac{q}{m}-1} = \frac{q}{m}\frac{\tau}{t_\infty}\left(\frac{n_{1e}}{N}\right)^{\frac{1}{q}} z^{1-\frac{m}{q}}. \tag{2.116}$$

The necessary condition for the supersaturation decrease (or its stabilization) at large t is therefore $q \le m$, which imposes the upper limit for the flux index at a given growth law.

Using (2.116) in (2.80), inserting the obtained result for $\varphi(z)$ into (2.85) and integrating it with the boundary condition $\psi(z_{nucl}) = \psi_{nucl}$ (in the Gaussian approximation of the double exponential distribution), we get the result of [50] for the leading asymptotes of the spectrum dispersion

$$\psi(z) \cong \psi_{nucl} + \frac{1}{m(2-m)} z^{2-m} + \left(\frac{N}{n_{1e}}\right)^{1/q} \frac{t_\infty}{\tau} \frac{2}{q(m/q - m + 1)} z^{\frac{m}{q}-m+1}, \tag{2.117}$$

This formula readily shows that the fluctuation-induced spreading occurs under the condition $q < q_c = m/(m-1)$. In particular, at a constant flux ($q = 1$), the spectrum width asymptotically increases with z with the leading dependence $\sqrt{z}$ regardless of m. The curve $q_{\min} = m/(m+1)$ corresponds to the lowest limit for the flux index where the spectrum width increases slower than the mean size itself, which validates (2.93).

As pointed out by Kukushkin and Osipov [66], the decrease of supersaturation does not guarantee that the ripening stage occurs. Indeed, the Ostwald ripening process under the material influx requires that the growth of critical size in terms of variable ρ, obtained from (2.9) at $\zeta \to 0$,

$$\rho_c(t) = \left[\frac{(d-1)a}{d\zeta(t)}\right]^{\frac{d}{m}}, \tag{2.118}$$

is faster than the growth of the mean size $z(t)$. Only in this case the critical size will overtake the spectrum and start the ripening. The mean size z_{OR}, relating to the beginning of the Ostwald ripening stage, can be therefore defined as the solution to the equation

$$\rho_c = z - \sqrt{2\psi}, \tag{2.119}$$

where ψ is given by (2.117). The equation for z_{OR} is readily obtained from (2.115)–(2.119) in the form

$$z^{\frac{d}{q}-\frac{d}{m}-1} = \left(\frac{q}{mC\omega}\right)^{d/m} - \in(z). \tag{2.120}$$

with the constant $C \equiv [(d-1)a]/d$, $\omega \equiv (N/n_{1e})^{1/q}(t_\infty/\tau)$ and the function $\in(z)$ summarizing the corrections for the spectrum width that are weaker than the left hand side at large z.

Since ω is usually much smaller than one (see the corresponding results of Sect. 2.6), the solution to (2.120) must be very large. Such a solution exists only when the left hand side increases with z, i.e., when the Kukushkin–Osipov inequality holds:

$$q < q_{\max} = \frac{1}{1/m + 1/d}. \tag{2.121}$$

Obviously, the value of $q_{\max}$ corresponds to the critical flux index below which one would observe the Ostwald ripening of d-dimensional nuclei with the growth index m. This critical index obeys the inequality $q_{\max} > q_{\min} = m/(m+1)$, where $q_{\min}$ relates to the spectrum spreading scaling as z. Since the critical flux indices for the Ostwald ripening lie within the region of spreading slower than z, the fluctuation-induced effects do not change the main power-law asymptote of the left boundary of size spectrum. The correction defined by (2.120) leads, however, to an earlier onset of the Ostwald ripening when it occurs, and can significantly change the corresponding initial condition.

The results concerning the asymptotic growth stage of growth are summarized in Fig. 2.22, where the spreading curves are complemented with the critical ripening curves. The growth diagram in the (m, q) plane maps out the regions of time-invariant homogeneous distributions (the region above the curve $q_c = m/(m-1)$ but below the curve $q = m$) and the spectrum spreading at different rates [the region between

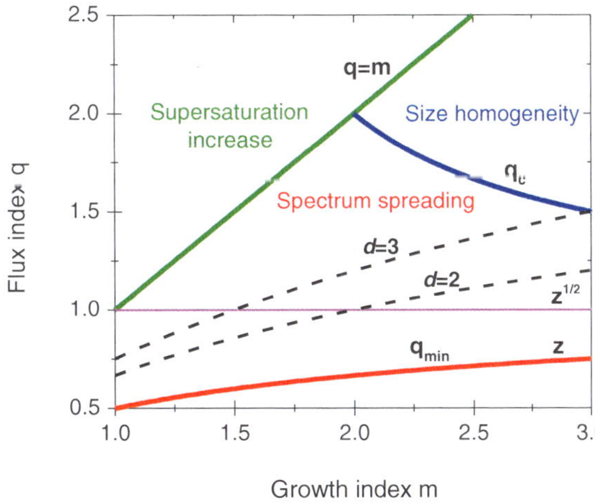

Fig. 2.22 Critical curves on the (m, q) plane. The line $q = m$ separates the domain of supersaturation decrease (*below the line*). The region of spectrum spreading lies below the curve q_c. Above the curve q_{min}, the spectrum broadens slower than z. *Horizontal line* at $q = 1$ relates to the spreading as $z^{1/2}$. *Dashed lines* are the critical ripening curves for 2D ($d = 2$) and 3D ($d = 3$) nuclei; the Ostwald ripening stage occurs below the curves

the curves $q_c = m/(m-1)$ and $q_{min} = m/(m+1)$]. The dotted lines show the critical ripening curves at $d = 2, 3$; the Ostwald ripening stage is present below the curves. As discussed already, the ripening curves are higher than q_{min} but lower than q_{max}, so that all the physical systems undergoing the Ostwald ripening under the material influx are affected by fluctuations.

In the case of $q = 1$, the Kukushkin–Osipov condition is reduced to

$$\frac{1}{d} + \frac{1}{m} < 1. \tag{2.122}$$

It is noteworthy that the ripening stage can never be observed under a constant material influx for $m \leq 2$ (2D islands growing in the ballistic regime) in 2D case and $m \leq 3/2$ (3D droplets growing in the diffusion regime) in 3D case, since these points relate to the crossing of corresponding ripening curves with the horizontal line $q = 1$. Among all physical systems listed in Table 2.2, the only exclusive system undergoing the Ostwald ripening under a constant monomer source is 3D droplets growing in the ballistic regime ($d = m = 3$).

2.8 Ostwald Ripening

Let us now turn to the analysis of the Ostwald ripening stage, which can occur only at $\zeta \to 0$ where, from (2.9), $\rho_c = [(d-1)a/d\zeta]^{d/m}$. In the foregoing analysis, we use the original approach developed by Lifshitz and Slezov [86, 87] for the late Ostwald ripening stage, some generalizations of the Lifshitz–Slezov theory [90–92] and the numerical data obtained in [106]. We consider here the case of an isolated system where the total number of monomers ("mass") is conserved, i.e., upon the termination of material influx. Following Lifshitz and Slezov [86, 87], we search for the asymptotic solution to the deterministic kinetic equation in the case of arbitrary growth index m

$$\frac{\partial g(\rho,t)}{\partial t} = -\frac{\zeta}{\tau}\frac{\partial}{\partial \rho}[(1-U)g(\rho,t)] \tag{2.123}$$

in the form

$$g(\rho,t) = \rho_c^{\alpha}(t) F\left(\frac{\rho}{\rho_c(t)}\right); \quad \alpha = -1-m. \tag{2.124}$$

The function U in (2.123) is given by (2.8) at $\zeta \to 0$. The power index α is chosen in such a way that the total number of monomers in the nuclei is time independent:

$$\begin{aligned}\int_0^\infty d\rho\rho^m f(\rho,t) &= \frac{1}{\rho_c^{m+1}(t)}\int_0^\infty d\rho\rho^m F\left(\frac{\rho}{\rho_c(t)}\right)\\ &= \int_0^\infty dx x^m F(x) = const, \quad x = \frac{\rho}{\rho_c(t)}.\end{aligned} \tag{2.125}$$

At $\zeta \to 0$ and $\Phi \to const$, such a form of distribution automatically provides the mass conservation at the asymptotic Ostwald ripening stage. Substitution of (2.124) into (2.123), upon the transformation to the new independent variables x and t and separation of variables, results in

$$\rho_c(t) = \left[\frac{(d-1)(m+d)a}{d^2C}\frac{t}{\tau}\right]^{\frac{d}{m+d}}; \tag{2.126}$$

$$\left[x + C(1-x^{-m/d})\right]\frac{dF}{dx} = -\left[C\frac{m}{d}x^{-m/d-1} + m + 1\right]F. \tag{2.127}$$

Here, C is the integration constant.

It is evident that the solution to (2.127) has the power law asymptote $F(x) \sim x^{-m-1}$ at $x \to \infty$ so that the integral in the mass conservation given by (2.125) becomes discontinuous at large x. Therefore, the Lifshitz–Slezov solution (2.124) is consistent only with the finite functions $F(x)$ such that $F(x) = 0$ at $x > x_{max}$, with x_{max} being related to the maximum nucleus present in the system at a given moment of time. Direct integration of (2.127) with accounting for the singularity of subintegral function at $x = 0$ yields

$$F(x) = C_1 x^{m/d} \exp\left[\int_0^x dx_1 \frac{-[(m+1+m/d)x_1 + Cm/d]}{x_1\left[x_1 + C(1 - x_1^{-m/d})\right]}\right]. \tag{2.128}$$

The normalization amplitude C_1 can be obtained from the mass conservation, while the constant C is generally unknown. Therefore, (2.128) describes a family of C-dependent model solutions to the deterministic kinetic equation at the Ostwald ripening stage. In their first paper [86], Lifshits and Slezov presented the arguments for the well-defined choice of C. These arguments are definitely valid in the case of infinite initial spectra without a maximum nucleus.

Later on, the choice of C was discussed in a more complex situation of finite spectra having the maximum droplet of size $\rho_{max}(t)$ [90, 91]. It has been shown that the shape of such finite spectra remains time-invariant near the maximum droplet size: $F \sim [\rho - \rho_{max}(t)]^{\mu}$. The solution given by (2.124) should also feature this behavior at $\rho \to \rho_{max}(t)$. To ensure that (2.127) is satisfied at $x = x_{max}$, where $F = 0$ but $dF/dx \neq 0$, the polynomial $P(x_{max})$ in the left hand side must equal zero:

$$P(x_{max}) = x_{max} + C(1 - x_{max}^{-m/d}) = 0. \tag{2.129}$$

In the linear approximation in the neighborhood of x_{max}, the function $P(x)$ can be presented in the form $P(x) \sim [1 + (m/d)x_{max}^{-m/d-1}](x - x_{max})$. Since the solution at $x \to x_{max}$ must be of the form $F \sim (x - x_{max})^{\mu}$, one finds

$$\mu\left(1 + C\frac{m}{d}x_{max}^{-m/d-1}\right) = \frac{m}{d}x_{max}^{-m/d-1} - m - 1. \tag{2.130}$$

Solution to (2.129) and (2.130) enables one to express the parameters x_{max} and C through the power index μ defined by the initial condition at the beginning of the Ostwald ripening stage. The results for x_{max} and C are given by

$$x_{max} = \left(\frac{m(\mu+1) + d(m+\mu+1)}{d(m+\mu+1)}\right)^{d/m};$$

$$C = -\frac{(m(\mu+1) + d(m+\mu+1))}{m(\mu+1)}\left(\frac{m(\mu+1) + d(m+\mu+1)}{d(m+\mu+1)}\right)^{d/m}. \tag{2.131}$$

The Lifshitz–Slezov theory [86] follows from (53) at $\mu \to \infty$, corresponding to the maximum smooth linking of the solution with the region $x > x_{max}$ where $F(x) = 0$:

$$x_{max} = \left(\frac{m+d}{d}\right)^{d/m}; \quad C = -\left(\frac{m}{d}\right)^{\frac{d}{m}}\left(\frac{m+d}{m}\right)^{\frac{m+d}{m}}. \tag{2.132}$$

In particular, in the case of $d = 3$, $m = 3/2$ (3D droplets growing in the diffusion regime), the integration in the right hand side of (2.128) at $C = -27/4$ yields the classical spectrum [86]

$$F(x) = C_1 \frac{\sqrt{x}}{(3-2\sqrt{x})^{11/6}(3+\sqrt{x})^{7/6}} \exp\left(\frac{3\sqrt{x}}{2(2\sqrt{x}-3)}\right), \tag{2.133}$$

re-arranged in terms of invariant size $x = \rho/\rho_c = (i/i_c)^{2/3} = (r/r_c)^2$, where r is the nucleus radius. The cases of 2D islands ($d = 2$) growing in the diffusion ($m = 1, C = -27/4$) and ballistic ($m = 2, C = -4$) regimes are described by the distributions

$$F(x) = C_1 \frac{\sqrt{x}}{(1-2\sqrt{x}/3)^{28/9}(1+\sqrt{x}/3)^{17/9}} \exp\left(\frac{4\sqrt{x}}{3(2\sqrt{x}-3)}\right) \tag{2.134}$$

and

$$F(x) = C_1 \frac{x}{(1-x/2)^4} \exp\left(\frac{2x}{x-2}\right), \tag{2.135}$$

respectively.

We now present the results of numerical analysis of the Ostwald ripening stage based on the second kinetic equations (2.10) and (2.11) and the deterministic kinetic equation (2.123) with the material balance given by (2.13) and (2.14) [106]. The parameters of the initial distributions for the two 2D systems with $m = 1$ and 2 are the same as in Table 2.1. We turn off the material influx at time $t_0 = 0.3$ s to enable the Ostwald ripening process. Figure 2.23 show the time dependences of the mean invariant size, the critical size (a) and the corresponding size spectra $g(\rho, t)$ (b) at the beginning of the Ostwald ripening stage for the system with $m = 2$, obtained from (2.11). As is seen from Fig. 2.23a, when the influx is turned off at $t_{\max} = 0.3$ s, the mean and the critical size quickly match each other. Initially, the islands just stop growing, as demonstrated by the plateau-like sections of the curves in Fig. 2.23 (a) at $t < 2$ s. After that, both $< \rho >^t$ and $\rho_c(t)$ feature an almost identical time evolution. The size distributions presented in Fig. 2.23b rapidly broaden, however, the differences between the spectra obtained with the second order (2.10) and (2.11) and the deterministic (2.123) remain rather small. So, the first order kinetic equation provides a good quantitative description for the spectrum behavior at the beginning of the Ostwald ripening for the ballistic regime of 2D island growth.

Figure 2.24 demonstrates the corresponding spectra for the system with $m = 1$. Opposite to the previous case, the figure shows that the deterministic kinetic equation predicts a much narrower size spectrum which is distinctly different from the fluctuational distributions. The difference between the U-linearized (2.11) and the general (2.10) is again almost negligible. Therefore, the correct description of an early ripening process for 2D islands growing in the diffusion regime must include fluctuations. On the other hand, the fluctuation-induced broadening makes the initial condition to the asymptotic Ostwald ripening stage much closer to an infinitely

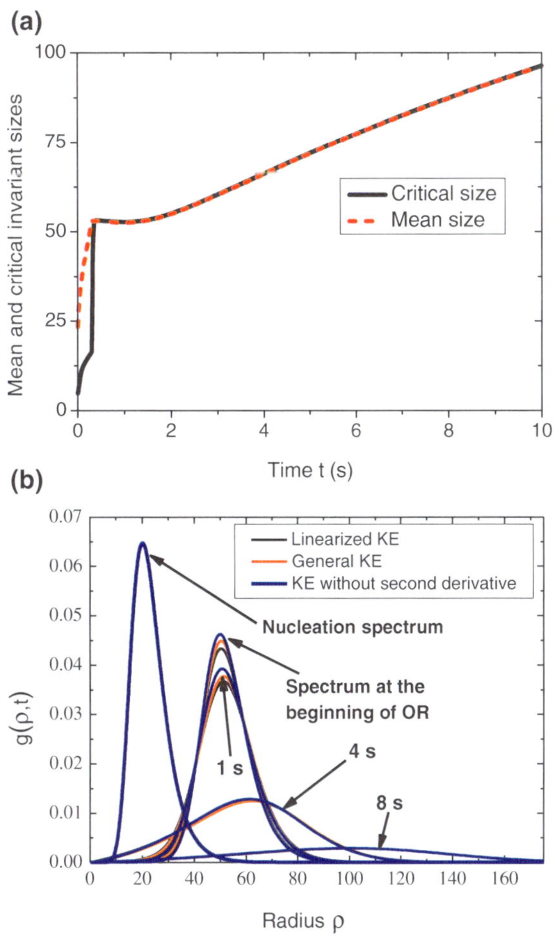

Fig. 2.23 Time evolution of the average and critical sizes (**a**) and the size spectra $g(\rho, t)$ (**b**) for the system with $m = 2$ at the beginning of the Ostwald ripening stage ($t = 0.3$), and $t = 1, 4$ and 8 s, calculated from different kinetic equations. Initial nucleation spectrum is shown for reference

smooth spectrum, the case where the Lifshitz–Slezov theory [i.e., the choice of constant C given by (2.132)] should be more easily justified [90–92].

We now analyze these arguments by the direct comparison of large time asymptotes of numerical solutions of the second order kinetic equation (2.11) and the deterministic (2.123) with the Lifshitz–Slezov distributions of 2D islands given by (2.134) and (2.135). Numerically solving the second order equation has been performed by applying the explicit scheme of the Crank–Nicolson finite difference method as described in Sect. 2.6, with only one modification. At the fast growth stage, the array pitch $\delta\rho$ was kept constant, since the spectra are well localized near the mean size. The situation at the Ostwald ripening stage is rather different, since the spectrum in

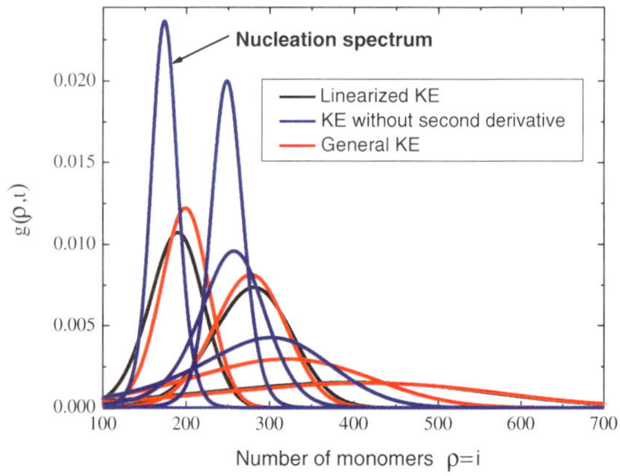

Fig. 2.24 Size spectra for the system with $m = 1$ at the beginning of the Ostwald ripening stage ($t = 0.3$), and $t = 1$, 4 and 10 s, obtained numerically from different kinetic equations

standard variables is no longer localized. To access the large time asymptotes of the spectra, we fix the number of points and increase the pitch accordingly. Because of the high degree of smoothness of the Ostwald ripening spectra, this procedure does not degenerate the accuracy of calculations and, on the other hand, radically decreases the computation time. For calculations with the first order (2.123) at the Ostwald ripening stage, the Crank–Nicolson method may produce artifact oscillations in the large time limit, which has been indeed observed in some of our computation runs. We therefore adopted a different numeric algorithm by considering an ensemble of ~50 000 nuclei having the known initial size distribution (computed at the previous growth stage). We then calculate the time evolution of each nucleus, whose growth rate is given by $d\rho/dt = 1 - U$, with the corresponding initial condition for the given nucleus. The resulting time-dependent spectrum describes the evolution of the ensemble of nuclei in the deterministic limit.

Figures 2.25 and 2.26 present the computed spectra, re-scaled in the conventional Lifshitz–Slezov variables r/r_c, with r being the radius of a 2D island and r_c being the critical radius. Figure 2.25 demonstrates a fairly good agreement of numerical results for the system $m = 2$ with the Lifshitz–Slezov asymptotic solution, almost regardless of the fluctuation term with second derivative. However, both numerical spectra do not exactly match the Lifshitz–Slezov shape. The spectra of the system with $m = 1$, shown in Figs. 2.26, seem to collapse to the function which is not exactly the Lifshitz–Slezov value, however, the integral discrepancy is small in both cases. The asymptotic spectrum obtained from the second order kinetic equation (Fig. 2.26b) is obviously farther from the Lifshitz–Slezov value than in the deterministic limit (Fig. 2.26a). We do not observe any significant difference in terms of time required to establish the asymptotic regime, but this time is very long in both cases considered.

To summarize, comparative analytical and numerical studies of 2D size distributions at the Ostwald ripening stage lead to the following conclusions:

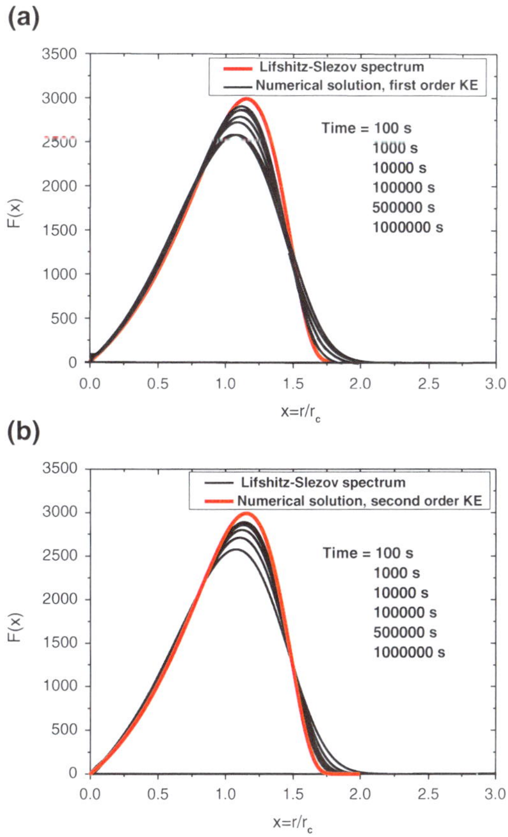

Fig. 2.25 Numerical distributions at the late Ostwald ripening stage for the system with $m = 2$, time t varying from 100 to 10^6 s, obtained with the first (**a**) and second (**b**) order kinetic equation and compared to the Lifshitz–Slezov spectrum in the conventional r/r_c variables

1. Keeping the fluctuation term increases significantly the spectrum broadening in the diffusion regime of 2D island growth but leads to an almost negligible effect in the ballistic regime.
2. The universal Lifshitz–Slezov shapes provide a reasonable approximation to the calculated spectra in the large time limit for the model systems considered.
3. However, numerical distributions collapse to the shapes that are not exactly the Lifshitz–Slezov values. The correspondence is much worse in the case of 2D islands with the growth index $m = 1$, with the discrepancy becoming larger when the kinetic fluctuations are included.

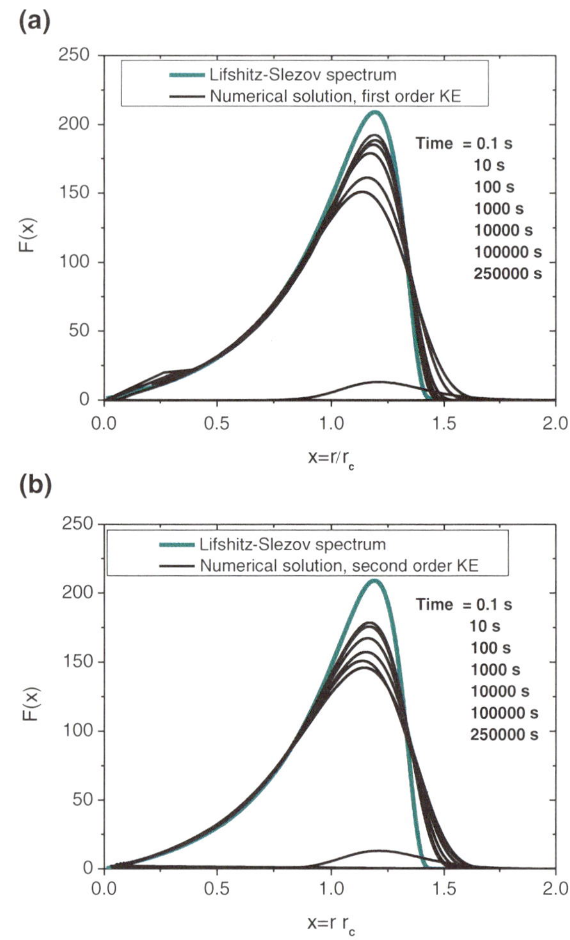

Fig. 2.26 Same as in Fig. 2.25 for the system $m = 1$, time t varying from 0.1 to 2.5×10^5 s

2.9 Tailoring the Size Distribution

The analyses of different stages of nucleation–condensation process presented above are useful for the formulation of several important rules that help to tailor the size distribution of self-induced nanostructures. The possibility of tuning the morphology of nanostructure ensembles by changing the technologically controlled external conditions such as the growth temperature, flux and total amount of material is of paramount importance for nanotechnology in general and epitaxy of nanostructures in particular. Such a kinetically controlled engineering is widely used, for example, for the fabrication of ensembles of coherent Stranski–Krastanow islands with the desired properties, as will be explored in more detail in Chap. 3. It is remarkable, however, that the kinetic tuning knobs for tailoring the size distribution are rather general and can be identified without knowing many details of a concrete material system. We now consider the nucleation stage, the shortest and possibly the most important step in forming the resulting size spectrum. Indeed, when the deposition process is fast and the structure is cooled down or overgrown immediately after the termination of monomer influx, the spectrum is not considerably affected by either the fluctuational broadening or the Ostwald ripening and remains almost the same as acquired at the nucleation stage.

The initial size nucleation spectrum might have rather different shapes and behaviors depending primarily on the rate of material influx and the total number of monomers introduced into the system. As we saw earlier in Sect. 2.3, the two major nucleation modes depend crucially on whether the maximum ideal supersaturation Φ_0 is larger or smaller than the critical supersaturation Φ_c. We can estimate Φ_c from the first (2.34): $F(\Phi_c) = (m+1)\ln Q$, where the kinetic control parameter is introduced by definition

$$Q = \frac{t_\infty}{\tau}. \tag{2.136}$$

Let us analyze the temperature and flux dependence of Q. In the regimes with a constant monomer source, the macroscopic pumping time is given by $t_\infty = n_{1e}/I$, where I is the monomer arrival rate (the flux). On the other hand, the results of Sect. 1.7 (see, e.g., (1.108) or (1.110)) show that the characteristic time of nucleus growth is always inversely proportional to the equilibrium monomer concentration: $\tau \propto 1/n_{1e}$. In thin films, liquid alloys and other condensed phases, the time τ is proportional to the characteristic diffusion time whose temperature dependence is given by (1.83). These considerations yield: $\tau \propto (1/n_{1e})\exp(T_D/T)$, where we introduce the diffusion temperature $T_D = E_D/k_B$ for brevity. The temperature dependence of n_{1e} is obtained from (1.22). Therefore, the leading T and I dependences of Q are determined by

$$Q(T, I) = \frac{I_0}{I}\exp\left(-\frac{T_D + 4T_c}{T}\right), \tag{2.137}$$

showing that the kinetic control parameter is inversely proportional to the flux and increases with the temperature as the Arrhenius exponent [25, 53]. Obviously, the critical temperature T_c is the entirely equilibrium value, while T_D has a kinetic origin showing simply that the diffusion becomes faster at elevated temperatures. At modest $(4T_c)/T$ of the order of one, the strong inequality $T_D/T >> 1$ shows that the temperature dependence of Q is driven by the growth kinetics rather than by the energetics.

Further, (1.58), (1.59) show that the constant a entering the nucleation barrier given by (1.67) is always proportional to the surface energy γ. In liquids and solids, the temperature dependence of γ is rather weak and, for modest variations in T, is well approximated by the linear function $\gamma = \Delta h - k_B T \Delta s$. Here, Δh and Δs are the specific enthalpy and entropy for the formation of interfacial boundary, respectively

[107]. Therefore, the temperature dependence of $a \propto \gamma/(k_B T)$ is given by

$$a(T) = \frac{T_A}{T} - B, \tag{2.138}$$

(where the constants T_A and B can be determined for any particular system), a typical equilibrium behavior. Using (1.67), (2.34), (2.137) and (2.138), we obtain the critical supersaturation in the form [53]

$$\Phi_c = \exp\left[\left(C_{md}\frac{(T_A/T - B)^d}{\ln(I_0/I) - (T_D + 4T_c)/T}\right)^{1/(d-1)}\right] - 1, \tag{2.139}$$

where $C_{md} = (d-1)^{d-1}/[d^d(m+1)]$. The critical supersaturation thus decreases with temperature and only slightly increases with flux in view of the strong inequality $I_0/I >> 1$. In most cases, the exponent in (2.139) is much weaker than that in (2.137), resulting in the typical Φ_c values of the order of one.

Neglecting the temperature dependence of Φ_c for a qualitative analysis, (2.42) yields the leading temperature and flux dependence of the nucleus density in the essentially super-critical pumping mode in the form

$$N_c \propto n_{1e} Q^{-m} = n_0 \left(\frac{I}{I_0}\right)^m \exp\left[\frac{mT_D + (4m-2)T_c}{T}\right]. \tag{2.140}$$

Estimating the maximum invariant size $z_{\max}$ (reached by the time t_0 when the ideal supersaturation is stabilized at $\Phi = \Phi_0$) from the material balance $(N_c/n_{1e})z_{\max}^m \cong \Phi_0$, we arrive at

$$z_{\max} = \Phi_0^{1/m}\left(\frac{N_c}{n_{1e}}\right)^{-1/m} \propto \Phi_0^{1/m}\frac{I_0}{I}\exp\left(-\frac{T_D + 4T_c}{T}\right). \tag{2.141}$$

These important formulas show that all super-critical morphologies (i.e., at Φ_0 well above Φ_c) are controlled primarily by the kinetic factors [53]. Indeed, the flux dependence of the density and the mean size is entirely kinetic, while their temperature dependence is much more sensitive to the kinetic parameters T_D and m compared to T_c. The island density increases with the flux as I^m, decreases with temperature as the Arrhenius exponent and is independent on Φ_0. The mean size decreases inversely proportional to the flux, increases with the growth temperature as the Arrhenius exponent and increases with the maximum ideal supersaturation as $\Phi_0^{1/m}$.

Let us now see how the situation changes when the flux is turned off at Φ_0 well below Φ_c according to the subcritical pumping regime. As follows from the corresponding (2.67), (2.70) and $[J(\Phi_0)\tau]/n_{1e} \propto \exp[-F(\Phi_0)]$, the nucleus density reached upon the completion of the nucleation stage behaves as [103]

$$N_{\infty} \propto n_{1e} \exp\left[-\frac{m}{m+1}F(\Phi_0)\right]. \tag{2.142}$$

For the maximum size this yields

$$z_{\max} \propto \exp\left[\frac{F(\Phi_0)}{m+1}\right]. \tag{2.143}$$

Using (1.67), (2.138), the leading temperature dependences of N_{∞} and $z_{\max}$ are obtained in the form

$$N_{\infty} \propto n_0 \exp\left[-\frac{2T_c}{T} - mC_{md}\frac{(T_A/T - B)^d}{\ln^{d-1}(\Phi_0 + 1)}\right]; \tag{2.144}$$

$$z_{\max} \propto \exp\left[C_{md}\frac{(T_A/T - B)^d}{\ln^{d-1}(\Phi_0 + 1)}\right]. \tag{2.145}$$

These equations contain only thermodynamic values, except for the growth index m which influences the spectrum shape even in the subcritical case. Opposite to the super-critical behavior, the density of nuclei exponentially increases and their size exponentially decreases with increasing the ideal supersaturation and the growth temperature. All subcritical structures are flux-independent. Therefore, the morphology of subcritical ensembles is controlled by thermodynamics. The density of subcritical islands is considerably lower and their fully relaxed size is larger than in the super-critical regime.

We now consider the numerical results of [53], obtained from (2.61), (2.62), (2.66), (2.138), (2.139) and (1.196) at $\kappa = \vartheta = 0$. The calculations were performed for 2D surface islands growing from the adatom sea in the ballistic regime ($d = m = 2$, $z = r/l_0$ is the island radius in the units of lattice spacing). The maximum size was obtained from the material balance at $\zeta = 0$. The model parameters were as follows: $n_0 = 6.25 \times 10^{14}\,\text{cm}^{-2}$, $T_c = 2200\,K$, $T_A = 4200\,K$, $B = 1$, $T_D = 10500\,K\,(E_D \cong 0.9\,\text{eV})$, $\nu = 10^{12}\,\text{s}^{-1}$ and $I_0 = 1.56 \times 10^{26}\,\text{cm}^{-2}\text{s}^{-1}$. At $T = 600$ K and $I = 2 \times 10^{10}\,\text{cm}^{-2}\text{s}^{-1}$, we get: $n_{1e} = 6.5 \times 10^{-4}$, $n_0 = 4.1 \times 10^{11}\,\text{cm}^{-2}$, $\Phi_c = 0.797$, $F(\Phi_c) = 15.4$ and $\Gamma_c = 12$. Calculation of the characteristic diffusion time by means of (1.83) on a square lattice and the growth rate by (1.90) yields $Q = 167$. The time t_{∞} required to reach the equilibrium concentration equals 20.5 s, and the critical time t_c is 37 s. In the super-critical growth regime, the surface density saturates at $N_c = 1.7 \times 10^8\,\text{cm}^{-2}$. The maximum size of islands at $\Phi_0 = 1.5\Phi_c$ amounts to 22 nm at $l_0 = 0.4$ nm. The dependences of the island size and density on the maximum ideal supersaturation Φ_0 at different temperatures and fluxes are shown in Figs. 2.27, 2.28, 2.29 and 2.30.

Figure 2.27 illustrates a typical temperature behavior of the island density at the fixed deposition rate. At higher temperatures, the nucleation under the material influx

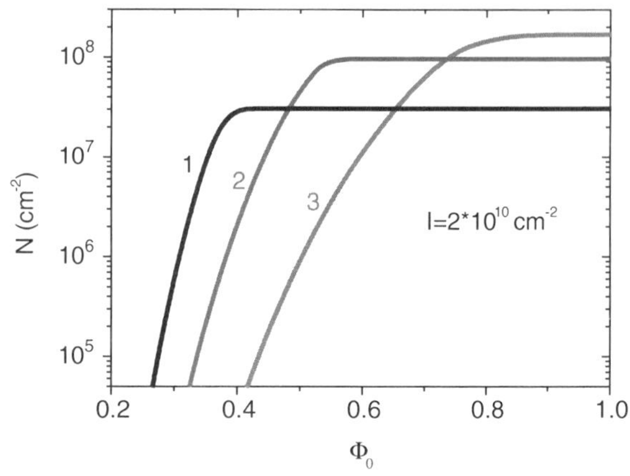

Fig. 2.27 Dependences of island density on ideal supersaturation at the constant $I = 2 \times 10^{10}\,\mathrm{cm}^{-2}\mathrm{s}^{-1}$ and three different temperatures: $T = 650$ (*1*), 625 (*2*) and 600 (*3*) K, for the model parameters described in the text. The critical supersaturation Φ_c amounts to 0.797, 0.531 and 0.387 at $T = 600$, 625 and 650 K, respectively

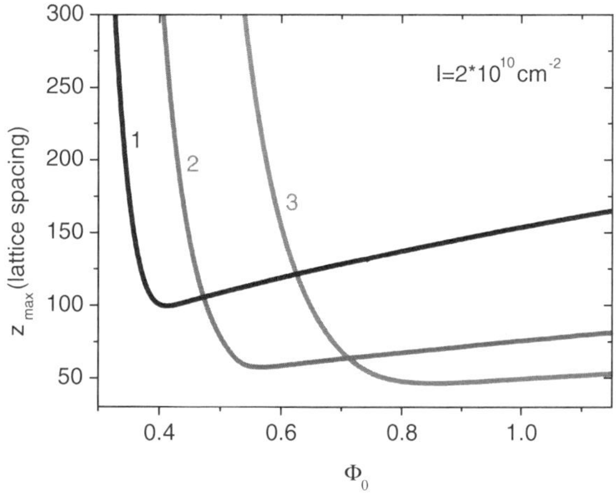

Fig. 2.28 Dependences of the maximum island radius on ideal supersaturation at the constant $I = 2 \times 10^{10}\,\mathrm{cm}^{-2}\mathrm{s}^{-1}$ and three different temperatures: $T = 650$ (*1*), 625 (*2*) and 600 (*3*) K

starts at a lower critical supersaturation and the super-critical density is lower. When the temperature is decreased by only 50°, the critical supersaturation is noticeably higher, while the supercritical density increases by the order of magnitude. Graphs in Fig. 2.28 show the dependence of the maximum island radius on ideal supersaturation at different temperatures. The radius of relaxed islands rapidly decreases with Φ_0 in the subcritical deposition range, reaches its minimum at the critical deposition thickness and then slowly increases in the super-critical deposition range. At a given $\Phi_0 > \Phi_c$, the maximum radius strongly increases with temperature. The dependences of island size on Φ_0 at fixed I and different T convert in the near-critical deposition range, indicating the transition from thermodynamically to kinetically controlled regime of nucleation in an open system.

As seen from the graphs in Fig. 2.29 at the fixed temperature, the island density saturates earlier at a lower flux, however, its super-critical value is lower. Increasing the flux makes the $N(\Phi_0)$ curves steeper; the critical concentration increases

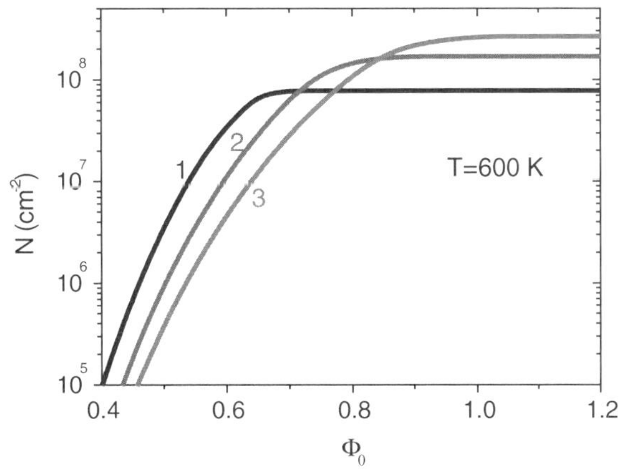

Fig. 2.29 Dependences of the island density on ideal supersaturation at the constant $T = 600$ K and three different fluxes: $I = 0.81$ (*1*), 2.0 (*2*) and 3.3 (*3*) $\times 10^{10}$ cm^{-2}s^{-1}. The critical supersaturation Φ_c amounts to 0.646, 0.797 and 0.916 at $I = 0.81$, 2 and 3.3 $\times$ 10^{10} cm^{-2}s^{-1}, respectively

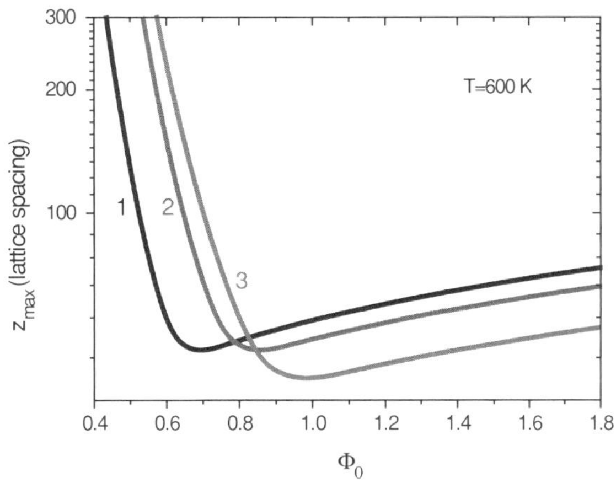

Fig. 2.30 Dependences of the maximum island radius on ideal supersaturation at the constant $T = 600$ K and three different fluxes: $I =$ 0.81 (*1*), 2.0 (*2*) and 3.3 (*3*) $\times 10^{10}$ cm^{-2}s^{-1}

and the super-critical density becomes noticeably higher. Figure 2.30 shows that the maximum radius is almost flux-independent in the subcritical deposition range and rapidly decreases with increasing the flux in the super-critical range. The size and density diagrams at the fixed temperature and different fluxes split in the near-critical deposition range. This splitting is caused by the transition to kinetically controlled regime of nucleation at a larger deposition thickness, where the morphology becomes flux-dependent. Of course, the time to reach the maximum size is much longer in the subcritical deposition range, because the nucleation and growth processes are very slow. Initially broad and essentially asymmetric size spectrum might be affected by the Ostwald ripening if the exposition time is too long.

These results provide useful tools for tuning the nanostructure morphology to the desired properties by changing the control parameters T, I and Φ_0. For example, if a dense ensemble of surface islands is required, the deposition thickness should be larger than critical and, very importantly, the growth process should be performed

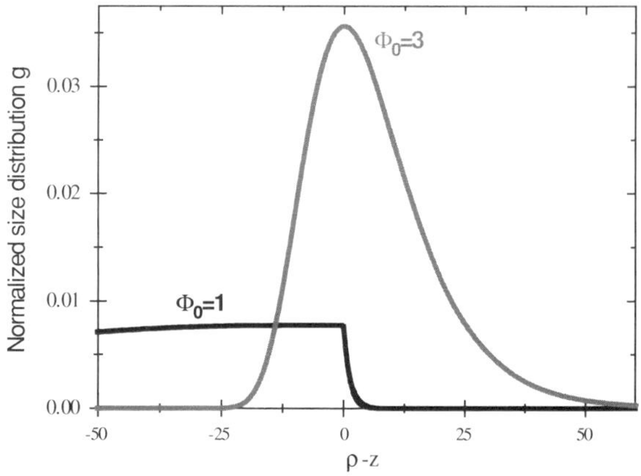

Fig. 2.31 "Frozen" size distributions of Fig. 2.7 showing the double exponential supercritical shape at $\Phi_0 = 3$ and an almost rectangular subcritical shape upon termination of the material influx at $\Phi_0 = 1$

at the lowest possible temperature with the highest possible flux. The density can hardly be increased by depositing more material far above Φ_c, unless the secondary nucleation is initiated due to a suppression of growth of larger islands. Such a process will lead, however, to a bimodal size distribution [23]. Increasing the deposition thickness always leads to a larger size. If a low density ensemble of nanostructures is needed, the deposition should be stopped well below the critical thickness, where a certain time might pass before the nucleation starts. In the super-critical deposition range, the lower density would correspond to a higher temperature and a lower flux.

Let us now analyze the distribution width, an important parameter that should be minimized in many self-induced nanostructure systems to enhance their properties. For example, the size homogeneity of the Stranski–Krastanow quantum dots is important for reducing the inhomogeneous broadening of quantum dot lasers [108]). As is seen from Fig. 2.7, a fully completed spectrum is generally much wider in the subcritical than in the supercritical nucleation mode. However, nothing prevents us from turning off the influx soon after the nucleation is started. This would result in an almost rectangular nucleation spectrum at $\Phi_0 < \Phi_c$, Fig. 2.31 showing the corresponding transformation.

As for the more usual super-critical nucleation mode, the double exponential distribution given by (2.37) is characterized by the half-width $\Delta x \cong \sqrt{2}/c$. The spectrum gets narrower at larger c. From (2.29), $c \propto \tau/t_\infty = 1/Q$ and, using (2.137), we get

$$c \propto \frac{I}{I_0} \exp\left(\frac{T_D + 4T_c}{T}\right). \tag{2.146}$$

This expression leads to a quite simple conclusion: the narrowest nucleation spectrum of self-induced nanostructures is expected at a higher flux and a lower temperature. As we saw earlier, such conditions also lead to a higher density. Time evolution of the size spectrum (in terms of the invariant size) at the follow-up growth stages can only result in its broadening due to the kinetic fluctuations or the Ostwald ripening.

The latter is rapidly switched on when the influx is terminated. To avoid the unwanted ripening, epitaxial samples should be cooled down or overgrown quickly after the deposition is stopped, a rather obvious recommendation.

To conclude this section, we briefly formulate the rules established:

- Below the critical supersaturation, the morphology of nanostructure ensembles is controlled by thermodynamics. The island density exponentially increases and the size exponentially decreases with the amount of deposited material and temperature. None of subcritical characteristics depends on the deposition flux.
- Above the critical supersaturation, the morphology is controlled by the growth kinetics. The density exponentially decreases with temperature, increases as a power function of flux and is independent of the amount of deposited material. The size increases with temperature and the amount of material and decreases with flux.
- The super-critical size spectra are generally narrower than the subcritical ones. The supercritical spectra are double-exponential, while the subcritical ones are triangle-like, converting to the rectangular shape when the influx is stopped at an early stage.
- The width of super-critical spectra is narrower at higher fluxes and lower temperatures.

2.10 Kolmogorov–Johnson–Mehl–Avrami Model

The final stage of growth process in dense systems such as 3D liquid alloys and 2D surface layers is the formation of a continuous, macroscopic solid phase. This process starts when the growing nuclei merge, as shown in Fig. 1.32d. At the beginning, the coalescence proceeds via the pair coagulation of particles that can be described by the Smoluchowski type equation [109]. As the crystallized fraction of the available volume increases, the coalescence necessarily involves interactions between many particles. At this stage, the description in terms of size and number of nucleus is no longer possible. Rather, the kinetics of mosaic-like structure should be treated in terms of time-dependent probabilistic values such as the crystallized fraction of the volume (surface), the surface area (perimeter) of the crystallization front, etc. A very useful crystallization model was formulated long ago independently by Kolmogorov [93] and Johnson and Mehl [94]. The original work by Johnson and Mehl was firstly misunderstood and criticized by Avrami [95, 96], however, this criticism had led to a better understanding of the model essentials.

Kolmogorov [93] formulated the crystallization model that enables a calculation of the crystallized fraction (or the filling factor) of 3D volume or 2D surface under the following assumptions:

(1) Poissonian nucleation process with a time dependent nucleation rate $J(t)$ which is not a function of spatial coordinate,

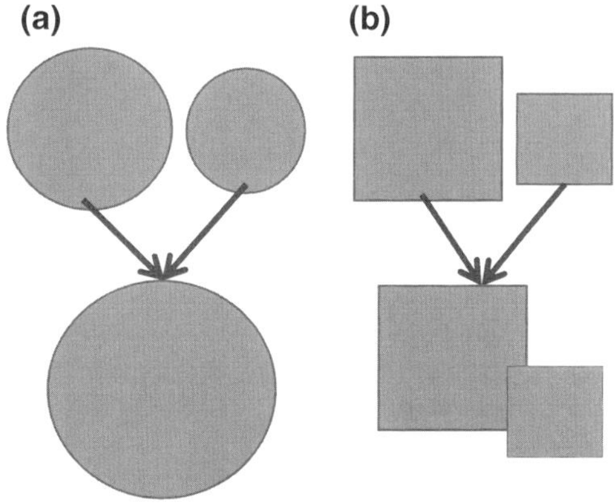

Fig. 2.32 Liquid-like (**a**) and solid-like (**b**) coalescence

(2) All particles nucleate with zero size and grow with a fixed convex shape and orientation that can be characterized by the radius r such that the nucleus volume (surface area) equals Cr^d with C as the corresponding shape constant;
(3) The growth rate of differently sized nuclei in terms of radius r, $v(t) = dr/dt$, is a function of time only and depends neither on r nor on spatial coordinate;
(4) Absence of boundaries;
(5) Solid-like coalescence such that the joint boundaries of merged particles just stop growing, while the free boundaries continue growing in the usual way (Fig. 2.32).

The liquid-like coalescence of droplets, also shown in Fig. 2.32, cannot be described by the same model. However, the solid-like case might be applied for useful estimates of the kinetics of droplet coalescence [7].

Under the above assumptions, the Kolmogorov-Johnson-Mehl-Avrami formula for the volume filling factor $\theta(t)$ in space dimension d can be proved rigorously and is given by the surprisingly simple expression:

$$\theta(t) = 1 - \exp[-\theta_0(t)]; \tag{2.147}$$

$$\theta_0(t) = C \int_0^t dt' J(t') r^d(t', t); \quad r(t', t) = \int_{t'}^t dt'' v(t''). \tag{2.148}$$

The last formula defines the current radius of a nucleus having evolved at time t'. The extended volume $\theta_0(t)$ in (2.147) is none but the filling factor in absence of coalescence effect, the value that would hold if all the merged nuclei were again separated. As long as $\theta_0 << 1$, the real filling factor equals θ_0 in view of $\exp(-\theta_0) \cong 1 - \theta_0$. In the large time limit where $\theta_0 \to \infty$, the filling factor tends to one showing that the entire available volume gets crystallized.

The functions $J(t)$ and $v(t)$ are external with respect to the model and should be obtained from separate considerations. The remarkable feature of the model is that the filling factor can be determined without knowing the details of many-body interactions. Indeed, the functions $J(t)$ and $v(t)$ characterize the nucleation and regular growth stages at a low filling factor, where the nuclei are entirely independent. That is why the corresponding expressions for the nucleation rate of Sect. 1.11 are directly applicable [65]. Furthermore, a simple formula for the growth rate $v(t) = \sigma^{1/d}\zeta(t)/\tau$ can be recalled when 3D or 2D nuclei grow in the ballistic regime such that $m = d$ and the invariant size $\rho = i^{1/d}$ equals the radius expressed in the units of lattice spacing (see Table 1.2). Of course, the model applies only if the critical size is small enough [this correlates with the assumption made in (2.18)] and does not account for the statistical randomness in the growth rates induced by kinetic fluctuations.

The Kolmogorov-Johnson-Mehl-Avrami theory is not only capable of describing the time evolution of the filling factor, but also provides explicitly the area of crystal surface per unit volume (in 3D space) or the perimeter of 2D crystal per unit surface area. These characteristics are given by

$$p(t) = \frac{1}{v(t)} \frac{d\theta_0(t)}{dt} \exp[-\theta_0(t)] \tag{2.149}$$

Obviously, the interfacial boundary increases at the stage of independent growth and decreases as the nuclei merge. The characteristic time t_{coal} at which the crystal starts reaching continuity can therefore be determined as the point of maximum of $p(t)$: $p'(t_{coal}) = 0$.

Let us now consider some important particular cases of (2.147)–(2.149):

1. Time independent nucleation and growth rates: $J = const$, $v = const$. This case is observed when the supersaturation is kept constant, for example, during the stationary vapor–liquid–solid growth of nanowires where the droplet is gradually refilled from the vapor phase [46, 47]. The general formulas are reduced to

$$\theta(t) = 1 - \exp[-(\omega t)^{d+1}]; \quad \omega = \left(\frac{CIv^d}{d+1}\right)^{1/(d+1)}; \tag{2.150}$$

$$p(t) = \frac{(d+1)\omega}{v}(\omega t)^d \exp[-(\omega t)^{d+1}]. \tag{2.151}$$

   The quantity ω gives the characteristic rate of crystallization, which is effectively completed at $\omega t \sim 1$ corresponding to the characteristic crystallization time $1/\omega$. At a time-independent supersaturation, the Kolmogorov exponent contains time to the fourth power for 3D nuclei and to the third power for 2D nuclei.
2. Instantaneous nucleation and a constant growth rate: $J(t) = N\delta(t+\varepsilon)$; $v = const$, where N is the nucleus density and ε is an infinitely small positive constant. The δ—like approximation for $J(t)$ is relevant for a short nucleation stage under the material influx which looks almost instantaneous on a time scale of interest, as described in Sect. 2.2. In this case, the filling factor and the area of crystallization front take the form

$$\theta(t) = 1 - \exp[-(\omega t)^d]; \quad \omega = (CN)^{1/d} v; \tag{2.152}$$

$$p(t) = \frac{d\omega}{v}(\omega t)^{d-1} \exp[-(\omega t)^d], \tag{2.153}$$

yielding the Kolmogorov exponent with the power index d.

3. More realistically, whenever the nucleation stage is short scale and the instantaneous approximation for $J(t)$ applies, the supersaturation should decrease with time. In the ballistic growth regime ($m = d$) and under a constant material influx ($q = 1$), (2.115) for the mean size $z(t)$ can be put in the form

$$r(t) = \left(\frac{\omega t}{CN}\right)^{1/d}, \tag{2.154}$$

where $\omega = \sigma I$ and I is the flux. This equation is identical to (2.115) in view of $Cr^d = \sigma i = \sigma z^d$ and $t_\infty = n_{1e}/I$. Therefore, the extended volume is the linear function of time: $\theta_0(t) = \omega t$. The general formulas are reduced to

$$\theta(t) = 1 - \exp(-\omega t), \quad \omega = \sigma I; \tag{2.155}$$

$$p(t) = dCN^{1/d}(\omega t)^{(d-1)/d} \exp(-\omega t). \tag{2.156}$$

Our examples show that, in most realistic systems, the power index k in the Kolmogorov exponent $\exp[-(\omega t)^k]$ lies between 1 and 4 in 3D space and between 1 and 3 in 2D space. Figures 2.33 and 2.34 illustrate the typical time dependences of $\theta(t)$ and $p(t)$in the cases considered.

Amongst the necessary conditions justifying the Kolmogorov formula, the strongest assumption (3) can be relaxed. In Sect. 2.1, we have shown that the growth rate can be made size-independent in terms of the invariant size even if dr/dt depends on r. Introducing the invariant size ρ, we consider a hypothetical nucleus with "radius" ρ growing at a size-independent rate $\zeta(t)/\tau$ rather than the same nucleus in terms of its real radius. This allows one to re-scale the Kolmogorov-Johnson-Mehl-Avrami model and to put (2.147)–(2.148) in a more general form

$$\theta(t) = 1 - \exp\left[-\sigma \int_0^t dt' J(t')\rho^m(t', t)\right], \tag{2.157}$$

where $\rho(t', t)$ is the invariant size defined by (2.47). The ideal filling factor θ_0 in this expression is given by $\theta_0(t) = \sigma n_{1e} G(t) = \theta_{1e} G(t)$, where $G(t)$ is defined by (2.46). All the previous results remain valid upon the substitution of space dimension d to the growth index m. The power indices of the Kolmogorov exponent are again ranged between 1 and 4, however, the fractional numbers are not excluded. For example, the growth index $m = 3/2$ (corresponding to 3D crystals growing in the diffusion

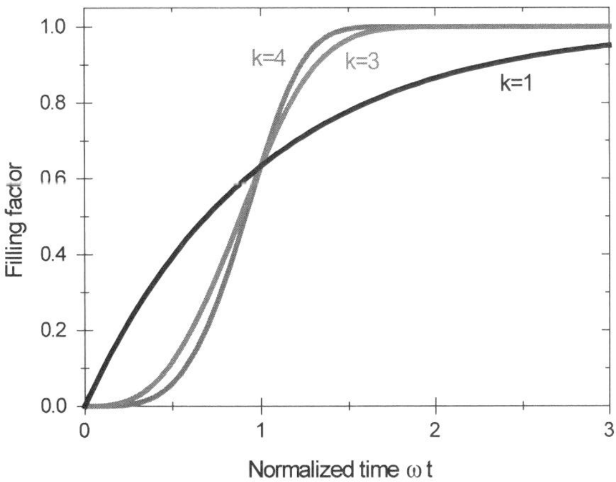

Fig. 2.33 Filling factors in 3D space with different indices $k = 4, 3$ and 1

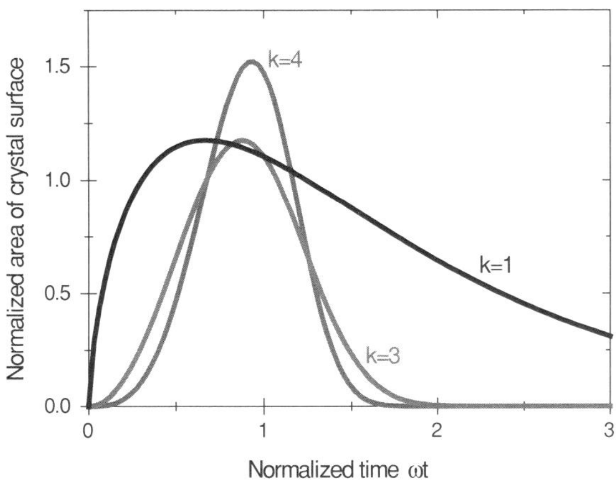

Fig. 2.34 Normalized areas of crystal surface in 3D space $(vp)/\omega$ at $k = 4, 3$ and $p/(cN^{1/3})$ at $k = 1$

regime from a 3D environment or 3D surface islands growing in the ballistic regime from a 2D environment) yields the Kolmogorov indices $k = 5/2, 3/2$ and 1 in the typical cases considered above.

2.11 Three-Dimensional Thin Films

Study of the growth kinetics of 3D thin films is a rather complex problem, and no complete theoretical description is achieved to this end. Many works have used a continuum approach based on various types of non-linear equations for the adatom diffusion on a curved and time dependent surface $h(\vec{r}, t)$, where $\vec{r}$ is the 2D radius-vector in the substrate plane [110–113]. If the surface morphology is controlled by adsorption-desorption processes, the 3D relief can be described by the Kardar-Parisi-Zhang equation [114]. Some papers have been aimed at developing a time-dependent generalization of the BET model for multilayer adsorbates with accounting for the interlayer diffusion and the attractive lateral interactions [31, 115]. In the so-called "pyramidal" model [116], only the growth kinetics of the first surface layer is considered, while the vertical growth is assumed to proceed at a constant velocity which is proportional to the deposition flux.

When a 3D film surface is formed due to the lateral growth of 2D layers that emerge randomly on the underlying layers, a useful approach proposed by Kashchiev [97] is based on the hierarchical sequence of the 2D Kolmogorov-Johnson-Mehl-Avrami models in each layer. We will now show how this approach helps to access the mean thickness and the surface roughness of a growing film. Let us consider a 3D surface without vacancies and overhangs shown in Fig. 1.20, where the coverage of the surface by the layer i at time t equals $\theta_i(t)$. As in Sect. 1.6, the time dependent probabilities to find the film thickness at the height of i MLs at time t, $p_0(t) = 1 - \theta_1(t)$, $p_1(t) = \theta_1(t) - \theta_2(t), \ldots\ p_i(t) = \theta_{i-1}(t) - \theta_i(t)$ are normalized to one. The time dependent mean thickness of the film $h(t)$ and its surface roughness $R_h(t)$, measured in the units of the monolayer height, are defined as follows

$$h(t) = \sum_{i=0}^{\infty} ip_i(t); \quad R_h^2(t) = \psi(t) = \sum_{i=0}^{\infty} [i - h(t)]^2 p_i(t) = \sum_{i=0}^{\infty} i^2 p_i(t) - h^2(t) \tag{2.158}$$

We now assume that the extended areas of the layers i on the fully completed underlying layers $i-1$ are the known functions of time, $\theta_0^i(\omega_i t) \equiv g_i(\omega_i t)$, with ω_i as the characteristic frequencies of lateral growth in the ith layer. The coverage of the surface by the layer i would be given by the Kolmogorov formula

$$y_i(t) = \theta_i(t)|_{\theta_{i-1}=1} = 1 - \exp[-g_i(\omega_i t)] \tag{2.159}$$

if the lower layer was fully completed. To take account for the fact that the upper layer can form only where an underlying layer is already built, we use the known integral relationships [97]

$$\theta_{i+1}(t) = \int_0^t dt' F_{i+1}(t-t')\theta_i(t'), \quad i \geq 1. \tag{2.160}$$

At $i = 1$ we have simply

$$\theta_1(t) = 1 - \exp\left[-g_1(\omega_1 t)\right], \tag{2.161}$$

because the substrate surface is fully built from the very beginning. The kernels $F_{i+1}(t)$ in (2.160) are determined by

$$F_{i+1}(t) = -\frac{d}{dt}\exp[-g_{i+1}(\omega_i t)], \quad i \geq 1. \tag{2.162}$$

In the two-layer BET-type kinetic model of heteroepitaxy [31, 97], we assume that the lateral growth kinetics is identical in all layers beginning from $i = 2$, while the very first surface layer is different (see Sect. 1.6 for the corresponding equilibrium results). Without any concretization of the form of $g_l(\omega_l t)$, we can write down quite generally:

$$g_1 \neq g_2 = g_3 = g_4 = \ldots \equiv g_*; \quad \omega_1 \neq \omega_2 = \omega_3 = \omega_4 = \ldots \equiv \omega_*. \tag{2.163}$$

Using the Laplace transformation of the probabilities $p_i(t) = \theta_{i-1}(t) - \theta_i(t)$,

$$\tilde{p}_i(\omega) = L\{p_i(t)\} = \int_0^\infty d\omega p_i(t)\exp(-\omega t) \tag{2.164}$$

and the known properties $L\{1\} = 1/\omega, L\{df/dt\} = \omega L\{f\} - f(0), L\{\int_0^t dt' f(t-t')g(t')\} = L\{f\}L\{g\}$, (2.160) and (2.161) in the approximation given by (2.163) yield the Laplace images of the form

$$\tilde{p}_i(\omega) = \frac{[1-\varphi_1(\omega)][1-\varphi_*(\omega)]^{i-1}\varphi_*(\omega)}{\omega}. \tag{2.165}$$

Here, the functions $\varphi_1(\omega)$ and $\varphi_*(\omega)$ are defined by

$$\varphi_1(\omega) \equiv \omega\int_0^\infty dt\exp\left[-\omega t - g_1(\omega_1 t)\right]; \quad \varphi_*(\omega) \equiv \omega\int_0^\infty dt\exp\left[-\omega t - g_*(\omega_* t)\right] \tag{2.166}$$

The reverse Laplace transformation of probabilities (2.165) and substitution of the result into (2.158) give the mean thickness and the surface roughness of 3D film:

$$h(t) = \int_0^t dt'\Psi_*(t-t')\theta_1(t'); \tag{2.167}$$

$$R_h^2(t) = 2\int_0^t dt'\Psi_*(t-t')h(t') - h^2(t) - h(t). \tag{2.168}$$

Here,

$$\Psi_*(t) = L^{-1}\{1/\varphi_*(\omega)\} = \int_{\lambda-i\infty}^{\lambda+i\infty} \frac{d\omega\exp(\omega t)}{\varphi_*(\omega)} \tag{2.169}$$

is the Laplace original of the function $1/\varphi_*(\omega)$ and λ is a real constant which is greater than all the singularity points of $1/\varphi_*(\omega)$. Therefore, the time dependences $h(t)$ and $R_h^2(t)$ can be readily calculated by means of (2.161) and (2.166) to (2.169) for the given Kolmogorov functions $g_1(\omega_1 t)$ and $g_*(\omega_* t)$ describing the lateral growth kinetics of the first and the upper layers.

Let us first consider the linear time dependence $g_*(\omega_* t) = \omega_* t$, where the exact solutions are easily obtained at arbitrary $g_1(\omega_1 t)$. In this case, the second (2.166) yields $\varphi_*(\omega) = \omega/(\omega + \omega_*)$, and, after simple calculations, (2.167)–(2.169) give the results of the form

$$h(t) = 1 - \exp\left[-g_1(\omega_1 t)\right] - \nu A_1(\omega_1 t) + \omega_* t; \tag{2.170}$$

$$\begin{aligned} R_h^2(t) = {} & [\theta_1(\omega_1 t) - 2\nu A_1(\omega_1 t) + 2\omega_* t] \exp\left[-g_1(\omega_1 t)\right] + \nu A_1(\omega_1 t) \\ & \times [\nu A_1(\omega_1 t)(2B_1(\omega_1 t) - 1) - 1] + \omega_* t. \end{aligned} \tag{2.171}$$

Here, $\nu = \omega_*/\omega_1$, and the functions $A_1(x)$ and $B_1(x)$ are given by

$$A_1(x) = \int_0^x dx' \exp\left[-g_1(x')\right]; \quad B_1(x) = \frac{1}{A_1^2(x)} \int_0^x dx' x' \exp\left[-g_1(x')\right]. \tag{2.172}$$

From (2.170) and (2.171), the characteristics of the first layer are forgotten after the time moment $t_1 \sim 1/\omega_1$ when it gets overgrown by the upper layers. Starting from this time, the mean film thickness scales linearly with time, while the height dispersion, or the squared roughness, equals the mean thickness. Such a behavior is typical for the Poissonian surface relief, as discussed in Sect. 1.9. Of course, the purely Poissonian case given by (1.146) follows from (2.165) at $\varphi_1(\omega) = \varphi_*(\omega) = \omega/(\omega + \omega_*)$ and $\omega_* = V$, where V is the deposition rate in ML/s.

In the case of arbitrary $g_*(\omega_* t)$, the large time asymptotes of the mean thickness and surface roughness at $\omega_* t >> 1$ are obtained from the Taylor expansion of $\varphi_*(\omega)$ (given by the second (2.166)) in ω/ω_* at $\omega/\omega_* << 1$:

$$\varphi_*\left(\frac{\omega}{\omega_*}\right) = \frac{\omega}{\omega_*} \int_0^\infty dx \exp\left[-\frac{\omega}{\omega_*} x - g_*(x)\right] \cong \frac{\omega}{\omega_*}\left[A_*(\infty) - \frac{\omega}{\omega_*} A_*^2(\infty) B_*(\infty)\right]; \tag{2.173}$$

$$\frac{1}{\varphi_*(\omega/\omega_*)} \cong B_*(\infty) + \frac{\omega_*}{\omega A_*(\infty)}. \tag{2.174}$$

Here, the constants $A_*(\infty)$ and $B_*(\infty)$ are defined as

$$A_*(\infty) = \int_0^\infty dx \exp\left[-g_*(x)\right]; \quad B_*(\infty) = \frac{1}{A_*^2(\infty)} \int_0^\infty dx x \exp\left[-g_*(x)\right]. \quad (2.175)$$

Calculating the Laplace original of the function (2.174) by means of (2.169), using the result in (2.167) and (2.168) and assuming also that $\omega_1 t >> 1$, the leading asymptotic terms of the mean thickness and surface roughness scales linearly with time as

$$h(t) \to \frac{\omega_* t}{A_*(\infty)}; \quad R_h^2(t) \to [2B_*(\infty) - 1] \frac{\omega_* t}{A_*(\infty)}. \quad (2.176)$$

Now, the asymptotic ratio of the surface roughness to the mean thickness is given by the planarity parameter

$$q_\infty = \lim_{t\to\infty} \left[\frac{R_h^2(t)}{h(t)}\right] = 2B_*(\infty) - 1 \quad (2.177)$$

As discussed in the previous section, the extended area in the Kolmogorov exponent is usually given by $g(x) = x^k$ with the index k ranging between 1 and 3 in 2D case. With this power law dependence, the second (2.175) and (2.177) yield

$$q_\infty(k) = \frac{2k\Gamma(2/k)}{\Gamma^2(1/k)} - 1, \quad (2.178)$$

where $\Gamma(\xi)$ are the gamma-functions. The asymptotic planarity rapidly decreases with increasing k, as demonstrated by Fig. 2.35, with $q_\infty(1) = 1$, $q_\infty(2) = 0.274$ and $q_\infty(3) = 0.132$ in the typical cases considered. Therefore, the surface of multilayer film becomes smoother towards larger k indicating that a time-independent nucleation is better than an instantaneous one and a higher growth index m is preferred for obtaining a better planarity.

The multilayer film growth considered above is typical for a low temperature deposition with a low diffusivity, where the adatoms prefer to nucleate in the upper layer rather than migrate to the corners of the underlying layer and attach to the monoatomic step there. In the Frank–van der Merwe layer-by-layer growth mode which occurs at higher surface temperatures, a better adatom diffusivity suppress the usually unwanted nucleation in the upper layers before the lower layer is fully formed. This leads to an abrupt interface where the surface roughness is a periodic function of time. For example, if a 2D layer is growing in the ballistic regime, (1.222) for $z(t)$ and (1.223) at $m = 2$ yield the extended covered area in the form $\theta_0(t) \cong (1 - \alpha_{des})Vt$, with V as the deposition rate in ML/s. Considering only two neighboring layers, the corresponding probabilities to find the lower layer either empty or filled are given by $p_0(t) = 1 - \theta(t)$, $p_1(t) = \theta(t)$, where $\theta(t) = 1 - \exp[-(1 - \alpha_{des})Vt]$. The surface roughness in such a layer-by-layer growth is therefore given by

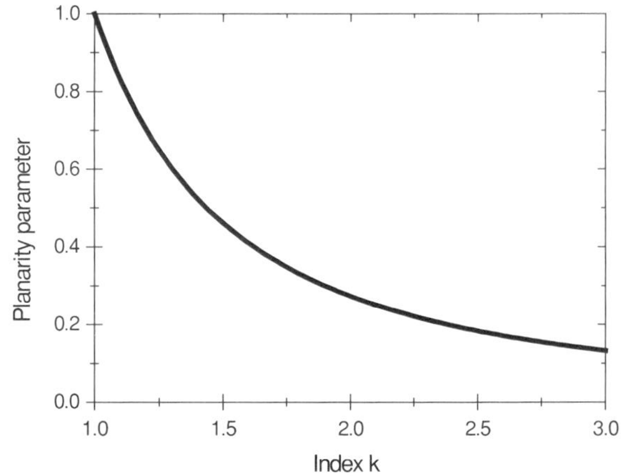

Fig. 2.35 Asymptotic planarity parameter as a function of the Kolmogorov index k

$$R_h(t) = \sqrt{(1 - e^{-\omega t})e^{-\omega t}} \tag{2.179}$$

with $\omega = (1 - \alpha_{des})V$ as the effective growth rate. Since the intensity of electron diffraction in a given direction will reach its maximum for a flat layer and minimum at the maximum roughness, one oscillation of the diffraction pattern relates exactly to the time required to form one monolayer. This property is used for the *in situ* analysis of the growth rate by the reflection high energy electron diffraction (RHEED) in molecular beam epitaxy (MBE) [62, 63].

2.12 Growth Rate of a Crystal Facet of Arbitrary Size

Systems of 2D surface islands considered up to now have been assumed as being grown on a substrate which is so large that the boundary effect can be safely neglected. Taking an example of circular 2D islands growing in the ballistic regime at a constant supersaturation $m = d = 2, J = const, v = const$, (2.150) yields $\omega = (\pi J v^2/3)^{1/3}$. In the case of layer-by-layer growth, the vertical growth rate $V_L = dL/dt$ (ML/s) of an infinitely large crystal facet is given by the same expression:

$$V_L = \left(\frac{\pi J v^2}{3}\right)^{1/3}. \tag{2.180}$$

The facet is growing in the so-called polynuclear mode [7, 80, 81], where many 2D islands emerge and coalesce to form a continuous monolayer.

If a circular facet has a finite radius R, the characteristic time between two consecutive nucleation events on its surface equals $t_n = 1/(\pi R^2 J)$, while the time required for a nucleus to cover the entire facet is given by $t_g = R/v$. The polynuclear growth is observed when $t_g >> t_n$. In the opposite case of $t_g << t_n$, the vertical growth is limited by the waiting time between the two consecutive nucleation acts: $V_L = 1/t_n$, or

$$V_L = \pi R^2 J, \tag{2.181}$$

while the lateral growth occurs almost instantaneously compared to the time scale t_n. This formula is the particular case of (1.205) for the growth of crystal facet in 2D mononuclear mode (see Sect. 2.12), where only one island succeeds in nucleation in a given layer and spreads over the entire facet well before the next nucleus is formed.

The boundary effect becomes significant at $t_g \sim t_n$, which naturally leads to the dimensionless control parameter

$$\alpha \equiv \frac{t_g}{t_n} = \frac{\pi R^3 J}{v}. \tag{2.182}$$

Equations (2.180) and (2.181) can be put together in the form

$$V_L = \begin{cases} \pi R^2 J, \alpha << 1 \\ \left(\pi J v^2/3\right)^{1/3}, \alpha >> 1 \end{cases}. \tag{2.183}$$

The vertical growth rate is v-independent in the mononuclear mode and becomes R-independent in the polynuclear mode. Generally, it is a function of three variables: $V_L = V_L(R, J, v)$, or

$$V_L = \pi R^2 J F(\alpha), \tag{2.184}$$

where $F(\alpha)$ describes the transition from the mononuclear to the polynuclear mode. From (2.183), $F(\alpha)$ should obey the following asymptotic behaviors: $F(\alpha) \to 1$ at $\alpha \to 0$ and $F(\alpha) \to 1/(3\alpha^2)^{1/3}$ at $\alpha \to \infty$. Finding the function $F(\alpha)$ in the general case is very important in modeling the growth of small facets such as the top facets of nanowires with the radius ranging from ten to hundreds of nanometers [45, 80, 81, 117].

Kashchiev [7, 81] proposed a simple interpolation formula matching the two known asymptotes at $\alpha \to 0$ and $\alpha \to \infty$: $F(\alpha) = 1/[1 + (3\alpha^2)^{1/3}]$. This reduces (2.184) to

$$V_L = \frac{\pi R^2 J}{1 + 3^{1/3}\left(\pi R^3 J/v\right)^{2/3}}. \tag{2.185}$$

It is evident that the vertical growth rate increases with R so that a smaller facet always grows slower than an infinitely large one at otherwise identical conditions. This property follows directly from the limit cases given by (2.183) in view of $V_L \to 0$ at $R \to 0$, and should be considered as a fundamental limitation on the growth rate of nanoscale facets. We note, however, than the vapor–liquid–solid growth of nanowires mediated by 2D nucleation at the liquid–solid interface [46] will be suppressed by the Gibbs–Thomson effect (see Sect. 1.3) even earlier, at a certain positive critical radius. This size-dependent effect will be considered in detail in Chap. 4.

In order to include the boundary effect, Dubrovskii and Sibirev [80] developed a size-dependent generalization of the Kolmogorov-Johnson-Mehl-Avrami model. The simplest way to introduce boundaries into the Kolmogorov scheme is presenting the 2D coverage in the form

$$\theta(t) = 1 - \exp\left[- \int_0^t dt' J(t') S[r(t', t), R] \right]. \tag{2.186}$$

The effective crystallized area of a facet with the radius R is defined as the averaged value

$$S(r, R) = \frac{1}{\pi R^2} 2\pi \int_0^R d\rho \rho S(r, \rho, R) \tag{2.187}$$

of the transformed areas $S(r, \rho, R)$ covered by the time moment t by a nucleus that has emerged at time t' at the distance ρ from the facet center. The transformed area depends on t' and t through the nucleus radius $r(t', t)$ defined by (2.148).

Three possible geometrical ways of positioning the nucleus inside the facet area πR^2 at different values of r, R and ρ are illustrated in Fig. 2.36. In case I, the facet is much larger than the nucleus so that the transformed area simply equals the nucleus area πr^2 lying completely inside the facet surface. In case III, the available facet area is so small that the nucleus would grow much larger than the facet but is stopped by the facet boundary, the mononuclear mode with $S(r, \rho, R) = \pi R^2$. The intermediate case II corresponds to the nucleus whose boundary is only partly at the facet boundary, whereby the resulting transformed area acquires a non-circular shape. The corresponding expressions for $S(r, \rho, R)$ are given by

$$S(r, \rho, R) = \begin{cases} \pi r^2, r \leq R - \rho \\ R^2[\theta/2 - \sin(\theta/2)] + r^2[\varphi/2 - \sin(\varphi/2)], R - \rho < r < R + \rho \\ \pi R^2, r \geq R + \rho \end{cases}, \tag{2.188}$$

where $\cos(\theta/2) = (R^2 + \rho^2 - r^2)/(2R\rho)$ and $\cos(\varphi/2) = (r^2 + \rho^2 - R^2)/(2r\rho)$.

Equations (2.188) and (2.187) are not of friendly form for further calculations due to the presence of trigonometrical functions. The detailed analysis of [80] shows that the averaged transformed area $S(r, R)$, obtained numerically upon integration of (2.187) with $S(r, \rho, R)$ given by (2.188), can be approximated with a high accuracy by the polynomial

$$S(y, R) = \begin{cases} \pi R^2[y^2 - (3/8)y^3 - (1/16)y^4 + (1/32)y^5], y \leq 2 \\ \pi R^2, y > 2 \end{cases}, \tag{2.189}$$

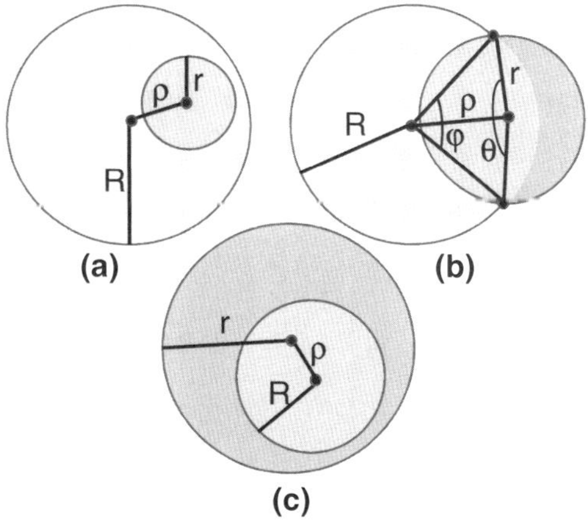

Fig. 2.36 Three different nucleus positions showing relevant geometrical parameters: **a** nucleus inside the facet area, **b** nucleus partly at the facet boundary, **c** nucleus spreading over the entire facet area. *Grey color* corresponds to the facet area, *yellow parts* show the nuclei (the transformed area); *blue circles* represent the nuclei that would form at the same supersaturation on an infinitely large substrate

where $y = r/R$. Substitution of (2.189) into (2.186) and integration of the extended area in the Kolmogorov exponent at $J = const$ and $v = const$ yields the generalized Kolmogorov-Johnson-Mehl-Avrami formula of the form

$$g(t) = g(y) = 1 - \exp[-\alpha f(y)]. \tag{2.190}$$

Here, $y(t) = r(0,t)/R = (vt)/R$, α is the control parameter defined by (2.182), and the function $f(y)$ is given by

$$f(y) = \begin{cases} (1/3)y^3 - (3/32)y^4 - (1/80)y^5 + (1/192)y^6, \; y \le 2 \\ y - 0.9, \; y > 2 \end{cases}. \tag{2.191}$$

The growth rate of a crystal facet of arbitrary size is now readily obtained in the form

$$V_L = \frac{\pi R^2 J}{\alpha y_*(\alpha)}, \tag{2.192}$$

where $y_*(\alpha)$ is the solution to the transcendent equation

$$\alpha f(y) = 1. \tag{2.193}$$

The latter determines the characteristic time of monolayer formation according to (2.190). The obtained solution again has the form (2.184), but the function $F(\alpha) = 1/[\alpha y_*(\alpha)]$ is more complex than the Kashchiev's interpolation. The correct asymptotic behaviors of this $F(\alpha)$ is rather obvious. Indeed, at small $\alpha << 1$ the solution to (2.193) should correspond to $f(y) \cong y >> 1$, yielding $y_* \cong 1/\alpha$, $F(\alpha) \cong 1$ and therefore $V_L \cong \pi R^2 J$ in the mononuclear mode. In the opposite limit case of $\alpha >> 1$, the corresponding asymptote of $f(y)$ at small y is given by

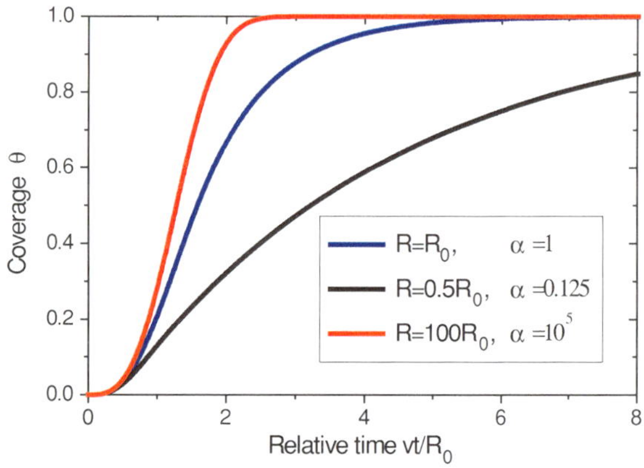

Fig. 2.37 Time dependences of facet coverage θ on the relative time vt/R_0 with the characteristic radius $R_0 = (v/\pi I)^{1/3}$ at three different facet radii R corresponding to different values of α

$f(y) \cong y^3/3$. This leads to the solution to (2.193) of the form $y_* \cong (3/\alpha)^{1/3}$, yielding $F(\alpha) \cong 1/(3\alpha^2)^{1/3}$ and reducing (2.192) to (2.180) in the polynuclear mode.

Typical time dependencies of the coverage θ obtained from (2.190) and (2.191) at fixed I, v and different values of facet radius R are presented in Fig. 2.37. The curve at $\alpha = 10^5$ corresponds to the usual Kolmogorov exponent in the polynuclear growth mode of 2D islands in the ballistic regime. The curve at $\alpha = 1$ shows an intermediate regime, while the curve at $\alpha = 0.125$ presents the situation where the boundary effects dominate. The figure demonstrates that, at a given supersaturation, the thinner rods will grow vertically much slower than the wider ones owing to the size-dependent effect illustrated in Fig. 2.36. When the facet surface shrinks, the nucleus cannot get larger than the facet surface, which is why a large portion of the material influx is lost for growth. When the island is fed from the droplet seated on top of the rod, an almost instantaneous lateral extension of the island decreases the droplet supersaturation considerably. Therefore, a much longer time is required to refill the droplet to enable the next nucleation. This effect has a major impact on the nucleation statistics where rather than being random Poissonian, the two nucleation events becomes anti-correlated in time [118].

Figure 2.38 presents the comparison of the normalized vertical growth rates $(R/v)V_L = \alpha F(\alpha)$ obtained from the Kashchiev's and the generalized Kolmogorov-Johnson-Mehl-Avrami models. It is seen that both models give very close results; however, the growth rate in the Kolmogorov case is always slightly higher than in the Kashchiev's interpolation. Finally, all the results of this section remain valid for an arbitrary island shape upon the substitution of the shape constant π to C in the corresponding expressions.

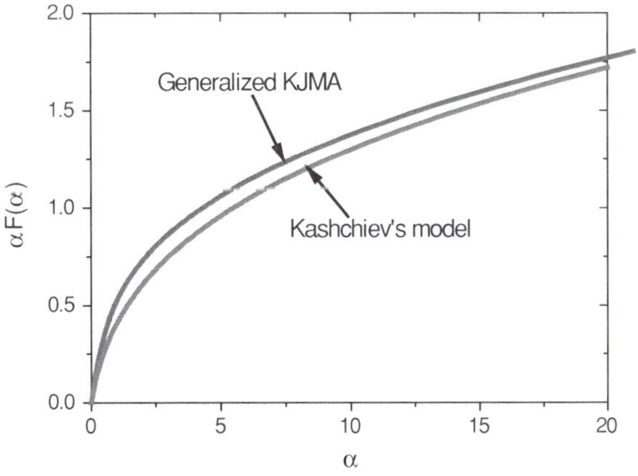

Fig. 2.38 Normalized vertical growth rate $(R/v)V_L$ obtained from the Kashchiev's and the generalized Kolmogorov-Johnson-Mehl-Avrami models

2.13 Irreversible Growth

To this end, we have considered only condensing systems with evaporation, where the quasi-chemical reaction scheme $A_i + A_1 \Leftrightarrow A_{i+1}$ contains the attachment rates $W_i^+ = n_1 k_i^+$ and the evaporation rates W_i^-. In some systems (but only within a restricted time interval), evaporation processes can be neglected. This situation applies, for example, at the initial stage of thin film nucleation at low temperatures [9, 241], and generally in the case of condensation at a very high supersaturation, where the critical size tends to one. It is clear that macroscopic models for the island formation energy and the growth rates do not apply in such systems. Let us now consider the irreversible homogeneous growth with monomolecular reactions

$$A_i + A_1 \xrightarrow{k_i^+} A_{i+1}, \quad i = 1, 2, 3 \ldots, \tag{2.194}$$

where A_1 denotes a monomer, A_i is the cluster containing i monomers and k_i^+ is the attachment rate constant. For example, if the growth process is diffusion-induced and the characteristic diffusion length is large, (1.103) yields $k_i^+ = D \cong const.$ More generally, we can write $k_i^+ = D\alpha_i$, with D as the diffusion coefficient and α_i as the corresponding capture number. The rate equations for the time-dependent concentrations $n_i(t)$ discussed in Sect. 1.8 are now considerably simplified to:

$$\frac{dn_1}{dt} = P - \frac{n_1}{\tau_A} - 2k_1^+ n_1^2 - n_1 \sum_{i=2}^{\infty} k_i^+ n_i; \tag{2.195}$$

$$\frac{dn_i}{dt} = k_{i-1}^+ n_1 n_{i-1} - k_i^+ n_1 n_i, \quad i = 2, 3.4 \ldots . \tag{2.196}$$

Here, P is the monomer influx (e.g., the deposition rate) and τ_A is the characteristic lifetime of monomers in the system (e.g., before desorption from a substrate). The total concentration of all clusters (the cluster density) and the total number of monomers in all clusters at time t are given by

$$N(t) = \sum_{i=2}^{\infty} n_i(t); \quad G(t) = \sum_{i=2}^{\infty} i n_i(t). \tag{2.197}$$

The mean number of monomers in clusters can be therefore defined as

$$i_*(t) = \frac{G(t)}{N(t)}. \tag{2.198}$$

From (2.196), the cluster density N changes in time only due to dimerization:

$$\frac{dN}{dt} = k_1^+ n_1^2. \tag{2.199}$$

The total number of monomers in the system, $n_1 + G$, is changed only by the monomer sources,

$$\frac{dn_1}{dt} + \frac{dG}{dt} = P - \frac{n_1}{\tau_A}, \tag{2.200}$$

and is conserved in an isolated system with zero sources:

$$n_1(t) + G(t) = n_{tot} = const. \tag{2.201}$$

This model system is much simpler for the analysis than the real systems with evaporation. Indeed, at $P = 0$ and $\tau_A \to \infty$, all terms in the right hand sides of (2.195) and (2.196) are proportional to n_1, because all the attachment processes require monomers (while evaporation does not). Expressing the rate constants in dimensionless form as $k_i^+ \equiv k^+ \alpha_i$, with k^+ being the characteristic growth rate, and normalizing all concentrations to the total concentration of monomers in the system, $f_i \equiv n_i/n_{tot}$, we introduce the new time-dependent variable z by definition

$$\frac{dz}{dt} = n_{tot} k^+ f_1; \quad z(t=0) = 0. \tag{2.202}$$

This substitution leads to major simplification, because all the kinetic equations become linear in terms of z:

$$\frac{df_1}{dz} = -2\alpha_1 f_1 - \sum_{i=2}^{\infty} \alpha_i f_i;$$

$$\frac{df_i}{dz} = \alpha_{i-1} f_{i-1} - \alpha_i f_i \tag{2.203}$$

The zero moment of this re-normalized distribution equals one:

$$f_1(t) + \bar{G}(t) = 1, \tag{2.204}$$

where $\bar{G} = G/n_{tot}$. Obviously, the monomer concentration f_1 can now only decrease with time, and the entire growth process stops when f_1 becomes zero. This feature is distinctly different from the real condensing systems where the monomer concentration cannot be smaller than its equilibrium value.

Kinetic equations of irreversible growth can be integrated exactly in some important particular cases. As in Sect. 1.9, we first examine the case of i-linear rate constants in the form

$$\alpha_i = a + i - 1, \quad i = 1, 2, 3 \ldots, \tag{2.205}$$

where $a > 0$ is the dimerization rate. Looking at Table 1.2, this model corresponds to an ensemble of 2D surface islands fed from a 3D vapor phase, while the limit case $a \to \infty$ approximately corresponds to 2D islands fed from a 2D adatom sea in the diffusion regime. Other scenarios include 1D chains of monomers grown from a 3D vapor or liquid phase. In this case, the growth rate is either proportional to the length i (if the monomers are captured along the entire length of the chain, then migrate along the chain and finally incorporate to the edges), or is i-independent (if the monomers are collected only at the edges). The initial condition is chosen in monodispersive form, i.e., no clusters are present at $t = 0$:

$$f_1(z = 0) = 1; \quad f_i(z = 0) = 0, \quad i = 2, 3, 4 \ldots. \tag{2.206}$$

Inserting this into the first (2.203), using (2.204) and $\bar{N} \equiv N/n_{tot} = \sum_{i=2}^{\infty} f_i$, we obtain

$$\frac{df_1}{dz} = (1 - 2a)f_1 - (a - 1)\bar{N} - 1. \tag{2.207}$$

On the other hand, $d\bar{N}/dz = af_1$ from (2.199). Differentiating (2.207), we therefore obtain the closed equation for the monomer concentration

$$\frac{d^2 f_1}{dz^2} = (1 - 2a)\frac{df_1}{dz} - a(a - 1)f_1; \quad f_1(z = 0) = 1, \quad \left.\frac{df_1}{dz}\right|_{z=0} = -2a, \tag{2.208}$$

where the second initial condition follows from (2.207) at $z = 0$. The solution is given by

$$f_1(z) = (1 + a)e^{-az} - ae^{(1-a)z}. \tag{2.209}$$

The cluster density is found simply by integration of this $f_1(z)$:

$$\bar{N}(z) = a\int_0^z dz' f_1(z') = (1+a)\left(1-e^{-az}\right) - \frac{a^2}{(1-a)}\left(e^{(1-a)z}-1\right). \tag{2.210}$$

The growth process is stopped when f_1 becomes zero at

$$z_{\max} = \ln\left(\frac{1+a}{a}\right). \tag{2.211}$$

Inserting the obtained $f_1(z)$ into (2.202), the relationship between z and t is obtained in the form

$$n_{tot}k^+t = \int_0^z \frac{dz' e^{az'}}{1+a-ae^{z'}} = \frac{1}{a(1+a)}\int_1^{e^{az}} \frac{d\xi}{1-[a/(1+a)]\xi^{1/a}}. \tag{2.212}$$

In particular, at $a = 1$ (2.212) yields: $\exp(z) = 2 - \exp(-n_{tot}k^+t)$, and $f_1 = 2\exp(-z) - 1$ becomes $f_1 = 2/[2-\exp(-n_{tot}k^+t)] - 1$ in terms of time. The value of $z_{\max} = \ln 2$ thus relates to the infinite growth time in this case.

Exact solution for the size distribution $f_i(z)$ can be then obtained by using the generating function

$$f(x,z) = \sum_{i=0}^{\infty} f_{i+1}(z)x^i. \tag{2.213}$$

Differentiating this with respect to z and using $df_i/dz = (a+i-2)f_{i-1} - (a+i-1)f_i$ for all $i \geq 2$, and also $af_1 + df_1/dz = -a\exp[(1-a)z]$, we arrive at the first order partial differential equation for the generating function

$$\frac{\partial f(x,z)}{\partial z} - (x-1)x\frac{\partial f(x,z)}{\partial x} = (x-1)af(x,z) - ae^{(1-a)z}. \tag{2.214}$$

Writing the equivalent system of ordinary differential equations in the canonical form

$$\frac{dz}{1} = -\frac{dx}{(x-1)x} = \frac{df}{a(x-1)f - ae^{(1-a)z}}, \tag{2.215}$$

we proceed to finding the first integrals. Integration of the first of these equations readily yields

$$\psi_1 = z - \ln\left(\frac{x}{1-x}\right). \tag{2.216}$$

Expressing x as $x = 1/[\exp(\psi_1 - z) + 1]$ and using the initial condition $f(x, z = 0) = 1$, integration of the second (2.215) gives

$$f = e^{-U(z)} \left[1 - a \int_0^z dz' e^{(1-a)z' + U(z')} \right];$$

$$U(z) = a \int_0^z \frac{dz' e^{\psi_1 - z'}}{e^{\psi_1 - z'} + 1} = a \ln \left(\frac{e^{\psi_1} + 1}{e^{\psi_1 - z} + 1} \right). \tag{2.217}$$

Performing the integration and then expressing ψ_1 through x and z from (2.126), the final result for the generating function is obtained in the form

$$f(x, z) = \frac{e^{-az}}{\left[1 - (1 - e^{-z})x\right]^a} - \frac{ae^{(1-a)z}}{(1-a)} \frac{\left\{1 - [1 - (1 - e^{-z})x]^{1-a}\right\}}{x}. \tag{2.218}$$

In order to find the explicit form of the size distribution, we expand the generating function in x by using the known formula

$$\frac{1}{(1-y)^a} = \sum_{i=0}^{\infty} \frac{\Gamma(a+i)}{\Gamma(a) i!} y^i, \tag{2.219}$$

where $\Gamma(\xi)$ denotes the gamma-function. Integration of this expansion from 0 to y yields

$$\frac{1}{(1-a)} \left[1 - (1-y)^{1-a}\right] = \sum_{i=1}^{\infty} \frac{\Gamma(a+i-1)}{\Gamma(a) i!} y^i. \tag{2.220}$$

These expressions ease the series expansion of (2.218) in $y = (1 - e^{-z})x$. The exact solution for the size distribution is therefore obtained in the form

$$f_i(z) = \frac{\Gamma(a+i-1)}{\Gamma(a)(i-1)!} (1 - e^{-z})^{i-1} e^{-az} \left[1 - \frac{a}{i}\left(e^z - 1\right)\right], \quad i = 1, 2, 3 \ldots, \tag{2.221}$$

the expression which also applies at $i = 1$.

Let us now analyze the behavior of the obtained solutions depending on the parameter a. As follows from (2.211), the case of $a >> 1$ corresponds to small $z_{max} \cong 1/a$, so that all physical $z < z_{max}$ are much smaller than one. The factors $1 - \exp(-z)$ in (2.221) can be than changed to z, while the asymptotic behavior of the gamma-function ratio at large a is given by

$$\frac{\Gamma(a+i-1)}{\Gamma(a)} = a^{i-1}(1 + 1/a)(1 + 2/a)\ldots(1 + (i-2)/a) \cong a^{i-1}. \tag{2.222}$$

We therefore obtain a quasi-Poissonian distribution

$$f_i(az) = e^{-az} \frac{(az)^{i-1}}{(i-1)!} \left(1 - \frac{az}{i}\right), \tag{2.223}$$

perturbed by the non-linear monomer sink. We note that $az < 1$ at any time and approximately equals one only at $z = z_{\max}$, which is why the distribution is always centered at $i = 2$ in the large time limit. This property shows the importance of the material balance given by (2.203). Indeed, we saw in Sect. 1.9 that the solutions to (2.204) at $\alpha_i = const$ would be exactly Poissonian with an infinitely increasing mean size at a constant monomer concentration (stabilized by external fluxes for example). This is however not the case in an isolated system, because monomers are rapidly consumed by the growing clusters, which prevents the Poissonian mean size az from increasing above one.

In the opposite limit case of $a << 1$, the value of $z_{\max} \cong -\ln(a)$ is much larger than one, and (2.221) matches with the Polya size spectrum (see Sect. 1.9) at the right tail ($i >> a\beta$):

$$f_i(\beta) = \frac{\Gamma(a+i-1)}{\Gamma(a)(i-1)!} \frac{\beta^{i-1}}{(1+\beta)^{a+i-1}}, \tag{2.224}$$

with $\beta = \exp(z) - 1$ and the "effective" mean size $a\beta$. It is interesting to consider the continuum form of this distribution. At $\beta >> 1$, the term $[\beta/(1+\beta)]^i$ can be reduced to the exponential by

$$\left(\frac{\beta}{1+\beta}\right)^i = \left(1 + \frac{1}{i}\frac{i}{\beta}\right)^{-i} \cong \lim_{n\to\infty}\left(1 + \frac{1}{n}\frac{i}{\beta}\right)^{-n} = \exp\left(-\frac{i}{\beta}\right). \tag{2.225}$$

By applying the Stirling formula, the first gamma-factor in (2.224) is reduced to i^{a-1}. The large i interpolation of the Polya spectrum is therefore given by

$$f(i,\beta) = \frac{i^{a-1}}{\Gamma(a)\beta^a} \exp\left(-\frac{i}{\beta}\right), \tag{2.226}$$

which is normalized to one and preserves the mean size

$$\int_0^\infty di i f(i,\beta) = \frac{\beta}{\Gamma(a)} \int_0^\infty dx x^a e^{-x} = a\beta. \tag{2.227}$$

By introducing the invariant size $\rho = \ln i$ (because the growth rate di/dt is proportional to i at large i), and using the normalization condition $f(\rho)d\rho = f(i)di$, we arrive at the time-invariant shape

$$f_a(x) = \frac{1}{\Gamma(a)} \exp\left[ax - e^x\right], \tag{2.228}$$

with $x = \rho - \ln(\beta) \cong \rho - z$. This distribution is reduced to the double exponential shape at $a = 1$ (with the reversed sign compared to (2.37), where x was defined as $z - \rho$), but is very different from it at large and small a. Figure 2.39a shows how the

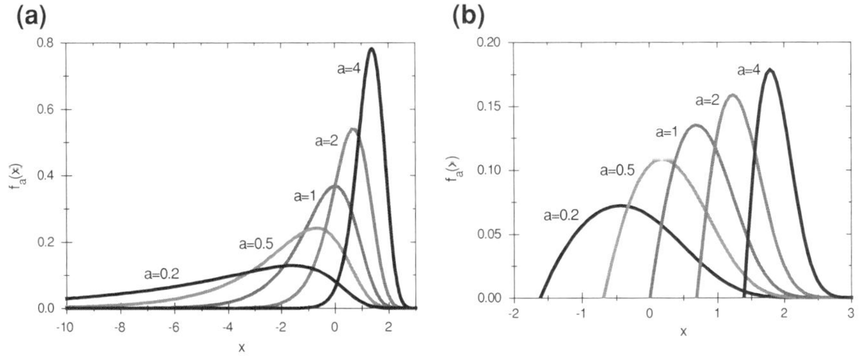

Fig. 2.39 Distribution shapes $f_a(x)$ at different a without (**a**) and with (**b**) accounting for the influence of monomer consumption on the size spectra

narrow and almost symmetrical spectrum at large a transforms through the double exponential shape at $a = 1$ to the essentially asymmetrical shape with a long and flat left tail and an abrupt right tail at small a. However, our irreversible system is described by the Polya distribution only when x is noticeably larger than $\ln a$. Re-arranging the exact solution (2.221) at large i in terms of the invariant variable $x = \rho - z$, one obtains the exact continuum distribution

$$f_a(x) = \frac{1}{\Gamma(a)} \exp\left[ax - e^x\right]\left[1 - ae^{-x}\right]. \tag{2.229}$$

This spectrum tends to zero exactly at the point of maximum of the non-perturbed distribution (2.228), as shown in Fig. 2.39b. The only difference between the two spectra in Fig. 2.39 is in (2.209) for the monomer concentration, whose influence is disregarded in Fig. 2.39a. This shows again the importance of monomer consumption, the non-linear process that suppresses the growth of smaller clusters and narrows the resulting spectrum.

At small a, the continuum Polya distribution in terms of the number of monomers i does not have any maximum. This property persists also for the exact solution given by (2.221). Figure 2.40a shows the i-distribution at $a = 0.1$, which remains a decreasing function of i most of the time, with monomers being the most representative in the spectra. The situation changes only at the very final growth stage where the monomer concentration is bound to vanish, and the spectrum maximum is eventually transferred from 1 to 2 (this discrete spectrum has physical sense only at the integer points, the lines are given as guide to the eye only). When the dimerization process is suppressed, the spectrum is always essentially asymmetric with a very long right tail. However, the mean size

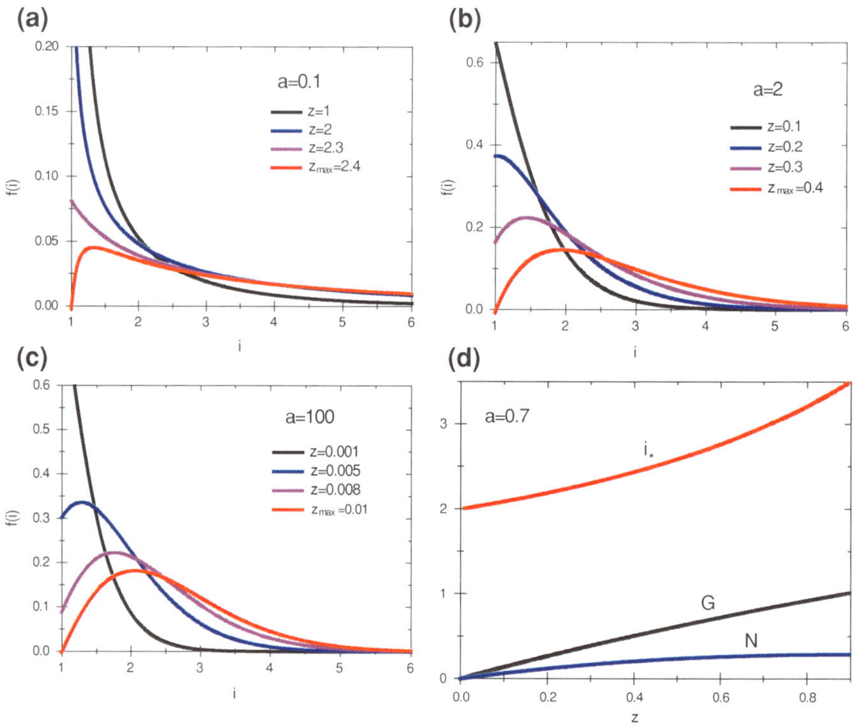

Fig. 2.40 Time evolution of size spectra at $a = 0.1$ (**a**), 2 (**b**) and 100 (**c**); z-dependences of the total number of monomers in clusters G, the total concentration of clusters N, and the mean size $i_* = G/N$ at $a = 0.7$ (**d**)

$$i_* = \frac{1-(1+a)e^{-az}+ae^{(1-a)z}}{(1+a)(1-e^{-az})-[a^2/(1-a)][e^{(1-a)z}-1]} \to \frac{1-a}{1-a^a(1+a)^{1-a}} \tag{2.230}$$

becomes larger toward smaller a. While this result might seem counter-intuitive at a glance, it is actually well explained by the fact that, at smaller a, clusters have more time to grow before the monomers are exhausted.

The spectrum shape changes significantly at $a = 2$ (Fig. 2.40b), acquiring a more centered shape with the maximum at $i = 2$ established much earlier than in the previous case. The spectrum at $a = 100$ (Fig. 3.40c) exactly matches the quasi-Poissonian shape given by (2.223). Its maximum evolves from $i = 1$ to $i = 2$ quite fast and stays there until the asymptotic stage, where the concentration at $i = 3$ is still slightly smaller than that of dimers. The mean size given by (2.230) tends to e at large a. The typical z-dependences of the total number of monomers in the clusters, the total concentration of clusters, and the mean cluster size are presented in Fig. 2.40d. Of course, the i_* value does not have physical sense at $z = 0$ for the mono-disperse initial condition, where we divide zero by zero.

To summarize the results, in a closed system with no monomer sources, the size-linear rate constants $k_i^+ = k^+(a+i-1)$ yield the size spectra with a maximum at either 1 or 2 for all times regardless of a. The final concentration of dimers is always larger than that of trimers, with the difference becoming smaller toward larger a. The spectrum narrows at larger a, while the mean number of monomers in the clusters increases toward smaller a. Since our model gives two limiting cases of a linearly increasing (at small a) and a size-independent (at $a \to \infty$) growth rates, it is anticipated that the maximum at $i = 2$ in the large time limit will persist also for a power law dependence $k_i^+ = k^+(a+i-1)^\alpha$ with $0 < \alpha < 1$. This property is indeed confirmed by numerical simulations.

In order to describe size distributions having a maximum at $i > 2$, we consider the rate constants of the form

$$k_1^+ = k^+ b; \quad k_i^+ = k^+, i = 2, 3, 4 \ldots. \tag{2.231}$$

This model assumes an identical attachment rate for all clusters with $i \geq 2$ but a different rate of dimerization at $i = 1$, with b as the measure of this asymmetry. Such a form of the rate constants is expected in the case of linearly aligned clusters (in a highly supersaturated vapor or liquid environment) that attach monomers from each end of the linear chain. With these rate constants, (2.203) and (2.204) become

$$\frac{df_1}{dz} = -2bf_1 - \bar{N}; \tag{2.232}$$

$$\frac{df_2}{dz} = bf_1 - f_2; \tag{2.233}$$

$$\frac{df_i}{dz} = f_{i-1} - f_i, \quad i = 3, 4, 5 \ldots, \tag{2.234}$$

with z given by (2.202). From (2.199), the total concentration of all clusters with $i \geq 2$ is given by

$$\frac{d\bar{N}}{dz} = bf_1. \tag{2.235}$$

Considering the monodispersive initial condition $f_i(0) = \delta_{i1}$, (2.232) and (2.235) again enables separating out the equation for the monomer concentration in the form

$$\frac{d^2 f_1}{dz^2} = -2b\frac{df_1}{dz} - bf_1; \quad f_1(z=0) = 1, \quad \left.\frac{df_1}{dz}\right|_{z=0} = -2b. \tag{2.236}$$

Solution to this equation yields the z-dependent monomer concentration

$$f_1(z) = \frac{1}{2c}\left[(c-1)e^{b(c-1)z} + (c+1)e^{-b(c+1)z}\right], \quad c = \sqrt{1 - 1/b}. \tag{2.237}$$

The coefficient c is real at $b > 1$ and imaginary at $b < 1$, so that $c = jq$ with $j^2 = -1$ and real $q = \sqrt{1/b - 1}$ in the latter case. At $b < 1$, (2.237) re-writes as

$$f_1(z) = e^{-bz}\left[\cos(bqz) - \frac{\sin(bqz)}{q}\right], \quad q = \sqrt{1/b - 1}. \tag{2.238}$$

In the particular case of $b = 1$, solution to (2.236) becomes

$$f_1(z) = e^{-z}(1 - z). \tag{2.239}$$

The maximum values of z where the entire growth process stops $[f_1(z_{\max}) = 0]$ are therefore given by

$$z_{\max} = \frac{1}{2bc}\ln\left(\frac{1+c}{1-c}\right), \quad b > 1;$$

$$z_{\max} = 1, \quad b = 1;$$

$$z_{\max} = \frac{1}{bq}\arctan(q), \quad b < 1. \tag{2.240}$$

The total concentration of clusters is readily obtained by integrating (2.235):

$$\bar{N}(z) = \frac{e^{-bz}}{c}\sinh(bcz); \quad b > 1, \tag{2.241}$$

$$\bar{N}(z) = \frac{e^{-bz}}{q}\sin(bqz); \quad b < 1. \tag{2.242}$$

We now introduce the generating function for the cluster concentrations at $i \geq 2$ by definition

$$f(x,z) = \sum_{i=0}^{\infty} f_{i+2}(z)x^i. \tag{2.243}$$

Differentiating this with respect to z and using (2.233), (2.234), one obtains the simple ordinary differential equation

$$\frac{df(x,z)}{dz} = (x-1)f(x,z) + bf_1(z); \quad f(z=0,x) = 0, \tag{2.244}$$

where x is only a parameter. Integrating this with arbitrary $f_1(z)$, we arrive at

$$f(x,z) = b\int_0^z dy f_1(z-y)e^{(x-1)y}. \tag{2.245}$$

Series expansion of the exponent under the integral yields the Poissonian-like exact solutions for concentrations that depend only on the particular form of $f_1(z)$:

$$f_{i+2}(z) = \frac{b}{i!}\int_{o}^{z} dy f_1(z-y)y^i e^{-y}, \quad i = 0, 1, 2 \ldots. \tag{2.246}$$

Alternatively, (2.236) can be integrated directly with the exponential $f_1(z)$ given by (2.237) and then expanded in x, the final result will be of course identical to (2.246).

In particular, integration of (2.245) at $b = 1$ with $f_1(z)$ given by (2.239) readily yields

$$f(x, z) = e^{-z}\left[\left(\frac{1}{x} - \frac{1}{x^2}\right)\left(e^{xz} - 1\right) + \frac{z}{x}\right]. \tag{2.247}$$

The Taylor expansion of $\exp(xz)$ therefore leads to the result

$$f_i(z) = e^{-z}\frac{z^{i-1}}{(i-1)!}\left(1 - \frac{z}{i}\right), \quad i = 1, 2, 3 \ldots, \quad b = 1, \tag{2.248}$$

which is exactly identical to (2.223) at $a = 1$. This result is anticipated, because the two models given by (2.205) at $a >> 1$ and (2.231) at $b = 1$ yield the same size-independent rate constant for all clusters including monomers.

The solutions at $b \neq 1$, obtained by integration of (2.246) with $f_1(z)$ given by (2.239), have a more complex form:

$$f_{i+2}(z) = \frac{be^{-bz}z^{i+1}}{2c}\left[(c-1)e^{bcz}\gamma^*_{i+1}[(1-b+bc)z] + (c+1)e^{-bcz}\gamma^*_{i+1}[(1-b-bc)z]\right]. \tag{2.249}$$

Here,

$$\gamma^*_{i+1}(y) \equiv \frac{1}{y^{i+1}i!}\int_0^y d\xi\xi^i e^{-\xi} = \frac{\gamma_{i+1}(y)}{y^{i+1}i!} \tag{2.250}$$

denotes the normalized incomplete gamma-function of the integer order $i+1$ of the argument y. This argument takes the real values $(1-b\pm bc)z$ at $b > 1$ and the complex conjugate values $(1-b\pm jbq)z$ at $b < 1$. The analysis of (2.249) is simple at $b >> 1$. Indeed, the coefficient c can be approximated by $c = \sqrt{1-1/b} \cong 1 - 1/(2b)$ at $b >> 1$, where the first (2.240) yields

$$z_{\max} \cong \frac{\ln(4b)}{2b} << 1, \quad b >> 1. \tag{2.251}$$

Using the known asymptotic representation of $\gamma^*_{i+1}(y)$

$$\gamma_{i+1}^*(y) = e^{-y}\left[\frac{1}{(i+1)!} + \frac{y}{(i+2)!} + \frac{y^2}{(i+3)!} + \dots\right],$$

in (2.249), we note that the approximation

$$f_{i+2}(z) \cong be^{-z}\frac{z^{i+1}}{(i+1)!}, \quad b >> 1 \tag{2.252}$$

applies uniformly for all z. Thus, the cluster size spectrum has the Poissonian form at large b, however with small $z << 1$ at any time. This property does not allow for the shift of the spectrum maximum farther than to $i = 2$, as we saw earlier in Fig. 2.40.

At $b < 1$, the value of $z_{\max}$ is given by the corresponding (2.240) and becomes much larger than one at $b << 1$ in view of $q = \sqrt{1/b - 1} \cong 1/\sqrt{b}$:

$$z_{\max} \cong \frac{\pi}{2\sqrt{b}} - 1 >> 1, \quad b << 1. \tag{2.253}$$

General solution given by (2.249) now contains complex arguments in the gamma-functions. In order to express the concentrations at $b < 1$ through real functions, we present $(1 - b \pm jbq)z$ in the form

$$(1-b)z \pm jbqz = re^{\pm j\varphi}, \quad r = \sqrt{1-b}z; \quad \tan\varphi = \sqrt{\frac{b}{1-b}}. \tag{2.254}$$

We then perform the integration under the integral of the incomplete gamma-function $\gamma_{i+1}(y)$ in (2.250) along the two rays of the complex plane with the constant polar angles $\pm\varphi$, as shown in Fig. 2.41. This is possible since the result of integration of an analytical function does not depend on the integration path from the

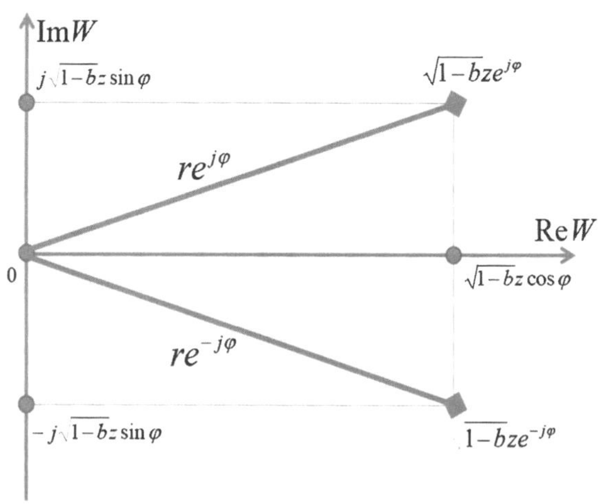

Fig. 2.41 Integration paths in the complex plane W used for obtaining (2.255)

Cauchy theorem. We thus obtain the following useful representation for the incomplete gamma-function of a complex variable:

$$\gamma_{i+1}(\sqrt{1-b}ze^{\pm j\varphi}) = e^{\pm j\varphi(i+1)} \int_0^{\sqrt{1-b}z} dr r^i e^{-r(\cos\varphi \pm j\sin\varphi)}. \tag{2.255}$$

Using this in (2.249), after some straightforward calculations we find the final result

$$f_{i+2}(z) = \frac{be^{-bz}}{i!(1-b)^{i+1}} \left[\left(\cos(bqz) - \frac{\sin(bqz)}{q} \right) C_i(z) + \left(\sin(bqz) + \frac{\cos(bqz)}{q} \right) S_i(z) \right], \quad b < 1. \tag{2.256}$$

The functions $C_i(z)$ and $S_i(z)$ are defined as

$$C_i(z) = \int_0^{(1-b)z} dy y^i e^{-y} \cos\left(\sqrt{\frac{b}{1-b}} y\right);$$

$$S_i(z) = \int_0^{(1-b)z} dy y^i e^{-y} \sin\left(\sqrt{\frac{b}{1-b}} y\right). \tag{2.257}$$

At small enough $b << 1$, the values of concentrations at $z = z_{\max}$ can be estimated from (2.256) at $f_1(z_{\max}) = 0$ and

$$S_i(z \to \infty) = i!(1-b)^{\frac{i+1}{2}} \sin\left[(i+1) \arctan\left(\sqrt{\frac{b}{1-b}}\right)\right]. \tag{2.258}$$

This gives

$$f_{i+2}(z_{\max}) \cong \frac{b \exp\left(-\sqrt{\frac{b}{1-b}} \arctan\left(\sqrt{\frac{1-b}{b}}\right)\right)}{(1-b)^{\frac{i+1}{2}}} \left(\sqrt{1-b} + \frac{b}{\sqrt{1-b}}\right) \sin\left[(i+1)\arctan\left(\sqrt{\frac{b}{1-b}}\right)\right]. \tag{2.259}$$

These asymptotic stationary concentrations for the first $i = 2, 3, \ldots 8$ are shown in Fig. 2.42 as the functions of b. It is seen that the spectrum maximum shifts toward larger i at smaller b so that the maximum concentration is reached at $i = 6$ with $b = 0.15$ and at $i = 8$ with $b = 0.05$. Therefore, the size spectrum of a closed irreversible system of clusters with rate constants given by (2.231) can reach its maximum at large i only if b is small compared to one. This property has not been seen for all other models discussed in this section. The effect is physically explained by the fact that a low rate of dimerization allows for a longer growth time of larger

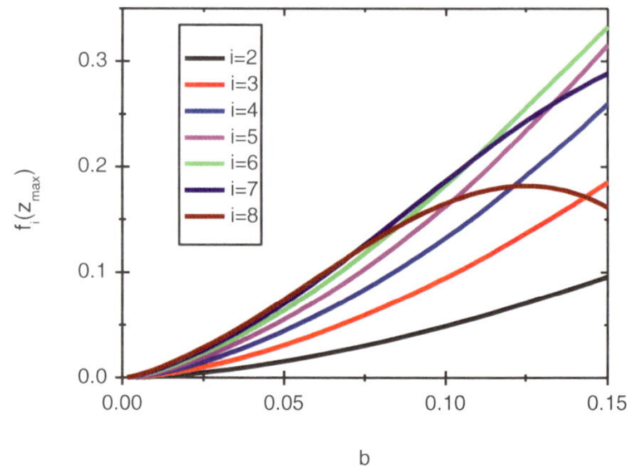

Fig. 2.42 Stationary concentrations $f_i(z_{max})$ versus b for the first seven i-mers

clusters before the monomer concentration is exhausted and the entire growth process ceases.

2.14 Formation of Peptide Chains

In this section, we discuss one interesting application of the irreversible growth model with size-independent growth rate constants, namely, a theoretical description of the salt-induced formation of peptides [72]. Peptides or proteins are linear polymers of α-amino acids performing a wide variety of biological functions in living organisms, such as catalysis of metabolic reactions, DNA replication and transport of biological molecules. While the structural and functional features of many proteins were comprehensively studied over the last decades, it is still of great interest if such molecules could be formed from simple inorganic compounds under abiogenic conditions. The salt-induced peptide formation reaction was discovered as relevant for the very first steps of chemical evolution. It has then been found that L-glutamic amino acid (L-Glu) nucleation in aqueous solutions in the presence of K^+ ions has the greater than an order of magnitude catalytic effect than the same concentration of Na^+ [73]. These data present the surprising evidence that the biological molecules might have had a greater probability to emerge not in a sodium-rich ocean, as commonly believed, but rather in excess potassium solutions [73].

The reaction of the peptide bond formation between amino acids that leads to elongation of the peptide chain involves dehydration process which is very difficult to carry out in a homogeneous aqueous environment. Particularly, 1,1'-carbonyldiimidazole (CDI) is commonly used as an efficient dehydrating agent to overcome the thermodynamic barrier of peptide bond formation in aqueous solutions and produce oligopeptides in water via a carboxyanhydride intermediate. This enables a considerable enhancement of the peptide growth rates in water solutions.

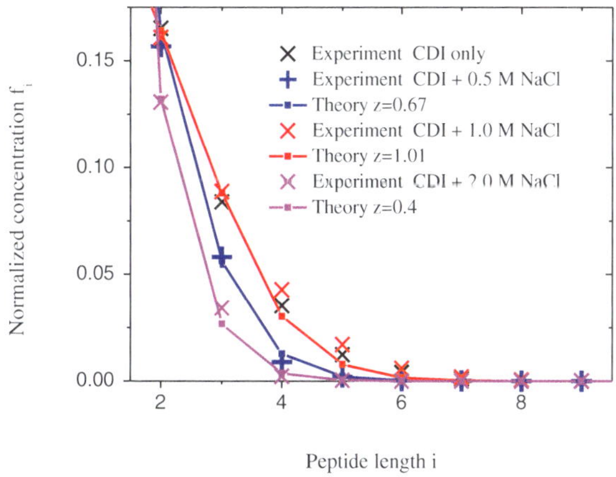

Fig. 2.43 Normalized concentrations of peptides containing *i* amino acid monomers in reaction mixtures with different concentrations of NaCl: experimental data (*symbols*) and theoretical fits (*lines + symbols*). The fit at $z = 1.01$ is the same for the CDI only and the CDI + 1.0 M NaCl data

Chemical experiments were as follows [72]. Amino acid polymerization reactions were performed with L-Glu and CDI as a dehydrating agent. Reaction mixtures contained 3 mM L-Glu in 5 ml distilled water and different amounts (5, 10 or 20 mM) of KCl or NaCl salt. The mixtures were diluted to 10 ml, yielding the final salt concentrations of 0.5, 1.0 and 2.0 M, and 6 mM CDI was added into each mixture. After that, the mixtures were incubated at room temperature for 24 hours. In order to access the resulting concentrations of differently sized peptides in the reaction mixtures, each mixture was analyzed using high pressure liquid chromatography and mass spectrometry. Analysis of the acquired chromatograms led to quantification of the relative concentrations of differently sized polypeptides in the reaction mixtures. The measured absorbance at 214 nm originates mainly from the peptide bonds in polypeptides, while the absorbance by the L-Glu side chain is commonly believed negligible. Taking into account that the polypeptide containing i amino acid monomers has $i - 1$ peptide bonds, each peak on the chromatogram corresponding to a given length was integrated and divided by the appropriate number of absorbing peptide bonds.

The obtained values were normalized by using the mass conservation $n_1 + \sum_{i=2}^{\infty} in_i = n_{tot}$. Here, n_1 is the concentration of free amino acid monomers, n_i are the concentrations of peptides of length i and n_{tot} is the total concentration of amino acid in a given solution. With the measured concentrations of i-mers and the known n_{tot}, this circumvents the uncertainty in the free monomer concentration, not seen on the chromatograms. The normalized concentrations of $i-$mers, $f_i = n_i/n_{tot}$, detected at different concentrations of NaCl and KCl, are shown by symbols in Figs. 2.43 and 2.44, respectively. Symbols in Fig. 2.45 correspond to the ratios of differently sized peptides grown in water solutions with the same concentrations of KCl and NaCl for the same time and under otherwise identical conditions. It is clearly seen that f_i^{KCl} are noticeably larger than f_i^{NaCl} in the entire domain studied. As follows from Fig. 2.43, NaCl does not show any catalytic effect in these conditions: the polymer length is usually smaller with NaCl than just with CDI. The ratio

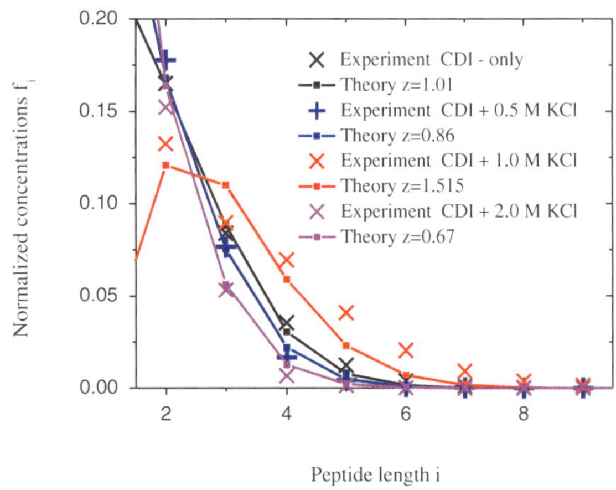

Fig. 2.44 Same as Fig. 2.43 but with KCl

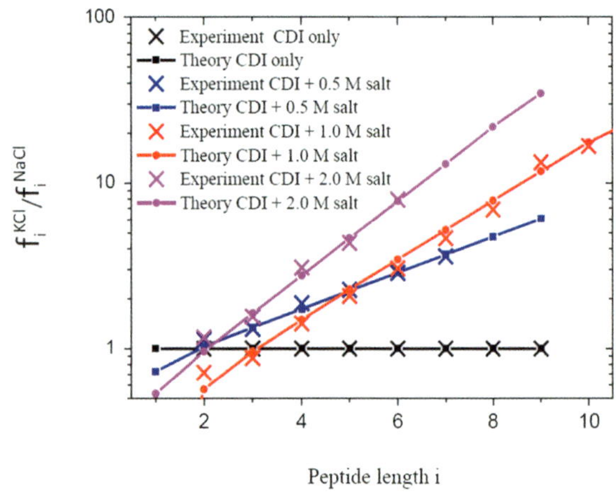

Fig. 2.45 Ratios f_i^{KCl}/f_i^{NaCl} versus i at different concentrations of KCl and NaCl: experiment (*symbols*) and theory (*lines + symbols*)

$f_i^{\rm KCl}/f_i^{\rm NaCl}$ increases super-linearly with length for a given salt concentration [73] and increases with the salt concentration for long peptides with $i > 5$. Lines with symbols in Figs. 2.43, 2.44 and 2.45 are theoretical fits that will be discussed shortly.

We now consider the reaction scheme shown in Fig. 2.46. As commonly assumed, an amino acid monomer should get activated by CDI before it can be attached to any i-mer, as shown in Fig. 2.46a. Therefore, not all the monomers participate in the reaction chain producing i-mers (Fig. 2.46b), but only the activated monomers. We also neglect the decay of i-mers which should not occur in these experimental conditions. With these assumptions, the reaction chain is described in terms of the irreversible growth model (2.194), with k_i^+ as the rate constant of linear polymerization. The rate equations for the normalized concentrations describing this process are

$$\frac{1}{n_{tot}}\frac{df_1}{dt} = -2k_1^+ f_1^2 - f_1 \sum_{i=2}^{\infty} k_i^+ f_i;$$
$$\frac{1}{n_{tot}}\frac{df_i}{dt} = k_{i-1}^+ f_1 f_{i-1} - k_i^+ f_1 f_i, \quad i = 2, 3.4\ldots, \tag{2.260}$$

with the initial conditions $f_i(0) = \delta_{i1}$. For the diffusion-controlled growth of linear polymers that can attach monomers only from their ends, the rate constants should be i-independent (at least for small $i \leq 10$ considered here). The dimerization rate is however different, since only a fraction of monomers is activated by CDI. These considerations yield the rate constants of the form

$$k_1^+ = k^+ b; \quad k_i^+ \cong k^+, i = 2, 3, 4\ldots, \tag{2.261}$$

where $b \leq 1$. This is identical to (2.231). Linearizing (2.260) with rate constants (2.261) by introducing a new variable z instead of time by definition

$$k^+ n_{tot} t = \int_0^z \frac{dz}{f_1(z)}, \tag{2.262}$$

the solutions for $f_1(z)$ and $f_i(z)$ are therefore given by (2.239), (2.248) at $b = 1$ and (2.238), (2.256) at $b < 1$.

We now turn to the analysis of the experimental size spectra. First we assume that b is close to 1/2 in the system considered. This follows from a thermodynamic study of the reversible activation of amino acid with CDI (not discussed here) as well as from analyzing the best numerical fits. It is also clear that the b value should not be strongly

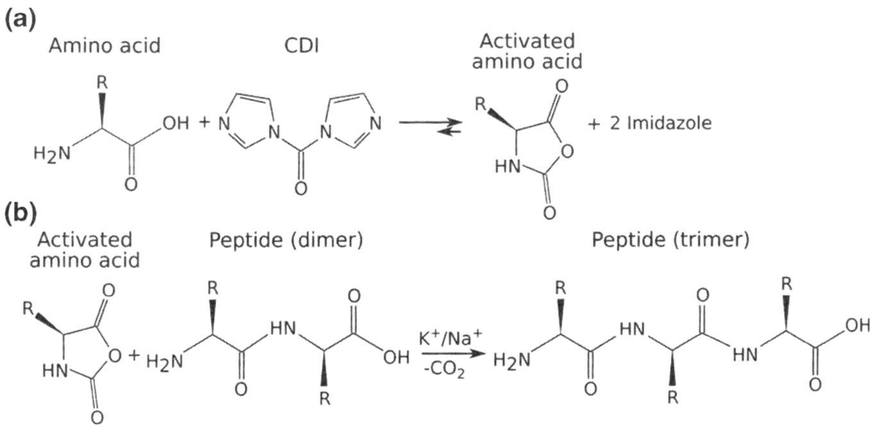

Fig. 2.46 Reaction scheme: **a** Reversible activation of a monomer with CDI; **b** Irreversible reaction chain producing peptides A_i ($i = 1, 2, 3, \ldots$)

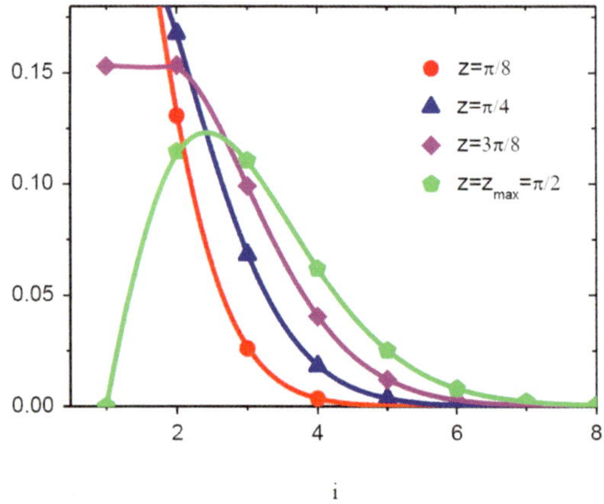

Fig. 2.47 Evolution of $f_i(z)$ at $b = 1/2$, obtained from (2.238) and (2.256)

Table 2.2 Fitting values of z, the estimated $k^+n_{tot}t$ values and the ratios of KCl to NaCl catalyzed rate constants at different salt concentrations

Fitting parameters	CDI only	CDI + 0.5 M salt	CDI +1 M salt	CDI + 2 M salt
z[KCl]	1.01	0.86	1.52	0.67
z [NaCl]		0.67	1.01	0.4
$k^+n_{tot}t$[KCl]	2.06	1.51	8.51	1.01
$k^+n_{tot}t$[NaCl]		1.01	2.06	0.50
k^+_{KCl}/k^+_{NaCl}	1	1.50	4.13	2.02

affected by the presence of salt. Adopting $b = 1/2$ uniformly for all the incubation experiments, the sole parameter that determines the spectrum shape is z, related to the incubation time and the polymerization rate constant k^+ through (2.262). In terms of z, the time evolution of the size spectrum has the universal form shown in Fig. 2.47, whatever $k^+n_{tot}t$ is. It is seen that the spectrum is purely decreasing at small z. The peak at $i = 2$ is developed at $z \cong 3\pi/8$, and remains at $i = 2$ throughout the follow-up growth steps including the steady state at $z_{\max}$, where the concentration of monomers tends toward zero and the entire growth process ceases.

Best fits of the experimental size spectra and their ratio with (2.238), (2.256) and (2.257) at $b = 1/2$ and adjustable z are shown in Figs. 2.43, 2.44 and 2.45. The fits reproduce very well the data, with only one discrepancy in Fig. 2.44 for 1 M of KCl, i.e., the longest polymer chains. This curve was fitted with z very close to $z_{\max}$, corresponding to only a trace amount of monomers. However, this is still insufficient to reproduce the enhanced length of the polymer chains. Using (2.262), we are able to deduce the $k^+n_{tot}t$ for each sample from the fitting values of z, yielding the ratios of KCl to NaCl-catalyzed polymerization rates at different salt concentrations. These data are summarized in Table 2.2.

From this analysis, the k^+_{KCl}/k^+_{NaCl} ratio appears to be systematically larger than one in the entire domain of salt concentrations from 0.5 to 2 M. This explains why the KCl-catalyzed peptides are statistically longer than the NaCl-catalyzed ones. We note, however, that the dependence of this effect on the salt concentration is not monotonic, with the maximum ratio of $\sim$4 reached at 1 M. This is consistent with the longest peptides being the most representative at 1 M of KCl in Fig. 2.44, while the NaCl curve at 1 M in Fig. 2.43 is almost the same as with CDI only. Combined with the non-linear spectrum shapes, the deduced $k^+_{\mathrm{KCl}}/k^+_{\mathrm{NaCl}}$ ratios reproduce well the superlinear increase of the concentration ratios with the peptide length, shown in Fig. 2.45. Qualitatively, for the quasi-Poissonian distributions (2.248) at $b = 1$ and small enough z, the $f_i^{\mathrm{KCl}}/f_i^{\mathrm{NaCl}}$ ratio scales as $(k^+_{\mathrm{KCl}}/k^+_{\mathrm{NaCl}})^{i-1}$. The general solutions at $b < 1$, given by (2.256), preserve this property, which explains the exponential dependence of the concentration ratio with i.

To sum up the results of this section, a qualitative analysis of the size spectra of polypeptides grown in water solutions in presence of KCl and NaCl can be performed within the irreversible growth model in a closed system. Our rate equations assume a size-independent polymerization rate k^+ for all i-mers, while the dimerization rate is smaller ($b \leq 1$). This non-linear growth model has an interesting analytical solution given by (2.238) and (2.256). Taking a plausible estimate of $b = 1/2$, we have obtained good one-parametric fits for all the experimental spectra. It has been shown that the value of k^+ is systematically larger in reaction mixtures with KCl than with NaCl, which is why the peptides elongate much faster in potassium-rich solutions. This leads to the exponential increase of the ratio between the peptide concentrations in potassium and sodium solutions with length. While our study is not extensive, it gives a simple scenario of the salt-induced linear polymerization of amino acids, which might be usefull in approaching schemes of chemical evolution. Of course, the microscopic origin of the k^+ dependence on the concentration of a given salt in water requires a separate study. Usually, the rate constant is expected to be proportional to the corresponding diffusion coefficient in water D and the capture number σ, $k^+ = D\alpha$, where the last quantity depends on the microscopic kinetics of the reactions shown in Fig. 2.46. Therefore, k^+_{KCl} should be greater than k^+_{NaCl} either due to an enhanced salt-induced diffusivity of the peptide monomers in presence of KCl or due to a faster salt-induced dimerization.

2.15 Open Irreversible Systems

Let us now discuss how the kinetics of irreversible growth changes in open systems with monomer influx ans sink. We again consider homogeneous growth of clusters via the monomolecular Scheme (2.194). To be more specific, we write down (2.195) and (2.196) in the particular case of surface islands growing by surface diffusion of adatoms with the diffusion coefficient D and the characteristic lifetime τ on a substrate subjected to a time-independent vapour flux I:

$$\frac{dn_1}{dt} = I - \frac{n_1}{\tau} - 2D\alpha_1 n_1^2 - Dn_1 \sum_{i=2}^{\infty} \alpha_i n_i; \tag{2.263}$$

$$\frac{dn_i}{dt} = Dn_1(\alpha_{i-1}n_{i-1} - \alpha_i n_i), \quad i = 2, 3, 4 \ldots . \tag{2.264}$$

Here, α_i represent the corresponding capture numbers. The surface concentration of islands $N = \sum_{i=2}^{\infty} n_i$ is given by

$$\frac{dN}{dt} = D\alpha_1 n_1^2. \tag{2.265}$$

In many important systems, the capture numbers of all islands can be treated as size-independent in the first approximation [20, 241]:

$$\alpha_i = \alpha_*, \quad i = 2, 3, 4 \ldots \tag{2.266}$$

More generally, (2.263) for the adatom concentration can be approximated by

$$\frac{dn_1}{dt} = I - \frac{n_1}{\tau} - 2D\alpha_1 n_1^2 - D\alpha_* n_1 N \tag{2.267}$$

with α_* being the averaged value of the capture numbers over the ensemble of differently sized islands:

$$\alpha_* = <\alpha_i> = \frac{\sum_{i=2}^{\infty} \alpha_i n_i}{\sum_{i=2}^{\infty} n_i}. \tag{2.268}$$

At $\alpha_* \cong const$, (2.267) and (2.265) constitute the closed system for finding $n_1(t)$ and $N(t)$, which can be reduced to one second-order differential equation for $n_1(t)$. This equation, however, cannot be linearized by introducing the z variable by definition

$$\frac{dz}{dt} = D\alpha_* n_1; \quad z(t = 0) = 0, \tag{2.269}$$

because the arrival rate I in (2.267) is not proportional to n_1. In the general case, the adatom concentration should be obtained numerically at the given sources. After that, it can be inverted as a function of z, while the size distribution of islands in terms of z is again given by the equation which is identical to (2.246). Indeed, (2.264) at $\alpha_i = \alpha_*$ in view of (2.269) take the form

$$\frac{dn_2}{dz} = bn_1 - n_2;$$

$$\frac{dn_i}{dz} = n_{i-1} - n_i, \quad i = 3, 4, 5 \ldots \tag{2.270}$$

with $b = \alpha_1/\alpha_*$. This is exactly identical to (2.233) and (2.234), although the adatom concentration is now driven by external fluxes. Therefore, the $n_i(z)$ are given by

$$n_{i+2}(z) = \frac{b}{i!} \int_0^z dy n_1(z-y) y^i e^{-y}, \quad i = 0, 1, 2 \ldots . \tag{2.271}$$

Let us now discuss some important particular cases of the time dependences of the adatom concentration and the island density. If the desorption rate is small ($\tau \to \infty$), as usually occurs at low substrate temperatures, an asymptotic growth stage is quickly reached where the adatom concentration tends toward zero and the island density increases with t. In this regime of complete condensation, (2.267) at $n_1 \to 0$ yields

$$D\alpha_* n_1 N \cong I. \tag{2.272}$$

Using this in (2.265), we obtain the asymptotic behavior of the island density:

$$N = \left(\frac{3\alpha_1 I^2 t}{D\alpha_*^2} \right)^{1/3}, \tag{2.273}$$

i.e., N scales with time as $t^{1/3}$. If the results of growth experiments are compared at a fixed amount of deposited material $n_{tot} = It_0$ with different growth times $t_0 \propto 1/I$, the island density obey the scaling behavior

$$N \propto (I/D)^{1/3}. \tag{2.274}$$

From (2.273), (2.272) and (2.269), the adatom concentration scales with t and z as

$$n_1 = \frac{C}{t^{1/3}} = \frac{A}{z^{1/2}}, \tag{2.275}$$

with $C = (I/3D^2\alpha_*\alpha_1)^{1/3}$ and $A = (I/2\alpha_1 D)^{1/2}$. Using this in (2.271), the island concentrations at the asymptotic stage are described by

$$n_{i+2}(z) = \frac{Ab}{i!} \int_0^z \frac{dy y^i e^{-y}}{(z-y)^{1/2}}. \tag{2.276}$$

In the opposite case of the incomplete condensation regime at high substrate temperatures, the maximum adatom concentration $n_1 \cong I\tau$ is reached due to the balance between the adsorption–desorption processes and then decreases slowly with time. At the beginning of growth, (2.269) yields $z \cong D\alpha_* I\tau t$, as it would be in a system with a constant adatom concentration. At $n_1 \cong I\tau$, (2.271) are simplified to

$$n_{i+2}(z) \cong \frac{I\tau b}{i!}\gamma_{i+1}(z). \tag{2.277}$$

However, this small time interpolation become incorrect when n_1 starts decreasing, and the system will be finally rendered to the asymptotic stage described by (2.273), (2.275) and (2.276).

Assuming that $\alpha_1 \cong \alpha_* \cong 1$, (2.265), (2.267) and (2.269) can be put in dimensionless form by introducing the adatom diffusion length $\lambda = \sqrt{D\tau}$, the normalized adatom concentration $f_1 = \lambda^2 n_1$, surface density $F = \lambda^2 N$, and dimensionless time $x = t/\tau$:

$$\frac{df_1}{dx} = v - f_1 - 2f_1^2 - f_1 F; \tag{2.278}$$

$$\frac{dF}{dx} = f_1^2; \tag{2.279}$$

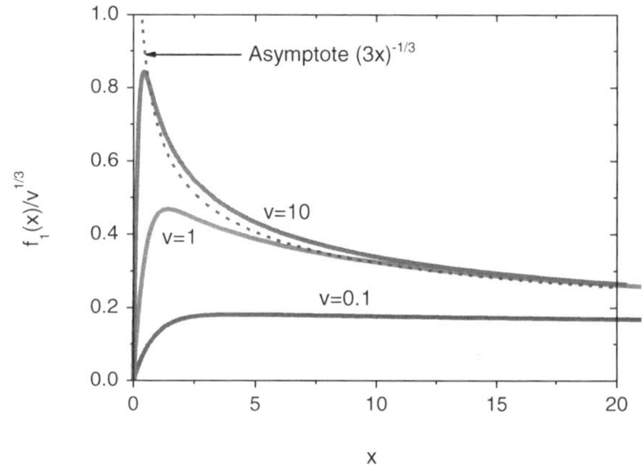

Fig. 2.48 Graphs of the normalized adatom concentration $f_1/v^{1/3}$ versus dimensionless time x at different v, compared to the large time asymptote

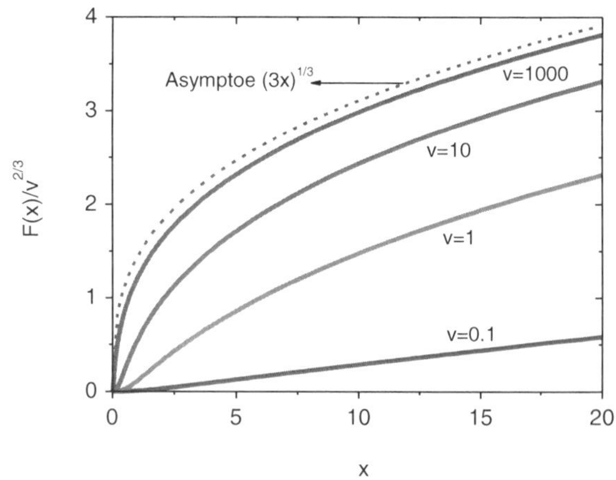

Fig. 2.49 Graphs of the normalized island density $F/v^{2/3}$ versus dimensionless time x at different v, compared to the large time asymptote

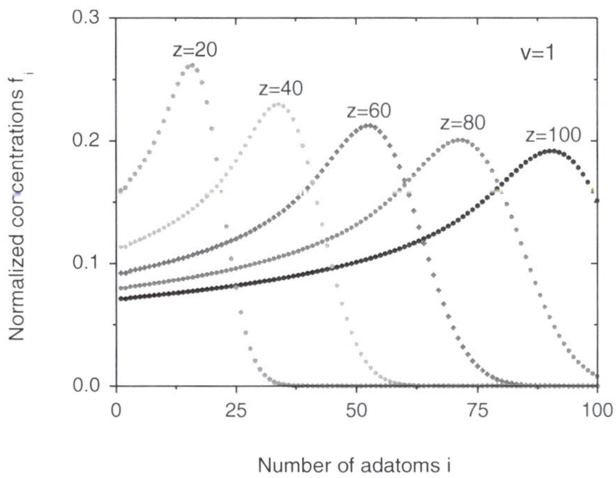

Fig. 2.50 Normalized concentrations f_i at fixed growth conditions ($v = 1$) and different z relating to different growth time

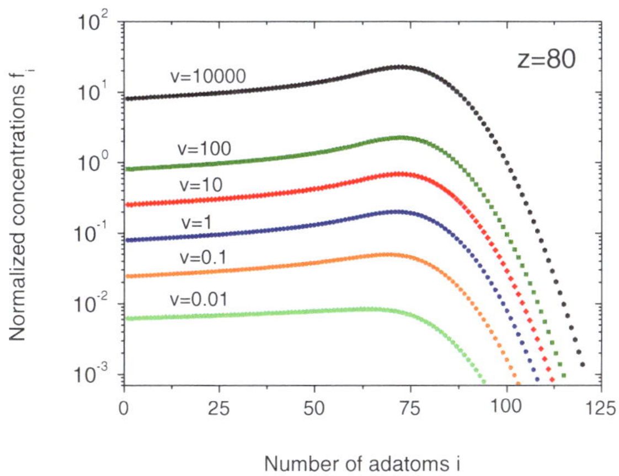

Fig. 2.51 Normalized concentrations f_i at a fixed growth time (relating to $z = 80$) and different growth conditions (v)

$$\frac{dz}{dx} = f_1, \tag{2.280}$$

with the initial conditions $f_1(0) = 0$, $F(0) = 0$, and $z(0) = 0$. Size spectra are obtained from the (2.271) for $f_i(z)$, with $b = 1$. In these variables, the system evolution depends on only one control parameter

$$v = \lambda^2 I\tau = DI\tau^2, \tag{2.281}$$

which equals the ratio between the characteristic diffusion area λ^2 and the adsorption area $(I\tau)^{-1}$. The value of v always decreases at higher surface temperatures.

Figures 2.48 and 2.49 show the normalized adatom concentration $f_1/v^{1/3}$ and island density $F/v^{2/3}$ as functions of the dimensionless time x at different growth conditions (v), obtained from numerical solutions of (2.278) and (2.279) and compared to their large time asymptotes $f_1(x)/v^{1/3} = (3x)^{-1/3}$ and $F(x)/v^{2/3} = (3x)^{1/3}$. Discrete concentrations of differently sized islands, obtained from (2.280) and (2.271), are shown in Figs. 2.50 and 2.51, relating to different growth times and deposition conditions, respectively. It is seen that the size distributions always have essentially asymmetric shapes, with an abrupt right tail and a much slower regression toward smaller i. It is noteworthy that these size distributions are universal and depend on the sole parameter v, while fitting a particular growth experiment just needs re-scaling of the results with adjustable values of λ and τ.

Chapter 3
Self-Induced Islands in Lattice Mismatched Systems

In the previous chapters, we deliberately gave an account of nucleation theory without connecting it with a concrete material system or growth experiment in most cases. Now, the reader will see some applications of the theory to real data. This chapter deals with the analysis of growth properties and morphology of ensembles of self-induced islands that form spontaneously in lattice mismatched material systems. One of the most known processes of this type is the Stranski-Krastanow growth resulting in the formation of 3D islands. The island size can be made so small that the band structure of a semiconductor material gets modified compared to bulk. Whenever a typical island size is compared to or smaller than the de Broile wavelength in bulk semiconductor material, the energy spectrum is no longer continuous. Rather, the density of states acquires a delta-like shape with maxima at the discrete energy levels and in this sense the island becomes an artificial atom. These quantum dots have many important advantages for applications in optoelectronics. In particular, the energy of optical transitions becomes size-dependent and increases when the size is confined. The properties of quantum dot devices are less affected by heating, while the threshold current of a quantum dot laser is much lower compared to its quantum well competitors. These advantages were first understood theoretically in the early 1980s when there was no easy way to produce semiconductor particles with the required morphology. In 1986, Goldstein and co-authors reported on a spontaneous formation of 3D InAs islands on a GaAs substrate [127], which is now a classical quantum dot system. It was soon realized that, in principle, any material system with a large enough lattice mismatch has an energetic tendency for 3D islanding in a certain thickness interval. Spontaneous formation of 3D islands is very attractive as it requires neither preliminary nor after growth surface treatment by applying a sophisticated lithography. On the other hand, the self-induced growth is rather difficult to control because the islands just "emerge themselves". It was therefore necessary to establish reproducible growth techniques where the island morphology (size, shape, surface density) could be tuned by technologically accessible external conditions such as the surface temperature, material fluxes and deposition time. This is exactly where theoretical approaches based on nucleation theory come into

V. G. Dubrovskii, *Nucleation Theory and Growth of Nanostructures*, NanoScience and Technology, DOI: 10.1007/978-3-642-39660-1_3,

play, because the nucleation process has an essentially spontaneous character and the corresponding models deal with size distributions of self-induced objects in metastable surroundings.

A very important feature of epitaxial nanostructures is their small footprint in contact with the substrate. In a thin film case, lattice mismatch would necessarily lead to the formation of misfit dislocations at a certain thickness, which is why it persists as one of the most challenging bottlenecks for heterogeneous integration of dissimilar semiconductor materials. In particular, the integration of optoelectronic and electronic integrated circuits is highly desirable because it will not only enable a vast range of otherwise unattainable capabilities, but also reduce power consumption, weight and size of personal electronics. While wafer bonding is not readily applicable for finished CMOS circuits so far, the integration requires direct growth of high quality single-crystalline optical materials directly on silicon. Due to a large lattice mismatch between most III-V materials and silicon, the dislocation-free growth is possible only in a nanostructure form. Nanostructures such as quantum dots, nanoneedles and nanowires give much wider opportunities for the band gap engineering as well as for the selection of material–substrate combinations. Therefore, this chapter deals not only with the traditional Stranski-Krastanow quantum dots that have been attracting much attention for at least 25 years already and are very well documented, but also with only recently emerged nanoneedles. Many results presented in the chapter will be used in the foregoing nanowire parts of the book. Interestingly, while lattice mismatch is considered as a major problem in the conventional crystal growth, it is absolutely necessary to observe most of the growth effects considered here. The self-induced nucleation of islands is essentially stress-driven, because it enables a very efficient relaxation of elastic stress at free sidewalls. On the other hand, these islands are useful only before the formation of dislocations or when the defects remain localized at the interface (in the case of submicron-sized nanoneedles). Relevant growth models should therefore carefully account for different mechanisms of stress relaxation.

We start the chapter with a brief (and simplified) description of quantum confinement in semiconductors. The energy spectra and the density of states in quantum wells, quantum wires and quantum dots are discussed in Sect. 3.1. In Sect. 3.2, we give a qualitative description of the Stranski-Krastanow growth, along with some examples for the InAs/GaAs and Si/Ge material systems. We briefly discuss relevant theoretical models of coherent strained islands, in particular, the model by Shchukin and co-authors [148] where a non-trivial interplay of the elastic relaxation, surface and edge energies was first considered in connection with thermodynamics of the Stranski-Krastanow growth. Sect. 3.3 deals with the elastic relaxation in differently shaped nanostructures depending on their aspect ratio. We present some useful analytical approximations for the elastic relaxation, our own semi-analytical approach, and a comparison between them. In Sect. 3.4, we discuss a nucleation model where the formation energy is treated as a function of two variables: the size and the aspect ratio. The results of this section might be useful for understanding the nanostructure growth in general. We show how the energetically preferred aspect ratio depends on the lattice mismatch and supersaturation and how the lattice mismatched growth may switch between different modes depending on the material constants and deposition conditions.

Section 3.5 presents a model for the free energy of coherent island formation that was originally developed by Osipov and co-authors in [24]. The model is of great importance for what follows because it connects the Stranski-Krastanow surface energetics with the general nucleation theory considered in Chaps. 1 and 2. It is shown that the stress-driven islanding can be formulated in terms of the first order phase transition where the wetting layer thickness plays a role of system metastability. Using the results of Chaps. 1 and 2, we then construct the nucleation distribution in Sect. 3.6. The critical thickness for 2D-3D growth transformation is considered in detail. We emphasize an important difference between the equilibrium and critical thickness: while the former is entirely determined by thermodynamic parameters, the latter is a kinetic value that approximately equals the maximum wetting layer thickness under the material flux. As a consequence, the islands may also form at a subcritical deposition thickness (the so-called subcritical quantum dots), usually with a much lower density. The time evolution of the mean island size is studied in Sect. 3.7. Section 3.8 gives the general modeling scheme and a numerical example. In Sect. 3.9, we consider the influence of surface steps on the island nucleation process, showing that the substrate vicinity can be used as an additional tuning knob to tailor the island morphology. Section 3.10 is dedicated to the subcritical islands that usually require a certain exposition of the surface and behave differently with changing the growth conditions. Following the kinetic approach described in Chap. 2, Sect. 3.11 gives a systematic account of the kinetically controlled engineering of quantum dot ensembles. It is shown how the mean size, surface density and other morphological characteristics can be tuned by the growth temperature, flux, deposition thickness and exposition time.

Section 3.12 presents some experimental data on the MBE growth, morphology and optical properties of InAs/GaAs quantum dot ensembles, and their comparison with theoretical predictions. We consider the temperature and deposition rate dependence of the mean lateral size and surface density of InAs islands, the system evolution during the exposition under arsenic flux, and the subcritical quantum dots grown on singular and vicinal substrates. In Sect. 3.13, we compare theoretical results with some experimental data on the Si/Ge islands. The temperature dependence of the mean size and density, and the threshold growth behavior of Si/Ge islands in presence of antimony flux are discussed. Overall, it is shown that the kinetic growth theory is capable of describing the major effects, and is in a good quantitative correlation with the experimental data in most cases. The kinetic approach provides a powerful tool to predict and control the morphology of self-induced nanostructures and is by no means limited to the particular systems studied here.

Section 3.14 gives an overview of the MOCVD growth experiments, structural and optical characterization of GaAs nanoneedles and InGaAs/GaAs nanopillars obtained in Chang-Hasnain group [185, 196, 222]. These highly anisotropic structures (with a typical aspect ratio of the order of ten) emerge as the Volmer-Weber islands and then grow in the core-shell mode maintaining the aspect ratio acquired at the nucleation stage. The GaAs nanoneedles can be obtained both on silicon and sapphire substrates. Although nanoneedles start with tiny islands, they can be grown to the micron-scale dimensions. Surprisingly, the bulk material of nanoneedles is single crystalline wurtzite regardless of the substrate used. We present very important data of extensive TEM characterization, revealing the crystal structure and the dislocations that are localized in the transition region close to the interface due to special growth geometry. The limit-breaking 160 nm thick dislocation-free GaAs core around $In_{0.2}Ga_{0.8}As$ shell is shown, against only 10 nm critical thickness in 2D thin film case. In Sect. 3.15, we present the modeling of the time evolution of GaAs nanoneedle length, and compare the results with the experimental data. We briefly discuss some other nanoneedle properties from the viewpoint of the kinetic growth theory.

Finally, Sect. 3.16 is devoted to the growth modeling of cobalt nanoparticles deposited onto $CaF_2(111)$ buffer layers on Si(111), a metal-on-insulator system which is interesting for its magnetic properties. Much higher surface energy of Co than that of an insulator results in the Volmer-Weber growth mode. Since magnetic properties of Co are expected to be strongly dependent on the surface morphology of islands, it is paramount to identify the kinetic tools enabling a precise control over the island size, shape and density. We show that the kinetic approach initially developed for the Stranski-Krastanow semiconductor islands is also useful for understanding rather peculiar growth properties in this material system.

3.1 Size Quantization in Semiconductor Nanostructures

Semiconductor nanostructures with reduced dimensionality such as 2D quantum wells, 1D quantum wires and 0D quantum dots feature unique fundamental properties enabling a radical improvement of device performance. An exciting breakthrough from bulk semiconductors toward nano-heterostructures that revolutionized high speed electronics and optoelectronics is documented, e.g., in [119]. In particular, the opportunity to tune the operating wavelength by changing the island size, a lower threshold current and an enhanced temperature stability of quantum dot lasers have recently been the major achievement [40, 41]. Although the detailed analysis of optical and transport properties as well as applications of nanostructures is beyond the scope of this book, we now present a semi-quantitative account of remarkable physical properties originating from the quantum confinement in semiconductors.

Energy levels of electrons in free space are determined as the eigenvalues E of the stationary Schrödinger equation for the Ψ-function

$$\left(-\frac{\hbar^2}{2m}\Delta + U\right)\Psi = E\Psi, \tag{3.1}$$

where $\hbar$ is the Plank constant, m is the mass, Δ is the Laplace operator and $U(\vec{r})$ is the potential. The function $\Psi(\vec{r})\Psi^*(\vec{r})$ gives the probability density for an electron to have the coordinate $\vec{r}$ and should be therefore normalized to one. In the absence of potential field ($U = 0$), the electrons in 3D space are described by the de Broile waves

$$\Psi(\vec{r}) = \exp(i\vec{k}\vec{r}). \tag{3.2}$$

The wave vector is related to the momentum $\vec{p}$ as $\vec{k} = \vec{p}/\hbar$, while the continuum energy spectrum is given by the classical dispersion relationship

$$E = \frac{\hbar^2}{2m}k^2. \tag{3.3}$$

Electrons and holes in bulk semiconductors near the corresponding band edges E_0 can be described by the same approximation, where the mass is changed to the effective mass to account for the quasi-classical motion in the inherent periodic potential of a crystal lattice. The dispersion law is now modified to [120]

$$E = E_0 + \frac{\hbar^2}{2m}k^2. \tag{3.4}$$

Here, we assume the spatial isotropy of the effective mass.

Let us now consider an idealized model for a quantum well with width L in x direction and infinitely large potential barriers at $x = 0$ and $x = L$, as shown in Fig. 3.1a. Substitution of the function $\Psi(\vec{r}) = \Psi_x(x)\exp[i(k_y y + k_z z)]$ into the Schrödinger equation $\Delta\Psi = -k^2\Psi$ leads to the separation of variables and yields

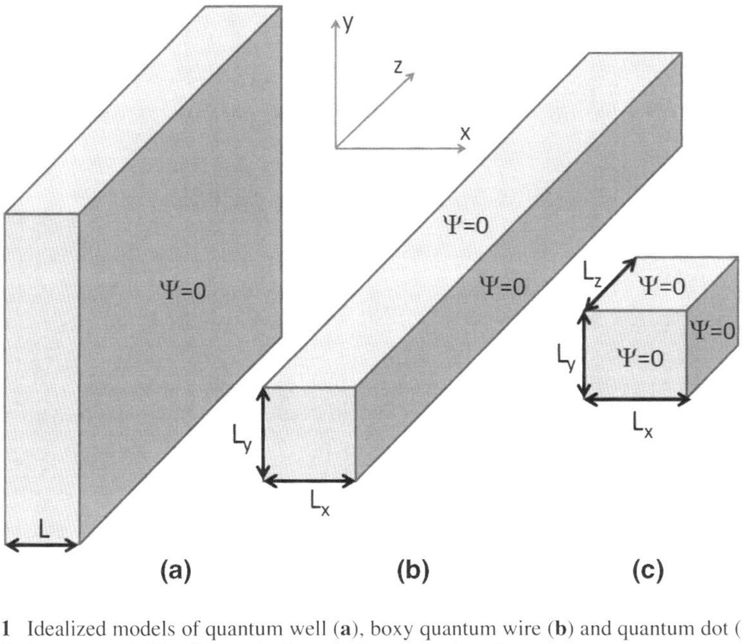

Fig. 3.1 Idealized models of quantum well (**a**), boxy quantum wire (**b**) and quantum dot (**c**) with infinitely large potential barriers at the boundaries where $\Psi = 0$

$$\frac{d^2\Psi_x(x)}{dx^2} = -k_x^2\Psi_x(x);\ \Psi_x(0) = \Psi_x(L) = 0. \tag{3.5}$$

Solutions to (3.5) are given by

$$\Psi_x(x) = A\sin(\pi nx/L) \tag{3.6}$$

with integer $n = 1, 2, 3\ldots$, where the constant $A = -i/\sqrt{2L}$ is easily determined from the normalization condition. This leads to the quantization of energy levels by the boundary conditions:

$$\Psi(\vec{r}) = A\sin\left(\frac{\pi nx}{L}\right)\exp[i(k_y y + k_z z)]; \tag{3.7}$$

$$E_n = \frac{1}{2m}\left(\frac{\pi\hbar n}{L}\right)^2. \tag{3.8}$$

Whatever is the discrete energy spectrum for the electron motion in x direction, the dispersion (3.4) is modified to

$$E = \sum_n E_n + \frac{\hbar^2}{2m}(k_y^2 + k_z^2). \tag{3.9}$$

The quasi-continuum term in x direction is thus replaced by the discrete energy levels in the quantum well.

Similarly, in the idealized case of a rectangular, infinitely long quantum wire with the dimensions L_x and L_y in x and y directions, respectively, and infinitely large potential barriers at the boundaries (Fig. 3.1b), the corresponding wave functions and energy levels are given by

$$\Psi(\vec{r}) = A\sin\left(\frac{\pi nx}{L_x}\right)\sin\left(\frac{\pi ly}{L_y}\right)\exp(ik_z z); \tag{3.10}$$

$$E_{nl} = \frac{(\pi\hbar)^2}{2m}\left[\left(\frac{n}{L_x}\right)^2 + \left(\frac{l}{L_y}\right)^2\right]. \tag{3.11}$$

Whenever the motion of carriers is confined in two dimensions and remains quasi-classical along the quantum wire axis, the dispersion law is modified to

$$E = \sum_{n,l} E_{nl} + \frac{\hbar^2}{2m}k_z^2. \tag{3.12}$$

For a quantum dot in the form of rectangular box with sides L_x, L_y, L_z and an infinitely large potential barrier at the boundaries (Fig. 3.1c), the carriers are confined in all the three directions. This yields purely discretized Ψ—function and the corresponding energy spectrum of the form

$$\Psi(\vec{r}) = A \sin\left(\frac{\pi n x}{L_x}\right) \sin\left(\frac{\pi l y}{L_y}\right) \sin\left(\frac{\pi q z}{L_z}\right) \tag{3.13}$$

$$E_{nlq} = \frac{(\pi\hbar)^2}{2m}\left[\left(\frac{n}{L_x}\right)^2 + \left(\frac{l}{L_y}\right)^2 + \left(\frac{q}{L_z}\right)^2\right] \tag{3.14}$$

When the motion of carriers is quantized in all directions, the total energy simply equals the sum over all discrete energy levels

$$E = \sum_{n,l,q} E_{nlq}, \tag{3.15}$$

and no quasi-classical terms are left in the entire spectrum.

One of the most important characteristics of semiconductors is the so-called density of states $\rho(E) = dN/dE$ showing how many energy states N are occupied within the band in the energy interval from E to $E + dE$ [41, 120]. In bulk semiconductors and in surface nanostructures such as quantum wells, wires and dots, it is reasonable to normalize the density of states to the volume V and to the surface area S, respectively:

$$\rho_V(E) = \frac{1}{V}\frac{dN}{dE}; \tag{3.16}$$

$$\rho_S(E) = \frac{1}{S}\frac{dN}{dE}. \tag{3.17}$$

We first consider the density of states in bulk semiconductors. Presenting (3.16) in the equivalent form

$$\rho_V(E) = \frac{1}{V}\frac{dN}{dk}\left(\frac{dE}{dk}\right)^{-1}, \tag{3.18}$$

the last term in the right hand side is given by

$$\frac{dE}{dk} = \frac{\hbar^2 k}{m} \tag{3.19}$$

from the dispersion curve defined by (3.4). To calculate the elementary volume of reciprocal space ω_k relating to the state with the wave vector $\vec{k}$ in the crystal, we note that the allowed state for the projection k_j in any of three directions $(j = x, y, z)$ must correspond to an integer number of the de Broile waves along the crystal length L_j. This corresponds to the periodic boundary conditions for Ψ (absolutely homogeneous 3D model) and yields $k_j = (2\pi n)/L_j$ with an integer n. Therefore, ω_k is given by $\omega_k = (2\pi)^3/(L_x L_y L_z)$, or

$$\omega_k = \frac{(2\pi)^3}{V}, \tag{3.20}$$

where $V = L_x L_y L_z$ is the crystal volume.

For an isotropic cubic crystal, the reciprocal space is spherical. The sphere of volume $\Omega_k = (4\pi/3)k^3$ accommodates Ω_k/ω_k allowed states with different wave vectors and twice more electron states due to the spin degeneracy:

$$N = \frac{2\Omega_k}{\omega_k}, \tag{3.21}$$

yielding $N = (Vk^3)/(3\pi^2)$. Using this together with (3.18), (3.19) and expressing k as $k = \hbar\sqrt{2m(E - E_0)}$ from (3.4), we arrive at the well-known expression [41]

$$\rho_V(E) = \frac{\sqrt{2}}{\pi^2}\left(\frac{m}{\hbar^2}\right)^{3/2}\sqrt{E - E_0}. \tag{3.22}$$

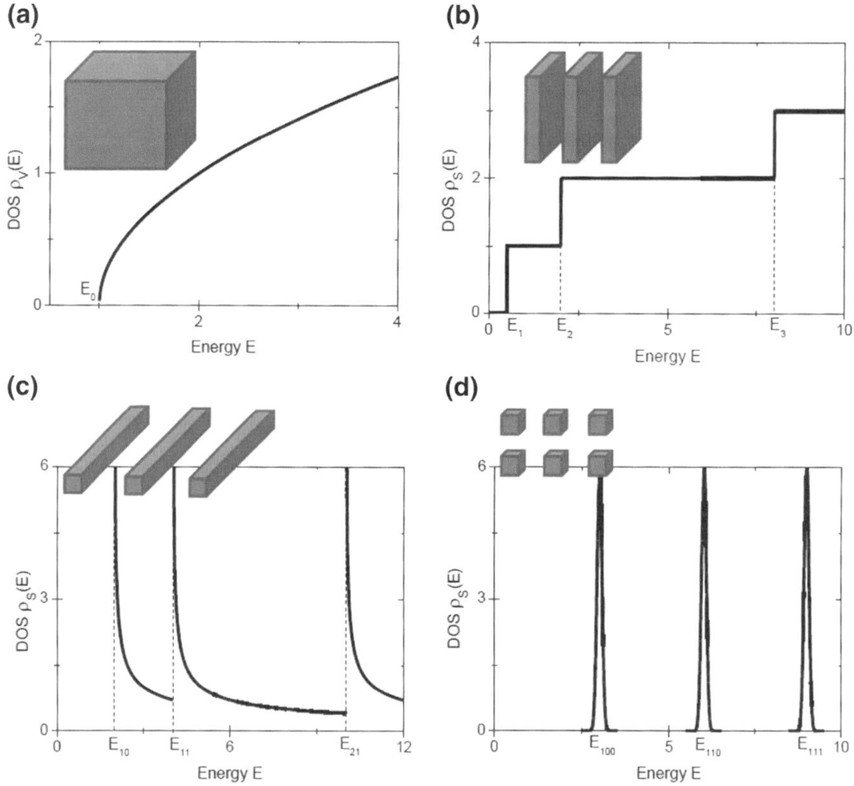

Fig. 3.2 Density of states in bulk semiconductors (**a**), 2D quantum wells (**b**), 1D quantum wires (**c**) and 0D quantum dots (**d**)

Therefore, the density of states in bulk semiconductors scales as the square root of energy counted from the band edge and tends to zero when the energy approaches E_0, as shown in Fig. 3.2a. Of course, the obtained expression is valid only near the bang edge, where the effective mass approximation is directly applicable. We also note that, if zero boundary conditions are imposed for Ψ at L_j instead of periodic, the resulting density of states would be 8 times larger [120].

In the case of quantum well, (3.7) yields (3.19), where the 3D wave vector is replaced to the 2D one, $k^2 = k_y^2 + k_z^2$. Equation (3.20) for 2D reciprocal space is changed to $\omega_k = (2\pi)^2/S$, where S is the surface area of the quantum well. The area of circle of the radius k equals $\Omega_k = \pi k^2$. Applying (3.21) in 2D case, one obtains: $N = (Sk^2)/(2\pi)$, which is distinctly different from the bulk form. From (3.9), (3.19) and

$$\rho_S(E) = \frac{1}{S}\frac{dN}{dk}\left(\frac{dE}{dk}\right)^{-1}, \tag{3.23}$$

we obtain the density of states in a 2D quantum well [41]

$$\rho_S^{2D}(E) = \frac{m}{\pi\hbar^2}\sum_n \Theta(E - E_n). \tag{3.24}$$

Here, $\Theta(E - E_n)$ is the step function that equals 1 at $E \geq E_n$ and zero otherwise. The summation is performed over the total number of discrete levels present in the quantum well. The latter depends on the width of the well L and the height of potential barrier at the heterointerface. Equation (3.7) can be applied for estimating the lower levels of a thick quantum well. For a thin well separated by a low barrier, only one ground level may survive in the well, simplifying (3.24) to $\rho_S^{2D} = \rho_0\Theta(E - E_1)$ with $\rho_0 \equiv m/(\pi\hbar^2)$. In the general case, every new energy level in the well adds the step of a constant height ρ_0 to the surface density of states, as shown in Fig. 3.2b.

For 1D quantum wires, (3.11) leads to (3.19) along the wire axis at $k = k_z$. The number of states with a given k now equals $\omega_k = (2\pi)/L_z$, where L_z is the wire length. The total number of states with different k is given by $\Omega_k = 2k$, with the corresponding number of electron states $N = (2L_z k)/\pi$. Using (3.12) and (3.23), we arrive at

$$\rho_S^{1D}(E) = n_{QWR}\frac{\sqrt{2m}}{\pi\hbar}\sum_{n,l}\frac{\Theta(E - E_{nl})}{\sqrt{E - E_{nl}}}, \tag{3.25}$$

where $n_{QWR} = L_z/S$ is the density of quantum wires. Above each discrete state E_{nl} in a given subband, the density of states decreases inversely proportional to the square root of energy thus reaching infinity as the energy approaches E_{nl}, as shown in Fig. 3.2c. As for the ensembles of 0D quantum dots, the size quantization effect occurs in all three directions. Since no quasi-classical motion of charge carriers is present in this case, the density of states is given by the sum of delta-functions

$$\rho_S^{0D}(E) = 2n_{QD} \sum_{n,l,q} \delta(E - E_{nlq}), \tag{3.26}$$

as shown in Fig. 3.2d. This equation has clear meaning that each quantum dot can accommodate two levels with different spin orientations whereas the surface density of such states is set by the density of quantum dots themselves.

It is seen that, compared to bulk semiconductors and even quantum well heterostructures, quantum wires and especially quantum dots can provide much more favorable density of states for optoelectronic applications, semiconductor lasers in particular. Indeed, an important feature of the bulk density of states given by (3.22) is that only few electrons and holes can occupy states near the corresponding band edge. However, it is such electron-hole pairs that mostly participate in optical transitions. In the ultimate case of 3D quantum confinement, the only allowed energy states correspond to the atomic states of the quantum dot. The density of the charge carriers accumulated at the energy of working optical transition is thus drastically enhanced at the expense of the high energy parasitic states of quasi-continuum. This results in a much steeper dependence of the optical gain on the current density, leading to a significant reduction in the threshold current of quantum dot lasers as compared to the double heterostructure or quantum well competitors [41]. The atomic-like energy spectrum also leads to enhanced temperature stability of optoelectronic devices based on quantum dots.

Taking the case where only one electron and one hole discrete levels are accommodated in a nanostructure as an example, the energy of optical transition increases with respect to the bandgap as

$$E_{opt} = E_g + E_{e1} + E_{h1} \equiv E_g + \Delta E(L). \tag{3.27}$$

The size-dependent addict $\Delta E(L)$ increases toward smaller L. This property, illustrated in Fig. 3.3, gives a unique possibility for tuning the working wavelength of optoelectronic devices by changing the size L. In other words, one can perform the engineering of the effective bang gap in a given material system, which is impossible in the bulk case. Of course, the size quantization effect is significant only when $\Delta E(L)$ is comparable with E_g. Taking (3.7) at $n = 1$ for the estimate, we arrive at the condition $(1/2m)(\pi\hbar/L)^2 \sim E_g$, or

$$L \sim \lambda_B, \tag{3.28}$$

where $\lambda_B = (2\pi\hbar)/\sqrt{2mE_g}$ is the characteristic de Broile wavelength of electrons in a given bulk semiconductor. To observe the size quantization, the nanostructure size at least in one dimension should be thus made comparable or smaller than λ_B, the quantity that equals approximately 20 nm in the particular case of GaAs.

It should be noted that the well-known (3.22), (3.24)–(3.26) for the density of states in a bulk semiconductor, quantum wells, quantum wires and quantum dots, respectively, are obtained under the assumption of either purely continuous band (for 3D) or purely discrete energy spectrum (for all the structures with reduced

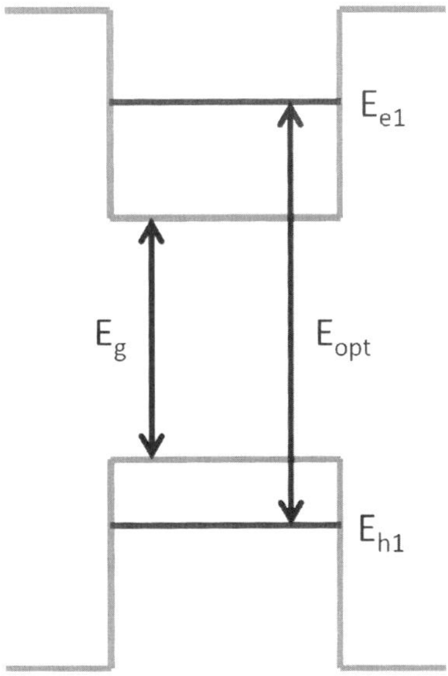

Fig. 3.3 Size quantization effect in a type II nano-heterostructure

dimensionality). Therefore, these expressions themselves do not allow for any accurate criterion to define the critical midway points for a transition between different dimensionality regions, for example, for 3D-2D transition when L_x changes from infinity to zero. An interesting approach to describe such a transition was proposed by Ren [120], who introduced the concept of size-dependent energy-level dispersion. In this theory, the expression for the density of states per unit volume is presented as $\rho_V(L_x, L_y, L_z, E) = (2/L_xL_yL_z)\sum_{nlq}F\left((E - E_{nlq})/D\right)$, where $F(x)$ is a certain delta-like function reaching its maximum at the peak energies E_{nlq} (corresponding to $x = 0$) and $D = D(L_x, L_y, L_z)$ is the size-dependend dispersion. This expression yields (3.26) for $D = 0$ and enables one to describe the smooth dimensionality transitions under some reasonable assumptions on the dependence of D on the dimensions L_x, L_y and L_z.

In real self-induced systems, the size distribution of quantum dots is always a Gaussian-like with a certain dispersion originating from random character of nucleation and growth. The effect of size distribution on the quantum dot density of states is illustrated in Fig. 3.4 [41]. At the known size distribution $f(L)$ and the dependence of quantum energy level on the island size $E(L)$, the density of states can be readily obtained as shown in the figure. Since the total number of states is independent of the size distribution, a broader distribution would result in a wider density of states and in lowering its maximum value. The maximum of density of states always corresponds to the most probable size over the ensemble of islands. If the density of states is narrower than the thermal energy k_BT, the total number of parasitic states and its temperature variation is reduced. Otherwise, the effect of quantum confinement on the characteristics of optoelectronic devices becomes questionable, which is why the elaboration of size-uniform ensembles of islands is of particular importance for applications.

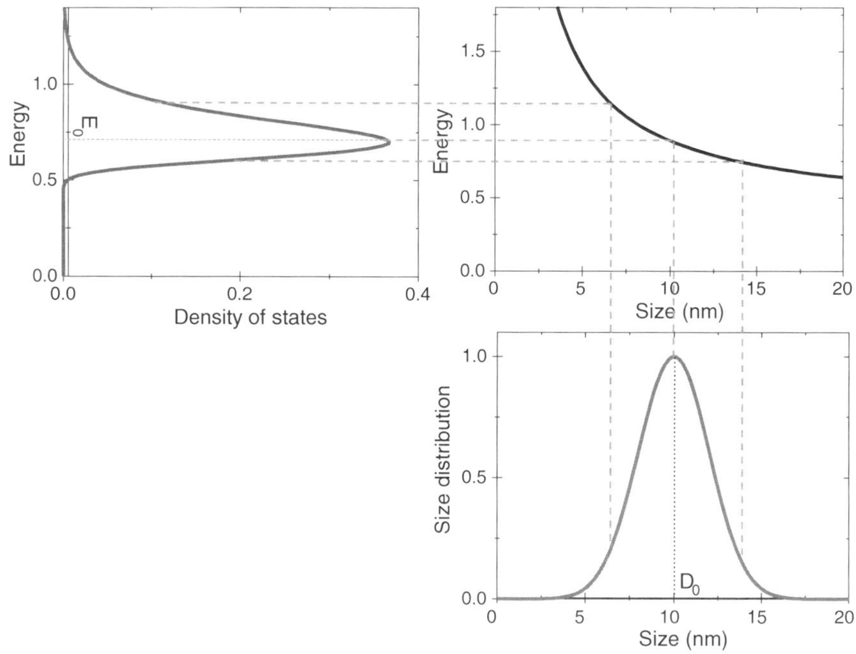

Fig. 3.4 Transformation of size distribution of 3D islands (*bottom part*) into the density of states (*top left part*) via the dependence of quantization energy on the island size (*top right part*) [41]

3.2 Stranski-Krastanow Growth

Advantages of quantum confinement in quantum dots for the creation of a low threshold and temperature-insensitive laser were well realized as early as the 1970s [121], and then evolved by Arakawa and Sakaki in 1982 [122]. However, no technological methods were known at that time to fabricate artificial quantum dots of reasonable quality. Since the 1980s, an extensive search for such methods has resulted in a number of reproducible *in situ* and *ex situ* quantum dot synthesis techniques. The corresponding overview can be found, for example, in [41]. While *ex situ* methods require a post-growth surface lithography, *in situ* methods use the formation of nanometer scale objects during the epitaxy process. Such a direct formation of quantum dots can be seeded, for example, by a preliminary substrate patterning or self-faceting of particular high-index surfaces. However, the most common method for quantum dot synthesis is based on the self-induced formation of 3D nanometer islands in lattice mismatched heteroepitaxial systems via the Stranski-Krastanow mode. We will now briefly consider the main features of the Stranski-Krastanow growth induced by the lattice mismatch.

It is well known that, when a 2D layer of the uniform thickness h and the surface area S is deposited onto a dissimilar substrate with the lattice mismatch ε_0, the elastic strain energy scales linearly with h as [40–42, 123]

$$W_{2D} = \frac{E}{1-\nu}\varepsilon_0^2 Sh \tag{3.29}$$

with E as the Young's modulus and ν as the Poisson's ratio of deposited material. The quantity $w_{2D} = (E\Omega\varepsilon_0^2)/(1-\nu)$ thus presents the elastic energy per atom in a 2D layer, with Ω as the elementary volume. At small h, a uniformly strained layer adopts the crystal lattice of the substrate in the (xy) plane. Such a layer is often called pseudomorphic, or wetting layer. Since the elastic energy increases with h, it must be relaxed via the introduction of misfit dislocations. This enables matching the two crystals with different lattice constants by plastic deformation. The critical thickness h_d corresponding to the onset of misfit dislocations can be obtained, e.g., from the Matthews-Blakeslee model [124]. A detailed analysis of plastic deformations induced by the lattice mismatch will be presented in Chap. 5. For now, it is enough to say that, for a given lattice mismatch, formation of dislocations is energetically preferred when the layer thickness exceeds $h_d(\varepsilon_0)$, and that $h_d(\varepsilon_0)$ rapidly decreases with ε_0. That is why lattice mismatch persists as one of the most challenging bottlenecks for growing high quality 2D layers on dissimilar substrates [125].

In a thin film case, plastic deformation is the only possible way to relax the infinitely increasing strain energy of the wetting layer. The situation changes drastically if the deposited material can segregate into 3D islands having free lateral surfaces where the corresponding components of stress tensor must equal zero [24, 25, 40–42, 123, 126]. Free surfaces enable a decrease of the elastic energy in a 3D island with respect to the elastic energy of a planar layer of the same volume. Whatever is the island shape, its strain energy can be presented as

$$W(\eta) = W_{2D} z(\eta), \tag{3.30}$$

where $\eta \equiv H/(2R)$ is the aspect ratio of an island with height H and base radius R. The η -dependent factor $z(\eta)$ measures the effect of strain relaxation at free sidewalls. Different approximations for the elastic relaxation $z(\eta)$ will be discussed in the next section. Quite clearly, the relaxation must equal one at $\eta \to 0$ (i.e., in a 2D thin film case) and scale as $1/\eta$ at $\eta \to \infty$, since no additional strain energy is generated when the island reaches a height of the order of R and completely recovers its strain-free state. Therefore, $z(\eta)$ obeys the following asymptotic behaviors:

$$z(\eta) \overset{\eta \to 0}{\longrightarrow} 1; \quad z(\eta) \overset{\eta \to \infty}{\longrightarrow} \text{const}/\eta. \tag{3.31}$$

Relaxation of elastic stress on free lateral surfaces reduces the total energy of the system. On the other hand, the stress-driven nucleation of 3D islands requires the formation of their lateral facets and also some work that should be done against the wetting force in a 2D layer. Therefore, one can anticipate that the Stranski-Krastanow islands become energetically preferred only after the thickness of 2D wetting layer exceeds a certain quasi-equilibrium value h_{eq}. Since the process involves a competition between an energetically favorable process of elastic stress relaxation and an energetically unfavorable faceting in the presence of wetting force, the island nucleation should have a barrier character. In other words, the wetting layer should acquire a certain value of metastability (associated with the elastic energy increasing linearly with the layer thickness as deposition proceeds) before the nucleation starts.

Stress-driven 3D islanding was first reported in 1985 by Goldstein and co-authors in the case of InAs islands grown on the GaAs(100) surface by molecular beam epitaxy (MBE) [127]. For this system, the lattice mismatch parameter equals approximately 7 %. It was soon realized that spontaneous formation of coherent strained islands within a certain interval of deposition thickness is a rather general phenomenon. In particular, quantum dot ensembles have been synthesized *in situ* in the Ge/Si [128] and InAs/Si [129] systems on Si substrates; InGaAs/GaAs [130], InGaAs/AlGaAs [131], InAlAs/AlGaAs [132, 133], InAs/InGaAs/GaAs [134] and InP/InGaP [135, 136] on GaAs substrates; InAs/InGa(Al)As [137] and InAs/InP [138, 139] on InP susbtrates; and GaInP/GaP, InAs/GaP and InP/GaP [140, 141] on GaP substrates. In principle, any heteroepitaxial system with a high enough lattice mismatch (typically, at $\varepsilon_0 \geq 2\,\%$) has a thermodynamic tendency for islanding. Very importantly, small enough islands remain dislocation-free. Typical lateral size of the order of 10–30 nm, island height of the order of several nm and surface density of the order of $10^{10}\,\mathrm{cm}^{-2}$ appear to be very advantageous for numerous applications. Most of the Stranski-Krastanow islands have the shape of full or truncated pyramids with a low aspect ratio of the order of 0.1–0.4 (see Sect. 3.4), so that their small vertical dimension leads to a pronounced quantum confinement.

In most experiments, the nucleation of coherent strained islands occurs under the deposition flux. In the case of MBE, the 2D-3D growth transformation is well recorded by *in situ* RHEED measurements. At the transition point, relating to the so-called critical thickness h_c, the RHEED pattern is transformed from streaks to spots originating from the electron diffraction from the lateral facets of 3D islands [40, 41]. It is noteworthy that the growth transformation occurs almost instantaneously, showing that the nucleation stage is much shorter than the follow-up growth stage as well as the time required to deposit a wetting layer of the critical thickness. As pointed out, e.g., in [24, 25, 142–144] and discussed in more detail later on, the critical thickness of 2D-3D growth transformation under the flux is generally larger than the equilibrium wetting layer thickness where such a transition becomes energetically preferred. The condition $h_c > h_{eq}$ reflects the barrier character of stress-driven island formation and is quite analogous to the nucleation in an open system considered in Chap. 2. Moreover, it might be anticipated that, while the value of h_{eq} is set primarily by the surface energetics and the lattice mismatch, the h_c can also be affected by the growth conditions such as the substrate temperature and material fluxes. In the simplest case of a single component deposition or III-V deposition under group V-rich conditions (where the growth kinetics is limited by the transport of group III element), we can therefore write down quite generally:

$$h_c = f\left[\{E_i\}, \varepsilon_0, T, V\right] h_{eq}(\{E_i\}, \varepsilon_0).$$

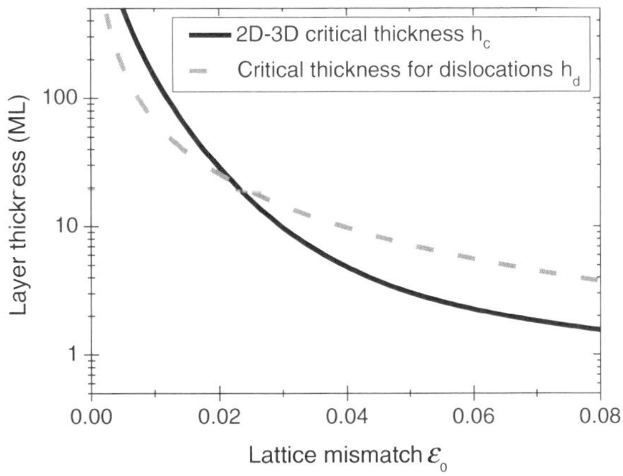

Fig. 3.5 Critical thicknesses for 2D-3D transformation and misfit dislocations

Here, $\{E_i\}$ denotes the entire set of relevant energetic parameters (elastic constants, surface energies of the substrate and relevant island facets, wetting energy etc.), T is the surface temperature and V is the deposition rate.

Experiments in different material systems show that the critical thickness of 2D-3D transformation rapidly decreases with increasing the lattice mismatch and slightly decreases with increasing the growth temperature, while the dependence on the deposition rate is rather weak [40, 41, 145, 146]. In particular, h_c equals approximately 4.6 ML at 600 °C for the Ge/Si(100) system with $\varepsilon_0 \cong 4\,\%$, and 1.7–1.8 ML for the InAs/GaAs(100) system with $\varepsilon_0 \cong 7\,\%$, at typical growth temperatures 450–500 °C. The dependence $h_c(\varepsilon_0)$ is different from $h_d(\varepsilon_0)$, as they originate from rather different stress relaxation scenarios. Qualitatively, both critical thicknesses decrease with ε_0 and cross at some point which is close to $\varepsilon_0 = 2\,\%$, as shown in Fig. 3.5. At a lower lattice mismatch, the Stranski-Krastanow islands are not observed, because $h_d < h_c$ and elastic stress is always released by the misfit dislocations. At $h_d > h_c$, the coherent islands form within the interval from h_c to h_d, where 3D islanding is energetically preferred to dislocations. This interval widens toward higher lattice mismatch and extends from ~1.7 to ~4 ML in the case of InAs/GaAs system. In fact, the quantum dot region extends even wider, from h_{eq} to h_d, but the formation of islands between h_{eq} and h_c often requires a surface exposition and usually results in a much lower island density.

A typical plan-view transmission electron microscopy (TEM) image of InAs quantum dots grown by MBE on a singular GaAs(100) substrate is presented in Fig. 3.6. The quantum dots were obtained after the deposition of 2 MLs of InAs at the substrate temperature $T = 485\,°\mathrm{C}$ with the InAs deposition rate $V = 0.03\,\mathrm{ML/s}$. The structure was covered by a low temperature GaAs cap immediately after the InAs deposition, i.e., with no further exposure to arsenic flux. Under these conditions, the average lateral size of islands equals 19 nm and the surface density is 1.3×10^{10} cm^{-2}. As will be discussed in detail later on, the average size, density and size distribution can be tuned by changing the deposition conditions (for example, V) even

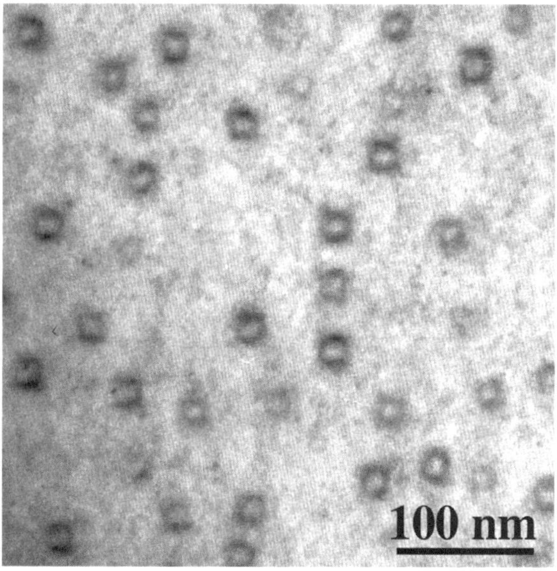

Fig. 3.6 Plan-view TEM image of InAs quantum dots obtained in the many-beam conditions with the incident beam directed along the [001] zone axis [144]

if the temperature and the deposition thickness are fixed. This reveals an essentially kinetic character of island nucleation providing additional tuning knobs to tailor the size distribution, as explained in Sect. 2.9. The image in Fig. 3.6 demonstrates that the islands have a square base, which is a usual geometry for III-V materials grown on the (100)-oriented substrates.

To estimate the equilibrium thickness after which a strained wetting layer becomes metastable, we use a simplified version of the Müller-Kern model [147]. Let us consider a 3D island consisting of i atoms (or III-V pairs), seated on top of a wetting layer with thickness h. As discussed above, the driving force for the island formation is the relaxation of elastic stress: the island nucleates because the elastic energy per atom in the wetting layer is larger than in the island [40–42]. Of course, the quantitative measure of such a relaxation strongly depends on the island shape and aspect ratio. We now assume for simplicity that the elastic energy in the island is much smaller than that in the wetting layer, the approximation valid only for very tall islands. Then, the elastic energy gain is given by the 2D elastic energy of i atoms, $\Delta G_{elastic} = (E\varepsilon_0^2\Omega i)/(1-\nu)$. On the other hand, the wetting force acts against the islanding, because the initial surface is stable against faceting. The wetting energy can be generally introduced as $\Psi_0 = \gamma_s - \gamma_d - \gamma_{s-d}$, where γ_s is the surface energy of the substrate, γ_d is the surface energy of the deposit and γ_{s-d} is the interfacial energy between them [24, 25, 147]. According to the Müller-Kern model [147], the energy of attractive deposit-substrate interactions decays exponentially with distance from the substrate h and equals $-\Psi_0 \exp(-h/k_0h_0)$, where h_0 is the height of a ML and k_0 is the relaxation coefficient which is of the order of one for most semiconductor systems. For sufficiently tall islands one can assume that the energy of interaction with the substrate is much weaker in the islands than on the surface of a wetting layer. This yields the expression $\Delta G_{wetting} = (\Psi_0/h_0)\exp(-h/k_0h_0)\Omega i$ for the free energy spent to overcome the wetting force in the island formation from $i$atoms on the surface of wetting layer.

Clearly, the wetting layer is stable at $\Delta G_{wetting} > \Delta G_{elastic}$ and metastable otherwise. The thickness h_{eq} relating to the condition $\Delta G_{wetting} = \Delta G_{elastic}$ can be therefore called the equilibrium wetting layer thickness:

$$h_{eq} = k_0h_0\ln\left(\frac{\Psi_0}{h_0E\varepsilon_0^2/(1-\nu)}\right). \tag{3.32}$$

This expression shows the decrease of h_{eq} with the lattice mismatch and its increase with increasing the wetting energy. The wetting layer is stable at $h < h_{eq}$ and has a thermodynamic tendency for the transformation of $h - h_{eq}$ MLs into 3D islands at $h > h_{eq}$. At very large ε_0, the equilibrium thickness obtained from (3.32) may become less than 1 monolayer, revealing the transition from the Stranski-Krastanow to the Volmer-Weber growth mode. The latter case has been observed, for example, in the InAs/Si system with $\varepsilon_0 = 11.6\,\%$.

As regards theoretical approaches of quantum dot formation in lattice mismatched systems, they can be divided into equilibrium [148–158] and kinetic [23–25, 142–144, 159–163] models. In a typical equilibrium approach, the system consisting of a substrate, a wetting layer and an ensemble of coherent strained islands is treated thermodynamically under the constraint of a constant temperature and deposition thickness (desorption is neglected). It is assumed that, after all relaxation processes are finished, the system will evolve to a thermodynamically preferred ordered state with the minimum free energy. In the simplest case of size homogeneous islands, such a state is characterized by the optimal size R_{opt} relating to the minimum energy. After obtaining R_{opt}, the free energy is compared to that of disordered state without islands but with the same amount of deposited material (i.e., the uniform wetting layer). If the energy difference between the ordered and disordered states is negative and the size R_{opt} is finite, the ensemble of islands is said to be stable. At a positive energy difference for any R, islanding is energetically suppressed; however, a positive minimum may be associated with a metastable quantum dot ensemble. If the energy difference decreases with R toward $R_{opt} \to \infty$, the system undergoes the Ostwald ripening.

A pioneering step in understanding of strain-induced formation of coherent 3D islands was taken in 1995 by Shchukin and co-authors [148]. The Shchukin's model has reflected several important features such as the elastic stress relaxation in 3D islands as the dominant driving force for islanding, the role of surface energies and the repulsive elastic interactions between the islands in their self-ordering. We now briefly describe this model in a simplified case of a dilute ensemble of islands. Consider the initial state of a heteroepitaxial system as being a flat wetting layer with the uniform thickness h, and the final ordered state corresponding to a size homogeneous ensemble of pyramidal islands having the square base with side R and making the contact angle β to the substrate. For simplicity, we also ignore a residual wetting layer thus letting the islands to consume all material from the substrate. In this case, the surface density of islands is given by $N = (6h \cot an\beta)/R^3$. Neglecting the entropy contributions, the variation of internal energy upon the formation of one island can be presented as

$$\Delta E_{isl} = \Delta E_{surf} + \Delta E_{elas} + \Delta E_{edges}. \tag{3.33}$$

The first term accounts for the surface energy change: the island formation eliminates the substrate area R^2 with the surface energy $\gamma(0)$ of a (100) surface and creates four equivalent facets of the surface area $R^2/\cos\beta$ with the surface energy $\gamma(\beta)$. The facet type is assumed to be dictated by the lowest of the local minimua of the deposit surface energy as a function of crystallographic orientation. This yields

$$\Delta E_{surf} = \left[\frac{\gamma(\beta)}{\cos\beta} - \gamma(0)\right] R^2. \tag{3.34}$$

Without strain-induced renormalization of the surface energy, ΔE_{surf} must be positive, because the wetting layer is initially stable against faceting. The second term gives the change of elastic energy and contains three contributions:

$$\Delta E_{elas} = -f_1(\beta)\lambda\varepsilon_0^2 R^3 - f_2(\beta)\tau\varepsilon_0 R^2 - f_3(\beta)\frac{\tau^2}{\lambda} R \ln\left(\frac{R}{2\pi a}\right). \tag{3.35}$$

The first contribution into (3.35) gives the decrease of volume elastic energy, where the function $f_1(\beta)$ summarizes the shape factor and the elastic stress relaxation on free sidewalls, with $\lambda \equiv E/(1-\nu)$. The second contribution accounts for the strain-induced renormalization of facet surface energy with τ as the corresponding surface stress tensor. The third contribution originates from the singularity of surface stress tensor at the corners of a pyramidal island, with a as the lattice spacing. All three contributions are negative, showing that the island formation decreases both the volume and surface elastic energy of the system. Finally, the third, positive term in (3.34) stands for the short-range edge energy which is proportional to the base perimeter

$$\Delta E_{edges} = f_4(\beta)\chi R \tag{3.36}$$

with χ as the specific edge energy. The functions $f_k(\beta)$ in (3.35), (3.36) depend on the crystallographic orientation of lateral facets.

Summing up all contributions and calculating the formation energy per unit surface area as $\Delta E = N\Delta E_{isl}$, the resulting expression for ΔE can be put in the form

$$\Delta E(l) = E_0\left[\frac{\alpha}{e^{1/2}l} - \frac{\ln(e^{1/2}l)}{l^2}\right]. \tag{3.37}$$

Here, l is the normalized lateral size expressed in the units of the characteristic size

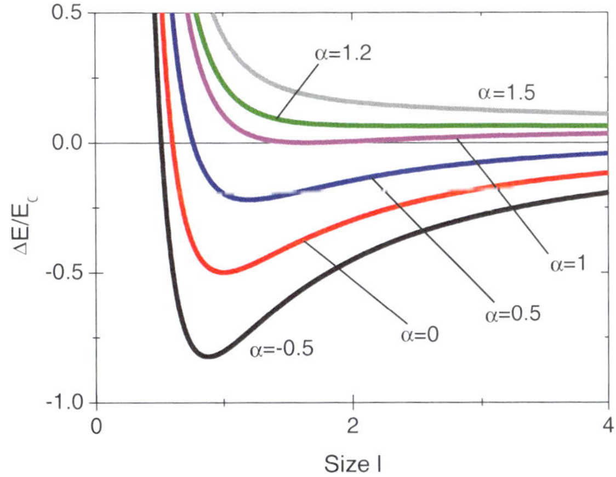

Fig. 3.7 Graphs of the Shchukin's driving force $\Delta E'(l)$ at different α

$$R_0 = 2\pi a \exp\left[\frac{f_4(\beta)\chi\lambda}{f_3(\beta)\tau^2} + \frac{1}{2}\right]. \tag{3.38}$$

The control parameter α is determined by

$$\alpha = \left[\frac{\gamma(\beta)}{\cos\beta} - \gamma(0) - f_2(\beta)\tau\varepsilon_0\right]\frac{e^{1/2}\lambda R_0}{f_3(\beta)\tau^2}. \tag{3.39}$$

We do not write here the expression for the factor E_0 as inessential for the analysis. In view of strain-induced renormalization of the surface energy, the value of α can be of either signs.

Graphs of the function $\Delta E' \equiv \Delta E/E_0$, obtained from (3.39) at different α, are shown in Fig. 3.7. It is seen that the formation energy has a negative minimum at $l = l_{opt}$ when $\alpha < 1$. In this case, islanding is more energetically favorable than a flat wetting layer. At $1 < \alpha < 2e^{-1/2}$ the minimum of $\Delta E'(l)$ becomes positive and disappears completely at $\alpha > 2e^{-1/2}$, showing that the islanding is suppressed on energetic grounds in this domain of material parameters. Thus, the Shchukin's model attributes the islanding effect to a decrease of internal energy (containing the elastic, surface and edge contributions) per unit area upon the formation of an ensemble of 3D islands, where the surface stress τ and the logarithmic term at the island corners play a crucial role. The tendency for islanding becomes more pronounced and the optimal size smaller at lower α , i.e., where the surface stress increases. Although the model does not account for entropy corrections, ignores the influence of metastability as well as the residual wetting layer, neither of these effects breaks the possibility for the stress-driven islanding within a plausible range of material constants.

Kinetic approach to the self-assembly of 3D islands induced by the lattice mismatch enables, in principle, a complete description of the time evolution of island size distribution in a given system (characterized by the lattice mismatch, elastic constants, surface energies etc.) and under particular deposition conditions (temperature, fluxes, deposition time). It is capable of describing the dependences of system morphology not only on the equilibrium parameters (temperature and deposition thickness) but also on the essentially kinetic factors. As such, the experimentally observed dependence of quantum dot size and density on the deposition rate in the InAs/GaAs [166–168] and Ge/Si [169] systems can be described only by considering the growth kinetics. Let us now qualitatively discuss the main kinetic growth steps in the Stranski-Krastanow mode. We consider the case of MBE at a growth temperature T with a constant flux V (Ge or group III element which limits the growth of III-V compounds under group V-rich conditions) which is turned off at time t_0. Assuming the effective absence of desorption on the time scale of interest, the deposition thickness is given by $H(t) = Vt$ at $t \leq t_0$ and $H = H_0 = Vt_0$ at $t > t_0$, with H_0 as the total deposition thickness. At any time, $H(t) = h(t) + g(t)$, where h is the wetting layer thickness and $g \cong NCR^3$ is the volume of islands per unit surface area. The last expression corresponds to the mono-disperse approximation of size distribution, N is the surface density of islands, R is their size and C is the shape constant. After the deposition of H_0 MLs, the structure can be either immediately cooled down or covered with a cap layer, or exposed at the growth temperature T during the exposition time Δt before capping. Assuming that the low-temperature cap does not affect significantly the island ensemble morphology, the entire process is controlled by four external parameters: T, V, H_0 and Δt. In the case of III-V quantum dots, the fluxes ratio may also influence the resulting morphology, as it changes the surface energetics as well as the growth kinetics. Surface vicinity can also be used as an additional knob to tune the island structure [170–173].

As discussed above, the driving force for the 2D-3D transformation is the elastic stress accumulated in the wetting layer, whose energy scales linearly with h. One can therefore assume that the measure of wetting layer metastability is given by

$$\zeta = h/h_{eq} - 1. \tag{3.40}$$

This quantity, often called "superstress" [24, 25, 144], is quite analogous to supersaturation of vapor; the island formation is possible only at $\zeta > 0$. Therefore, the very first stage of the Stranski-Krastanow growth is the formation of 2D wetting layer of equilibrium thickness h_{eq} (Fig. 3.8a). Further increase of thickness within the interval $h_{eq} < h < h_c$ develops the metastability which is insufficient, however, to onset the island nucleation under the deposition flux. The effective thickness H equals the wetting layer thickness h at $h < h_c$, because the total volume of islands remains negligibly small at this stage (Fig. 3.8b). The nucleation starts when the wetting layer reaches the critical thickness h_c . At this moment, the electron diffraction pattern changes from streaks to spots (Fig. 3.8c). Since the nucleation stage is short scale, the transformation occurs almost instantaneously. By the analogy with a usual nucleation from supersaturated vapor studied in Chap. 2, the critical thickness h_c should correspond to the minimum nucleation barrier $F(h_c) = \min$ and the maximum nucleation rate: $J(h_c) = \max$. Since the island emerge with a very small size, the critical wetting layer thickness is only slightly smaller than the equivalent deposition thickness so that $H_c \cong h_c$ in the first approximation.

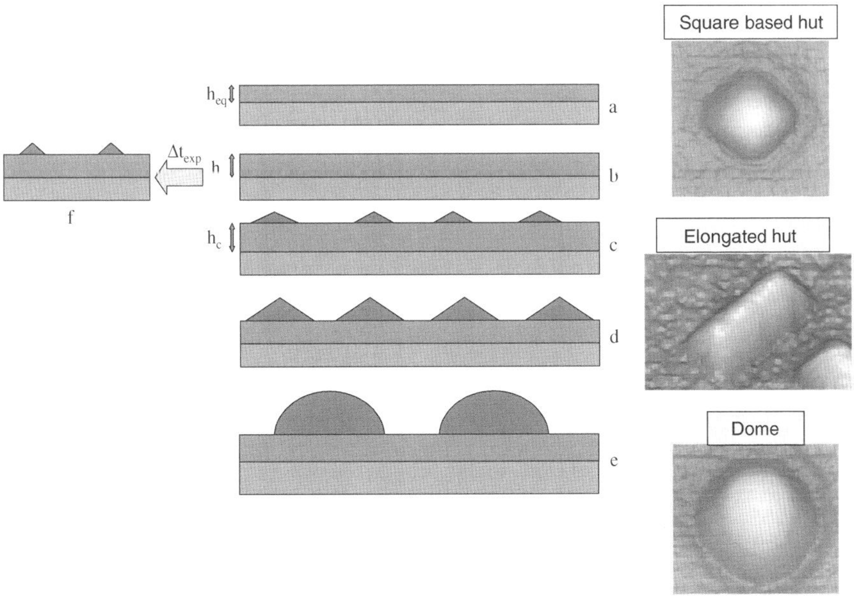

Fig. 3.8 Qualitative representation of different stages of quantum dot formation via the Stranski-Krastanow mode: **a**—equilibrium wetting layer; **b**—2D growth between h_{eq} and h_c where the wetting layer is metastable; **c**—nucleation stage; **d**—regular growth; **e**—shape transformation; the inserts showing differently shaped Ge islands on Si(100) surface [176]. Figure (**f**) illustrates the formation of subcritical quantum dots after a certain exposition time

Geometrical shape of islands depends on the crystallography of a given material system and might be influenced by the deposition conditions such as the surface temperature [174]. *In situ* RHEED analysis, and *ex situ* TEM and atomic force microscopy (AFM) imaging reveal that initially InAs quantum dots tend to emerge as full or truncated pyramids with a square base [40–42]. The so-called "hut" clusters in the Ge/Si system may be either square or elongated depending on the growth conditions. Very often, these two shapes are observed simultaneously. The nucleation of hut clusters is detected by the RHEED reflexes originating from electrons scattered from the (105) crystal planes [175]. The aspect ratio of hut clusters amounts to 0.134, and their typical lateral size is not larger than 25 nm. Above this size, the hut-to-dome shape transformation occurs to further relax the elastic stress (Fig. 3.8d). Dome islands are restricted by the high-index planes so that the resulting surface looks almost as a spherical cap, while the aspect ratio becomes larger than in the hut phase. Due to the initial size inhomogeneity or the strain-induced limitation on the growth rate of larger islands [23], differently sized hut and dome islands often coexist. Longer exposition time may lead to the Ostwald ripening process, where smaller hut clusters are consumed by the growing domes [175]. In a dense array, one should also take into account the dipole-dipole elastic interactions between the neighboring islands [148]. This process may prevent larger islands from growing and, in certain cases, narrow the resulting size-distribution and lead to spatial ordering. Of course, increasing the size of quantum dots will necessarily lead to the formation of misfit dislocations at their base at a certain critical dimension, where the transport and optical properties will be strongly impeded. For example, misfit dislocations are usually seen in the dome Ge islands [175]. Therefore, several methods have been developed to suppress the unwanted dome phase by using growth surfactants or employing specific growth conditions.

The above behavior would be typical for the Stranski-Krastanow growth under a constant deposition flux. As discussed, the wetting layer gets superstressed above the equilibrium thickness. While the island nucleation is onset only at $h = h_c$ under the flux, nothing prevents the formation of quantum dots between h_{eq} and h_c after a certain waiting time achieved, e.g., by exposing the surface to arsenic flux in the case of InAs deposition. The exposition is required simply because the level of metastability is not sufficient to start the nucleation immediately. Such quantum dots are often called subcritical since they form below the critical thickness h_c, the process schematized in Fig. 3.8f. Formation of subcritical quantum dots has some specific features leading to distinctly different morphological properties [143, 144, 177–179], as will be discussed in detail later on. Most importantly, their density is much lower and the growth rate much slower than at the super-critical deposition thickness. These features might be useful for the single photon sources for example, because the low density dots do not cross-talk.

While there is no doubt that the formation of coherent 3D islands is driven by the elastic stress originating from the lattice mismatch, the kinetic growth mechanisms can be rather different depending on the material constants, deposition conditions and the growth time. At a low enough coverage, the surface diffusion should be more important than the direct impingement from vapor. Than, two mechanisms are possible illustrated in Fig. 3.9:

(a) Growth from a metastable "adatom sea" on the surface of the wetting layer [23, 175, 180];
(b) Growth at the expense of the wetting layer itself by solid diffusion [24, 25, 129].

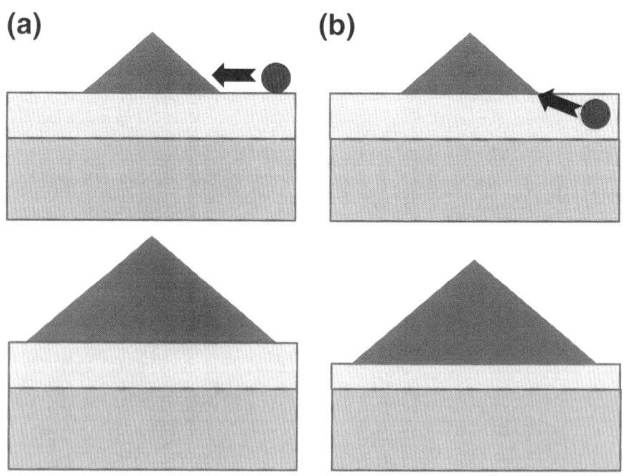

Fig. 3.9 The Stranski-Krastanow growth from the adatom sea at a constant wetting layer thickness (**a**) and by solid diffusion where the wetting layer thickness decreases (**b**)

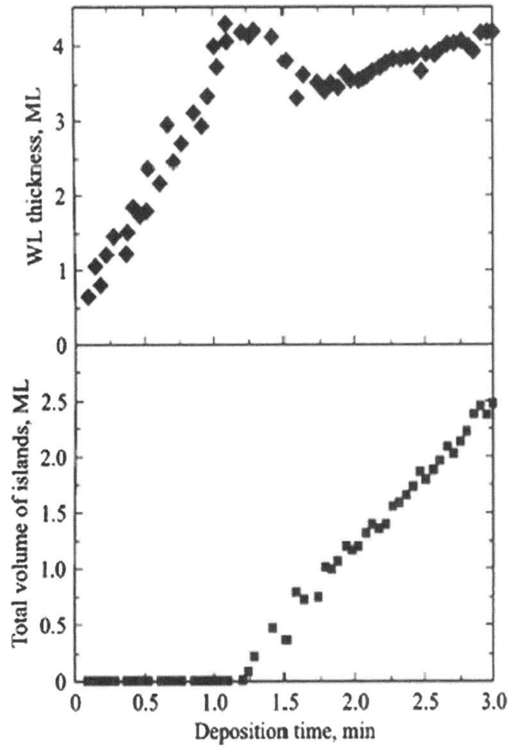

Fig. 3.10 Wetting layer thickness (*top*) and the total volume of islands per unit substrate area (*bottom*) in the case of CVD grown Ge/Si quantum dots, as revealed by *in situ* ellipsometry [24]

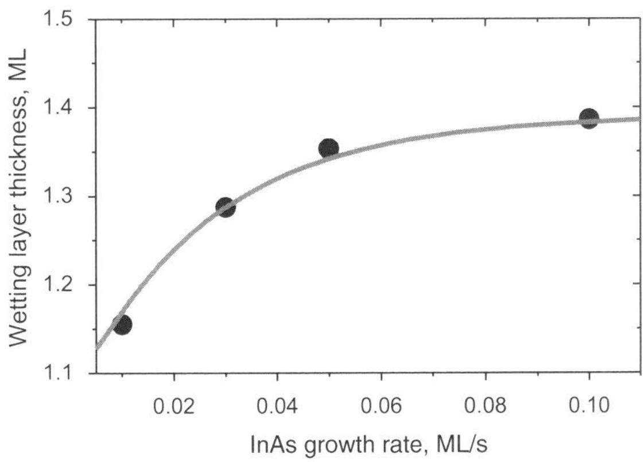

Fig. 3.11 Thickness of residual wetting layer in the 2 ML ensembles of InAs/GaAs quantum dots grown by MBE at 485 °C

Obviously, the wetting layer thickness would not change in the former case and decrease with time in the latter case. With a sufficient exposition time, $H_0 - h_{eq}$ of H_0 MLs deposited will be distributed in 3D islands, and the wetting layer thickness will decrease to the equilibrium value h_{eq}. Quite clearly, a combination of both kinetic mechanisms is not excluded.

Figures 3.10 and 3.11 show some experimental data evidencing the decrease of wetting layer thickness at the growth stage. Figure 3.10 presents the real time ellipsometric measurements of Ge islands on the Si(100) surface grown by CVD at 500 °C [24]. It is seen that ~1 ML of Ge is consumed by the islands at the initial step. This observation is in agreement with the data of [181] obtained by the Rutherford back scattering and AFM. Follow-up increase of h is most probably explained by the strain-induced potential barriers for the diffusion into large islands that limits the growth rates and often leads to a secondary nucleation [23]. Data of [182] shown in Fig. 3.11 regards the 2 ML ensembles of InAs/GaAs quantum dots grown by MBE at fixed $T = 485\,^{\circ}\mathrm{C}$ with different deposition rates of InAs and at zero exposition time. The results were accessed from the photoluminescence (PL) measurements of the peak position corresponding to the optical recombination in the InAs wetting layer. Its width was then obtained by applying the thin quantum well approximation with the parameters of InAs/GaAs. The residual wetting layer is noticeably thinner than the critical thickness (1.7–1.8 ML) for all $V = 0.01 - 0.1$ ML/s, although the growth process may not be fully completed at zero exposition. One can thus conclude that the equilibrium thickness in the InAs/GaAs system is not larger than 1.1 ML. In our growth modeling, we will therefore assume that the islands are fed from the wetting layer by solid-like diffusion (whose rate can be drastically enhanced by the elastic strain), while the adatom diffusion leads to a 2D growth between the islands, as in the standard thin film case.

3.3 Elastic Relaxation in Nanostructures

In this section, we present several important models for the elastic stress relaxation in differently shaped nanostructures on lattice mismatched substrates, and a semi-analytical treatment that enables the calculation of elastic strain energy for a given geometry [183]. These results will be relevant in what follows not only for the Stranski-Krastanow quantum dots, but also for the growth and structural modeling of other nanostructures such as nanowires [184] and nanoneedles [185, 186]. As regards the two asymptotes given by (3.31), the simplest formula for $z(\eta)$ can be chosen in the form

$$z(\eta) = \frac{1}{1 + A\eta}. \tag{3.41}$$

Such an expression was found previously in [187] by fitting the results of finite element calculations in the case of a mismatching layer seated on top of a dissimilar cylinder. The value of A depends on the Poisson's ratio but neither on the Young's modulus nor mismatch, and equals approximately 27.4 at $v = 1/3$.

The Ratsch-Zangwill formula [188], which is a good approximation for rectangular quantum dots, is given by

$$z(\eta) = \frac{1 - \exp(-\Lambda\eta)}{\Lambda\eta}, \tag{3.42}$$

with the relaxation coefficient $\Lambda \cong 3\pi$ for a cubic material. The Gill-Cocks expression [189], obtained beyond the assumption of identical elastic constants of the island and the substrate, in the case of an isotropic full cone takes the form

$$z(\eta) = \frac{1 + 6\eta^2}{1 + 2b\eta(1 + 6\eta^2) + (16 - 10k)\eta^2}. \tag{3.43}$$

Here, the coefficient $k \equiv [2(1-2v)]/[3(1-v)]$ and the coefficient $b \equiv 1.059[E(1-v_s^2)]/[E_s(1-v)]$ contains the elastic constants of the substrate E_s and v_s. The Glas fit [190] to the finite element calculations of the elastic relaxation of a uniformly strained cylinder standing on an infinite foreign substrate, writes down as

$$z(\eta) = \frac{p_1}{1 + p_2\eta} + (1 - p_1)\exp(-p_3\eta). \tag{3.44}$$

The p_k are v -dependent fitting coefficients such that $p_1 = 0.557, p_2 = 10.15$ and $p_3 = 9.35$ for a cubic material at $v = 1/3$. It is noteworthy that all the existing models show that the stress release in laterally confined nanostructures is very efficient: the strain-free state is already achieved when the height reaches the base radius, while the upper part of the island is fully relaxed.

Let us now formulate the general set of equations of linear elasticity theory in a simplified case of axi-symmetrical islands [183]. Typical 2D isotropic geometries describing nanowires (cylinder, truncated cone, reverse truncated cone), nanoneedles (cone) and quantum dots (cone, truncated cone) are shown in Fig. 3.12. For a 2D isotropic nanoisland of any shape, the displacement field is angular independent. In the case of circular plate under a radial force, the 2D radial displacement is presented as $u_r = B/r + Cr$, with r being the distance from the origin, while $u_\theta = 0$ [191–194].

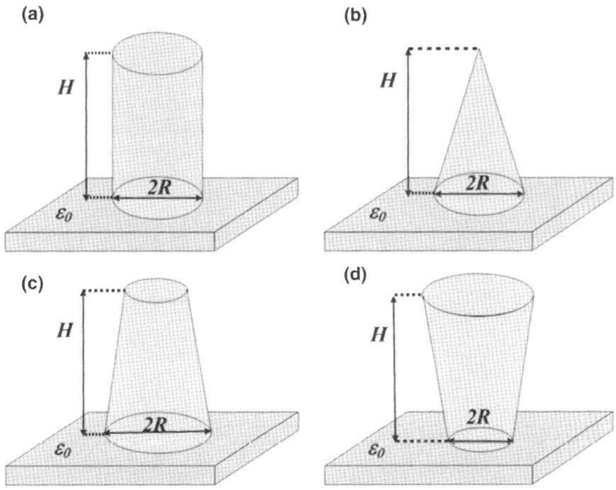

Fig. 3.12 Model nanostructure geometries: cylinder (**a**), full cone (**b**), truncated cone (**c**) and reverse truncated cone (**d**)

If there is no hole in the center, the displacement should remain finite when the radius tends to zero, yielding $B = 0$ and $u_r = Cr$. In a 3D case without a volume force, the displacement field caused by a 2D stress force acting upon the island-substrate interface takes the form: $u_r = Brf(z)$. The function $f(z)$ describes the relaxation with the distance from the interface z; its particular choice will be considered below.

In the cylindrical coordinates r, θ and z, the tangential component u_θ vanishes in view of 2D isotropy, while u_r, u_z are independent of θ. The displacement strains ε_{ij} are determined through u_i and their derivatives as follows

$$\varepsilon_{rr} = \frac{\partial u_r}{\partial r};\ \varepsilon_{\theta\theta} = \frac{u_r}{r};\ \varepsilon_{zz} = \frac{\partial u_z}{\partial z};\ \varepsilon_{rz} = \frac{1}{2}\left(\frac{\partial u_r}{\partial z} + \frac{\partial u_z}{\partial r}\right);\ \varepsilon_{r\theta} = \varepsilon_{z\theta} = 0. \quad (3.45)$$

According to the Hooke's law [191, 192], the elastic stress components are given by

$$\sigma_{ij} = 2\mu\varepsilon_{ij} + \lambda\delta_{ij}\varepsilon_{kk}, \quad (3.46)$$

with the elastic constants $\mu = E/[2(1+v)]$ and $\lambda = (Ev)/[(1-2v)(1+v)]$. Furthermore, the stress components must satisfy the following equations of equilibrium [191]

$$\frac{\partial\sigma_{rr}}{\partial r} + \frac{\partial\sigma_{rz}}{\partial z} + \frac{\sigma_{rr} - \sigma_{\theta\theta}}{r} = 0;\ \frac{\partial\sigma_{rz}}{\partial r} + \frac{\partial\sigma_{zz}}{\partial z} + \frac{\sigma_{rz}}{r} = 0. \quad (3.47)$$

Combination of (3.45)–(3.47) results in the well-known equations for the displacement field given by [192]

$$\frac{1}{1-2v}\frac{\partial g}{\partial r} + \Delta u_r = 0;\ \frac{1}{1-2v}\frac{\partial g}{\partial z} + \Delta u_z = 0; \quad (3.48)$$

$$\Delta g = 0. \quad (3.49)$$

Here,

$$g = \frac{\partial u_r}{\partial r} + \frac{u_r}{r} + \frac{\partial u_z}{\partial z} \quad (3.50)$$

and $\Delta = \frac{\partial^2}{\partial r^2} + \frac{1}{r}\frac{\partial}{\partial r} + \frac{\partial^2}{\partial z^2}$ is the angular independent 3D Laplace operator.

In order to obtain an analytical approximation for the z dependence of displacement fields, we introduce the normalized variables by definition

$$(\rho, \varsigma) = (r/R, z/H),\ (u_\rho, u_\varsigma) = (u_r, u_z)/R.$$

Re-scaling (3.49) and (3.50) for g in terms of these variables results in

$$\frac{1}{4\eta^2}\frac{\partial^2 g}{\partial\varsigma^2} + \left(\frac{\partial^2 g}{\partial\rho^2} + \frac{1}{\rho}\frac{\partial g}{\partial\rho}\right) = 0;\ g = \left(\frac{\partial u_\rho}{\partial\rho} + \frac{u_\rho}{\rho}\right) + \frac{1}{2\eta}\frac{\partial u_\varsigma}{\partial\varsigma}. \quad (3.51)$$

For tall enough NSs with $2\eta >> 1$, these equations are simplified to

$$\frac{\partial^2 g}{\partial r^2} + \frac{1}{r}\frac{\partial g}{\partial r} = 0; \, g = \frac{\partial u_r}{\partial r} + \frac{u_r}{r}. \tag{3.52}$$

In the same approximation, (3.48) are reduced to

$$\frac{1}{1-2v}\frac{\partial g}{\partial r} + \frac{\partial^2 u_r}{\partial r^2} + \frac{1}{r}\frac{\partial u_r}{\partial r} - \frac{u_r}{r^2} = 0; \, \frac{1}{1-2v}\frac{\partial g}{\partial z} + \frac{\partial^2 u_z}{\partial r^2} + \frac{1}{r}\frac{\partial u_z}{\partial r} = 0. \tag{3.53}$$

Obviously, any z-independent g is the solution to the first (3.53).

Let us now consider $g(z)$ in the form $g(z) = const \times (1-2v)\exp(-\alpha z/R)$, where α is a constant. The first (3.53) is satisfied when

$$u_r = Br\exp(-\alpha z/R) \tag{3.54}$$

with arbitrary constant B, i.e., when the decrease of the radial displacement with z is exponential. The second (3.53) is satisfied with

$$u_z = \frac{1}{4}\frac{(C\alpha r^2 + SR^2)}{R}\exp(-\alpha z/R), \tag{3.55}$$

with arbitrary constants C and S. Below we will use (3.54) and (3.55) for any η, assuming that the exponential approximation is also good for flat islands. This assumption will be then checked for validity by the direct comparison with numerical calculations.

Since the displacement fields in both radial and vertical directions decay exponentially with z, the boundary conditions of free lateral surfaces, $\sigma_{rr}(r = R(z), z) = 0$ and $\sigma_{rz}(r = R(z), z) = 0$ (where $R(z)$ describes the side facets) cannot be exactly satisfied. Instead, we proceed by using the minimization of total elastic energy of a given nanostructure to obtain the unknown coefficients α, C and S [193]. Likewise in 193, it can be shown that the minimum elastic energy approach can be used to best reflect the exact displacement fields that satisfy the above boundary conditions.

The non-zero components of strain field are now readily obtained from (3.45):

$$\varepsilon_{rr} = B\exp(-\alpha z/R); \, \varepsilon_{\theta\theta} = B\exp(-\alpha z/R);$$

$$\varepsilon_{zz} = -\frac{\alpha}{4}\frac{(C\alpha r^2 + SR^2)}{R^2}\exp(-\alpha z/R);$$

$$\varepsilon_{rz} = -\frac{1}{4}\frac{r\alpha(2B - C)}{R}\exp(-\alpha z/R). \tag{3.56}$$

The elastic stress components are easily calculated from (3.46):

$$\sigma_{\rm rr} = \frac{1}{4}\frac{(8\lambda BR^2 - \lambda C\alpha^2 r^2 - \lambda\alpha SR^2 + 8\mu BR^2)}{R^2}\exp(-\alpha z/R);$$

$$\sigma_{\theta\theta} = \frac{1}{4}\frac{(8\lambda BR^2 - \lambda C\alpha^2 r^2 - \lambda\alpha SR^2 + 8\mu BR^2)}{R^2}\exp(-\alpha z/R);$$

$$\sigma_{zz} = \frac{1}{4}\frac{(8\lambda BR^2 - \lambda C\alpha^2 r^2 - \lambda\alpha SR^2 - 2\mu\alpha^2 Cr^2 - 2\mu\alpha SR^2)}{R^2}\exp(-\alpha z/R);$$

$$\sigma_{rz} = -\frac{1}{2}\frac{\mu r\alpha(2B - C)}{R}\exp(-\alpha z/R). \tag{3.57}$$

The elastic energy density is given by [191]

$$w(r, z) = \frac{1}{2}\varepsilon_{ij}\sigma_{ij} \tag{3.58}$$

Thus, integrating the density over the entire island volume gives the total elastic energy stored in the island:

$$W = \int d^3xw\,(r, z)\,. \tag{3.59}$$

For a material with the lattice constant l grown on a substrate with the lattice constant l_0, the lattice mismatch parameter is defined as $\varepsilon_0 = (l - l_0)/l_0$. For a rigid substrate, the constant B must equal ε_0, and the total elastic energy $W_{tot} = W$. The constants α, C and S are then obtained by the minimization of total energy W given by (3.59) for a given island geometry.

For an island seated on an elastic substrate, the total elastic energy contains the contribution from the surface. Elastic energy of the substrate is induced by the radial surface traction force $\vec{f}(\vec{r})$ which is singular at the island edge and is highly dependent on the contact angle β (the latter relates to the aspect ratio as $\beta = \arctan(2\eta)$ in the case of full cone geometry). Generally, the elastic energy of the substrate is given by

$$W_{sub} = \frac{1}{2}\int d\vec{r}\int d\vec{r}' f_i(\vec{r})G_{ij}(\vec{r} - \vec{r}')f_j(\vec{r}'), \tag{3.60}$$

where the Green's tensor on the surface is defined as [192,195–197]

$$G_{ij}(\vec{r} - \vec{r}') = \frac{1+\nu}{\pi E}\left[\frac{(1-\nu)\delta_{ij}}{|\vec{r} - \vec{r}'|} + \frac{\nu(\vec{r} - \vec{r}')_i(\vec{r} - \vec{r}')_j}{|\vec{r} - \vec{r}'|^3}\right]. \tag{3.61}$$

Integration in (3.60) is performed over the entire substrate area outside the island base. In an isotropic case, the radial surface traction can be well approximated by the numerical fit of Gill [197]

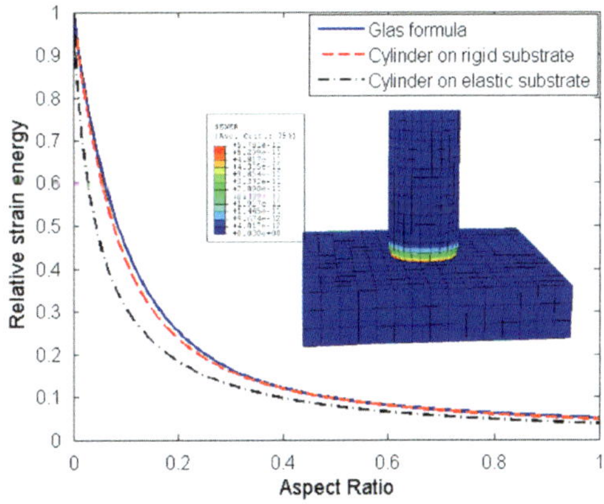

Fig. 3.13 Relative strain energy $z(\eta)$ in a cylindrical nanowire on a rigid (*dashed line*) and elastic (*dash-dotted line*) substrates, compared to the Glas formula given by (3.44); the insert showing how the strain is localized near the interface

$$f_r(r) = c\left[\left(1-\frac{r}{R}\right)^{-\chi} - \left(1-\frac{r}{R}\right)^{-\delta}\right]. \tag{3.62}$$

Here, $\chi = (1/2)\tanh(a\beta)$ is the index describing the singularity at the island edge, $a = 1.41$ for a cubic island having the identical elastic constants with the substrate, and $\delta = 0.114/\chi + 0.638$ is the exponent introduced to ensure that $f_r(0) = 0$.

With the surface traction given by (3.62), the strain energy of the substrate is readily obtained from (3.60) and (3.61) in the form

$$W_{sub} = \frac{E_s\varepsilon_0^2}{(1-\nu_s^2)}\frac{\pi R^3}{3}\frac{\rho}{\phi}. \tag{3.63}$$

Here, E_s and ν_s are the elastic constants of the substrate; other parameters are defined as

$$\rho = \frac{6}{(1-\chi)(2-\chi)(3-\chi)} - \frac{6}{(1+\delta)(2+\delta)(3+\delta)},\ \phi = 1.059 + 41.25\chi^3, \tag{3.64}$$

The work done by the surface traction relaxes the strain at the island base from the mismatch strain ε_0 to a lower value of $\varepsilon_0 - \varepsilon$. In the case of elastic substrate, the radial component of displacement field should be therefore changed to $u_r = (\varepsilon_0 - \varepsilon)r\exp(-\alpha z/R)$. The total elastic energy of the system is given by: $W_{tot} = W + W_{sub}$. The constant ε is obtained by minimizing the total energy, i.e., from the condition $\partial W_{tot}/\partial\varepsilon = 0$ at $\varepsilon = \varepsilon_*$. Substitution of this ε_* into the total energy W_{tot} and global minimization of the latter enables, as above, to find the unknown parameters α, C and S for a given island geometry. In particular, for a cylindrical island with the aspect ratio of 5 we obtain the following numbers: $\alpha = 5.23$, $C = 0.103$ and $S = 0.0336$.

Relative strain energies $z(\eta) = W(\eta)/W_{2D}$ for the four geometries shown in Fig. 3.12 were computed at different aspect ratios by the integration of strain energy density as given by (3.58) and (3.59). In the case of elastic substrate, the strain energy

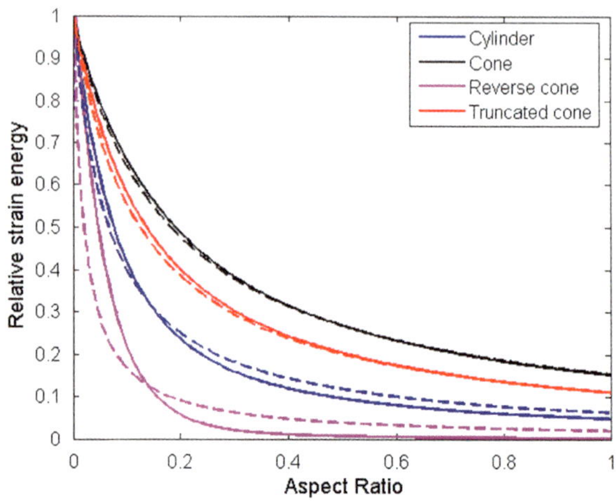

Fig. 3.14 Relative strain energy in the cylinder, cone, truncated cone and reverse truncated cone (*solid lines*) and their fits by the formula $z(\eta) = 1/(1 + A\eta)$ with $A =$ 15, 5.5, 8 and 50, respectively (*dotted lines*)

of the latter was included as given by (3.63) and (3.64). The minimization of the total strain energy was performed as described above for each η. Figure 3.13 shows the resulting elastic relaxation in the cylindrical island geometry, computed in the case of a cubic material having identical elastic constants E and v with the substrate. The computations were performed separately for the rigid and elastic substrates. As expected, both curves rapidly decay as the aspect ratio increases, so that about 80 % of the 2D elastic energy is relieved already at $\eta = 0.2$. As discussed, the effect of elastic relaxation at free lateral surface is huge. As is seen from Fig. 3.13, the relaxation on the elastic substrate is faster than on the rigid one, which is indeed a natural and anticipated result. It is also seen that the Glas fit given by (3.44) with coefficients $p_1 = 0.557$, $p_2 = 10.15$ and $p_3 = 9.35$ is excellent for a rigid substrate.

As discussed above, the simplest possible expression for the elastic relaxation is given by (3.41). Due to its simplicity, such formula makes it possible to obtain many important physical characteristics such as stress-driven nucleation barriers in an analytical, physically transparent form. It is therefore interesting to perform the calculations of the elastic relaxation for different island geometries and fit the results by (3.41). In the case of rigid substrate, the best correspondence is obtained with the fitting coefficient $A = 5.5$ for the full cone, 8 for the truncated cone (with the 70° contact angle at the base), 15 for the cylinder and 50 for the reverse truncated cone (with the 110° contact angle at the base), Fig. 3.14 showing the corresponding curves. As follows from the figure, the simple fits are excellent for the full and truncated cones, the typical model geometries for the Stranski-Krastanow quantum dots in different material systems. The full cone is also good model geometry for GaAs nanoneedles on silicon and sapphire substrates [185, 186]. The correspondence becomes worse for the cylinder and reverse truncated cone, with the numerical relaxation decreasing slower in the beginning and becoming steeper towards larger η than that given by (3.41). We can also conclude that the elastic relaxation depends dramatically on the

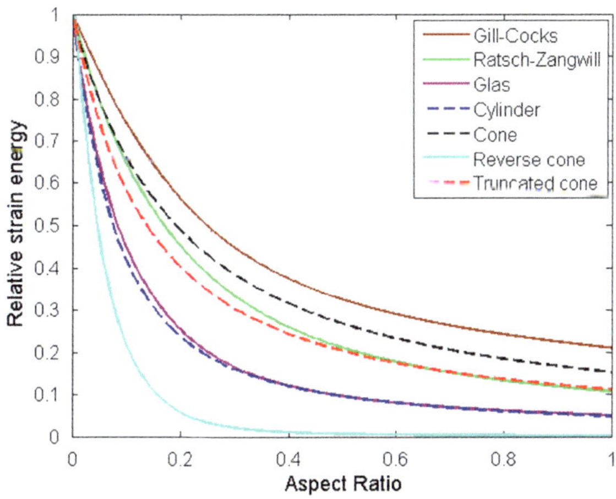

Fig. 3.15 Relative strain energy in different nanostructure geometries (*dashed lines*), compared to the Glas formula for the cylinder at $p_1 = 0.557, p_2 = 10.15$ and $p_3 = 9.35$, the Ratsch-Zangwill formula for the rectangular island at $\Lambda = 3\pi$ and the Gill-Cocks formula for the full cone at $E = E_s$, all calculated for cubic materials

contact angle at the nanostructure base, with the fitting coefficient A increasing by more than 6 times when the contact angle changes from 70° to 110°.

Figure 3.15 presents the comparison of our results obtained for different island geometries with the known analytical approximations for the elastic relaxation, in the case of a cubic material and identical elastic constants of the substrate and the island. All calculations were performed on a rigid substrate. It is seen that the Gill-Cocks expression (3.43) predicts a slower decay than our results for the identical full cone geometry. The Ratsch-Zangwill formula (3.42) for a rectangular island is close to our result for the truncated cone; however, the decay of our curve at small $\eta < 0.5$ is noticeably steeper. The Glas formula (3.44) for the mismatching cylinder is very close to our curve, as we saw earlier in Fig. 3.13.

Semi-analytical solutions for the strain fields presented above can be directly compared to the results of finite element calculations of the elastic energy density, Fig. 3.16 showing the corresponding results. The elastic strain density at the center of a cylindrical wire (i.e., at $r = 0$) was computed with the following parameters: $E = 90$ GPa, $\nu = 0.3$ and $R = 20$ nm, with different lattice mismatches from 0.01 to 0.07. The corresponding finite element calculations (shown by dots in Fig. 3.16) show excellent correspondence with the analytical results. Overall, the data presented in Fig. 3.16 again demonstrate a very efficient strain relaxation in a 20 nm wide cylinder, with the strain-free state completely recovered already at $z = 10$ nm.

3.4 Growth Scenarios and Preferred Aspect Ratio

In this section, we consider the nucleation barriers in lattice mismatched systems. In contrast to the conventional one-parametric case studied in Chaps. 1 and 2, where the formation energy depends only on the number of monomers in the nucleus,

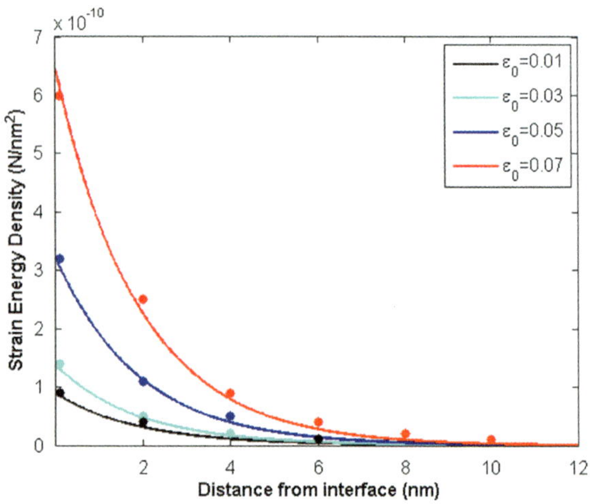

Fig. 3.16 Strain energy density of a cylindrical 20 nm radius nanowire as a function of z at $\varepsilon_0 = 0.01, 0.03, 0.05$ and 0.07 (*lines*). Results of corresponding finite elements calculations are shown by *dots*

the 3D island formation energy is generally a function of two variables: the size and the aspect ratio [24, 199]. The instability in the size variable is responsible for irreversible growth, while the shape-related variable is stable. This leads to a saddle-like character of the formation energy which is quite analogous to the previously studied case of non-isothermal nucleation [198]. The minimum nucleation barrier at the saddle point would normally correspond to the energetically preferred aspect ratio, which determination is very important for quantum dots [40–42, 175, 189] and nanoneedles [185, 186, 200]. It is necessary, however, to compare 3D nucleation barrier to a 2D one to access the preferred growth scenario.

The model geometries used below in this section are shown in Fig. 3.17: we consider a cylinder (a), a hexagonal prism (b), a rectangular prism (c) and pyramidal islands with a circle, regular hexagonal or square cross-sections (d). Tapered sidewalls of pyramidal islands are assumed to be composed of the vertical facets of elementary length dl, separated by the horizontal steps of width dr (d). The steps are arranged regularly so that $dl/dr = L/R = 2\eta$, where L is the height, R is the base dimension and $\eta \equiv L/(2R)$ is the aspect ratio. For the geometries considered, the volume, the sidewall surface area and the base surface area are given by: $V = k_1 C_1 R^3 \eta$, $S_F = k_2 C_2 R^2 \eta$ and $S_B = (C_1/2)R^2$, respectively. Geometrical coefficients are defined as follows: $C_1 = 2\pi$, $C_2 = 4\pi$ for a cylinder; $C_1 = 3\sqrt{3}$, $C_2 = 12$ for a hexagonal prism and $C_1 = 8$, $C_2 = 16$ for a rectangular prism. The coefficients k_1, k_2 equal one for straight islands and 1/3, 1/2, respectively, for pyramids, because the volume of a regular pyramid is three times smaller than that of the equivalent prism and the sidewall area of stepwise tapered pyramid is just two times smaller.

The cylinder and hexagonal prism are typical model geometries for straight nanorods and nanowires, the hexagonal pyramid with stepwise sidewalls is the experimentally observed nanoneedle geometry [185, 186] while the rectangular prism, the conical pyramid and the pyramid with a square base are good models for quantum

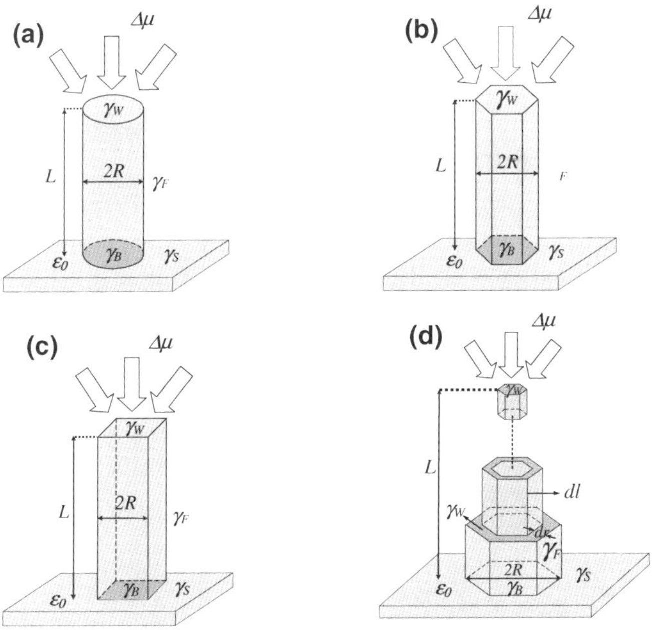

Fig. 3.17 Model island geometries, figure (**d**) represents the pyramidal island with stepwise tapered sidewalls

dots. Generalization to the case of truncated pyramids of different shapes is straightforward. It is noteworthy that the stepwise geometry allows one to use the same value of sidewall surface energy at different aspect ratios, which would be impossible for a full pyramid with crystallographic sidewalls due to the dependence of the sidewall surface energy on their orientation.

As usual in nucleation theory, the formation energy of a 3D island can be written down as

$$\Delta G^{3D}(R, \eta) = -\frac{\Delta\mu(\eta)}{\Omega} k_1 C_1 R^3 \eta + \gamma_F k_2 C_2 R^2 \eta + \Delta\gamma \frac{C_1}{2} R^2. \tag{3.65}$$

Here, $\Delta\mu(\eta)$ is the difference of chemical potentials between the metastable surface phase and the island, Ω is the elementary volume in the solid phase, γ_F is the surface energy of side facets (formed due to the nucleation), $\Delta\gamma \equiv \gamma_W + \gamma_B - \gamma_S$, γ_W is the surface energy of the top facet (formed due to the nucleation), γ_B is the surface energy of the solid-solid interface at the island base (also formed due to the nucleation) and γ_S is the surface energy of pre-existing substrate surface (eliminated by the nucleation). The first term in (3.65) gives the decrease in chemical potential due to the vapor-solid phase transition, the second term describes an energetically unfavorable formation of sidewalls and the third term stands for the change in the in-plane surface energy caused by the island formation. Due to the lattice mismatch between the substrate and the island material, the volume term $\Delta\mu(\eta) = \Delta\mu_0 - w(\eta)$ contains two contributions: the chemical vapor-solid energy difference $\Delta\mu_0$ and the η-dependent strain-induced elastic energy per atom in the island $w(\eta)$. For the latter, we use the general expression $w(\eta) = w_{2D}z(\eta)$, where the elastic relaxation $z(\eta)$ was studied in detail in Sect. 3.3.

Owing to the η-dependence of the volume, the sidewall area and the elastic relaxation, the free energy of 3D island formation is generally a function of two independent variables, R and η. Since ΔG^{3D} has a maximum in R at $\Delta\mu(\eta) > 0$ [denoted below as $\Delta G_*^{3D}(\eta)$], the variable R is thermodynamically unstable and supercritical islands will grow infinitely. The nucleation barrier is now determined by the minimum of $\Delta G_*^{3D}(\eta)$ in η. The minimum point η_* relates to the saddle point of the formation enthalpy [denoted as $\Delta G_*^{3D}(\eta_*) \equiv \Delta G_{\min}^{3D}$], as demonstrated in Fig. 3.18. The variable η is therefore thermodynamically stable and the islands tend to maintain the energetically preferred value of η_*, at least at the initial stage of growth. Maximizing the formation energy given by (3.65) in R, the nucleation barrier can be put in the form

$$\Delta G_*^{3D}(\eta) = \Delta G_0 \frac{(\eta + a)^3}{\eta^2\,[\chi - bz(\eta)]^2}, \tag{3.66}$$

where $\Delta G_0 \equiv (k_2^2/27k_1^2)C_2\gamma_F h_0^2$ (with h_0 as the height of a monolayer) is a material-related constant.

The function given by (3.66) contains three important physical parameters:

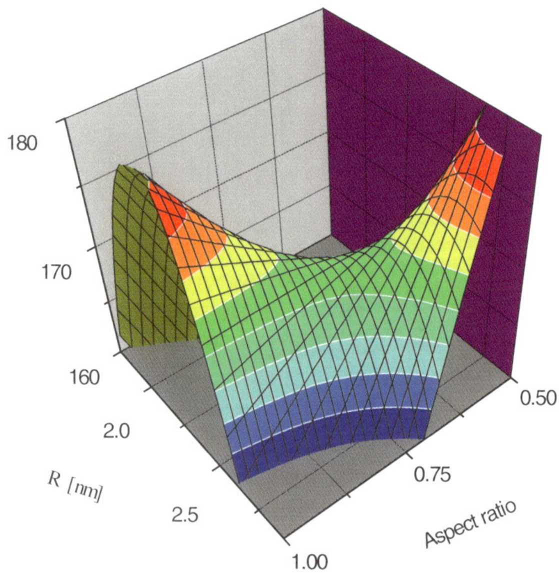

Fig. 3.18 Shape of the formation energy of a 3D island at $\chi = 0.1$, $a = 0$ and $b = 0.23$, with $z(\eta) = 1/(1 + 3\pi \times \eta)$

$$\chi = \frac{C_1}{2k_2C_2}\frac{h_0\Delta\mu_0}{\Omega\gamma_F};\ a = \frac{C_1}{2k_2C_2}\frac{\Delta\gamma}{\gamma_F};\ b = \frac{C_1}{2k_2C_2}\frac{h_0 w_{2D}}{\Omega\gamma_F}. \tag{3.67}$$

The supersaturation coefficient χ equals the chemical potential of a metastable phase with respect to the solid state, expressed in the units of the lateral surface energy per elementary area. The surface energy coefficient a is negative in the wetting ($\Delta\gamma < 0$) and positive in the non-wetting ($\Delta\gamma > 0$) case; homoepitaxy with $\gamma_W = \gamma_S$ and $\gamma_B = 0$ relates to $a = \Delta\gamma = 0$. The coefficient b equals the normalized elastic energy of a 2D layer, scaling with the lattice mismatch as ε_0^2. The necessary condition for 3D nucleation is given by the inequality $\chi > bz(\eta)$, i.e., the supersaturation must be larger than the elastic energy in the island with the aspect ratio η. Since the island height is physically limited by the height of a ML, the minimum value of the aspect ratio is obtained from the equation $\eta = h_0/2R_*^{3D}(\eta)$, where $R_*^{3D}(\eta)$ is the critical radius at which the 3D formation enthalpy reaches its maximum given by (3.66). With the above definitions for the parameters, the minimum aspect ratio is readily obtained in the form: $\eta_{\min}^{3D} = (3k_1/2)(\chi - b) - a$.

The case of 2D nucleation follows from (3.65) upon putting $\eta = h/2R$, $z(\eta) = z(0) = 1$ and $k_1 = k_2 = 1$:

$$\Delta G^{2D}(R) = -[\Delta\mu_0 - w_{2D} - (\Omega/h_0)\Delta\gamma]\frac{C_1}{2}\frac{h_0}{\Omega}R^2 + \gamma_F\frac{C_2}{2}h_0R. \tag{3.68}$$

The correction for the change in the in-plane surface energy $\Delta\gamma$ now enters the volume term, while the surface term contains a positive energy of the island boundary of a monolayer height. Maximizing (3.68) in R, we find the critical radius for 2D nucleation, R_*^{2D}, the corresponding aspect ratio $\eta_*^{2D} = h_0/2R_*^{2D} = 2k_2(\chi - b - a)$ and the nucleation barrier

$$\Delta G_*^{2D} = \Delta G_0\frac{\lambda}{\chi - b - a}, \tag{3.69}$$

where $\lambda \equiv (27k_1^2)/(16k_2^2)$. The condition for 2D nucleation relates to a positive denominator in (3.69): $\chi > b + a$. When 2D and 3D nucleation can occur simultaneously, the preferred island shape is found by minimizing the 3D nucleation barrier in η and comparing it to the 2D barrier. If the minimum 3D barrier is higher than the 2D one, the preferred system configuration would be a 2D thin film. When the minimum 3D barrier at $\eta = \eta_*$ becomes lower than in the 2D case, the situation is reverse and the island adopts a 3D shape with the preferred aspect ratio η_*. To access the preferred thermodynamic configuration, we therefore need to minimize the function $f(\eta) \equiv [\Delta G_*^{3D}(\eta) - \Delta G_*^{2D}]/\Delta G_0$ within the interval from $\eta_{\min} = \max\{\eta_*^{2D}; \eta_{\min}^{3D}\}$ to ∞ and to find the point of minimum η_*. The case with $f(\eta_*) > 0$ relates to 2D growth and the case with $f(\eta_*) < 0$ to 3D islands. From (3.66) and (3.69), the function $f(\eta)$ is given by

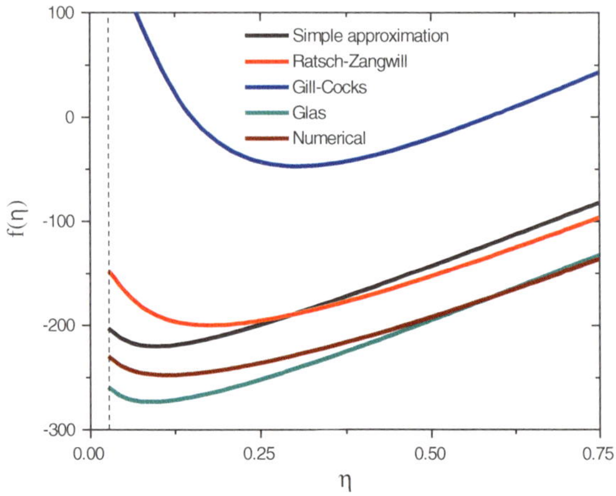

Fig. 3.19 Dependences $f(\eta)$ at fixed $\chi = 0.6, b = 0.58, a = 0$, obtained with different models for $z(\eta)$. The *vertical line* corresponds to the minimum aspect ratio in 2D case

$$f(\eta) = \frac{(\eta + a)^3}{\eta^2 [\chi - bz(\eta)]^2} - \frac{\lambda}{\chi - b - a}. \tag{3.70}$$

Analytical properties of $f(\eta)$ strongly depend on the parameters χ, a and b, as well as on the particular form of elastic relaxation $z(\eta)$.

The preferred growth scenario is thus determined by the sign of minimum of the function $f(\eta)$ at $\eta = \eta_*$ at a given set of parameters χ, a and b and the elastic relaxation $z(\eta)$. In the case of homoepitaxy relating to $\Delta\gamma = 0$, $\varepsilon_0 = 0$ and therefore $a = b = 0$, the function $f(\eta) = \eta/\chi^2 - \lambda/\chi$ reaches its minimum $(2k_2 - \lambda)/\chi$ at $\eta_*^{2D} = 2k_2\chi > \eta_{\min}^{3D}$. This minimum equals $5/(16\chi)$ for prisms and $(1/4\chi)$ for pyramids, i.e., is positive for any χ. Therefore, homoepitaxial films should always grow in 2D form, which is indeed the natural and well known result. For further analysis, we note that the relevant values of supersaturation coefficient χ at typical $\Omega/h \sim 0.1\ \text{nm}^2$, $\gamma_F \sim 1\ \text{J/m}^2$ and $\Delta\mu_0 \sim$ few hundreds meV are of the order of one. Possible behaviors of $f(\eta)$ at $\chi = 0.6$ are shown in Figs. 3.19 and 3.20. At a given b, larger a always favors 3D growth, as demonstrated by Fig. 3.19 at $b = 0.58$ (approximately relating to the parameters of InAs/GaAs system) and $a = 0$. The negative minimum of $f(\eta)$ is reached at the typical quantum dot values of the preferred η_* from 0.1 to 0.35, depending on the approximation used for $z(\eta)$. The decrease in a to the negative value of -0.12 (the wetting case) leads to almost linear dependences $f(\eta)$, reaching their positive minima at the left edge, where 2D nucleation is always favorable (Fig. 3.20). The curves in Figs. 3.19 and 3.20 were obtained with different approximations for the elastic relaxation discussed in Sect. 3.3: simple formula (3.41) at $A = 3\pi$, the Ratsch-Zangwill formula at $\Lambda = 3\pi$, the Gill-Cocks expression with the identical elastic constants for the substrate and the island, and the Glas fit at $p_1 = 0.557, p_2 = 10.15$ and $p_3 = 9.35$. The numerical curve represents our own finite element calculations for a cylindrical geometry.

Fig. 3.20 Same as in Fig. 3.18 at $a = -0.12$

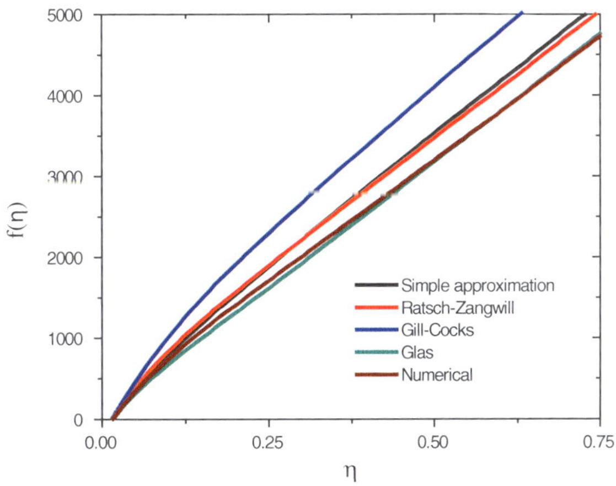

Minimization of $f(\eta)$ at the given elastic relaxation and solution of the parametric equation $f[\eta_*(\chi, a, b); \chi, a, b] = 0$ allows one to obtain the critical curves $b(\chi)$ at different a, separating the domains of 2D and 3D growth in the (χ, b) plane. If $z(\eta)$ is given by the simplified (3.41), the function $g(\eta) \equiv \Delta G_*^{3D}/\Delta G_0$ can be presented in the form

$$g(\eta) = \frac{1}{A\chi^2} y(x); \; y(x) = \frac{(x+C)^3(x+1)^2}{x^2(x+1-B)^2} \tag{3.71}$$

with $x \equiv A\eta$, $C = Aa$ as the surface energy coefficient and $B \equiv b/\chi = w_{2D}/\Delta\mu_0$ as the elastic term. Minimization of $y(x)$ in x results in the cubic equation for the preferred aspect ratio as a function of B and C. Solutions to this equation can be easily analyzed in some particular cases. In the non-wetting case $C > 0$ and with negligible elastic contribution $B \to 0$, (3.71) is reduced to $y(x) = (x+C)^3/x^2$, yielding $x_* = 2C$ and $y(x_*) = (27/4)C$. Since the condition for 3D growth is now given simply by $x > B - 1$, the interesting range in x at $B >> 1$ (where the 2D elastic energy is much larger than $\Delta\mu_0$) is $x >> 1$, which reduces (3.71) for $y(x)$ to $y(x) = (x+C)^3/(x-B)^2$. Such an approximation is always applicable to the growth of nanowires or nanoneedles with large enough aspect ratios. The minimum nucleation barrier is now reached at $x_* = 3B + 2C$, where $y(x_*) = (27/4)(B + C)$. Since $B \propto \varepsilon_0^2$, the obtained expressions demonstrate the tendency for a quadratic increase in the preferred aspect ratio and the corresponding nucleation barrier with the lattice mismatch.

Typical growth diagrams obtained from (3.70) and (3.71) are presented in Fig. 3.21. The critical curve $b(\chi)$ (2D growth below the curve and 3D growth above the curve) at $a = 0$ is almost linear. Since $a = 0$ corresponds to the nucleation on a similar substrate, this case describes the Stranski-Krastanow islands, where a 2D wetting layer is formed on a foreign substrate prior to 3D growth. The critical curve at $a = 0$ shows simply that the stress-driven 3D nucleation is always favored at a larger lattice mismatch and lower supersaturation. For a fixed supersaturation (determined

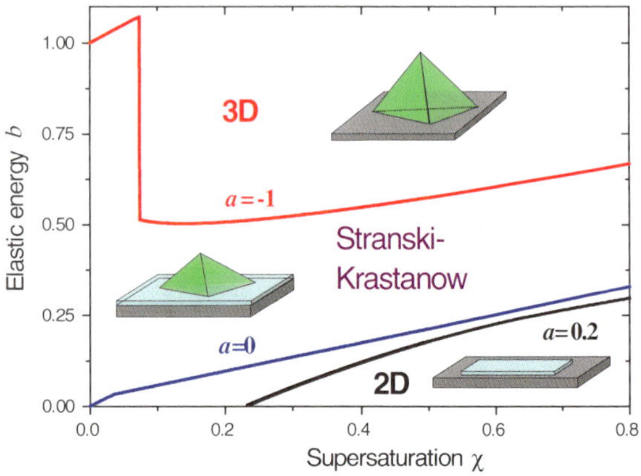

Fig. 3.21 Growth diagram in (χ, b) plane at $A = 3\pi$ for a pyramidal geometry at different a. The wetting curve at $a = -1$ separates the domain of the Volmer-Weber (3D) growth above the curve and the Frank–van der Merve (2D) growth below the curve. After the formation of a wetting layer, the region between the curves at $a = -1$ and $a = 0$ becomes preferential for 3D growth, as in the Stranski-Krastanow mode

by the epitaxy technique and the deposition conditions used) 3D growth starts only if the mismatch is larger than a certain critical value, below which the stress-driven islanding is energetically suppressed. In the wetting case with $a = -1$, the critical curve is constructed from two branches relating to the minima of nucleation barrier at $\eta_{\min}$ (2D growth) and $\eta_* > \eta_{\min}$ (3D growth), with the vertical line connecting the branches. This extends the region of 2D growth, particularly at a smaller lattice mismatch. The obtained behaviors of the critical curves naturally lead to a simple interpretation of the Stranski-Krastanow growth mode within the frame of our nucleation model. Whenever an initially wetting heteroepitaxial system is within the 2D region of Fig. 3.21, the first few monolayers are bound to grow in 2D form. Upon the completion of such a wetting layer, the in-plane surface constant a becomes zero (as in the case of homoepitaxy), which transforms the region between the curves at $a = 0$ and $a = -1$ to the preferred 3D growth. It is noteworthy that such an explanation of the Stranski-Krastanow growth transformation is different from a more usual approach with 3D growth being favored when the elastic energy stored in the wetting layer compensates the surface energy and the wetting force [147]. The critical curve at $a = 0.2$ corresponds to the non-wetting case, where the Volmer-Weber growth region is considerably extended and can be observed at much smaller or even no lattice mismatch. The sketch shown in Fig. 3.21 demonstrates schematically the shapes of crystals emerging in different regions.

It should be noted, however, that real growth scenarios could be much more complex than those discussed hereinabove. This is illustrated in Fig. 3.22, where we plot the curve $\eta_*^{3D}(\chi)$ for 3D islands and the linearly increasing 2D aspect ratio $\eta_*^{2D}(\chi)$. At low enough supersaturations, the preferred aspect ratio rapidly decreases with χ traveling along the curve $\eta_*^{3D}(\chi)$. Since 2D growth is blocked as long as supersaturation is smaller than $b + a$, the islands emerge only in a 3D form below point 1 in Fig. 3.22. 3D growth would be preferred between points 1 and 2 (where $f < 0$) until the curve $\eta_*^{3D}(\chi)$ hits the minimum 3D aspect ratio $\eta_{\min}^{3D}(\chi)$ relating to the monolayer height. In our example, the latter is larger than the 2D aspect ratio

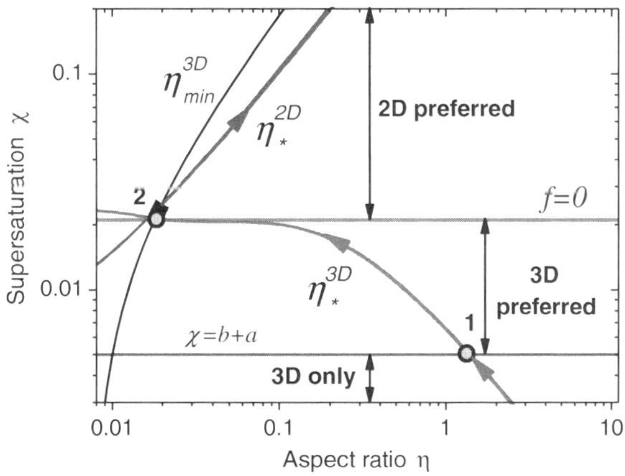

Fig. 3.22 Evolution of the preferred aspect ratio and the growth mode with increasing the supersaturation χ (shown by arrows). The curve $\eta_*(\chi) \equiv \eta_*^{3D}(\chi)$ is obtained by minimizing (3.71) at $b = 0.025$, $a = -0.02$, $A = 3\pi$ for a pyramidal geometry ($k_1 = 1/3$, $k_2 = 1/2$)

$\eta^{2D}_{\min}(\chi)$ at the crossing point. Above the horizontal line $\chi = \chi_0$ corresponding to $f = 0$, the function $f(\chi)$ is positive and therefore 2D growth is preferred. As discussed above, the aspect ratio of 2D islands equals the maximum of $\eta^{3D}_{\min}$ and $\eta^{2D}_{\min}$ at a given supersaturation, as shown in Fig. 3.22. The "2D only" region is formally absent, because the condition for 3D growth $\chi > bz(\eta_*)$ can always be satisfied at $\chi > 0$ with sufficiently large η_*. Realistically, 3D growth becomes completely disabled as the system moves far above point 2 in Fig. 3.22. Whenever the 2D and 3D nucleation are thermodynamically allowed simultaneously (this occurs in the "preferred" domains), the observed island shape is determined by the relative heights of the nucleation barriers. In particular, point 2 in Fig. 3.22 corresponds to 50 % probabilities of 2D and 3D nucleation, where the formation of mixed 2D-3D morphologies with wetting platelets is most anticipated.

Figure 3.23 shows the preferred aspect ratio plotted against $B = w_{2D}/\Delta\mu_0$. The curve relates to the case of $C = 0$, where the point of minimum of $y(x)$ given by (3.71) is readily obtained as $x_* = -1 + (3/2)B + (1/2)\sqrt{9B^2 - 8B} \cong 3B - 5/3$. The last expression (which is a good approximation at large enough B) again shows the tendency for a quadratic increase in the preferred aspect ratio with ε_0. Since the strain-related coefficient B contains an unidentified quantity $\Delta\mu_0$, the graph in Fig. 3.23 is fitted to some of the experimentally observed aspect ratios in different material systems [186, 189, 201–204] with the known lattice mismatch and elastic constants [205]. These data are summarized in Table 3.1. When required, the aspect ratio was obtained from the known orientation of the lateral facets under the assumption of full pyramid geometry.

Theoretical curve is fitted to the average of experimental values for a given system. The fit allows us to deduce the unknown supersaturations $\Delta\mu_0$. The obtained values, summarized in the right column, relate to a reasonable range of several hundreds of meV in all cases. General tendency for a higher supersaturation required to produce taller islands seems relevant, because taller islands have larger surface energy and therefore need larger $\Delta\mu_0$ to decrease the corresponding nucleation barrier. Finally we note that the preferred aspect ratio increases considerably with the lattice mis-

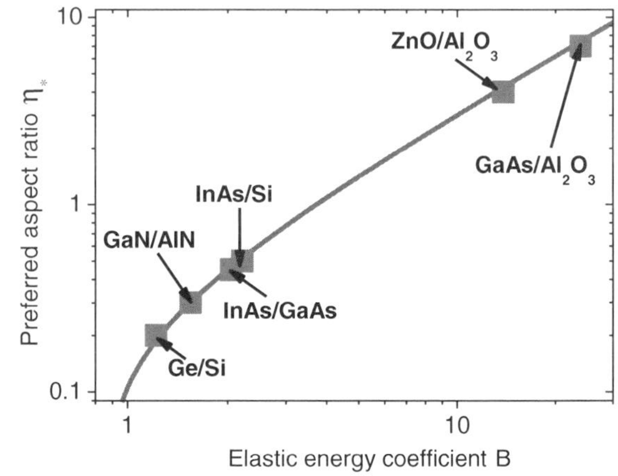

Fig. 3.23 Average experimental (*squares*) and theoretical dependences $\eta_*(B)$, fitted by the approximation of (3.71) at $A = 3\pi$ and $C = 0$ with the parameters summarized in Table 3.1

Table 3.1 Lattice constants and preferred aspect ratios in different material systems

Material system	Lattice mismatch ε_0	$E/(1-\nu)$ GPa	Reference	Preferred aspect ratio η_* Experiment	Theory	$\Delta\mu_0$ meV
Ge/Si	0.04	139	[201, 202]	0.10–0.33	0.2	160
GaN/AlN	0.04	279	[189]	0.31	0.3	260
InAs/GaAs	0.07	79	[201, 202]	0.23–0.71	0.45	300
InAs/Si	0.12	79	[203]	0.5	0.5	900
ZnO/Al_2O_3	0.22	209	[204]	3.0–5.0	4.0	470
$GaAs/Al_2O_3$	0.46	124	[186]	5.2–9.5	7.8	950

match, reaching record values of order of ten for GaAs nanoneedles on sapphire substrate with the extreme lattice mismatch of 46 %.

To sum up the results of this section, the two-parametric nucleation model presented here is capable of describing the nucleation barriers and the preferred aspect ratios of 3D crystal islands in different material systems depending on the lattice mismatch, surface energies and deposition conditions. With these considerations, the physics of self-induced nucleation of Stranski-Krastanow quantum dots, pyramidal nanoneedles and straight nanowires in lattice mismatched systems has many features in common. A wide range of the observed aspect ratios should be mainly due to different lattice mismatches. We note, however, that the model applies only to the initial stage of nucleation, where the preferred aspect ratio must correspond to the saddle point of the formation enthalpy. Whether this aspect ratio is maintained at the follow-up growth stage depends on many energetic and kinetic factors, so that an adequate theoretical description of the growth stage requires relevant modification of two-dimensional nucleation theory. Also, large lattice mismatches always enhance strain accommodation by the interface dislocations. As nanostructures extend in lateral direction, the formation of misfit dislocations at their foot should become favorable, and relevant growth theory must include this possibility. This important question will be considered in more detail in Chap. 5.

3.5 Formation Energy of Stranski-Krastanow Islands

As shown above, the saddle-like shape of formation energy as a function of island size and shape ensures that the shape-dependent variable self-stabilizes during phase transition. In this sense, the concept of the preferred aspect ratio [24, 123] is quite analogous to that used in most of earlier models [25, 40, 148], where the island shape is assumed as being dictated by a cusped local minimum of the surface energy as a function of facet orientation. Clearly, these two concepts have different physics behind them. For instance, while the aspect ratio depends on the growth conditions in the two-dimensional nucleation model, it remains an essentially equilibrium value when driven by the surface energy functional. However, both approaches indicate the possibility to consider the nucleation and growth of the Stranski-Krastanow islands under the constraint of a fixed shape in the first approximation. Whatever is the mechanism of maintaining the energetically preferred aspect ratio, the islands should have identical shape and orientation at a given set of deposition conditions, at least throughout a short-scale nucleation stage. Obviously, such an assumption does not apply during the entire growth stage at a longer duration, for example, when the hut-to-dome shape transformation of Si/Ge islands occurs.

Following Kukushkin and Osipov [24], we will use the following simplified expression for the free energy of island formation [25]:

$$\Delta G(i) = \Delta G_{elas} + \Delta G_{surf} + \Delta G_{wetting}. \tag{3.72}$$

Here, we assume that the island forms from a metastable layer. Therefore, ΔG_{elas} is the difference of elastic energies of i atoms (or III-V pairs) in the island and wetting layer, and ΔG_{surf} the difference of surface energies of the island and the wetting layer area covered by the island. Additionally, we introduce the Muller-Kern term $\Delta G_{wetting}$ standing for the difference in energy of deposit-substrate interactions of i atoms in the island and wetting layer [147]. The island is assumed as being the full pyramid with a square base with side R and contact angle β, yielding the aspect ratio $\eta = (\tan\beta)/2$. The model geometry is shown in Fig. 3.24. The relationship between i and R is given by $i = (R/\alpha l_0)^3$, where l_0 is the 2D lattice spacing and $\alpha = (6h_0 \cot an\beta/l_0)^{1/3}$ is the geometrical factor. For a cubic material with a square 2D surface lattice, the first term in (3.72) is given by

$$\Delta G_{elas} = -(1 - z(\beta))\, \lambda \varepsilon_0^2 l_0^2 h_0 i \tag{3.73}$$

with $\lambda \equiv E/(1 - \nu)$ as the elastic modulus of deposited material and $l_0^2 h_0$ as the elementary volume in the solid. The elastic contribution into the overall free energy

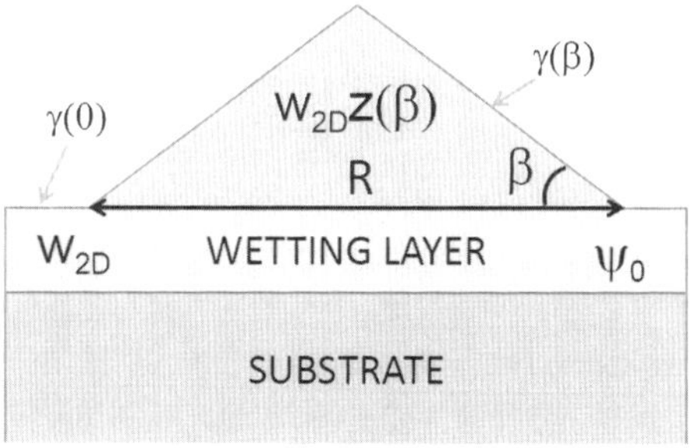

Fig. 3.24 Model geometry of the Stranski-Krastanow island and the parameters used

of island formation is always negative and increases by the absolute value for taller islands.

The surface energy contribution is determined by the difference

$$\Delta G_{surf} = \left[\frac{\gamma(\beta)}{\cos\beta} - \gamma(0)\right] R^2, \tag{3.74}$$

which is exactly equivalent to (3.34). As discussed, below we assume that differently sized island have the same contact angle and consequently identical elastic relaxation $z(\beta)$. It is also assumed that, in spite of strain-induced renormalization of the surface energy [148], the surface energy term is positive, i.e., the formation of additional surface area of lateral facets remains energetically unfavorable even under strain.

The third term in (3.72) describes the work done against the wetting force as i atoms are transferred from the wetting layer surface (with height h) to the island. In line with the Müller-Kern model [147], it is generally given by [24]

$$\Delta G_{wetting} = [(\Psi_0/h_0)\sum_H \exp(-H/k_0h_0) - (\Psi_0/h_0)\exp(-h/k_0h_0)]l_0^2h_0i, \tag{3.75}$$

where the summation is performed over all local heights within the island volume. Since the relaxation of wetting force is exponential for most semiconductor materials with $k_0 \sim 1$, one can neglect the first term in the first approximation. This implies that the absolute value of attractive deposit-substrate interactions in the wetting layer is much higher than in the island. For a cuboid island, this property can be easily justified by summing up the geometrical progression for sufficiently tall islands [24]. Therefore, (3.75) can be simplified to

$$\Delta G_{wetting} \cong -G_{wetting}^{WL} = \frac{\Psi_0}{h_0}\exp\left(-\frac{h}{k_0h_0}\right)l_0^2h_0i. \tag{3.76}$$

Combining all contributions together and expressing ΔG in the units of thermal energy at the substrate temperature T, the normalized island formation energy $F = \Delta G/(k_BT)$ takes the form [24, 25]

$$F = \frac{(\gamma(\beta)/\cos\theta - \gamma(0))}{k_B T} R^2 - \frac{1}{k_B T}\left[(1 - z(\beta))\lambda\varepsilon_0^2 - \frac{\Psi_0}{h_0}\exp\left(-\frac{h}{k_0 h_0}\right)\right] l_0^2 h_0 i. \tag{3.77}$$

The expression in the square brackets represents the effective difference of chemical potentials of atoms in the wetting layer and in the island (we remind that the kinetic growth mechanism considered is the stress-driven diffusion from the wetting layer to island). As usual in nucleation theory, negative difference of chemical potentials relates to a stable wetting layer with thickness $h < h_{eq}$. Positive difference corresponds to a metastable wetting layer with thickness $h > h_{eq}$, where the 2D-3D growth transformation becomes energetically preferred. The equilibrium thickness is therefore given by

$$h_{eq} = h_0 k_0 \ln\left(\frac{\Psi_0}{h_0(1 - z(\beta))\lambda\varepsilon_0^2}\right). \tag{3.78}$$

This important quantity is determined by the ratio between the characteristic wetting and elastic energies. Island formation at $h > h_{eq}$ is possible because the gain in the elastic energy is higher than the energy required to overcome the wetting force and to form the lateral facets. Since the first, surface tern in (3.77) is always positive, the stress-driven formation of 3D islands has a character of fluctuational surpassing the nucleation barrier. The height of the barrier decreases with increasing the wetting layer thickness and the elastic stress accommodated in it.

Introducing the superstress ζ by (3.40), expressing R through i and linearizing (3.77) in ζ (this is well justified only if the wetting layer thickness is not far above h_{eq}), the dependence of formation energy on the number of atoms in the island takes the form

$$F(i) = Ai^{2/3} - B\zeta i. \tag{3.79}$$

The constants A and B are defined as

$$A = \frac{[\gamma(\beta)/\cos\beta - \gamma(0)]}{k_B T}\alpha^2 l_0^2;\ B = \ln\left(\frac{\Psi_0}{h_0(1 - z(\beta))\lambda\varepsilon_0^2}\right)\frac{(1 - z(\beta))\lambda\varepsilon_0^2 l_0^2 h_0}{k_B T} \tag{3.80}$$

The surface constant A is given by the ratio of surface energy increase upon the island nucleation to thermal energy, while the volume constant B is the product of elastic energy gain in thermal units and the logarithm of ratio between the wetting and elastic energy. It is seen that (3.79) is almost exactly equivalent to the conventional (3.66) for the free energy of 3D island formation ($d = 3$), where the difference of chemical potentials $\Delta\mu/(k_B T) = \ln(\zeta + 1)$ is changed to $B\zeta$. Since the surface and volume terms have different signs at $\Delta\mu > 0$ and scale differently with i, (3.79) yields the ζ-dependent nucleation barrier. This enables to describe the Stranski-Krastanow process within the frame of the standard nucleation theory considered in Chaps. 1 and 2. All major notions of self-induced nucleation are directly applicable, although both "phases" considered are in solid state. In order to bridge directly the general theory with 3D islanding, we use the following links:

- Flat wetting layer as a metastable phase;
- Superstress ζ as the measure of metastability (an analogue of supersaturation);
- Excessive "chemical potential" $B\zeta$ as the driving force of "phase transition";
- Coherent strained 3D islands as the nuclei of a new phase, described by the size distribution $n(i, t)$;
- Stress-driven diffusion from the wetting layer to islands as the dominant kinetic pathway that determines the island growth rate;
- Deposition flux as the external material source that determines the technologically controlled ideal supersaturation.

In particular, maximizing (3.79) in i yields the following expressions for the number of atoms in the critical island, the nucleation barrier and the second derivative of formation energy at the critical size:

$$i_c = \left(\frac{2A}{3B\zeta}\right)^3; F = \frac{4}{27}\frac{A^3}{B^2\zeta^2}; -F''(i_c) = \frac{9}{8}\frac{B^4\zeta^4}{A^3}. \tag{3.81}$$

Let us now consider the parameters of Ge/Si(100) material system [24]: $\varepsilon_0 = 0.042$, $h_0 = 0.145\,\text{nm}$, $l_0 = 0.395\,\text{nm}$, $\lambda = 1.27 \times 10^{11}\,\text{N/m}^2$, $\gamma(0) = \gamma_{(100)} = 0.8\,\text{J/m}^2$, $\Psi_0 = 0.45\,\text{J/m}^2$. Taking for estimates $\gamma(\beta) \cong \gamma(0)$, $\beta = 20°$ (to adjust the experimentally observed aspect ratio of 0.18) and calculating the elastic relaxation by means of the Ratsch-Zangwill formula, we obtain: $A = 2.59$ and $B = 0.617$ at $T = 470\,°\text{C}$. At $k_0 = 0.8$, the equilibrium wetting layer thickness amounts to 3.0 ML. Figure 3.25 shows the island formation energy for three different values of wetting layer thickness. It is seen that the nucleation barrier decreases approximately from 20 to 12 and the number of atoms in the critical nucleus from 110 to 50 as the thickness increases from 4.75 to 5.25 ML. At $h = 5\,\text{ML}$, the critical lateral size is about 2.5 nm.

For modest variation in the substrate temperature where the material constants do not vary significantly with T, (3.81) for the nucleation barrier can be put in the form

$$F(T, \zeta) = \frac{T_e}{T\zeta^2}, \tag{3.82}$$

showing explicitly the dependence of F on temperature and superstress. The characteristic quasi-equilibrium temperature T_e contains all the material constants of a concrete heteroepitaxial system, its lattice mismatch parameter and the shape factor in the form

$$T_e = \frac{16}{3}\frac{[\gamma(\beta)/\cos\beta - \gamma(0)]^3 \cot an^2\beta}{k_B\left[(1 - z(\beta))\lambda\varepsilon_0^2\right]^2 \ln^2\left[\Psi_0/(h_0(1 - z(\beta))\lambda\varepsilon_0^2)\right]}. \tag{3.83}$$

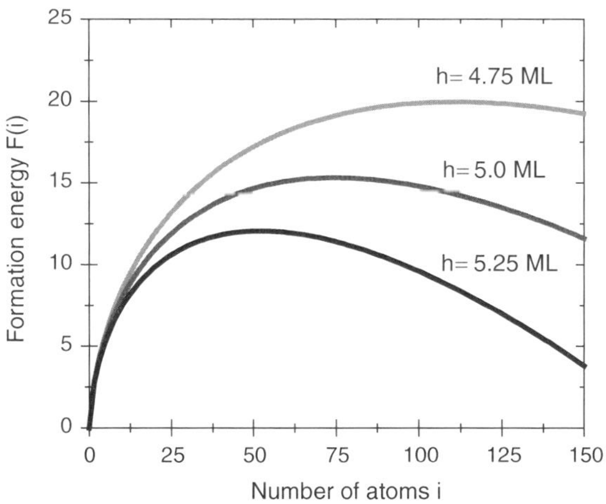

Fig. 3.25 Formation energy for the model parameters of Ge/Si islands

This parameter entirely determines the system at a given set of deposition conditions (which may affect the surface energies and the contact angle). Nucleation of islands is easier at lower T_e because the nucleation barrier gets smaller. Equation (3.83) shows that T_e strongly decreases with increasing the lattice mismatch (the leading dependence is $T_e \propto \varepsilon_0^{-4}$) and the contact angle.

The next step of our nucleation scheme is considering the island growth rate. With neglect of direct impingement from the vapor flux onto a small enough island, the diffusion flux from the wetting layer into a super-critical island can be written down as [24, 25]

$$\frac{di}{dt} = \frac{D}{l_0^2}\frac{\nabla\mu}{k_B T}4R. \tag{3.84}$$

Here, D is the coefficient of stress-driven diffusion from the wetting layer to island that arises due to the gradient of surface chemical potential $\nabla\mu$. In the case of composite islands such as III-V (InAs) quantum dots, we consider the diffusion of group III material (In) limiting the growth process under group V (As) reach conditions. For small enough islands, the diffusion through the island base boundary of perimeter $4R$ should be more important than the R^2 terms. Elastic stress field around the island is a very complex function of coordinates which is highly dependent on the island shape [40, 206–208]. Therefore, $\nabla\mu$ cannot be calculated in absolute fashion in the general case. We can, however, estimate the chemical potential gradient as $\nabla\mu \cong \Delta\mu/(\nu l_0)$, where $\Delta\mu$ is the difference of chemical potentials in the wetting layer and island and ν is the cutoff parameter for the elastic stress field (a quantity of the order of 10 lattice spacing). Equation (3.79) with neglect of surface energy term yields $\Delta\mu/(k_B T) = B\zeta$ for essentially supercritical islands. This simple approximation reduces (3.84) to

$$\frac{di}{dt} = \frac{4D}{l_0^2}\frac{\alpha B\zeta}{\nu}i^{1/3}, \tag{3.85}$$

where we used the relationship between R and i. For the equilibrium wetting layer at $\zeta = 0$, the island growth rate must equal zero. Using the standard formula $di/dt = W_i^+ - W_i^-$ with the corresponding attachment and detachment raters and making use of $W_i^- = W_i^+(\zeta = 0)$, we obtain the attachment rate in the form

$$W^+(i) = \frac{4D}{l_0^2}\frac{\alpha B(\zeta + 1)}{v} i^{1/3}. \tag{3.86}$$

Standard Zeldovich expression given by (1.185) of Sect. 1.8 should be written down as

$$J = \sqrt{\frac{2\pi}{|F''(i_c)|}}\frac{1}{l_0^2} W^+(i_c)\exp(-F), \tag{3.87}$$

because the surface density of a metastable wetting layer equals $1/l_0^2$. Using (3.81) for the nucleation barrier, critical size and formation energy width as well as (3.86) at $i = i_c$, one obtains the nucleation rate of the Stranski-Krastanow islands at a given superstress:

$$J(\zeta) = \frac{a}{\tau l_0^2}\zeta(\zeta + 1)\exp\left[-F(\zeta)\right], \tag{3.88}$$

where the constant $a = (3B)/(4\sqrt{\pi A})$ is of the order of one. As we saw earlier in Sect. 2.1, the size distribution function has a particularly simple form in terms of the invariant size for which the island growth rate becomes size-independent. Since 3D island is fed from the wetting layer predominantly through the base boundary, the growth proceeds in the diffusion regime such that $di/dt \propto i^{1/3}$. From Table 1.2, the invariant size is given by

$$\rho = i^{2/3} = \left(\frac{R}{\alpha l_0}\right)^2, \tag{3.89}$$

i.e., equals the normalized island base area. Using (3.86), the corresponding growth rate can be put in the conventional form

$$\frac{d\rho}{dt} = \frac{\zeta}{\tau} \tag{3.90}$$

with the characteristic time

$$\tau = \frac{3l_0^2 v}{8\alpha BD}. \tag{3.91}$$

This parameter determines the growth rate of supercritical islands: they grow faster at smaller τ. Finally, introducing the distribution over invariant sizes $f(\rho, t)$ by (2.4), the size-invariant stationary distribution is obtained from the general (2.12) in the form

$$f_s(\zeta) = \frac{\tau}{\zeta}J(\zeta) = \frac{a}{l_0^2}(\zeta + 1)\exp\left[-F(\zeta)\right]. \tag{3.92}$$

As usual in nucleation theory, the leading dependence of the nucleation rate and stationary size distribution on superstress is given by the Zeldovich exponent.

3.6 Nucleation Stage and Critical Thickness

The obtained analogy with classical nucleation theory allows one to describe the Stranski-Krastanow growth as the first order phase transition from a metastable wetting layer into coherent strained 3D islands. All the results presented in Chap. 2 are directly applicable. Such a description is most advantageous, as it enables to find the size distribution as a function of the material constants, deposition conditions and growth time. In particular, the nucleation distribution is obtained from the first order kinetic equation

$$\frac{\partial f}{\partial t} = -\frac{\zeta}{\tau}\frac{\partial f}{\partial \rho} \tag{3.93}$$

with the initial and boundary conditions of the form

$$f(\rho, t=0) = 0; f(\rho = 0, t) = f_s\left(\zeta(t)\right). \tag{3.94}$$

Time $t = 0$ now relates to the moment when the wetting layer reaches its equilibrium thickness. Kinetic equation should be coupled with the material balance

$$h_{eq} + h_0 \int_0^t dt' V(t') = h + h_0 l_0^2 \int_0^\infty d\rho \rho^{3/2} f(\rho, t). \tag{3.95}$$

Clearly, the left hand side gives the deposition thickness at time t, whereas the right hand size is the sum of the mean wetting layer thickness and the volume of all islands per unit surface area. The last equation shows simply that, with neglect of desorption, all material deposited must be distributed either in the wetting layer or in the islands.

Without repeating all the steps of the general Γ—procedure described in detail in Sect. 2.2, the distribution over invariant sizes ρ at the nucleation stage under a constant material flux $V(t) = V$ is easily obtained from (3.92) to (3.95) in the standard double exponential form given by (3.38):

$$f(x) = f_s(\Phi_*) \exp\left(cx - e^{cx}\right). \tag{3.96}$$

Here, $\Phi_* = H_*/h_{eq} - 1$ is the maximum "ideal superstress", H_* is the deposition thickness at time t_* where the wetting layer thickness reaches its maximum $[h(t_*) = h_* \equiv h_c]$, and $x(\rho, t) = z(t) - \rho$. The parameter c is given by

$$c = \frac{\Gamma}{\Phi_*^2}\frac{\tau}{t_{eq}}, \tag{3.97}$$

where $t_{eq} = h_{eq}/(Vh_0)$ is the time required to grow the equilibrium wetting layer with a given deposition rate and

$$\Gamma = -\Phi_* \frac{dF}{d\zeta}\bigg|_{\zeta=\Phi_*} = 2F(\Phi_*) >> 1 \tag{3.98}$$

is the large parameter of nucleation theory. The most representative size z relates to the islands having emerged at time t_* and is given by (2.20). The nucleation rate depends on $z(t)$ by means of

$$J(z) = \frac{\Phi_*}{\tau} f_s(\Phi_*) \exp(cz - e^{cz}). \tag{3.99}$$

The surface density of islands is given by

$$n(z) = N\left(1 - \exp(-e^{cz})\right),\ N = \frac{f_s(\Phi_*)}{c}, \tag{3.100}$$

where the second expression determines the maximum density reached after the nucleation is completed. As above, the assymmetric double exponential dependences given by (3.96) and (3.99) can be substituted by symmetric Gaussians in the first approximation, where the most representative size equals the mean size. The characteristic duration of short-scale nucleation stage is given by

$$\Delta t = \frac{2\sqrt{2}}{\Gamma} t_*. \tag{3.101}$$

This expression shows again the time scale hierarchy: at $\Gamma >> 1$, the nucleation stage turns out to be much shorter than the macroscopic deposition time $t_* = H_c/(Vh_0)$ required for growing the wetting layer of critical thickness.

In view of (3.89), the most representative lateral size of islands $R_*(t)$ is related to $z(t)$ as

$$R_*(t) = \alpha l_0 \sqrt{z(t)}. \tag{3.102}$$

Using the normalization condition $g(R)dR = f(\rho)d\rho$ and (3.89), the distribution over linear base dimensions R is obtained from (3.96) in the form

$$g(R,t) = \frac{2R}{(\alpha l_0)^2}\varphi(R,t);\ \varphi(R,t) = cN\exp\left(cx(R,t) - e^{cx(R,t)}\right) \tag{3.103}$$

with

$$x(R,t) = \frac{R_*^2(t) - R^2}{(\alpha l_0)^2}. \tag{3.104}$$

These expressions enable to plot the size distribution over lateral sizes that are more easily determined from the statistical analysis of plan view electronic micrographs.

We emphasize that the double exponential distributions are precise only at the nucleation stage, while the follow-up growth can result in broadening due to kinetic fluctuations and the Ostwald ripening (in the regimes with growth interruption), as described in Chap. 2.

We now turn to the analysis of critical thickness for 2D-3D transition under the material flux, the quantity which can be precisely determined in MBE growth of quantum dots by the *in situ* RHEED diagnostics [40, 41]. As described in Sect. 2.2 in the general case [see (2.33)], the critical superstress Φ_* can be obtained in the self-consistent manner by equalizing the stationary distribution $f_s(\Phi_*)$ given by (3.92) and resulting from the material balance. In the case of Stranski-Krastanow islands, the general formula is reduced to [25]

$$f_s(\Phi_*) = \frac{4}{3\sqrt{\pi}} \frac{h_{eq}}{h_0 l_0^2} \frac{\Phi_*}{\Gamma} c^{5/2}, \tag{3.105}$$

where c and Γ are defined by (3.97) and (3.98), respectively. Using also (3.81) for the nucleation barrier at $\zeta = \Phi_*$, the transcendent equation for Φ_* has the form

$$34 \frac{h_{eq}}{d_0} \frac{B}{A^{5/2} \Phi_*^2 (\Phi_* + 1)} \left[F(\Phi_*) \frac{\tau}{t_{eq}} \right]^{5/2} \exp\left[F(\Phi_*)\right] = 1. \tag{3.106}$$

This expression is very important for understanding the physics of self-induced formation of the Stranski-Krastanow islands as well as for tuning their morphology to the desired properties. Indeed, (3.106) contains thermodynamic characteristics of a given material system (the nucleation barrier) along with the two characteristic kinetic times, which bridges the surface energetics with the growth kinetics. The ratio of macroscopic time of equilibrium wetting layer formation to the microscopic time of island growth has been called the kinetic control parameter [25, 144]:

$$Q \equiv \frac{t_{eq}}{\tau} >> 1. \tag{3.107}$$

This large quantity can be changed by several orders of magnitude by tuning the deposition conditions such as the temperature and deposition flux. Equation (3.106) contains the ratio of two very large quantities: $\exp[F(\Phi_*)]$ and $Q^{5/2}$. Since $F(\Phi_*) >> 1$, the factor $F(\Phi_*)$ can also contribute into the result for the critical thickness, while other multiplying factors can be neglected with a logarithmic accuracy.

Therefore, the critical superstress can be well estimated as the solution to the simplified equation

$$F(\Phi_*) + (5/2) \ln F(\Phi_*) = (5/2) \ln Q, \tag{3.108}$$

where the right hand side is known in the particular growth experiment. Solving this by iterations, the result for the nucleation barrier is obtained in the form

$$F(\Phi_*) = (5/2)u(Q) \tag{3.109}$$

The function *u(Q)* is determined by the following recurrent equations:

$$u^{(n+1)}(Q) = \ln Q - \ln[(5/2)u^{(n)}(Q)],\, n = 0, 1, 2...(u^{(0)} \equiv 2/5), \tag{3.110}$$

yielding $F^{(1)}(\Phi_*) = (5/2)\ln Q$ in the first approximation. This result is exactly equivalent to the first (2.34) at $m = 3/2$. For example, for the parameters of Ge/Si hut clusters considered in Sect. 3.5 at $\Phi_* = 0.72$ and $Q = 4.9 \times 10^3$, $F^{(1)}(\Phi_*) = 21.2$, (3.109) and (3.110) yield $F(\Phi_*) = 14.5$, while the exact solution of (3.106) gives $F(\Phi_*) = 12.0$. Hence, the simple approximation $(5/2)\ln Q$ is inaccurate and may be used only for qualitative analysis, while more precise expressions (3.106) or (3.109), (3.110) should be used when the material parameters are known.

Very importantly, the obtained relationship between the nucleation barrier at the critical thickness and the growth conditions allows us to express all the characteristics of nucleation process through only two parameters: the quasi-equilibrium temperature T_e (the surface energetics) and the kinetic control parameter Q (the growth conditions). In particular, using (3.81), (3.97), (3.98), (3.101) and (3.109), we get:

$$\Phi_* = \sqrt{\frac{2}{5}\frac{T_e}{Tu(Q)}}; \tag{3.111}$$

$$c = \frac{5u(Q)}{\Phi_*^2 Q}; \tag{3.112}$$

$$\Delta t = \frac{0.57}{u(Q)} t_*; \tag{3.113}$$

$$N = \frac{4}{l_0^2}\frac{h_{eq}}{h_0}\frac{T}{T_e}\left(\frac{u(Q)}{Q}\right)^{3/2}. \tag{3.114}$$

The time t_* relating to the maximum wetting layer thickness at which the island nucleation rate is maximum equals $\Phi_* t_{eq}$, whereas the time to grow the critical wetting layer from the beginning of deposition equals $(\Phi_* + 1)t_{eq}$. It should be emphasized, however, that the obtained expressions are valid only when the Stranski-Krastanow islands nucleate under a material influx. This condition reflects the most usual case of supercritical deposition thickness with the flux turned off at $t_0 > t_* + \Delta t$ or $H > h_* + V\Delta t$ after the island nucleation is effectively completed. Other possible regimes will be considered later.

Using (3.107) and (3.91) along with the definition for t_{eq} and (3.78) for h_{eq}, the explicit expression for the kinetic control parameter can be presented as

$$Q = \frac{8}{3}\alpha k_0 \ln^2 \left[\frac{\Psi_0}{(1 - z(\beta))\lambda\varepsilon_0^2 d_0} \right] \frac{(1 - z(\beta))\lambda\varepsilon_0^2 h_0}{k_B T} \frac{D(T)}{\nu V}. \quad (3.115)$$

Assuming the Arrhenius-like dependence for the diffusion coefficient in the wetting layer, $D(T) \propto \exp(-E_D/k_B T)$ with E_D as the characteristic activation energy for the stress-driven diffusion, the leading temperature and flux dependence of Q is given by

$$Q \propto \frac{1}{VT} \exp\left(-\frac{E_D}{k_B T}\right). \quad (3.116)$$

Inserting this into (3.114) one obtains the leading temperature and flux dependence of the island density in the form

$$N \propto V^{3/2} T^{5/2} \exp\left(\frac{3E_D}{2k_B T}\right), \quad (3.117)$$

where the characteristic diffusion temperature $T_D = E_D/k_B$ is always much larger than T. Such a dependence holds provided that the island shape is temperature independent, which is not always the case throughout the entire growth process.

However, islands should evolve with a fixed energetically preferred shape at the short scale nucleation stage, as discussed in Sect. 3.4 and supported by experiment [40, 41, 175]. Upon the completion of initial nucleation, the island density may change only if the growth of larger islands is somehow suppressed. Under a material flux, a slower growth of larger islands would result in the increasing wetting layer thickness and may cause the secondary nucleation, where the size distribution becomes bimodal [23]. As in the general case described in Sect. 2.9, (3.117) show the major tendency: the island density in the supercritical deposition mode tend to increase with the growth rate and decrease with the temperature. Such a dependence has been observed experimentally in many material systems, for example, for the InAs/GaAs [144, 166–168] and Ge/Si [169, 174, 209] quantum dots. Physical explanation of this general tendency during the Stranski-Krastanow growth is the following. At a given temperature, an increased deposition rate creates more nucleation centers on the surface and consequently increases the density. At a given deposition rate and deposition thickness, the solid diffusion in the wetting layer is faster at elevated temperatures. The islands get larger, while their density must decrease because the total volume of islands is fixed by the deposition thickness. It should be emphasized that the dependence given by (3.117) has an essentially kinetic origin and is much less dependent on the system thermodynamics.

Let us now consider the critical thickness for 2D-3D transformation. As discussed qualitatively in Sect. 3.2, the total volume of islands remains relatively small throughout the nucleation stage. In particular, the maximum wetting layer thickness h_c almost equals the critical deposition thickness H_*. The shortness of nucleation stage also explains why the transformation of RHEED pattern from streaks to spots seems almost instantaneous compared to the entire growth process. Indeed, in view of the time scale hierarchy given by the strong inequality $\Delta t << t_{eq}+t_*$, the nucleation of the Stranski-Krastanow islands proceeds in such a way that

$$\Delta H << H_* = h_{eq} + Vt_*. \tag{3.118}$$

As in the general case considered in Chap. 2, the time scale hierarchy is physically explained by an extremely high sensitivity of the nucleation rate to the wetting layer thickness. In turn, a very strong system non-linearity leads to quite narrow size distribution, which is of major advantage for applications of self-induced quantum dots. Using (3.111) and $\Phi_* \cong h_c/h_{eq} - 1$, we are able to deduce the following analytical formula for the critical thickness [25]:

$$h_c \cong h_{eq}\left[1 + \left(\frac{2}{5}\frac{T_e}{Tu(Q)}\right)^{1/2}\right]. \tag{3.119}$$

Here, h_{eq} is the thermodynamic equilibrium thickness which is defined by (3.78) that contains only energetic parameters and depends on temperature only weakly through the wetting energy. The bracket term gives the difference between the critical and equilibrium thickness, which is not clearly understood sometimes. At $h < h_{eq}$, the stress-driven islanding is impossible, because the surface and wetting energy dominate over the elastic stress energy. At $h_{eq} < h < h_c$, the elastic energy relaxation outweighs the surface energy and the wetting layer becomes metastable. However, this metastability is insufficient to onset the islanding immediately, and a certain exposition time is required to observe it (see Fig. 3.8). An intense nucleation starts only near h_c, which has a kinetic origin.

Based on (3.119), one can summarize the following most important points regarding the critical thickness:

- The critical thickness is proportional to the equilibrium thickness which decreases with the lattice mismatch and is almost temperature independent.
- The critical thickness equals the maximum wetting layer thickness in a given growth experiment; the maximum is reached due to a kinetic balance between the material influx from a vapor phase and its sink due to the island nucleation.
- In spite of its kinetic nature, the critical thickness only logarithmically depends on the deposition conditions through the function $U(Q) \cong \ln Q$. In particular, the flux dependence of the critical thickness is only logarithmical. An almost flux-independent critical thickness has been confirmed experimentally, as reviewed in [40, 41].
- The critical thickness becomes closer to the equilibrium one with increasing the lattice mismatch, scaling approximately as $h_c = h_{eq}(1 + const/\varepsilon_0^2)$ with neglect of logarithmic corrections.
- In a given material system, the critical thickness decreases with increasing the surface temperature approximately as $h_c = h_{eq}(1 + const/\sqrt{T})$.

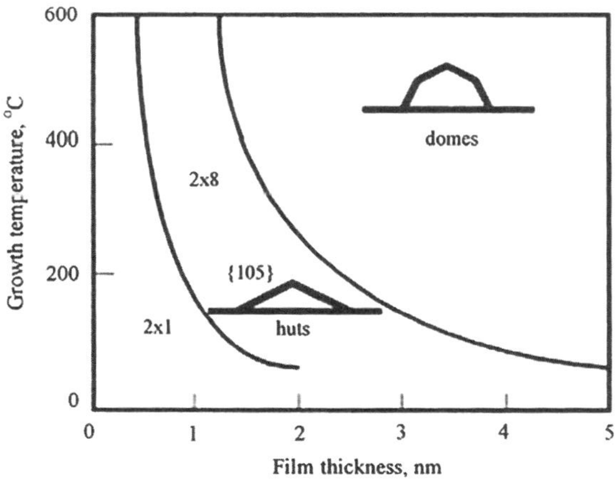

Fig. 3.26 Diagram of surface states in the Ge/Si(100) heteroepitaxial system [175]

In particular, Fig. 3.26 shows the phase diagram in the Ge/Si system from [175], where the critical thickness discussed now relates to the hut islands. The corresponding critical curve clearly shows the decrease of the critical thickness with increasing the surface temperature. Dependence on the lattice mismatch is demonstrated in Fig. 3.5 for the parameters of the InGaAs/GaAs system, where the range of 1.7–1.8 ML relates to the typical growth temperatures in the case of pure InAs quantum dots. We emphasize once again that optically bright quantum dots cannot be too large because of misfit dislocations emerging at their base. Therefore, useful deposition range cannot far exceed the critical thickness. On the other hand, the narrower size distributions are obtained in faster growth regimes and an immediate overgrowth by a wide band gap layer, where the resulting spectrum is less affected by the fluctuation-induced broadening, the Ostwald ripening and the re-evaporation of material. In such regimes, the size distribution is well described by the nucleation double exponential shape, as discussed in Chap. 2 and further in the next section. Larger islands are required to increase the operational wavelength of optical devices, however, the crystal quality and the size homogeneity might be sacrificed.

3.7 Growth of Stranski-Krastanow Islands

As shown in Sect. 2.6, the normalized total volume of islands at the growth stage is well described within the mono-dispersive approximation

$$G(t) = \frac{h_0 l_0^2}{h_{eq}} N z^{3/2}(t), \tag{3.120}$$

showing simply that the total volume equals the product of the island density and the volume of individual island with the average size over the ensemble. Writing the material balance in the standard form $\Phi = \zeta + G$ following from (3.95) and making use of $\zeta = \tau dz/dt$, one obtains the closed differential equation for the most representative invariant size $z(t)$ in the form

$$\tau\frac{dz}{dt} + \frac{h_0 l_0^2}{h_{eq}} N z^{3/2}(t) = \Phi(t); \tag{3.121}$$

$$\Phi(t) = \begin{cases} t/t_{eq}, 0 \le t \le t_0 \\ t_0/t_{eq} \equiv \Phi_0, t > t_0 \end{cases} \tag{3.122}$$

The ideal superstress Φ increases linearly with time before t_0 and stabilizes at t_0 when the deposition flux is turned off. The time t_0 corresponds to the deposition thickness H_0.

Introducing the characteristic time t_g and lateral size R_g by

$$t_g = \frac{2}{3\Phi_0^{1/3}}\left(\frac{h_{eq}}{h_0 l_0^2 N}\right)^{2/3}\tau; \tag{3.123}$$

$$R_g = \left[\frac{6(H_0 - h_{eq})\cot an\beta}{N}\right]^{1/3}, \tag{3.124}$$

and measuring the lateral size and time in the units of R_g and t_g, respectively, (3.121) can be put in the form

$$\begin{cases} 3r\frac{dr}{dx} + r^3 = \frac{\Phi(\varepsilon x)}{\Phi_0} \\ r(x=0) = 0 \end{cases} \tag{3.125}$$

Here, $r = R_*/R_g$, R_* is the most representative lateral size, and $x = (t - t_*)/t_g$; the initial condition assumes an infinitely small critical size at the critical wetting layer thickness. Obviously, the time t_g determines the characteristic duration of the size relaxation stage during which the island size grows from zero to the maximum size R_g that can be achieved at a given deposition thickness H_0. The small parameter

$$\varepsilon = \frac{t_g}{3t_0} < \frac{t_g}{3t_*} << 1 \tag{3.126}$$

reflects the timescale hierarchy: the growth stage is always much longer than the nucleation stage and, on the other hand, much shorter than the deposition time required to form the wetting layer of the critical thickness. The time scale hierarchy follows directly from (3.98), (3.101), (3.109) and (3.123) at $U(Q) >> 1$:

$$t_g = \left(\frac{h_c - h_{eq}}{H_0 - h_{eq}}\right)^{1/3} \frac{0.47}{u^{1/3}(Q)} t_* = \left(\frac{h_c - h_{eq}}{H_0 - h_{eq}}\right)^{1/3} 0.82u^{2/3}(Q)\Delta t, \qquad (3.127)$$

leading to the timescale hierarchy $\Delta t << t_R << t_*$.

Equation (3.125) can be integrated analytically if $\Phi/\Phi_0 = 1$, i.e., when the deposition is stopped soon after the critical thickness. In this approximation, the time evolution of the mean size is given by (2.106) at $m = 3/2$:

$$\frac{t - t_*}{t_g} = \ln\left[\frac{(1 + r + r^2)^{1/2}}{1 - r}\right] - \sqrt{3}\arctan\left(\frac{2r + 1}{\sqrt{3}}\right) + \frac{\pi}{2\sqrt{3}} \equiv f(r). \quad (3.128)$$

The universal function in the right hand side does not contain any model parameters. Therefore, all quantum dots would obey exactly identical growth law in terms of the normalized variables $(t - t_*)/t_g$ and R_*/R_g, with only t_g and R_g depending on the material system and the growth conditions. It is seen that the mean size saturates at $t - t_* \sim 3t_g$, showing again that t_g is the characteristic size relaxation time. After this time, the islands almost reach the size R_g. As follows from (3.124), the relaxed size scales with the island density as $N^{-1/3}$ and increases with the deposition thickness as $(H_0 - h_{eq})^{1/3}$. The obtained dependences are quite obvious and can be well understood simply from the material balance. Less obviously, using (3.117), the leading temperature, flux and deposition thickness dependence of the relaxed size are given by [25, 144, 174]

$$R_g \propto \frac{\exp(-E_D/2k_BT)}{V^{1/2}T}(H_0 - h_{eq})^{1/3}. \qquad (3.129)$$

As follows from the general nucleation kinetics described in Sect. 2.9, the island size increases with the temperature as the Arrhenius exponent and decreases as the power law of the flux; the particular power exponents correspond to the growth index $m = 3/2$.

As the deposition thickness increases, the simplified (3.128) becomes incorrect and the time evolution should be obtained from the general (3.125). One can also use the perturbation theory in the small parameter ε, because deposition process described by the right hand side of (3.125) is much slower than the island growth [210]. Typical time dependences of the mean size are presented in Fig. 3.27 for the typical parameters of the Ge/Si system: $h_0 = 0.145$ nm, $l_0 = 0.395$ nm, $h_{eq} =$ 3 ML, $\beta = 20°$, $\mathrm{N} = 2 \times 10^{10}\,\mathrm{cm}^{-2}$, $\varepsilon_0 = 0.042$, at $\tau = 0.03$ s, with the deposition thickness $H_0 = 7.1$ ML [210]. It is seen that, in this particular case, the analytical approximation gives largely overestimated mean size for all times, while the first approximation of the perturbation expansion in ε is the excellent approximation to the numerical solution of (3.125). The exact solution demonstrates a more linear increase of the mean size with time than the steep analytical dependence.

The size distribution of coherent islands is obtained by means of the corresponding expressions of Sect. 2.6–2.8 at $m = 3/2$ at different stages (growth under the material

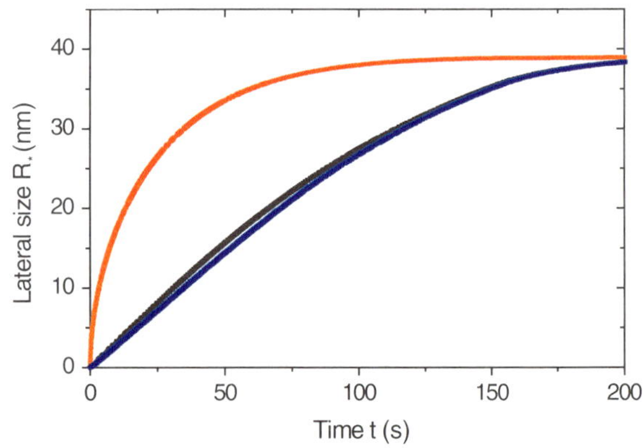

Fig. 3.27 Mean size of Ge hut islands on Si(100): *red line*—analytical solution given by (3.128), *blue line*—the first approximation of the perturbation expansion in ε, *black line*—numerical solution of (3.125). Zero time relates to the critical wetting layer thickness

influx, upon the growth interruption and the Ostwald ripening), or by numerically solving the system of (2.11)–(2.14). In particular, the fluctuation-induced broadening of the size spectrum under the flux is described by (2.117), where the equilibrium monomer concentration n_{1e} is changed to $1/l_0^2$ for the diffusion-induced Stranski-Krastanow growth from the wetting layer. Under a constant flux, the dispersion contains two terms proportional to $z^{1/2}$ and z, showing that the distribution width scales approximately as $z^{1/2}$. In the regimes with a small enough deposition thickness and a short exposition, the distribution over invariant size remains double exponential in the first approximation. When expressed in terms of the lateral size, the spectrum is no-longer time invariant. In particular, its relative width $\Delta R/R_*$ decreases with time. In view of

$$cz(t) \cong \frac{c\Phi_*}{\tau}(t - t_*) = \Gamma\frac{(t - t_*)}{t_*}, \tag{3.130}$$

the spectrum width upon the completion of nucleation stage equals $\Delta z = (\Gamma\Delta t)/(ct_*)$. For the relative distribution width after the growth stage, this yields

$$\frac{\Delta R}{R_g} \cong \frac{0.4}{u^{2/3}(Q)} \tag{3.131}$$

in the Gaussian approximation. At large enough $u(Q) >> 1$, the strong inequality following from (3.109), the dispersion remains relatively small. This important property is explained by an extremely steep exponential dependence of the nucleation rate on the wetting layer thickness that leads to a very fast nucleation under the material influx. Despite a random character of self-induced nucleation process, the Stranski-Krastanow islands have a narrow size distribution in optimal growth regimes, which is of paramount importance for applications. As discussed in the next section, typical values of $\Delta R/R_g$ are in the range of 7–10 %.

3.8 Modeling of Size Distribution

In this section, we present a general scheme for modeling the size distribution of the Stranski-Krastanow quantum dots at a supercritical deposition thickness, and a numerical example. Calculations involve the following steps:

1. *Modeling of the growth thermodynamics*. The input parameters include: the elastic modulus of deposited material $\lambda \equiv E/(1-\nu)$, the lattice mismatch parameter ε_0, the wetting energy Ψ_0, the contact angle β, the surface energies of deposit in the substrate plane $\gamma(0)$ and for lateral facets, the lattice parameters (the 2D lattice spacing l_0 and the height of a ML h_0 in the case of a cubic material), and the relaxation coefficient of attractive interactions with the substrate k_0. Using the results of Sect. 3.3, the elastic relaxation $z(\beta)$ is determined for a given β and particular island shape. With these input parameters, the equilibrium wetting layer thickness is calculated by means of (3.78). The constants A and B entering the island formation energy are obtained from (3.80), and the quasi-equilibrium temperature T_e from (3.83).
2. *Determination of the kinetic control parameter*. We first input the deposition rate V, which is a technologically controlled value. After that, the temperature dependent diffusion coefficient of the stress-driven diffusion $D(T)$ at a given surface temperature and the cutoff parameter of the elastic stress field v should be estimated. The kinetic control parameter Q can then be obtained from (3.115). Among other material parameters, the diffusion coefficient is less known, and can be increased by several orders of magnitude by the elastic stress. Therefore, the kinetic control parameter Q serves as a fitting parameter (in most cases) that can be deduced, for example, from the measured density of quantum dots or the critical thickness by (3.114) or (3.119), respectively.
3. *Modeling of the nucleation stage*. Numerically solving transcendent (3.106) or using approximate (3.109), we obtain the maximum superstress Φ_*, the corresponding nucleation barrier $F(\Phi_*)$ and Γ. The critical thickness of 2D-3D transformation is calculated by means of (3.119). The size distribution constant c, the characteristic duration of the nucleation stage Δt, and the maximum surface density of islands N are obtained from (3.112) – (3.114), respectively. The time dependences of the nucleation rate and the island density are given by (3.99) and (3.100). The distribution over the invariant size is obtained from (3.96), and the lateral size distribution is given by (3.103) and (3.104). The time dependence of the most representative size at the nucleation is given by (3.130).
4. *Island growth stage*. For a given deposition thickness H_0, the time evolution of the mean lateral size is obtained by numerically solving (3.125). Approximate (3.129) can be used when H_0 is close to h_c. The dimensional characteristics are calculated by scaling the results with the characteristic growth time t_g and the relaxed size R_g given by (3.123) and (3.124), respectively. The wetting layer thickness at the growth stage is obtained from $\zeta = \tau dz/dt$; at $\Phi(t) \cong \Phi_*$, where the formula

$$\zeta(t) \cong \Phi_*[1 - r^3(t)] \tag{3.132}$$

is a good approximation. The island size distribution at the growth stage is calculated by means of the general expressions of section of Sects. 2.6–2.8, depending on the growth regime. The double exponential nucleation distribution resulting in (3.103) and (3.104) with the mean lateral size $R_*(t)$ can be used when the exposition time is not too long.

Let us now consider the numerical example in the case of Ge/Si(100) hut clusters grown by MBE at the surface temperature $T = 470\,°\mathrm{C}$ with the deposition rate $V/h_0 = 0.07$ ML/s and the deposition thickness $H_0/h_0 = 6.2$ ML. In calculations, we use the energetic parameters of Sect. 3.5, for which the free energy of island formation is shown in Fig. 3.24. The equilibrium thickness of Ge layer equals 3.0 ML. At $A = 2.59$ and $B = 0.617$, the quasi-equilibrium temperature amounts to $T_e = (4A^3/27B^2)T = 5020$ K. Assuming the value of $D(T) = 6 \times 10^{-13}\,\mathrm{cm}^2/\mathrm{s}$ for the diffusion coefficient at 470 °C and $\nu = 10$ yields the kinetic control parameter $Q = 4.9 \times 10^3$. Numerical solution of (3.112) yields the maximum superstress $\Phi_* = 0.71$, with the corresponding nucleation barrier $F(\Phi_*) = 12$ and $\Gamma = 2F(\Phi_*) = 24$. The maximum wetting layer thickness h_c equals approximately 5.1 ML. The duration of nucleation stage Δt amounts to 4 s, with the corresponding $\Delta H = 0.3$ ML. Therefore, the island nucleation becomes effective at $H = h_c - \Delta H/2 \cong 4.95$ ML and is almost completed at $H = h_c + \Delta H/2 \cong 5.25$ ML. The time required to grow the equilibrium and the critical wetting layer amounts to $t_{eq} = 43$ s and $t_{eq} + t_* = 74$ s, yielding $t_* = 31$ s. The characteristic growth time $\tau = t_{eq}/Q = 8.8 \times 10^{-3}$ s. The value of constant c equals to 8.5×10^{-3}. The surface density of hut islands N after the nucleation stage is $3.6 \times 10^{10}\,\mathrm{cm}^{-2}$. The characteristic time of the growth stage $t_g = 9.0$ s, and the islands reach their maximum mean size $R_g = 24$ nm after $3t_g = 27$ s after the deposition of the critical thickness. If the structure is "frozen" at $H_0 = 6.2$ ML, the corresponding mean size $R(t_0) = 20.5$ nm is smaller than R_g.

The time dependences of the wetting layer thickness, surface density and mean lateral size normalized to their maximum values h_c, N and R_g, respectively, are presented in Fig. 3.28. These curves are typical for the Stranski-Krastanow growth. In particular, the time scale hierarchy is clearly seen: the maximum island density is achieved at the shortest nucleation stage and stays constant at the much longer growth stage where the mean lateral size changes from zero to 20.5 nm. In turn, the growth stage is noticeably shorter than the deposition of the critical wetting layer thickness. The corresponding lateral size distributions *F(R, t)* at four different times are shown in Fig. 3.29. These are obtained from (3.103) and (3.104). It is seen that the island growth is much faster at the beginning and saturates towards larger deposition time. In terms of lateral size, the relative dispersion slightly decreases as growth proceeds and reaches the value of $\Delta R/R_g = 7\,\%$ when the Ge flux is terminated.

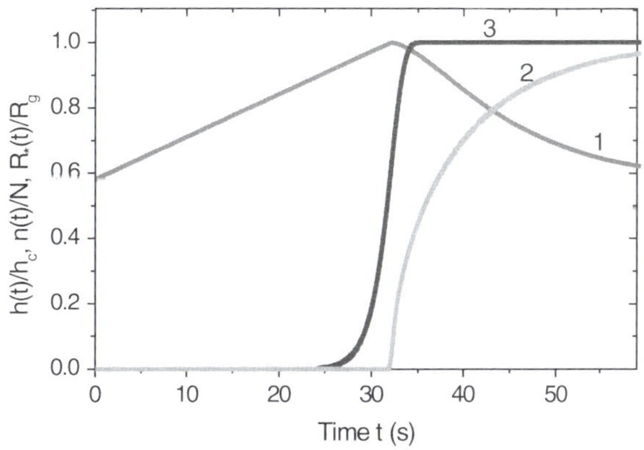

Fig. 3.28 Normalized wetting layer thickness (*1*), mean lateral size (*2*) and surface density (*3*) for the model parameters of Ge hut clusters

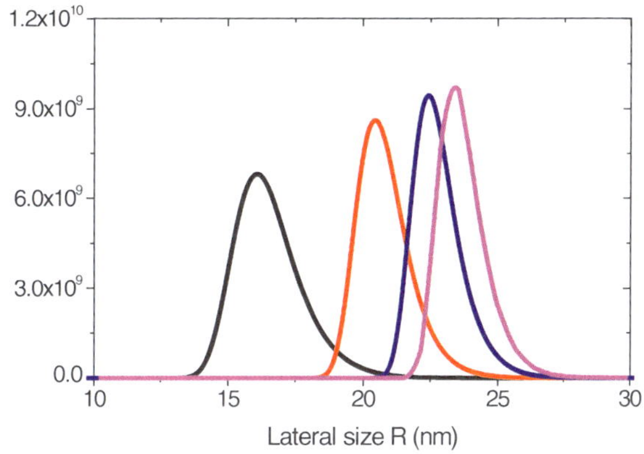

Fig. 3.29 Lateral size distributions of Ge hut clusters $g(R, t)$ at four different times $\beta = 39$, 46, 52 and 59 s

3.9 Role of Surface Steps

The models of stress-driven island formation described above relate to the case of an ideal singular substrate without surface steps. Since real semiconductor substrates always contain a certain number of surface steps, the step density was assumed as being so low that the nucleation events occurred on the flat terraces rather than at the steps or kinks. As already mentioned in Sect. 3.2, the misorientation (or miscut) angle of a vicinal substrate can be used as an additional control parameter to tune the morphology of island ensembles [41, 170–172, 179]. When the step density increases, the islands tend to nucleate near the steps, while the nucleation on the terraces becomes ineffective. In this case, the island will decorate the steps, which can be used for spatial ordering of quantum dots. Following [211], we now briefly consider the stress-driven nucleation of 3D islands at a linear step to see what effect the substrate vicinity has on the nucleation probability.

The model geometry is schematized in Fig. 3.30. As in Sect. 3.3, we consider a pyramidal island (with a square base), whose lateral surfaces are composed of

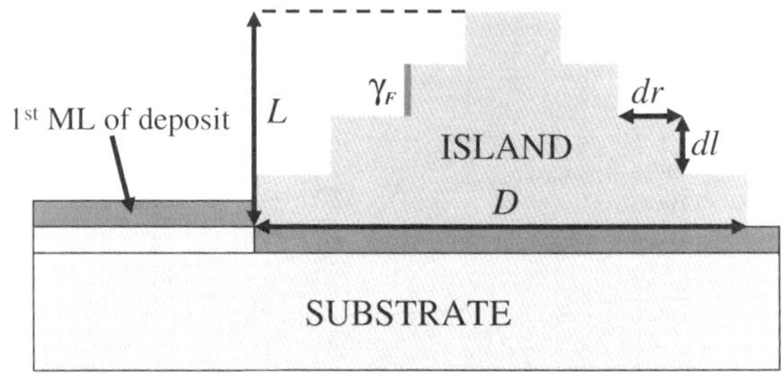

Fig. 3.30 The model geometry of island formed at the linear monoatomic step of a vicinal surface on a 1ML wetting layer; $D = 2R$—base dimension, L—height, γ_F—surface energy of lateral surfaces

regularly arranged steps of elementary height dl and length dr, respectively, such that $dl/dr = L/R = 2\eta$, where $\eta = L/D$ is the aspect ratio. Generalization to other geometries (full and truncated pyramids with flat crystallographic side facets) is rather straightforward. For simplicity, we assume that the islands may form after the growth of 1 ML thick wetting layer. As discussed above, 1 ML is very close to our estimate for the equilibrium wetting layer thickness of the InAs/GaAs(100) quantum dots. However, the assumption $h_{eq} = h_0$ is not crucial and all the expressions can be easily re-formulated for an arbitrary equilibrium thickness. We consider a vicinal surface with a given density of linear steps defined by the misorientation angle. The island is in contact with the monoatomic step of a wetting layer of the elementary height h_0. The island volume equals $D^3\eta/3 = \Omega i$, where Ω is the elementary volume and i is the number of atoms (or III-V pairs) in the island. This yields the relationship $D = (3\Omega i/\eta)^{1/3}$. In the geometry shown in Fig. 3.30, the total area of vertical facets S_F decreases by $2Dh_0$ with respect to a singular substrate, because the formation of two monoatomic steps of length D is economized by the step. Therefore, $S_F = 2D^2\eta - 2Dh_0 = 2\eta^{1/3}(3\Omega i)^{2/3} - 2h_0(3\Omega i/\eta)^{1/3}$. Horizontal facets of the sidewalls do not contribute to the total change of the surface energy for the geometry considered. Indeed, in the Stranski-Krastanow growth, their contribution cancels with the surface energy of pre-existing wetting layer area covered by the island.

The island formation thus leads to the increase of the surface energy by the value $\gamma_F S_F$. On the other hand, it leads to a relaxation of the elastic stress induced by the lattice mismatch. As discussed in Sect. 3.2, the elastic relaxation is larger for higher aspect ratios. As in Sect. 3.5, we assume that the dominant kinetic mechanism of the island growth is stress-driven diffusion from the wetting layer. The difference of chemical potentials in the wetting layer and islands can be then put as $-f(h)[1 - z(\eta)]w_{2D}$. As above, $w_{2D} = (E\Omega\varepsilon_0^2)/(1 - v)$ is the elastic energy per atom in a uniformly strained 2D wetting layer, while $z(\eta)$ is the elastic relaxation showing which percentage of the elastic energy is relaxed at the free island sidewalls. The function $f(h)$ describes the dependence of the chemical potential difference on the wetting layer, in particular, within the frame of the Müller-Kern model [147]. As already discussed in Sect. 3.5, $f(h_{eq})$ must equal zero, therefore the series expansion of $f(h)$ near h_{eq} should start from a linear term: $f(h) \cong c(h - h_{eq})/h_{eq} = c\zeta$. Whenever the island is in contact with the surface step, there exist the lattice mismatch between the island and the wetting layer along their contact line. The corresponding elastic energy can be described by the Ratsch-Zangwill formula given by (3.42) at $\eta = 1 : E_{lateral} = (w_{2D}/\Omega)(D^2 h_0)F(1)$, where $F(\eta) = (1/\Lambda\beta)[1 - \exp(-\Lambda\beta)]$ and $\Lambda \cong 3\pi$.

Summing up all the contributions, we arrive at the following expression for the free energy of coherent island formation:

$$\Delta G = \tilde{A}i^{2/3} - \tilde{B}i - \tilde{C}i^{1/3}. \tag{3.133}$$

The η-dependent coefficients here are given by

$$\tilde{A} = \tilde{A}_0 + \tilde{A}_1; \tilde{A}_0 = 2\eta^{1/3}(3\Omega)^{2/3}\gamma_F; \tilde{A}_1 = [(3^{2/3}h_0)/(\Omega^{1/3}\eta^{2/3})]F(1)w_{2D};$$

$$\tilde{B} = f(h)[1 - z(\eta)]w_{2D};$$

$$\tilde{C} = 2h_0(3\Omega/\eta)^{1/3}\gamma_F. \tag{3.134}$$

Obviously, the values of $\tilde{A}$, $\tilde{B}$ and $\tilde{C}$ are positive or zero for any material parameters of a given heteroepitaxial system. The coefficient $\tilde{B}$ presents the volume part of the free energy associated with the decrease of the elastic stress in the island, which is always negative at $h > h_{eq}$ (in our model, when the deposition thickness exceeds 1 ML). The coefficient $\tilde{A}$ contains the standard surface energy term $\tilde{A}_0$, whereas $\tilde{A}_1$ describes the additional elastic energy on a vicinal substrate. The coefficient $\tilde{C}$, which is non-zero only on a vicinal substrate, gives a negative correction for the surface energy of the island eliminated by the step. The case of nucleation on a planar surface follows from (3.133) and (3.134) at $\tilde{A}_1 = \tilde{C} = 0$. Upon normalization to the $k_B T$ and using the approximation $f(h) = c\zeta$, the corresponding formation energy is equivalent to (3.79), re-written for a slightly different geometry.

Differentiating (3.133) with respect to $x = i^{1/3}$ and looking at the extrema, it is seen that the formation energy reaches its minimum and maximum at the points $x_{\max} = (\tilde{A} + \Delta)/3\tilde{B}$ to $x_{\min} = (\tilde{A} - \Delta)/3\tilde{B}$, respectively, where $\Delta = \sqrt{\tilde{A}^2 - 3\tilde{B}\tilde{C}}$. The values of the maximum and minimum of ΔG equal

$$\Delta G_{\min} = (\tilde{A} - \Delta)(2\tilde{A}^2 - \tilde{A}\Delta - 9\tilde{B}\tilde{C} - \Delta^2)/(27\tilde{B}^2);$$
$$\Delta G_{\max} = (\tilde{A} + \Delta)(2\tilde{A}^2 + \tilde{A}\Delta - 9\tilde{B}\tilde{C} - \Delta^2)/(27\tilde{B}^2). \tag{3.135}$$

As in the case of heterogeneous nucleation considered in Sect. 1.5, the maximum corresponds to the critical size, while the nucleation barrier ΔG_* equals the difference between the maximum and minimum:

$$\Delta G_* = \frac{4(\tilde{A}^2 - 3\tilde{B}\tilde{C})^{3/2}}{27\tilde{B}^2}. \tag{3.136}$$

In the case of a singular substrate without steps ($\tilde{C} = 0$), the minimum disappears, and the nucleation barrier is reduced to the corresponding (3.81). As seen from Fig. 3.31,

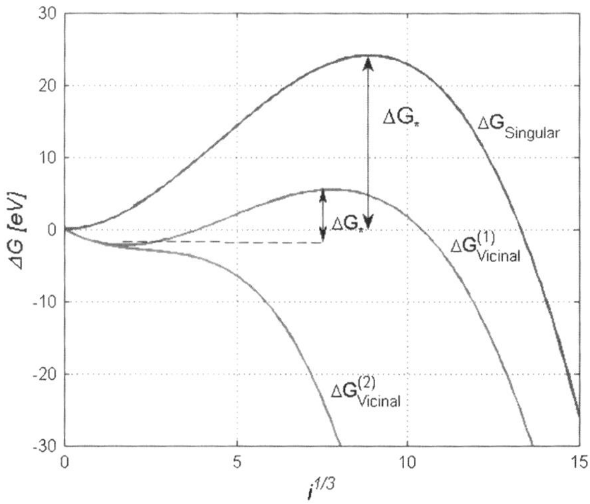

Fig. 3.31 Island formation energy on a singular ($\Delta G_{Singular}$) and vicinal ($\Delta G_{Vicinal}$) substrate as a function of $i^{1/3}$. The *curves* are obtained from (3.133), (3.134) at $\gamma_F = 0.8\,\mathrm{J/m^2}$, $E/(1-\nu) = 8\times 10^{10}\,\mathrm{J/m^3}$, $h_0 = 0.3\,\mathrm{nm}$, $\varepsilon_0 = 0.07$, $f(h) = 2$ (for $\Delta G_{Singular}$ and $\Delta G^{(1)}_{Vicinal}$) and 4 (for $\Delta G^{(2)}_{Vicinal}$) at the fixed elastic relaxation $z(\eta) = 1/2$

surface step leads to the modification of the formation energy and decreases the nucleation barrier. The latter disappears completely at $\tilde{A}^2 < 3\tilde{B}\tilde{C}$, showing that the nucleation at the surface step might proceed without any barrier, e.g., immediately after the wetting layer reaches its equilibrium thickness.

3.10 Subcritical Quantum Dots

In several promising applications of quantum dot heterostructures such as single photon emitters and detectors, the low density ensembles of coherent strained islands are required. The position controlled fabrication of low density quantum dots can be achieved, e.g., by top down etching of InAs/GaAs quantum well heterostructures into microcolumns [212, 213], or site-controlled nucleation of InAs/GaAs islands defined by the surface stressors [214]. Such methods necessarily involve advanced processing or growth techniques. Without a position control, low density InAs/GaAs quantum dots can be obtained, for example, with a very low deposition rate of InAs [215]. A strong decrease of surface density of supercritical islands with decreasing flux is well explained by (3.117). A simple alternative method for synthesis of the low density Stranski-Krasrtanow islands is the deposition of subcritical amount of material (which is below the critical thickness for the 2D-3D transformation under the flux). Subcritical islands were observed long ago, for example, in [130, 177, 178]. Subcritical MBE techniques involving a certain exposition under the arsenic flux [144, 216] or growth on vicinal GaAs(100) substrates [179] enable the fabrication of InAs quantum dots with the density down to $1\times 10^7\,\mathrm{cm^{-2}}$.

As already discussed in Sect. 3.2 and illustrated in Fig. 3.8f, the formation of subcritical islands on a singular substrate without steps can occur only when the wetting layer thickness exceeds its equilibrium value. For the parameters of the InAs/GaAs(100) system, our estimates give $h_{eq} \cong 1.0 - 1.1$ ML [144, 216]. The critical thickness h_c approximately equals the maximum wetting layer thickness at which the islands nucleate at the highest rate; its value is close to 1.7 ML [41, 217]. Between h_{eq} and h_c, the wetting layer is thermodynamically unstable, however, the level of metastability is too low to start the nucleation immediately. The formation of islands therefore requires a certain waiting time Δt_{exp}. General analysis of Sect. 2.9 shows that the morphology of subcritical islands depends rather differently on the deposition thickness, temperature and flux compared to the supercritical regime. Without repeating the detailed procedure of Sect. 2.9 in the particular case of the Stranski-Krastanow growth (the corresponding analysis can be found in [143]), we now only qualitatively consider the behavior of the surface density and maximum size of subcritical islands. We assume a small enough deposition thickness $H_0 < h_c$ and long enough Δt_{exp} such that the nucleation proceeds in the subcritical nucleation mode and is effectively completed before the surface is cooled down or overgrown by a barrier cap layer [216].

With neglect of desorption, the material balance on the substrate surface yields

$$H(t) = h(t) + g(t), \tag{3.137}$$

where $H(t)$ is the deposition thickness, $h(t)$ is the average wetting layer thickness (including the surface adatoms in the general case) and $g(t)$ is the total volume of islands per unit surface area. If the islands are small and the ensemble is dilute, they are fed mainly through the base perimeter. The island growth rate di/dt is proportional to the base dimension $R \propto i^{1/3}$. Therefore, the invariant size $\rho = i^{2/3} = R^2$ is proportional to the island base area. According to (2.46) and (2.47),

$$g(t) = const \int_0^t dt' J(t') \rho^{3/2}(t', t); \tag{3.138}$$

$$\rho(t', t) = \int_{t'}^{t} d\tau v(\tau), \tag{3.139}$$

where $J(t)$ is the nucleation rate, $V(t)$ is the growth rate such that $d\rho/dt = v(t)$ and $\rho(t', t)$ is the invariant size of islands having emerged at time t'. In the case of subcritical deposition with $H_0 < h_c$, the maximum nucleation rate is reached at $h = H_0$, i.e., when the flux is terminated but the total volume of islands is negligibly small. Due to extremely steep dependence of the nucleation rate on the wetting layer thickness, the value of $J(H_0)$ is much smaller than $J(H_c)$, which is why the resulting island density is much smaller and the nucleation process is much slower than in the supercritical case.

The second important difference is that the morphology of subcritical islands should not depend on the deposition rate. As typical for the regimes with an instantaneous generation of metastability (see Sect. 2.9), the nucleation stage is much longer than the deposition time $t_0 = H_0/V$ and, at long enough exposition, the time t_0 should not affect the resulting structure. The characteristic duration of the nucleation stage Δt can be estimated from (3.138) with

$$J(t) = J(H_c)f\left(\frac{t-t_0}{\Delta t}\right) \propto v\exp[-F(H_0)f\left(\frac{t-t_0}{\Delta t}\right), \tag{3.140}$$

where $f[(t-t_0)/\Delta t]$ is a steep function of its argument and $F(H_0)$ is the nucleation barrier at the deposition thickness. For the latter, we use (3.82) at $\zeta = \Phi_0 = H_0/h_{eq} - 1 : F(H) \cong T_e/[T(H_0/h_{eq}-1)^2]$, with T_e as the quasi-equilibrium temperature characterizing the energetics of a given heteroepitaxial system. The nucleation will stop when the island volume (expressed in the numbers of MLs) reaches the value of the order of one:

$$g(t) \sim v^{5/2}\,(\Delta t)^{5/2}\,e^{-F(H_0)} \sim 1. \tag{3.141}$$

Therefore, Δt scales as

$$\Delta t \propto \frac{1}{v}\exp\left[\frac{2F(H_0)}{5}\right]. \tag{3.142}$$

The surface density of subcritical quantum dots is obtained from

$$N = \int_0^\infty dtJ(t) \sim ve^{-F(H_0)}\Delta t \sim \exp\left[-\frac{3}{5}\frac{T_e}{T(H_0/h_{eq}-1)^2}\right], \tag{3.143}$$

where only the leading exponential dependence on the deposition thickness and temperature is shown. The maximum lateral size of islands can be obtained from (3.137) after the end of the growth stage, at $H = H_0$, $h = h_{eq}$ and $g \propto NR_g^3$:

$$R_g \sim (H_0 - h_{eq})^{1/3}N^{-1/3}, \tag{3.144}$$

showing that the relaxed size is smaller at a higher density. Obviously, (3.143) and (3.144) are just the particular case of the general (2.142), (2.143) at $m = 3/2$.

Opposite to the supercritical deposition regime, the density of subcritical quantum dots exponentially increases and their maximum size exponentially decreases with the deposition thickness and the substrate temperature. As pointed out in Sect. 2.9, such a behavior is typical for the thermodynamically controlled nucleation process. The density is obtained by the integration of the Zeldovich nucleation rate over the nucleation time, while the nucleation rate is enhanced at a higher metastability and temperature. Combining the results of this section with earlier conclusions of Sects. 3.6 and 3.7, Table 3.2 summarizes qualitative temperature, flux and deposition thickness dependences of the island size and density in the particular case of InAs/GaAs system. The transition from the thermodynamically to kinetically con-

Table 3.2 Qualitative behavior of InAs quantum dot morphology at different deposition thickness

Deposition thickness of InAs	Behavior of island density and mean size at increased temperature	Behavior of island density and mean size at increased flux	Growth regime
Smaller than 1.5 ML	Density increases Size decreases	Independent	Thermodynamic
1.5–1.7 ML	–	–	Transition
Larger than 1.7 ML	Density decreases Size increases	Density increases Size decreases	Kinetic

trolled growth regime relates to an intermediate interval of the deposition thickness near the critical value, approximately between 1.5 and 1.7 ML. Finally, we emphasize that the mean size of subcritical quantum dots can be significantly decreased with respect to R_g by applying a shorter deposition time.

3.11 Kinetically Controlled Engineering of Quantum Dot Ensembles

Based on the above considerations and following the general scheme for tailoring the size distribution of self-induced islands described in Sect. 2.9, we now give several examples of the control over the size and density of quantum dots by changing the MBE growth conditions. These results will be compared to the experimental data in the next two sections. First, Figs. 3.32 and 3.33 present theoretical dependences of the density and relaxed lateral size of InAs/GaAs quantum dots on the deposition thickness at different temperatures and fixed deposition rate $V = 0.1$ ML/s [143]. These graphs were obtained for the following parameters: $T_e = 5000$ K, $E_D/k_B = 6000$ K, $h_{eq} = 1.05$ ML, $h_0 = 0.303$ nm, $\Omega = l_0^2 h_0 = 0.0452\,\text{nm}^3$ and $\beta = 30°$. As we saw earlier (see Figs. 2.27 and 2.28 of Sect. 2.9 and Table 3.2), the temperature dependences convert near the critical thickness of about 1.7 ML: the density increases with temperature below and decreases above the critical thickness, while the relaxed size demonstrates the opposite behavior. At a given temperature, the density rapidly increases with H_0 below h_c and saturates above h_c. The distinct difference between the conventional 2D nucleation from surface adatoms and the Stranski-Krastanow growth from a wetting layer is that the critical thickness in the latter case is almost temperature independent. This is explained by a strong temperature dependence of the equilibrium adatom concentration and the temperature-independent density of the wetting layer for quantum dots.

Next, Figs. 3.34 and 3.35 show the same dependences at the fixed temperature $T = 450\,°$C and different deposition rates, calculated for the same parameters of

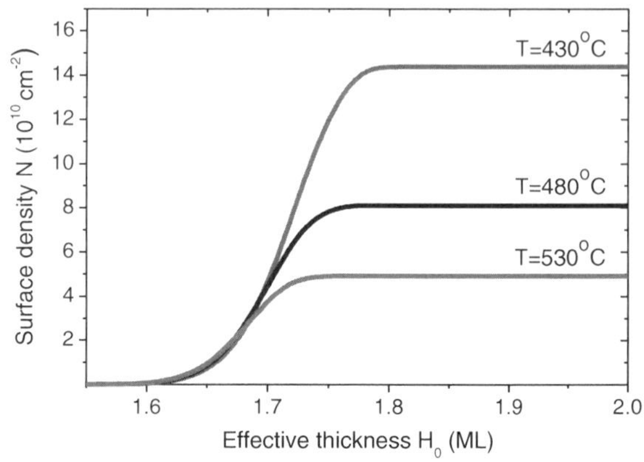

Fig. 3.32 Surface density of InAs/GaAs quantum dots depending on the deposition thickness at three different temperatures and fixed flux $V = 0.1$ ML/s

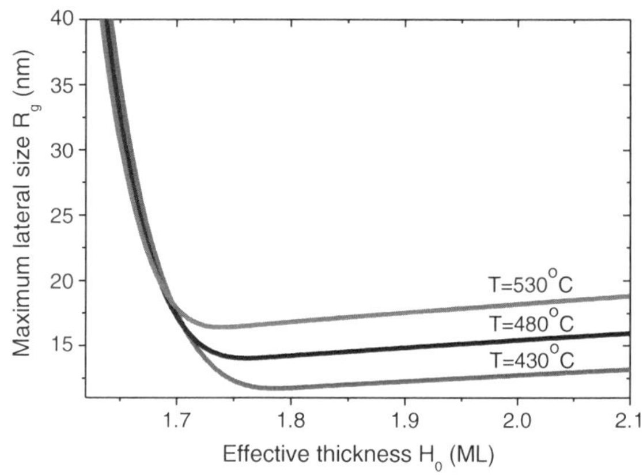

Fig. 3.33 Relaxed mean size of InAs/GaAs quantum dots depending on the deposition thickness at three different temperatures

InAs/GaAs. Following the general laws given in Table 3.2, the flux dependences split near the critical thickness (which is almost flux independent), because the subcritical structures are controlled by the surface thermodynamics, while the morphology of supercritical islands depends primarily on the growth kinetics. It is noteworthy that the flux dependence at $H_0 > h_c$ is very stong so that the density can be changed by the order of magnitude for the same deposition thickness. We point out again that the relaxed radius R_g plotted in Figs. 3.33 and 3.35 gives the absolute maximum dimension that can be achieved at the given growth conditions. The complete size relaxation requires a long exposition time, particularly for the subcritical islands, where the size distribution acquires essentially asymmetrical shape shown in Fig. 2.31.

Let us now see how the morphology of an ensemble of the Stranski-Krastanow islands can be tuned by changing only two essentially non-equilibrium parameters: the deposition rate V and the exposition time Δt after the termination of flux. We again consider the InAs/GaAs(100) system. If the structure is cooled down or overgrown immediately after the In flux is turned off, the mean lateral size $R(t_0)$ would be smaller than R_g. Since the nucleation stage in the supercritical deposition range is

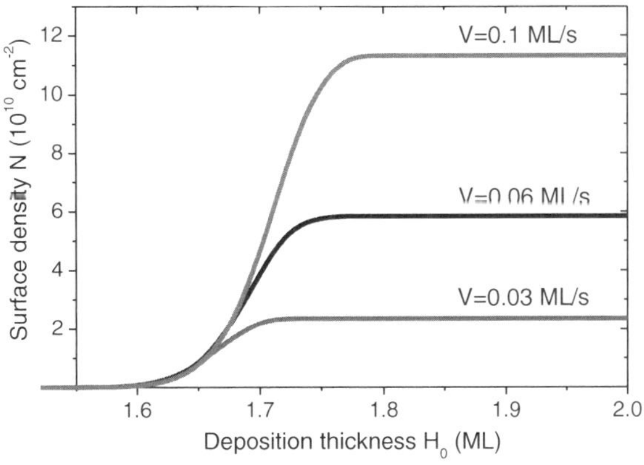

Fig. 3.34 Same as in Fig. 3.32 at three different fluxes and fixed temperature $T = 450\,^{\circ}$C

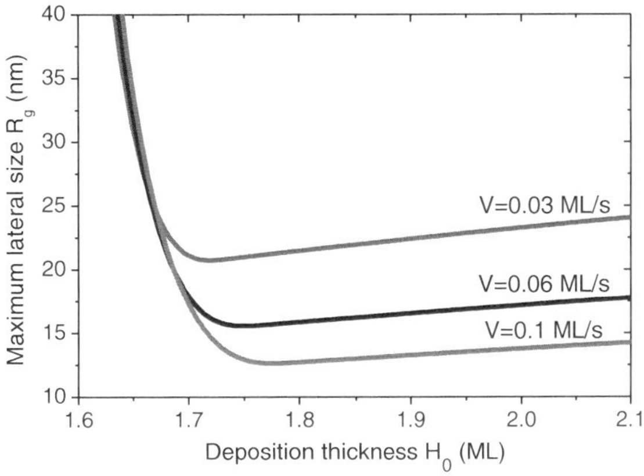

Fig. 3.35 Same as in Fig. 3.33 at three different fluxes and fixed temperature $T = 450\,^{\circ}$C

much faster than the island growth, one can anticipate that increasing the exposition time would result in the increase of the lateral size without noticeable change of the density. This can be analyzed by means of (3.125) or its approximate solution given by (3.128) (which becomes more precise as H_0 approaches h_c) [142]. The surface temperature T, the deposition rate V and the deposition thickness H_0 are assumed as being constant in a given growth run. We then calculate the maximum mean size R_g and the characteristic growth time t_g by means of (3.124) and (3.123), respectively, at $N = const$, because the nucleation stage is already completed. If the In flux is terminated at time $t_0 = H_0/V$, the current time t equals the sum of the deposition and the exposition times: $t = t_0 + \Delta t$. We thus obtain the relationship between R and Δt at given t_0. The minimum size at a given deposition thickness equals $R(t_0) = R(\Delta t = 0)$, while the maximum size is R_g. Changing the exposition therefore enables to change the size from $R(t_0)$ to R_g without affecting the island density.

Table 3.3 Theoretical characteristics of InAs/GaAs quantum dots at different InAs fluxes

V (ML/s)	Q	h_c (ML)	ΔH (ML)	Δt (s)	t_g (s)	N (cm^{-2})	$R(t_0)$ (nm)	R_g (nm)
0.01	10^4	1.70	0.040	4.0	14.6	7.8×10^9	26.6	32.4
0.03	3.33×10^3	1.73	0.048	1.6	5.3	3.3×10^{10}	15.0	20.0
0.05	2×10^3	1.75	0.052	1.0	3.3	6.5×10^{10}	11.4	16.0
0.07	1.42×10^3	1.76	0.056	0.8	2.5	1.0×10^{11}	9.4	13.8

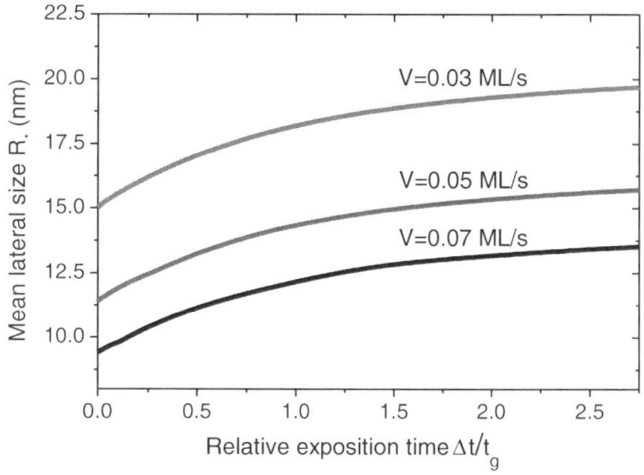

Fig. 3.36 Mean lateral size of InAs islands versus the normalized exposition time at three different fluxes

Table 3.3 summarizes theoretical characteristics of InAs/GaAs quantum dots obtained with the following parameters [142]: $h_{eq} = 1.05\,\text{ML}$, $T_e = 5800\,\text{K}$, $Q = Q_0 = 10^4$ at $T = 430\,°\text{C}$ and $V_0 = 0.01\,\text{ML/s}$ and $\beta = 30°$. The calculations were performed at the fixed temperature of 430 °C and the deposition thickness $H_0 = 1.9\,\text{ML}$. The deposition rate of InAs was varied from 0.01 to 0.07 ML/s. The density of quantum dots was obtained by (3.114), the critical thickness by (3.119), the duration of nucleation stage by (3.113), the maximum size by (3.124) and the growth time by (3.123), within the approximation $u(Q) = \ln Q$. The kinetic control parameter at different fluxes equals $Q = Q_0(V_0/V)$, which follows from (3.116) at a constant temperature. The thickness interval of the nucleation stage equals $\Delta H = V\Delta t$. The time evolution of the mean size was obtained from (3.129). Figure 3.36 shows theoretical dependences of the mean lateral size on the exposition time at three different fluxes. It is seen that, depending on the deposition rate and exposition time, one can obtain the structures with identical size but different density or *vice versa*, which would be impossible if the process was controlled entirely by the surface energetics.

3.12 Theory and Experiment: InAs/GaAs System

In this section, we present some relevant experimental data on the optical and structural properties of InAs/GaAs quantum dots grown by MBE at different conditions, and compare them with theoretical predictions [144, 182, 216, 218, 219]. Let us first consider the photoluminescence (PL) properties of 2 ML ensembles of InAs quantum dots grown at different temperatures and deposition rates without exposition. A brief description of the experimental techniques is as follows. Growth experiments were carried out in an EP1203 solid source MBE machine on semi-insulating singular GaAs (100) substrates. The deposition thickness of the InAs layer was fixed to 2.0 ML for all samples. The InAs growth rate V was varied from 0.01 to 0.1 ML/s. Two series of samples were grown, for which the substrate temperature T during the deposition of the quantum dot layer was kept at 440 and 485 °C, respectively. The As shutter was opened through the whole growth run. For each sample, the quantum dot layer was covered by a low-temperature GaAs cap immediately after the In deposition. The active region was confined by two short-period superlattices of GaAs/$Al_{0.3}Ga_{0.7}As$ (25/25Å, 10 pairs) in order to prevent the carrier escape from the active region into the substrate and surface areas. These superlattices and the GaAs buffer layer were grown at 600 °C. PL measurements were carried out in a standard lock-in configuration. The excitation was provided by a 514.5 nm Ar^+ laser.

The measured dependence of the PL peak position on the InAs growth rate at two different substrate temperatures is presented in Fig. 3.37. A typical PL spectrum is shown in the insert. The two PL peaks are associated with the quantum dot ground and excited states, respectively. Figure 3.37 demonstrates that the PL peak position is gradually shifted towards a shorter wavelength range with increasing the growth rate for both series of different substrate temperature applied. At the same growth rate of InAs, the PL peak is always higher at 485 °C than at 440 °C. Therefore, the characteristic size of islands increases with increasing the temperature and

Fig. 3.37 Room temperature PL peak position from 2 ML InAs/GaAs quantum dots depending on the InAs growth rate at two different substrate temperatures. Room temperature PL spectrum shown in the insert relates to the sample pointed by the *arrow*

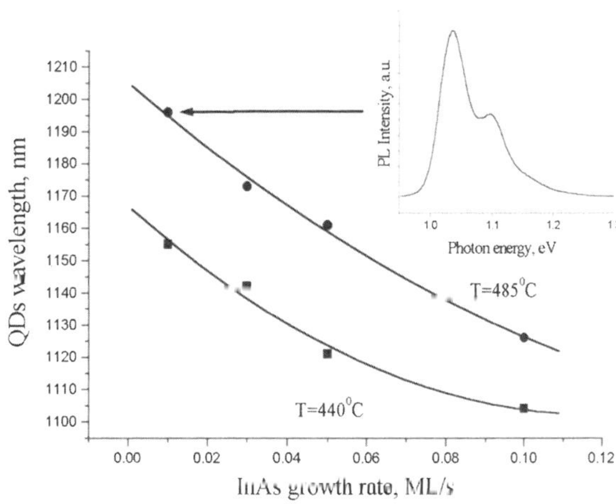

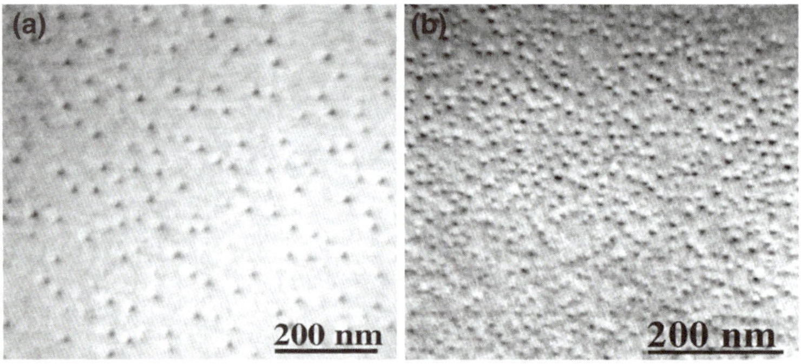

Fig. 3.38 Plan-view TEM images of 2 ML InAs quantum dots on the GaAs(100) surface grown by MBE at different conditions: $T = 485\,°C$, $V = 0.03$ ML/s (**a**) and $T = 440\,°C$, $V = 0.05$ ML/s (**b**)

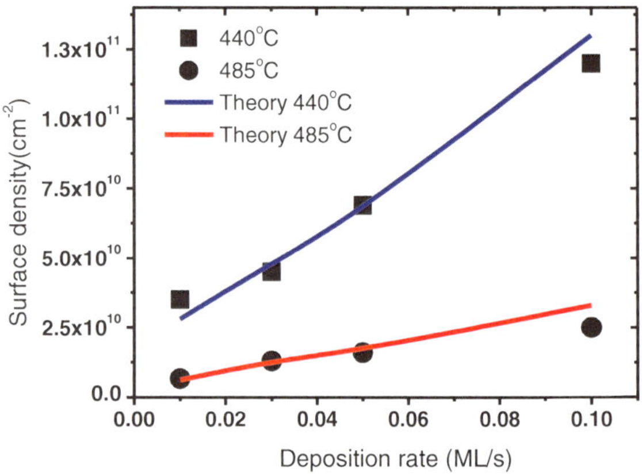

Fig. 3.39 Experimental and theoretical dependences of the density of 2 ML InAs quantum dots on the deposition rate at two different substrate temperatures

decreasing the growth rate. Such a behavior of optical properties of InAs quantum dot ensembles follows the general tendencies imposed by the kinetic limitations on the island formation process and is in qualitative agreement with the theoretical predictions concerning the supercritical islands summarized in Table 3.2. The overall wavelength shift of about 100 nm at the fixed deposition thickness is quite substantial.

Quantitative information on the morphology of different samples was obtained by the statistical analysis of TEM images. TEM measurements were carried out applying a Philips EM420 electron microscope. Typical TEM plan-view images of 2 ML quantum dot ensembles at different temperatures are shown in Fig. 3.38. Concerning the used diffraction contrast technique, the islands have a square base with sides parallel to the crystallographic $< 110 >$ directions. It is clearly seen that the islands grown at lower temperature and higher flux are much denser. Experimental dependences of the island density and mean size on the deposition rate at two different temperatures are plotted in Figs. 3.39 and 3.40. It is seen that all ensembles grown at 485 °C are more dilute and have larger mean size than those grown at 440 °C at

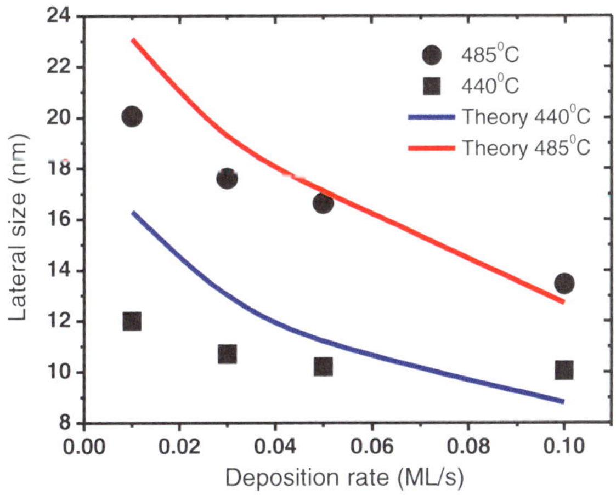

Fig. 3.40 Same as in Fig. 3.39 for the mean lateral size

the same deposition rate. At a given temperature, the island density increases and the size decreases with the deposition rate, with the density reaching the maximum value of $\sim 1.2 \times 10^{11}$ cm^{-2} at $T = 440\,°C$ and $V = 0.1$ ML/s.

Theoretical fits to the experimental data were obtained for the following model parameters of an InAs/GaAs(100) system: $\alpha(\beta) = 1.82 = \text{const}$, $E_D/k_B = 4700\,\text{K} = \text{const}$, while the value of the quasi-equilibrium temperature T_e was varied in such a way that the calculated value of the equilibrium thickness equals 1.0 ML and the critical thickness varies within the range of 1.7–1.8 ML. The growth rate of InAs was varied from 0.01 to 0.1 ML/s for the two chosen temperatures of 440 and 485 °C, respectively, at the deposition thickness fixed to 2.0 ML. The simulation results for the main characteristics of island ensembles are presented in Table 3.4, where $R(t_0)$ is the mean lateral size right after the deposition of 2MLs of InAs. It is seen that the density curves fit very well the experimental data, while theoretical mean size is slightly overestimated towards lower fluxes. However, overall quantitative correlation between theory and experiment is fairly good. The variation in

Table 3.4 Theoretical characteristics of 2 ML InAs/GaAs quantum dot ensembles grown at different conditions

V (ML/s)	T(°C)	T_e (K)	h_c (ML)	N (10^{10} cm^{-2})	$R(t_0)$(nm)	R_g (nm)
0.01	440	4630	1.70	2.8	16.3	18.5
0.03	440	7630	1.75	4.8	12.7	15.9
0.05	440	9900	1.77	6.7	11.0	14.7
0.1	440	11400	1.80	13	8.8	12.1
0.01	485	4590	1.68	0.62	23.1	27.0
0.03	485	8110	1.74	1.3	18.9	23.9
0.05	485	11300	1.77	1.7	17.1	20.2
0.1	485	13700	1.79	3.3	12.7	17.6

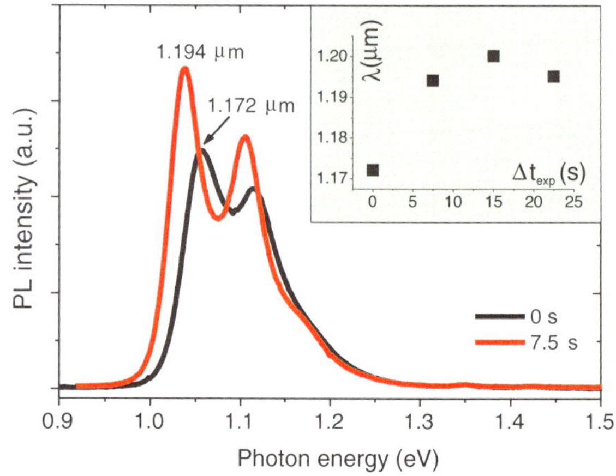

Fig. 3.41 Room temperature PL spectra from InAs quantum dots grown with zero and 7.5 s exposition; the insert showing the measured dependence of the PL peak position on the exposition time

the lateral size from 10 to 20 nm at a fixed deposition thickness clearly shows the importance of kinetic growth effects.

We now consider experimental data on the dependence of optical properties of the InAs/GaAs quantum dots on the exposition time under As_4 flux. Four samples were grown by MBE with the fixed deposition rate of 0.03 ML/s and deposition thickness of 1.9 ML, at a substrate temperature of 485 °C. Upon the termination of In deposition, the samples were exposed to the As_4 flux for four different times: $\Delta t_{exp} = 0$; 7.5; 15 and 22.5 s, respectively, at the same substrate temperature of 485 °C. All samples were then covered by a 5 nm thick GaAs cap layer. For optical measurements, the active quantum dot region was confined from both sides by two short-period superlattices of $GaAs/Al_{0.25}Ga_{0.75}As$ (25/25Å, 5 pairs). For the buffer layer, cap layer and superlattices, the growth temperature was 600 °C. Figure 3.41 presents PL spectra from the samples grown with $\Delta t_{exp} = 0$ and 7.5 s. The insert shows the experimentally measured dependence of the peak wavelength on the deposition time.

It is seen that the width of spectrum corresponding to the optical recombination in quantum dots narrows from 80 to 33 meV after 7.5 s exposition, the effect associated with narrowing the size distribution. This is qualitatively explained by our theoretical results (see 3.131 and Fig. 3.29) showing the decrease of the relative distribution width with time. At a fixed aspect ratio less than one, the minimum size (that equals the height) responsible for the size quantization is directly related to a larger lateral dimension. The peak position λ is shifted towards a longer wavelength as Δt_{exp} increases from 0 to 15 s, as it should be when the islands get larger. A slight decrease of the wavelength at 22.5 s exposition might be explained by the effect of misfit dislocations which emerge in larger islands and excludes them from optical recombination. Such an explanation is supported by the measured decrease in the PL intensity from the sample exposed to the As flux for 22.5 s.

In Sect. 3.10, we discussed the possibility of obtaining the so-called subcritical quantum dots in the deposition range between the equilibrium (~1–1.1 ML) and critical (~1.7–1.8 ML) thickness upon a certain exposition under the As flux.

To investigate that, two series of InAs structures were grown on planar GaAs(100) substrates, where the effective thickness of the InAs layer was fixed to 1.5 and 1.6 ML, respectively. The growth temperature was varied from 420 to 485 °C within both series. The InAs growth rate was fixed to 0.03 ML/s for all samples. After the deposition of the InAs layer, the surface was exposed under the As_4 flux with the exposition time varied from 1.5 to 2.5 min for different samples. The active region was confined by two short-period superlattices of $GaAs/Al_{0.25}Ga_{0.75}As$ (25/25Å, 5 pairs), grown at 585 °C. The surface structure was controlled *in-situ* by a RHEED system composed of a high sensitivity video camera, a videotape recorder, and a computer. It has been found that all RHEED patterns remain streaky during the deposition of InAs. The exposition of samples under the As_4 flux after switching off the In source leads to the transition of RHEED patterns from streaky to spotty for all the samples except for the 1.5 ML sample grown at 485 °C (for this sample, the formation of islands was not observed even after 90 s exposition). The moment of this transition after the waiting time is interpreted as the beginning of island nucleation. Further exposition of samples leads to the appearance of additional RHEED patterns corresponding to the diffraction from the island facet planes. The moment of their appearance can be associated with reaching the quasistationary size of islands. The optical and structural properties of subcritical quantum dot ensembles were subsequently studied by PL and TEM.

As regards the data on the characteristic waiting time before the nucleation starts (summarized in Table 3.5), it is seen that the island formation is observed (except for only one sample) both for 1.6 and 1.5 ML subcritical deposition thickness. The waiting times for 1.5 ML samples are noticeably longer than for 1.6 ML ones, which is clearly explained by a higher metastability of the wetting layer in the latter case. It is also seen that the temperature dependence of the waiting time shows a distinct tendency: at a fixed deposition thickness, the island nucleation starts earlier at a lower temperature. This effect should be associated with a larger number of nucleation centers that emerge when the surface diffusion is suppressed at a lower temperature.

TEM studies were performed using a Philips EM2420 electron microscope with an accelerating voltage of 100 kV. TEM contrast originating from coherent strained inclusions was observed in all the samples except for 2 a. PL studies confirmed an excellent optical quality of these islands which is typical for the conventional InAs/GaAs quantum dots. Plan-view TEM images of subcritical InAs quantum dots

Table 3.5 Growth parameters and waiting times before the nucleation of subcritical InAs quantum dots

Sample No	Deposition thickness (ML)	T(°C)	Waiting time (s)
1 a	1.6	485	13
1 b	1.6	450	4
1 c	1.6	420	2
2 a	1.5	485	>90
2 b	1.5	450	20
2 c	1.5	420	15

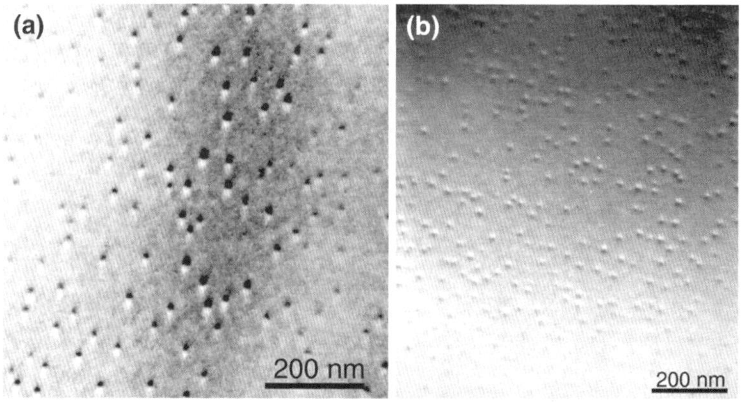

Fig. 3.42 Plan-view TEM images of subcritical InAs quantum dots grown by MBE at different conditions: $T = 485\,°C$, $H_0 = 1.6\,ML$ (**a**) and $T = 450\,°C$, $H_0 = 1.5\,ML$ (**b**)

at 1.6 and 1.5 ML deposition thickness are shown in Fig. 3.42. Statistical analysis of TEM images of samples 1 a–c and 2 b, 2 c revealed that all the islands had a density of the order of $10^{10}\,cm^{-2}$. This value is of the order but always lower than the density of supercritical islands obtained at exactly identical MBE conditions with the deposition thickness of 2 ML without any exposition under the As_4 flux. The density in 1.5 ML samples is always lower than in 1.6 ML ones. Experimental dependences of the surface density on the deposition thickness within the range from 1.5 to 2.0 ML at two different temperatures are presented in Fig. 3.43. It is seen that the curves at a given temperature are increasing, while the temperature behavior is more complex. The density of 2 ML ensemble is larger at 420 °C than at 450 °C for 1.6 and 2 ML samples, whereas at 1.5 ML thickness the situation is reversed and the density becomes higher at 450 °C. As discussed is Sect. 3.11, such a different temperature dependence occurs because the island morphology is controlled by the growth kinetics above 1.6 ML and by the surface thermodynamics for thinner layers, with the conversion somewhere close to 1.6 ML.

To further decrease the density of InAs quantum dots, several 1.3–1.5 ML structures were grown by MBE on the off-cut GaAs(100) substrates with different misorientation angles at the surface temperature $T = 485\,°C$ and deposition rate $V = 0.05\,ML/s$ [179, 219], with no exposition. The active regions were confined from both sides by GaAs/AlGaAs superlattices for optical measurements. Surface density of InAs islands was obtained by looking at their PL spectra. Below we present the data on the samples grown on 5° off-cut substrates. For the low density samples with 1.3 and 1.4 ML of InAs deposited, it was possible to estimate the density by simply shining the laser light over an extended area of the sample and imaging the photoluminescence on a CCD camera (Figs. 3.44a–c). The scale of the image was calibrated based on lithographically defined markers on the sample surface.

The sample with 1.5 ML of InAs (Fig. 3.44d) showed too high concentration of quantum dots that could not be determined by simple imaging but required some

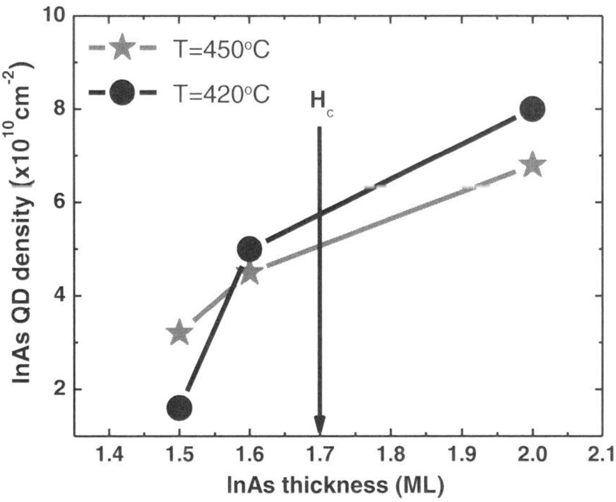

Fig. 3.43 Dependences of InAs island density on the deposition thickness at fixed $V = 0.3\,\text{ML/s}$ and two different growth temperatures

spectroscopic measurements. The density on this sample was therefore counted in the following way. A region of the sample was scanned in discrete steps and a spectrum was recorded at each point. In this way, the PL intensity maps were obtained at different wavelengths. A Matlab script was let to find the location of peaks on each map in order to have a list of different optical transitions of different dots. To determine the number of dots, it was then necessary to divide these transitions in groups that belong to the same dot. Different transitions were attributed to the same dot if they were located within a certain distance d and if their wavelengths differ by less than $\Delta\lambda$. The choice of d and $\Delta\lambda$ was made based on the spatial resolution of the system and the confinement energy of the dots. In this way, the density of dots emitting in the range between 880 and 940 nm was found to be $3 \times 10^8\,\text{cm}^{-2} \pm 30\,\%$. The optical method was then checked for validity by atomic force microscopy (AFM) of the uncovered samples and TEM measurements of the selected covered samples.

Experimental data shown in Fig. 3.45 reveal a very low island density of $1.2 \times 10^7\,\text{cm}^{-2} \pm 6\,\%$ and $2.5 \times 10^7\,\text{cm}^{-2} \pm 6\,\%$ at 1.3 and 1.4 ML deposition thickness, respectively, followed by $\sim$12 times increase at 1.5 ML at otherwise identical conditions. Since no exposition under arsenic flux was applied in this set of experiments, the islands started nucleating in the kinetic growth mode, where the double-exponential formula for the surface density given (3.100) is directly applicable. The data are therefore fitted by this expression, re-written in terms of the deposition thickness H as

$$n(H) = N\left[1 - \exp\left(-\exp\left(\frac{\Gamma(H - h_c)}{h_c - h_{eq}}\right)\right)\right]. \tag{3.145}$$

Here, h_c is the critical thicknesss, h_{eq} is the equilibrium thickness, Γ is twice the nucleation barrier at the critical thickness and N is the density of supercritical dots upon the completion of the nucleation stage. Using the typical values of $h_{eq} = 1\,\text{ML}$, $h_c = 1.7\,\text{ML}$, $N = 4 \times 10^{10}\,\text{cm}^{-2}$, the curve at $\Gamma = 16$ fits well the experimental

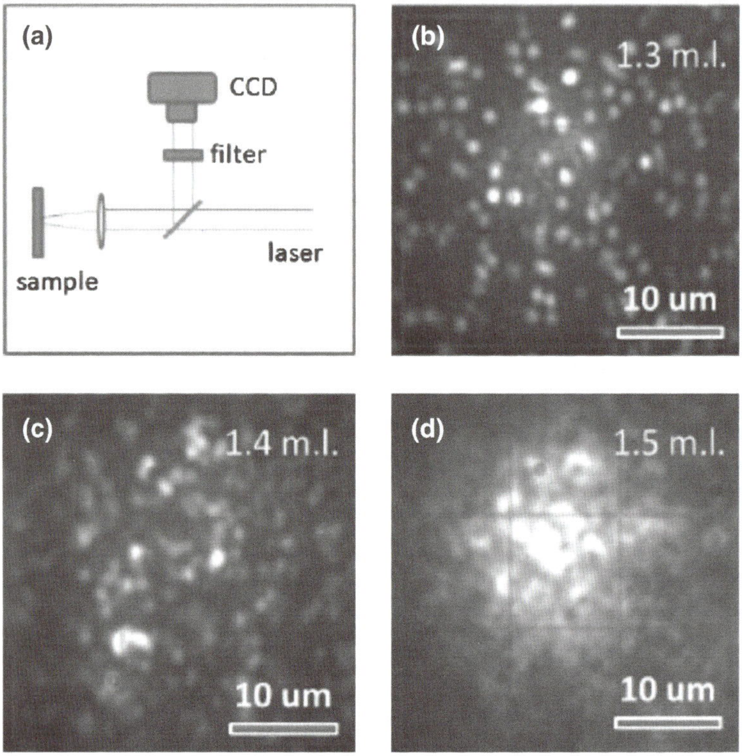

Fig. 3.44 Schematics of PL measurements (**a**) and PL images of 1.3 (**b**), 1.4 (**c**) and 1.5 (**d**) ML InAs quantum dots on 5° misoriented GaAs(100) substrates

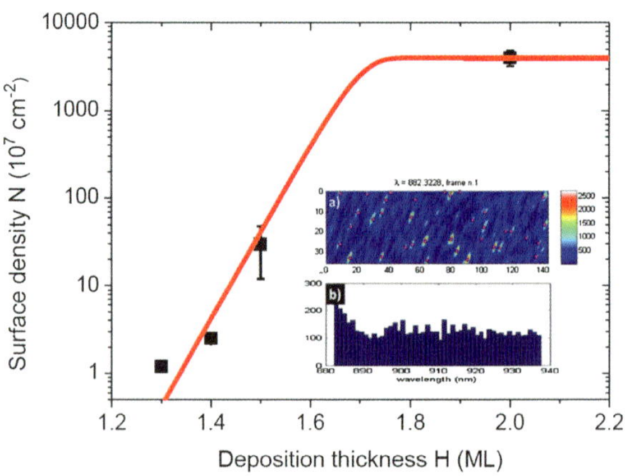

Fig. 3.45 **a** Experimental (*symbols*) and theoretical (*line*) dependences of the density of InAs quantum dots on 5° misoriented GaAs(100) substrates on the deposition thickness. The insert shows the PL map and quantum dot counts used for the measurements of 1.5 ML sample

data. Thus, the island formation well below the critical thickness is not exceptional and enables a substantial decrease of density which might be useful for particular applications.

3.13 Theory and Experiment: Ge/Si System

We now consider some experimental results on the growth kinetics and morphology of Ge islands obtained by MBE on singular Si(100) substrates, and their interpretation within theoretical models described above [144, 174, 220]. As in the case of InAs/GaAs system, Ge islands can be obtained below as well as above the critical thickness of 2D-3D growth transformation under Ge flux at a given temperature. In particular, Fig. 3.46 shows typical AFM images of the two samples grown using a Riber SIVA MBE machine. In sample A, 0.75 nm (or 5.2 ML) thick Ge layer was deposited onto a 100 nm Si buffer at $T = 600\,°C$. Sample B was different only by a lower Ge deposition thickness of 0.55 nm ((3.8 ML). We note that the critical thickness h_c in the Si/Ge system equals approximately 0.66 nm (4.6 ML) at 600 °C [175], so that the deposition thickness of Ge is supercritical in sample A and subcritical in sample B. The surface morphology was controlled *in situ* by RHEED. After growth, both samples were immediately cooled down to the room temperature and subsequently studied by AFM using a Digital Instruments Inc. setup.

As revealed by the characteristic transformation of RHEED pattern from streaky to spotty at approximately 4.6 ML of deposited Ge, the island formation in sample A proceeds via the conventional Stranski-Krastanow mode. The nucleation stage is completed soon after the critical thickness and is followed by the independent growth of islands. Typically for a high surface temperature applied [175], only dome islands are observed in the AFM image of Fig. 3.46a. Their mean lateral size amounts to 73 nm, while the mean height is systematically more than 10 nm. The surface density equals $5 \times 10^9\ cm^{-2}$. In sample B, the RHEED transformation from streaky to spotty was not observed. However, the AFM imaging revealed the existence of smaller

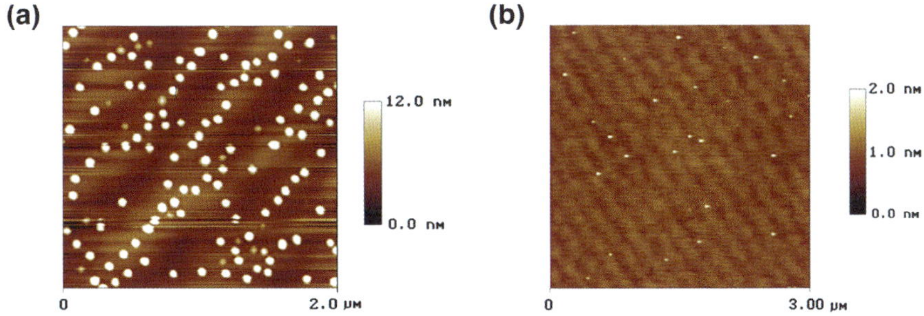

Fig. 3.46 AFM images of Ge islands in samples A (**a**, scan area $= 2 \times 2\,\mu m$) and B (**b**, scan area $= 3 \times 3\,\mu m$), with the colour contrast showing the island height

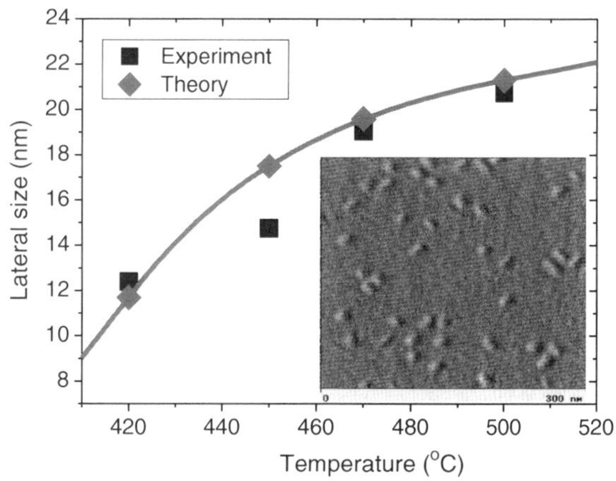

Fig. 3.47 Experimental and theoretical dependence of the mean lateral size of square base Ge islands on the surface temperature, the insert showing plan-view AFM image at 450 °C with the scan area of 300 × 300 nm

islands with the mean lateral size of 34 nm, height of 3 nm and a much lower density of $3.1 \times 10^8\ \mathrm{cm}^{-2}$. The island formation starts in sample B because its thickness is larger than equilibrium (∼3 ML). The 16 times decrease of density in the subcritical sample is well explained by the double-exponential dependence given by (3.145). We note that the lateral size in sample B does not reach its maximum value since exposition is not applied.

The temperature dependence of the morphology of supercritical Ge islands was investigated in the following experiment. Four samples of Ge islands on Si(100) surface were grown using a Riber SIVA MBE machine. The substrate temperature was set at 420, 450, 470 and 500 °C, respectively. The growth rate of Ge was fixed to 0.035 ML/s for all samples. As the moment of growth interruption for each sample, a deposition thickness of 6.2 ML was chosen. The corresponding deposition time amounted to 177 s. Afterwards the samples were studied by the plan-view and cross-sectional AFM imaging. A typical AFM image of Ge islands at a growth temperature of 450 °C is presented in the insert to Fig. 3.47. It was found that the islands have a pyramidal shape with an approximately square base only at 420 °C. At higher temperatures, the ensembles contain both square and rectangular base islands. Rectangular islands are elongated in [100] direction, with the R_x/R_y ratio ranging from 1.74 to 2.64 depending on the temperature. For all the samples, the fraction of square base islands remains predominant. The aspect ratio of both types of islands increases with temperature. In particular, it ranges from 0.09 at 420 °C to 0.24 at 500 °C for square base islands. This shows that the island shape is not exactly a crystallographically defined full pyramid, because such a variation in the aspect ratio is strictly speaking a crystallographic nonsense. Experimental data on the island morphology at different growth temperatures are summarized in Table 3.6, whereas the "effective" square base $R = (R_x + R_y)/2$ is plotted against temperature in Fig. 3.47. It is seen that the mean size increases approximately from 12 to 21 nm as the surface temperature rises from 420 to 500 °C.

Table 3.6 Morphological characteristics of Ge islands

T(°C)	Density (10^{10}cm^{-2})	Height (nm)	Mean lateral size (nm)			
			Square base islands		Elongated islands	
			R_x(nm)	R_y (nm)	R_x (nm)	R_y (nm)
420	5.6	1.1	11.8	13.0	–	–
450	3.9	2.5	14.2	15.3	24.7	11.8
470	1.1	3.0	20.2	17.9	33.1	19.0
500	1.5	5.0	20.2	21.3	53.1	20.1

Table 3.7 Theoretical characteristics of 6.2 ML ensembles of Ge islands

(T°C)	β (deg)	N(10^{10}cm^{-2})	R (nm)	R_g (nm)
420	13	5.7	11.7	25
450	19	4.0	17.5	26
470	19	3.3	19.6	28
500	26	1.7	21.3	31

Experimental points in Fig. 3.47 are fitted by the corresponding theoretical expressions following the general scheme of Sect. 3.8, with the material parameters of the Ge/Si system given there. The characteristic diffusion temperature E_D/k_B in the Ge wetting layer was set at 7750 K in all cases. The value of the contact angle β was varied to adjust the experimental aspect ratios at different temperatures in the model pyramidal geometry. Theoretical characteristics of the island formation process and the morphology at $H = 6.2$ ML are presented in Table 3.7. The surface density is predicted to decrease gradually from 5.7×10^{10}cm^{-2} to 1.7×10^{10}cm^{-2} as the temperature increases from 420 °C to 500 °C, while the mean size is expected to increase from 12 to 21 nm. The size $R_*(t)$ frozen right after the deposition of 6.2 ML of Ge is considerably smaller than the maximum size R_g, because the growth termination followed by an immediate cooling does not allow a complete size relaxation. It is seen that theoretical and experimental values for the mean size and density are in a fairly good agreement with each other, except for the discrepancy at 470 °C for the density and at 450 °C for the size. These could be associated with geometrical effects (a fraction of elongated islands is not taken into account either in measurements or calculations), or a spatial inhomogeneity of the samples.

As already mentioned, misfit dislocations nucleating at the base of dome Ge clusters impede material quality of Si/Ge superlattices, therefore the dome phase is usually unwanted. One of the possible methods to suppress the formation of domes is growth under an antimony flux [221]. Due to its surfactant properties, Sb is known to suppress the surface diffusion of Ge, leading to a decreased growth rate and a much narrower size distribution. Let us now consider some experimental data on the influence of Sb flux on the morphology of Ge islands and their qualitative interpretation within the frame of kinetic theory [220]. Experimental procedure is the following. The samples were grown on singular Si(100) substrates by MBE using

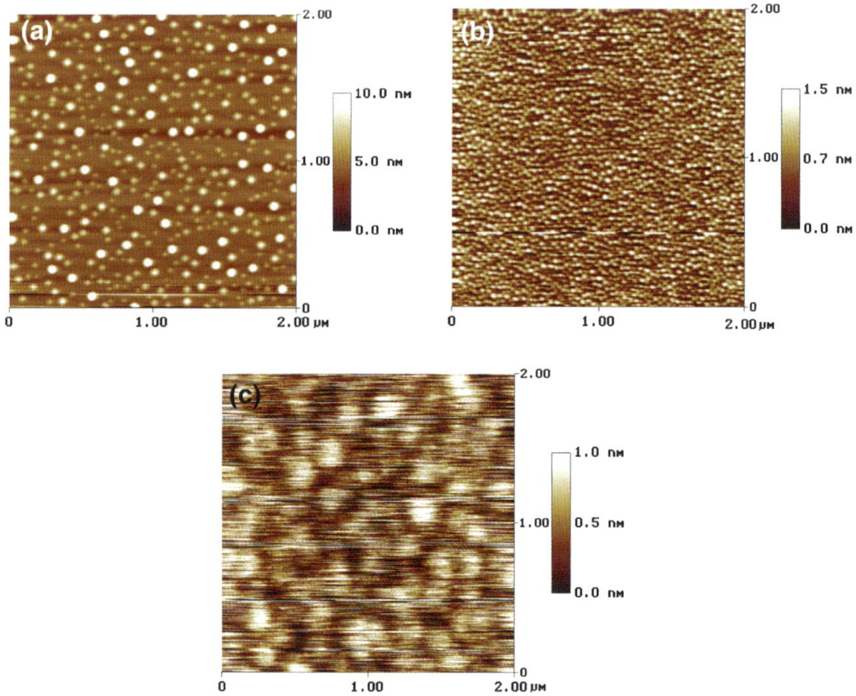

Fig. 3.48 AFM images of samples 1 (**a**), 4 (**b**) and 5 (**c**) described in Table 3.8, scan area $= 2\times2\,\mu$m

a Riber SIVA machine. Before Ge deposition, all the substrates were covered by a 100 nm thick buffer layer of Si. After that, a Ge layer with the equivalent thickness of 0.8 nm (5.5 ML) was grown at a substrate temperature of 550 °C with a deposition rate of 0.016 nm/s. Deposition of Ge was performed under a Sb_4 flux, regulated by the temperature of a Sb effusion cell that was varied between 450 and 550 °C. The surface morphology was controlled *in situ* by RHEED. After Ge deposition, all the samples were immediately cooled down to room temperature and subsequently analyzed by AFM using a Digital Instruments Inc. setup.

For the first four samples grown with the Sb cell temperature of 450, 475, 500 and 525 °C, respectively, the transformation of RHEED pattern from streaky to spotty was in place, revealing a transition from a strained wetting layer to 3D islands. For sample 1 5 grown at the maximum Sb cell temperature of 550 °C, the RHEED pattern stayed predominantly streaky with some widening of the primary reflexes. Typical AFM images of samples obtained with different Sb cell temperatures are shown in Fig. 3.48. The results of their statistical analysis are summarized in Table 3.8. Figure 3.48a demonstrates a bimodal size distribution over lateral size and height in sample No. 1. This reflects both hut and dome shapes of islands coexisting with each other, as it would be without any Sb deposition at this substrate temperature [175]. Therefore, the influence of Sb source is almost negligible at 450 °C. Sample No. 1 is

Table 3.8 Description of samples and results of AFM analysis of hut Ge islands

No	T Sb cell (°C)	Results of AFM characterization					
		Dome-islands			Hut-islands		
		Lateral size (nm)	Height (nm)	Density (10^{10}cm^{-2})	Lateral size (nm)	Height (nm)	Density (10^{10} cm^{-2})
1	450	64.0	9.3	0.24	39.3	3.38	1.1
2	475				38.5	2.90	1.8
3	500				35.1	2.99	3.4
4	525				29.6	1.15	6.7
5	550	–	–	–	–	–	–

the only one where the dome phase is present. As seen from Fig. 3.48b which shows the AFM image of sample No. 4 grown at the Sb cell temperature of 525 °C, the island size distribution is much narrower and their characteristic size much smaller than in sample 1. However, further increase of Sb cell temperature up to 550 °C has a tremendous effect on the surface structure: Fig. 3.48c reveals no islands at all. Rather, the surface just becomes roughened with a typical height less than 1 nm, which is consistent with what was observed by *in situ* RHEED diagnostics.

Plotting out the data of Table 3.8 in Fig. 3.49a, it is seen that the increase of Sb cell temperature leads to a remarkable threshold behavior of surface morphology. At a low temperature hut clusters coexist with domes, then the dome phase disappears while the hut clusters become denser and smaller as the Sb cell temperature increases up to 525 °C. When the temperature is ramped up to the threshold value of 550 °C, hut clusters also disappear. Such a non-monotonous dependence has not been predicted so far and requires an additional theoretical examination.

Let us now consider simplified theoretical expressions for the dependences of island density and size on the kinetic control parameter Q following from (3.114) and (3.124) in the approximation $u(Q) \cong \ln Q$ and under the assumption of a fixed island shape:

$$N = N_0 \frac{T}{T_e} \left(\frac{\ln Q}{Q} \right)^{3/2}; R_g = R_0 \left(\frac{H_0 - h_{eq}}{N} \right)^{1/3}. \tag{3.146}$$

If the factors N_0 and R_0 are independent of the Sb flux, the main parameters influencing the density and size are the quasi-equilibrium temperature T_e and the kinetic control parameter Q. The latter depends on the system energetics and the surface diffusion barrier E_D according to (3.115) and (3.116). We now assume that the major effect of Sb flux is the suppression of surface diffusion, i.e., an increase of E_D at higher Sb flux (which of course increases at a higher Sb cell temperature). If the surface energies are less affected by the presence of Sb, the quasi-equilibrium temperature might be assumed as being constant in the first approximation. At a constant surface temperature and Ge deposition rate, we can therefore write

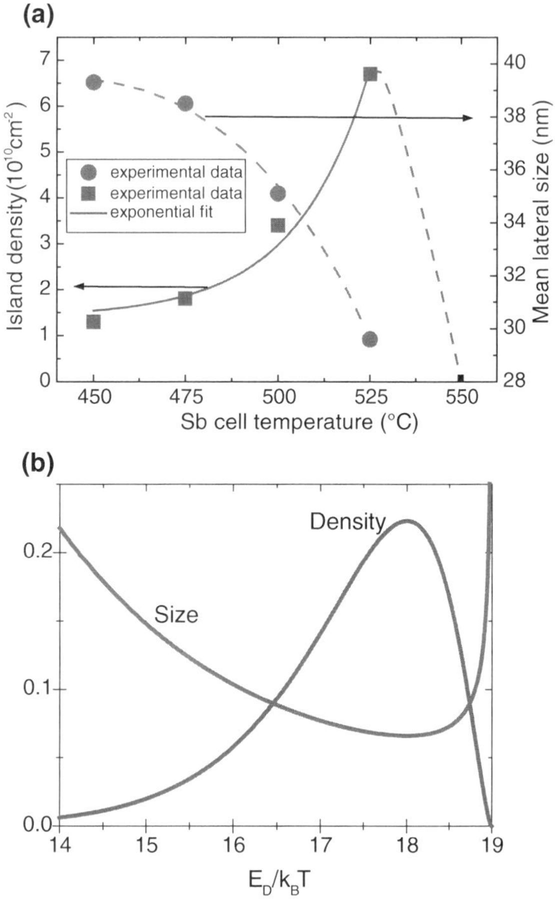

Fig. 3.49 Experimental dependences of mean size and density of Ge hut islands on the Sb cell temperature (**a**) and their qualitative theoretical behavior (**b**)

$$N \propto \left(\frac{\ln Q}{Q}\right)^{3/2}; R_g \propto \left(\frac{Q}{\ln Q}\right)^{1/2}; Q \propto \exp\left(-\frac{E_D}{k_B T}\right), \tag{3.147}$$

where only E_D increases with the Sb flux.

In our previous analyses, we always assumed that Q was very large so that the logarithmic dependence in these equations could be neglected. Decreasing Q would than result in the increase of density, for example, at a lower substrate temperature where diffusivity becomes smaller. However, without any thermodynamic restrictions imposed on the nucleation process, island formation cannot occur if the material is not supplied from the surface at a very low diffusivity. This effect could be seen from (3.147): considering N and R_g as functions of Q, the density reaches its maximum and the size minimum at $Q = e$. Therefore, decreasing the diffusivity should first lead to an increasing density but, after exceeding a certain threshold value, the island formation is completely suppressed on kinetic grounds. The non-monotonous behaviors of island density and size on $E_D/(k_B T)$, shown in Fig. 3.49b, qualitatively correlates with the experimental data. However, any quantitative analysis based on such expressions as (3.147) is hardly possible, because the major assumption of classical nucleation theory (large enough nucleation barrier) becomes violated.

3.14 III–V Nanoneedles and Nanopillars

Unconventional growth mechanism that yields the formation of highly anisotropic, vertically elongated needle-like structures on lattice mismatched substrates has been reported by Chang-Hasnain group for several material systems including different III-V compounds on silicon and GaAs on sapphire [185, 186, 222]. Similarly to the self-induced Stranski-Krastanow quantum dots, III-V nanoneedles and nanopillars enable monolithic integration of dissimilar semiconductors. Due to their large aspect ratio of the order of ten, these 3D structures show great promise to overcome large mismatch of lattice constants and thermal expansion coefficients between III-V compounds and silicon. This offers several otherwise unattainable properties (even by using the Stranski-Krastanow quantum dots) that are of paramount importance for the heterogeneous integration of optoelectronic and electronic integrated circuits. In particular, due to a very efficient stress relief on free sidewalls and in the bottom region, the bulk material of nanoneedles is pure and single crystalline. It is also possible to fabricate high quality core-shell heterostructures whose width far exceeds the thin film limit for dislocation formation. Concerning the applications, room-temperature operation of light-emitting diodes and avalanche photodiodes grown on silicon has been demonstrated [223, 224]. The nanostructures consist of (Al,In)GaAs core-shell heterostructures and can be monolithically grown on single-crystalline silicon and sapphire at low temperature (around 400 °C), via MOCVD on silicon substrates and processed by standard fabrication techniques. Furthermore, room-temperature operation of nanopillar-based lasers on silicon by optical pulsed pumping has been shown [225]. These exciting results motivate detailed studies of this unique growth mechanism that enables high-quality growth of III-V structures on silicon.

We now briefly discuss the most important experimental data on the nanoneedle and nanopillar growth mechanism and crystal structure. Let us first consider the GaAs nanoneedles [185]. They were grown without catalyst at ~400 °C on both GaAs and Si substrates, using a low-pressure MOCVD reactor. The substrates were deoxidized and mechanically treated to initiate surface roughness to catalyze 3D GaAs growth. The group III and V sources were triethylgallium and tertiarybutylarsine, which have relatively low decomposition temperatures (300 and 380 °C, respectively). Figure 3.50a shows nanoneedles grown on a GaAs (100) substrate. The nanoneedles are found to grow along the degenerate $< 111 >$B orientations. Nanoneedles enveloping each other during growth are observed, showing that the growth proceeds in a core-shell mode. With increasing or decreasing the growth time, the nanoneedle length is correspondingly increased or decreased without changing its shape or nanoneedle tip dimension. Typical SEM images of nanoneedles grown

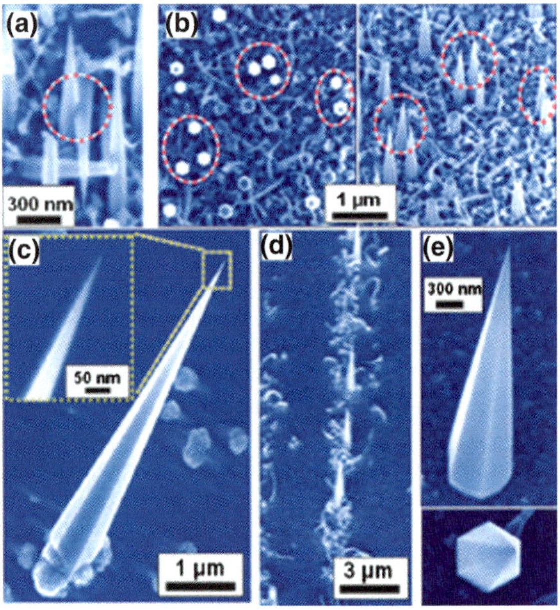

Fig. 3.50 SEM images of GaAs nanoneedles. **a** Two needles which enveloped each other during growth; **b** Nanoneedles grown on a GaAs(111)B substrate, viewed top down (*left*) and tilted by 30° (*right*) indicating the uniformity and alignment of the [0001] nanoneedle growth axis to the $<111>$B substrate directions. **c** Nanoneedle viewed near to its side, with the extremely sharp tip shown in the inset. **d** A linear array of nanoneedles made by mechanically roughening only a very thin line. **e** GaAs nanoneedle grown on 4° off-cut Si (111) substrate, with views of 30° tilted and top down [185]

on a GaAs (111)B substrate (viewed both normal and 30° tilted to the substrate) are shown in Fig. 3.50b. The white hexagonal shapes in the first image indicate well-aligned vertical, sharp nanoneedles with a length of 2–3 μm, which takes ~40 min of growth. The nanoneedle sidewalls align to the $< 211 >$ zinc blende substrate directions. Figure 3.50c shows a zoomed-in SEM image of a typical nanoneedle tip viewed nearly perpendicular to the growth axis. A linear array of nanoneedles is also attained on GaAs as shown in Fig. 3.50d.

The GaAs nanoneedles can be also grown on roughened Si substrates using the same growth conditions and exhibit the same characteristics, despite the 4 % lattice mismatch between GaAs and Si. Figure 3.50e shows 30° tilted and top-down views of a 4 μm long nanoneedle grown on a 4° off-cut Si(111) substrate. The typical nanoneedle density is $\sim 10^7\,\text{cm}^{-2}$ on GaAs substrates and $\sim 5 \times 10^5\,\text{cm}^{-2}$ on Si substrates in the roughened areas.

Further, in [186], it was shown that the GaAs nanoneedles can be grown directly on sapphire substrates without any preliminary surface roughening. The growth details are as follows. The sapphire substrate was first cleaned with acetone, methanol, and water, for 3 min for each step. The growth was carried out in an EMCORE D75 MOCVD reactor. The growth started with an *in-situ* pre-growth annealing process for 3 min. Annealing temperature of 600 °C was used. After the annealing, the temperature was brought down to the growth temperature, between 385 and 415 °C in 3 min and then stayed at this temperature for 2 more minutes for temperature stabilizing. TEGa was then introduced to the reactor to begin the nanoneedle growth. The

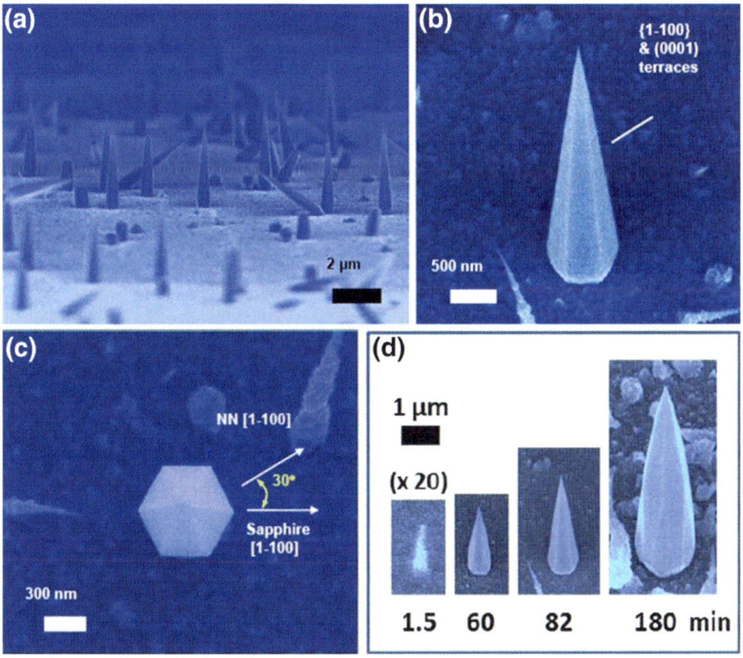

Fig. 3.51 **a** Side-view SEM image of as-grown nanoneedles on sapphire. **b** 30°-tilt SEM image of a nanoneedle. The sharp tip, submicron base, and smooth sidewall facets are seen. The sidewalls were made of $\{1\bar{1}00\}$ and (0001) terraces. **c** Top-down view of a nanoneedle showing a hexagonal cross section. The in-plane orientation of the nanoneedle shows a 30° rotation with respect to the sapphire substrate. **d** 30°-tilt SEM images of nanoneedless with different growth times. The sharp tip and almost identical shape are seen for all the needles shown here [186]

TEGa mole fraction was kept constant at 1.12×10^{-5} in a 12 l/min hydrogen carrier gas flow. The TBA mole fraction was 5.42×10^{-4} hence the V/III ratio was 48.

Some results of SEM characterization are presented in Fig. 3.51. Figure 3.51a shows the sideview of the spontaneously grown nanoneedles. The epitaxial growth feature of these needles is manifested by their alignment with the (0001) crystal orientation. Figure 3.51b shows a 30°-tilt image of a GaAs nanoneedle with 82-minute growth time. The sharp tip, submicron-wide base and smooth sidewall facets are observed. The nanoneedle taper angle is typically 9–11° for the growth temperature of 400 °C. Figure 3.51c shows a top-down SEM image of a nanoneedle. The hexagonal cross section corresponding to the 6 sidewall facets is clearly seen. The needle in-plane orientation is rotated by 30° with respect to [1–100] axis of the sapphire substrate. The nanoneedle facet orientation is assigned via the TEM analysis, which will be shown later. The 30° in-plane rotation is attributed to the initial bonding of As atoms with Al on sapphire, causing Ga atoms to align with Al thereafter. The lattice mismatch corresponding to this alignment is determined to be 46 % with a compressive strain. Nanoneedles with different growth times are shown in Fig. 3.51d.

Fig. 3.52 An HRTEM image showing that the nanoneedle sidewalls are composed of atomic steps which result in tapering. The tapered sidewall is determined consisting of $(1\bar{1}00)$ and (0001) terraces [186]

ranging from 1.5 to 180 min. It is seen that the nanoneedle shape and aspect ratio are maintained almost constant during growth.

Next, Fig. 3.52 shows a HRTEM image of a nanoneedle sidewall. Single atomic steps are clearly seen on the side of the nanoneedle which results in the taper. The taper sidewall facets are determined as consisting of $(1\bar{1}00)$ and (0001) terraces, as labeled in Fig. 3.52. This justifies the model geometry of a nanoneedle presented in Fig. 3.17d and used in calculations of Sect. 3.4. Such a stepwise tapered geometry also explains why nanoneedle aspect ratio can vary with the growth conditions.

Thus, the GaAs/sapphire system in this particular surface alignment relates to an extremely large lattice mismatch of 46 %. The GaAs nanoneedles are formed spontaneously on sapphire and have identical physical properties to those on a silicon substrate. It was hypothesized that the GaAs nanoneedles on sapphire were initiated as nanoclusters whose nucleation was driven by a large lattice mismatch between GaAs and the substrate [186]. Indeed, the model described in Sect. 3.4 attributes the observed nanoneedle geometry to the increase in the preferred aspect ratio with the lattice mismatch. Results shown in Fig. 3.22 and Table 3.1 give the correct value for the experimentally observed aspect ratio of the GaAs needles on sapphire, while the saddle shape of the formation energy qualitatively explains why the aspect ratio acquired at the nucleation stage is maintained throughout the follow-up growth steps. However, it is clear that, as the nanoneedle extends laterally, the misfit dislocations must develop at their base (for both silicon and sapphire substrates but at different base dimension.) Below in this section we discuss why these dislocations tend to stay in the bottom region and do not propagate upward as growth proceeds.

Before discussing the elastic stress relaxation and misfit dislocations, we point out a surprising result than holds for all GaAs nanoneedles growing perpendicular to the (111) plane (in cubic notation) regardless of the substrate used: the bulk material of a nanoneedle is single crystalline wurtzite (WZ)! (That is why we have used the WZ notations for crystallographic directions in Figs. 3.50, 3.51, and 3.52). It is well known that all III-V materials, except for nitrides, have stable zincblende (ZB) crystal phase

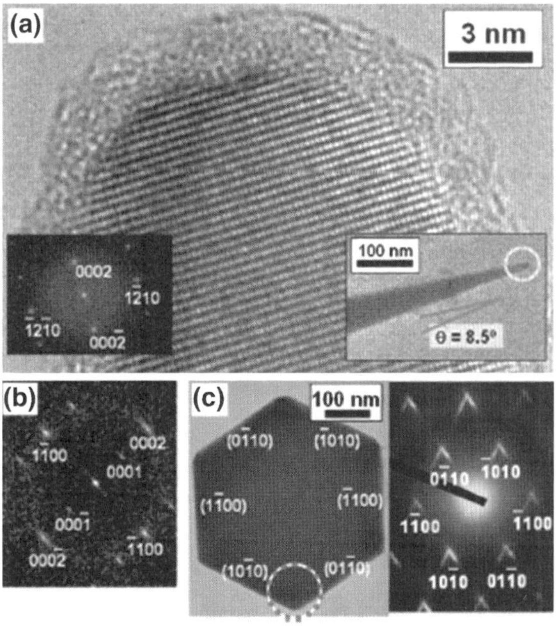

Fig. 3.53 HRTEM images of GaAs nanoneedles. **a** [1–100] zone axis HRTEM image of an as-grown GaAs nanoneedle. The insets show the zoomed-out view, and also the image FFT. **b** FFT from another nanoneedle on its [1–210] zone axis with a distinct WZ pattern. The (1–100) spacing is 3.45 Å. **c** Top-down [0001] TEM image of a nanoneedle. The image to the right shows a SAED pattern from the circled area, with distinct wurtzite {1–100} spots matching the expected unique wurtzite 3.45 Å spacing. The chevron spot shape is due to electron scattering from the two sidewalls contained in the circled area [185]

under bulk form [226]. Taking the example of GaAs, calculations give the difference in bulk cohesive ehergies of 24 meV per III-V pair at zero ambient pressure in favor of ZB phase [226]. This value is indirectly supported by experimental results of [227]. Cubic zinc blende phase becomes unstable only at a huge pressure higher than 14 GPa, while stable hexagonal WZ phase is observed at a pressure of the order of 100 GPa [227]. Under normal conditions, WZ phase never forms either in III-V thin films or the Stranski-Krastanow islands. However, GaAs nanoneedles [185, 186] and different III-V nanowires [45, 184, 228–234] grown on (111) oriented substrates systematically adopt hexagonal WZ phase. This remarkable phenomenon has been observed for most III-V compounds and epitaxial techniques, both for Au-assisted, selective area and self-induced growths, on III-V (111)B, Si(111) and sapphire (for nanoneedles) substrates. Leaving the detailed discussion of WZ-ZB crystallographic polytypism for Chap. 6 (specially dedicated to the crystal structure), we just point out that the WZ phase formation is physically explained by a lower surface energy of relevant sidewall facets and some peculiarities of growth. The latter are different for self-induced nanoneedles (extending radially during growth) and metal-catalyzed cylindrical nanowires (whose radius stays constant in most cases).

Experimental data on the crystal structure of GaAs nanoneedles are presented in Fig. 3.53. Figure 3.53a shows a high resolution TEM (HRTEM) image on the [1–100] zone axis of a nanoneedle, along with its corresponding fast Fourier transform (FFT). The tip in the image comes to an atomically sharp point of just 2–4 nm wide, which is one of the sharpest self-assembled semiconductor tips reported. The material remains single crystalline WZ all the way up until the tip. There is a surrounding 2 nm oxide layer which forms due to exposure to air. Figure 3.53b presents a FFT from another nanoneedle on the [1–210] zone axis, showing the distinct WZ pattern, free of any ZB segments. The *c* and *a* axes for these nanoneedles were determined to be 6.52 and 3.98 Å, respectively, within ±0.5 %. The *c/a* ratio is 1.638, which is close to the ideal hexagonal *c/a* ratio of 1.633 and in close agreement with the x-ray diffraction analysis of WZ GaAs in powder form created through high-pressure treatments [227]. Figure 3.53c shows a TEM image of a nanoneedle oriented along its [0001] growth axis with crystallographic directions labeled. A selected area electron diffraction (SAED) pattern was recorded in the area indicated by the circle. The SAED chevron shape is due to scattering from the two nanoneedle side facets. The interplanar spacings uniquely match those of WZ GaAs. This 3.45 Å {1–100} pattern spacing is distinct from the similar looking 2.00 Å {2–20} spacing in ZB GaAs(111).

Next, we discuss very important data of [222] on the relaxation mechanisms of strain induced by the lattice mismatch of two different origins: (i) at the InGaAs/Si interface of $In_{0.2}Ga_{0.8}As$ nanopillars grown on Si(111) substrates and (ii) at the $In_{0.2}Ga_{0.8}As$/GaAs interface of the core-shell nanoneedle heterostructure. Details of growth technique can be found in [222]. The evolution of the $In_{0.2}Ga_{0.8}As$/GaAs core-shell nanopillars grown on silicon is illustrated by the SEM images displayed in Fig. 3.54a, which show that the nanopillar length scales with growth time. Initially, the nanostructure grows into a hexagonal pyramid with an extremely sharp tip: the facet-to-facet taper angle is as small as 5°. Growth then continues in a core-shell manner, with the sharpness well preserved. Vertical growth stops beyond a certain point (in this case after about 25 min), while radial growth continues, transforming the originally sharp needle into a hexagonal frustum. This growth mechanism is schematically illustrated in Fig. 3.54b, and is similar to that described above for GaAs nanoneedles, except for the abrupt stop to vertical growth. In Fig. 3.54b, an inverse-cone shape root is shown, as well as a polycrystalline "wetting" layer surrounding the pillars. The root shape will be discussed in more detail later on.

A statistical study was performed in which more than 50 nanostructures were measured for each of the six selected growth durations. Figure 3.54c, d show the average base diameter and height as a function of growth time, respectively. The base diameter increases linearly with time. In particular, the base diameter can scale up to 1.5 μm while maintaining excellent crystal quality, a distinct difference from a dimension-limited nanowire growth [126, 184, 234, 235] that will be discussed in Chap. 5. The average length also shows a linear dependence on time at the early stage of growth when the nanostructure is still sharp. The length saturates when the growth time reaches ~40 min, transforming the needles into pillars, as seen in Fig. 3.54d.

Extensive HRTEM studies were carried out to investigate the nanopillar crystal structure including the bottom region and the buried root. It was chosen to expose $(2\bar{1}\bar{1}0)$ so that the distinction between WZ and ZB phases can be clearly seen. Figure 3.55a, b show a schematic diagram and a typical cross-sectional TEM image of an $In_{0.2}Ga_{0.8}As$/GaAs nanopillar grown on Si(111), respectively. The nanopillar can be clearly seen to directly grow on silicon with a 'footprint' much smaller than the base diameter of the nanopillar, which is 720 nm in this particular lamella. In more

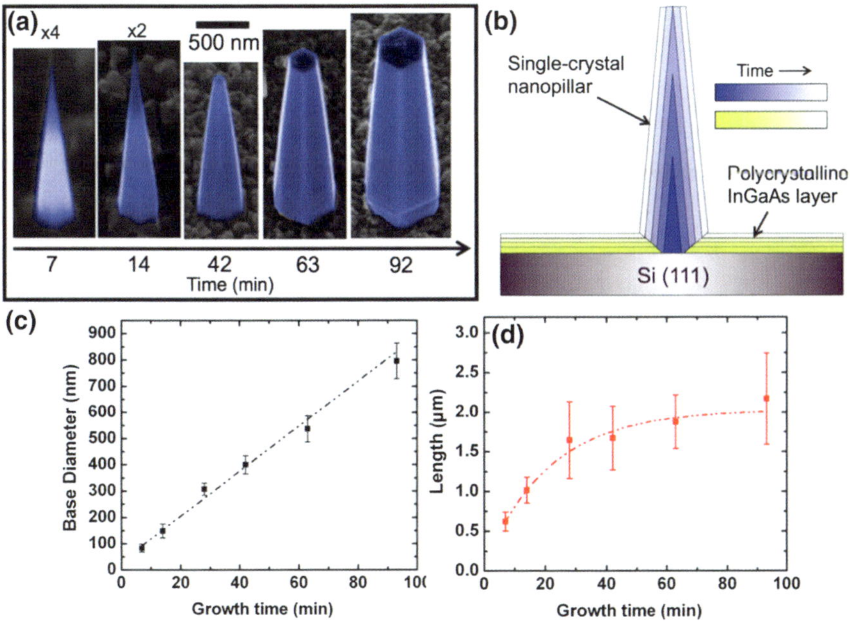

Fig. 3.54 **a** Time evolution of the nanostructure from a sharp nanoneedle to a blunt nanopillar **b** Schematic illustration of the core-shell growth mode of the nanopillar structure. **c** The nanopillar base diameter increases linearly with time. **d** The nanopillar length scales linearly with time initially, and then saturates at around 25 min of growth

than ten nanopillars being studied, the footprint diameter is found to range from 70 to 130 nm. The pillar has an initial section that tapers upwards with an increasing diameter at about 45°, as shown in Fig. 3.55c. The tapered region has a thickness approximately equal to that of the "wetting" layer, about 280 nm, and consists of polytypes and defects, as seen in Fig. 3.55c. A range of angles from about 45–60° were observed for various TEM samples. The formation of such reverse cone is due to the formation of a wetting layer which "wraps" around the bottom of the nanopillar during growth and masks the subsequent growth, as illustrated in Fig. 3.54b. The presence of the reverse cone is attributed as the key for stress relaxation in the core-shell growth mode, as will be discussed in more detail shortly.

The surface layer surrounding the pillar root is examined with high resolution scanning transmission electron microscopy (HRSTEM), as shown in Fig. 3.55d. This material actually covers the entire substrate and forms a rough, continuous layer. The STEM image reveals that this material is polycrystalline with many domains, showing short-term ZB crystallinity with random orientations. The origin of this film can be attributed to the coalescence of ZB phase islands that nucleate randomly during the initial growth stage at low temperature. In contrast to the defective surface layer, the bulk material of the nanopillar is pure and single crystalline. Above the tapered transition region, no noticeable defects or polytypic regimes can be observed, as

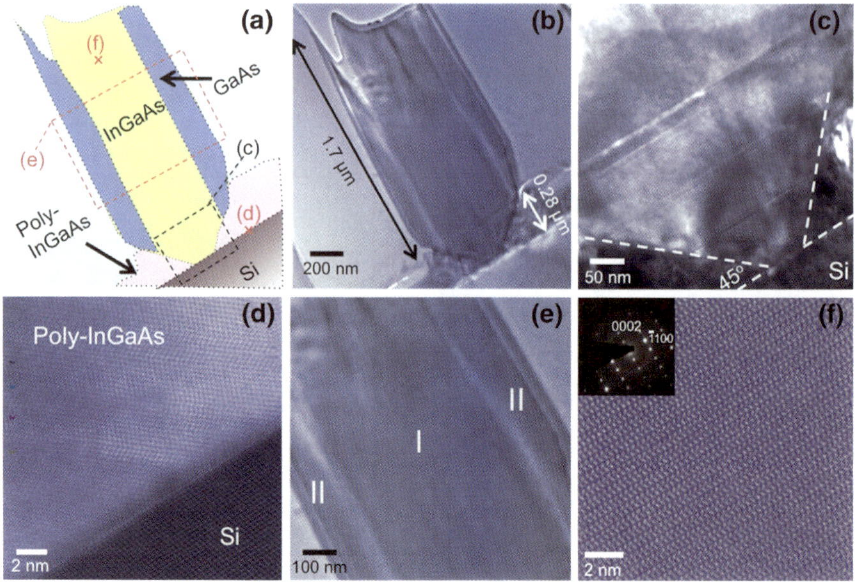

Fig. 3.55 **a** Schematic diagram showing an InGaAs/GaAs core-shell nanopillar grown on Si. Areas examined in **c**–**f** are labeled. **b** Cross-sectional TEM image of an InGaAs/GaAs core-shell nanopillar with base diameter ~740 nm. Horizontally-terminated stacking faults are well confined to the bottommost 280 nm of the tapered region. The bulk material above the tapered region consists of high quality single-crystal WZ. **c** Magnified view of the tapered root. Stacking faults arise in order to relax misfit stress. **d** An HRSTEM image of polycrystalline InGaAs. Short-term crystallinity can clearly be seen. **e** Magnified view of the bulk structure. Region *I*: InGaAs core; Region *II*: GaAs shell. Neither stacking faults nor dislocations are observed. **f** An HRTEM image of bulk $In_{0.2}Ga_{0.2}As$

seen in Fig. 3.55b, e. This indicates that most of the stress is relieved in the bottom transition region and that the crystal above is essentially stress-free. In this particular TEM sample, the single-crystal bulk material extends 1.7 µm above the root and would continue all the way up to the tip of the originally 5-µm-long nanopillar, except for the fact that the tip was milled away inadvertently during sample preparation. Figure 3.55f shows an HRTEM image of $In_{0.2}Ga_{0.2}As$ in the bulk along $[1\bar{2}10]$. The lattice displays a characteristic zig-zag configuration, attesting to the WZ nature of the crystal. Excellent crystal quality is confirmed by the very clear diffraction pattern taken along $[1\bar{2}10]$ in the inset of Fig. 3.55f.

Unlike the single crystalline bulk material, imperfections are present in the bottommost transition region of the nanopillar, as seen in Fig. 3.55c. Inverse tapering and horizontally terminated stacking disorders and defects in this region should be the key to this high-quality mismatched growth. Figure 3.56a–c show the roots of nanopillars with different base diameters. We note that reverse cone taper is only observed in Fig. 3.56c in which the nanopillar base is 740 nm in diameter. As mentioned before, the footprint of InGaAs on silicon is typically 70 ~ 130 nm, which is

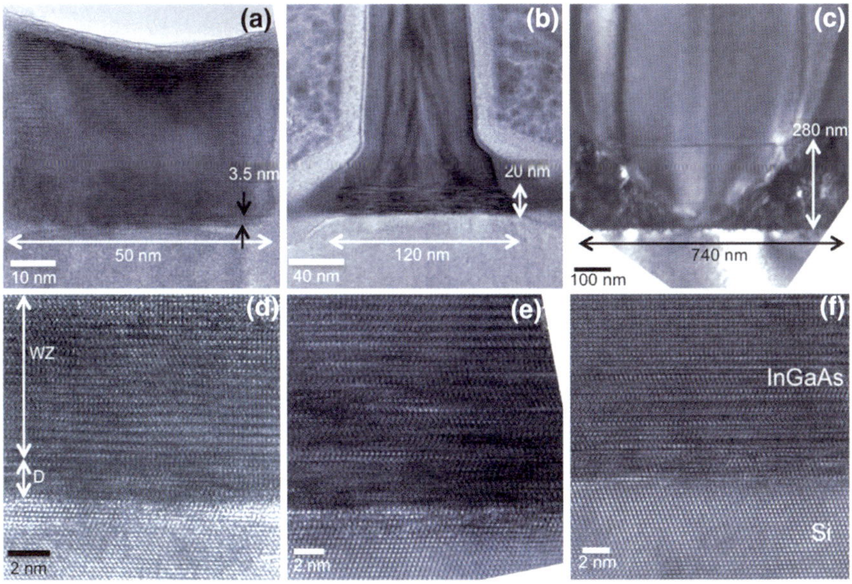

Fig. 3.56 TEM images of InGaAs/GaAs nanopillars with base diameters of (**a**) 50 nm (**b**) 120 nm and (**c**) 740 nm. The nanostructures are free of defects except the bottommost defective region. As base diameter increases, the thickness of defective region increases from 3.4 to 280 nm, relaxing the 6 % misfit stress between InGaAs and silicon. **d**–**f** show the exact InGaAs/Si interfaces of nanopillars in (**a**)–(**c**), respectively. InGaAs stems directly on Si without any amorphous material in between. In the smallest nanopillar, the defective region (labeled 'D') is only 3.5 nm in thickness, as seen in (**d**)

likely to be the average distance between islands during initial nucleation. When the nanopillar size is less than 130 nm, the surrounding polycrystalline grains are yet to get close enough and mask the nanopillar growth. Therefore, the reverse cone is not observed in Fig. 3.56a, b. Nevertheless, the same phenomenon is observed in all the three cases: all stacking faults and defects are well confined within the bottommost region. In particular, when the base diameter is 50 nm, stacking defects only extend 3.5 nm, or ~12 monolayers, above silicon, as seen in Fig. 3.56a, d. As nanopillar base diameter increases, the disordered region gets thicker to accommodate the extra misfit stress, as seen in Fig. 3.56b, c. We note that all defects propagate laterally rather than vertically along [0001]. In other WZ crystals like GaN, threading dislocations propagating along the growth direction are usually observed. Epitaxial lateral overgrowth is then developed to promote lateral growth so as to bend the dislocations from propagating upward into the active region [236]. The growth mechanism described here, on the other hand, is a pure core-shell growth mode in which growth occurs only in the lateral direction, except the very tip of the structure. Misfit defects therefore propagate horizontally and terminate at the sidewalls. Hence, the crystal structure remote from the substrate stabilizes into a single pure WZ phase, which is energetically preferred due to a lower number of dangling bonds on the WZ sidewalls,

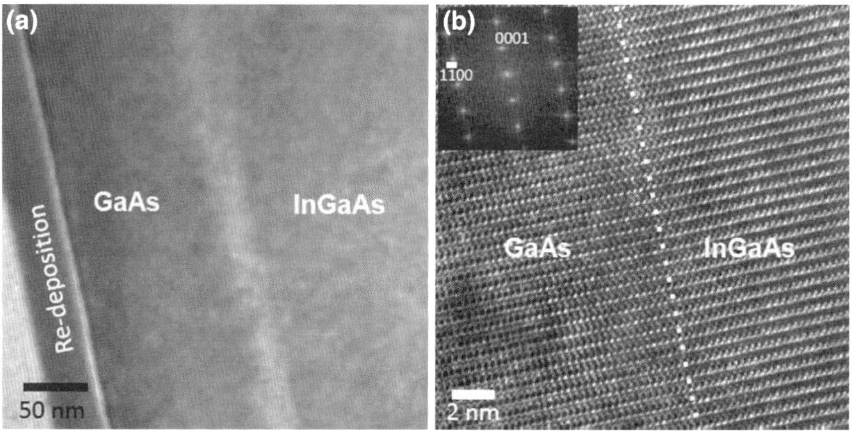

Fig. 3.57 **a** TEM image of the InGaAs/GaAs interface showing an effective absence of crystal defects. To the left of GaAs is a re-deposition layer induced during focus ion beam milling. **b** HRTEM image of the InGaAs/GaAs interface. GaAs grows seamlessly on top of InGaAs. Dotted line is an eye guide for the InGaAs/GaAs interface. The inset shows the corresponding fast Fourier transform. Clear pattern reveals an excellent quality of the material

as will be explained in Chap. 6. In addition to stacking disorders, reverse tapering is crucial in stress relaxation. The inverse-cone taper serves to limit the footprint area while the base expands, somewhat similar to the critical diameter observed in nanowires below which misfit strain can be relaxed elastically [126, 183, 234, 235] (see Chap. 5 for the details). Although the footprint (>70 nm) has already exceeded the theoretical critical value, small contact area and outward tapering still facilitate elastic strain relaxation. As seen in Fig. 3.56c, stacking disorders only appear infrequently in the tapered transition region, suggesting that the special taper geometry is the dominant mechanism in stress relaxation.

To realize electrical devices, it is essential to study how the III-V material is connected to the substrate nanoscopically. Figure 3.56d–f show the HRTEM images of the exact InGaAs/Si interfaces of the nanopillars shown in Fig. 3.56a–c, respectively. We note that InGaAs always grows directly on silicon without any amorphous material in between. This guarantees direct electrical conduction from III-V to silicon.

Understanding the mismatched growth of the GaAs shell on the $In_{0.2}Ga_{0.2}As$ core is of critical importance for heterojunction device engineering. With a 2 % lattice mismatch in WZ structure (instead of 1.4 % in ZB), the critical thickness of 2D $In_{0.2}Ga_{0.8}As$ layer on GaAs(100) has been reported to be less than 10 nm [237]. In core-shell nanopillars, however, 160 nm GaAs can be grown on $In_{0.2}Ga_{0.8}As$ with no misfit dislocations or stacking faults nucleating at the interface, as shown in Fig. 3.57a. HRTEM image in Fig. 3.57b, taken along [11$\bar{2}$0] zone axis, shows that GaAs grows seamlessly on InGaAs, maintaining the characteristic zig-zag wurtzite lattice arrangement. In over 20 TEM samples examined, no noticeable dislocations can be observed at the interface along the entire pillar. This limit-breaking coherent

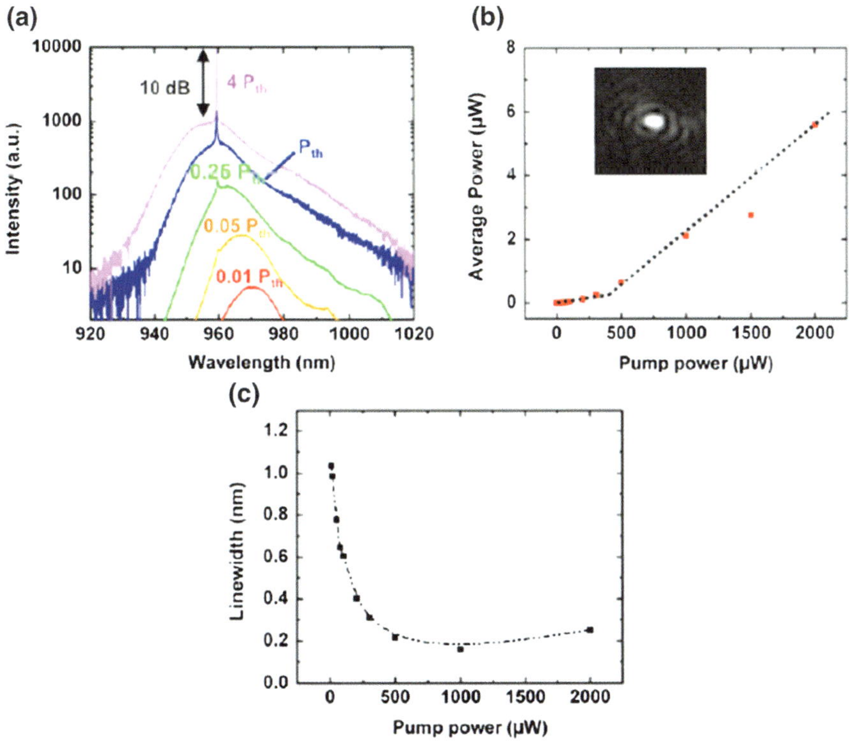

Fig. 3.58 **a** CW emission spectra at different pump powers. The cavity mode peak is observed when the pump power is as low as 0.25 times the threshold pump power (0.25 P_{th}). A sideband suppression ratio of 10 dB is observed at 4 P_{th}. **b** Emission power plotted as a function of pump power. Threshold behavior is observed and the maximum output power is close to 6 μW. The inset shows clear speckle patterns in the near-field. **c** Linewidth narrowing is observed with increasing pump power. The slight increase in linewidth at 2000 μW is due to heating

growth comes as a consequence of the core-shell growth mode; the surface area of the shell layer increases almost linearly with its thickness. A large surface area facilitates the elastic relaxation of stress induced at the $In_{0.2}Ga_{0.8}As/GaAs$ interface; detailed theoretical modeling is given in Chap. 5. This unique stress relaxation mechanism enables the growth of mismatched layers to thicknesses far beyond the conventional thin film limit and could lead to device structures with potentially unprecedented functionalities.

The excellent crystal quality described above gives rise to remarkable optical properties of heterostructured nanoneedles. Continuous wave (CW) operation is achieved in nanopillar lasers grown on silicon under optical pumping at 4 K [222]. A 785-nm diode laser was used as the CW pumping source. Figure 3.58a shows the emission spectra under various pump powers. At low pump levels, spontaneous emission is observed with a peak wavelength at 970 nm and a 3-dB bandwidth of approximately

10 nm. As the excitation power increases, we observe the emergence of a cavity mode at 960 nm, which finally evolves into laser oscillation. Clear threshold behavior in the pump power dependence of light output power and the near-field pattern are illustrated in the inset of Fig. 3.58b. Lasing action is also indicated by the prominent linewidth narrowing, as seen in Fig. 3.58c. At a pump power of 1000 μW, the laser linewidth is as narrow as 0.2 nm, which is comparable with the narrowest linewidths observed for existing nanocavity lasers [225, 238–240]. The laser assumes a helically-propagating mode and, hence, the output emitted from the top surface is relatively low. Nevertheless, more than 5.5 μW of CW optical power was collected, which is among the highest reported for nanolasers. This underscores the potential usefulness of integrating these nanopillar lasers onto silicon for various applications.

3.15 Growth Kinetics of GaAs Nanoneedles

We now present some results concerning theoretical modeling of the MOCVD growth of GaAs nanoneedles on silicon and sapphire substrates [200], while the analysis of stress relaxation and WZ crystal structure will be given in Chaps. 5 and 6. Growth model of a single GaAs nanoneedle is illustrated in Fig. 3.59. As suggested by the experimental results presented in the previous section, the nanoneedle is assumed to have a pyramidal shape with length L, base dimension D and a regular hexagonal cross-section. The aspect ratio $\eta = L/D$ and the corresponding taper angle θ are assumed as being constant at any time t. The growth under As-rich condition is assumed as being Ga-limited. The nanoneedle grows due to (i) direct impingement onto the sidewalls of surface area $S_W = (3/2\eta)\sqrt{1+3/(16\eta^2)}L^2$; and (ii) migration of Ga adatoms from the diffusion ring of width λ, from a diffusion area $S_D = (3\sqrt{3}/2)\left[(R+\lambda)^2 - R^2\right]$ around the needle base (see Fig. 3.59). The quantity λ is essentially the effective diffusion length of Ga on the surface. Re-evaporation processes are neglected, which is reasonable at a low surface temperature. The total change of nanoneedle volume, $\Omega_{tot} = (\sqrt{3}/8\eta^2)L^3$, per unit time is given by

$$d\Omega_{tot}/dt = \chi_W V S_W + \chi_S V S_D. \tag{3.148}$$

The first term represents the volume of GaAs pairs adsorbed by the sidewalls, the second stands for the volume of GaAs pairs originating from Ga adatoms migrating from the diffusion ring on the substrate surface. The quantity V is the arrival rate in nm/s, χ_W, χ_S are the pyrolisis efficiencies at the corresponding surfaces, thus the $\chi_S V$ term gives the effective deposition rate.

Using the above definitions, (3.148) can be put in the dimensionless form

$$\frac{dx}{dh} = \frac{ax^2 + 4\eta x + 4\eta^2}{x^2}. \tag{3.149}$$

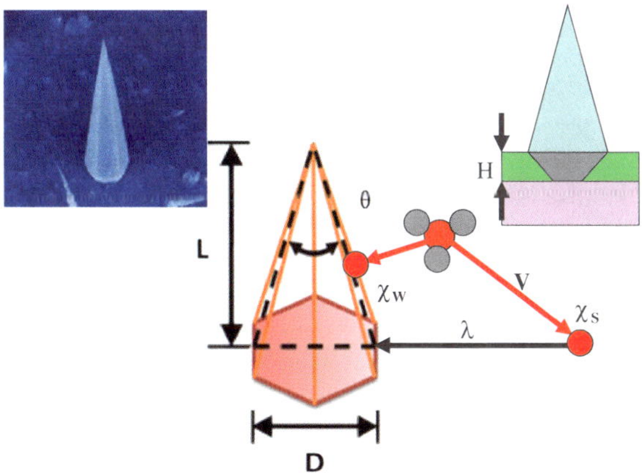

Fig. 3.59 Nanoneedle growth model with the parameters described in the text. *Left insert* shows the SEM image of a GaAs nanoneedle on sapphire. *Right insert* schematize the surface layer of height $H = V_S t$ burying the needle root

Here, $x = L/\lambda$ is the normalized length, $h = (\chi_S V t)/\lambda$ is the normalized deposition thickness and $a = (\chi_W/\chi_S)\sqrt{(16\eta^2/3) - 1} - V_S/(\chi_S V)$. The quantity V_S represents the vertical growth rate of a surface layer (see Fig. 3.59). Integration of (3.149) with the initial condition $x(h = 0) = 0$ readily gives the reverse dependence $h(x)$:

$$h = \frac{x}{a} - \frac{2\eta}{a^2}\ln\left(\frac{ax^2 + 4\eta x + 4\eta^2}{4\eta^2}\right) - \frac{2\eta(a-2)}{a^2\sqrt{a-1}}\left[\arctan\left(\frac{ax + 2\eta}{2\eta\sqrt{a-1}}\right) - \arctan\left(\frac{1}{\sqrt{a-1}}\right)\right] \tag{3.150}$$

The asymptotes at small and large lengths follow directly from (3.149):

$$L \cong \lambda^{2/3}(12\eta^2\chi_S V t)^{1/3},\ L << \lambda;\ L \cong \left[\sqrt{(16\eta^2/3) - 1}\,\chi_W V - V_S\right]t,\ L >> \lambda. \tag{3.151}$$

The first asymptote shows that the nanoneedle grows primarily by the surface diffusion as long as L is much smaller than λ, with the length scaling with time as $t^{1/3}$. At $L >> \lambda$, the diffusion-induced contribution disappears and the length becomes proportional to t.

We now turn to the description of growth experiments and the comparison of theoretical and experimental length-time dependences. The MOCVD growth procedure is the following. The silicon or sapphire substrates were first cleaned with acetone, methanol, and water. The growth was carried out in an Emcore D75 MOCVD reactor using TBA and TEGa as group V and III sources, respectively. The growth temperature was between 385 and 415 °C. The TBA and TEGa mole fractions were kept constant at 5.42×10^{-4} and 1.12×10^{-5}, respectively, in a 12 l/min hydrogen carrier gas flow.

Figure 3.60 shows 30° tilted SEM images of GaAs nanoneedles obtained on the Si(111) substrates at $T = 400$ °C after different growth times. The corresponding images of GaAs nanoneedles grown on sapphire at the same temperature are shown in Fig. 3.51d. Table 3.9 summarizes the details of nanoneedle growth evolution at $T = 400$ °C. The data on the length and diameter are the average values of 15–20

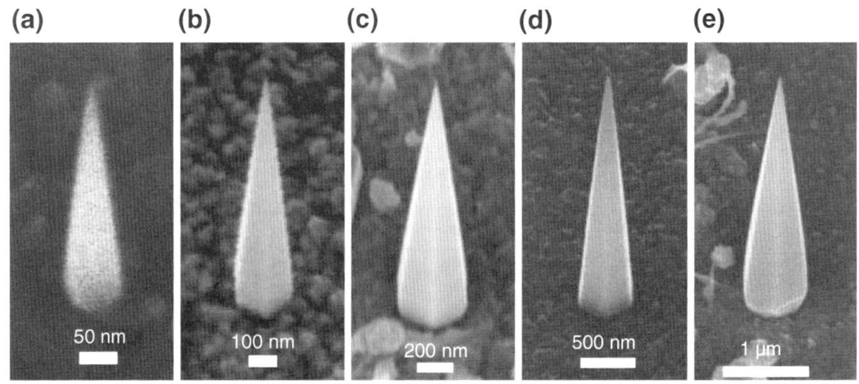

Fig. 3.60 GaAs nanoneedles on Si(111) substrate grown at $T = 400\,^{\circ}\mathrm{C}$ for $t = 6$ (**a**), 15 (**b**), 38 (**c**), 60 (**d**) and 90 (**e**) min

Table 3.9 Time evolution of GaAs nanoneedles grown at 400 °C on different substrates

Silicon substrate			Sapphire substrate		
t (min)	L (nm)	D (nm)	t (min)	L (nm)	D (nm)
6	450	70	1.5	120	20
15	1060	194	15	870	194
30	1820	303	60	3000	601
38	2480	381	82	4200	850
60	4000	600	180	8200	1340
90	5200	740			

Table 3.10 Growth parameters of GaAs nanoneedles

Substrate	Aspect ratio	V (nm/min)	λ (nm)
Silicon	6.7	3.5	80
Sapphire	5.7	2.9	400

needles from the same sample (except for the smallest nanoneedles where the results are averaged over only 5 needles). In agreement with the experimental data and the model of Sect. 3.4, the aspect ratios for a given growth run are indeed approximately constant. The linear fits to the data presented in Table 3.9 yields the mean values of $\eta = 6.7$ on the silicon and 5.7 on the sapphire substrate (in these particular experiments).

Experimental length-time curves were fitted by (3.150) with the above aspect ratios, $\chi_S = \chi_W = 1$ and $V_S = 0$. Theoretical $L(t)$ dependence contains two fitting parameters: the arrival rate V and the diffusion length on the substrate surface λ. From the best fits shown in Fig. 3.61, we deduce the parameters summarized in Table 3.10. Investigation of parameter robustness shows that the fits are less sensitive to λ than to

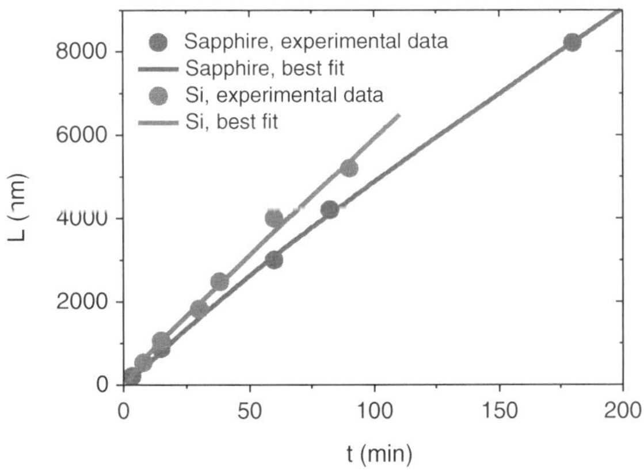

Fig. 3.61 Time evolution of GaAs nanoneedle length on Si and sapphire substrates

V, so that 20 % variation in λ at fixed V does not change significantly the curves shown in Fig. 3.61. This is well understood intuitively, because most of the time nanoneedles grow due to the sidewall impingement, while the diffusion-induced contribution is important only at the very beginning of growth. The obtained value of 80 nm for the effective diffusion length of Ga on Si substrate is consistent with previously published data on the group III element diffusion during the nanowire growth (from tens to one hundred nanometers) [184], while the 400 nm value on sapphire substrate is noticeably larger. A smaller diffusion length on the Si substrate is probably explained by its preliminary roughening. As for the arrival rate, the obtained values of 2.9–3.5 nm/min are rather small, yielding only 174–210 nm equivalent heights of a 2D layer growing in 1 h. The nanoneedles grow much faster, with average vertical growth rates of 46 nm/min on sapphire and 58 nm/min on Si, because their developed lateral surfaces absorb surrounding vapors very efficiently.

As regards the temperature dependence, Fig. 3.62 presents the results of [186] for GaAs nanoneedles on sapphire. Fig. 3.62a shows a nanoneedle with 415 °C growth temperature. The growth time was 60 min for all samples. At this 15 °C higher growth temperature, 6 more facets near the nanoneedle root, with 30° rotation to the upper 6 main $(1\bar{1}00)$ and (0001) terraces, are observed. These new set of facets are the $(1\bar{2}10)$ facets. No sharp tips are seen at this growth temperature. Instead, all the needles show a flat c-plane top surface, somewhat similar to the terminated vertical growth of nanopillars discussed in the previous section. For a lower growth temperature of 385 °C as shown in Fig. 3.62b, the sharp nanoneedle feature is maintained. The needle diameter, however, becomes 31 % smaller than at 400 °C, while the needle length is about the same for these two growth temperatures. Therefore the taper angle of the 385 °C nanoneedle is reduced to only 7°. On the other hand, although the 415 °C growth did not result in sharp nanoneedless, a taper angle can still be defined and measured as ~17°. The taper angle versus the growth temperature is shown in Fig. 3.62c. The taper angle decreases with decreasing growth temperature. Figure 3.62c also shows the nanoneedle density as a function of growth temperature.

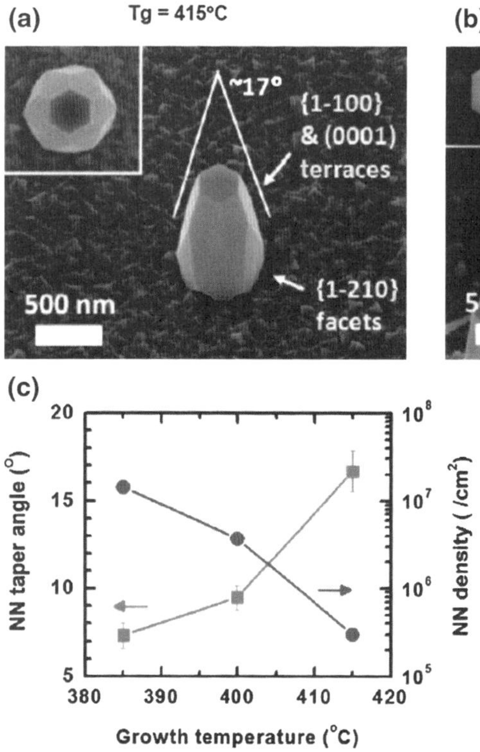

Fig. 3.62 **a** 30° tilt view of a nanoneedle grown at 415 °C. *Inset* is the top-down view. **b** 30° tilt view and the top-down view (*inset*) of a nanoneeedle grown at 385°. The nanoneedle shape still shows 6 tapered facets with sharp tip, which is similar to the nanoneedles grown at 400 °C but with a smaller taper angle. **c** Nanoneedle taper angle and nanoneedle density as a function of growth temperature

The density is larger than 10^7 cm^{-2} for the 385 °C growth but drops by nearly two orders of magnitude when the temperature is 415 °C, which is only 30 °C higher.

Let us now discuss the above experimental findings from the viewpoint of kinetic theory of 3D island formation. We note that the Stranski-Krastanow growth theory can be easily re-formulated for the Volmer-Weber islands by simply putting the equilibrium thickness to less than 1 ML, with preserving all the major results. Therefore, the observed decrease of the nanoneedle surface density with increasing the growth temperature is qualitatively explained by the corresponding results of Sect. 3.11. At higher temperatures, a kinetically established nucleation barrier is always larger due to the Arrhenius-like temperature dependence of the surface diffusivity (we assume here that the islands are mainly fed from the surface at the short scale nucleation stage). As for the temperature dependence of the taper angle, Fig. 3.22 of Sect. 3.4 shows that the aspect ratio always decreases as supersaturation increases. In the case of low temperature MOCVD, increase of substrate temperature should result in a higher arrival rate due to better pyrolysis efficiency at the surface. Thus, the supersaturation value is larger at higher temperatures, which explains the dependence shown in Fig. 3.62c.

Finally we note that, when the contribution from the surface diffusion becomes much smaller than the direct sidewall collection, the nanoneedles are mainly fed through their sidewall surface. At a constant vapor supersaturation (which should be maintained in a steady state MOCVD process), this yields the growth law of the form $di/dt \propto i^{2/3}$, where i is the number of GaAs pairs in the nanoneedle. Recalling the results of Sect. 2.4, the nanoneedle growth from vapor corresponds to the unique case with the growth indices $\alpha = 2/3$, $m = 3$ where the fluctuation-induced spreading of the size distribution is absent (see (2.77)). The invariant size equals $i^{1/3}$, which is proportional to a linear dimension (e.g., the nanoneedle length). One can thus conclude that the length distribution of self-induced nanoneedles is entirely determined by their random nucleation and should be maintained throughout the follow-up growth steps. This property guarantees a narrow length (and diameter) distribution of nanoneedles with a constant dispersion, which is important for applications involving the ensembles of nanoneedles.

3.16 Growth Properties of Co Nanoislands on CaF_2/Si(111)

Nucleation and growth of different metals on insulating surfaces have been of interest for a long time, particularly in connection with magnetic properties. Much higher surface energy of metals than that of insulators usually results in the Volmer-Weber growth mode whereby three-dimensional (3D) metal islands nucleate on the surface directly, without forming any wetting layer in between. The most known systems of this type are noble metals on alkali halides [241, 242]. It has been found that, although single metal adatoms easily desorb from the halide surface above room temperature, they can also form the irreversibly growing supercritical nuclei by joining other adatoms or small clusters via the surface diffusion process. Considerable attention has been paid to growing metals on oxides such as TiO_2 [243], MgO [244], NiO [245], and $SrTiO_3$ [246]. Much less is known about the growth properties of metals on the alkaline-earth fluorides. Growth of Fe, Co and Ag on the CaF_2(111) surface via a rather specific defect-induced nucleation mode was reported in [247], however, no epitaxial relationship to the substrate was shown under these particular growth conditions. Epitaxial growth of the α-Fe(110) on the CaF_2(111) surface was reported in [248], where the epitaxial relations were established by x-ray diffraction. Later on, the epitaxial growth of Co nanoparticles on the CaF_2(111) and CaF_2(110) surfaces was confirmed by *in situ* RHEED diagnostics [249]. Since Co segregates into an ensemble of 3D islands on insulating surfaces and forms a two-dimensional continuous layer on metal surfaces, there is a drastic difference between magnetic properties of the same amount of Co deposited onto an insulator or metal [250]. Ferromagnetic metals on insulators could exhibit antiferromagnetic ordering, which is of particular interest in connection with physics and applications of the exchange bias effect. Since magnetic properties of Co on insulating substrates are expected to be strongly dependent on the surface morphology of 3D islands, it is paramount to achieve a precise control over the size distribution of self-induced Co nanoparticles during growth.

In this section, we discuss the growth and structural properties of Co nanoparticles obtained by MBE on atomically clean $CaF_2(111)$ surfaces [251]. We show that deposition of Co onto atomically flat $CaF_2(111)$ surface results in the formation of ensembles of epitaxial 3D nanoislands whose density and size can be tuned in a wide range by changing the substrate temperature and the deposition time. We then explain several important growth features observed in this system by applying our kinetic growth theory.

MBE growth of Co/CaF_2/Si(111) samples was carried out in a dedicated ultra-high vacuum (UHV) system. Slightly off-cut silicon substrates (with misorientation angles of less than 15 angular minutes) were first cleaned by the conventional Shiraki chemical treatment and then flash-annealed in UHV chamber at 1200 °C to remove the oxide. Distinct 7×7 patterns characteristic of clean Si(111) surface were observed by RHEED. Growth of CaF_2 buffer layer on Si was performed from an effusion cell with a graphite crucible loaded with pieces of CaF_2 crystals. The CaF_2 deposition rate was calibrated to 2–3 nm/min by using an Inficon quartz thickness monitor. Cobalt was then deposited on top of CaF_2 buffer layer from an e-beam source where the target was a Co rod with 6 mm diameter. The cobalt flux was usually kept at 0.2–0.3 nm/min. The substrate temperature during the cobalt deposition was varied from room temperature to 700 °C, while the Co deposition thickness was changed between 0.1 and 45 nm. Here and below, we define the Co deposition thickness H (or the Co exposure) as the thickness of an imaginary flat Co layer that would form in absence of desorption and nucleation. This value is equivalent to the flux measured by a quartz thickness monitor (mass per unit time per unit area divided by the Co density) multiplied by the growth time. In some experiments, a low temperature Co seeding layer with less than 1 ML coverage was grown at a low T (between room temperature and 200 °C) before depositing Co at elevated temperatures in order to increase the number of nucleation sites. The AFM images were obtained with an ambient-air NT-MDT microscope operated in the semi-contact mode. SEM studies were performed on a JSM 7001F (JEOL) microscope operating in the secondary electron regime.

It has been found that the most regular surface morphology of the buffer CaF_2 layers on silicon is obtained by applying the two-step growth procedure. At the first, low-temperature stage (3 ML deposited at 250 °C), a thin fluorite layer uniformly covers Si surface without changing its step structure. The second, high-temperature stage (20 ML deposited at 770 °C) yields the formation of relatively wide terraces with the average width close to that of the initial Si(111) substrate. The RHEED patterns during and after growth show narrow streaks confirming a high crystalline quality and flatness of the surface. It is worth mentioning that the crystallographic axis of CaF_2 and Si are co-oriented during the low-temperature stage, which corresponds to the so-called A-type epitaxial relations at the CaF_2/Si(111) interface. This relation transforms into the B-type one (rotated by 180°) right after the increase of temperature

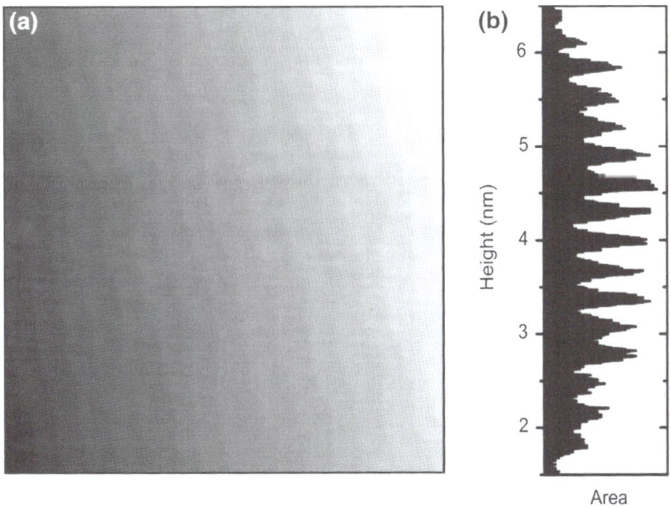

Fig. 3.63 **a** AFM image of CaF_2 buffer layer grown on silicon by the two-step growth procedure (size 1400 nm × 1400 nm × 6 nm). **b** The calculated height histogram showing the terrace-related peaks that correspond to ∼0.31 nm step height

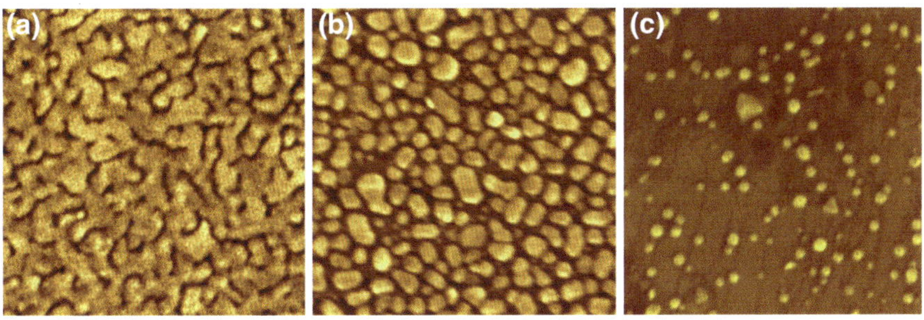

Fig. 3.64 SEM images of Co islands grown at 100 °C (**a**), 300 °C (**b**), and 500 °C (**c**) with a fixed deposition thickness of 20 nm. The image size is 800 nm × 800 nm

above 500 °C. The AFM studies prove that the two-stage growth procedure with the temperature ramping results in a uniform CaF_2 layer with smooth monoatomic steps at the surface, as shown in Fig. 3.63.

We now consider experimental data on the temperature and Co exposure dependence of the island morphology. Figure 3.64 shows SEM images of three samples grown with a fixed deposition thickness of 20 nm at 100, 300 and 500 °C, respectively. It is clearly seen that the total surface area occupied by Co is drastically decreasing with temperature. To measure the total volume of Co on the surface, the medium energy ions scattering measurements were carried out. It has been found that 100 % of cobalt remains on the surface at 100 °C (the complete condensation regime [241]), decreasing to a 30 % fraction at 300 °C and less than 10 % fraction at 500 °C (the incomplete condensation regime). These values give the sticking coefficient integrated over the entire growth time. More precisely, two different processes should be distinguished. The first one involves sticking of Co on CaF_2, a heterogeneous metal-insulator process which sticking coefficient is expected to be strongly temperature dependent. The second process is the direct impingement of Co on Co with almost 100 % sticking probability in our temperature window. At the nucleation stage where the Co coverage is very low, the Co-CaF_2 sticking is always dominant. As the coverage increases with time, more and more of the arriving Co atoms directly impinge the islands surface and stick to it with 100 % probability.

Experimental dependences of the island height L on the deposition thickness H at different temperatures shown in Fig. 3.65 feature a qualitatively dissimilar behavior. At 100 °C, the island height increases sub-linearly with H and can be well fitted by the power law dependence $L = aH^{1/3}$ with $a = 5$ nm$^{2/3}$, while the dependence at 500 °C is linear: $L = bH$ with $b = 1.05$. The observed difference can be explained by different growth mechanisms of islands. Indeed, the surface diffusion length of Co should be lower at higher temperatures due to re-evaporation. Therefore, the islands grow primarily by the direct impingement of Co atoms onto their surface from the vapor phase. For 3D island with a time-independent shape, the growth rate *di/dt* (atoms per unit time) is given by

$$\frac{di}{dt} = IC_1L^2, \tag{3.152}$$

where I is the arrival rate and C_1 is the shape constant such that C_1L^2 is the island cross-section intercepted by the molecular beam. Since $i = (C_2L^3)/\Omega$, where C_2 is the shape constant such that C_2L^3 is the island volume and Ω is the elementary volume in the solid phase, and the Co deposition thickness $H = I\Omega t$ by definition, integration of (3.152) leads to a linear dependence of the island height on the Co exposure

$$L = \frac{C_1}{3C_2}H, \tag{3.153}$$

i.e., the high temperature growth at 500 °C shown in Fig. 3.65a.

Islands grown at 100 °C should be mainly fed from the surface, because the Co diffusion length at this low temperature is much larger. Possible mechanisms of mass transport into the islands include the surface diffusion of Co from a planar terraces and the diffusion along the surface step. Since islands nucleate and grow by consuming the surface adatoms, the surface supersaturation rapidly tends to zero, as discussed in Chap. 2. Whatever is the growth mechanism, the mass conservation at $\zeta \to 0$ yields the material balance of the form

$$H \cong H_{eq} + \Omega Ni. \tag{3.154}$$

Here, H_{eq} denotes the residual equilibrium Co coverage of the surface. The last term in the right hand side gives the total volume of islands per unit area in the monodispersive approximation of the island size distribution, with i as the number of Co

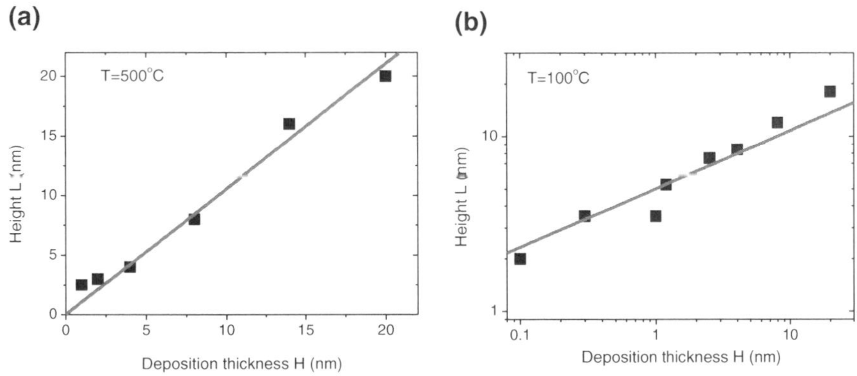

Fig. 3.65 Measured dependences of the island height on the Co deposition thickness at 500 °C (**a**) and 100 °C (**b**), fitted by the power law dependences discussed in the text

atoms in the island of the mean size and N as the island density. Assuming that $H >> H_{eq}$ (this inequality should always pertain for the Volmer-Weber growth) and using the relationship between i and L, we obtain

$$L = \left(\frac{H}{C_2 N}\right)^{1/3}, \quad (3.155)$$

which explains the experimental dependence shown in Fig. 3.65b.

Figure 3.66 shows the measured dependences of the island density on the Co deposition thickness at different T. It is seen that the island density is dramatically affected by the growth temperature. At a fixed $H = 20$ nm, the maximum density of islands grown at 500 °C is two orders of magnitude lower than at 100 °C. Lines in Fig. 3.66 represent theoretical fits obtained from the double exponential formula (3.145), written in the form

$$n(H) = N\left[1 - \exp\left(-e^{C(H-H_*)}\right)\right]. \quad (3.156)$$

Here, N is the maximum density acquired upon the completion of nucleation stage, H_* is the critical deposition thickness relating to a maximum nucleation rate, and C the parameter which is inversely proportional to the size distribution width (in terms of the invariant size). The maximum density decreases with the temperature as the Arrhenius exponent

$$N = N_0 \exp\left(\frac{E_D}{k_B T}\right). \quad (3.157)$$

with N_0 being a constant and E_D the quantity of the order of activation energy for their surface diffusion. Since the size distribution width increases due to thermal fluctuations, the C value must decrease with the temperature.

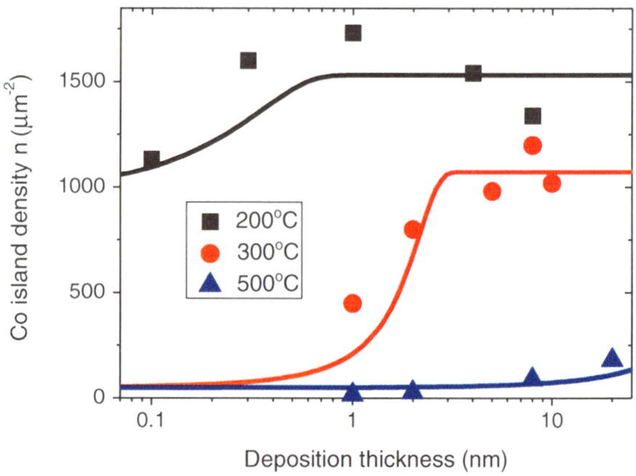

Fig. 3.66 Experimental (*symbols*) and theoretical (*curves*) dependences of the island density on the Co deposition thickness at three different temperatures

The fits in Fig. 3.66 are obtained from (3.156) and (3.157) at $E_D = 0.055$ eV, $N_0 = 200\,\mu\text{m}^{-2}$, and $C = 5.76 - T/135$ K. The critical thickness H_* is set to zero at 200 °C, 1.6 nm at 300 °C and 54 nm at 500 °C (we note that most of Co desorbs at elevated temperatures so that these fitting values do not necessarily contradict with the Volmer-Weber growth mode). It is seen that theoretical curves represent reasonably well the experimental data at $T = 300$ and 500 °C. However, we cannot describe a slight decrease of density with H observed at 200 °C. This effect is most probably explained by a partial coalescence of a dense ensemble of islands at lower temperatures. As follows from Fig. 3.66, the nucleation stage is faster at lower temperatures, having been fully completed at the deposition thickness of 1 nm at 200 °C, 3 nm at 300 °C, and not yet completed after the deposition of 20 nm of Co at 500 °C.

Careful analysis of SEM and AFM images reveals that the spatial distribution of cobalt islands on the CaF_2 surface is highly influenced by the fluorite surface steps. For a migrating adatom, the probability to be trapped at a surface step is usually higher than on a terrace because the step sites have more dangling bonds. Nucleation therefore occurs at the step much more often, provided that the adatom diffusion length is large enough to reach the step and its kinetic energy is low enough to feel the difference between the at-the-step and on-the-terrace bonding energies [251]. It has been noticed that nucleation always starts at the steps in the entire temperature window studied (100–600 °C). This is most clearly seen at temperatures below 300 °C, where the distance between the neighboring Co islands is much less than the width of CaF_2 terraces at the initial growth stage. Below 1 nm of Co deposited, the islands tend to form chains along the steps, with a smaller fraction of on-the-terrace islands (Fig. 3.67a). As the deposition thickness increases, more islands emerge on the terraces so that the island distribution transforms to a spatially uniform, as shown in Fig. 3.67b.

In order to better understand the growth mechanisms of Co islands, the height distributions were obtained from the statistical analyses of AFM images of the samples grown at different temperatures and deposition times. Typical height histograms at 100 °C (2 nm of Co deposited) and 300 °C (20 nm of Co deposited) are shown

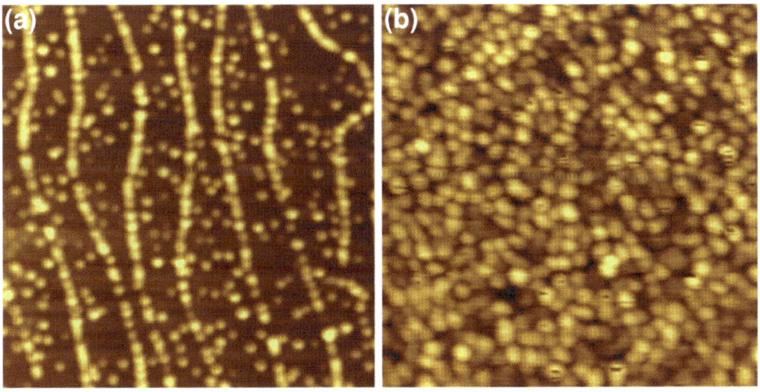

Fig. 3.67 AFM images of Co islands at 100 °C after deposition of 1 nm (**a**) and 8 nm (**b**) of Co. The image size is 620 nm × 620 nm × 9 nm

in Fig. 3.68. To model the experimentally observed shapes, we use the universal double-exponential distribution over invariant sizes ρ

$$g(\rho, z) = cN \exp\left(C\,(z-\rho) - e^{C(z-\rho)}\right). \tag{3.158}$$

As usual, $z(t)$ is the time-dependent most representative size, and C is the same constant as in (3.156). Let us now consider possible dependences of ρ on L. As discussed above, when the island is primarily fed by the direct impingement from vapor through its surface, the growth rate di/dt is proportional to $L^2 \propto i^{2/3}$, as given by (3.152). In the ballistic regime of the surface diffusion, the growth rate is proportional to the perimeter of the island base: $di/dt \propto L \propto i^{1/3}$. When growth is induced by the diffusion along the surface step (the step flow growth), the growth rate di/dt is size-independent. We can therefore write down quite generally:

$$di/dt \propto L^n \tag{3.159}$$

with $n = 0, 1, 2$ for the step diffusion growth, growth by the ballistic surface diffusion and by the direct impingement, respectively.

Since the growth rate $d\rho/dt$ in terms of invariant size must be L—independent, we obtain

$$\rho \propto L^k \tag{3.160}$$

with $k = 3 - n = 3, 2, 1$ for the step flow growth, surface diffusion and direct impingement, respectively. Distributions expressed in terms of different size-related variables should preserve the number of islands (see Sect. 2.1), yielding $f(L, t)dL = g(\rho, t)d\rho$. Therefore, (3.158) and (3.160) yield the height distributions of the form

$$f(L, t) = A_k L^{k-1} \exp\left(C_k(L_0^k(t) - L^k) - e^{C_k(L_0^k(t)-L^k)}\right). \tag{3.161}$$

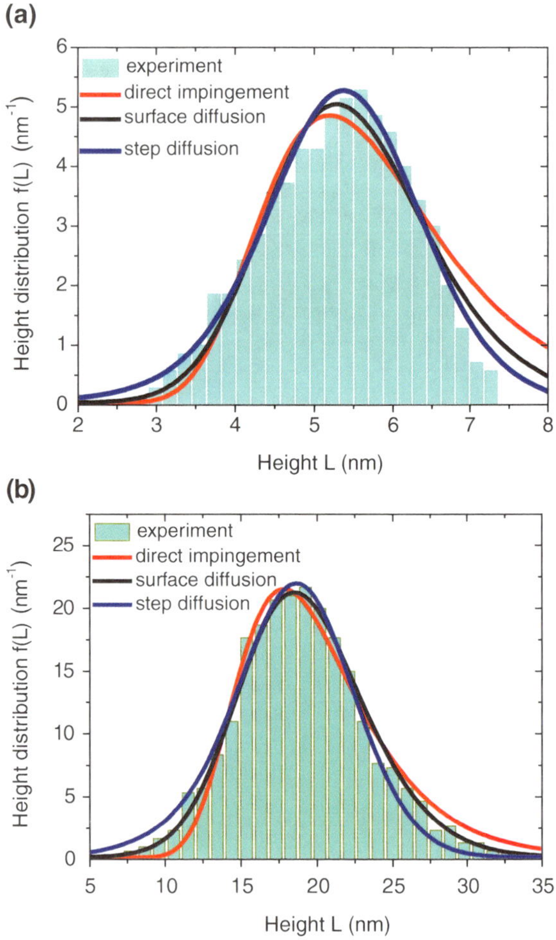

Fig. 3.68 Experimental height distributions of Co islands at $T = 100\,^{\circ}\mathrm{C}$, $H = 2\,\mathrm{nm}$ (**a**) and $300\,^{\circ}\mathrm{C}$, $H = 20\,\mathrm{nm}$ (**b**), fitted by different distributions with the following parameters (index 1—direct impingement, 2—surface diffusion, 3—step flow growth): $A_1 = 13.2$, $c_1 = 0.91$, $H_0^1 = 5.2\,\mathrm{nm}$, $A_2 = 2.64$; $c_2 = 0.1$, $H_0^2 = 26\,\mathrm{nm}$; $A_3 = 0.53$, $c_3 = 0.0132$, $H_0^3 = 125\,\mathrm{nm}$ in (**a**), and $A_1 = 58.5$, $c_1 = 0.258$, $H_0^1 = 17.8\,\mathrm{nm}$; $A_2 = 3.25$, $c_2 = 0.0071$, $H_0^2 = 308\,\mathrm{nm}$; $A_3 = 0.19$, $c_3 = 0.00029$, $H_0^3 = 4700\,\mathrm{nm}$ in (**b**)

Here, $L_0(t)$ are the peak values of the island heights, A_k are the normalization constants, C_k are the coefficients that determine the corresponding distribution widths. The dependences given by (3.161) are shown by lines in Fig. 3.68 at different k with fitting parameters listed in the figure caption. It is seen that the experimental histogram at $100\,^{\circ}\mathrm{C}$ is better fitted with the step diffusion mode. Histogram at an intermediate growth temperature of $300\,^{\circ}\mathrm{C}$ is little better fitted by the two-dimensional ballistic diffusion model, which seems reasonable because the Co exposure is ten times longer and the islands are much larger than after the deposition of 2 nm of Co at $100\,^{\circ}\mathrm{C}$. However, a combination of contributions from different growth mechanisms is not excluded in this case.

Finally, we consider the influence of low temperature seeding layer on the resulting morphology of Co islands. As discussed above, a weak sticking of Co to the CaF_2 surface suppresses its nucleation and growth at elevated temperatures (above $300\,^{\circ}\mathrm{C}$), where most of the material is lost because of the incomplete condensation regime. On the other hand, high temperature growth is believed to significantly improve the crystallinity of Co islands. Furthermore, the growth well above the hexagonal-to-cubic (HCP-FCC) crystallographic transition temperature for Co (450 °C [252]) should favor single crystalline cubic lattice. Another challenge is to produce densely packed linear chains of high temperature islands aligned along the surface steps. This is difficult because of a large diffusion length at intermediate temperatures, where the distance between the neighboring nucleation sites at the same step becomes comparable to the step spacing. A 0.02–0.5 nm seeding layer of Co was therefore grown on CaF_2 at a low temperature (from room temperature up to 200 °C) prior to the high temperature Co deposition at 300–700 °C. It has been found that the two-step growth procedure enables producing perfectly aligned islands decorating the steps, with the density being much higher and the size much larger than in the case of low temperature growth.

These features are clearly seen from comparing the morphologies shown in Figs. 3.69a and 3.67a. The sample obtained by the two-step procedure contains much less on-the-terrace islands between the chains, while its better crystallinity is confirmed by RHEED. As the Co exposure increases from 13 to 30 nm, the islands get larger and the spatial ordering is lost (Fig. 3.69b). As seen from Fig. 3.69b after 30 nm exposure and Fig. 3.69c after 45 nm exposure, the island coalescence process leads to a significant decrease of density at this late growth stage only when the growth temperature at the second stage is between 500 and 700 °C. At lower temperatures, the on-the-terrace to at-the-step density ratio increases drastically while at higher temperatures the CaF_2 buffer layer is no longer stable. The suppression of on-the-terrace nucleation at elevated temperatures can be qualitatively explained by the exponentially decreasing temperature dependences of on-the-terrace nucleation rate and surface density of supercritical islands discussed in Sects. 2.9 and 3.11. (We point out again that the Stranski-Krastanow growth theory can be re-formulated for the Volmer-Weber islands simply by changing the metastable wetting layer to the adatom sea with submonolayer coverage). Indeed, small on-the-terrace Co islands nucleated at the low-temperature step may either decompose to adatoms or migrate as a whole to much more stable at-the-step islands when the surface temperature is increased. After that, no new islands nucleate between the steps. Rather, a certain percentage of deposited Co diffuses toward at-the-step islands and contributes to their growth while the rest of Co re-evaporates.

As for the coalescence mechanism, a high surface mobility of Co at elevated temperatures does not seem to favor the solid-like coalescence discussed in Sect. 2.10 (although islands themselves are of course solid during growth). This is clearly seen from Fig. 3.69: if the coalescence had a solid-like character, linear chains of islands with small inter-island spacing in Fig. 3.69a would be transformed into continuous stripes that are not present in Fig. 3.69b. Instead, the islands get larger in size simultaneously with the gaps between them. This is most probably due to an inter-diffusion of adatoms along the island surfaces and bases, a process that tends to decrease the base perimeter after two or more islands merge. This complex coalescence process requires an additional study. In any case, the coalescence is not of the

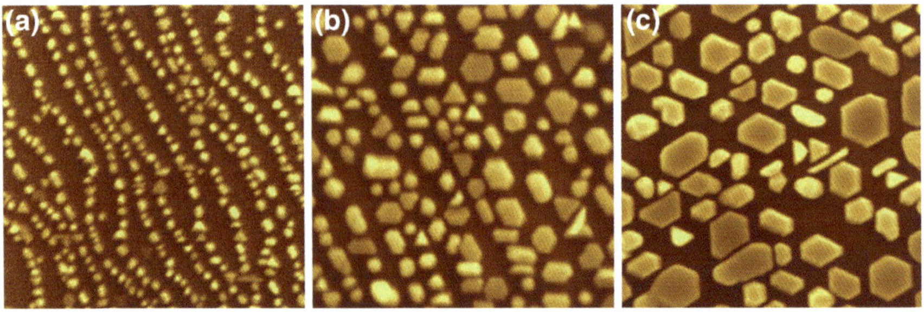

Fig. 3.69 SEM images of Co islands grown with a 0.1 nm seeding layer and different exposures at the main stage: **a** 13 nm, **b** 30 nm, **c** 45 nm

Kolmogorov type. Indeed, statistical analysis of AFM images of samples at different growth stages confirms a linear dependence of the island mean height on the Co deposition thickness. This is in agreement with the high temperature growth model by the direct impingement given by (3.153). Within the instantaneous nucleation approximation, such a growth yields the Kolmogorov exponent with index $\delta = 2$. The step flow and the surface diffusion growths at the nucleation stage would result in the indices $\delta = 2/3$ and 1, respectively. However, the measured dependence of the surface coverage θ on the deposition thickness H is best fitted by the formula $\theta(H) = 1 - \exp(-\nu H^{\delta})$ with $\delta = 0.825$ and $\nu = 0.0348\,\mathrm{nm}^{-0.825}$, showing a large discrepancy with the Kolmogorov values.

Chapter 4
Vapor–Liquid–Solid Growth of Nanowires

This chapter concerns growth modeling of semiconductor nanowires obtained via the vapor–liquid–solid growth mechanism by modern epitaxy techniques. In this mechanism, the growth species are supplied from vapor to a metallic catalyst particle that seats on the nanowire top and remains liquid during growth. Nanowires of different materials, particularly semiconductor nanowires, have recently drawn much attention for their unique physical properties and a variety of promising applications. It is the nanowire growth process that primarily determines their morphology, crystal structure, optical, transport and other physical properties unattainable in two-dimensional layers and even quantum dots. The complex character of nanowire growth involving different interfacial boundaries, position- and size-dependent nucleation, physical chemistry of metal-semiconductor alloys, non-stationary effects and many other non-trivial phenomena offers an exciting field for both fundamental and applied research.

In Sect. 4.1, we give a brief overview of the history and major concepts of the vapor–liquid–solid growth, starting from the seminal works by Wagner and Ellis and Givargizov till nowadays. A rapid progress in the bottom-up wire synthesis at the nanoscale is described, started in the 2000s by the research groups led by Liber, Samuelson, Young and some other researches. We discuss epitaxy techniques, the role of metal catalyst droplet, state of metallic particle during growth, typical nanowire shapes, material transport pathways, and advantages of this catalytic growth for size-, shape-, position- and composition-controlled fabrication of nanowires and nanowire heterostructures. We also formulate some general conditions for the vapor–liquid–solid growth and list several important problems of nanowire modeling that will be addressed in detail in the foregoing analysis. Not all modern nanowires are fabricated through the vapor–liquid–solid process. We therefore discuss alternative growth mechanisms in Sect. 4.2, including the so-called vapor-solid-solid growth from a frozen particle below the eutectic point, selective area techniques of nanowire fabrication, and self-induced GaN nanowires on silicon substrates.

Section 4.3 starts the modeling part. We first discuss the role of size-dependent effects in nanowire growth, namely, the curvature-induced Gibbs-Thomson effect and

V. G. Dubrovskii, *Nucleation Theory and Growth of Nanostructures*, NanoScience and Technology, DOI: 10.1007/978-3-642-39660-1_4,

the mononuclear character of the nucleation-mediated vertical growth of nanowires. The Wagner-Ellis and Givargizov-Chernov models are considered and compared with relevant experimental data. In Sect. 4.4, we consider the role of surface energies and in particular the dropet stability on the nanowire top. Whenever a liquid droplet wets the growth front but not the sidewalls and the growth interface is planar, the monolayer islands will tend to nucleate at the triple phase line where the vapor, liquid and solid phases meet. In Sect. 4.5 we show that the Glas condition for the triple phase line nucleation is actually equivalent to the Nebol'sin-Shchetinin condition for the droplet stability. Section 4.6 presents very interesting results obtained by IBM and Cambridge groups through *in-situ* growth monitoring, revealing non-planar growth interfaces with truncated edges and an oscillating degree of truncation.

In Sect. 4.7 we describe the adsorption-induced growth which is typical for some chemical epitaxies or self-catalyzed formation of III-V nanowires. Relevant theoretical models are discussed, with an emphasis on the self-consistency. This is required to equalize the material transport and the nucleation-limited growth under the droplet in the steady state. In most modern epitaxy techniques, a metallic catalyst particle also acts as a material collector directing the adatom diffusion fluxes to the wire top. Section 4.8 presents the diffusion growth model with the most general form of the boundary conditions. These conditions account for neighboring wires and non-equivalence of the substrate and sidewall crystal planes. Simplified growth models are considered in Sect. 4.9 in some important particular cases such as a single nanowire, small and large diffusion lengths on different surfaces. Section 4.10 presents a self-consistent approach in the general case with the diffusion-induced contributions included.

Section 4.11 concerns typical length-radius dependences of group IV and III-V nanowires. It is shown how the length-radius correlations are modified from increasing to decreasing through an intermediate regime with a maximum, and which parameters are responsible for this modification. A complex interplay of the size-dependent limitations on the growth rate and the diffusion transport is considered, leading to more complex length-radius dependences in some cases. Many experimental length-radius curves are fitted with the self-consistent growth equation. In Sect. 4.12, we analyze the temperature conditions for the vapor–liquid–solid growth of different nanowires through the temperature dependences of desorption rates and the effective diffusion lengths. Section 4.13 presents the results reported by LPN CNRS group on the growth kinetics of a single InPAs nanowire with intentionally modulated composition. This gives a unique possibility to directly check the validity of the growth equation and to deduce a number of important physical parameters.

On of the most important features of the diffusion growth equation is its nonlinearlity. Consequently, the nanowire growth kinetics might be very different at different time domains, and strongly depends on the nanowire radius. We consider some non-linear effects of nanowires growth in Sect. 4.14, such as negative growth through the reversed diffusion flux from top to base, and a continuing growth from a pre-existing stem. Theoretical predictions are compared with relevant experimental data. Finally, Sect. 4.15 describes a surprising effect of narrowing the length distribution of Au-catalyzed Ge nanowires. It is shown that the effect originates from the change in the direction of the flow of adatoms on the nanowire sidewalls during growth, due to the droplet activity dependence on the radius of the seed particle.

This chapter deals only with straight nanowires without any radial growth, and does not consider morphological transformations. We also ignore strain induced by the lattice mismatch and a complex problem of a time-dependent polytypism in III-V nanowires. These questions will be considered in Chaps. 5 and 6.

4.1 Vapor–Liquid–Solid Growth Mechanism

Au-catalyzed growth of freestanding Si "whiskers" with sub-micrometer radii from vapors of $SiCl_4$ and H_2 deposited onto a Si(111) substrate at a temperature of around 1000 °C was first demonstrated by Wagner and Ellis in 1964 [46]. It was shown that the whisker growth originated from the droplet and continued in the $< 111 >$ direction perpendicular to the substrate surface. The whiskers elongated at a constant radius that was determined by the size of initial droplet, while the growth on the non-activated surface was suppressed. This new method for fabrication of high quality, high stiffness 1D crystals with a precise position and shape control immediately attracted much attention [47, 253–257]. In particular, an outstanding contribution into the fundamentals of metal-catalyzed growth of highly anisotropic crystals was made in the 1970s by Givargizov, with the results being summarized in his book [47]. Later on toward the 1990s, the interest in synthesis and physical studies of 1D crystals somehow declined. The bottom-up approach was re-visited starting from the early 2000s, primarily by the research groups led by Lieber [258–269], Samuelson [270–279], and Yang [280–286], who basically implied the same "vapor–liquid–solid" (VLS) growth method at the nanoscale. Their tremendous achievements gave rise to the exponential increase in the number of publications [Fig. 4.1a] and many other researchers joining the field in the 2000s [44, 45, 48, 49, 230–232, 235, 287–302].

Using the bottom-up VLS approach, nanowires with a typical radius of 10–100 nm and length up to several μm can be obtained by using modern epitaxy techniques such as MOCVD and MBE. Nanowire materials include elementary semiconductors Si and Ge [258–260, 281, 298], III-V [264, 265, 270–279], II-VI [282, 301] compounds and oxides [302]. VLS semiconductor nanowires are interesting for their unique physical properties, some of which are listed below:

(i) The growth mechanisms including a complex physical chemistry of liquid and solid alloys, 1D diffusion transport, specific features of nucleation in confined nano-volumes, droplet configurations, different non-stationary and non-linear growth effects;
(ii) The morphology including the control over length and radius uniformity, fabrication of spatially ordered nanowire arrays, and tunable nanowire shapes;

(iii) The crystal structure featuring surprising ZB-WZ polytypism of III-V nanowires, rotational twinning, developing methods to control the crystal phase during growth;
(iv) Dramatically enhanced role of surface effects leading, in particular, to a very efficient relaxation of elastic stress on free sidewalls and enabling coherent growth of epitaxial nanowires on lattice mismatched substrates;
(v) Synthesis methods and physical properties of axial and core-shell heterostructures in nanowires, including elastic relaxation and structural defects;
(vi) The doping processes and the formation of p-n junctions in nanowires;
(vii) Transport and optical properties, including those in nanowire heterostructures, with a variety of possible dimensions, morphologies and crystal structures and a possibility to realize single electron transport and single photon emission and detection.

Most of the basic properties that are directly related to the growth mechanism, morphology, crystal structure, and strain relaxation will be discussed in Chaps. 4–6.

As for applications, semiconductor nanowires show great promise as the key elements for field effect transistors [260, 267, 277, 283], high electron mobility transistors [264], resonant tunneling diodes [273], single electron memory devices [279], high efficiency light emitters [265, 274], solar cells [281], sensors [262, 266] and many other functional nanodevices. One of the most important directions in the entire nanowire physics and technology is the on-chip integration of photonic and electronic circuits [261, 286]. For example, Fig. 4.1b from [286] shows the layout of a future nanophotonic circuit for chemical analysis where almost all functional elements (power supplies, light sources and detectors, waveguides, sensors) are made of nanowires. A detailed overview of nanowire applications can be found, e.g., in [184]. It is obvious that, in the bottom-up approach, the physical properties of nanowires are mainly determined by their synthesis process, which is why the VLS growth should be very well understood to optimize these properties for different applications. Fundamental and technological achievements in synthesis processes of vertical 1D nanostructures are summarized in several recent reviews, for example, in [184, 303–305].

In simple terms, the VLS method utilizes the catalytic effect of a metal droplet which is liquid during growth. The droplet promotes the vertical nanowire growth and defines the nanowire position and radius, while the nanowire length is roughly proportional to the deposition time. Let us first discuss qualitatively the physical nature of the VLS growth during chemical vapor deposition (CVD) according to Wagner and Ellis [46] and Givargizov [47]. The process is illustrated in Fig. 4.2 In the 1960–1970s, typical CVD growth temperatures of Si from $SiCl_4$ and H_2 on silicon substrates were above 1100 °C. By decreasing the substrate temperature to 1000 °C or lower, one could therefore almost completely avoid the surface growth. The (111) substrate orientation was chosen for two reasons: the lowest rate of 2D growth and, more importantly, the preferred $< 111 >$ wire growth orientation where the wires are perpendicular to the (111) substrate. The Si(111) substrates were activated by Au droplets, which can be accessed by simply depositing a thin solid Au layer and

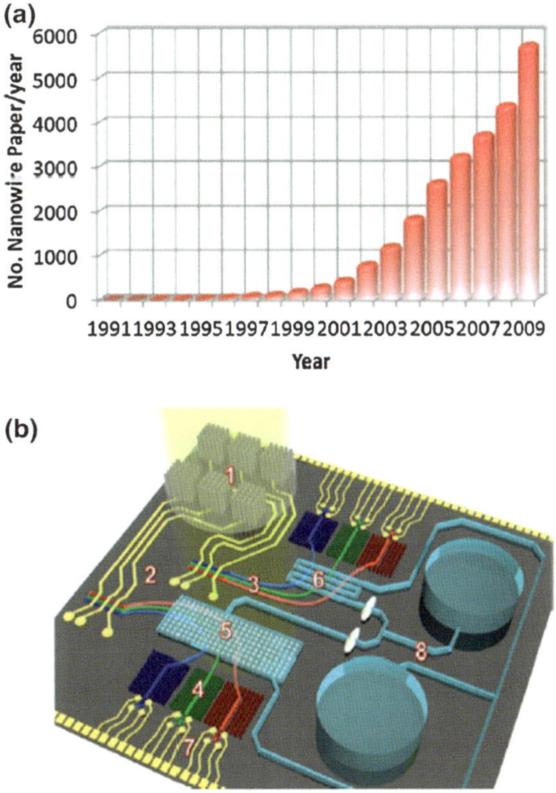

Fig. 4.1 **a** Increase in the number of publications on nanowire-related topics from 1991 to 2009. **b** Illustration of future nanophotonic circuits for chemical analysis that will require the integration of different nanowire-based devices, such as nanowire solar cells as on-chip power supplies (*1*), laser diodes and LEDs as light sources (*2*), nanoribbons (*3*) and 2D photonic bandgap nanowire arrays (*4*); filters and waveguides to select and route input and output signals; sample analysis chambers (*5*) or nanoribbon/sample intersection for absorption analysis (*6*); nanowire based photodetectors (*7*) and a microfluidic system for liquid sample transport (*8*) [286]

then annealing it before turning on a Si source. The annealing temperature as well as the VLS growth temperature must be higher than the eutectic melting temperature of the corresponding metal-semiconductor alloy (Au-Si in our example). Surprisingly, while bulk Au and Si themselves melt at 1064 °C and 1410 °C, respectively, the melting temperature of the Au-Si alloy is as low as 363 °C [306]. This value gives the lowest possible VLS growth temperature disregarding the size-dependent depression of the melting point. Since most of liquid metals do not wet semiconductor surfaces, the alloy segregates into an ensemble of spherical cap droplets.

When Si precursor is supplied to the substrate, the pyrolysis reaction $SiCl_4 + 2H_2 \rightarrow Si + 4HCl$ takes place only at the droplet surface. Silicon either dissolves in the droplet or re-evaporates. However, more Si arrives from vapor than desorbs, thus creating a certain Si concentration in the Au-Si alloy (C_L) which is above its equilibrium concentration (C_E) at a given temperature(T). On the other hand, C_L is lower than the concentration (C_V) at which the arrival rate from vapor would be equalized by desorption to the vapor phase (which is at a different temperature). Liquid alloy in the droplet is rendered into a supersaturated value by the vapor influx and crystallizes at the liquid–solid interface under the droplet. The composition of

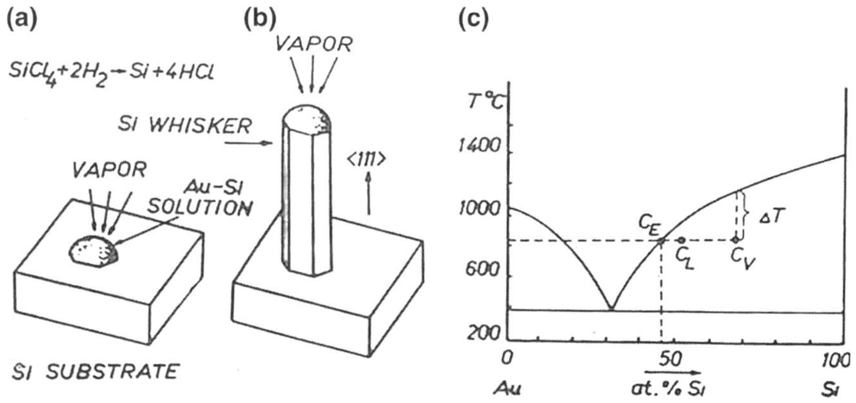

Fig. 4.2 The Wagner-Ellis sketch schematizing the VLS growth of Si "whiskers" by the Au-assisted CVD [46]: **a** Au-Si droplet on the substrate surface; **b** whisker growth under the droplet; **c** Au-Si phase diagram showing the melting point at 363 °C and the equilibrium silicon concentration C_E at the liquidus (*solid line*) at a given surface temperature T (*dashed line*). Vapor flux renders the Si concentration in the alloy into a supersaturated value C_L; $\Delta\mu_{LS} = \mu(C_L) - \mu(C_E)$ is the difference of Si chemical potentials in the liquid and solid states, and $\Delta\mu_{VS} = \mu(C_V) - \mu(C_E)$ is the difference of Si chemical potentials in the vapor and solid states. Temperature $T + \Delta T$ corresponds to the point at which the entire VLS system would be in equilibrium

the precipitated solid phase corresponds to that of the phase boundary and thus is pure Si according to the diagram. This crystallization usually proceeds via the nucleation and lateral extension of 2D islands, as in the conventional liquid phase epitaxy but in a laterally confined area. This leads to the nucleation-mediated vertical growth of a cylindrical solid rod under the droplet, whose radius is defined by the initial droplet size. In a steady state, the arrival of Si from vapor is exactly equalized by its sink due to the crystallization. Silicon is thus transformed from the vapor to the solid state via an intermediate liquid state by two consecutive phase transitions: the vapor-liquid condensation and the liquid–solid crystallization, which explains the term "vapor–liquid–solid" growth. The mechanism described above can be realized only when the vapor-solid difference of chemical potentials $\Delta\mu_{VS}$ is larger than the liquid–solid one $\Delta\mu_{LS}$ (the necessary condition for a positive vapor-liquid flux) and the latter is positive (the necessary condition for the liquid–solid flux). Here and below, we define the effective vapor (Φ) and liquid (ζ) supersaturations with respect to the equilibrium state in the droplet (which is usually at the same chemical potential as the solid state in the wire) as

$$\Phi = \exp\left(\frac{\Delta\mu_{VS}}{k_B T}\right) - 1; \zeta = \exp\left(\frac{\Delta\mu_{LS}}{k_B T}\right) - 1. \tag{4.1}$$

The necessary condition for the VLS wire growth is therefore given by

$$\Phi > \zeta > 0. \tag{4.2}$$

Modern MOCVD and MBE techniques for the VLS nanowire synthesis use very similar growth strategy, illustrated in Fig. 4.3 in the case of Au-catalyzed MBE of GaAs wires on GaAs(111)B substrates. Although Au is the most common catalyst for Si, Ge and III-V nanowires, some other metals can also be utilized as discussed later on. The growth procedure usually includes four steps:

1. Substrate de-oxidation and growth of a GaAs buffer layer in the MBE growth chamber.
2. Deposition of $\sim$1 nm thin Au film in an ultra-high vacuum chamber or a metallization chamber which is placed inside the vacuum part of MBE machine but is separated from the growth chamber in order to avoid the unwanted Au contamination.
3. Anneal of Au above the eutectic melting temperature with Ga (usually around 600 °C) in the MBE growth chamber in order to form the equilibrium droplets. This simple procedure always results in a broad size distribution, with typical examples shown in Fig. 4.4. Under the identical annealing conditions, larger equivalent thickness of Au layer leads to a large mean size and a wider size distribution.
4. MBE deposition of GaAs above the eutectic temperature resulting in the VLS nanowire growth. The growth process can be performed under different conditions such as the substrate temperature (usually in the range of 500–600° for GaAs), Ga deposition rate, As_2 or As_4 flux, and deposition thickness. Together with the initial size distribution of Au droplets, these conditions influence the resulting nanowire morphology: radii, lengths, surface density and shapes.

Although modern VLS techniques have many features in common with the early days Au-catalyzed CVD of Si wires, there are also important differences. First, the use of low temperature precursors enables a considerable decrease of typical MOCVD growth temperatures down to 400–500 °C in the case of III–Vs. At these temperatures, adatoms have a high surface diffusivity and can migrate from the substrate surface and the nanowire sidewalls to the droplet seated at the nanowire top. In the case of MBE where typical growth temperatures (usually $<$620 °C in the case of GaAs nanowires) are well below the evaporation temperatures for group III metals, the VLS growth just cannot be explained without considering the surface diffusion effects. Indeed, at an almost 100 % sticking probability to the substrate surface and possibly a higher desorption rate from the droplet, the vertical nanowire elongation rate cannot be much larger than that of the quasi-2D surface layer (see Fig. 4.3) growing concomitantly with the wires. As will be discussed in detail later on, the diffusion-induced contribution into the overall growth rate [47] is the essential feature of the entire VLS process. It plays the dominant role in the nanowire formation in most cases. In particular, the wire length can up to ten times exceed the deposition thickness during MBE [184]. Formation of 1D crystals by the surface diffusion was initially considered in connection with the dislocation-induced growth where the screw dislocation directed the diffusion flux to the top [307–309]. Historically such a growth was observed even earlier than the VLS one. While no screw dislocations are present in the VLS wires, many of the concepts and theoretical models developed

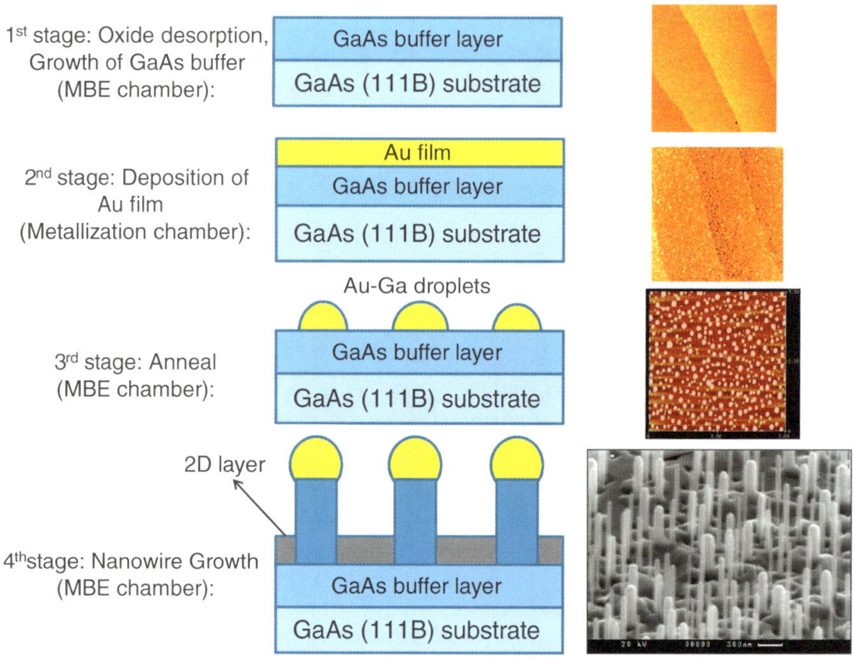

Fig. 4.3 Au-assisted VLS growth of GaAs nanowires on GaAs(111)B substrate by MBE, AFM images show the GaAs buffer (1st stage), the surface after Au deposition (2nd stage), 3D droplets after annealing (3rd stage), and SEM image shows the vertically aligned GaAs wires (4th stage)

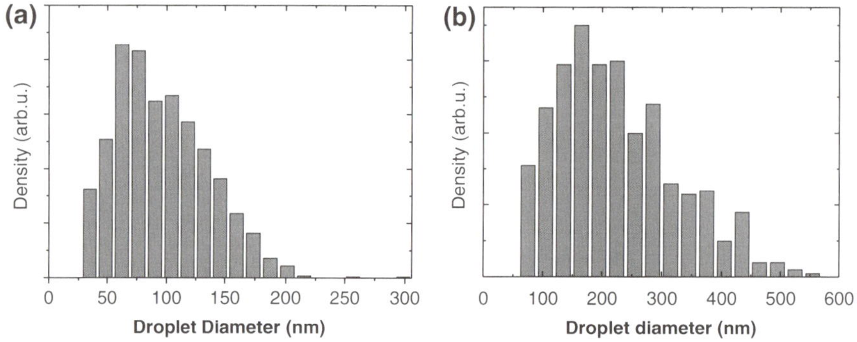

Fig. 4.4 Diameter distributions of Au-Ga droplets on the GaAs(111)B substrate after 5 min anneal of 1 nm (**a**) and 2.5 nm (**b**) Au layer in MBE growth chamber at 630 °C [310]

earlier are now widely used in studies of the VLS growth with surface diffusion effects. The role of metal droplet as a purely chemical catalyst should be therefore reconsidered: it also acts as a material collector that directs the adatom diffusion fluxes to the nanowire top.

The diffusion flux is directed to the top provided that the liquid chemical potential (μ_L) is smaller than that of surface adatoms far away from the wire top (μ_A): $\Delta\mu_{LS} < \Delta\mu_{AS}$. This inequality should hold for at least one of the two adatom systems on the substrate surface or on the nanowire sidewalls. Introducing the effective adatom supersaturation $\eta = \exp(\Delta\mu_{AS}/k_BT) - 1$, the condition for the positive diffusion flux to the top is given by

$$\eta > \zeta. \tag{4.3}$$

The adatom supersaturation is influenced by desorption processes and the surface growth. If the formation of parasitic surface layer is suppressed, the adatoms far away from the wire base or top are at the chemical potential of the vapor phase. This yields $\eta = \Phi$ and consequently $\Phi > \zeta$, which is equivalent to the condition for the adsorption-limited growth given by (4.2). When the surface growth occurs, the adatoms are at a lower chemical potential than the vapor phase. Therefore, the VLS growth usually proceeds under the condition $\Phi > \eta > \zeta$, where both vapor and diffusion fluxes to the droplet are positive. However, since surface ("s") and sidewall facet ("f") adatom systems have different supersaturations η_s and η_f, more complex growth scenarios are possible where the two adatom fluxes have different signs. A non-trivial interplay between different kinetic pathways during the VLS nanowire growth, combined with the nucleation kinetics at the liquid–solid interface under the droplet, will be studied in detail in what follows.

The second important feature of modern VLS nanowires is their reduced lateral dimension with typical radii of the order of 10 nm. As mentioned in Sect. 1.3, the Laplacian pressure $2\gamma_{LV}/R$ in the Au-Ga droplet of 10 nm radius is in the range of 0.14–0.23 GPa, which leads to a huge elevation of its chemical potential. For radii below 100 nm, the VLS growth usually proceeds via the mononuclear mode, whereby only one 2D island nucleate in each nanowire layer and then rapidly extends to fill the complete monolayer slice, as discussed in Sect. 2.12. In this regime, the nanowire growth rate is limited by the waiting time between two consecutive nucleation acts rather than by the lateral growth time of the island. Therefore, the formation of the nanowire monolayer occurs almost instantaneously. The nucleation may decrease the droplet surface area by ΔS_L. The corresponding correction $\gamma_{LV}\Delta S_L$ is of the order of other surface terms in the island formation energy as given by (1.206). Finally, the nucleation from a nano-scale droplet becomes essentially position-dependent. For instance, the nuclei may emerge in the center of the liquid–solid interface or at the triple phase line where some part of the island perimeter is in contact with the vapor phase [45]. These two positions are identical in terms of the volume energy but different in surface energies, leading to a dependence of the nucleation barrier on the position of the island. These examples show how complex is the individual nucleation process in nano-volumes compared to the conventional one where a metastable phase is infinite and homogeneous. On the other hand, the VLS growth in the steady state (i.e., at a time-independent droplet volume) contains one major simplification: the level of liquid supersaturation is maintained constant during growth. This is achieved through a dynamic balance between two non-equilibrium processes: material transport to the droplet and 2D nucleation from the droplet. Also, neither

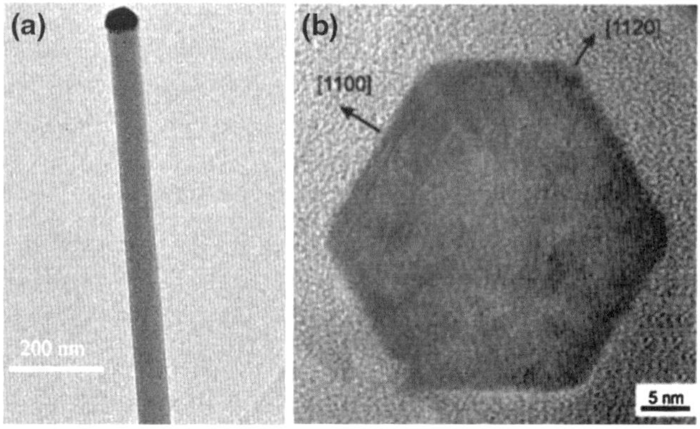

Fig. 4.5 **a** TEM image of a 34 nm radius GaAs nanowire obtained by the Au-catalyzed MOCVD on GaAs(111)B substrate [311]. **b** *Top-view* HRTEM image of a catalyst-free GaN nanowire obtained by MBE on Si(111) substrate covered by a 2 nm Si_xN_y thick amorphous interlayer, revealing that the vertical sidewalls are composed of m-planes (i.e. $(1\,\bar{1}\,00)$ planes) [312]

nucleus size distribution nor many-body effects such as coalescence are involved in the case of mononuclear growth.

Figure 4.5a shows a typical side-view TEM image of Au-catalyzed GaAs nanowire after growth [311]. It is seen that, under optimized conditions, the nanowire radius (34 nm in this case) does not vary along the length. The catalyst particle can be clearly seen at the wire top, and its round shape naturally leads to the assumption that it was liquid during growth. However, the liquid state of catalyst droplet is not always guaranteed and should be carefully checked taking into account the alloy composition [231] and the size-dependent depression of the melting temperature [303]. As for the sidewall geometry, the cubic (ZB) $< 111 >$ oriented nanowires of a constant radius are usually hexahedral, restricted by six equivalent low energy (110) planes or sometimes by the higher energy $(2\bar{1}1)$ planes. The hexahedral (WZ) $< 0001 >$ oriented wires are restricted by either $(1\bar{1}00)$ or $(11\bar{2}0)$ planes. Figure 4.5b shows the top-view HRTEM image of a WZ GaN nanowire obtained by the catalyst-free MBE [312] (the non-VLS self-induced GaN nanowires will be considered briefly in the next section and in more detail in Chap. 5), where the top facet is not hidden under the droplet. The image reveals that the vertical nanowire sidewalls are constructed of $(1\bar{1}00)$ planes in this case. The (110) and (211) ZB planes and their respective $(11\bar{2}0)$ and $(1\bar{1}00)$ WZ counterparts differ by 30° angular rotation. It should be noted here that, depending on the growth conditions and the nanowire material, more complex sidewall geometries can be observed, for example, the so-called twinning superlattice ZB nanowires composed of octahedral slices [44, 313, 317]. These nanowires are restricted by the alternating (111)A and (111)B microfacets and exhibit a sawtooth faceting as shown in Fig. 4.6.

Given the hexagonal shape of the top facet in hexahedral wires, the positioning of a liquid droplet on the nanowire top is not as straightforward as it would be if the

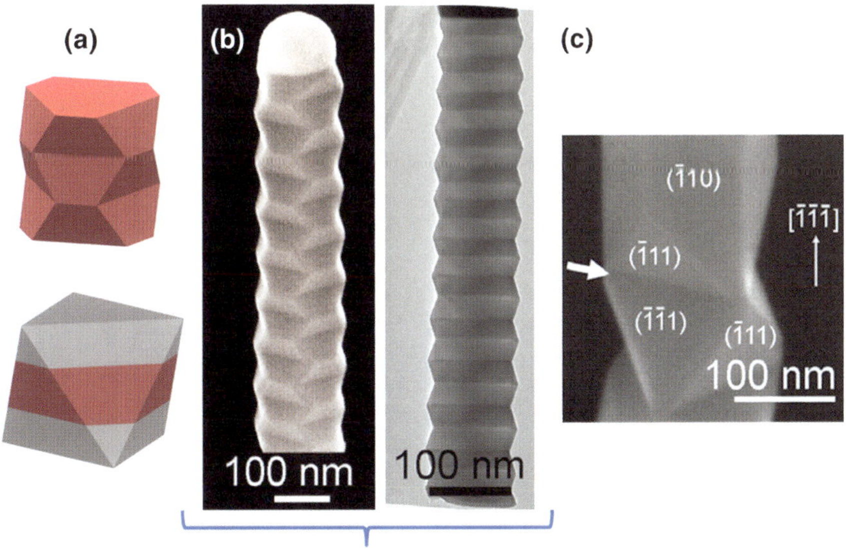

Fig. 4.6 **a** 3D model of a twinned nanowire with (111) microfacets. The octahedral slice is cut from the middle part of the full octahedron. **b** Examples of InAs nanowires from [234] exhibiting this facet structure are shown by SEM (left) and TEM (right). The height of the section varies with growth conditions up to the limit case of the full octahedron. **c** SEM image of an InAs nanowire containing a full octahedron, with some of the (111)-type facets labelled. The *thick arrow* indicates the twin plane following an octahedron. Above the octahedron, the side facets convert to the (110) type. These InAs nanowires were grown by MOCVD with aerosol Au nanoparticles as the growth seeds at temperatures above 400 °C [234]

wire had a cylindrical shape. This important problem was addressed by Krogstrup and co-authors in [318] in connection with the ZB-WZ polytypism of III-V nanowires that will be discussed later on. The program SURFACE EVOLVER [319] was used to calculate the equilibrium shapes of a droplet that is solely in contact with the (111) top facet, i.e., does not wet the sidewalls. In a cylindrical geometry, a spherical cap droplet with contact angle β would entirely cover the circular facet with the vapor–liquid–solid triple phase line being located at the top periphery, as shown in Fig. 1.7c of Chap. 1. The fraction χ of the triple phase line in contact with the cylindrical side surface would always equal one. The situation is much more complex on a hexagonal facet, as shown in Fig. 4.7. At a given equilibrium contact angle β_0 of the droplet with respect to the (111) plane, the local apparent contact angle β will depend on the angle ω between the middle of the side facet and the viewing position and the droplet volume. The χ value will also become a function of the droplet volume. Only the interval $0° \leq \omega \leq 30°$ can be considered because of the six fold symmetry of the top facet.

From Fig. 4.7a, the position $\omega = 0$ always favors the triple phase line in contact with the side facet, while the position $\omega = 30°$ at the corner favors the liquid-free sidewall. Figure 4.7a shows that there are two main configurations of the droplet: (b1)

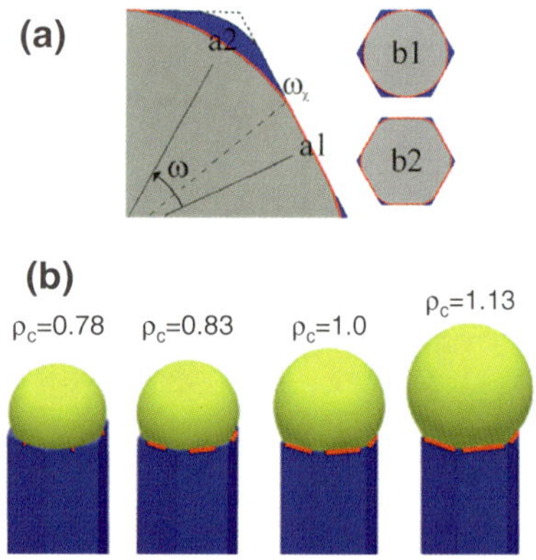

Fig. 4.7 **a** A top view illustration of a corner of the (111) top facet showing how the triple phase line separates the liquid and liquid-free regions of the facet. Position a1 with $\omega = 0°$ in the middle of the side facet is more easily covered by liquid than position a2 with $\omega = 30°$ at the corner. **b** Droplet shapes showing the increase of χ with the droplet volume at $\beta_0 = 110°$ [318]

the droplet is small enough to be an axisymmetric spherical cap with a circular triple phase line which means that $\beta = \beta_0$ all along the triple phase line and $\chi = 0$; (b2) parts of the droplet are pinned at the six edges of the wire, and the triple phase line acquires the shape of a hexagon with rounded corners. Figure 4.7b shows the droplet configurations calculated at fixed $\beta_0 = 110°$ [which is the estimate for the liquid Au on GaAs(111)] depending on the dimensionless length

$$\rho = k\rho_c, \tag{4.4}$$

introduced as a measure of the relative droplet volume. Here, $\rho_c(\beta_0) = V_c^{1/3}/(2R) = (\pi/24)^{1/3}[(1-\cos\beta_0)^{2/3}(2+\cos\beta_0)^{1/3}]/\sin\beta_0$ is the critical dimensionless length where the triple phase line is tangent to the edge of the side facets (corresponding to $\chi \to 0$), V_c is the critical droplet volume, R is the nanowire radius, and $k = (V/V_c)^{1/3}$ is the scaling factor. The χ value increases from zero at $\rho_c = 0.78$ to one at $\rho = 1.13$, and is about 0.8 at $\rho = 1$. In reality, the liquid will entirely cover the top facet even earlier, because the facet is expected to be rounded due to thermal and kinetic effects, as shown in Fig. 4.7a. One of the important conclusions here is that the triple phase line entirely fills the side facets at large enough droplet volumes, and in this sense the cylindrical wire geometry is a good approximation. However, the contact angle of the droplet has nothing to do with its equilibrium value on the (111) surface of the nanowire material, and changes with the position ω, because the geometry of pinned droplet is not exactly sapherical.

Let us now consider a very important question of the composition and state of metallic particle during growth of Au-catalyzed GaAs wires by MBE. In [231], three types of GaAs nanowires were grown on GaAs(111)B from Ga and As_2 molecular fluxes at a substrate temperature of 590 °C with a nominal Ga deposition rate of 0.2 nm/s. Wires A were grown for 45 min, and the Ga flux was stopped while arsenic was maintained for 2 min at 590 °C, before decreasing the substrate temperature. Wires B were also grown for 45 min, but Ga and As_2 fluxes were stopped simultaneously, and the substrate was cooled. Wires C were obtained with the same growth termination as wires B, but the growth time was increased to 160 min. Figure 4.8 shows TEM images of the wire tips, including the metallic particles. It is seen that the wire A exhibits a peculiar shape with a 26-nm-long constriction just before the metallic particle. This is explained by the consumption of residual Ga from the droplet under the arsenic flux, leading to a gradual diminishing of the droplet volume and consequently the wire radius. Wires B and C are not tapered at the top, because Ga and As_2 fluxes were turned off simultaneously. Selected area electron diffraction was used to investigate the structure of the metallic nanoparticles for each type of wire. The particle compositions were determined by energy dispersive X-ray spectroscopy (EDXS). Figures 4.8d–f show the diffraction patterns obtained on single wires of type A, B and C, respectively. These patterns contain spots related to GaAs diffraction and also reveal the crystalline structure of the metallic particles. The metallic crystal phases were assessed by the analysis of several nanowires of the same type. Each particle consisted of a single crystal. No As was found in these metallic phases. Wire A particle has a face centered cubic lattice with a lattice constant of 0.408 nm. The Ga and Au atomic concentrations are $c_{Ga} = 0.01$ and $c_{Au} = 0.99$. Wire B particle has an orthorhombic lattice with lattice constants $a = 0.640$ nm, $b = 0.627$ nm and $c = 0.345$ nm, and concentrations $c_{Ga} = 0.50$ and $c_{Au} = 0.50$. Wire C particle has the hexagonal lattice with $a = 0.772$ nm and $c = 0.875$ nm, and concentrations $c_{Ga} = 0.31$ and $c_{Au} = 0.69$.

It is remarkable that the EDXS data for three different types of solidified particles corresponds exactly to well defined solid compounds of the Au-Ga phase diagram at room temperature [306] which is shown in Fig. 4.9. Namely, the particle of wire A consists of almost pure Au, the type B particle contains the AuGa phase, and the type C particle contains the $\beta' Au_7Ga_2$ phase. According to the As-Au phase diagram [306], there is no stable As_xAu_{1-x} solid alloy under 636 °C. Consistently, no arsenic in found in the particles. The volumes of the purely Au particle in sample A and the volume of the GaAs constriction together with the numbers N_{Au} and N_{Ga} of Au and Ga atoms in these volumes were estimated, yielding a ratio N_{Au}/N_{Ga} of about 3. If we consider that these atoms were mixed in the particle before growth termination, the Ga concentration of the particle c_{Ga} should be close to 0.25 at that time. The Au-Ga phase diagram in Fig. 4.9 indicates that the $Au_{0.75}Ga_{0.25}$ alloy is liquid at the growth temperature of 590 °C. As regards the type B and type C particles, the As_2 and Ga fluxes were stopped simultaneously at the end of the growth. In this case, the metallic particles could not be emptied of their Ga, because As_2 vapor was not available to crystallize an extra volume of GaAs. However, the Ga concentration of 0.5 in type B wires is higher than the concentration of 0.25 which was deduced above for sample A before growth termination, and higher than the value of 0.31 for sample C. As discussed in [231] and further in [320], wires C, which are longer than A and B wires, have the same cylindrical shape within the first 3 μm of their length, but the last 4–5 μm of the wires are tapered. Their diameter decreases at a rate of about 25 nm per 1 μm length. This happens because the diffusion length of Ga adatoms on the wire sidewalls is of the order of 3 μm at 590 °C, which is why the diffusion flux of Ga to the top and consequently the Ga concentration in the droplet is decreased for wires longer than 3 μm. This explains why, although the growth

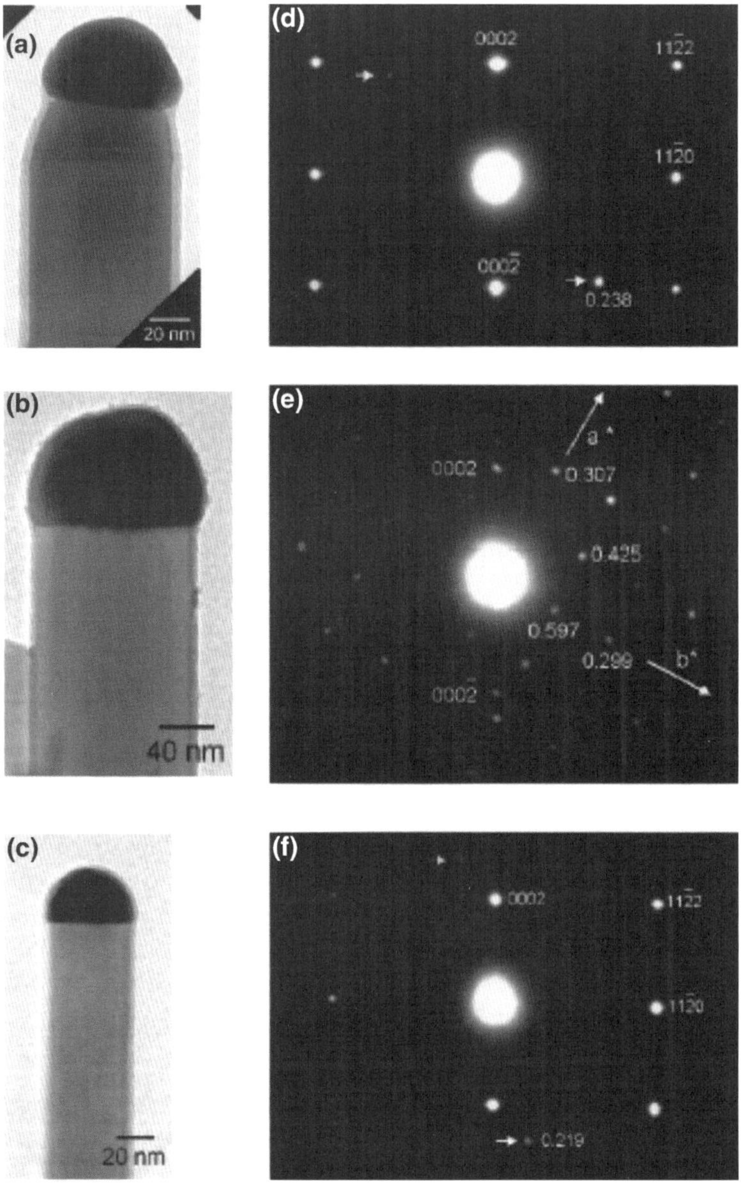

Fig. 4.8 TEM views showing the nanowire tips and the metallic nanoparticles of (**a**). sample A, (**b**) sample B, (**c**) sample C. Corresponding electron diffraction patterns; (**d**) sample A; (**e**) sample B; (**f**) sample C. The indexed spots correspond to GaAs. Inter-planar spacings in the metallic phases are given in nm for representative sets of diffraction spots shown by arrows in (**d**) and (**f**) [231]

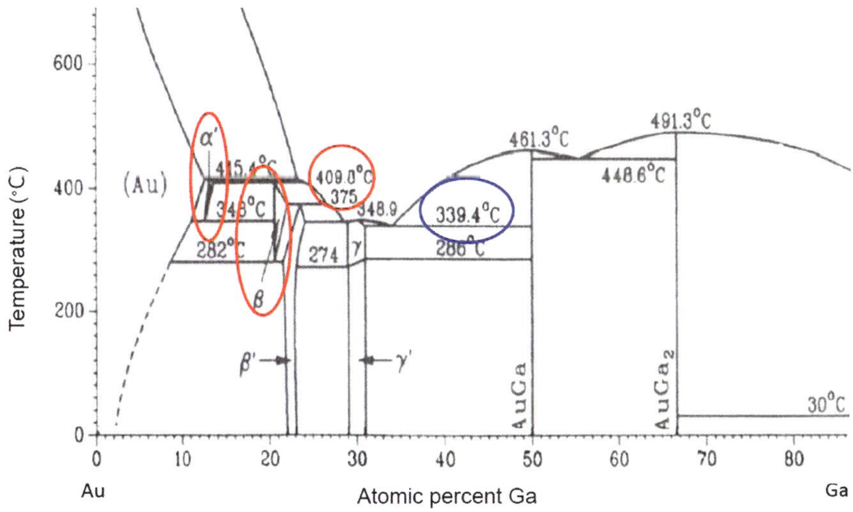

Fig. 4.9 Equilibrium Au-Ga phase diagram showing different solid compounds, liquidus lines, and the eutectic melting temperatures. Highlighted values correspond to the AuGa and the β' Au_7Ga_2 phases

termination procedure of samples B and C is the same, the metallic phase of wires C contains less Ga than that of wires B. As for the difference in composition between the particles of wires A and B, it might be due to their different lateral dimensions, because the Ga concentration and the corresponding supersaturation is generally a function of the wire radius.

In any case, the Ga concentration was systematically higher than 0.25 during growth. According to the Au-Ga phase diagram, this condition implies that the particles are liquid and the growth proceeds via the VLS mechanism. Further studies of metallic particles on the GaAs(111)B substrates by RHEED revealed solid state near 330 °C and melting at about 400 °C, while TEM observations at room temperature confirmed the Au_7Ga_2 solid phase in metallic particles [321, 322]. As discussed in [321, 322], the state of catalyst particle is very important for the temperature conditions for the Au-assisted growth of GaAs wires. The presence of As could modify the binary Au-Ga phase diagram. However, the role of As on the state of liquid particle must be negligible for the following reasons: (i) at 590 °C, there is no possibility of alloying As and Au; (ii) alloying As and Ga results in GaAs growth and in an eventual Ga-rich liquid phase with less than 3 % As. Finally, if for whatever reason the liquid particles are gradually emptied of their Ga, a critical concentration could be reached where a solid phase coexists with liquid. At 590 °C, this should happen for $c_{Ga} < 0.20$, and the particles might be completely solid at $c_{Ga} < 0.08$.

We can now formulate several general properties of metal catalysts that are suitable for facilitating the nanowire growth in the VLS approach:

1. Reasonably low melting temperature of alloy with a given semiconductor material (or a group III metal) to enable the liquid droplet state at the growth temperature.
2. Possibly simple phase diagram where the metal cannot form a stable solid alloy with a given semiconductor to avoid the nanowire contamination.
3. Efficient catalytic effect of metallic particle on the adsorption-desorption and diffusion processes with the droplet acting as a chemical catalyst or material collector under optimized growth conditions.
4. High liquid surface energy which is required for the stability of the non-wetting droplet configuration at the wire top as shown in Fig. 1.10 of Chap. 1.

The last requirement will be discussed in more detail later on.

Although a low surface energy group III metals such as Ga and In obviously do not satisfy the fourth condition of the list, there is an increasing interest in growing the so-called self-catalyzed III-V wires. Growth is promoted by the group III metals, e.g., Ga-catalyzed GaAs [48, 49, 293, 294, 323–325] or In-catalyzed InAs [325, 326] wires. The self-catalyzed approach is advantageous in several respects. First, the issue of unwanted Au contamination is safely avoided. Although equilibrium phase diagrams show no possibility of forming stable solid alloys of Au with Si and most III-V materials at typical growth temperatures, Au can still migrate from the droplet due to kinetic effects, for example, by diffusing onto the sidewalls. In particular, it has been shown in [356] in the case of Au-catalyzed Si nanowires that Au from the catalyst droplets wets the nanowire sidewalls, eventually consuming the droplets and terminating VLS growth. Gold diffusion from the smaller droplets to the larger ones (a process similar to the Ostwald ripening) leads to nanowire diameters that change during growth. Si nanowires shown in Fig. 4.10 were grown directly in a Hitachi ultra-high vacuum TEM H-9000 (UHVTEM) at 665 °C from disilane at 10^{-6} Torr. As the nanowires grew away from the surface they could be imaged in transmission mode. During growth the droplet labelled A grows larger, while the one labelled B decays and eventually disappears. These results clearly show that Au can diffuse from one catalyst droplet to another during growth. Au migrates over the surface as well as up and down the nanowire sidewalls Therefore, Au diffusion can fundamentally limits the length, shape and chemical purity of Si nanowires. Gold migration from Au-Si droplets was also observed in [327] by post-growth atom probe tomography (APT). Figure 4.11 shows a typical result of laser-assisted APT measurements of Au distribution in a Si wire. These wires were grown with Au catalyst on Si micropillars for APT measurements. Atomically resolved tomographic reconstruction clearly reveals the Au diffusion from the droplet during growth. Presence of Au clusters on the sidewalls of the Au-catalyzed Ge nanowires has also been confirmed by TEM imaging [328].

The second advantage is that self-catalyzed III-V nanowires can be monolithically grown on Si(111) substrates, where well-established silicon patterning techniques can be used to organize group III droplets in order to produce regular nanowire arrays. Third, group III-catalyzed III-V nanowires do not generally suffer from uncontrolled ZB-WZ polytypism (which is a common issue in the Au-catalyzed

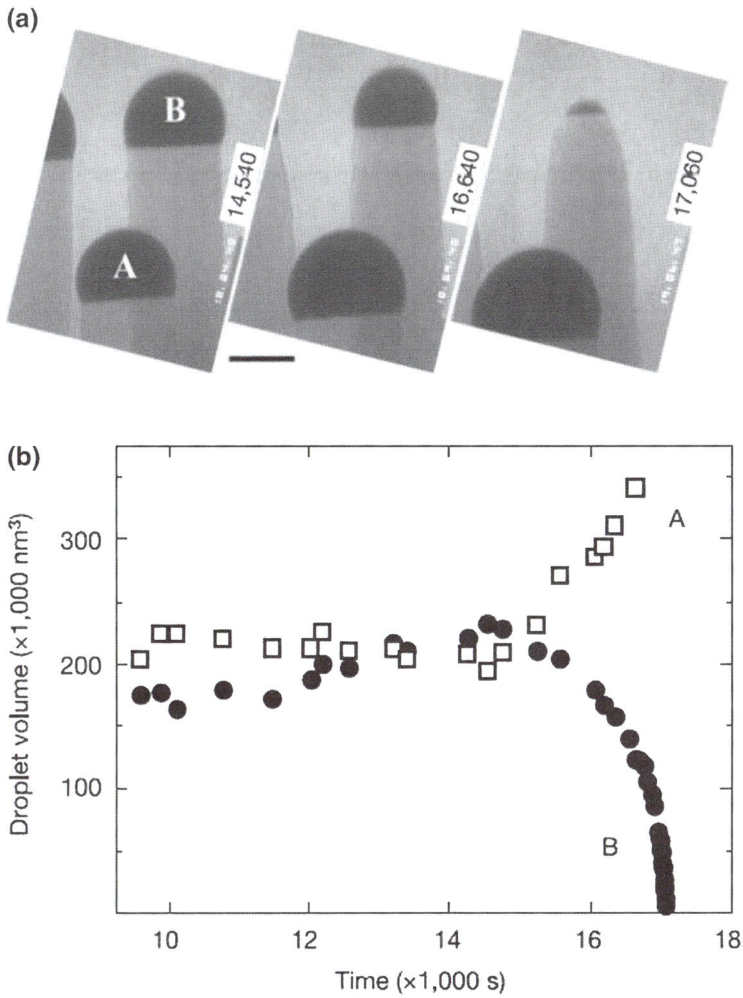

Fig. 4.10 In situ UHVTEM images recorded during the growth of Si nanowires at 655 °C in 10^{-6} Torr disilane. **a** Three images labelled by the time (in seconds) after the start of growth. *Scale bar* 50 nm. **b** Measured volume versus time for the two catalyst droplets labelled in **a** [356]

case). For example, most Ga-catalyzed GaAs nanowires adopt almost a pure ZB crystal phase under optimized conditions [48, 323].

Synthesis of self-catalyzed nanowires involves many features that are distinctly different from a more conventional growth with a foreign metal catalyst, and that are not completely understood so far. Figure. 4.12 shows a typical example of Ga-catalyzed GaAs wires obtained by MBE on Si(111) substrates. A typical MBE growth procedure in the self-catalyzed approach is briefly as follows [49]. Before introduction into the growth chamber, the substrates are chemically treated in HF (10 % in

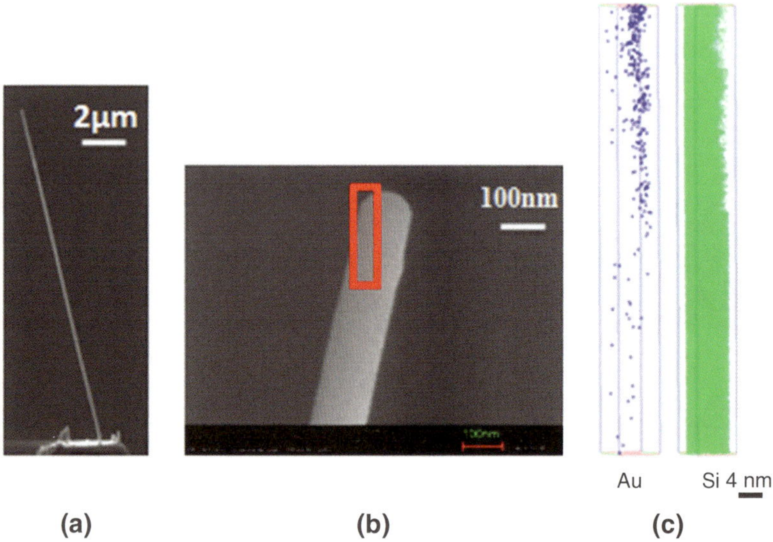

Fig. 4.11 **a** SEM image of an Au-catalyzed Si wire grown on a Si micropillar for APT measurements. **b** Magnified image showing the investigated region. **c** Atomically resolved 3D APT reconstruction of Au and Si distributions in the sidewall region under the droplet

water), and then rinsed in de-ionized water. Before the growth of nanowires, the substrate temperature is increased to the desired value, usually within the range 560–630 °C, and is kept constant during the whole growth run. In this temperature window, a native oxide layer on Si(111) substrate loses continuity and openings (i.e., the holes penetrating through the oxide toward the substrate surface) are formed. The surface density of openings depends on the temperature. Above 630 °C, the oxide layer is completely desorbed, as detected by the corresponding transformation of RHEED pattern. After the formation of openings, the Ga flux is supplied to the surface for some time (5 s in this particular case), while the As shutter is closed. This initiates the formation of Ga droplets in the openings. When the As flux is switched on (at a typical V/III fluxes ratio of $\sim$1), the wire growth is started after an incubation time. The latter is clearly detected by the transformation of the RHEED pattern and typically amounts to 2–10 s, depending on the temperature.

The droplets seated on the tops of the longest wires are clearly seen on the SEM image of Fig. 4.12a. However, some fraction of short nanowires does not seem to have the droplets on top. Regardless of the substrate temperature, the wire cross-section is a regular hexagon with six equivalent lateral facets, identified as being of (211) family in this case. Statistical analysis of the SEM images of different samples allowed evaluation of the substrate temperature dependence of the wire length and surface density, shown in Fig. 4.12b. It is seen that the average length of 1.8–1.9 μm, and the density of $\sim 1 \times 10^8\,\mathrm{cm}^{-2}$ remain almost constant in the lower temperature range, i.e., from 560 to 600 °C. At higher temperatures, the density increases up to

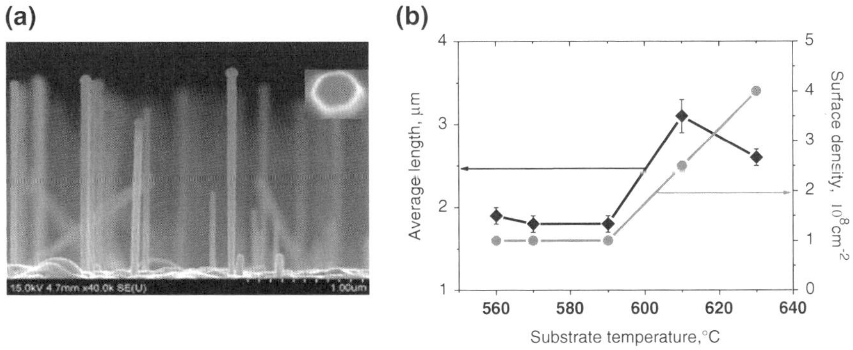

Fig. 4.12 **a** Lateral cross-section view SEM image of Ga-catalyzed GaAs nanowires on Si(111) substrate grown at 590 °C, the *inset* shows the hexagonal nanowire cross-section. **b** Substrate temperature dependences of the average length and surface density

$\sim 4 \times 10^8$ cm^{-2} at 630 °C, while the length behavior is non-monotonic, first increasing up to ~3 μm at 610 °C and then decreasing down to ~2.5 μm at 630 °C. Energy dispersive X-ray analysis after growth shows that the droplet consists of pure Ga. The average length of 1.8 μm corresponds to the average nanowire growth rate of 1.5 nm/s. At a higher temperature of 610 °C, the nanowire growth rate reaches its maximum at 2.5 nm/s. The observed growth rates are 5.4–8.9 times higher than the deposition rate of 0.28 nm/s, so that the nanowire length is much larger than the deposition thickness. This proves the diffusion-induced mode of nanowire growth with major contributions from the adatom migration to the top, as discussed in the foregoing sections.

Given a low melting temperature of group III metals used as catalysts for III-V nanowire growth (157 °C for In and only 30 °C for Ga), there is no question about the liquid state of the droplet and the VLS growth mechanism of self-catalyzed nanowires. The latter should now be re-formulated in the following way: initial droplets formed in the openings in the absence of an As flux must consist of pure liquid Ga. When the As deposition is started, the growth species are supplied to the droplet by direct impingement and, more importantly for Ga, by surface diffusion. The corresponding diffusion fluxes are driven by the difference in chemical potentials on the surface and in the droplet. Since stoichiometric GaAs cannot be formed in the liquid volume (because of its low As content), excessive Ga and As are transferred through Ga droplet to the solid surface (initially Si and then GaAs), where a supersaturated liquid crystallizes in the VLS fashion via 2D nucleation. After the steady state growth mode is achieved, the droplet size and shape remains constant due to a dynamic balance between the GaAs arrival to the droplet and the GaAs sink at the liquid–solid interface. The latter provides continuous building of nanowire bilayers.

It is obvious that this process should be extremely sensitive to the V/III fluxes ratio. Since Ga droplet serves as a reservoir of gallium for the arriving arsenic, the VLS growth of GaAs wires should be always As-limited (while it is usually Ga-limited in the Au-catalyzed case). Many experiments (see, e.g., [324] for a review) indeed show that the growth rate is proportional to the As flux. The As diffusivity on the sidewalls is much lower than that of Ga [329, 330], while the Ga diffusion flux should remain as a dominant kinetic pathway for gallium supply in most cases [331]. The steady state stoichiometric growth at a time-independent droplet volume can proceed only within a certain window of optimal V/III fluxes ratio, where the net atomic fluxes of As and Ga equal each other: $dN_{As}/dt = dN_{Ga}/dt$. The As influx may be significantly enhanced by its re-emission from the neighboring nanowires. The droplet would inflate with excessive Ga at a lower As flux ($dN_{As}/dt < dN_{Ga}/dt$) and shrink when the As supply is too large ($dN_{As}/dt > dN_{Ga}/dt$). Nanowires grown with an excessive Ga influx will extend radially. Even when the two influxes are equalized, a very low surface energy of group III metals (0.556 J/m^2 for In and 0.711 J/m^2 against 1.145 J/m^2 for Au at their respective melting temperatures [39]) should not favor the droplet stability at the wire top as discussed above. Some of the droplets would completely wet the sidewalls and finally disappear, which might be the explanation of the droplet-free short wires in Fig. 4.12a. This important issue will be considered in this chapter and further in Chap. 6 in connection with the crystal structure of GaAs nanowires.

Regardless of the epitaxy technique used for material deposition, modern VLS growth methods are characterized by the following general features:

- Appropriately optimized growth conditions usually enable close to 100 % yield of vertical wires.
- Position and surface density of VLS nanowires are determined by the properties of the initial ensemble of metallic droplets.
- The nanowire radius is determined by the size of initial droplet, while the nanowire length depends on the deposition conditions, growth time and the nanowire radius.
- VLS growth from the size-uniform and spatially ordered arrays of droplets results in the regular arrays of exactly identical nanowires.

Size, shape and position control of VLS nanowires is clearly of great relevance for applications. Such a control is impossible in the self-induced approach where the spatial position is random and the size distribution always has a certain width due to a random character of the nucleation process. Initial metallic droplets can be made identical in size and organized in a regular 2D lattice by applying various lithographical techniques such as nanoimprint or e-beam lithography. In a given growth run, the wire length and even the shape (in the case where the droplet volume changes in time due to kinetic effects or the radial extension is enabled) is dictated by the size of the initial droplet catalyzing its growth. Therefore, the nanowires grown from identical droplets would have identical length and geometry. Typical examples of regular arrays of Au-catalyzed III-V nanowires obtained by MOCVD with nanoimprint lithography [332] and by MBE with e-beam lithography for organizing the Au droplets are shown in Fig. 4.13.

As discussed in Sects. 3.3 and 3.4 (in the case of self-induced islands and nanoneedles) and will be further explored in Chap. 5 for the VLS nanowires, a small footprint of highly anisotropic vertical nanostructures in contact with a mismatching substrate and their developed strain-free sidewalls enable a radical increase of the

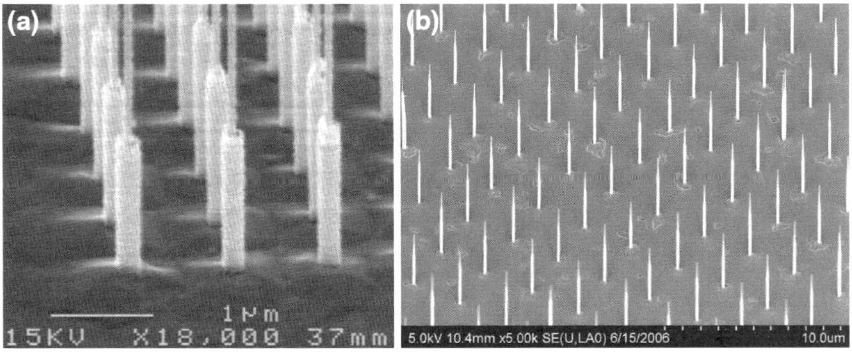

Fig. 4.13 Regular arrays of Au-catalyzed InAs (**a**) [332] and GaAs (**b**) nanowires

critical thickness for plastic deformation. Furthermore, it has been demonstrated that the formation of misfit dislocations in cylindrical nanowires grown on lattice mismatched substrates is completely suppressed provided that their lateral dimension is smaller than a certain critical value [187, 232, 235]. The same property applies for axial [187] and core-shell [333–337] heterostructures in nanowires, where the elastic stress relaxation is energetically preferred to the plastic deformation at a small enough radius of the wire or that of the core. Therefore, nanowires show great promise to enable a coherent growth of high quality heterostructures that are much less restricted by lattice mismatch than in the 2D thin film case and even in quantum dots. During the VLS nanowire synthesis, one can switch between different material compositions and regulate the rates of their axial and radial growths by tuning the deposition conditions. In this way, different types of heterostructures can be obtained within the VLS wires as illustrated in Fig. 4.14:

- Axial hererostructure where the two materials stack vertically.
- Radial, or core-shell, heterostructure where one material surrounds the other along the entire nanowire length.
- Quantum dot in nanowire where a low bandgap material forms an island inside the wire.
- Quantum dot formation on the nanowire sidewalls or on the top facet, a self-induced process that may proceed when the material is deposited onto the lateral facets of a lattice mismatched wire.

Typical examples of different nanowire heterostructures are also shown in Fig. 4.14. Quite clearly, the formation mechanisms and the resulting properties of these heterostructures are rather different from 2D thin films, in-plane quantum wires and quantum dots.

As regards theoretical modeling, this chapter is devoted solely to the nanowire growth. More delicate aspects of nanowire physics (elastic relaxation and plastic deformation in nanowire heterostructures, nanowire shape, doping process, nucleation statistics, self-induced GaN nanowires, non-stationary and collective effects) will be considered in Chap. 5, while Chap. 6 is devoted to the crystal structure

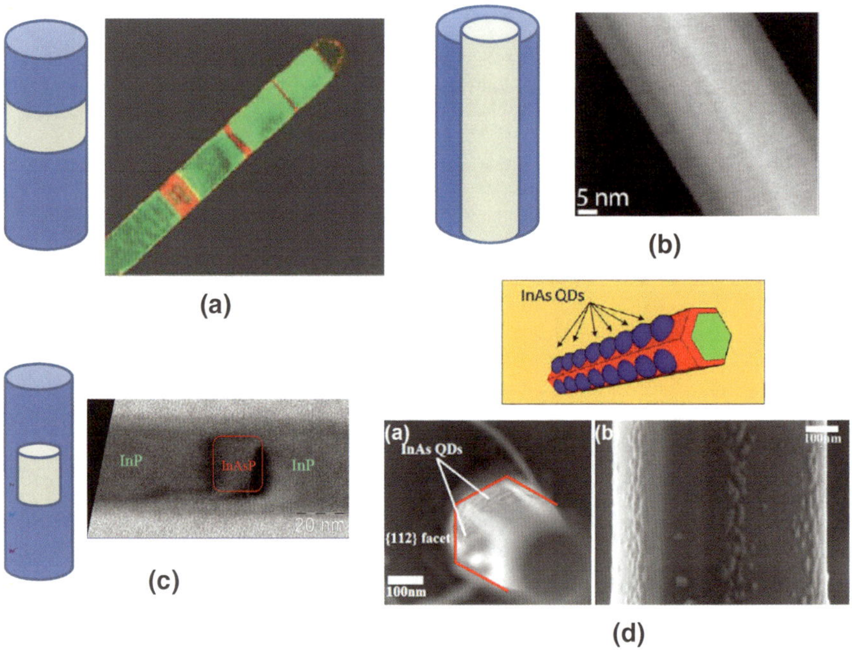

Fig. 4.14 Different types of nanowire heterostructures: axial heterostructure in the InAs/InP system [338] (**a**), core-shell InAs/InP nanowire [337] (**b**), $InAs_{0.35}P_{0.65}$ quantum dot inside an InAs nanowire [339] (**c**), and InAs quantum dots on the sidewalls of a GaAs nanowire [335, 340]

of the VLS III-V nanowires. However, these problems cannot be separated in most cases, because the resulting morphology and crystal phase are strongly affected by the growth process. Illustration of the main kinetic processes contributing into the steady state VLS growth is shown in Fig. 4.15. The VLS growth is driven by the vapor influx I creating the vapor supersaturation Φ. In the case of III-V compound nanowires, we assume that the growth is limited by the transport of group III element, which usually occurs under group V reach conditions. In the directional MBE or chemical beam epitaxy (CBE) methods, the flux is characterized by the incident angle α with respect to the substrate normal. In CVD techniques involving chemical reactions, the arrival rates depend on the substrate temperature via the pyrolysis efficiencies at different surfaces. The material directly impinges and desorbs from the droplet surface (processes 1 and 2 in the figure). If the arrival onto the substrate and nanowire sidewalls is disabled or the adatom diffusion lengths are negligibly small (such a situation is typical for a high temperature CVD where the droplet acts as a chemical catalyst), the VLS growth is adsorption-induced. The vapor influx renders the droplet into a metastable state with supersaturation ζ. Semiconductor material in the droplet precipitates to the liquid–solid interface under the droplet where 2D islands nucleate and then extend laterally to form a complete ML slice. This gives rise to the nucleation-mediated vertical growth rate V_L which is generally a function of the wire radius R and the deposition conditions such as temperature and flux.

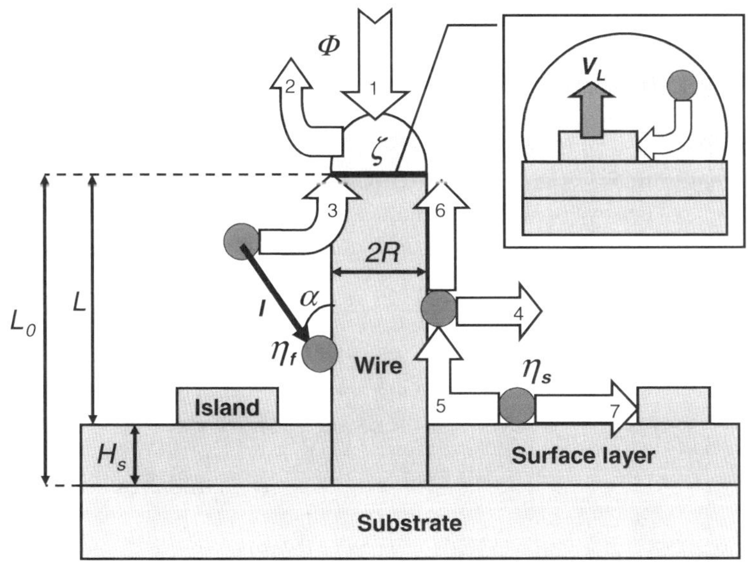

Fig. 4.15 Kinetic processes during the VLS nanowire growth: *1*—direct impingement onto the droplet surface; *2*—desorption from the droplet, *3*—diffusion flux from the sidewalls to the droplet, *4*—desorption from the sidewalls, *5*—diffusion from the substrate to the sidewalls, *6*—diffusion from the substrate along the sidewalls to the drop, *7*—nucleation on the surface. The insert shows the nucleation-mediated growth on crystal facet under the drop, resulting in the vertical growth rate $V_L = dL_0/dt$. The wire length $L = L_0 - H_s$, where H_s is the mean thickness of a 2D surface layer burying the wire bottom

As discussed above and will be explored in more detail later on, the adsorption-induced growth mode does not pertain for most of modern VLS nanowires whose growth is strongly affected by the surface diffusion of adatoms. Therefore, Fig. 4.15 shows the diffusion-induced kinetic pathways to the droplet, including the adatom diffusion from the sidewalls (process 3) which may be limited by either desorption (process 4) or incorporation to the sidewall surface steps. In the latter case, the wire would extend radially with time. With a high enough surface diffusivity, adatoms can also impinge the substrate surface, diffuse to the sidewalls and then to the wire top along the sidewalls (processes 5 and 6). Nucleation on the substrate surface results in a quasi-2D growth of the substrate (process 7), which is why the bottom part of a nanowire gets buried by the surface layer. When the surface diffusion processes are included, one has to consider the two adatom subsystems on the sidewal facets ("f") and on the main surface ("s"), characterized by the effective supersaturations η_f and η_s, respectively.

Adequate theoretical modeling of the VLS nanowire growth should therefore include:

- The mass transport equations describing different kinetic pathways to the droplet;
- Modeling of droplet configurations and stability depending on the material parameters and the growth conditions;

- Description of 2D nucleation process at the liquid–solid interface or at the triple phase line;
- Accounting for the size-dependent effects such as the Gibbs-Thomson effect, the mononuclear growth and the position-dependent nucleation rate;
- Self-consistency in the steady state VLS process implying that the nucleation-mediated vertical growth rate equalizes the total material flux into the droplet;
- Modeling of material and kinetic parameters entering the growth equations: chemical potentials in the droplet, surface energies of different interfaces, diffusion lengths etc.

Many of these problems were addressed right after whisker physics and technology emerged [46, 47, 256, 257, 307–309, 341–344], in particular, by Wagner and Ellis themselves [46, 341–343], and Givargizov and Chernov [47, 256, 257, 343]. In the 2000s, nanowire growth modeling was re-visited at the nanoscale. Important contributions into the understanding of fundamentals of VLS growth have been made by the Lund group [83, 229, 234, 276, 325, 326, 345–351], LPN CNRS [45, 118, 231, 290, 321, 322, 352–355], the IBM research group [44, 356–360], the MPI Halle group [287, 288, 361–363], the group at McMaster University (Canada) [296, 364–367], the Duisburg group [329, 330, 368, 369], the group at Niels Bohr Institute Copenhagen [318, 323, 370], the Delft-Eindhoven group [314, 371, 372], EPFL [324, 373], University of California at San Diego [374–376] and many other researchers [43, 81, 300, 377–403], including some works from our group at St. Petersburg [48, 49, 82, 84, 117, 184, 291, 311, 320, 404–414]. Theoretical approaches include kinetic modeling based on the transport equations [117, 290–292, 345–348, 366, 377, 379, 380, 408–410] and the nucleation theory [45, 47, 49, 81, 82, 84, 118, 184, 318, 311, 323], approaches based on energy balance considerations and growth thermodynamics [43, 44, 48, 256, 257, 342, 343, 400], molecular dynamic and kinetic Monte Carlo simulations [378, 389–391], and first principle modeling of growth, morphology and crystal structure [329, 330, 368, 369, 392]. In the following, we adopt the first approach where the mass transport equations are used to calculate the material influx to the droplet. Classical nucleation theory is used for the description of the layer-by-layer growth at the liquid–solid interface and is substantially modified to account for the finite droplet size and configuration.

4.2 Alternative Growth Mechanisms

Before continuing with the VLS nanowires, we only briefly discuss some alternative mechanisms that can also be used for fabrication of freestanding semiconductor nanowires. First, the so-called vapor-solid-solid (VSS) mechanism was introduced by Person *et al.* [229] and further developed by Dick *et al.* [415]. This mechanism suggests that the catalyst particle at the wire tip is solid during growth. In [229], through in situ TEM analysis and EDXS measurements of CBE grown GaAs nanowires, the authors observed the crystallinity of the Au-Ga particle and a low Ga concentration below the eutectic melt level. The analysis of [415] refers to the case of Au-catalyzed InAs nanowires grown by MOCVD below the eutectic temperature of Au-In alloy at 454.3 °C [306]. As discussed in Sect. 4.1, arsenic is present in the droplet only in trace amount and thus only the binary phase diagram can be considered in the first approximation. The basis of the VSS mechanism is consistent with the results obtained earlier by Kamins *et al.* [416], where the Ti-catalyzed Si nanowires were obtained at ~600 °C, which is much below the lowest melting temperature of Ti-Si of about 1300 °C [306]. While the concept of droplet acting as a pure chemical catalyst is more difficult in the VSS case, a low chemical potential in an Au-semiconductor alloy can still favor the diffusion-induced growth. At a high surface temperature (400 °C and above), the solid diffusion in the crystalline catalyst particle should be fast enough to promote the diffusion transport of semiconductor material to the topmost facet of the wire.

However, as discussed by Park and Prokes in [303], strictly based on the Gibbs-Thomson effect (see Sect. 1.3), one can expect the melting point depression to be about 60 °C for ~30 nm nanowire tips. The experimental values, measured during heating of 30 nm diameter Ge wires in a TEM, give even more significant effect of a 25 % melting point reduction [303]. The reason for this large size-dependent reduction of the melting temperature compared to theoretical value is not clear at this point. Taking into account the size-dependent liquidus depression in binary systems, the results of [229, 415, 416] should be re-interpreted [303]. In the case of the Ti-catalyzed Si wires described by Kamins *et al.* [416], the growth temperature is so far below the melting point that the liquidus depression factor does not seem applicable, and the catalyst particle should be indeed solid. As discussed in Sect. 4.1, we do not see any physical reason for the VSS mechanism in the case of the Au-assisted GaAs nanowires, at least for the temperatures above 500 °C. As for the Au-catalyzed InAs wires with a typical growth temperature of about 400 °C [292, 415], the liquidus depression factor seem to contradict with the claim of a solid metal alloy tip. However, the state of catalyst particle may strongly depend on the wire radius, and a coexistence of the liquid and solid phases is not excluded [385].

A detailed study of Au-assisted MBE of InAs nanowires [292] revealed the wire formation in a relatively narrow temperature window extending from 380 °C to 430 °C (at a fixed nominal deposition rate of 0.2 nm/s, and the V/III fluxes ratio of 3). The wire tip diameters were in the range 20–40 nm. To address the catalyst phase state issue, RHEED technique was used as an in situ probe to observe the behavior of the catalyst particles on the InAs surface. After Au deposition, the sample was annealed under As flux at 460 °C to induce Au alloying with the substrate constituents. Then, the sample was cooled down at a rate of ~10 °C min^{-1} and the RHEED patterns were recorded. At temperatures above 360 °C only a (2×2) streaky pattern typical for a (111)B InAs surface was observed (Fig. 4.16a). Around 360 ± 20 °C, additional diffraction spots appeared superimposed on the (2×2) pattern (Fig. 4.16b). The spots were ordered, indicating that a part of the electron beam was diffracted through the volume of a crystalline phase. These spots were situated outside the surface diffraction streaks and therefore cannot originate from InAs. As in the case of Au deposition on GaAs [322], the spots were attributed to the

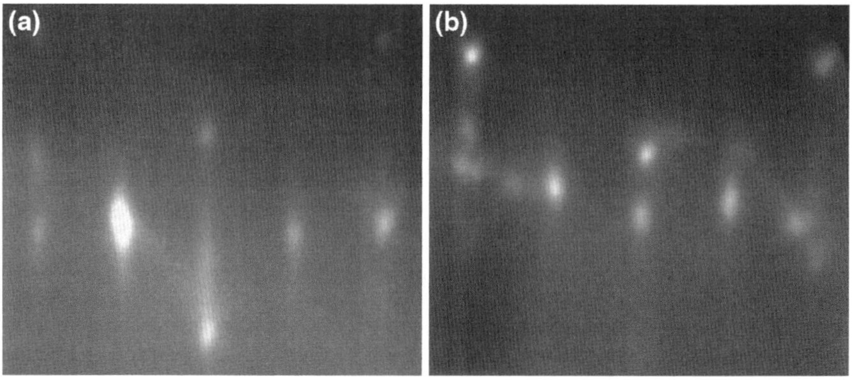

Fig. 4.16 RHEED patterns of Au deposited on a clean (111)B InAs surface: **a** at 410 °C after annealing; **b** at 360 °C after cooling. Additional diffraction spots originating from the catalyst particles are observed outside the surface streaks at 360 °C

diffraction by the solid metallic particles resulting from the interaction of Au with the sample surface. When the sample was heated again, the additional diffraction spots gradually disappeared at around 425 ± 20 °C and the picture returned to the (2 × 2) pattern. Several temperature cycles were repeated, resulting in reproducible observations. We believe that these observations correspond to the phase transitions of the metallic particles containing Au from the liquid to the solid state and vice versa. Based on these data, it can be concluded that the catalyst particles are liquid above 425 °C and solid below 360 °C. Between these limits a hysteresis region is observed, where the particle phase state is determined by the history of the sample. This 65 °C hysteresis most likely corresponds to supercooling of the liquid particle: there is a nucleation barrier which, under cooling below the eutectic point, prevents the formation of the catalyst solid phase within this range of temperatures. The enhancement of the melting point hysteresis has been reported in nanocrystals [417].

Similar experiments were performed for different Au deposition thicknesses, showing that the Au thickness has no significant influence on the phase transition temperature, at least for reasonably small amounts of Au thickness less than 5 nm. On the contrary, the change of the sample environment (for example, the incident As flux) can influence the hysteresis cycle. The experiments reported above were performed under the same As pressure as the one used for the nanowie growth (9×10^{-6} Torr). The temperature cycling performed without As flux revealed the 20 °C lowering of the liquid-to-solid transition temperature with respect to the experiment under As flux; the solid-to-liquid catalyst transition temperature was not affected. This could be explained by the different catalyst composition in the two cases. Indeed, the In composition of the solid catalyst phase may change if As is available to consume some of this In, which can re-condense as InAs on the substrate. Other conditions like the V/III ratio and time factor may also influence the hysteresis cycle. Because the hysteresis region overlaps with the wire growth temperature domain, no definite conclusion can be drawn concerning the catalyst phase state during the growth of InAs nanowires, but the liquid phase seems plausible. We also note that, after growth, the catalyst particles on top of the wires have a hemispherical shape with no defined facets. This also suggests that these particles were liquid during growth. Indeed, when the catalyst particle solidifies during the growth, the in situ TEM observations show that its surface becomes faceted [418].

Next, we mention the self-induced GaN nanowires that can be grown on silicon without any catalyst [312, 419–437]. MBE growth of GaN nanowires is usually performed on Si(111) substrates either covered with a lattice mismatched AlN inter-layer or an amorphous Si_xN_y layer. The self-induced approach commonly employs specific MBE growth conditions: a highly nitrogen-rich vapor phase (with V/III ratio larger than 4) is required and often combined with high substrate temperature (>750 °C). One of the examples is shown in Fig. 4.17, where the diffusion-induced growth model is illustrated [422]. No Ga droplet is detected at the nanowire top by electron microscopy imaging, so the self-catalytic effect cannot explain their formation according to the VLS mode. As will be discussed in more detail in Chap. 5, GaN never nucleates in the nanowire morphology [428, 429, 432, 434, 436]. Rather, the nanostructures emerge as the Volmer-Weber 3D nanoislands having a fixed shape and a complex free surface composed of high-index atomic planes. After that, the islands grow with a fixed aspect ratio (on an amorphous inter-layer) before their transformation to hexahedral nanowires with a flat top plane. Very importantly, self-induced GaN nanowires usually grow in both vertical and radial directions, as demonstrated by the histograms in Fig. 4.17c after 30 and 60 min of growth. Therefore, the self-induced formation of GaN nanowires is a rather complex growth process combining the features of a spontaneous 3D islanding (which may be strain-induced or driven by the non-wetting surface energetics) and the anisotropic nanowire elongation which becomes energetically preferred after a certain critical dimension. As shown in Fig. 4.17, the growth is controlled by the direct impingement onto the top facet and, more importantly, by the surface diffusion of adatoms. In absence of metal catalyst, the flat top facet with a low chemical potential acts as a material collector directing the diffusion flux to the top.

Finally, Fukui group [438–441], Hamano *et al.* [442], Poole *et al.* [443] and some other researches pioneered catalyst-free, selective area MOCVD growth of GaAs and other III-V nanowires, which is now widely used in different epitaxy techniques and material systems, including GaN [442–445]. In the selective area approach, the nanowire growth direction is defined by lithographically defined openings. Figure 4.17a shows schematic illustrations of a typical fabrication process of GaAs nanowires. After a 20 nm thick SiO_2 layer was deposited by plasma sputtering on GaAs (111)B substrates, periodic hexagonal opening patterns were formed by electron-beam lithography, and by wet-chemical etching techniques. Hexagonal openings were arranged in a triangular lattice (Fig. 4.17b), with the pitch a ranging from 0.4 to 3 mm. The actual shape of the opening tended to be circular because of the resolution limit in electron beam lithography and wet chemical etching. The diameter of nanowires appeared to be directly related to the opening diameter d_0, and a smaller d_0 resulted in a longer nanowire height.

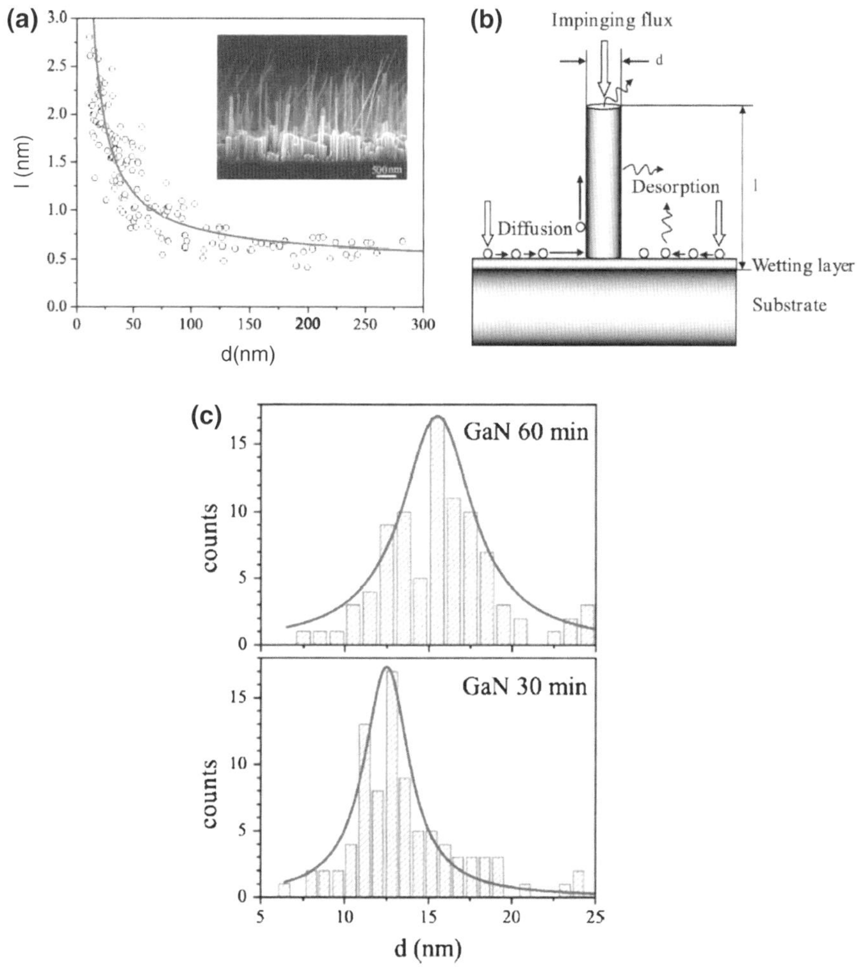

Fig. 4.17 Diffusion-like decreasing length-diameter dependence of the self-induced GaN nanowires on Si(111) shown in the insert (**a**), and the diffusion-induced growth model where the *top* facet acts as a material collector (**b**) [422], (**c**) Histograms showing the increase of GaN nanowire diameters with the growth time [422]

Selective area GaAs nanowire growth was carried out in a horizontal MOCVD system working at 0.1 atm. Trimethylgallium (TMGa) and 20 % arsine (AsH_3) diluted in H_2 were used as precursors. The partial pressure of TMG was 2.7×10^{-7} atm and that of AsH_3 was varied within the range from 2.5×10^{-4} to 1.7×310^{-3} atm. The growth temperature was 750 °C. In these conditions, the growth rate of GaAs on a planar (100) substrate is 0.21 and 0.03 nm/ s on (111)B, but the growth rate of nanowires is much higher, i.e., is controlled by the surface diffusion. The total growth time was 20 min. Fig. 4.18c show secondary electron microscopy (SEM) images for

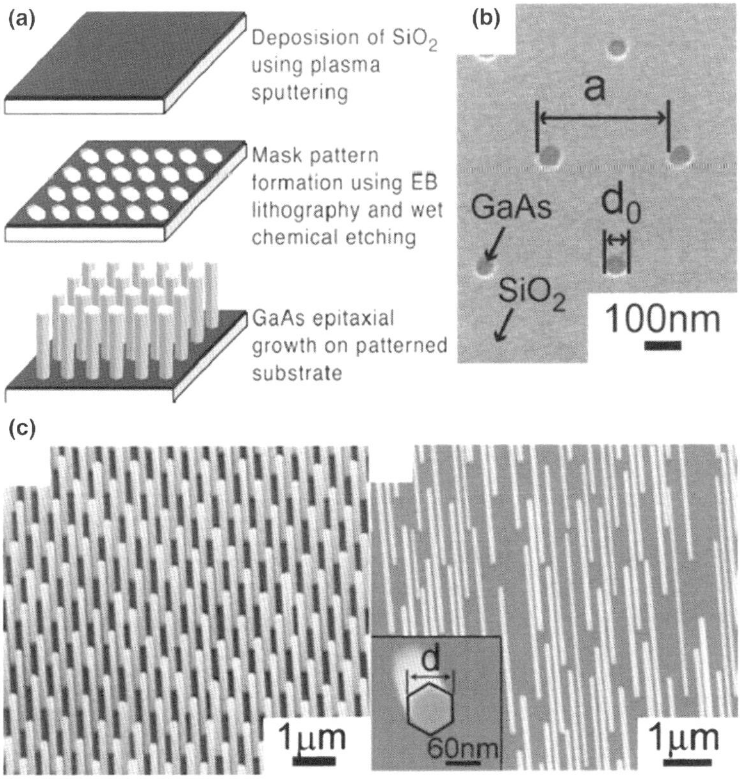

Fig. 4.18 **a** Schematic illustration of the fabrication process. **b** SEM image of the mask pattern on a GaAs (111)B substrate. The pitch a of the triangular lattice is 0.5 μm for this pattern [440]. **c** SEM images of GaAs nanowires for pattern opening d of (1) 200 nm and (2) 50 nm. The inset in (2) is an image from the *top* [440]

two typical samples of GaAs nanowires grown on a GaAs (111)B masked substrate at an AsH_3 partial pressure of 5.0×10^{-4} atm. The pattern period a is 1 mm, and the diameter d_0 is 200 and 50 nm for the samples shown in Fig. 4.18c. GaAs nanowires were grown in a (111)B vertical direction on the substrate. As shown in the inset in Fig. 4.18c, the nanowires have a hexagonal cross section with six (110) side facets. The diameter of the nanowires becomes smaller as the initial opening diameter d_0 is reduced. For the sample with $d_0 = 200$ nm, the diameter and the height of the nanowires are 200 nm and 2.8 μm, while at $d_0 = 50$ nm these are 60 nm and 5 μm, respectively. The observed increase of nanowire height with decreasing its diameter is a typical feature of the diffusion-induced growth as will be discussed in detail later on.

4.3 Role of Size-Dependent Effects

For many particular applications of semiconductor nanowires in nanoelectronics and nanophotonics, observation of quantum confinement effects, fabrication of defect-free nanowires on lattice mismatched substrates and lattice-mismatched heterostructures in nanowires, and phase perfection in III-V nanowires [184], it is desirable to grow them with possibly reduced lateral dimension. This is suppressed, however, by the curvature-induced Gibbs-Thomson effect in the droplet and the nanowire (discussed in Sect. 1.3), because small nanowires will emit more semiconductor species than adsorb. In 1965, Wagner and Ellis [341] wrote their formula for the minimum nanowire radius that can be obtained in the VLS growth at a given vapor supersaturation and surface temperature. In our notations, the Wagner-Ellis expression has the form

$$R_{\min} = \frac{2\Omega_L \gamma_{LV}}{k_B T \ln(\Phi + 1)}. \tag{4.5}$$

Here, Ω_L is the elementary volume in the liquid phase of the catalyst, γ_{LV} is the droplet surface energy, k_B is the Boltzmann constant, T is temperature and Φ is the vapor supersaturation with respect to the reference solid state in the wire. The solid state can be assumed as being at the same chemical potential as the infinitely large equilibrium liquid alloy (e.g., Au-Si): $\mu_S = \mu_L^{eq}$. The Wagner-Ellis formula is easily understood by considering the Gibbs-Thomson effect in a hemispherical droplet with the curvature $1/R$. Indeed, using the second (1.43) for the curvature-induced Gibbs-Thomson modification of the chemical potential in the liquid phase (at $\beta = \pi/2$), we obtain: $\mu_L = \mu_L^{\infty} + (2\Omega_L \gamma_{LV})/R$ for the chemical potential of the droplet resting on a cylindrical nanowire of arbitrary radius R. The effective difference of the chemical potentials in the vapor phase and in the droplet therefore equals $\Delta\mu_{eff} = \mu_V - \mu_L = \mu_V - \mu_L^{\infty} - (2\Omega_L \gamma_{LV})/R$. Dividing this by $k_B T$, assuming that $\mu_L^{\infty} \cong \mu_S = \mu_L^{eq}$ and using the definition $(\mu_V - \mu_L^{\infty})/k_B T = \ln(\Phi + 1)$, one can see that $R_{\min}$ given by (4.5) relates to $\Delta\mu_{eff} = 0$, i.e., no growth condition. While larger droplets can facilitate the VLS nanowire growth, all semiconductor material will evaporate from smaller droplets with $R < R_{\min}$, because shrinking the droplet size will only increase its curvature and the effective chemical potential. Expression (4.5) can be easily modified for an arbitrary contact angle of the droplet by changing the nanowire radius R to the droplet radius $R_d = R \sin\beta$.

Later on, Givargizov and Chernov [256] proposed their formula for the vertical growth rate of a cylindrical nanowire:

$$\frac{dL}{dt} = K\left[\ln(\Phi + 1) - \frac{2\Omega_s \gamma_{WV}}{k_B T R}\right]^2. \tag{4.6}$$

Here, K is a certain crystallization coefficient from the liquid alloy (nm/s), Ω_s is the elementary volume in the solid phase, γ_{WV} is the solid-vapor surface energy of nanowire sidewalls and R is the nanowire radius. Putting this *dL/dt* to zero, we obtain the minimum nanowire radius in the form

$$R_{\min} = \frac{2\Omega_s \gamma_{WV}}{k_B T \ln(\Phi + 1)}. \tag{4.7}$$

This formula is distinctly different from (4.5), because it contains the characteristics of the solid-vapor interface rather than that of the liquid-vapor surface. We note that Givargizov and Chernov never derived their expression theoretically. Rather, they arrived at it empirically by fitting the experimental data on the growth rates of Si nanowires from vapors $SiCl_4$ and H_2 at a temperature of $\sim$1000 °C [256], where the minimum radius was about 25 nm. It was then found by Givargizov [47] that this formula was suitable for describing the measured length-radius dependences of many other wires. The Givargizov-Chernov expression can be qualitatively explained by the Gibbs-Thomson effect in a cylindrical wire of radius R, while the quadratic dependence of the growth rate on the effective chemical potential, $dL/dt = K(\Delta\mu_{wf}/k_B T)^2$, remains purely empirical. The droplet is now considered as only a catalyst facilitating the chemical reaction, through which semiconductor material is transferred from vapor to solid state, and such a transfer is energetically forbidden when the wire is too narrow. However, (1.42) shows that, for a cylindrical curvature, the coefficient 2 in (4.7) is wrong and must be changed to 1. The difference between the expressions (4.5) and (4.7), as well as the kinetic justification of the Givargizov-Chernov formula with modified parameters, will be considered in Sect. 4.7.

The Givargizov-Chernov expression for the nanowire elongation rate is widely used for the analysis of the VLS growth up to now, e.g., in [376, 401, 448, 449]. A very simple and physically transparent form of this formula enables experimental determination of several important parameters. From (4.6), the dependence of $\sqrt{dL/dt}$ is L-independent and is a linearly decreasing function of $1/R$:

$$\sqrt{\frac{dL}{dt}} = \sqrt{K}\left[\ln(\Phi + 1) - \frac{2\Omega_s \gamma_{WV}}{K_B T R}\right]. \tag{4.8}$$

Measuring the nanowire lengths at different radii, calculating the growth rate dL/dt (or L/t) for a given growth time t, and plotting the data in the Givargizov-Chernov variables $\sqrt{dL/dt} = f(1/R)$, one can estimate the kinetic coefficient of crystallization K from the slope angle and the surface energy of the nanowire sidewalls at the known supersaturation Φ or *vice versa*. Obviously, extrapolation of the linear fit to zero gives the cutoff nanowire radius $R_{\min}$ below which the VLS growth is impossible. We also note that (4.6)–(4.8) are not restricted to the case of the VLS growth (because the droplet itself is not considered in the energetic balance) and are equally applicable to the VSS-grown or even catalyst-free wires.

Direct observation of the size effect on the growth of semiconductor Ge nanowires was reported in [376]. In these growth studies, the size-uniform growth seeds were prepared on Ge(111) substrates by either randomly dispersing Au colloids or lithographically patterning the spatially ordered arrays of Au dots. Ge nanowires were obtained by a two-step process where an $\sim$1.5 min nucleation step was performed at 366 °C and the temperature was ramped down in 3 min to 276 °C where nanowire elongation took place. Figure 4.19 shows field-emission SEM images of Ge nanowires grown on Ge(111) surfaces. The nanowire length decreases with decreasing diameter for a given growth time as shown in Fig. 4.19a for wires grown from solution-deposited Au nanoparticles. Similarly, for Ge nanowires grown from lithographically patterned Au dots, smaller diameter Au dots lead to shorter wires as demonstrated in Fig. 4.19b. The same diameter dependent growth rate is observed for nanowires grown in arrays with similar dot diameters and variable interwire spacing. For nanowire arrays with constant 1 μm interwire spacing, 100 μm spacing between arrays, and constant dot sizes, those nanowires from arrays of smaller Au dot diameters resulted in shorter nanowires than those of larger Au dot diameters as shown in Fig. 4.19c–j. Variation in only the interwire spacing for a constant Au dot size had a negligible effect on the growth rate.

Figure 4.20a shows the average wire length, L, as a function of diameter, $d = 2R$, for Ge nanowires grown from Au seeds at 276 °C for different elongation times in steps of 5 min. The nanowire lengths show a persistent decrease with diameter and their elongation is linear with time for a given diameter. For diameters less than 20 nm, measurements were taken from epitaxial (111) oriented wires grown from e-beam patterned Au film regions on the Ge(111) surface just outside the patterned Au dot arrays and grown simultaneously with the arrays. By fitting (4.8) to the experimental data (Fig. 4.20b), it was possible to determine values for $\Delta\mu = k_B T \ln(\Phi + 1)$, K, and the cutoff diameter, $d_c = 2R_{\min}$, below which the nanowire growth ceases. Here, both assignments of the growth velocity as (i) dL/dt or (ii) L/t resulted in a similar cutoff diameter $d_c \sim 3.5$ nm for all growth times used in Fig. 4.20a.

In the controlled experiments on Ge nanowire growth at elevated temperatures [376], the diameter-dependent growth rates were observed similar to those shown in Fig. 4.20. In particular, the data consistently show a linear dependence of $\sqrt{dL/dt}$ on $1/R$. Hence, nanowire growth at different temperatures can be understood within a common framework of the Gibbs-Thomson effect. The TEM experiments have established that the equilibrium Ge concentration in Au-Ge alloys increases with temperature. Thus a reduced supersaturation and consequently larger cutoff diameters are expected at higher growth temperatures. This behavior is indeed confirmed experimentally, as demonstrated by the data summarized in Table 4.1. Aside from the conventional growth parameters such as temperature and precursor pressure, the surface energy γ_{LV} is the key parameter for the synthesis process at the nanoscale. Manipulation of the nanowire sidewall energy by using different dopants and surfactants can enable the cutoff dimension to be tuned, thus extending the achievable radii close to the atomic scale.

The Gibbs-Thomson effect leads to increasing $L(R)$ curves with saturation at large R that corresponds to the limiting case of 2D liquid phase epitaxy. The mononuclear to polynuclear growth transformation discussed in Sect. 2.12 also results in an increasing length-radius correlation, because the growth rate of small crystal facet is always lower than of an infinitely large one. It might be therefore difficult to separate the two effects without a close examination of the region near $R_{\min}$, where the Givargizov-Chernov curve tends to zero (while the mononuclear growth rate starts

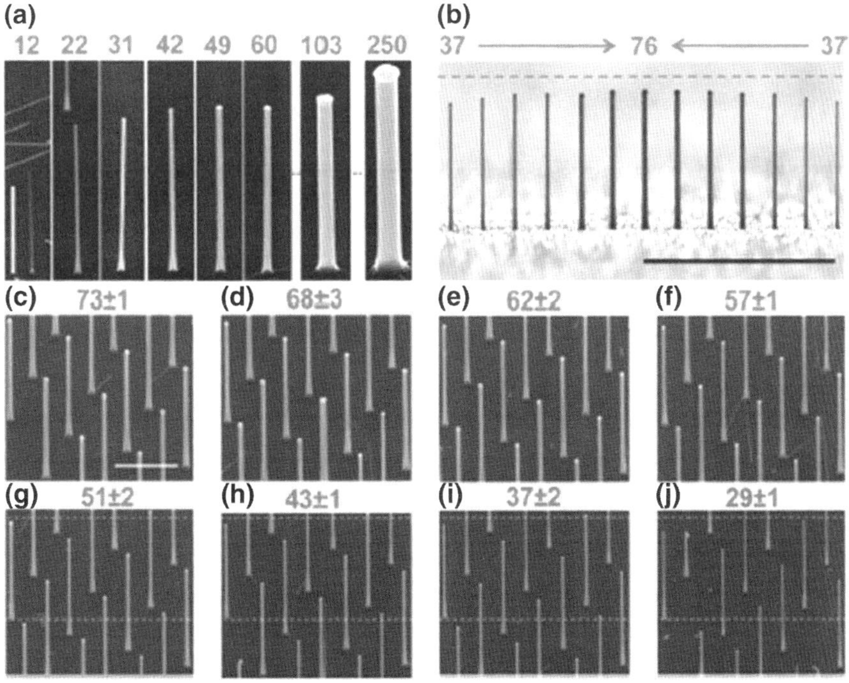

Fig. 4.19 Ge nanowires with controllable diameter. **a** SEM images of vertical nanowires grown for 15 min at 276 °C from randomly dispersed Au colloids of nominal diameter 10, 20, 30, 40, 50, 60, 100, and 250 nm. The resultant measured average nanowire diameter (nm) is indicated at the *top* of each SEM image. The images were taken at the same magnification and the angle-corrected scale bar (far left side, vertical bar) is 1.41 μm. **b** Linear array of Ge nanowires grown from lithographically patterned Au *dots*. The *dashed red line* highlights the systematic changes in nanowire length at different diameters. Scale bar is 3 μm. **c–j** Arrays of uniformly sized Ge nanowires grown from Au *dots* with 100, 90, 80, 70, 60, 50, 40, and 30 nm diameter, 20 nm thickness, and 1 μm spacing, grown for 10 min at 276 °C. The resulting average nanowire diameters and statistical standard deviations are indicated at the *top* of each SEM image. Scale bar is 1 μm. The *dashed white lines* in (**e–h**) serve as a guide-to-the-eye for comparing the left-most nanowire lengths [376]

from $R = 0$). In particular, lines in Fig. 4.21 show the fits to the experimental $V_L(R)$ dependence for Au-assisted, MBE grown GaAs nanowires ($T = 550\,°C$, GaAs deposition rate = 0.4 ML/s, GaAs deposition thickness = 200 nm [310]), obtained from different models [81]. Dotted line is the fit obtained from (4.7) with $K = 2.5$ ML/s, $2\gamma_{WV}\Omega_S/k_BT = 5.2$ nm and $\ln(\Phi+1) = 0.37$. Kashchiev [81] argued that this value of vapor supersaturation seems largely overestimated. If the liquid supersaturation in the droplet is close to Φ, (1.68) would give the value of 198 for the critical number of atoms i_c in a circular 2D nuclei ($d = 2$, $c_2 = \pi$) of monolayer height h (with the solid-liquid surface energy γ_{SL} of 0.2 J/m^2, which even seems underestimated). This number is too large: 2D nucleation typically occurs when $i_c < 50$ [7]. Solid line in Fig. 4.21 is obtained from (2.185) for the growth rate of a crystal facet of radius R,

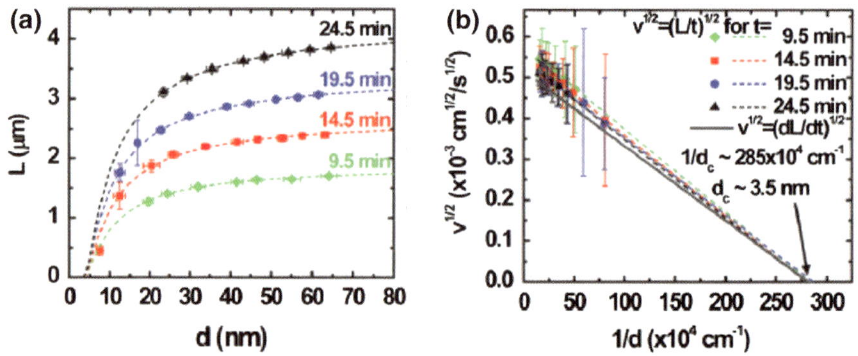

Fig. 4.20 Plots of Ge nanowire growth rate with and without doping illustrating applicability of the Gibbs-Thomson effect on their growth. **a** Plot of the nanowire length as function of diameter with a P[GeH4] = 0.6 Torr for different growth times at the low temperature step (276 °C) = 5, 10, 15, and 20 min, resulting in a linear increase of length with time for all diameters. **b** Plot of the square root of the growth velocity for the data shown in **a** calculated using L/t and dL/dt showing convergence of all data to the same cutoff diameter of ~3.55 nm

Table 4.1 Calculated critical diameters, supersaturation values, and crystallization coefficients for Ge nanowires grown at different temperatures

T (°C)	d_c (nm)	$\Delta\mu$ (meV)	$\ln(\Phi + 1)$	K (10^{-1} nm/s)
276	3.5	145	3.1	3.0
366	6.0	82	1.5	20.5
406	9.7	51	0.9	80.5
440	11.6	43	0.7	86.9

re-written in the form

$$V_L = \frac{a_1 R^2}{1 + a_2 R^2}, a_1 = \pi J; a_2 = 3^{1/3}(\pi J/v)^{2/3}. \quad (4.9)$$

Here, J is the 2D nucleation rate from a supersaturated alloy, and v is the lateral growth velocity of 2D islands. From fitting the experimental curve, the unknown parameters J and v are estimated at 42 $\mu m^{-2} s^{-1}$ and 34 nm/s, respectively. For comparison, the dashed curve in Fig. 4.21 depicts the $V_L(R)$ dependence obtained numerically in [80] within a self-consistent model where the two effects are taken into account simultaneously. It is seen that all the three curves fit well the data, although (4.9) does not predict any cut-off dimension.

It should be noted that the nucleation rate J in (4.9) can itself depend on the nanowire radius due to the Gibbs-Thomson suppression of the liquid–solid supersaturation. Even more importantly, the nucleation-mediated sink of semiconductor atoms from the droplet must exactly equal their arrival rate into the droplet by different kinetic pathways (adsorption-desorption processes in the simplest case without

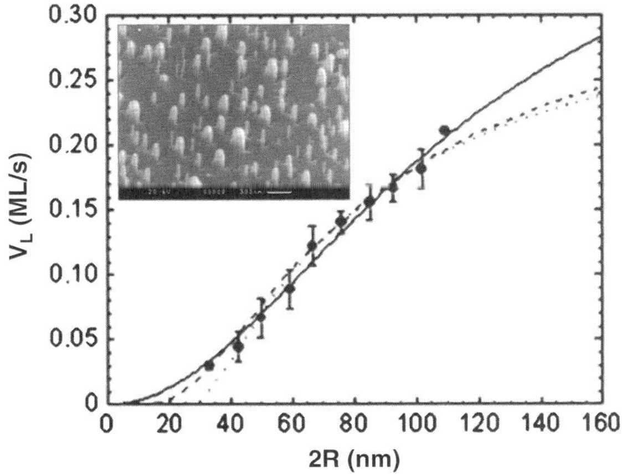

Fig. 4.21 Diameter dependence of the elongation rate $V_L = dL/dt$, self-consisten of Au-catalyzed GaAs nanowires shown in the insert: *Solid line*—(4.9) [81], *dotted line*—(4.6), *dashed line*—self-consisten model of [80]

surface diffusion) [80, 84, 320]. That is why the liquid supersaturation is in a way self-regulated: it should be lower in slower growing wires and increase with increasing the elongation rate. Otherwise the droplet will either inflate with excessive Ga or shrink by emptying Ga from the droplet into the wire. It will be shown in the foregoing sections that this self-consistency condition is very important for the resulting $L(R)$ dependence and may considerably modify the latter even when the Gibbs-Thomson effect is ineffective and the wires grow in a purely mononuclear mode.

Furthermore, as shown experimentally in [118, 360, 391] and theoretically in [45, 82, 118, 360, 389], liquid supersaturation ζ [that determines the nucleation rate $J(\zeta)$] is actually not constant during growth but oscillates with a period of monolayer formation cycle. At first, it rises approximately linearly due to material influx into the droplet until the island can nucleate. After the nucleation, semiconductor species are quickly removed from the droplet and supersaturation drops down. Since the monolayer formation time is much shorter than the refill stage, supersaturation exhibits a sawtooth behavior with almost instantaneous decrease, as shown schematically in Fig. 4.22. We have seen in Chaps. 1 and 2 that the nucleation rate is extremely sensitive to supersaturation and usually decreases by the order of magnitude even for modest variation in ζ, the behavior also schematized in Fig. 4.22. Therefore, even for a much higher growth rates of the order of 1 nm/s (that are actually typical for the Au-assisted MBE of GaAs nanowires under optimized conditions [117, 291]), theoretical analyses strongly suggest the dominant mononuclear growth regime for the nanowire radii up to ~100 nm [45, 46, 118]. The mononuclear growth is also revealed by in-situ monitoring of the VLS growth of group IV and GaP nanowires [360, 391] as discussed in Sect. 4.6. As regards (4.9), the value of J in a_1 should be taken at maximum ζ, while it should be reduced to $\zeta - \Delta\zeta$ in a_2, which may dramatically decrease the probability of polynucleation. Of course, the mononuclear growth mode can be broken even for small lateral dimensions at higher supersaturations relating to an extremely high growth rates. These are not achievable for MBE but are quite typical for chemical epitaxies. For example, the Au-catalyzed

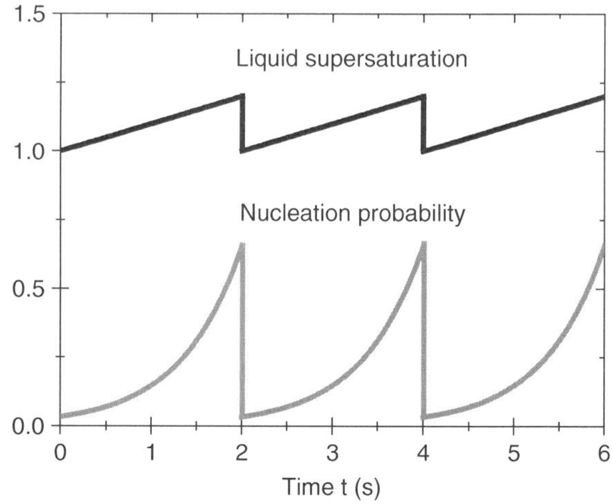

Fig. 4.22 Graphs schematizing the sawtooth time dependence of liquid supersaturation and nucleation rate with a growth cycle of 2 s

GaAs nanowires of [311] elongated at a rate of ~10 nm/s during MOCVD. In [450], the record elongation rate of 44 nm/s was reported for GaAs nanowires obtained in the novel hydride vapor phase epitaxy (HVPE) environment, enabling the fabrication of forty micrometer long rod-like nanowires in only 15 min!

4.4 Role of Surface Energies

As was shown in Sect. 1.4, the condition for the droplet stability on a planar top facet of a cylindrical nanowire is given by (1.49), which is now written for a given contact angle of the droplet β:

$$\gamma_{WV} < \gamma_{LV} \sin\beta + \gamma_{SL}^{l}. \tag{4.10}$$

In this case, the droplet wets the growth front but not the sidewalls, as depicted in Fig. 1.10. Whenever this inequality is reversed, the droplet wets the sidewalls and will slide down to acquire the wetting configuration shown in Fig. 1.11. We now show that the inequality given by (4.10) can be obtained from rather different considerations, namely, by comparing the free energies for the formation of one nanowire monolayer (or bilayer) under the constraint of a fixed amount of semiconductor material before and after the monolayer formation [451]. This analysis refers to the case of instantaneous monolayer nucleation in the mononuclear mode. As suggested in [45] and further developed in [82], the conventional time scale hierarchy of the nucleation process should be significantly modified at the nanoscale. Since the island nucleation and its lateral extension is always much faster than the waiting time between two consecutive nucleation events in the mononuclear mode, the refill stage that returns the droplet volume to its initial value after nucleation should also be aslow process.

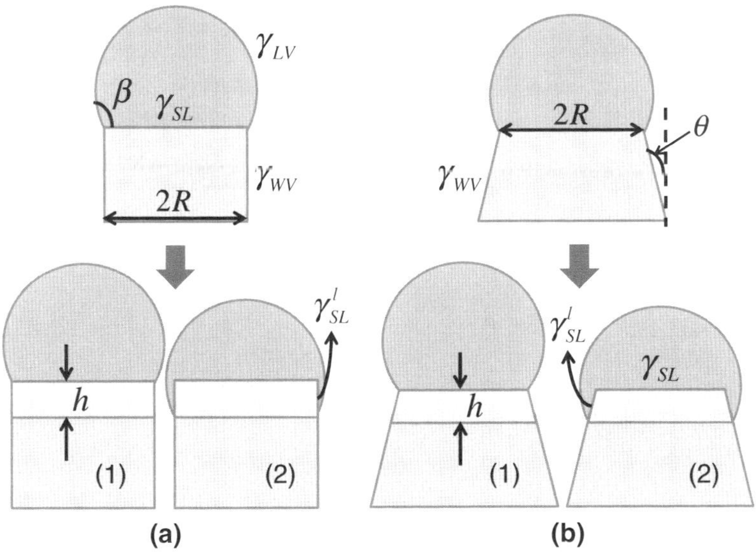

Fig. 4.23 Model geometries of the droplet-nanowire system before and after the monolayer formation for a cylindrical (**a**) and conical (**b**) nanowire: (*1*)—droplet remains on top, and (*2*)—liquid surrounds the monolayer

We can therefore assume that the entire monolayer forms at a fixed number of semiconductor atoms (or III-V pairs) in the system before and after one monolayer is removed from the catalyst.

With this consideration, we analyze a model system incorporating a spherical droplet with the contact angle β and the base radius R, and a cylindrical nanowire with a flat top facet, as shown in Fig. 4.23a. We also allow for 2D isotropic conical nanowire geometry with angle θ to the vertical direction (Fig. 4.23b) with $\theta > 0$ corresponding to the narrowing and $\theta < 0$ to the widening side facet. This geometry reflects possible sawtooth faceting of the nanowire sidewalls [44, 317] (see Fig. 4.6), which might occur with the monolayer period in the case of straight ZB III-V nanowires [45]. The cylindrical geometry is the particular case of the conical one at $\theta = 0$. The triple phase line in these geometries is the periphery of radius R. The surface energies of the wire-vapor, liquid-vapor and horizontal sold-liquid interfaces are denoted as γ_{WV},γ_{LV} and γ_{SL}, respectively. The droplet volume and surface area before the monolayer formation are given by

$$V_L = \frac{(1-\cos\beta)^2(2+\cos\beta)}{3\sin^3\beta}\pi R^3; \quad S_L = \frac{2\pi R^2}{1+\cos\beta}. \tag{4.11}$$

The formation energy of a monolayer of height h is the sum of the volume term $-i\Delta\mu/k_BT$, where $i = (\pi R^2 h)/\Omega_S$ (with Ω_S as the elementary volume in the solid phase) is the number of atoms in the monolayer, and the surface term. Clearly, the volume term is insensitive to the system configuration and cancels in the energy difference. For the analysis of the lowest energy configuration, it is therefore sufficient to compare the surface energies only. If the droplet remains on the top facet upon the monolayer formation, the liquid is outset by the solid, and the change of the droplet volume equals $\Delta V_L = -\Omega_L i$ (with Ω_L as the elementary volume in the liquid phase). Thus, we obtain: $\Delta V_L = -(\Omega_L/\Omega_S)\pi R^2 h$. For the geometry shown in Fig. 4.23b1, the surface energy change upon the monolayer formation can be written as

$$\Delta G_{surf} = \left(\frac{\gamma_{WV}}{\cos\theta} - \gamma_{SL}\tan\theta\right) 2\pi Rh + \gamma_{LV}\Delta S_L. \tag{4.12}$$

The first term in the right hand side gives the surface energy of newly formed side facet of monolayer height h, the second term stands for the surface energy of the eliminated (created) part of the horizontal solid-liquid interface under the droplet, and the third term gives the change of the liquid-vapor surface energy due to the variation of its surface area ΔS_L according to (1.206). In the non-wetting configuration shown in Figs. 4.23a1, b1, this variation is negative.

The problem is thus reduced to finding ΔS_L at a given ΔV_L. Assuming $h << R$, calculating the differential ΔV_L from the first (4.11) and using the geometrical relation $\Delta R = -h\tan\theta$, we relate the result to $-(\Omega_L/\Omega_S)\pi R^2 h$ and obtain

$$\frac{R\Delta\beta}{(1+\cos\beta)^2} = \frac{(1-\cos\beta)^2(2+\cos\beta)}{\sin^3\beta}(\tan\theta)h - \frac{\Omega_L}{\Omega_S}h. \tag{4.13}$$

On the other hand, the second (4.11) yields the variation ΔS_L in the form

$$\Delta S_L = 2\pi R\left[-\frac{2(\tan\theta)h}{1+\cos\beta} + \frac{R\sin\beta\Delta\beta}{(1+\cos\beta)^2}\right]. \tag{4.14}$$

Using (4.13) in (4.14) and excluding $\Delta\beta$, we find

$$\Delta S_L = -[(\Omega_L/\Omega_S)\sin\beta + \tan\theta\cos\beta]2\pi Rh. \tag{4.15}$$

Substitution of (4.15) into (4.12) yields the surface energy of the monolayer formation in the non-wetting configuration 1:

$$\Delta G^1_{surf} = \left[\frac{\gamma_{WV}}{\cos\theta} - \gamma_{LV}\frac{\Omega_L}{\Omega_S}\sin\beta - (\gamma_{SL} + \gamma_{LV}\cos\beta)\tan\theta\right]2\pi Rh \equiv \Gamma_1 2\pi Rh, \tag{4.16}$$

where Γ_1 plays the role of the effective surface energy for this configuration. This expression is similar to the result obtained by Ross *et al.* in [44] (with a minor correction of $\sin\theta$ to $\tan\theta$), but also accounts for the change of elementary volumes in phase transition. In the case of straight nanowire sidewalls ($\theta = 0$), (4.16) is reduced to

$$\Gamma_1 = \gamma_{WV} - \gamma_{LV}\frac{\Omega_L}{\Omega_S}\sin\beta, \tag{4.17}$$

which is further simplified to $\Gamma_1 = \gamma_{WV} - \gamma_{LV}\sin\beta$ at $\Omega_L = \Omega_S$ [45].

If the droplet surrounds the monolayer in the wetting growth mode as shown in Fig. 4.23a2, b2, the change of total volume surrounded by the liquid can only be due to the difference in the elementary volumes: $\Delta V_L = (\Omega_S - \Omega_L)i = -[(\Omega_L/\Omega_S) - 1]\pi R^2 h$, while $\Delta R = 0$. From similar considerations as above, the variation of the droplet surface area equals

$$\Delta S_L = -[(\Omega_L/\Omega_S) - 1]\sin\beta 2\pi Rh. \tag{4.18}$$

The wire-vapor surface energy in (4.12) should now be changed to the surface energy of the lateral solid-liquid interface γ_{SL}^l. Substitution of (4.18) into the modified (4.12) gives

$$\Delta G_{surf}^2 = \left[\frac{\gamma_{SL}}{\cos\theta} - \gamma_{LV}\left(\frac{\Omega_L}{\Omega_S} - 1\right)\sin\beta - \gamma_{SL}\tan\theta\right]2\pi Rh \equiv \Gamma_2 2\pi Rh, \tag{4.19}$$

where Γ_2 is the effective surface energy in the wetting VLS configuration. For a cylindrical wire, this is reduced to

$$\Delta G_{surf}^2 = \left[\gamma_{SL}^l - \gamma_{LV}\left(\frac{\Omega_L}{\Omega_S} - 1\right)\sin\beta\right]2\pi Rh \equiv \Gamma_2^{//} 2\pi Rh. \tag{4.20}$$

Comparing (4.16) and (4.19), the formation of a monolayer in the non-wetting configuration 1 is energetically preferred provided that

$$\gamma_{WV} - \gamma_{LV}\sin(\beta + \theta) < \gamma_{SL}^l, \tag{4.21}$$

while the wetting configuration 2 is preferred otherwise. This inequality gives the surface energetic condition for a conical nanowire. From Fig. 4.24, (4.21) yields a positive surface force F returning the triple phase line back to its original position at the top after a random downward shift. In the case of sawtooth ZB III-V nanowires, we need to compare the surface energies of the non-vertical (111)A and (111)B sidewall facets ($\theta_A = -\theta_B = 19.5°$) in contact with vapor with the vertical facets in the wetting configuration. In this case, comparison of (4.16) and (4.20) gives the condition for the droplet stability in the form

$$\frac{\gamma_j}{\cos\theta_j} - (\gamma_{SL} + \gamma_{LV}\cos\beta)\tan\theta_j - \gamma_{LV}\sin\beta < \gamma_{SL}^l, \tag{4.22}$$

with $j = A, B$ and $\gamma_j = \gamma_{WV}^j$. This inequality should hold for both facets since they alternate from each side to preserve a constant radius during growth. In the case of

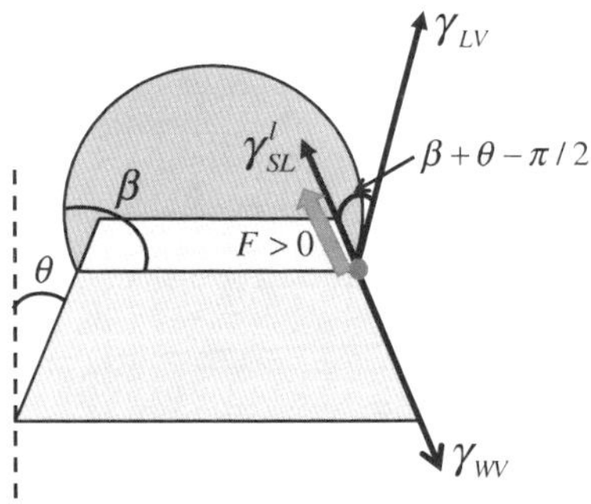

Fig. 4.24 Capillary force returning the triple phase line to the nanowire top in the non-wetting VLS configuration

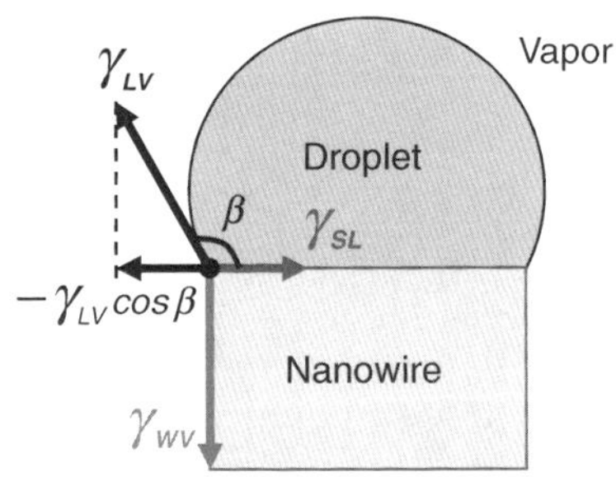

Fig. 4.25 Illustration of the balance of horizontal forces acting upon the triple phase line of a cylindrical nanowire

vertical nanowire sidewalls in both configurations, (4.21) and (4.22) are reduced to the Nebol'sin-Shetinin condition (4.10).

In their analysis of the stability of the VLS growth of Si nanowires catalyzed by different metals, Nebol'sin and Shetinin [43] also used the balance of horizontal forces at the triple phase line given by (1.55):

$$\gamma_{SL} = -\gamma_{LV}\cos\beta. \tag{4.23}$$

This condition, illustrated in Fig. 4.25, has been already discussed in Sect. 1.4, and does not seem to be rigorously justified from surface energy considerations (We note however, that it was obtained in [44] for facetted sidewalls as the condition for balancing ΔG^1_{surf} in the symmetrical case where the two facets j have equal and opposite angle and equal energy γ_{WV}). In any case, (4.23) might be used for rough estimates. Together with (4.10), (4.23) gives the droplet stability condition for a cylindrical nanowire:

$$\frac{\gamma_{WV}}{\gamma_{LV}} < \sin\beta - \cos\beta. \tag{4.24}$$

The right hand size reaches its maximum at $\sqrt{2}$ at $\beta = 135°$. For a given surface energy of the nanowire sidewalls γ_{WV}, this inequality imposes the lower limit for possible values of the droplet surface energy γ_{LV}, which depends on the catalyst metal and the alloy composition during growth.

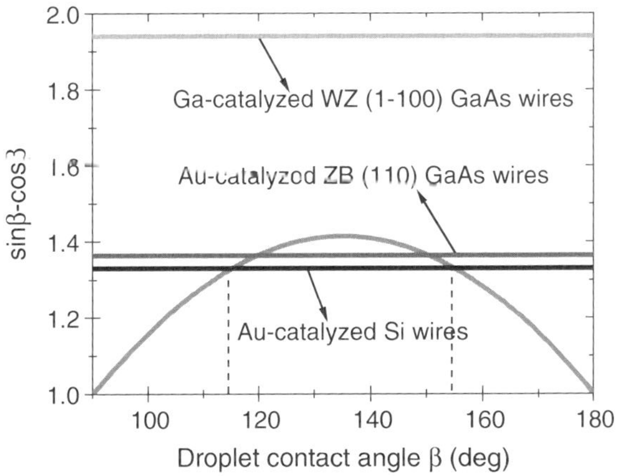

Fig. 4.26 Charts showing the estimated γ_{WV}/γ_{LV} ratios for different material combinations; stable VLS growth occurs below the *red curve* and is suppressed above the *curve*

Figure 4.26 shows the Nebol'sin-Shetinin chats, where the horizontal lines approximately correspond to the cases of Au-catalyzed <111> Si nanowires with {110} side facets ($\gamma_{WV} = 1.2\,\text{J/m}^2$, $\gamma_{LV} = 0.9\ \text{J/m}^2$ [44]), Au-catalyzed ZB GaAs nanowires with {110} side facets ($\gamma_{WV} = 1.5\,\text{J/m}^2$, $\gamma_{LV} = 0.67\,\text{J/m}^2$) and Ga-catalyzed WZ GaAs nanowires with {1–100} side facets ($\gamma_{WV} = 1.3\,\text{J/m}^2$, $\gamma_{LV} = 1.1\,\text{J/m}^2$) [453]. It is seen that the first two cases correspond to a plausible range of contact angles (from 115 to 155° for Si wires), while the Ga-catalyzed GaAs nanowires cannot grow in the standard non-wetting VLS configuration with a single plane top facet. The only possibility to preserve the droplet on top is forming a tapered growth interface that changes the entire surface energy balance. The data of [44] indeed confirm that Si nanowire growth is stable when catalyzed by a high surface energy metals such as gold, nickel, platinum, and iron. The VLS growth process with silver, zinc, and aluminum is unstable: the liquid droplet breaks down, and the wires branch and bend. No wire growth occurs with tin, bismuth, antimony, and some other elements. The data on the VLS wire growth with different catalyst metals are summarized in Table 4.2.

4.5 Triple Phase Line Nucleation

The concept of the triple phase line nucleation was introduced by Glas *et al.* [45] in connection with a surprising ZB-WZ polytypism of III-V nanowires. Glas *et al.* [45] considered the formation free energies of 2D nucleus either entirely surrounded by the liquid phase or at the triple phase line (TPL) (or trijunction) where a certain fraction of the island perimeter x is in contact with the vapor phase. Figure 4.27a shows the two positions of a nucleus, labelled "C" and "TL". Figure 4.27b illustrates the C and TL positions of an island in the form of equivalent triangle with side r on top of a hexahedral wire, with $x = 0$ and 1/3, respectively. Figure 4.28 from

Table 4.2 Surface energies γ_{LV}, γ_{WV}/γ_{LV} ratios, contact angles β and stability of Si wire growth ($\gamma_{WV} = 1.2\,\mathrm{J/m^2}$) in different metal-Si systems [44]

Metal	T (K)	γ_{LV} ($\mathrm{J/m^2}$)	γ_{WV}/γ_{LV}	β (deg)	Stability of VLS growth
Cu	1400	1.300	0.97	135	High
Au	1400	0.910	1.37	115	High
Ag	1300	0.900	1.33	–	Intermediate
Sn	1300	0.575	2.8	–	No wire growth
Pb	1300	0.400	3.15	–	No wire growth
Pt	2100	1.740	0.72	120	High
Zn	–	–	–	–	Low
Al	1300	0.820	1.46	–	Low
Sb	900	0.360	3.5	–	No wire growth
Ni	1850	1.750	0.72	120	High
Bi	800	0.355	3.38	–	No wire growth

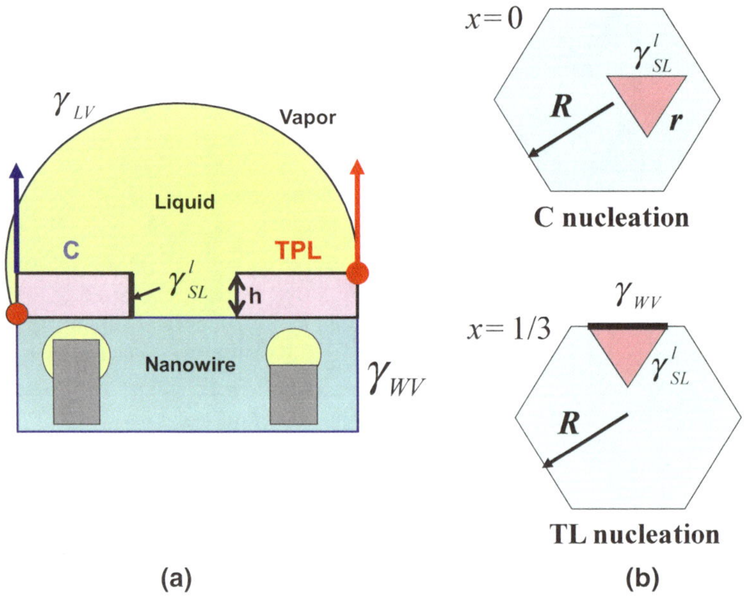

Fig. 4.27 **a** Illustration of possible positions of 2D island either entirely surrounded by the liquid (C) or at the triple phase line (TL). **b** Triangle island on top of a hexagonal top facet in different positions

[300] shows more complex geometries accounting for the non-vertical ZB facets of a hexagonal island ($x = 1/6$ in the TL position).

Whatever are the shapes of the nucleus and of the wire top facet, the free energy of its formation can be approximated by [82]

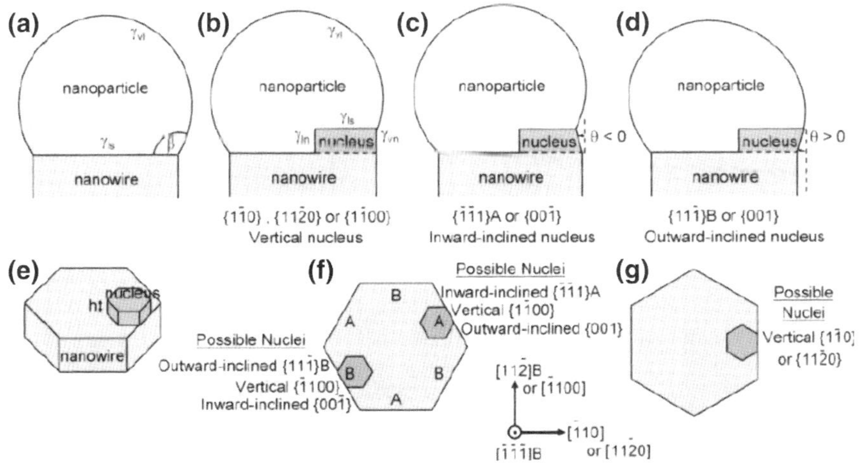

Fig. 4.28 Schematic illustrations of the nucleus, liquid nanoparticle, and nanowire, indicating key parameters. **a** β is the contact angle of the liquid nanoparticle. **b** γ_{ln}, γ_{vn}, γ_{ls}, and γ_{vn} are the energies of the liquid–nucleus sidewall, vapor–nucleus, liquid–nucleus *upper* face, and vapor–liquid interfaces, respectively. The external nucleus facet is **b** vertical with $\theta = 0$, **c** inward-inclined with $\theta < 0$, and **d** outward inclined with $\theta > 0$. **e** A hexagonal nucleus of bilayer height h. **f**, **g** Possible nuclei on the different pre-existing side-facets. The labels A and B denote two types of nonequivalent pre-existing side-facets [300]

$$\Delta G = -\left[k_B T \ln(\zeta_\infty + 1) - \frac{2\gamma_{LV}(\Omega_S - \Omega_L)\sin\beta}{R}\right]\frac{c_2 r^2 h}{\Omega_S} + \gamma_{eff} c_1 rh, \quad (4.25)$$

where

$$\gamma_{eff} = \left[\frac{\gamma_j}{\cos\theta_j} - (\gamma_{SL} + \gamma_{LV}\cos\beta)\tan\theta_j - \gamma_{LV}\sin\beta\right]x + \gamma_{SL}^{l}(1 - x). \quad (4.26)$$

Here, ζ_∞ is the liquid supersaturation at $R \to \infty$, r is the linear size of the nucleus and c_1, c_2 are the shape constants such that $c_2 r^2$ is the surface area and $c_1 r$ is the perimeter of the island. In these expressions, we assume vertical sidewalls for a nucleus away from the triple phase line. It is seen that the volume energy term is modified by the Gibbs-Thomson effect, but only with the difference of elementary volumes $\Omega_S - \Omega_L$ rather than with Ω_L alone, as considered, e.g., in [234]. This effect can be of either signs and cancels at $\Omega_S = \Omega_L$, because the curvature of the droplet surface does not change at $x = 0$ in this case. The $\Omega_S - \Omega_L$ term is ineffective in the energy difference $\Delta G_{TL} - \Delta G_C = \Delta G(x) - \Delta G(0)$, as we saw in the previous section for the complete monolayers.

Thus, we obtain:

$$\Delta G_{TL} - \Delta G_C = x\left(\gamma_j^{eff} - \gamma_{SL}^{l}\right)c_1 rh, \quad (4.27)$$

showing the condition for the preferred triple phase line nucleation is exactly equivalent to (4.22) for arbitrary facet orientation, and is reduced to (4.10) for vertical nanowire sidewalls [454]. This result is well understood intuitively, because the surface forces acting upon the triple phase line are the same for a nucleus and a complete monolayer slice. In fact, the absolute difference of the nucleation probabilities is given not by (4.27), but by $\Delta G^*_{TPL} - \Delta G^*_C$, where ΔG^* are the nucleation barriers at the critical radii r_c [obtained by maximizing (4.25)]. The corresponding expressions will be studied in detail later on.

The Glas condition given by (4.10) might be criticized for absence of any radius dependence, while it is clear that we should resume the preferred C nucleation at large enough nanowire radius R. This property is recovered by considering a purely geometrical effect [82] (sometimes called the entropic effect [387]) of nucleation. Indeed, the nucleation rate, $J \propto \exp(-\Delta G^*/k_B T)$, gives the probability of nucleation per unit time per unit area, and should be multiplied by the available nucleation area in the TL and C positions. It is clear that the nucleation in position C is possible on the whole facet without the ring of width r_c, while the nucleation in position TL is possible only within this ring of critical radius. Assuming circular geometry and $r_c << R$, we get the ratio of the TL and C nucleation-mediated growth rates

$$\frac{P_{TL}}{P_C} = \frac{2r_c}{R}\frac{J_{TL}}{J_C} \cong \frac{2r_c}{R}\exp\left[\frac{(\Delta G^*_C - \Delta G^*_{TL}}{k_B T}\right]. \tag{4.28}$$

This ratio tends to zero at $R \to \infty$ even if the exponent is large compared to one.

4.6 Non-planar Growth Interfaces

Analysis of Sects. 4.4 and 4.5 refers to a planar {111} growth interface, which is a common assumption in modeling the VLS growth and crystal structure of III-V nanowires. In this case, the triple phase line has a simple geometry and provides a unique location for the island nucleation. However, it has recently been shown by in situ growth monitoring for group IV [360, 391] and GaP [360] nanowires that the growth interface may have a more complex non-planar geometry with a truncated edge at the triple phase boundary. A similar geometry has been seen in computer simulations [388], and observed earlier in sapphire nanowires [455]. Moreover, the degree of truncation varies over the monlayer growth cycle. In [360], different nanowires were grown in an ultrahigh vacuum TEM. For Si and Ge nanowires, Si(111) substrates were flash cleaned in vacuum and 0.1–1 nm of Au, Cu, AuGa, or AuAl deposited by thermal evaporation. Substrates were transferred to the TEM, heated to form catalyst droplets (for VLS growth) or particles (for VSS growth), and then exposed to disilane or digermane gases. To examine group III-V nanowire growth, pregrown GaP nanowires on Si substrates were heated in situ during exposure to triethylgallium in an excess of phosphine.

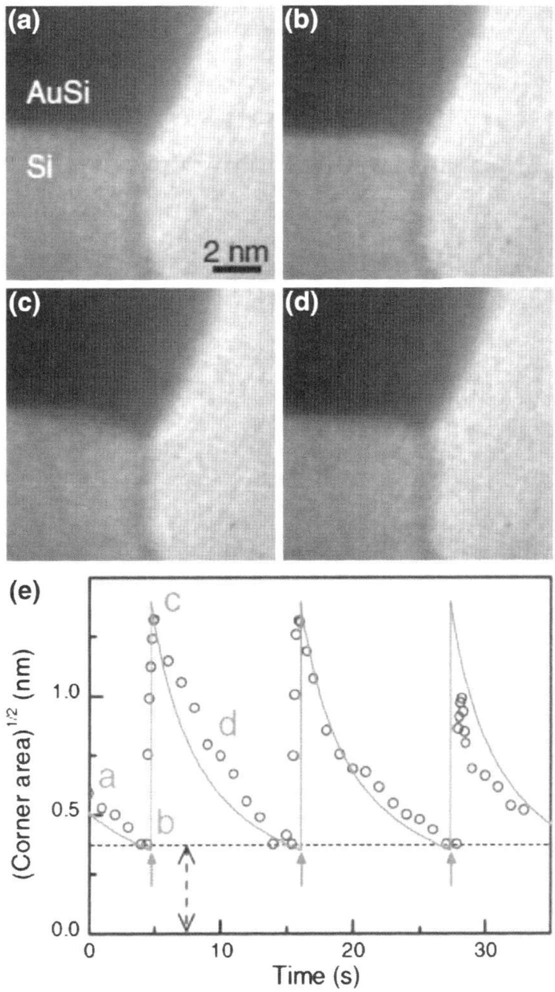

Fig. 4.29 Time evolution of nanowire edge morphology. **a–d** Bright-field TEM images extracted from a 30 images per second video showing the trijunction morphology during VLS Si/AuSi growth at 550 °C and a partial pressure of 4×10^{-7} Torr Si_2H_6. **e** The time variation of the truncated area. Measurements were made over three consecutive cycles, and the data point corresponding to **a–d** are labeled. A Si(111) bilayer forms after point b (shown with an *arrow* for each period) followed by a rapid increase of the truncated area. The horizontal *dotted line* and *dashed arrow* indicate that the amount of truncation remains above zero. The maximum truncated volume is 0.2 bilayers at the growth interface, assuming a trigonal hexagon cross section with a 2:1 ratio for the edge lengths and three truncated edges. The *solid line* is fitted with (4.30) with $a_0 = 0.5$ and $a_1 = 1.5$ [360]

Figures 4.29 and 4.30 show a selection of different nanowire-catalyst interfaces during growth. In each case at least one trijunction was observed where the growth interface and the nanowire sidewall do not meet at a sharp angle, but instead the edge appeared truncated, as shown schematically in Fig. 4.31. Furthermore, in all these cases the degree of truncated changes periodically during growth. For Si/AuSi nanowires in Fig. 4.29a–d, for example, the edge is most truncated in (c) and least truncated in (b). Surprisingly, no evidence was found that the catalyst-nanowire interface is ever a single plane during growth in any of the VLS systems studied. Figure 4.30a–f show truncated structures Si/AuAlSi, Si/AuGaSi, and Ge/AuGe that oscillate with different amounts of truncation. Note that the truncation is different on each side of the main facet. Oscillation also occurs during growth from a solid catalyst, VSS Si/Cu3Si (Fig. 4.30i, j).

We now present a model of [360] explaining why this truncated geometry occurs, and why it oscillates. We start with a 90° edge of length L and calculate the free energy of an extra facet formed by truncating this edge at angle θ to the main facet, as shown in Fig. 4.31. The free energy change ΔG of introducing the edge facet, expanded to the second order in the linear truncation size y, is

$$\Delta G = c_1 yL + \frac{1}{2} y^2 L \cot an\theta \, \Delta\mu + c_2 y^2 L. \tag{4.29}$$

Here, for the quadratic term, the contribution from supersaturation is separated out, where $\Delta\mu$ is the chemical potential of the growth species (say, Si) in the catalyst with respect to the solid phase (bulk Si).

For a sufficiently small edge facet, ΔG is dominated by the lowest order linear term with coefficient c_1. This reflects the capillary surface forces acting on the edge facet, i.e., the change in interfacial energies with facet size y, illustrated in Fig. 4.31c. Changing y also changes the size of the neighboring facets, as well as doing work against the droplet surface tension (Fig. 4.31c). Thus the value of c_1 is given simply by a geometrically weighted sum of the three crystal facet energies and the liquid surface tension [388, 389]. For $c_1 > 0$ the small facet is unstable and will shrink and disappear, while for $c_1 < 0$ an infinitesimal facet will always grow. In was argued in [456] for a classic two-phase problem of a crystal in a fluid that $c_1 < 0$ for any facet present in the equilibrium crystal shape. The quadratic terms in (4.29) determine the size to which the edge facet will grow when $c_1 < 0$. The first quadratic term describes how the free energy depends on the supersaturation in the catalyst, where $(y^2 L \cot an\theta)/2$ is the volume of the truncated region. All other quadratic contributions are lumped together in the last term, with c_2 independent of supersaturation. The value of y that minimizes ΔG (assuming $\theta = const$) is

$$y = \frac{y_0}{1 + \alpha \Delta\mu}, \tag{4.30}$$

where $y_0 = -c_1/2c_2$ is the degree of truncation for the catalyst at $\Delta\mu = 0$, and $\alpha = (\cot an\theta)/2c_2$. Equation (4.30) shows how the amount of truncation depends on the supersaturation $\Delta\mu$ of the growth species in the catalyst. Alternatively, to infer chemical potential from geometry, we could invert (4.30): $\Delta\mu = \alpha^{-1}(y_0 - y)/y$.

To calculate the time dependence of the truncation, we note that, as the growth species arrives from the gas phase, $\Delta\mu$ rises linearly until a step can nucleate [388, 389]. Mononuclear VLS growth is a particularly simple case since this step flows rapidly to complete the monolayer. This quickly reduces $\Delta\mu$ from its maximum to its minimum value as one monolayer of the growth species is removed from the catalyst. Assuming periodic step nucleation with period t_{ML}, and substituting this sawtooth $\Delta\mu(t)$ into (4.30), we obtain an explicit expression for the time variation of the truncation:

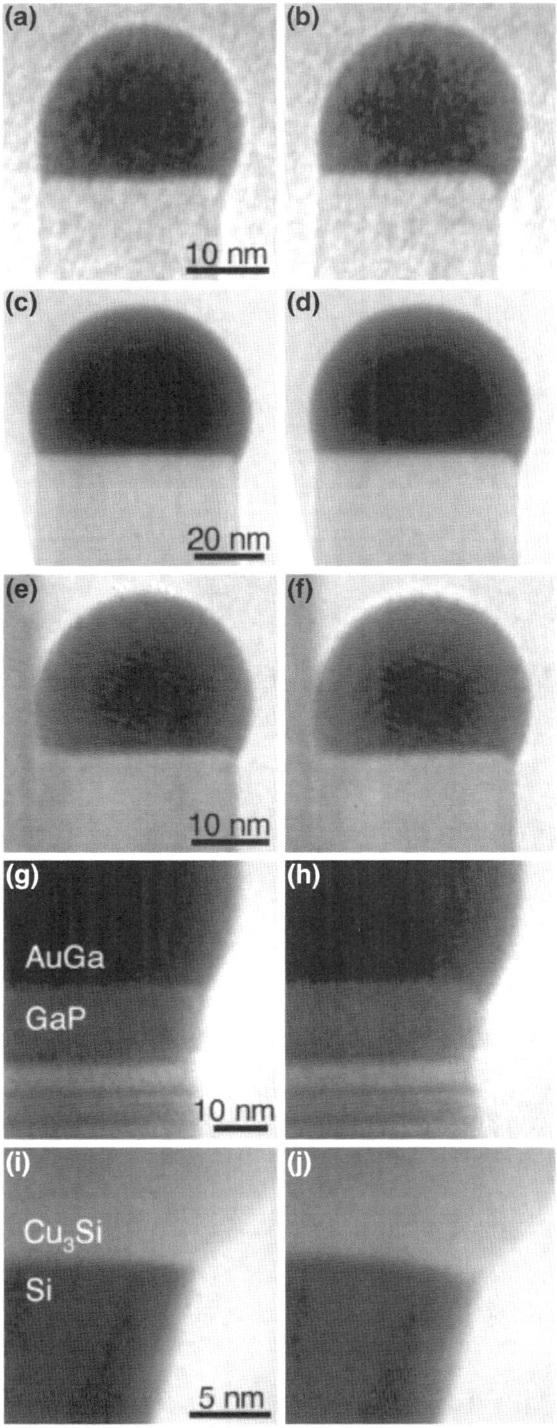

Fig. 4.30 Video image pairs showing approximately the minimum and maximum truncation in (**a**), (**b**) a VLS Si/AuAlSi wire of 23 nm diameter grown at 560 °C and 8×10^{-6} Torr Si_2H_6. For (**b**) the maximum truncated volume is 54 nm^3 or 1.2 bilayers. In (**a**) the minimum volume is 0.3 bilayers. (**c**), (**d**) VLS Si/AuGaSi at 500 °C and 1×10^5 Torr Si_2H_6. (**e**), (**f**) VLS Ge/AuGe at 410 °C and 4×10^{-6} Torr Ge_2H_6. (**g**), (**h**) VLS GaP/AuGa at 450 and 1×10^{-5} Torr $PH_3 + 2 \times 10^{-8}$ Torr $Ga(C_2H_5)_3$; (**i**),(**j**) VSS Si/Cu3Si at 530 °C and 2×10^{-7} Torr Si_2H_6 [360]

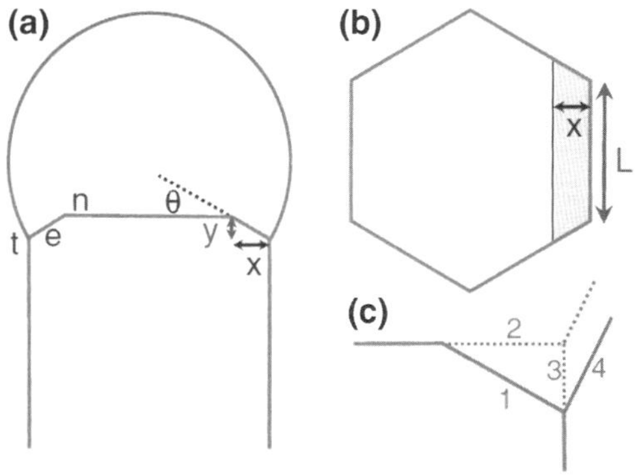

Fig. 4.31 Geometry of the catalyst-nanowire interface as viewed (**a**) from the side and (**b**) from above. The electron beam is parallel to the nanowire edge, so that the edge shows as a corner in the images, and the corner appears truncated. x and y are the width and height of the truncation, respectively, and $y = x \tan\theta$, where θ is the angle between the two facets. t is the trijunction, e the edge facet, and n the favored nucleation site on the main facet. Only one edge facet is shown in (**b**) for clarity. (**c**) Geometrical changes caused by truncating the corner. Contributions to the linear-order capillary term in (4.29) are (*1*) the area of the small facet, (*2*), (*3*) the decreased areas of the adjacent facets, and (*4*) the increase in the liquid-vapor interface length [360]

$$y = \frac{1}{a_0 + a_1(t/t_{ML})}. \tag{4.31}$$

Here $a_0 = y_{\max}^{-1}$ and $a_1 = y_{\min}^{-1} - y_{\max}^{-1}$ are determined directly from the measured extremal values of y, circumventing the uncertainties in c_2 and α. With two parameters a_0 and a_1 reflecting the maximum and minimum truncation, this simple prediction agrees well with the data in Fig. 4.29e. In particular, the model correctly predicts that, when an edge facet is present, it never shrinks to zero size and that the size varies nonlinearly over the cycle. The excellent fit supports the model and the relationship between truncation and catalyst supersaturation.

Whenever the main facet does not meet the trijunction (labeled "t" in Fig. 4.31a), steps on the main facet do not nucleate at the trijunction. Instead, the nucleation is expected where the main facet meets the edge facet, labeled "n". This spot is favorable because extending the facet must have lower energy than introducing a step on the terrace. Otherwise, the facet would decompose into steps. This is important for understanding ZB-WZ polytypism of III-V nanowires, where the concept of the triple phase line nucleation plays a central role [45, 48, 82, 83, 317, 234]. We note, however, that the necessary condition for growth of a truncated facet ($c_1 < 0$ in the above model) may be strongly dependent on the kinetic factors and the catalyst composition, and many post-growth images of III-V nanowires reveal a planar liquid–solid interface without truncation [45, 184]. To our knowledge, the oscillatory behavior of the truncated edge facet is more common for Si and Ge nanowires, while it has been observed only once for GaP nanowires [360] and under specific growth conditions implying a very low growth rate and therefore small $\Delta\mu$. One simple observation is that when $\Delta\mu$ is increased, (4.30) predicts $y \to 0$ and the truncation effect may disappear. Another possible growth scenario might be associated with non-equivalence of different sidewall facets of ZB nanowires. Images shown in Fig. 4.31 clearly suggest that there are some facets which are not truncated. While the triple phase line nucleation should be energetically suppressed on the truncated facet (its growth rate is lower than of the planar top facet), the layer may start at the trijunction of the vertical facet, rapidly fill the top facet and then complete the inclined facet through a slower step flow process. Thus the triple phase line nucleation in III-V nanowires is not excluded generally. In any case, the truncated geometry has far-reaching implications for understanding polytypism in III-V nanowires, deep-subeutectic growth in Ge and perhaps other systems, supersaturation kinetics, interface abruptness and requires further studies.

4.7 Adsorption-Induced Growth

In this section, we consider the VLS systems without surface diffusion. This case corresponds to the early days Wagner-Ellis [46, 341, 342] and Givargizov [47, 343] high-temperature VLS growths by chemical deposition, As-limited growth of self-catalyzed GaAs nanowires [457, 458], and generally to the systems with low adatom diffusivity. The material transport considerations in the steady state (at $R = const$ and $\beta = const$) yield the balance equation

$$\frac{\pi R^2}{\Omega_S}\frac{dL}{dt} = \pi R^2 f(\alpha, \beta) I - \frac{2\pi R^2}{1+\cos\beta} I_{des}. \tag{4.32}$$

Here, the left hand side gives the net flux of atoms contributing into the elongation of a cylindrical wire with radius R, which equals the difference between the adsorption and desorption terms in the right hand side. In the adsorption term, I is the arrival rate from a vapor phase per unit time per unit area, and $f(\alpha, \beta)$ in the case of MBE describes the amount of material collected by a spherical droplet sitting on top of a nanowire and submitted to a uniform parallel flux, as a function of the direction of the beam α and the contact angle of the droplet β. The desorption flux equals the product of the desorption rate I_{des} per unit time from unit area and the droplet surface area. In the case of MOCVD, the function $f(\alpha, \beta)$ simply equals $2/(1 + \cos\beta)$, because the droplet collects growth species through its entire surface area. Only the balance of the growth species is considered, assuming that the catalyst does not leave the droplet.

The function $f(\alpha, \beta)$ was calculated by Glas [354] by considering a spherical cap droplet submitted to a uniform flux at an angle α with respect to the vertical direction z (Fig. 4.32). The droplet is characterized by its center O, contact angle β and radius $R_d = R/\sin\beta$, where R is the radius of its base. We call x the direction of the beam projected in the base plane and y the orthogonal horizontal direction. When subjected

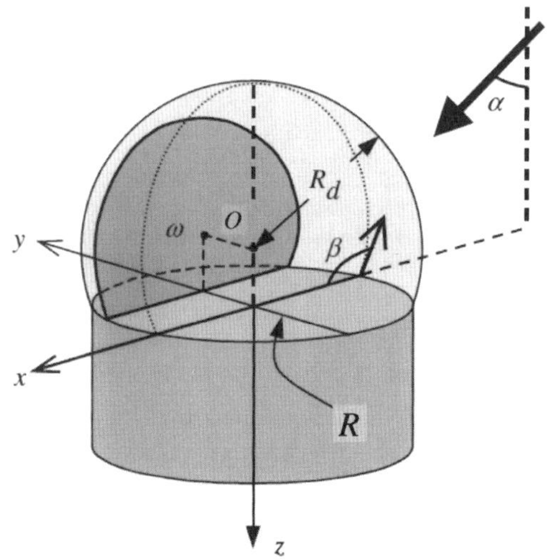

Fig. 4.32 Geometry of the droplet. An arbitrary vertical slice is indicated. The *thick arrow* marks the beam direction

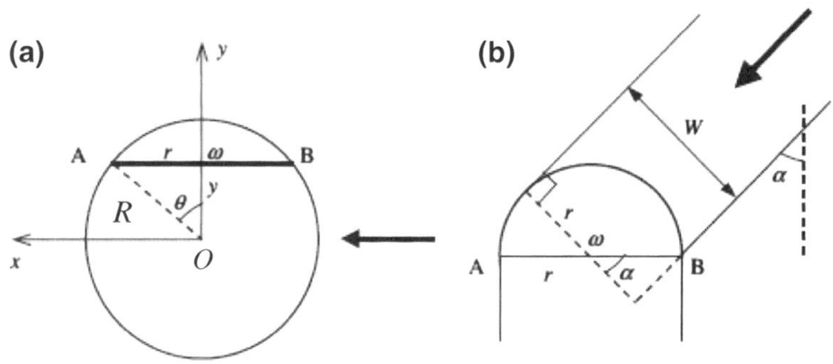

Fig. 4.33 Hemispherical droplet. **a** Section in its base plane, with projection of beam direction (*arrow*) and trace AB of a vertical slice. **b** section in the plane of slice AB. *Thick arrow* marks the beam direction

to a flux J, the droplet collects the latter at a rate χJS, where S is the area normal to the beam intercepted by the droplet and χ the collection efficiency (it is usually assumed that $\chi = 1$ for a liquid droplet). To calculate area S, we decompose the sphere in vertical slices of infinitesimal thickness, parallel to the beam direction and normal to the direction y. In calculations, three cases should be distinguished: (i) hemispherical droplet with $\beta = \pi/2$, (ii) $\beta < \pi/2$ and (iii) $\beta > \pi/2$.

Let us consider the simplest case of $\beta = \pi/2$, shown in Fig. 4.33. The position of a slice at distance y from the droplet center is alternatively measured by polar angle $\theta (0 < \theta < \pi)$ such that $y = R\cos\theta$ (Fig. 4.33a). This slice is a half-disk of radius $r = R\sin\theta$ (Fig. 4.32a) which, in the vertical plane (Fig. 4.33b), intercepts a beam width W, with

$$W = r + r\cos\alpha. \tag{4.33}$$

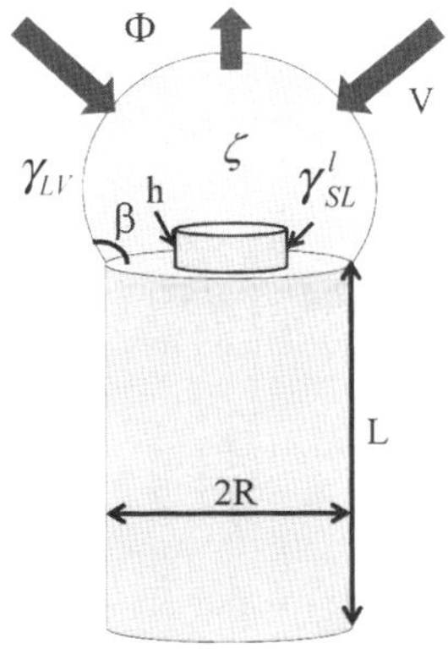

Fig. 4.34 Illustration of the main parameters of the adsorption-induced growth

Thus, the total area intercepted by the beam is

$$S = 2\int_{\theta=\pi/2}^{\theta=0} W dy = -2\int_{\pi/2}^{0} d\theta \sin^2\theta R^2(1+\cos\alpha) = \frac{1+\cos\alpha}{2}\pi R^2, \tag{4.34}$$

showing that $f(\alpha, \pi/2) = (1+\cos\alpha)/2$. Therefore, the droplet collection area is generally smaller than πR^2 and equals πR^2 only for the normal incidence $\alpha = 0$. Similar geometrical considerations provides explicitly the function $f(\alpha, \beta)$ also for $\beta < \pi/2$ and $\beta > \pi/2$. In a more usual case of $\beta > \pi/2$, the result of [354] is given by

$$f(\alpha,\beta) = \frac{1}{\sin^2\beta}\left[\frac{\pi}{2} - \text{acrsin}\left(\frac{\cos\beta}{\sin\alpha}\right) - \cos\beta\sqrt{\sin^2\beta - \cos^2\alpha} + \cos\alpha\sin^2\beta \arcsin\left(\frac{\sqrt{\sin^2\beta-\cos^2\alpha}}{\sin\alpha\cos\beta}\right)\right] \tag{4.35}$$

at $\beta < \alpha + \pi/2$ and

$$f(\alpha,\beta) = \frac{1}{\sin^2\beta} \tag{4.36}$$

at $\beta > \alpha + \pi/2$. In the latter case, the droplet collection area equals simply $\pi R_d^2 = \pi R^2/\sin^2\beta$, meaning that the beam nonwhere reaches the base of the droplet.

If the desorption rate I_{des} is R—independent or negligibly small, (4.32) would result in a radius-independent nanowire growth rate $dL/dt = A$, which can be indeed observed under particular growth conditions within a certain range of R[357]. In the general case, the desorption rate can be written as [320, 459]

$$I_{des} = \left(\frac{\zeta + 1}{\Phi + 1}\right) I, \tag{4.37}$$

where ζ is the droplet supersaturation and Φ is the vapor supersaturation with respect to the bulk solid phase (Fig. 4.34). This equation shows that the desorption rate equals the deposition rate I at $\zeta = \Phi$, where the VLS growth stops. The ζ value is modified by the Gibbs-Thomson effect in the droplet with curvature $\sin\beta/R$ as

$$\zeta + 1 = (\zeta_\infty + 1)\exp\left(\frac{2\Omega_L \gamma_{LV} \sin\beta}{k_B T R}\right). \tag{4.38}$$

Here, ζ_∞ is the supersaturation of the growth species in an infinitely large liquid alloy, Ω_L is the elementary volume in the liquid phase and γ_{LV} is the liquid-vapor surface energy of the droplet (Fig. 4.34). From (4.37) and (4.38), the adsorption-desorption growth rate is given by

$$\frac{dL}{dt} = V\frac{f(\alpha,\beta)}{\cos\alpha}\left[1 - \frac{(\zeta_\infty + 1)}{(\Phi + 1)g}\exp\left(\frac{2\Omega_L \gamma_{LV} \sin\beta}{k_B T R}\right)\right] = V\frac{f(\alpha,\beta)}{\cos\alpha}\left[1 - \frac{(\zeta + 1)}{(\Phi + 1)g}\right], \tag{4.39}$$

where $V = I\Omega_S \cos\alpha$ is the deposition rate in nm/s. The geometrical function $f(\alpha,\beta)$is defined above, cos α should be put to one in MOCVD, and

$$g = \frac{(1 + \cos\beta) f(\alpha,\beta)}{2} \tag{4.40}$$

is the geometrical factor that equals one in MOCVD. This growth rate becomes zero at the minimum radius

$$R_{\min} = \frac{2\Omega_L \gamma_{LV} \sin\beta}{k_B T\{\ln[(\Phi + 1)g] - \ln(\zeta_\infty + 1)\}}. \tag{4.41}$$

This expression generalizes the Wagner-Ellis formula (4.5) for arbitrary contact angle, directional flux and, more importantly, a non-zero liquid supersaturation ζ_∞ during growth.

As discussed by Givargizov [47] and further developed in [80, 84, 320, 459], the adsorption-induced VLS growth could be divided into 2 major modes. The first mode takes place when the vapor supersaturation is only slightly larger than the effective liquid one, a situation that always pertains near $R_{\min}$ and for all R at a modest Φ, i.e., at high temperatures. In this case the limiting stage of nanowire formation is the nucleation-mediated layer-by-layer growth at the liquid–solid interface under the droplet (or at the triple phase line). The elongation rate is quite low and rapidly decreases for smaller droplets. The second mode occurs when the vapor supersaturation is very high, and the wire formation is controlled entirely by the adsorption-desorption transport at the droplet surface. Looking at (4.39), at $(\Phi + 1)g >> \zeta + 1$ the liquid supersaturation does not affect the growth rate and we arrive at the R-independent wire length. For example, at $g = 1$ and $\Phi >> \zeta$ we get:

$L = [2Vt/(1+\cos\beta)][\Phi/(\Phi+1)] \cong 2Vt/(1+\cos\beta)$, where the last expression is valid at $\Phi >> 1$ with neglect of desorption.

Whenever $\zeta + 1$ is close to $(\Phi + 1)g$, the liquid supersaturation must be obtained by equalizing the resulting transport flux into the droplet to the nucleation-mediated sink of the growth species at the liquid–solid interface [84]. In the mononuclear mode which should always occur near $R_{\min}$, the nucleation-mediated growth rate writes down as

$$\frac{dL}{dt} = h\begin{cases}\pi R^2 J\\ 2\pi R r_c J\end{cases} \tag{4.42}$$

for the C and TPL nucleation positions, respectively, with J as the 2D nucleation rate from a supersaturated liquid alloy, r_c as the critical radius and h as the height of a monolayer. Applying (4.9) instead of (4.42) accounts for the mono- to polynuclear growth transformation at larger R. The nucleation rate is given by

$$J = J_0 \exp\left[-\frac{\Delta G^*}{k_B T}\right], \tag{4.43}$$

where $\Delta G^* = \Delta G(r_c)$ is the nucleation barrier. Maximizing (4.25) in r yields the supersaturation dependent critical size and nucleation barrier [82]:

$$r_c = \frac{c_1}{2c_2}\frac{\Omega_S \gamma_{eff}}{k_B T\left[\ln(\zeta_\infty + 1) - 2\gamma_{LV}(\Omega_S - \Omega_L)\sin\beta/(k_B T R)\right]}; \tag{4.44}$$

$$\Delta G^* = \frac{c_1^2}{4c_2}\frac{\Omega_S h \gamma_{eff}^2}{k_B T\left[\ln(\zeta_\infty + 1) - 2\gamma_{LV}(\Omega_S - \Omega_L)\sin\beta/(k_B T R)\right]} = \frac{c_1}{2} h \gamma_{eff} r_c. \tag{4.45}$$

These expressions are valid for both C and TPL nucleation with the corresponding γ_{eff} given by (4.26). They would be exactly equivalent to the standard formulas (1.67), (1.68) at $d = 2$ without the position dependence of $\gamma_{eff}(x)$ and the $1/R$ Gibbs-Thomson correction. These two factors thus reflect size-dependent modifications of nucleation theory in a confined space.

The 2D nucleation rate in our case can be thus put as

$$J = J_0 \exp\left[-\frac{A_*}{\ln(\zeta_\infty + 1) - (2(\Omega_S - \Omega_L)\gamma_{LV}\sin\beta)/(k_B T R)}\right];$$

$$A_* \equiv \frac{c_1^2}{4c_2}\Omega_S h\left(\frac{\gamma_{eff}}{k_B T}\right)^2 >> 1, \tag{4.46}$$

where the pre-factor J_0 only weakly depends on ζ_∞ and will be considered as a constant below in this section. The A_* value equals $a^2/4$ in (1.67) and elsewhere in Chaps. 1 and 2 for a 2D island. Comparing this to (4.38), the denominator under the exponent can be presented as $\ln(\zeta + 1) - (2\Omega_S\gamma_{LV}\sin\beta)/(k_B T R)$. Relating (4.45), (4.42) and (4.39), the self-consistent equation for the unknown liquid supersaturation

ζ (modified by the surface curvature) has the form

$$V\frac{f(\alpha,\beta)}{\cos\alpha}\left[1-\frac{(\zeta+1)}{(\Phi+1)g}\right]=\pi hR^2J_0\exp\left[-\frac{A_*}{\ln(\zeta+1)-(2\Omega_S\gamma_{LV}\sin\beta)/(k_BTR)}\right] \tag{4.47}$$

for the C nucleation. For the TPL nucleation, the factor πR^2 is changed to $2\pi Rr_c$ according to (4.42), while A_* and r_c are also different for the two different nucleation positions. We will continue the analysis for nucleation away from the triple phase line, which is in fact the Wagner-Ellis and Givargizov-Chernov case (all the results can be easily re-formulated for the TPL nucleation when required). It is noteworthy that, regardless of the nucleation position, our (4.47) is not consistent at $R = R_{\min}$: while the left hand side (material transport into the droplet) is zero, the right hand side (the nucleation from the droplet) is not. This might seem contradictory at a glance, but in fact means simply that the steady state elongation without material transport is not possible: the droplet must empty some growth species into the wire to reach its equilibrium concentration at a smaller volume.

We now use the Γ—method of Chap. 2 which enables expanding the function under exponent in the right hand side of (4.46) near $\zeta + 1 = (\Phi + 1)g$, where the growth rate is zero:

$$\frac{A_*}{\ln(\zeta+1)-(2\Omega_S\gamma_{LV}\sin\beta)/(k_BTR)}\cong\frac{A_*}{F}+\frac{A_*}{F^2}\left[1-\frac{(\zeta+1)}{(\Phi+1)g}\right]. \tag{4.48}$$

Here, we denote

$$F\equiv\ln[(\Phi+1)g]-\frac{2\Omega_S\gamma_{LV}\sin\beta}{k_BTR} \tag{4.49}$$

for brevity. Now (4.39) and (4.47) can be re-arranged as

$$\frac{dL}{dt}=\frac{Vf(\alpha,\beta)}{\cos\alpha}\frac{F^2}{A_*}u; \tag{4.50}$$

$$u\exp(u)=\left(\frac{R}{R_0}\right)^2\frac{A_*}{F^2}\exp\left(-\frac{A_*}{F}\right); \tag{4.51}$$

with

$$u\equiv\frac{A_*}{F^2}\left[1-\frac{(\zeta+1)}{(\Phi+1)g}\right] \tag{4.52}$$

and $R_0^2 \equiv [Vf(\alpha,\beta)]/(\pi hJ_0\cos\alpha)$ as the characteristic "nucleation" radius. At given material constants, temperature and flux, (4.51) should be solved first to find the unknown u, and then the growth rate is obtained as afunction of R from

(4.49) and (4.50). Very importantly, we excluded the unknown liquid supersaturation from the growth equation which is now related to the technologically controlled value of Φ.

The growth rate given by (4.50) tends to zero at $F = 0$ relating to

$$R_{\min} = \frac{2\Omega_S \gamma_{LV} \sin\beta}{k_B T \ln[(\Phi + 1)g]}. \tag{4.53}$$

This minimum radius is almost identical to the Wagner-Ellis formula (4.5) at $\beta \cong \pi/2$ with very minor changes: the elementary volume in the liquid is changed to that in the solid and the factor g accounts for the beam direction. The solution to (4.51) is the Lambert function of the right hand side, defined as the solution to equation $u(z)\exp[u(z)] = z$. Obviously, $u(z) \cong z$ at small z and $u(z) \cong \ln z - \ln(\ln z)$ at large z. We therefore find two asymptotic cases. The first one pertains just near $R_{\min}$, where the radius dependence through F is exponential

$$\frac{dL}{dt} \cong \pi h R^2 \exp\left(-\frac{A_*}{F}\right), \; u << 1. \tag{4.54}$$

The second one occurs when the right hand side of (4.51) is large compared to one, i.e., when the pre-exponent is much larger than the exponent. This limiting regime starts quickly at increasing R if the R_0 value is small enough, i.e., $(R_{\min}/R_0)^2 >> 1$. Neglecting the logarithmic dependence on R, we can write approximately

$$\frac{dL}{dt} \cong \frac{Vf(\alpha,\beta)}{\cos\alpha}\frac{< u >}{A_*}F^2 = K\left[\ln[(\Phi + 1)g] - \frac{2\Omega_S \gamma_{LV}\sin\beta}{k_B T R}\right]^2, \tag{4.55}$$

where $< u >$ is the averaged value of u over the measured nanowire radii. It is seen that this expression is of the Givargizov-Chernov type. Comparing (4.8) and (4.55), the surface energy of the nanowire sidewalls is changed to that of the liquid droplet, because the Gibbs-Thomson effect limiting the growth rate is associated with the droplet surface and has nothing to do with the curvature of cylindrical nanowire itself. That is why the factor 2 in the Gibbs-Thomson appears to be correct. Of course, the modified Givargizov-Chernov growth rate becomes zero at the Wagner-Ellis minimum radius given by (4.53). We also note that the crystallization coefficient K is inversely proportional to the large parameter A_* and therefore is expected to be much smaller than the deposition rate.

In what follows, we will often use the characteristic Gibbs-Thomson radii

$$R_{GT} = \frac{2\Omega_L \gamma_{LV}\sin\beta}{k_B T}; \; R^s_{GT} = \frac{2\Omega_S \gamma_{LV}\sin\beta}{k_B T} \tag{4.56}$$

that determine the influence of nanoscale size effect on the nanowire growth rate. Since $\Omega_L \cong \Omega_S$ for most systems of interest, R^s_{GT} is close to R_{GT} and equals R_{GT} when the elementary volumes in the liquid and solid phases equal each other. It is also useful to introduce the ratio of liquid to vapor activities by definition

$$\frac{\theta_l}{\theta_v} = \exp\left(\frac{\mu_L - \tilde{\mu}_V}{k_B T}\right) = \frac{\zeta + 1}{(\Phi + 1)g} \tag{4.57}$$

and

$$\frac{\theta_l^\infty}{\theta_v} = \exp\left(\frac{\mu_L^\infty - \tilde{\mu}_V}{k_B T}\right) = \frac{\zeta_\infty + 1}{(\Phi + 1)g}, \tag{4.58}$$

at $R \to \infty$. Here, the effective vapor chemical potential $\tilde{\mu}_V$ is corrected for geometrical effect in directional deposition methods. In these notations, (4.39), (4.46), (4.47) and (4.55) take the form

$$\left(\frac{dL}{dt}\right)_{AD} = V_{eff}\left(1 - \frac{\theta_l}{\theta_v}\right) = V_{eff}\left(1 - \frac{\theta_l^\infty}{\theta_v}\exp\left(\frac{R_{GT}}{R}\right)\right); \tag{4.59}$$

$$\frac{1}{h}\left(\frac{dL}{dt}\right)_{nucl} = \frac{\pi R^2 J}{1 + \Theta[3(\pi J/v)^2]^{1/3} R^2}; \tag{4.60}$$

$$J = J_0 \exp\left(-\frac{A_*}{\ln(\zeta_\infty + 1) - (R_{GT}^s - R_{GT})/R}\right); \tag{4.61}$$

$$\left(\frac{dL}{dt}\right)_{GC} = K\left[\ln[(\Phi + 1)g] - \frac{R_{GT}^s}{R}\right]^2, \tag{4.62}$$

with the effective deposition rate $V_{eff} = Vf(\alpha, \beta)/\cos\alpha$. Here, (4.60) accounts for the mono- to polynuclear growth transformation, with $v = dr/dt$ being the lateral growth velocity of 2D islands and Θ the parameter describing the supersaturation depression due to the sawtooth oscillations shown in Fig. 4.22. Equation (4.59) gives the adsorption-desorption growth rate with the Gibbs-Thomson effect and the unknown liquid supersaturation, used, e.g., in [348, 401, 409]. In the self-consistent approach, the supersaturation value should be obtained by equalizing (4.59) to the nucleation-mediated growth rate given by (4.60), with the nucleation rate defined in (4.61). The Givargizov-Chernov expression (4.62) contains three parameters: the effective vapor activity $(\Phi + 1)g$, the Gibbs-Thomson radius R_{GT}^s, and the kinetic coefficient K. General expressions (4.59) to (4.61) additionally depend on the surface energy coefficient A_* and the kinetic factors J_0, v and Θ.

The radius R_{GT} for different alloys can be estimated at given T from the known atomic volumes and surface energies of liquid Au, Si, In and Ga [23, 24] (for III-V materials, we consider only group III alloys with Au, because the concentration of group V elements in the drop is rather low as discussed above). Since the values of γ_{LV} are usually given at the melting temperature T_m, the surface energies at the growth temperature $\gamma_{LV}(T)$ are calculated by linear approximation with corresponding coefficients $d\gamma/dT$ from [39, 460]. All these data are given in Table 4.3 from

Table 4.3 Gibbs-Thomson radii for different alloys and material parameters

Material	Ω_L (nm^3)	T_m (°C)	$\gamma(T_m)$ (J/m^2)	T (°C)	$d\gamma/dT$ [10^{-3} J/(m$^2 \times K$)]	$\gamma_{pure}(T)$, (J/m^2)	$\gamma_{Au}(T)$ (J/m^2)	R_{GT} (nm)
Au	0.0189	1065	1.145		−0.2			
In	0.0261	157	0.556	470	−0.247	0.479	1.264	2.44–4.66
Ga	0.0191	30	0.711	585	−0.07	0.674	1.246	2.24–4.1
Si	0.0181	1410	0.775	500	−0.145	0.907	1.258	3.08–4.46

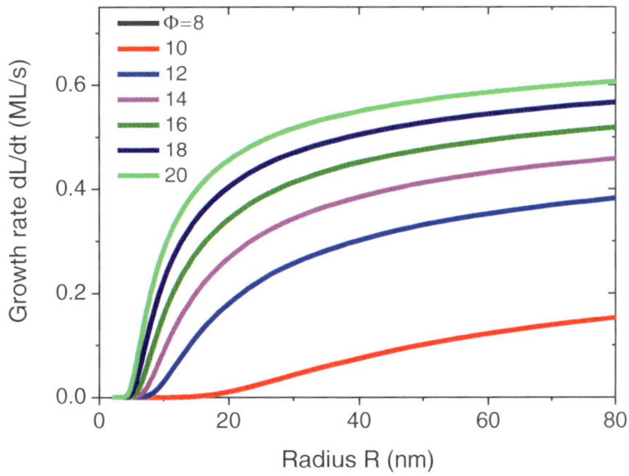

Fig. 4.35 Nanowire growth rate versus radius at different Φ

[409]. The last column presents relevant range of the GT radii at $\beta = \pi/2$ with the surface energy γ_{LV} and the elementary volume Ω_L between that of liquid Au (γ_{AU}) and liquid pure material (γ_{pure}), at typical growth temperatures for different materials. It is seen that the values of R_{GT} are between 2.2 and 4.7 nm in all the cases considered.

We therefore used the value of $R_{GT} = R^s_{GT} = 4$ nm in numerical calculations. The graphs of the growth rate versus radius shown in Fig. 4.35 were obtained from (4.59) to (4.61) for a hypothetical system with $A_* = 30, J_0 = 10^5\,\text{nm}^{-2}\,\text{s}^{-1}$, at $\alpha = 0, \beta = \pi/2$ (yielding $g = 1/2$), $\Theta \to \infty$ (the mononuclear growth), and different vapor supersaturations Φ in steps of 2. It is seen how the growth rate is suppressed and the minimum radius increased toward smaller Φ . Figure 4.36 shows the comparison between the growth rates obtained at $\Phi = 10$ and 20 within the self-consistent model, with the Givargizov-Chernov (4.62) with $K = 0.0610$ ML/s and 0.113 ML/s , respectively and (4.59) at $\theta_l^\infty/\theta_v = 0.38$ and 0.80, respectively. These parameters were chosen to match the three curves at large R. It is seen that the correspondence if fairly good at large vapor supersaturation but becomes worse at $\Phi = 10$, where the self-consistent curve lies much lower and increases very slowly near $R_{\min}$. The exponential (4.59) always starts with a larger slope and predicts a larger $R_{\min}$ than that given by the Wagner-Ellis formula.

To summarize the results of this section, we conclude that:

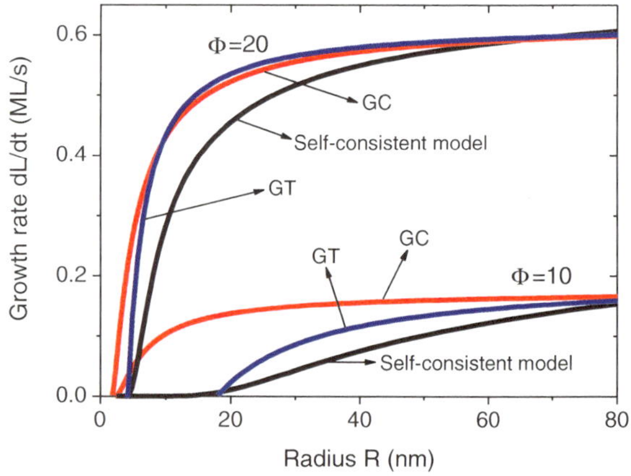

Fig. 4.36 Growth rates obtained from the self-consistent (4.59)–(4.61), the Givargizov-Chernov (GC) (4.62) and the Gibbs-Thomson (GT) (4.59) without self-consistency at $\Phi = 10$ and 20 with the parameters described in the text

1. The Givargizov-Chernov formula for the nanowire growth rate has the correct quadratic form in $1/R$ near $R_{\min}$ for large enough vapor supersaturations but contains incorrect surface energy γ_{WV}. This should be changed to the liquid-vapor energy γ_{LV}, the value which is actually measured from the Givargizov plots.
2. With this substitution, the coefficient 2 becomes correct, because the Gibbs-Thomson effect is due to the curvature of a spherical droplet rather than a cylindrical wire.
3. The Wagner-Ellis formula for the minimum nanowire radius turns out to be "almost correct", with only a small change of the elementary liquid volume to the solid one.
4. The exponential Gibbs-Thomson curve can fit well the measured length-radius dependences at high enough Φ, but does not feature a small slope behavior near $R_{\min}$ and predicts overestimated $R_{\min}$.
5. At $\Phi >> \zeta$ and $R >> R_{\min}$ the curvature-induced suppression of the VLS growth is absent, and nanowire growth should proceed at a radius-independent rate.

Comparison of theoretical and experimental length-radius dependences in different systems will be considered later on in a more general case with the diffusion-induced contributions.

4.8 Diffusion-Induced Growth

The models discussed above predict increasing length-radius dependences with the nanowire growth rate being suppressed at small R by the size-dependent effects. In most modern epitaxy techniques such as MOCVD [345, 348], MBE [117, 184, 287, 296, 411–413], CBE [350] and magnetron sputtering deposition [406], Au-catalyzed group IV and III-V semiconductor nanowires exhibit decreasing $L(R)$ correlation or at least decreasing sections of the $L(R)$ curves. One of the earliest examples of such curves in the case of Au-catalyzed Si and GaAs nanowires grown by MBE are shown in Figs. 4.37 and 4.38. These $L(R)$ curves are well fitted by the inverse dependence $L = (A + C/R)H$, where H is the equivalent deposition thickness and A, C are the radius-independent constants. In some cases, such as shown in Fig. 4.39, the measured nanowire length is better fitted by the inverse squared law $L = (A + C^2/R^2)H$. With only 1000 nm of GaAs deposited, the maximum length of the narrowest GaAs nanowires in Figs. 4.38 and 4.39 is 4700 and 3200 nm, i.e., exceeds the equivalent deposition thickness by 4.7 and 3.2 times, respectively. The $L_{\max}/H$ ratio can eventually reach 15 times under optimal conditions [184]. As mentioned in Sect. 4.1, a metallic droplet seated on the nanowire top acts as the material collector that directs the adatom (Ga, Si) diffusion fluxes from the substrate surface and the nanowire sidewalls to the top. These growth species first dissolve in the droplet and then precipitate to the liquid–solid interface, thus largely enhancing the resulting elongation rate. Of course, in the case of III-V nanowires, such a process proceeds only at group V (As) rich conditions, where every Ga adatom arriving to the droplet can find its As pair. The diffusion-induced growth is very advantageous because it enables growing longer nanowires at a lower material consumption. It has been found that the diffusion-induced growth of 1D nanostructures is a rather general phenomenon that occurs also in III-V nanowires obtained by selective area epitaxy [440], II–VI [301] and oxide nanowires [302], self-induced GaN nanowires [422, 430, 312], III-V nanoneedles [200] and carbon nanotubes [461, 462]. For this reason, early diffusion growth models [47] have been reconsidered and substantially developed by many authors, with [82, 117, 291, 345, 347, 348, 350, 364, 366, 377, 383, 401, 404] giving a non-exclusive list.

"Elementary" growth laws can be easily understood by considering a cylindrical wire of a fixed radius R that collects adatoms from either its top part or from the substrate surface. In the first case, we assume that adatoms are collected from the sidewall area $2\pi R\lambda_f$ (in the case of MOCVD), where λ_f is the adatom diffusion length on the sidewalls. Some growth species impinge directly onto the hemispherical surface area $2\pi R^2$ with an arrival rate I and desorb at a rate I_{des}. Writing the material balance in the steady state as $\Omega_S^{-1}\pi R^2 dL/dt = (I - I_{des})2\pi R^2 + I2\pi R\lambda_f$, we obtain: $L = 2H(1 - \theta_l/\theta_v + \lambda_f/R)$, which explains the inverse $1/R$ dependence. If adatoms are collected from the diffusion ring of width λ_s on the substrate surface and can migrate to the top without desorption or incorporation to sidewall steps, one can write: $\Omega_S^{-1}\pi R^2 dL/dt = (I - I_{des})2\pi R^2 + I\pi\lambda_s^2$ at $\lambda_s >> R$. This results in $L = 2H(1 - \theta_l/\theta_v) + H\lambda_s^2/R^2$, i.e. the $1/R^2$ length-radius correlation.

In a more precise approach, the diffusion-induced growth rate at a time-independent radius should be obtained from

$$\frac{\pi R^2}{\Omega_S}\left(\frac{dL}{dt}\right)_{diff} = -2\pi R D_f \left.\frac{dn_f}{dz}\right|_{z=L}. \tag{4.63}$$

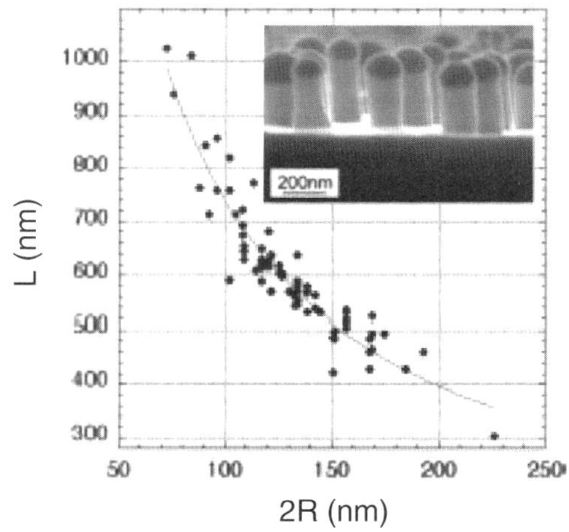

Fig. 4.37 Length-diameter dependence of Au-catalyzed Si nanowires grown by MBE on Si(111) substrate at $T = 525\,°\mathrm{C}$ with the deposition rate $V = 0.05\,\mathrm{nm/s}$ after 120 min of Si deposition (*dots*), fitted by $75000/(2R)$ (*line*)

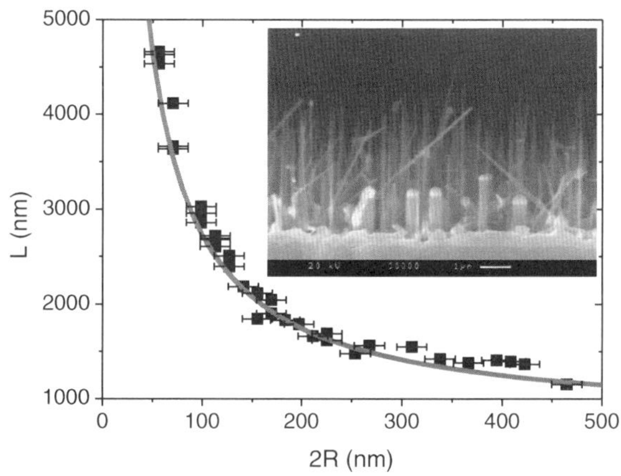

Fig. 4.38 Length-diameter dependence of Au-catalyzed GaAs nanowires grown by MBE on a GaAs(111)B substrate after annealing of a 2.5 nm Au layer at the growth temperature $T = 585\,°\mathrm{C}$, deposition rate $V = 1\,\mathrm{ML/s}$, and GaAs deposition thickness $H = 1000$ nm (*symbols*), the insert showing a typical SEM image. The *line* shows the fit by $L = (A + C/R)H$ at $A = 0.75$ and $C = 100\,\mathrm{nm}$ [291]

Here, the right hand side gives the total diffusion flux of adatoms through the circular triple phase line at the nanowire top (at vertical coordinate $z = L$), with D_f as the diffusion coefficient and n_f as the z—dependent adatom concentration on the sidewall facets, labelled "f". In order to calculate this flux in the general case, we should consider two steady state diffusion equations for the substrate (n_s) and sidewall (n_f) adatoms [409, 410]:

$$D_s \Delta n_s + I \cos\alpha - \frac{n_s}{\tau_s} = 0; \tag{4.64}$$

$$D_f \frac{d^2 n_f}{dz^2} + \omega I \sin\alpha - \frac{n_f}{\tau_f} = 0. \tag{4.65}$$

The first terms here describe the diffusion, the second stand for the adsorption from the flux I directed at an angle α to the substrate normal, and the third give the adatom

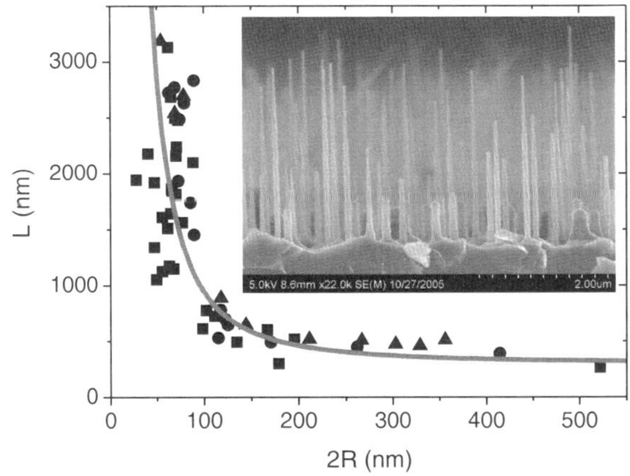

Fig. 4.39 Length-diameter dependence of Au-catalyzed GaAs nanowires grown by MBE on a GaAs(111)B substrate after annealing of a 2 nm Au layer at the growth temperature $T = 560\,^{\circ}\mathrm{C}$, deposition rate = 1 ML/s, and GaAs deposition thickness = 1000 nm (*squares*, *dots*, and *up triangles* correspond to different parts of the sample), the insert showing a typical SEM image [117]. The *line* shows the fit by $L = (A + C^2/R^2)H$ at $A = 0.3$ and $C = 40$ nm [117], the insert showing a typical morphology [287]

sinks. The Δ is the two-dimensional Laplace operator in the substrate plane (r), D_s is the adatom diffusion coefficient on the substrate surface, and τ_s, τ_f are the effective lifetimes on the substrate and sidewall surfaces, respectively. These lifetimes are limited by either desorption from the corresponding planes or incorporation to the surface steps. Adatoms incorporating to the sidewalls contribute into the radial growth, in which case the steady state model with $R = const$ becomes only approximate. The coefficient ω in (4.65) equals 1 in MOCVD and $1/\pi$ in MBE. The angle α does not enter the diffusion equations in the case of vapor deposition. Instead, the temperature-dependent pyrolysis efficiencies χ_f and χ_s can be introduced into the adsorption terms to account for different precursor decomposition probabilities at different surfaces.

General solutions to (4.64) and (4.65) are given by

$$n_s(r) = I\tau_s \cos\alpha + c_1 I_0(r/\lambda_s) + c_2 K_0(r/\lambda_s); \tag{4.66}$$

$$n_f(z) = \omega I \tau_f \sin\alpha + a_1 \cosh(z/\lambda_f) + a_2 \sinh(z/\lambda_f). \tag{4.67}$$

Here, $\lambda_s = \sqrt{D_s\tau_s}$ is the effective diffusion length of adatoms on the substrate surface and $\lambda_f = \sqrt{D_f\tau_f}$ is that on the sidewalls. The functions $I_m(\xi)$ and $K_m(\xi)$ here and below denote the modified Bessel functions of the order m in standard notations, the I increasing and K decreasing at $\xi \to \infty$. The physics of the growth process is now determined by the four boundary conditions that are required for finding the four coefficients c_1, c_2, a_1 and a_2.

For a single nanowire, the first boundary condition for the substrate adatoms writes down as

$$n_s(r \to \infty) = I\tau_s \cos\alpha, \tag{4.68}$$

meaning simply that the adatom concentration is constant far away from the nanowire. This yields $c_1 = 0$ to eliminate the increasing Bessel function I_0. For an array of nanowires with average spacing $2R_W \sim 1/\sqrt{N_W}$, with N_W as the surface density of wires, this condition is changed to [404]

$$\left.\frac{dn_s}{dr}\right|_{r=R_W} = 0, \tag{4.69}$$

implying zero diffusion flux at equal distance between the wires. Dubrovskii *et al.* [404, 409, 410] proposed using the conditions of continuity of adatom chemical potential and flux at the nanowire base ($r = R$), given by

$$\sigma_s n_s(R) = \sigma_f n_f(0); \tag{4.70}$$

$$D_s \left.\frac{dn_s}{dr}\right|_{r=R} = -D_f \left.\frac{dn_f}{dz}\right|_{z=0}. \tag{4.71}$$

The second boundary condition follows from the expression $\mu = k_B T \ln(\sigma n) + const$ for the chemical potential of both adatom subsystems "s" and "f" in the perfect gas approximation (see Chap. 1), with σ_s and σ_f being the elementary areas on the substrate and sidewall surfaces, respectively. The third condition is simply the continuity equation at the nanowire foot.

Johansson *et al.* [347, 348] put zero boundary condition at the nanowire top:

$$n_f(L) = 0, \tag{4.72}$$

which was also used by us [117, 404] and some other authors (see, e.g., the paper by Lu *et al.* [401]). With this condition, the diffusion flux is always directed to the top where the concentration is zero. Dubrovskii *et al.* [409, 410] then changed the fourth boundary condition to the continuity of chemical potentials at the triple phase line:

$$k_B T \ln\left[\sigma_f n_f(L)\right] = \mu_L^{\infty} + \frac{2\gamma_{LV}\Omega_L \sin\beta}{R}. \tag{4.73}$$

This regards for the Gibbs-Thomson effect of elevation of chemical potential in the droplet as discussed in the previous section. It has been shown that such a condition has rich implications for describing non-linear growth effects such as negative growth [410] and narrowing the length distribution of nanowires [413]. Equation (4.73) can be re-arranged in the equivalent form

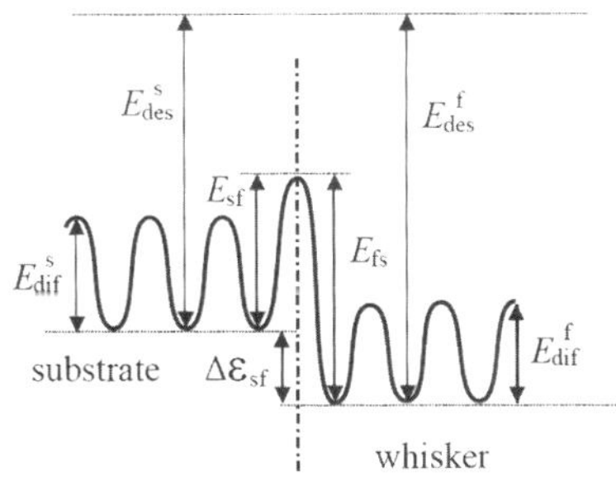

Fig. 4.40 Potential relief for adatom crossing the substrate-nanowire boundary, E_{des} are the desorption barriers, E_{diff} are the diffusion barriers and E_{sf}, E_{fs} are the barriers for the direct and reverse substrate-sidewall transitions

$$n_f(L) = \frac{\theta_l}{\sigma_f} = \frac{\theta_l^\infty}{\sigma_f} \exp\left(\frac{R_{GT}}{R}\right) \equiv \tilde{n}_f \tag{4.74}$$

by using the liquid activity and the Gibbs-Thomson radius defined in (4.58) and (4.56). Here, $\tilde{n}_f$ has a sense of quasi-equilibrium adatom concentration at the liquid-solid boundary, which is sensitive to the liquid supersaturation and the curvature of the droplet surface.

Recently, Hervieu [463] proposed more general boundary conditions at the substrate-nanowire boundary that account for different desorption and diffusion barriers on the substrate and sidewall surfaces [464], as shown in Fig. 4.40. In this case, the second and the third boundary conditions are modified to

$$D_s \frac{dn_s}{dr}|_{r=R} = k_{sf} n_s(R) - k_{fs} n_f(0); \tag{4.75}$$

$$D_f \frac{dn_f}{dz}|_{z=0} = k_{fs} n_f(0) - k_{sf} n_s(R). \tag{4.76}$$

Here, $k_{sf} = (\nu\sigma_s^{1/2})\exp(-E_{sf}/k_B T)$ is the velocity of the adatom transition from the substrate to the sidewalls (nm/s), and $k_{fs} = (\nu\sigma_f^{1/2})\exp(-E_{fs}/k_B T)$ is the velocity of the reverse transition, with ν as the corresponding frequency. Summing up these equations, we resume the continuity of the flux given by (4.71). However, the continuity of the adatom chemical potential at the nanowire base is not fulfilled in the general case, because the substrate and sidewall planes have different crystallographic orientations and may even belong to different materials. Furthermore, the condition at the nanowire top is changed to

$$D_f \frac{dn_f}{dz}|_{z=L} = k^+ n_f(L) - k^- C \tag{4.77}$$

from similar considerations. Here, the right hand side gives the resulting adatom flux, with k^+ and k^- as the velocities of the adatom transition into and from the droplet and C as the local concentration of semiconductor material in the droplet near the liquid–solid boundary.

We will proceed with the most general boundary conditions given by (4.69), (4.75)–(4.77), and then will see how the results of [410] [obtained with boundary conditions (4.68), (4.70), (4.71) and (4.73)] are modified. After some straightforward calculations, we calculate the four coefficients entering (4.66) and (4.67) and obtain exact solutions for adatom concentrations [465]

$$n_s(r) = I\tau_s \cos\alpha - a_2 \frac{D_f \lambda_s}{D_s \lambda_f} \frac{G(r/\lambda_s)}{G'(R/\lambda_s)}; \tag{4.78}$$

$$n_f(z) = \omega I \tau_f \sin\alpha + a_1 \cosh(z/\lambda_f) + a_2 \sinh(z/\lambda_f) \tag{4.79}$$

with

$$a_1 = \frac{I(\xi\tau_s\cos\alpha - \tau_f\sin\alpha)[\Gamma_d\cosh(L/\lambda_f) + \sinh(L/\lambda_f)] - \Gamma_f\left(1 + \frac{1}{\Gamma_s F}\right)(I\tau_f\sin\alpha - \tilde{n}_f)}{\sinh(L/\lambda_f) + \Gamma_d\cosh(L/\lambda_f) + \Gamma_f\left(1 + \frac{1}{\Gamma_s F}\right)\left[\Gamma_d\sinh(L/\lambda_f) + \cosh(L/\lambda_f)\right]};$$

$$a_2 = -\frac{I(\xi\tau_s\cos\alpha - \tau_f\sin\alpha)[\Gamma_d\sinh(L/\lambda_f) + \cosh(L/\lambda_f)] + I\tau_f\sin\alpha - \tilde{n}_f}{\sinh(L/\lambda_f) + \Gamma_d\cosh(L/\lambda_f) + \Gamma_f\left(1 + \frac{1}{\Gamma_s F}\right)\left[\Gamma_d\sinh(L/\lambda_f) + \cosh(L/\lambda_f)\right]}. \tag{4.80}$$

Here, we denote $\xi = k_{sf}/k_{fs} = (\sigma_s/\sigma_f)^{1/2}\exp\left(\Delta\varepsilon_{sf}/k_B T\right)$ with the energy barrier $\Delta\varepsilon_{sf}$ describing the difference between the activation energies for desorption and diffusion of the substrate and sidewall planes as shown in Fig. 4.40, $\Gamma_d = D_f/(\lambda_f k^+)$, and $\Gamma_s = D_s/(\lambda_s k_{sf})$. The quantity $\tilde{n}_f = k^- C/k^+$ corresponds to quasi-equilibrium concentration given by (4.74), at which the exchange flux $k^+ n_f(L) - k^- C$ equals zero. The radius-dependent functions G and G' are defined as

$$\begin{aligned} G(r/\lambda_s) &= K_1(R_W/\lambda_s)\, I_0(r/\lambda_s) + I_1(R_W/\lambda_s)\, K_0(r/\lambda_s); \\ G'(r/\lambda_s) &= \frac{dG}{d(r/\lambda_s)} = K_1(R_W/\lambda_s)\, I_1(r/\lambda_s) - I_1(R_W/\lambda_s)\, K_1(r/\lambda_s). \end{aligned} \tag{4.81}$$

The function F in (4.80) depends on the ratios R/λ_s and R_W/λ_s as

$$F \equiv -\frac{G'(R/\lambda_s)}{G(R/\lambda_s)} = \frac{I_1(R_W/\lambda_s)\, K_1(R/\lambda_s) - K_1(R_W/\lambda_s)\, I_1(R/\lambda_s)}{I_1(R_W/\lambda_s)\, K_0(R/\lambda_s) + K_1(R_W/\lambda_s)\, I_0(R/\lambda_s)}. \tag{4.82}$$

Differentiating (4.79) at $z = L$, the exact expression for the diffusion-induced elongation rate given by (4.63) is obtained in the form

$$\left(\frac{dL}{dt}\right)_{diff} = \frac{2\Omega_S}{R}\frac{D_f}{\lambda_f} \cdot \frac{I(\xi\tau_s\cos\alpha - \tau_f\sin\alpha) + \left[\cosh(L/\lambda_f) + \left(1 + \frac{1}{\Gamma_s F}\right)\Gamma_f\sinh(L/\lambda_f)\right](I\tau_f\sin\alpha - \tilde{n}_f)}{\sinh(L/\lambda_f) + \Gamma_d\cosh(L/\lambda_f) + \Gamma_f\left(1 + \frac{1}{\Gamma_s F}\right)\left[\Gamma_d\sinh(L/\lambda_f) + \cosh(L/\lambda_f)\right]}. \tag{4.83}$$

This formula can be considerably simplified if the barrier for the adatom transition from the sidewall to the droplet equals the diffusion barrier, i.e., $k^+ \sim D_f/\sigma_f^{1/2}$. Since both L and λ_f are much larger than the interatomic spacing $\sigma_f^{1/2}$ in all cases where the adatom diffusion flux is significant, we have $\Gamma_d \cosh(L/\lambda_f) \sim (\sigma_f^{1/2}/\lambda_f)\cosh(L/\lambda_f) << \sinh(L/\lambda_f)$ for all relevant L/λ_f. In such a diffusion regime of nanowire-droplet atomic exchange, (4.83) is reduced to the form obtained in [410]:

$$\frac{1}{V}\left(\frac{dL}{dt}\right)_{diff} = \frac{BU(L/\lambda_f) + C}{U'(L/\lambda_f)}, \tag{4.84}$$

where $V \equiv I\Omega_S \cos\alpha$ is the deposition rate (nm/s). The functions U and U' are defined as follows

$$U(L/\lambda_f) = \sinh(L/\lambda_f) + \frac{D_s\lambda_f}{D_f\lambda_s\xi}\frac{F}{(1+\Gamma_s F)}[\cosh(L/\lambda_f) - 1]; \tag{4.85}$$

$$U'(L/\lambda_f) = \frac{dU}{d(L/\lambda_f)} = \cosh(L/\lambda_f) + \frac{D_s\lambda_f}{D_f\lambda_s\xi}\frac{F}{(1+\Gamma_s F)}\sinh(L/\lambda_f), \tag{4.86}$$

with an obvious behavior at $L = 0 : U(0) = 0$ and $U'(0) = 1$. The radius-dependent coefficients B and C describe the diffusion fluxes originating from the sidewall and substrate adatoms, respectively:

$$B = \frac{2\lambda_f \omega \tan\alpha}{R}\left(1 - \frac{\theta_l}{\theta_f}\right); \tag{4.87}$$

$$C = \frac{2\lambda_s}{R}\frac{F}{(1+\Gamma_s F)}\left(1 - \frac{\theta_l}{\theta_s}\right). \tag{4.88}$$

Here, θ_l is the droplet activity defined by (4.74) and

$$\theta_f = \omega I \tau_f \sigma_f \sin\alpha; \tag{4.89}$$

$$\theta_s = I\tau_s\sigma_f\xi\cos\alpha = I\tau_s\left(\sigma_f\sigma_s\right)^{1/2}\exp\left(\frac{\Delta\varepsilon_{sf}}{k_BT}\right)\cos\alpha \tag{4.90}$$

are the effective activities of the sidewall and substrate adatoms, respectively.

In the diffusion regime of the nanowire-droplet exchange which should hold for Si, Ge, Ga, In and other group III adatoms with a high solubility in liquid Au, the Hervieu boundary conditions (4.69), (4.75)–(4.77) imply two differences compared to the Dubrovskii conditions (4.68), (4.70), (4.71) and (4.73). First, the denominator $1 + \Gamma_s F$ in (4.85), (4.86) and (4.88) was equal one in the corresponding expressions of [410]. Second, the renormalized coefficient $(D_s\lambda_f/D_f\lambda_s\xi)$ in (4.85), (4.86) and the effective activity of substrate adatoms in (4.90) were $(D_s\lambda_f\sigma_f/D_f\lambda_s\sigma_s)$ and $\theta_s = I\tau_s\sigma_s\cos\alpha$, respectively, in [410]. We now show that the first difference eventually disappears under the conventional conditions of the diffusion-induced growth. Indeed, in absence of an additional barrier for the transitions through the substrate-nanowire boundary, we have: $k_{sf} \sim D_s/\sigma_s^{1/2}$, yielding $\Gamma_s \sim \sigma_s^{1/2}/\lambda_s$.

Let us consider the case of a single nanowire with $R_W/\lambda_s >> 1$, where there is no overlap of the diffusion feeding rings on the substrate surface. From (4.82), the function F now equals $K_1(R/\lambda_s)/K_0(R/\lambda_s)$, the value which lies between 1 at $R/\lambda_s >> 1$ and $\lambda_s/[R\ln(\lambda_s/R)] >> 1$ at $R/\lambda_s << 1$. Therefore, the value of $\Gamma_s F$ is bound between $\Gamma_s \sim \sigma_s^{1/2}/\lambda_s$ and $\sigma_s^{1/2}/[R\ln(\lambda_s/R)]$. Both values are much smaller than one when the diffusion length and the wire radius are much larger than the lattice spacing. Equations (4.85), (4.86), (4.88) are now simplified to

$$U(L/\lambda_f) = \sinh(L/\lambda_f) + b\frac{\lambda_f}{\lambda_s}F[\cosh(L/\lambda_f) - 1]; \tag{4.91}$$

$$U'(L/\lambda_f) = \cosh(L/\lambda_f) + b\frac{\lambda_f}{\lambda_s}F\sinh(L/\lambda_f); \tag{4.92}$$

$$C = \frac{2\lambda_s F}{R}\left(1 - \frac{\theta_l}{\theta_s}\right), \tag{4.93}$$

where $b \equiv D_s/(D_f\xi)$ is a constant. As for the difference in θ_s, it equals the adatom coverage of the surface far away from the nanowire (obtained by balancing their arrival rate and the sink by desorption or surface growth) with the Dubrovskii boundary conditions, and is re-normalized by the ξ-factor (accounting for the difference between the substrate and sidewall planes) in the Hervieu model. The two results match at $\Delta\varepsilon_{sf} = 0$ and $\sigma_s = \sigma_f$, i.e., when the substrate and sidewall planes are exactly identical.

Very importantly, (4.93) and (4.87) show that the diffusion flux from the sidewalls is proportional to $1 - \theta_l/\theta_f$, the diffusion flux from the substrate is proportional to $1 - \theta_l/\theta_s$, and that the two fluxes are independent of each other. In other words, the substrate adatoms feel only the difference of chemical potentials between themselves and the droplet, but do not feel the sidewall adatoms. The sidewall diffusion flux is sensitive only to the difference of chemical potentials between the sidewall adatoms and the droplet, and is not affected by the substrate diffusion flux passing along the sidewalls. Since the droplet activity is generally a function of R, the magnitude and even the direction of the two diffusion fluxes can change with R and depend on the nanowire length. The corresponding diffusion fluxes are positive when the activity of adatoms is larger than the liquid activity and negative otherwise. This may give rise to interesting non-linear effects in nanowire growth that will be discussed in detail in the foregoing sections.

Combining the diffusion-induced contribution to the elongation rate given by (4.84) with the adsorption-desorption growth rate given by (4.59) and subtracting the growth rate V_s of the non-activated surface burying the nanowire root (see Fig. 4.15), we obtain:

$$\frac{1}{V}\frac{dL}{dt} = A + \frac{BU(L/\lambda_f) + C}{U'(L/\lambda_f)}. \tag{4.94}$$

Here,

$$A = \frac{f(\alpha, \beta)}{\cos\alpha}\left(1 - \frac{\theta_l}{\theta_v}\right) - \varepsilon \tag{4.95}$$

is the L- independent contribution from the adsorption-desorption processes and the surface growth, with $\varepsilon = V_s/V$. In the case of vapor deposition, all α factors should be put to one, and $f(\alpha, \beta) = 2/(1 + \cos\beta)$. This equation gives the most general form of the transport-limited nanowire elongation rate at $R = const$, which is widely used for modeling the nanowire length as a function of time, radius, and deposition conditions. The self-consistency discussed in the previous section implies that this growth rate equals the nucleation-mediated sink, which enables finding the unknown liquid activity θ_l as a function of R and deposition conditions in the general case. This is not required if $\theta_l/\theta_v << 1, \theta_l/\theta_s << 1$ and $\theta_l/\theta_f << 1$, the case where the liquid supersaturation does not affect the growth rate.

Equation (4.94) can be re-written in the non-dimensional form by introducing the dimensionless length $l \equiv L/\lambda_f$ and the deposition thickness $h = Vt/\lambda_f$:

$$\frac{dl}{dh} = A + \frac{BU(l) + C}{U'(l)}; \quad l(h = 0) = l_0. \tag{4.96}$$

The initial condition at $h = 0$ determines the length l_0 of a pre-existing nanowire, e.g., obtained at different growth conditions. If growth starts from the substrate, which is a usual case, we should put $l(h = 0) = 0$. It is noteworthy that the non-linear differential (4.96) with the function $U(l)$ given by (4.91) can be integrated analytically at arbitrary coefficients A, B and C, and initial condition l_0. The result is obtained in the form of reverse dependence $h(l)$ [410]:

$$h(l) = \frac{1}{B^2 - A^2}\left[B\ln\left(\frac{Y(l)}{Y(l_0)}\right) - A(l - l_0)\right] + A[\Psi(l) - \Psi(l_0)]. \tag{4.97}$$

The functions $Y(l)$ and $\Psi(l)$ are determined by

$$Y(l) = (A + B\nu F)(\cosh(l) - 1) + (B + A\nu F)\sinh(l) + A + C; \tag{4.98}$$

$$\Psi(l) = \frac{2(C - B\nu F)}{(B^2 - A^2)\sqrt{G}}\arctan\left[\frac{(A - C + 2B\nu F)\tanh(l/2) + B + A\nu F}{\sqrt{G}}\right], \tag{4.99}$$

where $G = A^2[1 - (\nu F)^2] - B^2 - C^2 + 2BC\nu F$ and $\nu = b(\lambda_f/\lambda_s)$.

4.9 Simplified Growth Equations

The diffusion term in (4.94) generally depends on the ratios L/λ_f, R/λ_s and R_W/λ_s. We now consider some important particular cases where the growth equation is simplified and, in particular, can be reduced to the $1/R$ and $1/R^2$ dependences shown in Figs. 4.37, 4.38 and 4.39. For a dilute nanowire ensemble with $R_W/\lambda_s >> 1$, the function F in (4.82) is reduced to the Blakely-Jackson factor [308]

$$F = \delta\left(\frac{R}{\lambda_S}\right) = \frac{K_1(R/\lambda_S)}{K_0(R/\lambda_S)}. \tag{4.100}$$

The single nanowire model is further simplified in the following situations:

1. Short nanowire with $L/\lambda_f << 1$, the $b\lambda_f/\lambda_s$ factor not too large. From (4.91) and (4.92), we get $U \cong 0$ and $U' \cong 1$, yielding

$$\frac{dL}{dH} = A + C, \tag{4.101}$$

with A defined by (4.95) and $H = Vt$. This result shows that a nanowire with a given radius can emerge from the substrate only at $A + C > 0$. The contribution from the substrate given by (4.93) at $R_W/\lambda_s >> 1$ takes the form

$$C = \frac{2\lambda_s}{R}\delta\left(\frac{R}{\lambda_s}\right)\left(1 - \frac{\theta_l}{\theta_s}\right). \tag{4.102}$$

At $R/\lambda_s >> 1$, i.e., small adatom diffusion length on the substrate surface, $\delta(R/\lambda_s) \cong 1$, yielding the classical $1/R$ radius dependence

$$\frac{dL}{dH} = A + \frac{2\lambda_s}{R}\left(1 - \frac{\theta_l}{\theta_s}\right) \tag{4.103}$$

with neglect of the radius dependence of θ_l. In the opposite case of $R/\lambda_s << 1$, i.e., large diffusion length, $\delta(R/\lambda_s) \cong \lambda_s/[R\ln(\lambda_s/R)]$. This yields the $1/R^2$ dependence

$$\frac{dL}{dH} = A + 2\left(\frac{\lambda_s}{R}\right)^2 \frac{1}{\ln(\lambda_s/R)}\left(1 - \frac{\theta_l}{\theta_s}\right) \tag{4.104}$$

with neglect of a weak logarithmic function at large values of argument. When a nanowire is mainly fed by the surface adatoms and the Gibbs-Thomson effect is ineffective, the nanowire length therefore depends on the radius as $1/R$ at small and as $1/R^2$ at large adatom diffusion length.

2. Short nanowire at $R/\lambda_s << 1$ and $\lambda_f \to \infty$. This interesting situation, studied in [413] and [452], holds when all adatoms collected from the substrate can reach the droplet regardless of the nanowire length. On the other hand, in view of

$R/\lambda_s << 1$, the surface collection area is much larger than the droplet surface. Neglecting the term A in (4.94), we get

$$\frac{dL}{dH} = \frac{BU + C}{dU/dl}. \tag{4.105}$$

with $l = L/\lambda_f$. Integrating this with initial condition $L(H = 0) = 0$, we obtain

$$U = \frac{C}{B}\left[\exp\left(\frac{BH}{\lambda_f}\right) - 1\right]. \tag{4.106}$$

According to (4.87) and (4.102), the B term is proportional to λ_f, while C does not depend on λ_f. The exponential term in (4.106) is therefore independent of λ_f, and C/B is inversely proportional to λ_f. The universal, λ_f—independent asymptote of the nanowire length is thus obtained by expanding (4.91) for U to the first order in $1/\lambda_f$. This corresponds to the substitution of $\sinh(L/\lambda_f)$ to L/λ_f and $\cosh(L/\lambda_f) - 1$ to $L^2/(2\lambda_f^2)$, where U is inversely proportional to λ_f. The result for L has the form

$$L + \frac{b}{2}\delta\left(\frac{R}{\lambda_s}\right)\frac{L^2}{\lambda_s} = \frac{\lambda_s}{a}\delta\left(\frac{R}{\lambda_s}\right)\frac{g_s}{g_f}\left[\exp\left(\frac{2ag_fH}{R}\right) - 1\right], \tag{4.107}$$

where $a = \omega \tan\alpha$ is the geometrical factor that equals one in MOCVD, and we write $g_f = 1 - \theta_l/\theta_f$ and $g_s = 1 - \theta_l/\theta_s$ for brevity. In view of $\lambda_s >> R$, where only the direct impingement can be neglected with respect to the diffusion flux, the asymptotic behavior of $\delta(R/\lambda_s)$ is: $\delta(R/\lambda_s) \cong \lambda_s/[R\ln(\lambda_s/R)] >> 1$. The L^2 term in the left hand side of (4.107) quickly becomes much larger than L at $L >> [2\ln(\lambda_s/R)/b]R$, i.e., starting already from the length of the order of R. With neglect of the L term, (4.107) becomes

$$L = \lambda_s\sqrt{\frac{2g_s}{abg_f}\left[\exp\left(\frac{2ag_fH}{R}\right) - 1\right]}. \tag{4.108}$$

Analysis of this approximate solution leads to the following conclusions:

(i) The formula

$$L \cong \frac{2\lambda_s}{R}\delta\left(\frac{R}{\lambda_s}\right)g_sH, \tag{4.109}$$

which is equivalent to (4.101), (4.102) at $A << C$, follows from (4.107) in the formal limit $H \to 0$. This gives the time-linear law of the wire elongation. We note, however, that such a behavior occurs only at the very beginning of growth, and that the steady state model itself may not be relevant at this incubation stage.

(ii) A more relevant solution at the initial stage is given by

$$L \cong 2\lambda_s \sqrt{\frac{g_s H}{bR}}. \tag{4.110}$$

which is valid as long as $2ag_f H/R << 1$. This gives an unusual, square root time dependence of the nanowire length and the $1/\sqrt{R}$ radius correlation.

(iii) The initial stage is followed by the exponential growth law

$$L \cong \lambda_s \sqrt{\frac{2g_s}{abg_f}} \exp\left(\frac{ag_f H}{R}\right), \tag{4.111}$$

which pertains as long as $L << \lambda_f$.

(iv) From (4.89) and (4.90), the adatom activities on the sidewall and substrate surfaces are related as

$$\frac{\theta_f}{\theta_s} = \frac{D_s}{D_f}\frac{\omega \tan\alpha}{\xi}\left(\frac{\lambda_f}{\lambda_s}\right)^2, \tag{4.112}$$

showing that the sidewall activity should be much larger than the substrate one at $\lambda_f >> \lambda_s$, apart from the systems with very large ξ. This is the most usual situation for MBE, because the substrate diffusion length is normally limited by the surface growth. One exception is the MBE systems with the direction of the beam close to the substrate normal ($\alpha \to 0$). In such setups, we have $g_f < 0 < g_s$, where nanowires can emerge from the substrate but then tend to a finite critical length

$$L \to \lambda_s \sqrt{\frac{2g_s}{ab|g_f|}} \tag{4.113}$$

rather than growing infinitely.

Figure 4.41 shows a typical time dependence of the nanowire length at $\lambda_f \to \infty$, obtained from (4.107) and (4.108) at $V = 0.17$ nm/s, b $= 1$, $\lambda_s = 200$ nm, $g_s = 1$, $g_f = 0.5$, $R = 35$ nm and $a = \sin 35°/\pi = 0.184$. The transition from $1/\sqrt{t}$ behavior at small t to the exponential law of elongation at large t is clearly seen. It is also seen that (4.108) is perfect approximation to a more general (4.107). Figure 4.42 illustrates the transformation from the limited growth at $g_f = -0.5$ through the critical curve given by (4.110) at $g_f = 0$ to the unlimited exponential growth at $g_f = 0.5$, for the same parameters.

For a dense forest of short nanowires with $L/\lambda_f << 1$ and a large diffusion length such that $R/\lambda_s << 1$, the asymptotes of the Bessel functions reduce (4.82) and (4.93) to

$$C = \frac{2\lambda_s^2}{R^2[-\ln(R/\lambda_s) + K_1(R_W/\lambda_s)/I_1(R_W/\lambda_s)]}. \tag{4.114}$$

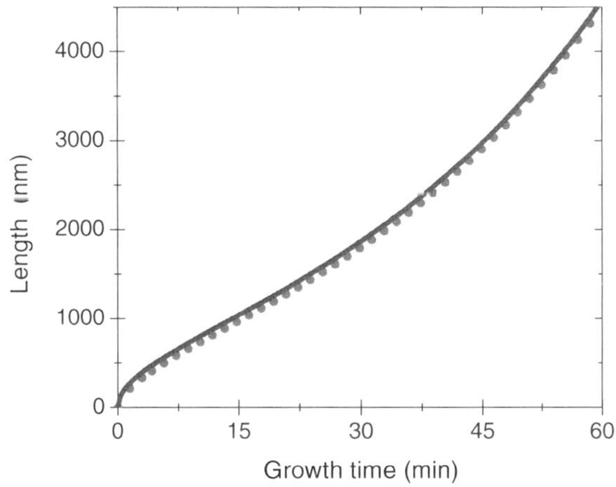

Fig. 4.41 Nanowire length versus time at $V = 0.17$ nm/s, $b = 1$, $\lambda_s = 200$ nm, $g_s = 1$, $g_f = 0.5$, $R = 35$ nm and $a = 0.184$, obtained from the asymptotic solution (4.107) (*dots*) and the approximate formula (4.108) (*line*)

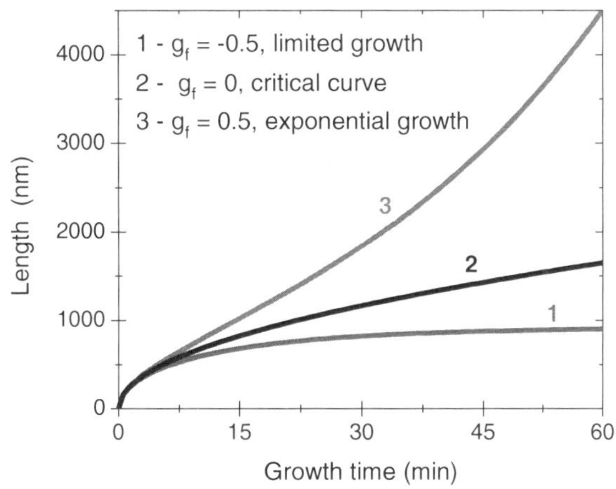

Fig. 4.42 Transformation from limited to infinite nanowire growth with increasing g_f

At $R_W/\lambda_s << 1$, i.e., when the diffusion length on the substrate is much larger than the inter-wire distance, the growth rate is limited entirely by R_W:

$$\frac{dL}{dH} \cong A + 4\left(\frac{R_W}{R}\right)^2. \tag{4.115}$$

This also gives the inverse $1/R^2$ length-radius correlation, with the growth rate rapidly decreasing with increasing the wire density because the neighboring wires compete for the surface adatoms. Figure 4.43 illustrates the effect by showing the dependence of the characteristic collection radius R_* defined as $(dL/dH)_{diff} = R_*/R$, for the model parameters given in the caption. It is seen that the competition of the neighboring wires for the substrate diffusion flux becomes significant at $\sim 2 \times 10^8$ cm^{-2}

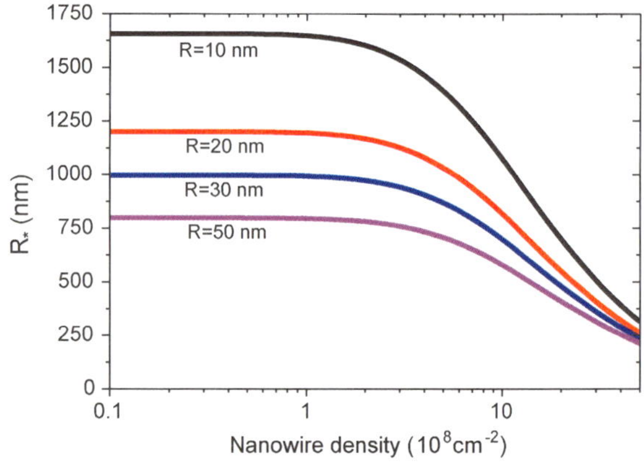

Fig. 4.43 Characteristic collection radius as a function of the nanowire density N_W, obtained from (4.84) for different $R = 10$, 20, 30, and 50 nm. Other parameters were fixed at: $\lambda_s = 200$ nm, $\lambda_f = 10000$ nm, $b(\lambda_f/\lambda_s) = 0.42$, $\theta_l = 0$, and $L = 1000$ nm [404]

density for $\lambda_s = 200$ nm. It is also clear that the elongation rate increases faster than $1/R$ with decreasing the radius in this example.

For long nanowires with $L/\lambda_f >> 1$, the droplets collect the adatoms only from the upper part of the wire of length λ_f, while the substrate adatoms cannot reach the top due to intensive re-evaporation or radial growth. This case relates to $U \cong U' \to \infty$, which reduces (4.94) to the classical $1/R$ dependence [47]

$$\frac{dL}{dH} = A + B = A + \frac{2\omega\lambda_f \tan\alpha}{R}\left(1 - \frac{\theta_l}{\theta_f}\right). \tag{4.116}$$

This shows the possibility for infinite growth only at $A + B > 0$. Of course, in real systems the droplet cools down far away from the substrate and finally freezes, which will stop the VLS growth at a certain length. Also, when the diffusion flux from the substrate becomes ineffective, the droplet will gradually shrink and the wire will taper. This effect can be described within a non-stationary growth model of [320].

Whenever the adatom diffusion length is limited entirely by desorption, i.e., no surface growth occurs, one can assume that adatoms far away from the nanowire base are at the chemical potential of the vapor. Putting $\theta_s \cong \theta_v$ and $\varepsilon \cong 0$, (4.101), (4.102) and (4.95) give the growth equation of the form

$$\frac{dL}{dH} = \left(1 - \frac{\theta_l^\infty}{\theta_v} e^{R_{GT}/R}\right)\left[\frac{f(\alpha,\beta)}{\cos\alpha} + \frac{2\lambda_s}{R}\frac{K_1(R/\lambda_s)}{K_0(R/\lambda_s)}\right]. \tag{4.117}$$

Equations (4.116), (4.95) and (4.74) at $\theta_f \cong \theta_v$ and $\varepsilon \cong 0$ yield

$$\frac{dL}{dH} = \left(1 - \frac{\theta_l^\infty}{\theta_v} e^{R_{GT}/R}\right)\left[\frac{f(\alpha,\beta)}{\cos\alpha} + \frac{2\omega\lambda_f \tan\alpha}{R}\right]. \tag{4.118}$$

In both equations, the geometrical factor $f(\alpha, \beta)/\cos\alpha = 2/(1+\cos\beta)$ in MOCVD and equals 2 at $\beta = \pi/2$. These growth rates become zero at the minimum radius $R_{\min} = R_{GT}\ln(\theta_v/\theta_l^{\infty})$, which is exactly identical to (4.41). This is not surprising, because the adatom fluxes are now driven by the same difference of activities $\theta_v - \theta_l$ as the vapor flux. If we assume that $\varepsilon \cong [f(\alpha,\beta)(1-\theta_s/\theta_v)]/\cos\alpha$, i.e., the surface layer growing at the same rate as the wire at $\theta_s = \theta_l$, (4.101),(4.102) and (4.95) give [82]

$$\frac{dL}{dH} = \left(1 - \frac{\theta_l^{\infty}}{\theta_v} e^{R_{GT}/R}\right)\left[\frac{f(\alpha,\beta)}{\cos\alpha}\frac{\theta_s}{\theta_v} + \frac{2\lambda_s}{R}\frac{K_1(R/\lambda_s)}{K_0(R/\lambda_s)}\right]. \tag{4.119}$$

This growth rate becomes zero at $R_{\min} = R_{GT}\ln(\theta_s/\theta_l^{\infty})$, where the activity of surface adatoms equal the liquid activity, modified by the Gibbs-Thomson effect. Simplified (4.117), (4.118) and (4.119) are often used for modeling the experimental length-radius correlations with one maximum [348, 409, 411]. These curves start from zero at $R = R_{\min}$, increase due to diminishing the curvature of the droplet surface, reach their maximum at a certain radius and then convert to decreasing as in the standard diffusion-induced mode.

4.10 Self-Consistent Growth Models

As in Sect. 4.7, the liquid supersaturation entering the diffusion growth equations is not an independent parameter, but should be obtained by balancing the transport-limited and the nucleation-mediated growth rate in the steady state [320]. This becomes essential when the liquid supersaturation is close to that of the adatom phases and therefore considerably affects the resulting diffusion fluxes to the droplet. From (4.94), (4.95), (4.87), (4.93), (4.60) and (4.46), we have the self-consistent equation for ζ_{∞}(or ζ) of the form

$$\left(\frac{dL}{dt}\right)_{AD} + \left(\frac{dL}{dt}\right)_{diff} = \left(\frac{dL}{dt}\right)_{nucl}, \tag{4.120}$$

with

$$\left(\frac{dL}{dt}\right)_{AD} = \frac{Vf(\alpha,\beta)}{\cos\alpha}\left(1 - \frac{\zeta+1}{g(\Phi+1)}\right); \tag{4.121}$$

$$\left(\frac{dL}{dt}\right)_{diff} = V\left[\frac{BU(L/\lambda_f) + C}{U'(L/\lambda_f)}\right]; \tag{4.122}$$

$$B = \frac{2\lambda_f \omega \tan\alpha}{R}\left(1 - \frac{\zeta + 1}{\eta_f + 1}\right); \quad C = \frac{2\lambda_s}{R} F\left(\frac{R}{\lambda_s}, \frac{R_W}{\lambda_s}\right)\left(1 - \frac{\zeta + 1}{\eta_s + 1}\right); \tag{4.123}$$

$$\left(\frac{dL}{dt}\right)_{nucl} = \frac{h\pi R^2 J}{1 + \Theta[3(\pi J/v)^2]^{1/3} R^2}; \tag{4.124}$$

$$J = J_0 \exp\left(-\frac{A_*}{\ln(\zeta + 1) - R_{GT}^s/R}\right);$$

$$\zeta + 1 = (\zeta_\infty + 1)\exp(R_{GT}/R). \tag{4.125}$$

Here, the geometrical function $f(\alpha, \beta)$is defined in (4.34)–(4.36), the g factor in (4.40), the functions U, U' are given by (4.85), (4.86), the F function by (4.82), the lateral growth velocity of a 2D island v is usually proportional to ζ_∞, the energetic parameter A_* is defined in (4.46), and the Gibbs-Thomson radii R_{GT} and R_{GT}^s in (4.56). In (4.123), we use the adatom supersaturations related to the corresponding activities as $\eta + 1 = \theta$, and the liquid supersaturation $\zeta + 1$ instead of θ_l, to more closely relate the notations to that of classical nucleation theory.

This system should be solved numerically at given growth conditions, material and kinetic constants. Unfortunately, some of the parameters, such as J_0, v, η_f, η_s and even the normalized liquid surface energy A_* and the vapor supersaturation Φ are not known with reasonable accuracy in most cases. Furthermore, there is no straightforward way to control and even to define those values in complex three-phase systems for variable temperatures, fluxes and other conditions. It is therefore interesting to consider simplified cases with a reduced number of fitting parameters. In particular, the case of vapor pressure deposition [$\cos\alpha = 1$, $\tan\alpha = 1$, $g = 1$, $\omega = 1$, $f(\alpha, \beta) = 2/(1 + \cos\beta)$] at $L/\lambda_f >> 1$ and $\eta_f \cong \Phi$ (no radial growth) corresponds to the transport-limited growth rate

$$\frac{dL}{dt} = \left(\frac{dL}{dt}\right)_{AD} + \left(\frac{dL}{dt}\right)_{diff} = 2V\left(\frac{1}{1 + \cos\beta} + \frac{\lambda_f}{R}\right)\frac{\Phi - \zeta}{\Phi + 1}. \tag{4.126}$$

Relating it to the nucleation-mediated growth rate in the mononuclear mode ($\Theta \to \infty$)

$$\frac{dL}{dt} = h\pi R^2 J(\zeta) \cong h\pi R^2 J(\Phi)\exp\left[-i_c(\Phi)\frac{\Phi - \zeta}{\Phi + 1}\right], \tag{4.127}$$

and following the procedure described in Sect. 4.7, we obtain the approximate solution

$$\frac{dL}{dt} = 2V\left(\frac{1}{1+\cos\beta} + \frac{\lambda_f}{R}\right)\frac{\left[\ln(\Phi+1) - R_{GT}^s/R\right]^2}{a}W[\varphi(R)]. \tag{4.128}$$

Here, W denotes the Lambert function, and $\varphi(R)$ is given by

$$\varphi(R) = \frac{A_*}{\left[\ln(\Phi+1) - R_{GT}^s/R\right]^2}\left(\frac{R}{R_0}\right)^2 \frac{1}{1/(1+\cos\beta) + \lambda_f/R}\exp\left[-\frac{A_*}{\ln(\Phi+1) - R_{GT}^s/R}\right], \tag{4.129}$$

with $R_0 = (2\mathrm{V}/\pi h J_0)^{1/2}$ as the characteristic radius. This growth rate equals zero at the Wagner-Ellis minimum radius given by (4.53), and contains the Givargizov-Chernov factor in the square brackets of (4.128). However, it accounts for the adatom diffusion flux and therefore can be used for modeling more complex, non-monotonic length-radius dependences.

Whenever the wire growth is controlled by the diffusion flux from the substrate at $R/\lambda_s << 1$, the transport-limited growth rate can be approximated as

$$\frac{dL}{dt} \cong VC = 2V\left(\frac{\lambda_s}{R}\right)^2 \frac{1}{\ln(\lambda_s/R)}\left(\frac{\eta_s - \zeta}{\eta_s + 1}\right). \tag{4.130}$$

In this case, the growth rate is controlled by the difference of activities of the surface adatoms and the liquid phase, $\eta_s - \zeta$. Neglecting a weak logarithmic correction and completely neglecting the Gibbs-Thomson effect for large enough droplets, i.e., putting $\zeta \cong \zeta_\infty$, we re-arrange this in the form ($\eta_s \equiv \eta$ for brevity):

$$\frac{dL}{dt} = V\left(\frac{\Lambda_s}{R}\right)^2\left(\frac{\eta - \zeta}{\eta + 1}\right), \tag{4.131}$$

with the effective substrate collection length Λ_s. Without the Gibbs-Thomson modification, the mononuclear nucleation-mediated growth rate is given by

$$\frac{dL}{dt} = h\pi R^2 J_0 \exp\left[-\frac{A_*}{\ln(\zeta+1)}\right] \cong h\pi R^2 J(\eta)\exp\left[-i_c(\eta)\left(\frac{\eta-\zeta}{\eta+1}\right)\right], \tag{4.132}$$

where we again used the Γ -method at η close to ζ. Now, the factor $J(\eta)$ does not contain any radius dependence, and the critical size simply equals $i_c(\eta) = A_*/\ln^2(\eta+1)$. (4.132) is written for nucleation away from the triple phase line. If nucleation is preferred at the triple phase line, the factor πR^2 should be changed to $2\pi R r_c(\eta)$, where r_c is the critical radius at $\zeta = \eta$.

Relating (4.131) to (4.132), we obtain:

$$xe^x = \left(\frac{R}{R_*}\right)^\delta \tag{4.133}$$

where $x = i_c(\eta)(\eta - \zeta)/(\eta + 1)$. The power exponent δ equals 4 for the C and 3 for the TPL nucleation. The characteristic nucleation radii here are defined as $R_* = [(V\Lambda_s^2)/(\pi h i_c(\eta) J(\eta))]^{1/4}$ for the C and $R_* = \left[(V\Lambda_s^2)/(2\pi h i_c(\eta) r_c(\eta) J(\eta))\right]^{1/3}$ for the TPL positions of the island. (4.133) has the solution $x = W[(R/R_*)^\delta]$ where W is the Lambert function. Substituting it into (4.131) and integration gives:

$$L = H\left(\frac{\Lambda}{R}\right)^2 W\left[\left(\frac{R}{R_*}\right)^\delta\right] \tag{4.134}$$

where $\Lambda \equiv \Lambda_s/\sqrt{i_c(\eta)}$ is a renormalized collection length. Equation (4.134) can be approximated in two limiting cases. First, the diffusion-limited growth regime follows from (4.134) for high adatom supersaturation and a fast nucleation process ($R/R_* >> 1$), which leads to a logarithmic asymptote of the Lambert function:

$$L \cong H\left(\frac{\Lambda}{R}\right)^2 \left[\delta \ln\left(\frac{R}{R_*}\right) - \ln\left(\delta \ln\left(\frac{R}{R_*}\right)\right)\right], R >> R_*. \tag{4.135}$$

Equation (4.135) describes the conventional $1/R^2$ behavior of the length for small R_*, neglecting a weak logarithmic dependence, as is typically observed in the diffusion-induced nanowire growth. In the second growth regime, nucleation-limited growth occurs for $R/R_* << 1$ where $W[(R/R_*)^\delta] \cong (R/R_*)^\delta$. The $L(R)$ relationship is then described by:

$$L \cong H\frac{\Lambda^2 R^{\delta-2}}{R_*^\delta}, R << R_* \tag{4.136}$$

This results in the unusual, increasing length dependence for large R_* where the supersaturation is low, and the decreased nucleation rate suppresses the nanowire growth of smaller radii. Equation (4.136) gives either R^2 or R behavior at small R depending on the nucleation position.

The non-monotonic behavior considered here has nothing to do with he Gibbs-Thomson effect (which is ineffective in large enough wires), and may pertain for much larger dimensions. Rather than by the droplet curvature, it is caused by the competition between the mononuclear 2D islanding and the diffusion transport from the surface. At very small radii, the VLS growth is limited just by the facet size because the nucleation probability is lower for a smaller facet. As the radius increases, the conventional R^{-2} diffusion becomes dominant and the growth rate of larger nanowires slows down. Depending on the numerical value of the characteristic radius R_*, one can therefore observe distinctly different $L(R)$correlations within a given range of radii. This is illustrated in Fig. 4.44 at $\delta = 4$, where curve 1 with $R_* = 14$ nm decreases starting from ~30 nm diameter, while curve 4 with $R_* = 125$ nm is always increasing.

We now consider experimental data of [466] on the Au-catalyzed, Be-doped InP nanowires that feature distinctly different length-radius correlations at different nominal doping levels, and model them by (4.134). MBE growth experiments

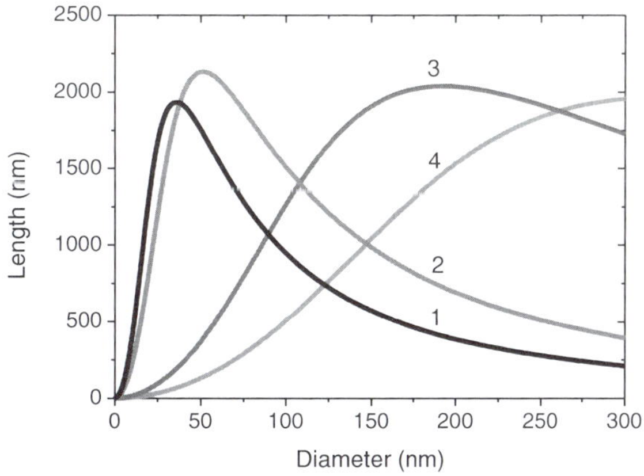

Fig. 4.44 Transition from the non-monotonic to purely increasing length-diameter dependences with increasing R_*, obtained from (4.134) at $\delta = 4$: *curve 1*—$R_* = 14$ nm, $\Lambda = 50$ nm; *curve 2*—$R_* = 40$ nm, $\Lambda = 75$ nm; *curve 3*—$R_* = 75$ nm, $\Lambda = 225$ nm; *curve 4*—$R_* = 125$ nm, $\Lambda = 450$ nm

Table 4.4 Description of samples

Sample	Structure (Growth duration)	Be doping concentration (cm^{-3})
A1	InP (30 min)	Undoped
A2	InP (30 min)	10^{18}
A3	InP (30 min)	2×10^{19}
B1	InAs (30 min) / InP (13 min)	Undoped
B2	InAs (30 min) / InP (13 min)	4×10^{19}

were as follows. InP (111)B wafers were activated by e-beam evaporation of 1 nm of Au. Wafers were then cleaved into 6 segments for separate growths as outlined in Table 4.4. Nanowires were grown by gas source MBE using an In effusion cell and group V dimers (As_2, P_2) derived from a hydride gas cracker, at 420 °C with a 2D equivalent growth rate of 0.139 nm/s. Samples A1–A3 consisted of InP nanowires with increasing Be dopant concentration, provided by a Be effusion cell. Nominal doping concentrations were based on Hall effect measurements on InP epilayers. Samples B1 and B2 were first comprised of a base InAs segment, followed by an InP segment. The V/III flux ratios were 5.6 and 3.6 for InP and InAs, respectively, in all samples.

Nanowire morphology was examined by SEM using a JEOL 7000F. Nanowire crystal structure was examined by a Philips CM12 TEM. Samples B1-B2 were examined by high angle annular dark field (HAADF) in a JEOL 2010F TEM. SEM analysis of the InP nanowires from samples A1–A3, shown in Fig. 4.45a–c, respectively, indicated a rod morphology with no observable tapering. Hence, the nanowire diameter was approximately equal to the Au droplet diameter. The nanowires grew vertical to the (111)B surface of the substrate with length up to 1.8 μm, and diameter ranging from 20 to 140 nm corresponding to the initial size distribution of the Au droplets. The size distribution of Au droplets at the end of the nanowires after growth was identical among samples A1–A3. Thus, Be doping had no effect on the Au particle

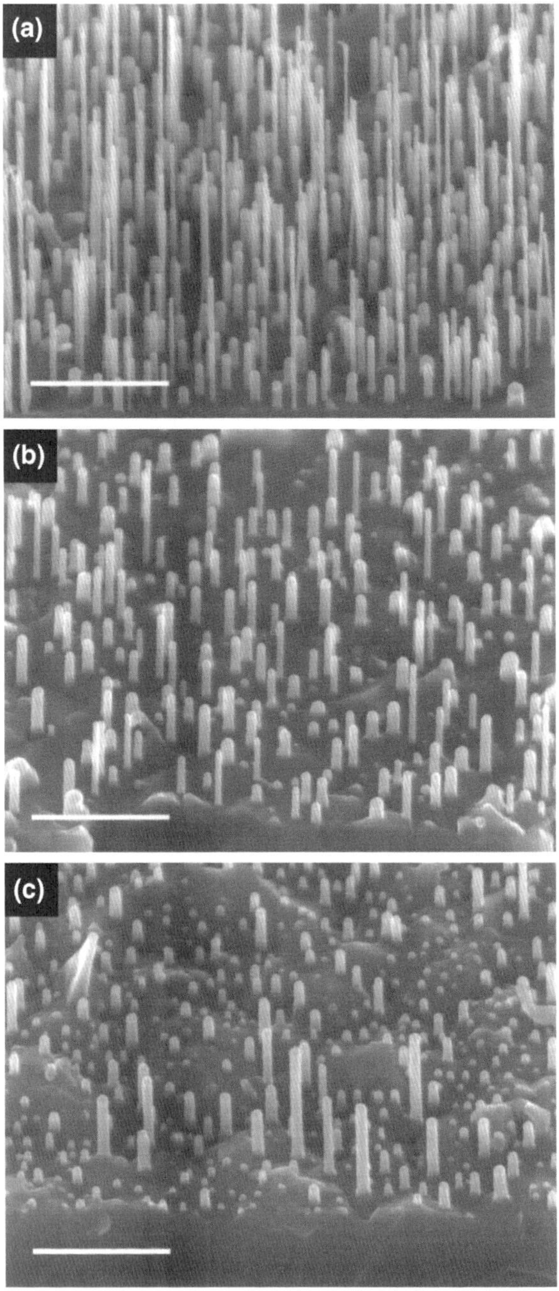

Fig. 4.45 Tilted view (45°) SEM images of (**a**) sample A1, undoped, (**b**) sample A2, Be doping $= 10^{18}\,\mathrm{cm}^{-3}$, and (**d**) sample A3, Be doping $= 2 \times 10^{19}\,\mathrm{cm}^{-3}$. Scale bars indicate 1 μm

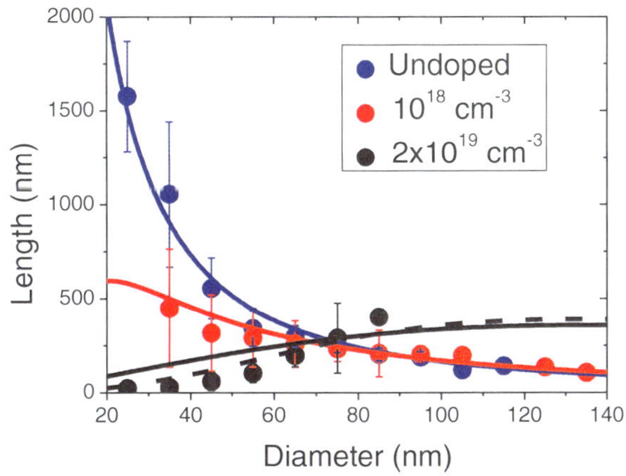

Fig. 4.46 Length versus diameter for different Be doped InP nanowires (sample A1: undoped, sample A2: $10^{18}\,\text{cm}^{-3}$, sample A3: $2 \times 10^{19}\,\text{cm}^{-3}$). Experimental data (*symbols*) are fitted by (4.134) at $H = 250\,\text{nm}$ (*lines*) with the parameters summarized in Table 4.5

size distribution. TEM analysis indicated a WZ crystal structure for all samples A1–A3 with the nanowire axis oriented along the (0001) growth direction, as commonly observed in InP nanowires [467]. Stacking faults were present in all samples A1–A3 with an increased occurrence in smaller diameter nanowires (~0.2 faults/nm) compared to larger diameter nanowires (~0.06 faults/nm). However, Be doping concentration did not appear to have any influence on the stacking fault density or crystal structure.

Next, the length versus diameter dependences of the InP nanowires were obtained from SEM images. Similar results were obtained by direct measurements from TEM images. The lengths are plotted against nanowire diameters in Fig. 4.46. Each symbol represents an average length measured from up to 35 nanowires within a 5 nm diameter range, and error bars show the standard deviation. As seen in Fig. 4.46, for undoped nanowires the length shows an inverse relationship with diameter. This decreasing $L(R)$ relationship indicates that growth is limited by adatom diffusion to the Au droplet from the nanowire sidewalls or surrounding substrate, as discussed previously. However, as the Be concentration increased, a suppression of the growth rate occurred for nanowire diameters below ~70 nm, resulting in an increasing length-diameter relationship saturating at a length of about 400 nm for the doping level of $2 \times 10^{19}\,\text{cm}^{-3}$.

A similar study was performed using the InAs/InP heterostructure samples B1 and B2. The contrast between the InP and InAs segments in HAADF imaging (Fig. 4.47) enabled the measurement of the top InP segment length, using the bottom InAs segment as a marker layer. Sample B1, which had a nominally undoped InP segment, featured the typical inverse $L(R)$ relationship, shown in Fig. 4.48, which is identical for undoped InP nanowires. In comparison, sample B2 with a Be doping concentration of $4 \times 10^{19}\,\text{cm}^{-3}$ showed the unusual increasing $L(R)$ relationship, also presented in Fig. 4.48. Thus, samples B1 and B2 showed that the unusual $L(R)$ behavior is not associated with differences in the initiation of growth, such as an incubation time.

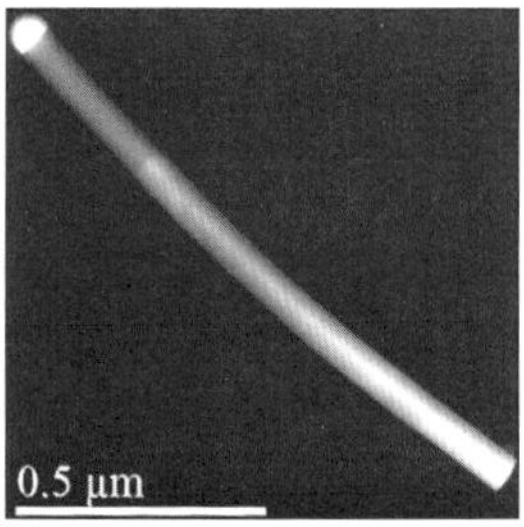

Fig. 4.47 HAADF image of InAs-InP nanowire from sample B1

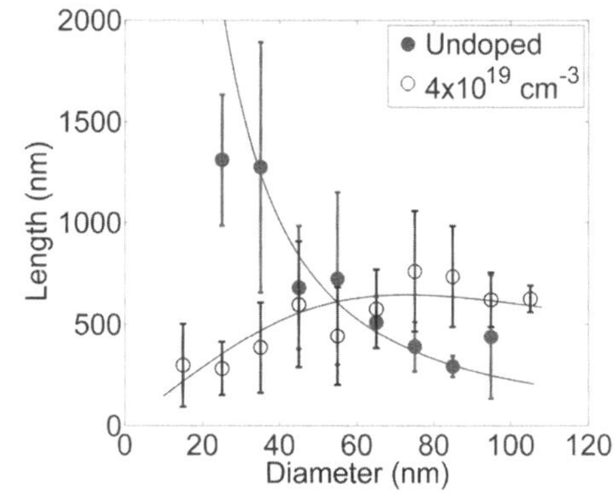

Fig. 4.48 InAs segment length versus diameter for different Be doped InAs/InP nanowires (sample B1: undoped, sample B2: $2 \times 10^{19}\,\mathrm{cm}^{-3}$). Experimental data (*symbols*) are fitted by (4.134) at $H = 108\,\mathrm{nm}$ (*lines*) with the parameters summarized in Table 4.5

The reproduction of doping effects in the heterostructures also excluded a change in adatom diffusion as a cause of the $L(R)$ behavior. Studies of Be-doped GaAs [468], Te-doped GaAs [469], Si-doped InAs [470], and Si-doped InP nanowires [471] indicated a reduction in length and an increase in tapering of the nanowires with doping. However, no $L(R)$ relationships were presented in these studies. These prior results suggested that dopants, present on the nanowire surface during growth, may influence diffusion length via surfactant effects or surface passivation, resulting in a preference for radial as compared to axial nanowire growth and a tapered nanowire morphology. However, such a tapered morphology was not observed in our A1–A3 nanowires. Furthermore, such a suppression of adatom diffusion length would be expected to produce a core-shell structure in nanowire heterostructures, but this was not observed in samples B1 or B2. Instead, the TEM analysis of the InAs-InP nanowires in samples B1 and B2 indicated a purely axial growth, with no visible tapering or core-shell (radial) growth regardless of doping concentration. The axial nature of the heterostructures and absence of radial growth excludes a reduction in diffusion length as the cause of the unusual $L(R)$ dependence in the InP nanowires.

In modeling these length-radius correlations, we disregard the Gibbs-Thomson effect since it is absent in the experimental $L(R)$ curves at small doping levels; the measured dependences are decreasing for all R. A small Be concentration at the droplet surface seems insufficient to cause such a significant increase in surface energy as to dramatically convert the diffusion-like $L(R)$ relationship. We hypothesize

Table 4.5 Fitting parameters at $\delta = 3$

Sample	Be doping level (cm^{-3})	R_*(nm)	Λ(nm)
A1	Undoped InP	1.5	14
A2	InP 10^{18}	11	22.5
A3	InP 2×10^{19}	72.5	115
B1	Undoped InAs	1.5	25
B2	InAs 4×10^{19}	39	126.5

that increasing Be doping level results in a decrease of the adatom supersaturation with respect to the droplet, a drastic decrease in the nucleation rate $I(\eta)$, and a rapidly increasing R_* in (4.134). The growth is then converted from the diffusion-induced to the nucleation-limited. Lines in Figs. 4.46 and 4.48 present theoretical fits to the experimental data obtained from (4.134) at $\delta = 3$, because all the nanowires have WZ crystal structure, and therefore nucleation is expected at the triple phase line. The fitting parameters are summarized in Table 4.5. It is seen that the fits are reasonable in all cases, however, sample A3 is better fitted with $\delta = 4$ (corresponding to nucleation away from the triple phase line), giving a quadratic increase of L at small R. This curve at $R_* = 51.5$ nm and $\Lambda = 82$ nm is shown by the dashed line in Fig. 4.46. In the model, R_* increases with Be concentration as discussed above, and the rapid increase of R_0 from 1.5 nm for undoped samples to 39–72.5 nm at $10^{19}\,cm^{-3}$ is attributed to a nucleation rate threshold dependence on supersaturation. The significant increase from $\Lambda = 14$ nm to $\Lambda = 126.5$ nm obtained from our fits may be the result of collective effects of the growth rate suppression, such as decreased competition among nanowires or a reduction in shadowing effects for adatom collection length.

4.11 Length—Radius Dependences

We saw earlier that the growth rate and the resulting length of the VLS nanowires can be radius-independent [357, 401], increasing due either to the Gibbs-Thomson-effect [376, 401, 448] or a suppressed nucleation rate on smaller facets [80, 466], decreasing in the purely diffusion-induced mode [287, 291, 401, 404], or exhibit a non-monotonic behavior with one maximum due to a competition between these effects [348, 345, 405, 409]. More examples will be given in this section, including even more complex S-like dependences with two extrema [412] (although such curves are quite rare). General expressions given in Sects. 4.8–4.10 enable modeling different types of $L(R)$ correlations but contain some unknown parameters. We also note that, in most cases such as shown in Figs. 4.20 and 4.21, the minimum radius at which the VLS growth stops is only an extrapolation point, while the experimental nanowire dimensions are much larger than R_{min}. Figure 4.44 shows that simple (4.134) with only two fitting parameters (the effective collection length Λ and the characteristic nucleation radius R_*) is capable of describing the diffusion-like decreasing, purely increasing and non-monotonic $L(R)$ curves with one maximum. This approach can be generalized. If the adsorption-desorption term given by (4.121) is much smaller than the diffusion-induced contribution into the overall elongation rate, the latter can be put in the form

$$\frac{dL}{dt} = V\left(\frac{\Lambda_a}{R}\right)^p\left(\frac{\eta-\zeta}{\eta+1}\right). \tag{4.137}$$

Here, Λ_a is the effective diffusion length of adatoms, ζ is the droplet supersaturation, and η is the supersaturation of the corresponding adatom phase. The power index p depends on the mechanisms of the diffusion transport and equals 1/2, 1 and 2 in the typical cases, as discussed in Sect. 4.9. In particular, (4.131) at $p = 2$ relates to the diffusion transport from the substrate at $R/\lambda_s << 1$, i.e., large adatom diffusion length compared to the nanowire radius.

Repeating step-by-step the procedure described in the previous section for obtaining (4.134), we arrive at the generalized self-consistent expression for the growth rate

$$\frac{dL}{dt} = V\left(\frac{\Lambda}{R}\right)^p W\left[\left(\frac{R}{R_*}\right)^{p+q}\right], \tag{4.138}$$

where W is the Lambert function. The effective collection length Λ and the characteristic radius R_* are now defined as

$$\Lambda \equiv \frac{\Lambda_a}{i_c^{1/p}(\eta)};\ \frac{1}{R_*^{p+q}} \equiv \frac{h}{V}\frac{J(\eta)}{\Lambda^p} \times \begin{cases} \pi, q = 2 \\ 2\pi r_c(\eta), q = 1 \end{cases}. \tag{4.139}$$

The index q equals 2 for the C and 1 for the TPL nucleation position. As above, the quantities $J(\eta), i_c(\eta)$ and $r_c(\eta) = (\sigma i_c(\eta)/C)^{1/2}$ denote the nucleation rate, the number of atoms in the critical nucleus and the critical radius, with σ as the elementary area of the corresponding plane and C as the shape constant. When the nucleation process is fast ($R/R_* >> 1$), the growth rate is limited by the diffusion transport:

$$\frac{dL}{dt} \cong const \times V\left(\frac{\Lambda}{R}\right)^p, R_* << R, \tag{4.140}$$

and decreases as $1/R^p$. When the nucleation is slow ($R/R_* << 1$), the liquid supersaturation gets higher and approaches that of adatoms. This quickly decreases the diffusion flux, and the VLS growth becomes nucleation limited. In this case, the growth rate increases with R as

$$\frac{dL}{dt} \cong V\frac{\Lambda^p R^q}{R_*^{p+q}}, R_* >> R, \tag{4.141}$$

i.e., as R for the TPL and as R^2 for the C nucleation positions. Generally, (4.138) gives a non-monotonic length-radius correlation starting from zero and reaching its

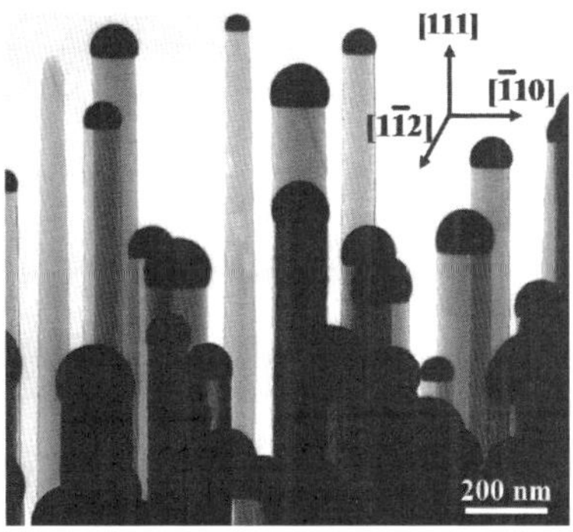

Fig. 4.49 A typical bright field TEM image obtained during the growth of Si nanowires at a temperature T of 575 °C and disilane pressure $P_{Si_2H_6} = 5 \times 10^{-6}$ Torr [357]

maximum at $R \sim R_*$. Equation (4.138) can be easily modified to yield a minimum radius (where the VLS growth ceases) by simply accounting for the fact that islands cannot nucleate on the facet whose radius is less than the size of the critical nucleus.

We now discuss typical experimental length-radius dependences of different nanowires. The simplest case is the radius-independent kinetics of the Au-catalyzed VLS Si nanowires, described by Kodambaka *et al.* [357]. Growth experiments were carried out in a UHV system composed of a TEM with in situ physical and chemical vapor deposition facilities. Phosphorous-doped Si(111) wafers with a miscut <0.5° were covered by 2–3 nm thick Au layer, heated prior to nanowire growth to form Au-Si eutectic droplets. A reasonable growth rate from Si_2H_6 can be obtained using temperatures in the range 500–660 °C, with typical growth times between 3 and 6 h. Figure 4.49 is a typical bright field TEM image acquired during Si wire growth. Note that all the wires grow in < 111 > directions, with the tips composed of the expected Si-Au eutectic droplets. Many wires taper visibly, because the catalyst droplet is shrinking over time.

Figure 4.50a is a typical plot of dL/dt versus nanowire diameter $d = 2R$, for wires grown at $P_{Si_2H_6} = 10^{-6}$ Torr and $T = 655$ °C. Although there is some scatter, there is no systematic dependence on d. Similar results were obtained from measurements of over 50 wires grown under different pressures and temperatures with diameters ranging from 30 to 150 nm. Figure 4.50b shows the growth rate plotted as a function of $P_{Si_2H_6}$ at $T = 575$ °C. It can be seen that dL/dt is proportional to $P_{Si_2H_6}$ over 3 orders of magnitude in pressure. The temperature dependence of dL/dt is shown in Fig. 4.50c. Assuming the Arrhenius behavior, an activation energy of 0.53 eV is obtained for the VLS growth of Si wires. Note that this activation energy is significantly lower than the activation energy of ~2 eV associated with the same process on a noncatalyzed Si surface [472, 473]. To consistently explain all of these observations, including the diameter independence of the growth rate, it is natural to assume that the sole rate-limiting step for wire growth is the thermally activated Au-catalyzed dissociative adsorption of disilane directly on the catalyst droplet [357]. Since the growth rate is found to be simply proportional to $P_{Si_2H_6}$, neither the reactive sticking probability nor the desorption rate depend on the chemical potential,

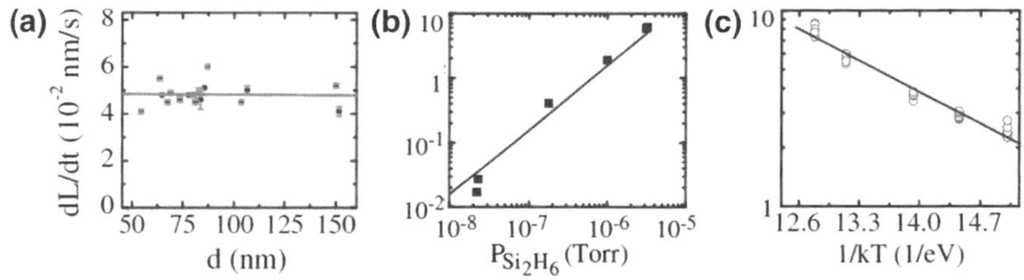

Fig. 4.50 **a** dL/dt versus diameter for an ensemble of Si wires grown at $P_{Si_2H_6} = 10^{-6}$ Torr and $T = 655$ °C. The *solid line* is a least-squares fit. The *error bars* represent the measurement uncertainties in growth rates. **b** Log-log plot of dL/dt of one particular wire grown at varying pressure at a fixed temperature $T = 575$ °C. The *solid line* is the best (*least-squares*) fit of slope 1. **c** Arrhenius plot of dL/dt versus T for 28 wires at $P_{Si_2H_6} = 10^{-6}$ Torr. Each data point represents an individual wire. The straight line is the least squares fit, with an activation energy of 0.53 ± 0.02 eV [357]

This excludes the Gibbs-Thomson effect and the increasing part of the $L(d)$ dependence, at least for the nanowire diameters studied in this work. The observed diameter independence also shows that only the molecules directly incident on the catalyst particle contribute significantly into the wire growth in this case. We note here that the radius-independent growth kinetics should be a common case for Ga-catalyzed GaAs nanowires and for group III-catalyzed III-V nanowires in general, because the VLS growth is usually controlled by the direct impingement of As.

A more complex behavior, already discussed in Sect. 4.3, is the size-dependent suppression of the growth rate at small R that yields an increasing $L(R)$ correlation. Two examples are shown in Figs. 4.51 and 4.52. The $L(R)$ curve in Fig. 4.51 corresponds to the Au-catalyzed GaAs nanowires grown by MBE under the conditions of sample No. 1 in Table 4.6. The dependence shown in Fig. 4.52 is the experimental $L(R)$ curve of Si/SiGe superlattice nanowire structure obtained by applying a hybrid pulsed laser ablation/chemical vapor deposition (PLA-CVD) process (sample No. 2 in Table 4.6) [474]. It is seen that both increasing $L(R)$ curves are reasonably fitted by (4.138) or by the equivalent formula

$$L = H\left(\frac{\Lambda}{R}\right)^p W\left[\left(\frac{R}{R_*}\right)^{p+q}\right] \tag{4.142}$$

which follows from (4.138) upon integration. In the case of CVD, the deposition thickness H is usually unknown, which is why we use $H\Lambda^p$ instead of Λ as the fitting parameter. The fitting values for each sample are also summarized in Table 4.6. The increasing $L(R)$ curves in Figs. 4.51 and 4.52 are best fitted with large values of the characteristic radius R_*, which reaches a maximum of 110 nm for sample 2 amongst all the samples. It is also seen that the two curves are best fitted with $p = 1$ and $q = 2$, i.e., the $1/R$-like diffusion transport and nucleation away from the triple phase line.

As the diffusion-induced contribution into the resulting growth rate increases, the $L(R)$ curves acquire a non-monotonic shape with a maximum. A typical example

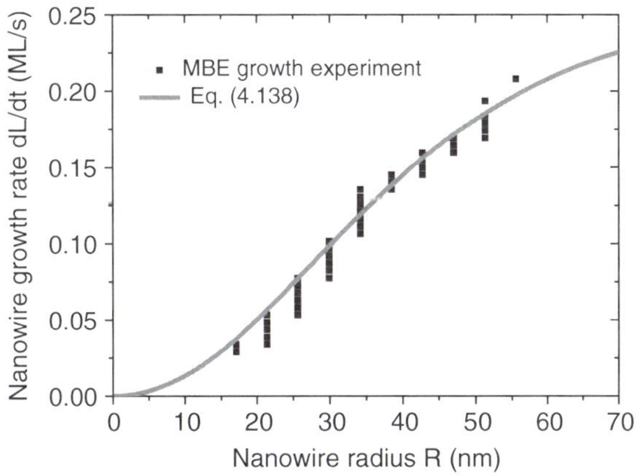

Fig. 4.51 dL/dt versus R for sample 1 (*symbols*), fitted by (4.138) (*line*) with the parameters given in Table 4.6

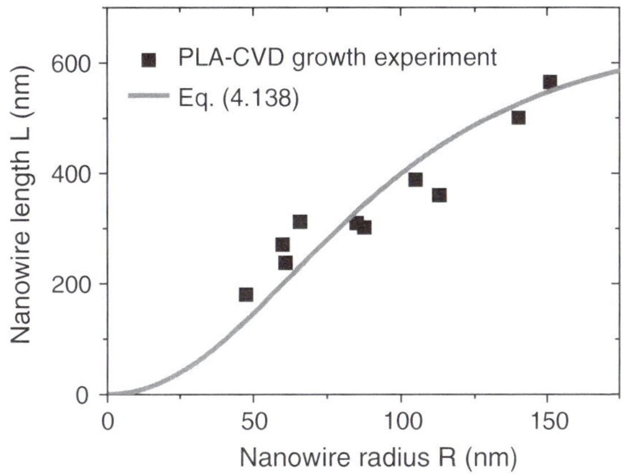

Fig. 4.52 L versus R for sample 2: experiment (*dots*) and theoretical fit by (4.138)

Table 4.6 Summary of experimental nanowire samples and fitting parameters with different types of $L(R)$ dependences

No	Material	Growth method	T (°C)	V (ML/s)	H (nm)	p	q	Λ (nm)	$H\Lambda^p$ (μm^{p+1})	R_* (nm)
1	GaAs	MBE [310]	550	0.4	200	1	2	37	–	48
2	SiGe	PLA-CVD [474]		–	–	1	2	–	85	110
3	InP	MOCVD [409]	470	–	–	2	2	–	1340	23
4	Si	MOCVD [409]	500	–	–	1	2	–	427	70
5	GaAs	MBE [409]	585	1	420	2	2	36	–	24
6	GaAs	MBE [291]	585	1	500	2	1	32	–	8

of such a curve is shown in Fig. 4.53 from [348]. These InAs nanowires were grown by the Au-seeded CBE from TMIn and precracked tertiarybutylarsine, at a temperature of 425 °C. The measured nanowire growth rate reaches its maximum at about 30 nm diameter. Solid line in Fig 4.52 is the fit obtained by Froberg *et al.*

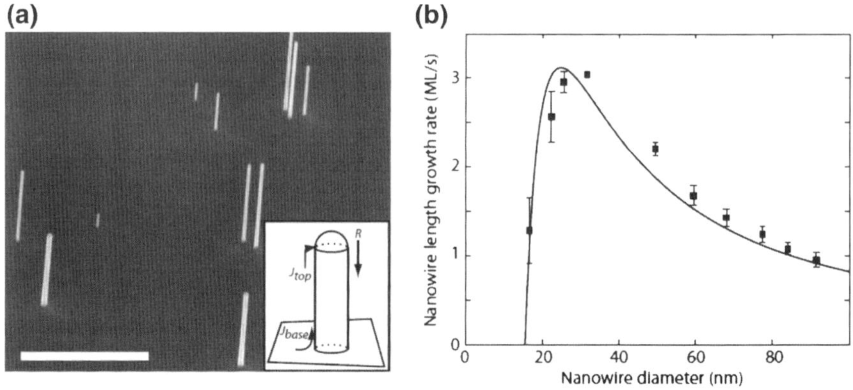

Fig. 4.53 **a** 30° tilted SEM image of InAs nanowires. Scale bar is 1 μm. The insert illustrates the model of material transport. **b** Experimental (*symbols*) and theoretical (*line*) growth rate versus diameter [348]

from the model which is similar to (4.101), (4.102), (4.95), at $I = 1\,\text{nm}^{-2}\,\text{s}^{-1}$, $\Delta\mu_{VL}^{\infty}/k_BT = \Delta\mu_{SL}^{\infty}/k_BT = 0.54$ and $\lambda_s = 130\,\text{nm}$. It is seen that the Gibbs-Thomson growth equation without self-consistency features a very steep radius dependence of the growth rate for small nanowire dimensions. As a result, the minimum diameter is predicted at ~16 nm.

Figure 4.54a shows a typical SEM image of InP nanowires grown on the GaAs(111)B substrates using colloidal Au nanodrops as catalysts in a MOCVD reactor under the VLS growth mode [409]. The mole fractions of the group V (tertiarybutylphosphine) and group III (trimethylindium) sources were 1.91×10^{-5} and 1.17×10^{-3}, respectively, in a 12 l/min hydrogen carrier gas flow yielding a V/III ratio of 61. The growth temperature was 470 °C and the growth time was 3 min. The size of Au drops ranges from 10 to 160 nm. Analysis of high-resolution SEM images was used to study the length-diameter dependences, presented in Fig. 4.54b. The growth parameters are summarized in line 3 of Table 4.6. The best fit of the experimental $L(R)$ curve by (4.138) is shown by solid line in Fig. 4.54b with the parameters summarized in line 3 of Table 4.6. Also for comparison, the dashed line presents the fit obtained from (4.116) and (4.95) at $R_{GT} = 4\,\text{nm}$, $\theta_v/\theta_l^{\infty} = 1, \theta_f/\theta_l^{\infty} = 1.4$ and $\lambda_f = 50\,\text{nm}$ [409]. It is seen that the fit by the self-consistent growth equation is excellent for all R. In contrast, the Gibbs-Thomxon fit represents correctly the maximum nanowire length but fails to describe the behavior at small $R < 15\,\text{nm}$. In particular, its rapid decrease toward smaller R predicts the incorrect minimum radius of 12 nm, while much smaller wires are observed experimentally.

Similar performance has been observed for Au-catalyzed Si nanowires grown on Si(111) substrates by MOCVD techniques [409]. The Au was deposited in-situ by controlling the evaporation flow. The evaporation pressure was 8×10^{-10} mbar and the deposition time was 4 min. Under these conditions, two Si substrates were prepared at 460 °C and 430 °C respectively. The corresponding SEM images of Au

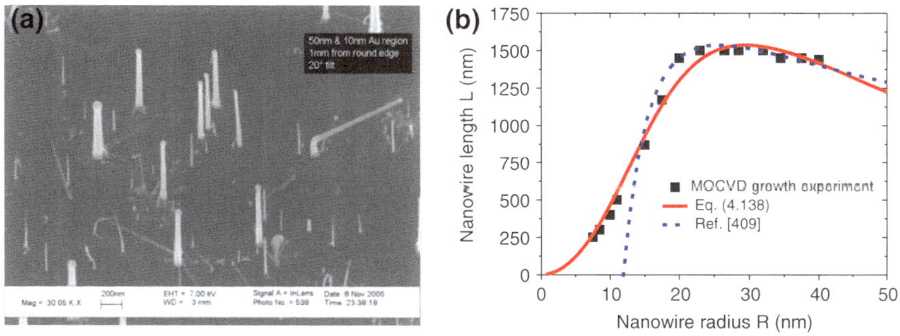

Fig. 4.54 **a** SEM image of MOCVD grown InP nanowires; **b** Experimental length-radius dependence (*black squares*) and theoretical fits obtained from (4.138) (*solid line*) and from the model of [409] (*dashed line*) with the parameters described in the text

droplets on the substrate surface before growth are shown in Figs. 4.55a and 4.56a. Obviously, a higher substrate temperature of 460 °C during Au deposition yields a much lower Au droplet density than at 430 °C. These two substrates were used for Si nanowire growth in a CVD chamber. The growth temperature and time for both samples were 500 °C and 30 min, respectively. The total pressure was 1.1 mbar and SiH_4 was diluted by H_2 with the ratio of 1:150. Figures 4.55 and 4.56b show SEM images of the Si nanowires in both conditions. Measurements from SEM images demonstrate that the length-diameter curve (Fig. 4.57) is almost independent of the number density of Au nanodroplets on the substrate surface, which suggests that the contribution from the diffusion of the substrate adatoms is negligible. Therefore, the data shown in Fig. 4.57 were fitted by (4.138) at $p = 1$, corresponding to the diffusion from the nanowire sidewalls. Other fitting parameters are given in line 4 of Table 4.6. This fit is shown by solid line in Fig. 4.57. The dashed line is the fit obtained from the Gibbs-Thomson (4.116), (4.95) at $R_{GT} = 4$ nm, $\theta_v/\theta_l^\infty = 1$, $\theta_f/\theta_l^\infty = 1.1$ and $\lambda_f = 90$ nm [409]. Given the scatter of the experimental data, both fits seem reasonable. However, the Gibbs-Thomson fit again shows too steep length-radius dependence at small R such that the minimum radius is predicted at an overestimated value of ~40 nm.

Figure 4.58a shows a typical SEM image of Au-catalyzed GaAs wires grown by MBE under the conditions summarized in line 5 of Table 4.6. The measured length-radius correlation shown in Fig. 4.58b is of the same type as discussed previously, i.e., a non-monotonic curve with a maximum. Solid line presents the best fit obtained from (4.138) with the parameters summarized in line 5 of Table 4.6. In particular, the index $p = 2$ corresponds to the diffusion flux from the substrate as the dominant contribution into the resulting elongation rate. The dashed line is the fit by (4.101), (4.102) and (4.95) at $R_{GT} = 3.5$ nm, $\theta_v/\theta_l^\infty = 2.2$, $\theta_s/\theta_l^\infty = 1.1$ and $\lambda_s = 95$ nm [409]. Comparison of the two theoretical fits leads to the same conclusion as before: the self-consistent (4.138) describes much better the experimental data at small $R < 30$ nm, although does not account for the Gibbs-Thomson effect.

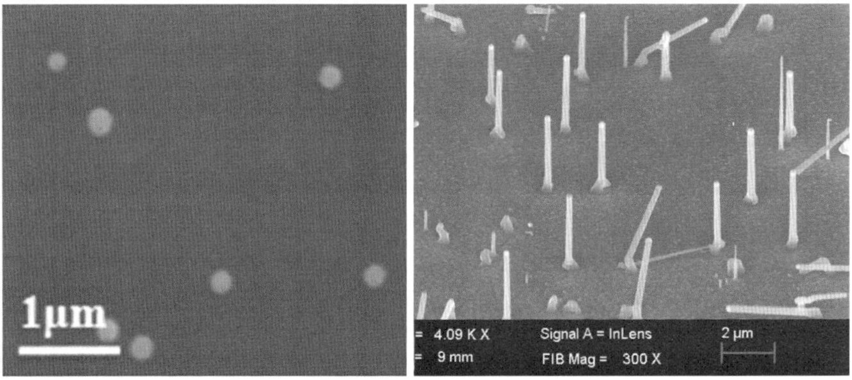

Fig. 4.55 SEM images of the Si surface with low droplet density (sample 1) before (*left*) and after (*right*) growth

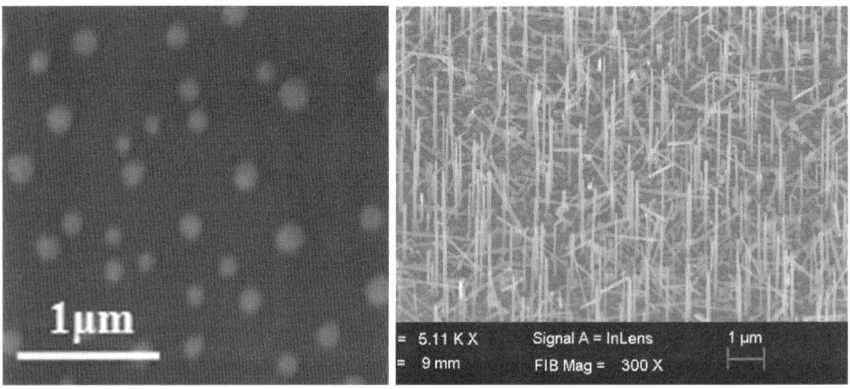

Fig. 4.56 SEM images of the Si surface with high droplet density (sample 2) before (*left*) and after (*right*) growth

The exponential Gibbs-Thomson correction to the diffusion transport equation predicts a rather overestimated minimum radius of ~17 nm, while 10 nm wide wires are observed in the SEM images.

Finally, Fig. 4.59 shows the diffusionlike length-radius dependence of the Au-assisted GaAs nanowires grown by MBE [291], which is decreasing all the way in the entire range of the measured dimensions. These wires were obtained under the conditions given in line 6 of Table 4.6. The fit is obtained from the self-consistent (4.138) with the parameters summarized in line 6 of Table 4.6. The indices $p = 2$ and $q = 1$ correspond to the substrate diffusion flux and nucleation at the triple phase line. The fitting curve at a small $R_* = 8$ nm reaches its maximum at $R = 10$ nm, which is below the experimentally observed dimensions. That is why the measured length-radius correlation is eventually seen as purely decreasing. This would not necessarily

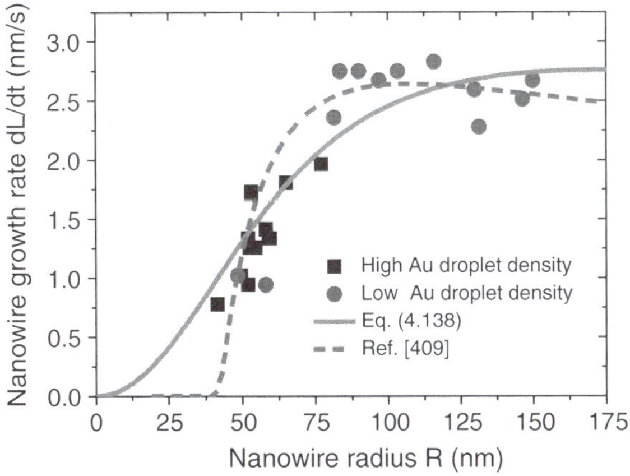

Fig. 4.57 Experimental length-radius dependence of MOCVD grown Si nanowires (*squares* for sample 2 and *circles* for sample 1), and theoretical fits obtained from (4.138) (*solid line*) and in [409]

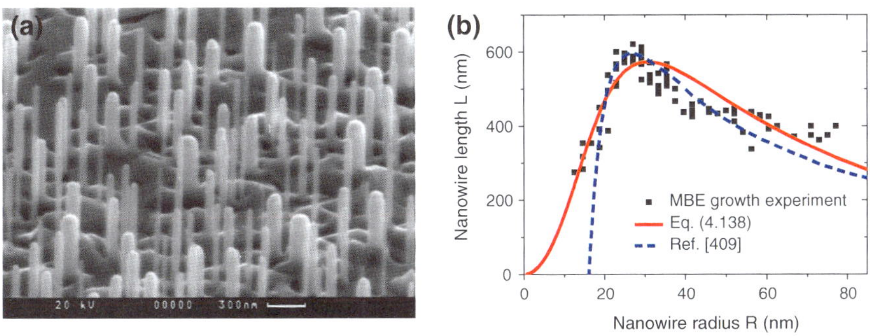

Fig. 4.58 **a** SEM image of MBE grown GaAs nanowires. **b** Experimental length-radius dependence (*symbols*) and theoretical fits obtained from (4.138) and from the model of [409]

be the case if smaller Au-Ga droplets were present in the size distribution of the initial particles.

Perhaps the most sophisticated radius-dependent growth kinetis has been reported in [412] in the case of InSb nanowire sections grown by Au-assisted CBE on InAs stems. InAs-InSb axial nanowire heterostructures were fabricated on InAs(111)B substrates in a Riber Compact-21 system, using trimethylindium (TMIn), tertiary-butylarsine (TBAs), and trisdimethylaminoantimony (TDMASb) as metal-organic precursors. A 0.5 nm thick Au film was first deposited by thermal evaporation in a separate evaporator chamber. The sample was then transferred into the CBE growth chamber, where it was annealed at 520 °C under the TBAs flux in order to generate the Au nanoparticles with diameters typically between 20 and 100 nm. The InAs segment was grown at a temperature of 435 °C, with the metal-organic line pressures

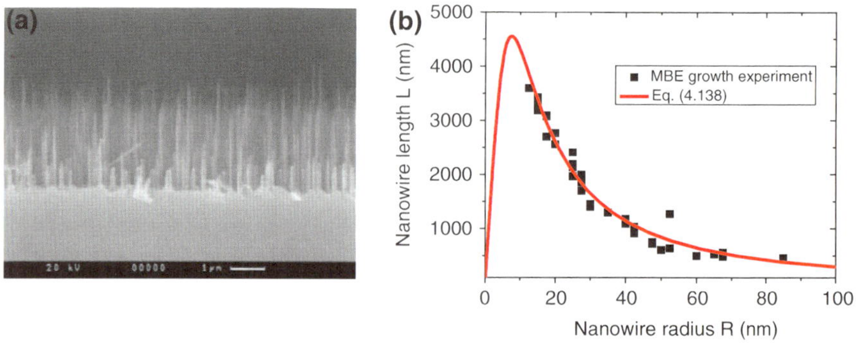

Fig. 4.59 SEM image (**a**) and length-radius correlation (**b**) of the Au-assisted GaAs nanowires grown by MBE [291], fitted by (4.138) with the parameters given in the text

of 0.3 and 1.0 Torr for TMIn and TBAs, respectively, and for a duration between 45 and 75 min. After that, the group V precursor was abruptly switched from TBAs to TDMASb in order to grow the InSb segment, while the temperature was lowered by 15°. Different metal-organic pressures, ranging between 0.3 and 0.45 Torr for TMIn and between 0.5 and 1.0 Torr for TDMASb, were used for different samples in order to investigate their effect on the growth rate. The growth duration of the InSb segment ranged between 45 and 60 min and was terminated by simultaneously stopping the TMIn and TDMASb fluxes and cooling the substrate down to the room temperature.

The samples were analyzed by SEM in a Zeiss Ultra Plus Field Emission Gun SEM. SEM image of a typical as grown sample is shown in Fig. 4.60. The InSb-InAs nanowires feature three important properties. First, nanowires growing in the vertical $< 111 >$B direction are composed of two axially stacked parts: a lower, smaller diameter InAs stem and an upper InSb segment of a larger diameter with a rather abrupt diameter change and almost uniform diameters of each part. The diameter typically increases by $18 \pm 2\,\%$ when passing from InAs to InSb part. Second, while the InAs stems form in WZ phase with the $< 0001 >$ direction along the growth axis, the InSb segments have pure ZB structure with the $< 111 >$ axis. Third, when the growth is switched from InAs to InSb, the hexagonal cross section defining the nanowire sidewalls rotates by 30° around the growth direction. This effect is demonstrated in Fig. 4.61. Consistent with earlier results [475], the InAs section is terminated by six equivalent $(1\ \bar{1}\ 00)$ WZ facets. After the WZ to ZB crystal phase transition and 30° angular rotation, the InSb section is defined by six (110) ZB facets, as schematized by the model geometries in Fig. 4.61b, c.

A statistical analysis of InSb morphology was then performed in different samples. The measured nanowires were chosen in such a way that the whole range of nanoparticle diameters was uniformly covered. The growth rate of InSb segment was obtained by dividing its length to the corresponding growth time. The nanoparticle diameter on the wire top almost equals twice the apothem of InSb segment

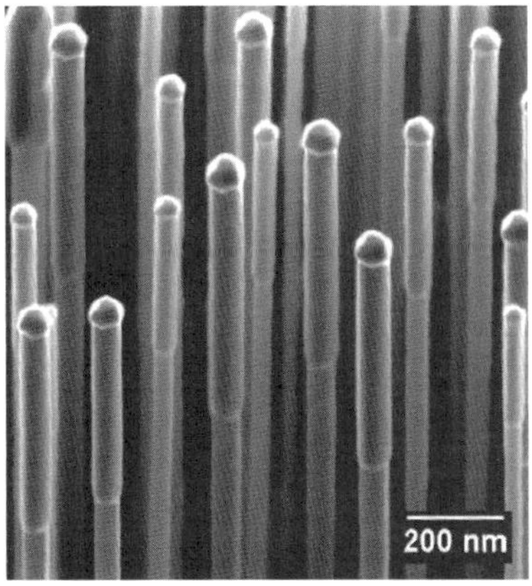

Fig. 4.60 45° tilted SEM image of as grown sample obtained at TMIn and TDMASb pressures of 0.45 and 0.75 Torr, respectively

$2R$ (see Fig. 4.61), called in what follows the diameter (d) for brevity. Experimental dependences of the growth rate of InSb segments on their diameter at different In/Sb pressure ratios are shown by symbols in Fig. 4.62. While the curve at $P_{In} = 0.45$ Torr is increasing all the way, the three curves at $P_{In} = 0.3$ Torr clearly exhibit an unusual, non-monotonic behavior: the growth rate first increases with the InSb diameter to a certain maximum value, decreases to a minimum and then is again converted to an increasing dependence. Such dependence cannot be explained within either the simplified self-consistent (4.138) or the expressions of Sect. 4.9 with the exponential Gibbs-Thomson corrections. The curves with two extrema can be, however, fitted within the general self-consistent approach given by (4.120)–(4.125) or its simplified version presented below.

The model is based on the material balance equation for the In-limited growth

$$\frac{\pi R^2}{\Omega_S}\frac{dL}{dt} = \chi_l I \pi R^2 \frac{(1+\cos\alpha)}{2} - C I_{eq}(\zeta_\infty + 1) 2\pi R^2 \exp\left(R_{GT}/R\right) + j_{diff}. \quad (4.143)$$

Here, the left hand side gives the change of the number of In atoms in the cylindrical nanowire with radius $R = d/2$ per unit time, with L as the total nanowire length (including the InAs stem). The first term in the right hand side describes the In flux that directly impinges the droplet making 90° contact angle to the top facet, with I as the arrival rate per unit area, χ_l as the pyrolysis efficiency of TMIn at the droplet surface and α as the incident angle of In beam. The second term stands for the desorption from the droplet, where I_{eq} is the equilibrium desorption rate per unit area of the liquid (same as equilibrium desorption rate from the solid InSb (111)B surface) and ζ_∞ is the supersaturation of bulk alloy in the droplet with respect to the solid state. The coefficient C in the desorption term of (4.143) describes possible deviation of chemical potential in the droplet surface layer from its value in the volume alloy. The exponential factor accounts for the Gibbs-Thomson effect as described above, with R_{GT} defined by the first (4.56) at $\beta = \pi/2$.

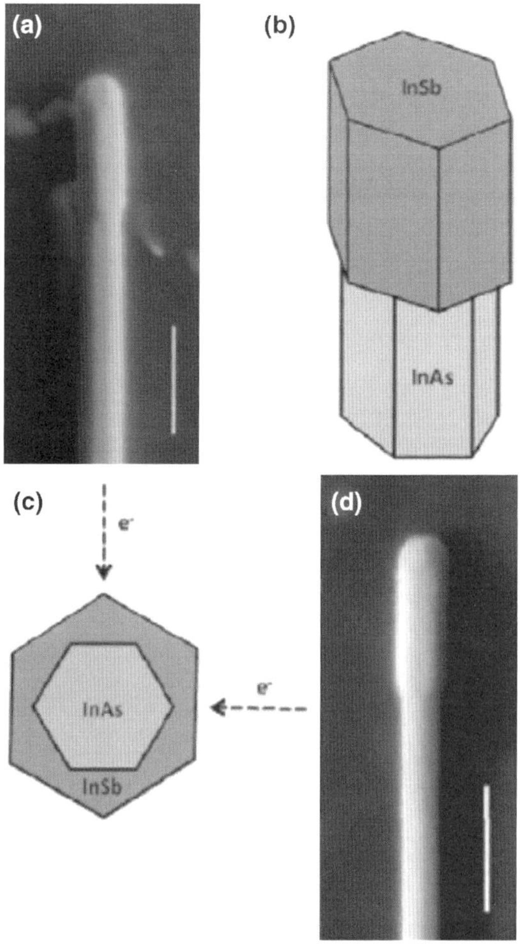

Fig. 4.61 SEM images of two InSb-InAs nanowires imaged along different projections: panel (**a**) and panel (**d**). Scale bars represent 200 nm. The facets present in the InAs and InSb segments confirm the rotation of the whole wire, as shown schematically in panel (**c**). Three-dimensional model of nanowire morphology is shown in panel (**b**)

The last term of (4.1), $j_{diff} = j^+_{diff} - j^-_{diff}$ collects contributions originating from the diffusion of sidewall and surface In adatoms, for which the exact solution is presented in Sect. 4.8. The j^+_{diff} term is the diffusion flux from the sidewalls to the droplet and j^-_{diff} is the reverse flux from the droplet. We will now derive a simplified version of the diffusion term which is capable of describing the observed diameter dependences with the two extrema [412]. In line with the approach of [82], we consider an idealized case where the surface diffusion is infinitely fast, while the desorption and nucleation on the sidewall facets can be neglected. The positive contribution to the diffusion-induced growth rate, $j^+_{diff} = \chi_s I\pi[(R + \lambda_s)^2 - R^2]\cos\alpha + \chi_f I2RL\sin\alpha$, contains two terms. The first term accounts for In atoms that firstly impinge the diffusion ring of width λ_s at the rate $\chi_s I\cos\alpha$ and then migrate to the droplet along the sidewalls, with χ_s as the pyrolysis efficiency of TMIn at the substrate surface.

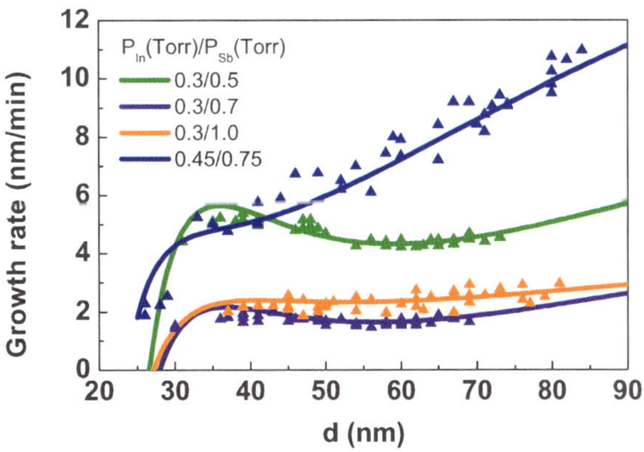

Fig. 4.62 Diameter dependences of the growth rate of InSb nanowire sections obtained at different pressure ratios: 0.3/0.5, 0.3/0.7, 0.3/1.0, and 0.45/0.75. *Dots* represent experimental data and *solid lines* are the fits obtained from (4.146) with the parameters given in Table 4.7

The second term describes In atoms impinging the sidewalls of the surface area $2RL\sin\alpha$ seen by the beam, with the pyrolysis efficiency χ_f. All these atoms participate in the wire growth if the effective diffusion length of In adatoms on the sidewalls is larger than the total length L (which is the sum of the lengths of InAs stem and InSb top). The reverse flux in such a simplified model should be obtained as $j^-_{diff} = I_{eq}(\zeta_\infty + 1)\exp(R_{GT}/R)\{\pi[(R+\lambda_s)^2 - R^2]\cos\alpha + 2RL\sin\alpha\}$, since $(\zeta_\infty + 1)\exp(R_{GT}/R)$ is the activity in the droplet driving the surface diffusion.

According to (4.124) and (4.125), the rate of nucleation-mediated nanowire growth from a supersaturated alloy in the mononuclear mode is given by

$$\left(\frac{dL}{dt}\right)_{nucl} = \pi R^2 J_0 \exp\left(-\frac{A_*}{\ln(\zeta+1) - 2\gamma_{LV}(\Omega_S - \Omega_L)/k_B TR}\right). \tag{4.144}$$

Let us now consider an idealized case of $J_0 \to \infty$, where the nucleation process is instantaneous. To preserve a finite value of the growth rate given by (4.144), we must put the denominator of the exponent to zero. This yields

$$(\zeta_\infty + 1)\exp(R_{GT}/R) = \exp(R^s_{GT}/R), \tag{4.145}$$

where R^s_{GT} is given by the second (4.56). Using (4.145) in (4.143) and in the above expressions for j^+_{diff} and j^-_{diff}, we arrive at

$$\begin{aligned}\frac{1}{V_{eq}}\frac{dL}{dt} &= \chi_l(\Phi+1)\frac{(1+\cos\alpha)}{2} - 2C\exp(R^s_{GT}/R)\\ &+ \left(\frac{2\lambda_s}{R} + \frac{\lambda_s^2}{R^2}\right)\left[\chi_s(\Phi+1)\cos\alpha - \exp(R^s_{GT}/R)\right]\\ &+ \frac{2L}{\pi R}\left[\chi_f(\Phi+1)\sin\alpha - \exp(R^s_{GT}/R)\right]\end{aligned} \tag{4.146}$$

Here, $V_{eq} = I_{eq}\Omega_s$ is the equilibrium desorption rate and $\Phi = I/I_{eq} - 1$ is the supersaturation in the vapor phase with respect to the solid state. As discussed in Sect. 4.8, this equation contains the $1/R$ diffusion term originating from the sidewall adatoms, while the diffusion of surface adatoms is described by the $1/R$ and $1/R^2$ terms scaling differently with the radius. Besides, the diffusion terms are modified to give correctly the direction of diffusion fluxes depending on the activity in the droplet. The use of approximation of instantaneous nucleation enables eliminating the unknown ζ_∞, thus avoiding a complex self-consistent procedure of equalizing the incoming material flux of In atoms by their nucleation-mediated sink.

In modeling the InSb growth rate, the total length L entering the right hand side of (4.146) should be adjusted to the averaged length of nanowires of a given diameter. However, since InSb tops are much shorter than InAs stems, it can be put to the length of the stem in the first approximation. The incident angle of In beam in the CBE system equals 40°. Under the assumption of 100 % pyrolysis efficiency at the droplet surface ($\chi_l = 1$), (4.146) contains five fitting parameters: the desorption rate V_{eq}, the vapor supersaturation Φ, the effective diffusion length λ_s and the two pyrolysis efficiencies at the substrate and sidewalls, χ_s and χ_f. The supersaturation Φ should increase with the TMIn pressure P_{In}. Lines in Fig. 4.62 show the best fits to the experimental diameter dependences of the InSb growth rate obtained from (4.146) with the parameters summarized in Table 4.7.

Theoretical fits reproduce quite well the non-monotonous behaviors of dL/dt with the two extrema, observed at P_{In}/P_{Sb} pressure ratios of 0.3/0.5, 0.3/0.7 and 0.3/1.0, as shown in Fig. 4.62. The difference between the two extrema progressively reduces as the P_{In}/P_{Sb} pressure ratio decreases. The increase of P_{In} from 0.3 to 0.45 Torr leads to the increasing length-diameter dependence. The non-monotonous curves are explained by the competition of the Gibbs-Thomson effect and the diffusion fluxes described by the $1/R$ and $1/R^2$ terms in (4.146). The minimum diameters lie within the interval from 25 to 32 nm for the entire range of TMIn and TDMASb pressures. As regards the values of fitting parameters given in Table 4.7, the vapor supersaturation indeed increases almost proportionally to P_{In}. The increase of P_{Sb} from 0.5 to 1 Torr at fixed $P_{In} = 0.3$ Torr leads to the decrease of In diffusion length from 106 to 95 nm. This is most probably explained by the known surfactant effect of Sb suppressing the surface diffusivity of group III metals [476]. Further decrease of λ_s to 66 nm at $P_{In}/P_{Sb} = 0.45/0.75$ might be associated with a faster surface nucleation at a larger material flux onto the surface. As the diffusion length decreases, the non-monotonous diameter dependence of the growth rate becomes less pronounced and finally disappears at 66 nm, because the wire growth becomes driven by the direct impingement and the Gibbs-Thomson effect. We also note that the obtained values of TMIn pyrolysis efficiencies are close for all samples (0.0068 for the sidewalls and 0.23–0.28 for the main surface), since they should be mainly determined by the surface temperature. However, the pyrolysis efficiency at the sidewalls is more than 30 times smaller than that at the main surface. This property was found to be essential for describing the non-monotonous dL/dt curves. It might be associated with several effects, amongst which we mention different chemical composition and atomic structures of the sidewalls and the surface layer, possible passivation of

Table 4.7 Fitting parameters at different TMIn and TDMASb pressures

P_{In} (Torr)	P_{Sb} (Torr)	V_{eq} (nm/min)	λ_s (nm)	Φ	χ_f	χ_s
0.3	0.5	1.80	106	18.1	0.0068	0.229
0.3	0.7	0.80	105	18.1	0.0068	0.230
0.3	1	0.54	95	18.1	0.0068	0.232
0.45	0.75	1.34	66	25.5	0.0068	0.28

sidewalls with group V element and organic compounds, and surface nucleation on the sidewalls.

To conclude the analysis of this section, we emphasis the following points:

1. Without surface diffusion, the length-radius dependence is an increasing function that saturates at a certain level where the size-dependent suppression of the growth rate becomes ineffective.
2. Purely diffusion-induced growth (limited just by the material transport) always yields decreasing length-radius dependences.
3. In most realistic systems, the observed length-radius correlation is dictated by a competition of the size-dependent and the diffusion-induced effects, leading to a non-monotonic $L(R)$ dependence with a maximum.
4. The exponential Gibbs-Thomson corrections of the type $\exp(R_{GT}/R)$ predict too steep decrease of the growth rate at small radii and too large cut-off radius compared to the experimentally observed values in most systems.
5. Self-consistent growth model (even without the Gibbs-Thomson correction to the liquid chemical potential) with the Lambert function enables perfect fits of the increasing, decreasing, and non-monotonic length-radius correlations with only two free parameters, the effective collection length and the characteristic nucleation radius.
6. S-like shape with two extrema is the most general form of the $L(R)$ dependence, which can be observed under particular growth conditions and over a wide range of nanowire dimensions.

4.12 Temperature Conditions for the VLS Growth

We now consider some experimental data regarding temperature dependence of the VLS nanowire growth rate. Typical temperature dependences of the growth rate of Au-catalyzed III-V nanowires obtained by MOCVD [255, 345, 346], MBE [292, 321, 322] and CBE [350] techniques can be described by a function with a maximum. This function rapidly tends to zero outside a certain temperature domain for a given material system. At otherwise identical deposition conditions and dropet dimensions, there is always a certain optimal temperature to grow the longest nanowires of a given material with a given catalyst.

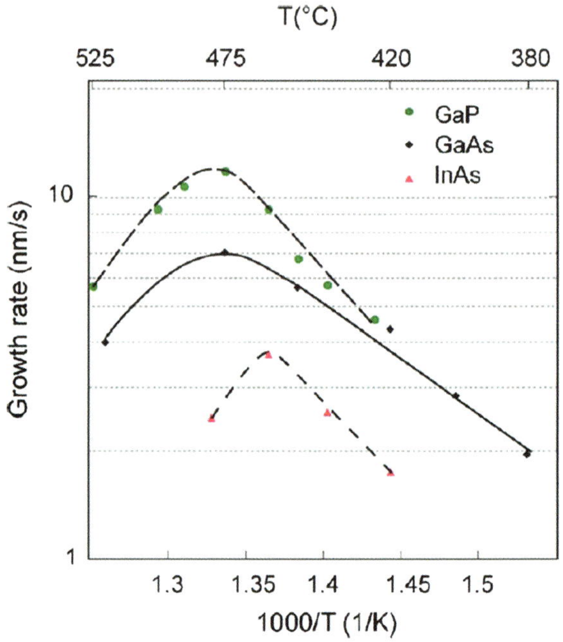

Fig. 4.63 Temperature dependence of the nanowire length growth rate for GaAs, GaP and InAs. The molar fraction of the TMGa and TMIn precursor was in all cases 1.2×10^{-5} [345]

In particular, Fig. 4.63 from [345] shows experimental temperature dependences of the growth rate of GaP, GaAs and InAs nanowires obtained by the Au-seeded MOCVD. These wires were grown on (111)B oriented substrates activated by aerosol Au nanoparticles with ~25 nm radius and $\sim 10^8$ cm^{-2} surface density. The substrates were annealed for 10 min in an atmosphere of $H_2/AsH_3(H_2/PH_3)$ at a pressure of 10 kPa and 580 °C for GaAs and InAs and 650 °C for GaP, in order to remove the oxide layer and to alloy Au with the substrate material. After anneal, the substrates were cooled down to the nanowire growth temperature T and the MOCVD process was started by supplying TMGa and TMIn, respectively, to the reaction cell. The curves given in Fig. 4.63 correspond to the fixed molar fractions of the TMGa and TMIn precursors of 1.2×10^{-5}. It is seen that the maximum growth rate of GaAs and GaP nanowires is observed near 475 °C, while for InAs nanowires it is shifted toward a lower temperature of ~455 °C. With the identical optimal temperature for GaAs and GaP, the growth rates of GaAs and GaP nanowires are different, showing that group V elements also affect the VLS process. The temperature domain for InAs is much narrower and the elongation rate is much lower than for GaAs. Overall, the difference between the InAs and GaAs nanowires is much more significant than between the GaAs and GaP ones. For all the three materials studied, the temperature dependence of the elongation rate has similar form, although over different temperature domains for different group III metals.

A comprehensive analysis of temperature conditions for the Au-assisted MBE growth of GaAs nanowires has been reported in [321] and [322]. The samples

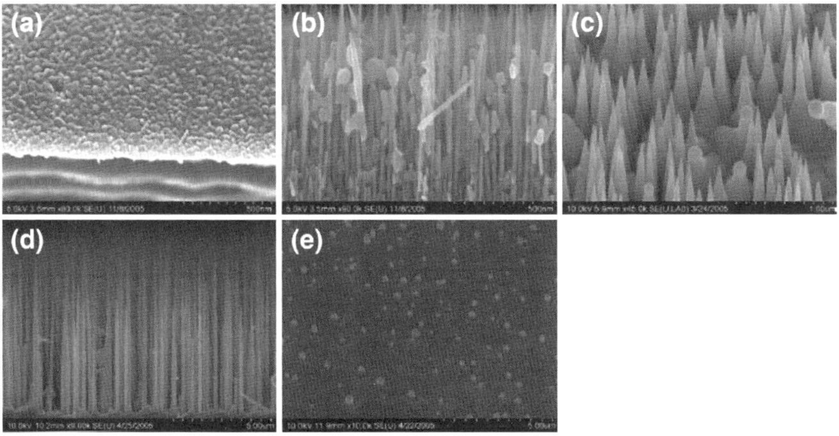

Fig. 4.64 SEM images of GaAs nanowires grown at different temperatures: **a** 300 °C, tilted view; **b** 370 °C, cross view; **c** 420 °C, tilted view; **d** 580 °C, cross view; **e** 630 °C, tilted view

were grown in a Riber 32 MBE equipped with a standard elemental source of Ga and a cracking cell providing As_2 dimers. The experiments were systematically started by depositing a GaAs buffer layer on a deoxidized GaAs (111)B substrate at 620 °C, followed by deposition of a 1 nm thick Au layer at room temperature in the ultrahigh vacuum system of the MBE machine. The samples were annealed for 5 min under As flux at 550 °C, then the substrate temperature was set to the desired growth temperature and the deposition of GaAs was started with the deposition rate $V = 0.2$ nm/s and the V/III fluxes ratio = 3. First, the range of temperatures was established where GaAs nanowires can be grown by this procedure. When growth was performed at 300 °C, i.e. below the particle solidification point, the particles were found to be buried under a rough GaAs layer, and no wire grew (Fig. 4.64a). At 370 °C, i.e., in the range corresponding to the hysteresis between solidification and melting of the metallic particles (as attested by RHEED), nanowires were produced, but their morphology suffers from parasitic crystallites which develop at their top or on the sidewalls (Fig. 4.64b). This parasitic growth should be attributed to instabilities of the particles which can easily switch from liquid to solid phase at this temperature. Growth at higher temperature (420–600 °C) produces regular nanowires, but their shape is changing from conical (Fig. 4.64c) to cylindrical (Fig. 4.64d) as the temperature is increased. Nanowires with a uniform radius from base to a certain critical length (3 μm at 580 °C) are usually obtained above 500 °C. Above 620 °C, the particles segregate at the GaAs surface with almost no nanowire formation (Fig. 4.64e). The averaged elongation rate of GaAs nanowires versus temperature is shown in Fig. 4.65.

InAs nanowires studied in [292] were grown by the Au-assisted MBE on InAs (111)B substrates. The equivalent deposition thickness of Au was 1 nm. Anneal at 430 °C resulted in the formation of 20–40 nm size droplets formed by thermal dewetting. The InAs deposition rate was fixed at 0.2 nm/s, the V/III fluxes ratio

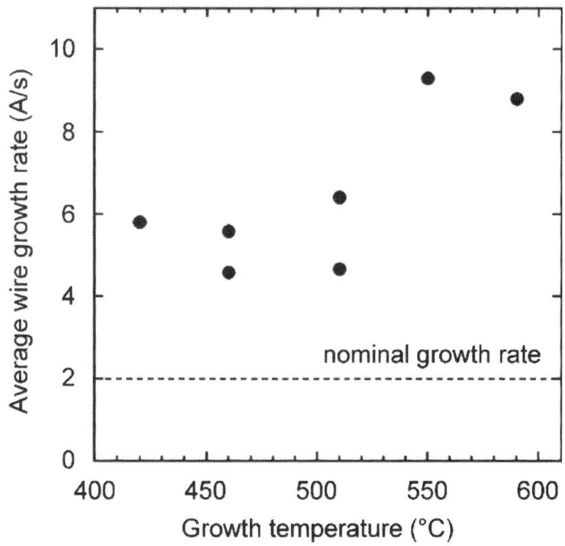

Fig. 4.65 Average GaAs nanowire growth rate versus growth temperature. The *dotted line* represents the nominal growth rate, i.e. the growth rate on a planar surface with no Au

was 3, while the growth temperature was varied. Statistical analysis of SEM images of different samples, with typical examples shown in Fig. 4.66, gave the temperature dependence of the nanowire elongation rate presented in Fig. 4.67. It is seen that the temperature window for the Au-assisted MBE of InAs nanowires under these conditions is much narrower than for GaAs, extending only from 380 to 430 °C. The maximum elongation rate of 1.75 nm/s is reached at 410 °C. It should be noted that the shape of low temperature InAs nanowires is not cylindrical. Nanowires grown at 390 °C (Fig. 4.66a) develop a pencil-like shape. The radius in the lower part of the wire is 2–3 times larger than that of the catalyst particle, and the nanowire end is tapered. This morphology demonstrates that at low temperatures, in addition to the vertical growth determined by the catalyst, the nucleation on the lateral facets also takes place. Radial growth should occur when the mean diffusion length of In on the lateral facets becomes smaller than the nanowire length. In this case, In adatoms cannot reach the catalyst and participate in the growth of the lateral surface. This phenomenon, also reported for MBE-grown InP [339] and GaAs [477] nanowires, will be considered in detail below.

To understand the temperature dependence of the vertical growth rate, we consider the diffusion-induced growth which is the dominant mechanism in all the cases presented in this section. For example, if the VLS growth is limited by diffusion from the substrate surface, the elongation rate can be approximated by (4.119). In this equation, the ratios of liquid to adatom activity θ_l^∞/θ_s and adatom to vapor activity θ_s/θ_v can be put in the equivalent form

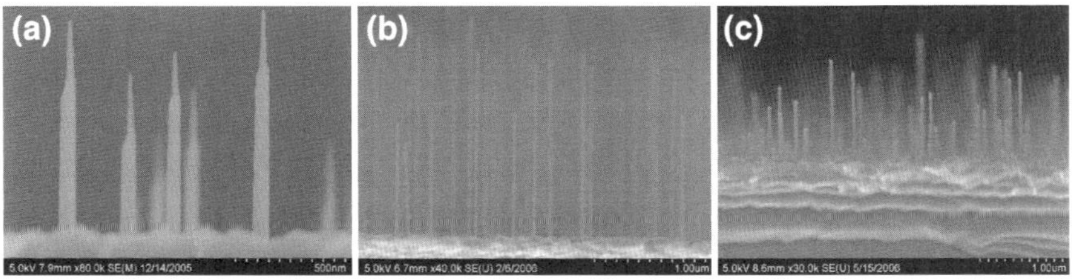

Fig. 4.66 InAs nanowires grown at **a** 390 °C, **b** 410 °C, and **c** 430 °C for 20 min. Scale bars correspond to 0.5 μm

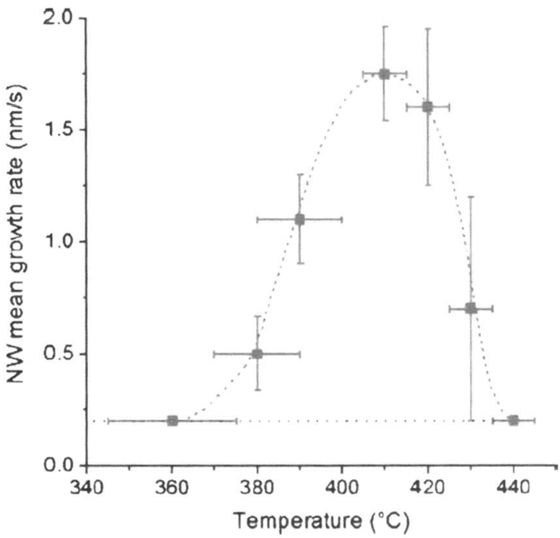

Fig. 4.67 InAs nanowire mean growth rate as a function of temperature. The nominal deposition rate of 0.2 nm/s is indicated with a *dotted line*. *Dashed line* represents a guide for the eyes

$$\frac{\theta_l^{\infty}}{\theta_s} = \exp\left(\frac{\Delta\mu_l^{\infty} - \Delta\mu_s}{k_B T}\right); \ \frac{\theta_s}{\theta_v} = \exp\left(\frac{\Delta\mu_s - \Delta\mu_v}{k_B T}\right). \tag{4.147}$$

Here, $\Delta\mu_v$ is the driving force for the wire growth, the difference of chemical potentials between the vapor and the bulk solid phase. The value of $\Delta\mu_v$ is determined by the surface temperature T and the deposition rate V as

$$\Delta\mu_v = k_B T \ln(V/V_0) = E_A + \tilde{\Lambda}_s - k_B T \ln(h\nu_A/V), \tag{4.148}$$

where V_0 is the equilibrium deposition rate at a given temperature, at which no growth occurs, E_A is the activation energy of adatom desorption, $\tilde{\Lambda}_s$ is the specific condensation heat of the adatom-substrate phase transition, ν_A is the vibration frequency, and h is the height of a monolayer [82]. Equation (4.148) shows that $\Delta\mu_v$ is a linearly decreasing function of temperature due to increased desorption rate at higher T. The $\Delta\mu_s$ in (4.147) denotes the difference between the surface adatoms far away from the nanowire and the bulk solid phase of the substrate. This value is generally lower than $\Delta\mu_v$ due to the surface nucleation. One can therefore write [82]

$$\Delta\mu_s = \Delta\mu_v - 2k_BT \ln\left(\lambda_s^0/\lambda_s\right). \tag{4.149}$$

Here, λ_s^0 is the diffusion length of surface adatoms at the equilibrium condition between a bare substrate and the vapor phase (limited by desorption), while λ_s is the effective diffusion length when the growth occurs (which implies $\lambda_s \le \lambda_s^0$).

The growth equation (4.119) now takes the form

$$\frac{dL}{dH} = \left[1 - \exp\left(\frac{\Delta\mu_l^\infty - \Delta\mu_s}{k_BT} + \frac{R_{GT}}{R}\right)\right]\left[\frac{1}{\sin^2\beta}\exp\left(\frac{\Delta\mu_s - \Delta\mu_v}{k_BT}\right) + \frac{2\lambda}{R}\frac{K_1(R/\lambda_s)}{K_0(R/\lambda_s)}\right], \tag{4.150}$$

where the characteristic Gibbs-Thomson radius R_{GT} is inversely proportional to T from the first (4.56). To access the temperature dependence of this nanowire growth rate, we need to determine the temperature behavior of the effective adatom diffusion length λ_s when surface growth occurs. This can be approximated by considering the probabilities ε_1 of adatom migration to the wire base, $\varepsilon_2 = V_s/V$ of adatom incorporation to a growing surface layer and $\varepsilon_3 = V_{des}/V$ of re-evaporation from the surface [84, 117]. Due to conservation of mass $\varepsilon_1 + \varepsilon_2 + \varepsilon_3 = 1$. Probability ε_1 must be proportional to the overall diffusion flux, which equals the sum of individual fluxes $j_{diff}(0)$ to the base of differently sized wires:

$$\varepsilon_1 = \frac{\Omega_s}{V} N_W < j_{diff}(0) > . \tag{4.151}$$

Here and below $< g >$ denotes the average value with normalized size distribution of droplets $f(R)$, i.e. $< g >= \int_0^\infty dRf(R)g(R)$, and N_W is the surface density of nanowires. From the results of Sect. 4.8, individual diffusion flux to the wire base at $L/\lambda_f << 1$ can be presented as

$$j_{diff}(0) = \pi R \frac{V}{\Omega_s} R_{diff}, \tag{4.152}$$

with R_{diff} as the characteristic diffusion radius. Inserting this into (4.151) and upon averaging we obtain

$$\varepsilon_1 = \pi N_W < RR_{diff} > \tag{4.153}$$

If λ_s is considerably smaller than the radii of wires and of two-dimensional islands arising on the surface, we can assume that probability ε_1 is proportional to the total perimeter of nanowire bases per unit surface area $P_W = 2\pi N_W < R >$. Similarly, the probability ε_2 is proportional to the time averaged perimeter of islands per unit area P_I, with accounting for the coalescence of islands into a continuous surface layer. Applying (2.156) for 2D circular islands, the time dependence of P_I can be approximated as $P_I(\tau) = \pi\sqrt{N_I}\tau e^{-\tau}$. Here, N_I is the surface density of islands and τ is the relative time measured in the units of time of layer formation. Averaging this P_I over τ, we arrive at

$$P_I \cong \pi\sqrt{N_I}\int_0^\infty dx x^{1/2}e^{-x} = \frac{\pi^{3/2}}{2}\sqrt{N_I} \tag{4.154}$$

Equation (2.140) of Sect. 2.9 at $m = 2$ show that N_I at small λ_s depends on temperature T and deposition flux V approximately as

$$N_I \propto V^2 \exp\left(\frac{3\tilde{\Lambda}_s + 2E_D^s}{k_B T}\right) \tag{4.155}$$

where E_D^s is the activation energy for adatom diffusion. Finally, the probability of desorption ε_3 is inversely proportional to the diffusion length of single adatom on a bare substrate λ_s^0. Temperature dependence of λ_s^0 is given by the conventional Arrhenius exponent, $\lambda_s^0 \propto \exp[(E_A^s - E_D^s)/2k_B T]$, where E_A^s is the activation energy for desorption. The expression ε_1 from the above considerations is as follows

$$\varepsilon_1 = \frac{P_W \lambda_s^0}{1 + (P_W + P_I)\lambda_s^0} \tag{4.156}$$

In the opposite case, when λ_s is much larger than the radii of wires and islands, the probability ε_1 is approximately proportional to surface density of wires N_W, the probability ε_2 is proportional to density of islands N_I and the probability ε_3 is proportional to $(2\pi\lambda_s^0)^{-2}$. Repeating the above described considerations, the expression for ε_1 is as follows

$$\varepsilon_1 = \frac{2\pi N_W (\lambda_s^0)^2}{1 + 2\pi(N_W + N_I)(\lambda_s^0)^2} \tag{4.157}$$

The flux and temperature dependence of the island density at large λ_s follows from (2.140) at $m = 1$

$$N_I \propto V \exp\left(\frac{\tilde{\Lambda}_s + E_D^s}{K_B T}\right). \tag{4.158}$$

As discussed in Sect. 4.9, the characteristic diffusion radius R_{diff} in (4.152) can be approximated as

$$R_{diff} = 2\lambda_s \frac{K_1(R/\lambda_s)}{K_0(R/\lambda_s)} \to \begin{cases} 2\lambda_s, \lambda_s << R \\ \frac{2\lambda_s^2}{R\ln(\lambda_s/R)}, \lambda_s >> R. \end{cases} \tag{4.159}$$

If $R_{diff} = 2\lambda_s$, (4.153) gives the result for λ_s in the form $\lambda_s = \varepsilon_1/P_W$. Comparing this to (4.156), one obtains the expression

$$\lambda_s = \frac{\lambda_s^0}{1 + (P_W + P_I)\lambda_s^0}. \tag{4.160}$$

In the case of $\lambda_s >> R$, if we neglect the logarithmic dependence on λ_s/R in (4.159), the formula for R_{diff} becomes $R_{diff} \cong 2\lambda_s^2/RR_*$, and (4.153) gives $\lambda_s^2 = \varepsilon_1/2\pi N_W$. Using (4.157) for ε_1, the result for λ_s is given by

$$\lambda_s = \frac{\lambda_s^0}{\sqrt{1 + 2\pi(N_W + N_1)(\lambda_s^0)^2}} \tag{4.161}$$

Using the temperature and flux dependence of N_I given by (4.155), (4.158) at small and large λ_s, respectively, (4.154) for P_I and the temperature dependence of λ_s^0 we finally arrive at

$$\lambda_s = \left[P_W + P_I(T_0, V_0)\frac{V}{V_0}\exp\left[F_s\left(\frac{T_0}{T} - 1\right)\right] + \frac{1}{\lambda_s^0(T_0)}\exp\left[-G_s\left(\frac{T_0}{T} - 1\right)\right]\right]^{-1}, \lambda_s << R; \tag{4.162}$$

$$\lambda_s = \left[2\pi N_W + 2\pi N_I(T_0, V_0)\frac{V}{V_0}\exp\left[F_s'\left(\frac{T_0}{T} - 1\right)\right] + \frac{1}{[\lambda_s^0(T_0)]^2}\exp\left[-2G_s\left(\frac{T_0}{T} - 1\right)\right]\right]^{-1/2}, \lambda_s >> R. \tag{4.163}$$

Here, the parameters $F_s \equiv \frac{(3/2)\tilde{\Lambda}_s + E_D^s}{k_B T_0}$, $F_s' \equiv \frac{\tilde{\Lambda}_s + E_D^s}{k_B T_0}$, and $G_s \equiv \frac{E_A^s - E_D^s}{2k_B T_0}$ are determined by the activation energies and by the specific condensation heat of surface atoms, $P_I(T_0, V_0)$ and $N_I(T_0, V_0)$ are the island perimeter and density at the reference temperature T_0 and flux V_0, and $\lambda_s^0(T_0)$ is the diffusion length at $T = T_0$. If nanowire growth is controlled by the diffusion from sidewall facets, the elongation rate is given by (4.116), (4.95), where the temperature dependence of the effective diffusion length on the sidewalls λ_f is determined by (4.162) or (4.163) at $P_W = N_W = 0$:

$$\lambda_f = \left[P_I(T_0, V_0)\frac{V}{V_0}\exp\left[F_f\left(\frac{T_0}{T} - 1\right)\right] + \frac{1}{\lambda_f^0(T_0)}\exp\left[-G_f\left(\frac{T_0}{T} - 1\right)\right]\right]^{-1}, \lambda_f << \rho; \tag{4.164}$$

$$\lambda_f = \left[2\pi N_I(T_0, V_0)\frac{V}{V_0}\exp\left[F_f'\left(\frac{T_0}{T} - 1\right)\right] + \frac{1}{[\lambda_f^0(T_0)]^2}\exp\left[-2G_f\left(\frac{T_0}{T} - 1\right)\right]\right]^{-1/2}, \lambda_f >> \rho, \tag{4.165}$$

Here, ρ denotes the characteristic spacing between the islands or steps on the wire sidewalls. All the parameters here have the same meaning as before, but are related to the sidewalls facets rather than to the substrate surface. We note that the temperature dependence of the liquid chemical potential $\Delta\mu_l^\infty$ in (4.150) is generally unknown and should be obtained in the self-consistent manner described in Sect. 4.10. When the wire is mainly fed by the surface adatoms, the probabilities of migration to the wire base ε_1 and escaping from the substrate $\varepsilon_1 + \varepsilon_3 = 1 - \varepsilon_2$ are obtained from $\varepsilon_1 = P_W\lambda_s, \varepsilon = \varepsilon_1 + \frac{\lambda_s}{\lambda_s^0}$ in the case of small λ_s and $\varepsilon_1 = 2\pi N_W\lambda_s^2, \varepsilon = \varepsilon_1 + \left(\frac{\lambda_s}{\lambda_s^0}\right)^2$ in

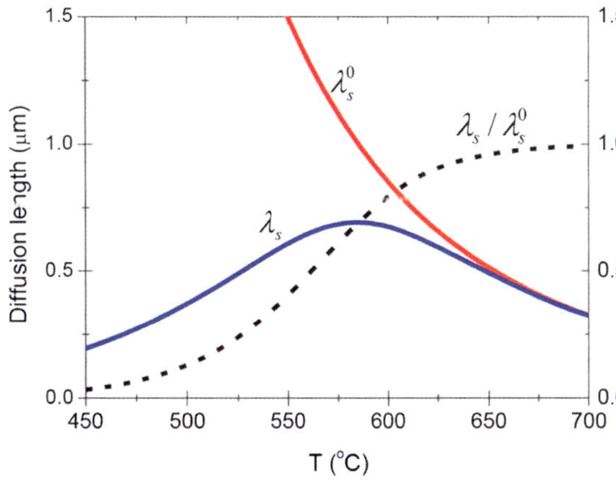

Fig. 4.68 Diffusion length of a single adatom on bare substrate λ_s^0 (limited by desorption), effective diffusion length and the ratio λ_s/λ_s^0 versus temperature for the parameters of GaAs system described in the text

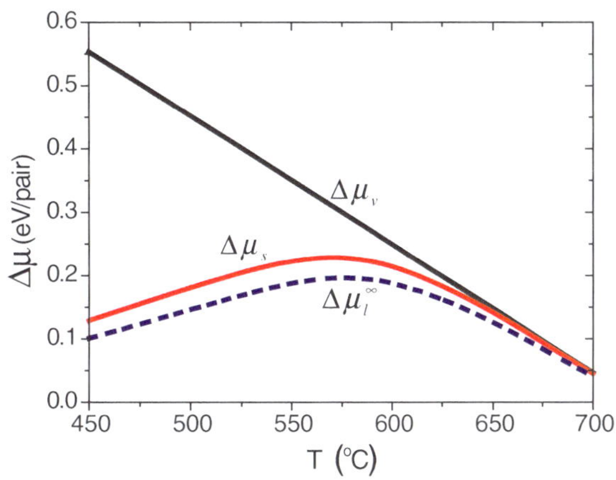

Fig. 4.69 Temperature dependences of $\Delta\mu_V$ and $\Delta\mu_s$ at fixed $V/h = 0.6$ ML/s for the parameters of GaAs described in the text. *Dotted line* is the liquid–solid chemical potential used in calculations

the case of large λ_s. Whenever the effective diffusion length on the sidewalls is limited by sidewall incorporation, wires extend radially concomitantly with elongation.

Figure 4.68 shows the temperature dependences of the substrate diffusion lengths λ_s^0, λ_s and their ratio, for the parameters approximately corresponding to the VLS growth of GaAs nanowires on GaAs(111)B substrate [82]: $E_A^s = 1.8$ eV, $\widetilde{\Lambda}_s = 0.22$ eV, $\nu_A = 10^{10} s^{-1}$, $\lambda_s^0 = 1.5\,\mu$m at $T = 550\,°$C, $E_D^s = 0.4$ eV (yielding $G_s = 9.9$) at a constant deposition rate $V/h = 0.6$ ML/s, As-rich conditions and a low nanowire surface density. It is seen that as long as λ_s is limited by desorption, it increases at decreasing T as the conventional diffusion length of a single adatom. This process is suppressed, however, by the increase of the surface density of islands N_I due to surface nucleation at low temperatures. When λ_s is limited by surface nucleation, the majority of adatoms are captured by surface islands before they reach the wire bases, resulting in the reduction of the effective diffusion length. Therefore, λ_s always has a maximum at a certain temperature, around 580 °C in our example. Figure 4.69

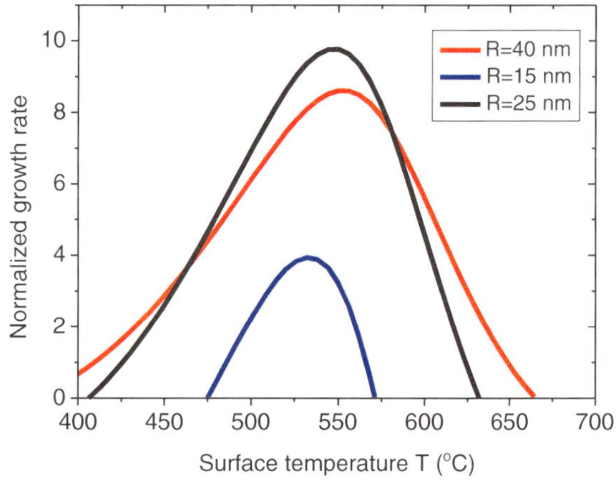

Fig. 4.70 Temperature dependences of normalized growth rate for the parameters described in the text and three different radii $R = 15$, 25 and 40 nm

depicts the linear decrease of $\Delta\mu_v$ with temperature that follows from (4.148) for the same model parameters. The chemical potential of the surface adatoms $\Delta\mu_s$ is obtained from (4.149) with the λ_s/λ_s^0 ratio shown in Fig. 4.68. The $\Delta\mu_s$ also has a maximum at ~570 °C in the case of GaAs. Finally, Fig. 4.70 shows the temperature dependent vertical growth rate *dL/dH*, obtained from (4.150) for the substrate parameters of Fig. 4.68, with a liquid–vapor surface energy $\gamma_{LV} = 1\,\mathrm{J/m^2}$, which is between the surface energies of pure liquid Ga and Au (0.72 and 1.14 J/m^2, respectively), $\Omega_L = 0.038\,\mathrm{nm}^3$, $\beta = 90°$ and $\Delta\mu_l^\infty$ shown by the dotted line in Fig. 4.69. It is seen that the temperature domains for the Au-assisted MBE of GaAs nanowires approximately correspond to the experimental window between 420 and 620 °C. However, theoretical curves strongly depend on the nanowire radius. For 15 nm radius nanowires, the temperature window is narrowed to ~ 480–570 °C. The maximum growth rate of ~10 V is expected at 560 °C for $R = 25$ nm.

Our results show that the centric temperature dependences of the nanowire elongation rates in the diffusion-induced VLS growth mode are explained by a non-monotonic temperature behavior of the adatom diffusion length, coupled with the Gibbs-Thomson effect and the increased desorption from the droplet surface at high *T*. At low temperatures, the effective diffusion length is always limited by surface nucleation, leading to the radial growth on the wire sidewalls. Similar dependences can be obtained from the self-consistent (4.138), where the growth rate is directly proportional to Λ^p (if the adatom diffusion is the dominant growth mechanism). In the nucleation-limited regime, the elongation rates and the entire temperature domains for the VLS nanowire growth are determined by the temperature-dependent nucleation probabilities. Other factors, such as the temperature-dependent pyrolysis efficiencies at different surfaces in MOCVD, may play an important role in the overall temperature behavior of the VLS process and lead to a more complex dependences than those described above.

4.13 Growth Chronology of a Single InPAs Nanowire

Most experimental studies evaluate the kinetics of nanowire growth from postgrowth measurements of final nanowire lengths at several given growth times. This is far from satisfactory since such measurements can hardly record the transient and nonlinear behaviors, and deal only with the averaged values over an ensemble of nanowires. More precise information can be derived from the use of markers consisting of thin layers of a chemically different material inserted in the nanowire [257, 478]. However, fine variations of growth rate are difficult to measure in this way and the insertion of the marker itself can alter the growth kinetics. A simple method to investigate finely the growth of a single nanowire was proposed by Harmand *et al.* [290]. This method relies on small periodic modulations of the incident vapor fluxes, produced on purpose. The amplitude and the time period of the flux modulation are fixed experimentally and chosen to produce small but detectable changes in the nanowire composition along the growth direction. The growth kinetics is determined by counting the number of monolayers formed during each time period of the modulation.

The method was implemented on $InP_{1-x}As_x$ nanowires grown at 420 °C by the Au-catalyzed MBE on InP (111)B substrates. The growth conditions produce nanowires of WZ structure which grow in the [0001] direction without stacking faults. As_4 and P_2 fluxes were evaporated from solid sources and a periodic modulation of the incident As_4:P_2 flux ratio was produced. The flux modulations may be obtained either by directly modulating the sources or indirectly. In [290], the second method was used: the modulation was obtained at constant source fluxes, by exploiting the geometry of the MBE growth chamber, namely, the fact that the molecular beams are not perfectly uniform on the wafer area. When the substrate is rotating, the nanowires situated close to its periphery experience a small modulation of the incoming fluxes during one substrate rotation. The flux modulation amplitude varies along the wafer radius, from 0 at the wafer center to about 15 % of the average flux at the edge of a 2 in wafer. The nanowires presented in the following were sampled at 5 mm from the edge.

Several important conditions must be met in order to transfer the flux modulation to an axial modulation of the nanowire composition. First, several monolayers must grow within one modulation time period. A time period of 3.6 s was therefore chosen, corresponding to the growth of about 10–30 monolayers. The corresponding length (3–10 nm) is sufficiently small to probe finely the growth kinetics. Second, one should avoid a complete damping of the flux modulation upon transferring atoms into the liquid droplet, which acts as a reservoir. In this respect, modulating the group V flux is expected to be more appropriate since P and As are much less soluble in the catalyst than group III metals. On the other hand, the modulation must be small enough to minimize the resulting perturbation on the growth kinetics.

The nanowires analyzed have uniform diameter, close to the size of the catalyst particle. This morphology indicates that nucleation at the sidewalls is negligible. The $InP_{1-x}As_x$ nanowires grown with modulated beams were removed from their substrate, deposited on a thin silicon nitride membrane and imaged by high angle

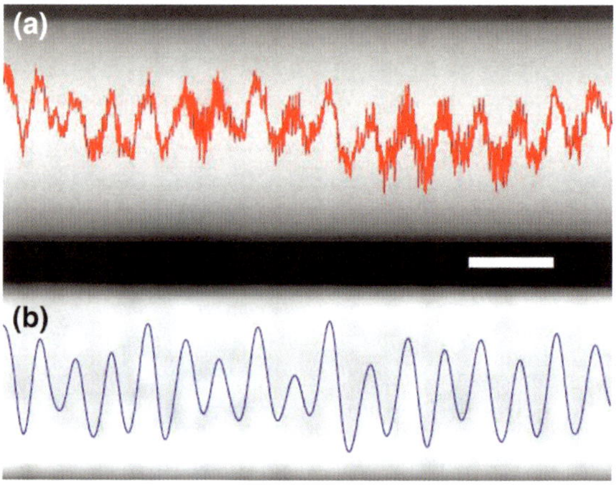

Fig. 4.71 **a** Modulation of composition in an $InP_{1-x}As_x$ nanowire segment grown with a modulated As_4:P_2 flux ratio. HAADF STEM image of the nanowire segment observed along the $< \bar{2}110 >$ zone axis. The scale bar represents 20 nm. The HAADF signal is integrated over the whole diameter and the profile is plotted along the growth axis. **b** FFT-filtered HAADF image and corresponding profile [290]

annular dark field (HAADF) STEM. The images were obtained in a 200 kV field emission gun JEOL 2200FS microscope. The image of a nanowire segment of 45 nm diameter is presented in Fig. 4.71. The faint contrast along the growth axis reveals that the composition of this nanowire has indeed been modulated. The modulation of the HAADF intensity profile integrated over the whole diameter is superimposed [Fig. 4.71a]. The FFT-filtered HAADF image with the corresponding intensity profile is also shown [Fig. 4.71b]. The oscillation amplitude is about 0.7 % of the total HAADF intensity. The corresponding variation Δx of the As concentration x was determined by EDXS to be about 0.03 around the mean value of 0.66.

The distance between two concentration extrema equals the increase of nanowire length during one time period. Therefore, it is straightforward to determine the instantaneous growth rate at each oscillation. The complete chronology of the nanowire growth is thus accessible. Counting the oscillations from a reference point deliberately introduced in the growth sequence allows one to date the exact time at which each oscillation was formed. This procedure was applied to a 21 nm diameter nanowire. The dependence with time of the nanowire length counted from the cleaved end is presented in Fig. 4.72. The measured length of the nanowire displays a nonlinear behavior which will be discussed below. Before that, we comment on two particular features. First, in this growth experiment, the flux modulation was stopped for 12 time periods. Consequently a short segment of the nanowire does not present regular composition oscillations. The corresponding region, indicated in Fig. 4.72, may serve as a time reference. Second, at some point during growth, the nominal As_4:P_2 ratio was changed abruptly from 1:5 to 3:3 while the total average group V flux was kept constant. Thanks to the former time reference, this event can be accurately dated, as shown in Fig. 4.72. The impact of this large change of flux ratio is strong and immediate: as seen from the change of slope in Fig. 4.72, the growth rate was suddenly reduced by a factor of 2.

We now turn to modeling the nanowire growth chronology. In order to reduce the number of unknown parameters, a simplified diffusion model was used to fit the data.

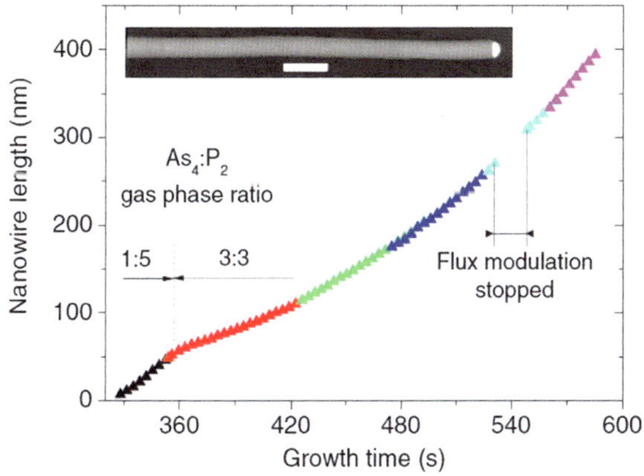

Fig. 4.72 Growth chronology of a single $InP_{1-x}As_x$ nanowire. The length of the nanowire shown in the inset (scale bar represents 50 nm), is measured from its cleaved edge and plotted as a function of the absolute growth time, as deduced from the composition oscillations along its axis. The different colors correspond to different overlapping HAADF images. During the growth experiment, the flux modulation was stopped for 12 time periods. The corresponding region, where composition oscillations are missing, serves as a time reference. At 355 s, the nominal As_4:P_2 flux ratio was changed abruptly from 1:5 to 3:3 [290]

This model follows from (4.94), (4.95) at $\varepsilon = 0$, (4.87), (4.93), (4.91), (4.92) and (4.100). The factor b in (4.91) and (4.92) is put to $(D_s\sigma_f)/(D_f\sigma_s)$ and $\xi = \sigma_s/\sigma_f$ in (4.90), corresponding to the Dubrovskii boundary conditions given by (4.69)–(4.72) and $R_W \to \infty$. In this case,

$$b\frac{\lambda_f}{\lambda_s} = \frac{D_s\sigma_f\lambda_f}{D_f\sigma_s\lambda_s} = \frac{\lambda_s}{\lambda_f}\frac{\theta_f}{\theta_s}\frac{\pi}{\tan\alpha} = \frac{\lambda_s}{\lambda_f}\frac{\theta_{ls}}{\theta_{lf}}\frac{\pi}{\tan\alpha}, \tag{4.166}$$

where we use the notations

$$\theta_{ls} \equiv \frac{\theta_l}{\theta_s} = \exp\left(-\frac{\Delta\mu_{sl}}{K_BT}\right);\, \theta_{lf} \equiv \frac{\theta_l}{\theta_f} = \exp\left(-\frac{\Delta\mu_{fl}}{K_BT}\right) \tag{4.167}$$

Here, $\Delta\mu_{sl} = \mu_s - \mu_l$ and $\Delta\mu_{fl} = \mu_f - \mu_l$ are the differences between the chemical potentials per III-V pair at the substrate surface far from the nanowire and on the sidewall at the maximum possible adatom coverage of the latter, and the liquid phase in the droplet. The elongation rate can now be put in the form

$$\frac{dL_0}{dt} = I\Omega_S\cos\alpha\left[A + \frac{\left\{\frac{2\lambda_f}{\pi R}(1-\theta_{lf})\tan\alpha\right\}U\left(\frac{L}{\lambda_f}\right) + \left\{\frac{2\lambda_s}{R}(1-\theta_{ls})\delta\left(\frac{R}{\lambda_s}\right)\right\}}{U'\left(\frac{L}{\lambda_f}\right)}\right]. \tag{4.168}$$

Here, L_0 is the total nanowire length measured from a fixed reference, namely, the initial substrate-wire interface, and L the length of the emerging part of the nanowire above the 2D surface layer of height H_s. If all atoms which do not participate in the nanowire elongation are incorporated at the substrate surface, the surface layer grows at the rate

$$\frac{dH_s}{dt} \cong I\Omega_s \cos\alpha - \pi R^2 N_w \frac{dL_0}{dt}, \tag{4.169}$$

where N_W is the surface density of nanowires. This equation follows from the material balance considerations in the lowest order in $\pi R^2 N_W << 1$. Obviously, $L_0 = L + H_s$ and $dL_0/dt = dL/dt + dH_s/dt$. Therefore, we obtain

$$\frac{dL}{dt} = \frac{dL_0}{dt}(1 + \pi R^2 N_W) - I\Omega_s \cos\alpha \tag{4.170}$$

The direct impingement factor A in (4.168) does not account for desorption from the droplet ($\theta_l/\theta_v \to 0$) and is very close to $1/(\sin^2\beta\cos\alpha)$in the present case, because the contact angle β during growth can be estimated to be about $120\,°$. The functions U and U' are given by (4.91) and (4.92) with $F = K_1(R/\lambda_s)/K_0(R/\lambda_s)$ and $(b\lambda_f)/\lambda_s$ defined by (4.166).

We now have four adjustable parameters for each As_4:P_2 ratio: $\Delta\mu_{sl}$, $\Delta\mu_{fl}$, λ_s and λ_f. In fact, the value of λ_f has little influence on the results, provided that it is larger than the final nanowire length. Therefore, it was chosen to be 2µm, a reasonable value since in these experimental conditions, nanowires up to this length can be obtained without radial growth. Concerning λ_s, it has been found that very small values (a few nm) are necessary to fit the experiment. The best fit is obtained with $\lambda_s =$ 2 nm. This very short length might be surprising at first sight, but can be understood as a limitation of adatom diffusion by atomic steps. Indeed, the surface layer between the wires has an ample surface roughness and the widths of the terraces between the atomic steps can be as small as a few nm. In these conditions, changing the group V flux is also expected to have little impact on the substrate surface diffusion. For that reason, the same value of λ_s was used for both growth stages. On the other hand, the abrupt change of As_4:P_2 flux ratio is expected to modify significantly the composition and the chemical potential of the catalyst droplet. Accordingly, two sets of chemical potential differences were adjusted, valid before and after this change.

The best fit for the first part of growth at As_4:P_2 flux ratio of 1:5 was obtained with $\Delta\mu_{fl} = 270$ meV and $\Delta\mu_{sl} = 18.2$ meV, indicating net flows of adatoms diffusing from the sidewall facets and from the substrate surface to the droplet. The contribution from the substrate surface is very small due to the combination of low λ_s and moderate $\Delta\mu_{sl}$. Very quickly, the contribution from the sidewalls dominates largely. After the change of As_4:P_2 flux ratio to 3:3, $\Delta\mu_{fl}$ must be reduced to 160 meV, which indicates that the flow from the sidewalls to the droplet is less efficient when P_2 is replaced by As_4. At the same time, $\Delta\mu_{sl}$ takes a negative value of -145 meV, showing that a significant part of the atoms intercepted by the nanowire now migrates to the substrate surface, to the detriment of nanowire growth. This illustrates the crucial role of group

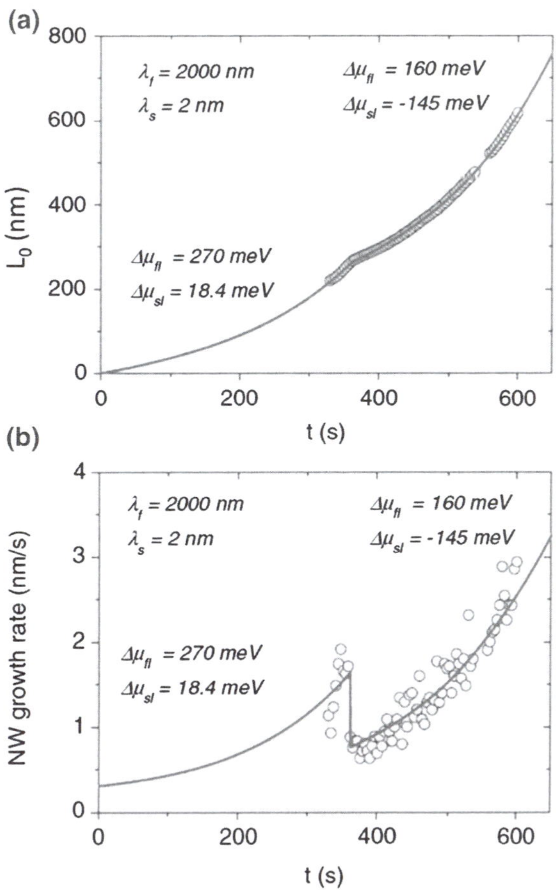

Fig. 4.73 **a** Experimental nanowire elongation and **b** axial growth rate (*circles*), fitted with the diffusion growth model given by (4.168) and (4.170) with four parameters $\Delta\mu_{sl}$, $\Delta\mu_{fl}$, λ_s and λ_f lines. Different chemical potential differences are used for each As_4:P_2 ratio, before and after $t = 355$ s [290]

V fluxes which can, via their interaction with the other constituents of the droplet, affect the chemical potential of In in the latter [355] and redirect part of the diffusion flow of In adatoms. With these parameters, the diffusion growth model describes fairly well the experimental dependence $L(t)$, and the abrupt change in the nanowire growth rate is well reproduced, as shown in Fig. 4.73.

Several important conclusions can be derived from these results. First, nanowire growth is mainly sustained by contributions A and B in (4.94). As the nanowire aspect ratio becomes significant, contribution A tends to be marginal as compared to B. The latter is responsible for the nonlinear variation of L: as the nanowire becomes longer, it collects more material on its sidewalls and consequently grows faster. During the first growth stage, almost all the In atoms intercepted by the nanowire sidewalls are transferred to the catalyst droplet. During the second stage, this transfer is less efficient as a consequence of the change of As_4:P_2 ratio which affects the droplet chemistry. Contribution C in (4.94) appears to be close to zero or even negative under these particular experimental conditions. This conclusion goes against the prevailing idea that collection of substrate adatoms is necessary to explain the beginning of the VLS growth. As highlighted in [479] and further developed in [290], the nanowire growth can be initiated by purely geometrical magnifying effect of droplet collection. Consider the catalyst droplet as a spherical cap sitting on the substrate with the contact angle β. The sidewalls are not yet formed ($B = 0$), and we assume that $C = 0$. The drop intercepts the vapor $IA\cos\alpha$ flux while the substrate surface receives the flux $I\cos\alpha$. Thus, if $A > 1$, more In is supplied to the liquid/solid interface via the drop as compared to the flux impinging on the rest of the substrate surface. This condition is satisfied as soon as $\beta > \pi/2 - \alpha$, which is very likely the case in these experiments. This geometrical effect is thus sufficient to explain how nanowire growth starts at $C = 0$.

The nanowire sidewall facets consist of $\{\bar{2}110\}$ WZ planes. At a given time, the chemical potential on these facets has a certain profile along z, with a maximum value, $\mu_{\max}$, situated at altitude z_* ($\mu_{\max}$ tends toward μ_f for an infinitely long nanowire). Let us call j_b and j_t the diffusion fluxes of adatoms at the bottom and at the top of the nanowire, counted positive if directed upward. At the first growth stage, j_b and j_t are positive and $\mu_{\max}$ is located at the bottom of the nanowire ($z_* = 0$). At the second stage, j_b is negative and in that case the diffusion equation on the sidewalls leads to

$$z_* = \frac{\lambda_f}{2}\ln\left[\frac{j_b\exp(L/\lambda_f) - j_t}{j_b\exp(-L/\lambda_f) - j_t}\right]. \tag{4.171}$$

As long as L is less than λ_f, z_* varies slowly with L and $z_* \cong Lj_b(j_b - j_t)$ at $L << \lambda_f$. For $L = 500\,\text{nm}$, we obtain $z_* = 114\,\text{nm}$. This value gives the effective collection length from the nanowire sidewalls, because z_* corresponds to the sidewall altitude where the In adatom concentration is highest. Diffusion flux from the bottom part of the wire at $z < z_*$ is directed downward and does not contribute to the nanowire elongation. This explains why the effective collection length obtained from fitting the experimental data can be much smaller than the total nanowire length. At nanowire lengths longer than those investigated here, the nucleation on the sidewall facets is likely to occur near z_* and radial growth can develop by step flow parallel to the facets.

4.14 Non-linear Effects in Nanowire Growth

Integration of the non-linear growth (4.94) is particularly simple at $A \to 0$, where the adsorption-desorption term is marginal compared to the diffusion-induced contributions B and C. Introducing the normalized length, $l \equiv L/\lambda_f$, and the deposition thickness, $h \equiv H/\lambda_f$, the equation for nanowire length at $A \to 0$ writes

$$\frac{dl}{dh} = \frac{BU(l) + C}{dU/dl};\ l(h = 0) = l_0. \tag{4.172}$$

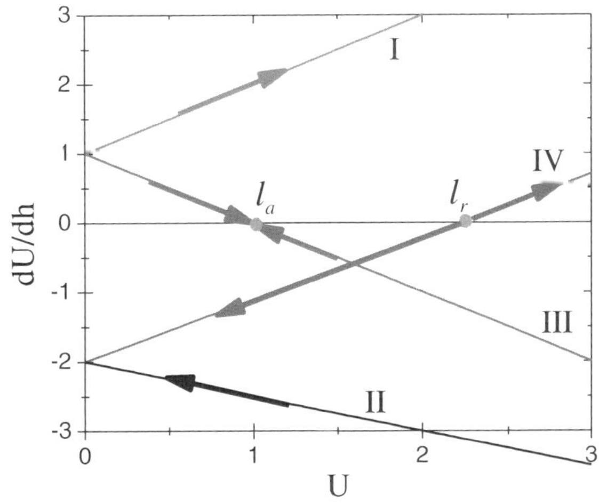

Fig. 4.74 Illustration of the growth modes: *I*—infinite growth, *II*—negative growth, *III*—averaging growth with attractive stationary point $l_* = l_a$, *IV*—continuing growth with repulsive stationary point $l_* = l_r$

The initial condition at $h = 0$ determines the length l_0 of pre-existing nanowire stem, e.g., obtained at different growth conditions as described below. Introduction of a new variable U converts (4.172) to

$$\frac{dU}{dh} = BU + C; \; U(h = 0) = U(l_0). \tag{4.173}$$

Solution to this equation is given by

$$h(l) = \frac{1}{B} \ln \left[\frac{BU(l) + C}{BU(l_0) + C} \right], \tag{4.174}$$

where $U(l)$, by its definition, increases with l and equals zero at $l = 0$.

We now analyze possible scenarios of nanowire growth following from simplified (4.173) in the diffusion-induced mode. Obviously, the sign of the right hand side of (4.173) determines the sign of derivative dl/dh: the growth rate is positive at $BU + C > 0$ and negative at $BU + C < 0$. The points where $BU(l_*) + C = 0$ are the stationary points with the corresponding critical lengths. Four possible behaviors of the linear function with sign alternating coefficients B and C are illustrated in Fig. 4.74, where the arrows indicate the time evolution of the non-linear system considered. We therefore arrive at four possible growth scenarios [410]:

(I) $B > 0, C > 0$: Nanowire of any initial length l_0 grows infinitely.
(II) $B < 0, C < 0$: Nanowire of any initial length l_0 decays.
(III) $B < 0, C > 0$: The critical length determined by $U(l_*) = C/|B|$ is the attractive stationary point such that the nanowires with initial length $l_0 < l_*$ grow to l_* and the nanowires with $l_0 > l_*$ diminish their length to l_* (averaging growth).

(IV) $B > 0, C < 0$: The critical length determined by $U(l_*) = |C|/B$ is the repulsive stationary point such that the nanowires with initial length $l_0 < l_*$ decay and the nanowires with $l_0 > l_*$ grow infinitely (continuing growth).

From (4.167), (4.168) and $\theta_l = \theta_l^\infty \exp(R_{GT}/R)$, the coefficients B and C change sign at critical radii R_B and R_C determined by

$$R_B = \frac{R_{GT}}{\ln(\theta_f/\theta_l^\infty)};\ R_C = \frac{R_{GT}}{\ln(\theta_s/\theta_l^\infty)}. \quad (4.175)$$

Here, θ_f/θ_l^∞ and θ_s/θ_l^∞ are the effective supersaturations of the sidewall and surface adatoms with respect to an infinitely large liquid phase. Diffusion flux of surface adatoms is positive at $R > R_C$ and negative at $R < R_C$; the condition for the nanowire growth from zero length in this model is therefore $R > R_C$. Diffusion flux of sidewall adatoms is positive at $R > R_B$ and negative at $R < R_B$; the infinite growth at $l \to \infty$ yields $R > R_B$. Four growth modes of nanowires can therefore be re-formulated in terms of their radius:

(I) Infinite growth at $R > \max(R_B, R_C)$;
(II) Negative growth at $R < \min(R_B, R_C)$;
(III) Averaging growth at $R_C < R < R_B$;
(IV) Continuing growth at $R_B < R < R_C$.

To be more specific, consider the case of Au-assisted MBE of GaAs nanowires on GaAs(111)B surface. We study the temperature dependences of the critical radii given by (4.175) with $\theta_f/\theta_l^\infty = (1/\pi)(V/h)(\sigma_f/\sigma_s)\tau_f \tan\alpha \exp(-\mu_L^\infty/k_BT)$ and $\theta_s/\theta_l^\infty = (V/h)\tau_s \exp(-\mu_L^\infty/k_BT)$. Here, V/h is the equivalent deposition rate in monolayers per second, for which we use the typical value of 1 ML/s. The ratio σ_f/σ_s amounts to $4/\sqrt{6}$ for the (110) sidewall and (111)B substrate surfaces of ZB GaAs. The adatom lifetime on the sidewalls, limited by the desorption, is as usual taken in the Arrhenius form $\tau_f = \nu_f^{-1} \exp\left(E_A^f/k_BT\right)$. We use the values of $E_A^f = 2$ eV for the activation energy for desorption (estimated from the data of [480]) and $\nu_f = 10^{11}$ s^{-1} for the corresponding vibration frequency. As discussed above, the adatom lifetime on the substrate is limited either by the desorption (at high temperatures) or by the surface nucleation (at low temperatures). As a result, τ_s has a maximum at a certain temperature, corresponding to the maximum diffusion length λ_s in Figs. 4.68 and 4.69, with the liquid chemical potential shown by the dotted line in Fig. 4.69. Finally, the value of $R_{GT} = 6.6$ nm corresponds to 40 % concentration of Ga in the Au-Ga alloy and the droplet contact angle $\beta = 120°$.

Theoretical curves for $R_B(T)$ and $R_C(T)$ with these parameters are presented in Fig. 4.75. The temperatures higher than 500 °C yield straight GaAs nanowires with uniform radius from base to top, for which the stationary growth model is directly applicable. As seen from Fig. 4.75, infinite nanowire growth (scenario I) cannot proceed at very high temperatures due to the negative diffusion from the wire top to the substrate. For the model parameters considered, the averaging growth to a finite length (scenario III) is never observed since $R_C > R_B$ within the entire temperature domain.

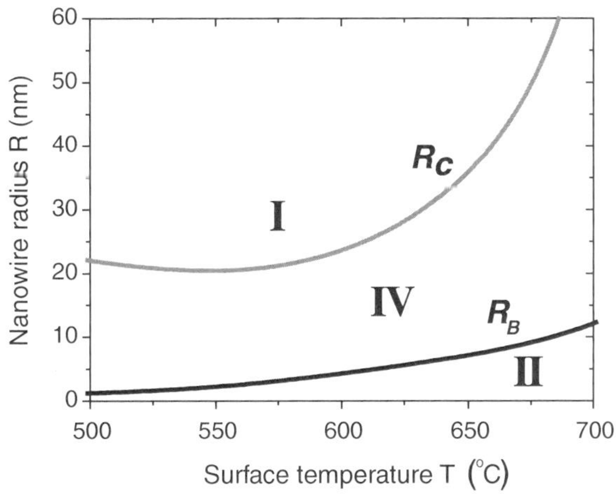

Fig. 4.75 Typical growth diagram of Au-assisted MBE of GaAs nanowires for the model parameters described in the text: zone *I*—infinite growth, zone *IV*—continuing growth, zone *II*—negative growth

The VLS growth is completely forbidden in zone II below the curve $R_B(T)$ (scenario II), which is, however, smaller than 10 nm for all reasonable temperatures. Non-trivial scenario IV can happen in zone IV between the curves $R_B(T)$ and $R_C(T)$, where the growth can not proceed from the droplet lying on the substrate surface but can continue from pre-existing nanowire stems with sufficiently large initial length $l_0 > l_*$. Figure 4.75 shows that the domain of nanowire radii exhibiting such a continuing growth extends at higher surface temperatures. As follows from the equation $U(l_*) = |C|/B$ and $U(l) \propto \exp(l)$ at $l >> 1$, the temperature dependence of dimensional critical length L_* at high T mimics that of $\lambda_f(T)$ with minor logarithmic corrections. As the diffusion length $\lambda_f(T)$ decreases with T, the critical length for the continuing growth becomes smaller at higher temperatures.

We note that negative growth via scenario II during MBE or MOCVD should always take place at zero deposition rate ($V = 0$), when the coefficients B and C become negative. Since the diffusion flux is directed from the nanowire top to the substrate, the decrease of nanowire length should be associated with the increase of surface layer thickness. It should also be mentioned that the growth scenarios in MOCVD technique strongly depend on the temperature behavior of precursor decomposition probabilities, which may be different at the substrate and sidewall surfaces. The resulting growth diagram will therefore differ from the MBE case. Finally, the detailed analysis of general growth (4.96) at $A \neq 0$ leads to six possible growth scenarios instead of four so that non-trivial cases of continuing and averaging growth could be observed within certain intervals of nanowire radii [481].

Obviously, scenario I corresponds to most of the growth experiments reported in the literature. However, experimental evidences of scenarios II and IV in the case of Au-assisted MBE of GaAs nanowires on GaAs(111)B substrates have also been found [410]. In growth experiments in Riber 32 MBE setup, substrate was deoxidised at 620 °C and a GaAs buffer layer was grown at 630 °C. A 70 nm AlGaAs marker layer was systematically inserted in this buffer layer in order to have an absolute

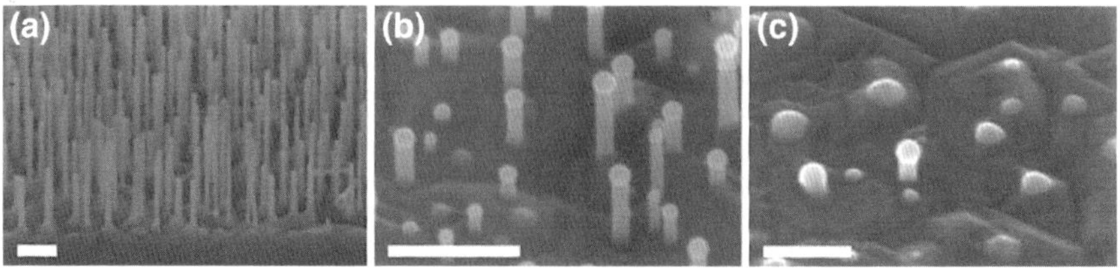

Fig. 4.76 GaAs nanowires after *in situ* anneal under As_4: **a** reference sample without annealing; **b** 4 min anneal; **c** 5 min anneal. Scale bars represent 200 nm

reference of vertical coordinate z. Then the substrate temperature was decreased and stabilized to the value desired for the wire growth. An amount of Au equivalent to a uniform layer of 1 nm was deposited on the substrate surface from an Au source installed in the MBE growth chamber. This procedure resulted in the formation of Au droplets alloyed with the substrate constituents which could activate the growth.

Scenario (II) is well illustrated by the following annealing experiments. GaAs nanowires were grown at 530 °C for 15 min with an As_4 beam equivalent pressure of 6×10^{-6} Torr and a Ga flux corresponding to a nominal growth rate V of 0.2 nm/s. This growth stage produced nanowires of 800 nm length and 30 nm diameter in average, as shown in Fig. 4.76a. Several samples were grown and subsequently annealed in the growth chamber under the As_4 flux at 630 °C and for various durations. After the growth, the resulting morphology of nanowire ensembles was investigated by SEM. The sample cross-sections were observed after cleaving and the contrast between GaAs and AlGaAs allowed us to locate the marker layer and then to determine the exact position where Au was deposited before initiating the nanowire growth. We could deduce the thickness of the 2D GaAs layer growing concomitantly with the nanowires. The results are presented in Fig. 4.76b, c. It is obvious from this figure that nanowires gradually decompose during the annealing. We also observed that the average thickness of GaAs 2D layer increased from 110 nm for non-annealed samples to 147 and 176 nm for 4 and 5 min annealing time. This clearly indicates that the nanowire decomposition did not result from re-evaporation of Ga which is not expected at 630 °C. We propose the following mechanism: nanowires dissolve in the catalyst droplets, As is most likely re-evaporated from the drop, but Ga migrates from the droplet to the surface where it can reincorporate the GaAs solid phase since As_4 flux is supplied. In other words, we observe the negative nanowire growth rate at $V = 0$ which is predicted in scenario (II) where adatom diffusion flow is reversed as compared to the standard situation of scenario (I).

Another series of MBE growth experiments evidence that scenario (IV) also occurs. Au catalyst was deposited on GaAs (111) B with the same procedure as described hereinabove. After Au deposition, we tried to grow GaAs nanowires with the following conditions: in a first sample (sample A) the growth was realized at 630 °C for 15 min. The SEM view shown in Fig. 4.77a reveals that nanowires do not grow from the substrate at this relatively high temperature and the segregation of the Au particles at the substrate surface is observed. This result is similar to what is shown in Fig. 4.64e. Then, two other samples were grown by using a two-temperature process: the first growth step was performed at 530 °C for $t_{LT} = 1.5$ min (sample B) or 15 min (sample C). At this temperature, nanowires are formed, as already observed in Fig. 4.76a. After this first step, the temperature was ramped to 630 °C within 2 min. During this ramping, the incident gallium and arsenic fluxes were maintained. Finally the growth was continued at 630 °C for 15 min for sample B and C. Therefore, this last growth step was similar for the three samples A, B and C. Figure 4.77b, c show SEM images of samples B and C. It is observed that no wires are present in sample B, the final figure being quite comparable to sample A. This means that the nanowires formed at the low-temperature-growth step of 1.5 min, did not continue to grow at high T: they were either decomposed like in annealing experiments, or buried by GaAs growing faster on the rest of the surface. Conversely, sample C shows nanowires longer than what is expected for the 15 min of growth at low temperature: most of the nanowires are longer than 2 μm (Fig. 4.76c). Obviously, in this sample, nanowire growth was possible at high T. Scenario (IV) is believed to be responsible for the different behaviours of sample B and C: the initial nanowire length is lower (sample B) or higher (sample C) than the critical length. The latter corresponds to the nanowire sidewall cross-section which intercepts enough incoming Ga flux to compensate for the flow of adatoms which otherwise diffuses from the droplet to the substrate surface, leading to nanowire decomposition.

Nanowires shown in Fig. 4.77c have rather complex geometry: some of them develop branches, and the majority of nanowires are tapered towards their top. Such a complex geometry can be explained by changes of the droplet size and position at elevated temperature. Most probably, the droplet first inflates with the temperature and then becomes unstable on the wire top so that it splits into several smaller droplets. Some of those give branches and other may migrate to the substrate. We note that similar MBE growth experiments carried out in EP1203 setup at slightly different conditions allowed us to obtain more regular shape of nanowires. The GaAs(111)B surfaces were activated by Au with the same procedure as described above. A low temperature growth was performed at $T = 550$ °C form atomic Ga and As_4 beams with a nominal growth rate of 0.3 nm/s for 1, 2 and 12 min. After the temperature ramping to 630 °C, the high temperature growth step was performed at a lower nominal growth rate of 0.15 nm/s for 48 min. Similarly to the previous case, nanowires did not grow for $t_{LT} = 1$ and 2 min. Nanowires obtained with $t_{LT} = 12$ min, having average initial length of 800 nm (Fig. 4.78a) continued to grow at 630 °C and reached more than 10 μm in length (Fig. 4.78b) after this high temperature step.

The continuing growth scenario via the two-step procedure with a temperature ramping thus allows for the high temperature growth of GaAs nanowires at 630 °C from pre-existing stems, which would be impossible if started from the substrate. This possibility is very important for controlling the crystal structure of GaAs nanowires, which tends to adopt pure ZB phase at high temperature corresponding to a low liquid supersaturation, as discussed in Chap. 6.

Fig. 4.77 Samples of GaAs grown with a low temperature (530 °C) step of variable duration t_{LT}, followed by a high temperature step (630 °C) of 15 min. **a** sample A with $t_{LT} = 0$; **b** sample B with $t_{LT} = 1.5$ min; **c** sample C with $t_{LT} = 15$ min. Scale bar is 400 nm

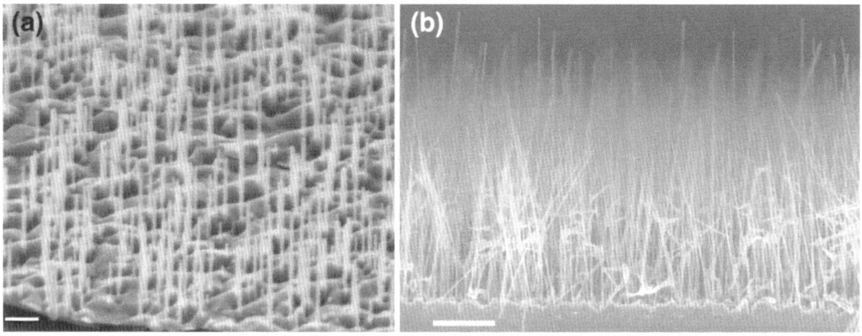

Fig. 4.78 SEM images of GaAs nanowires grown with a low temperature step for 12 min before (**a**) and after (**b**) the high temperature growth. Scale bars are 200 nm in (**a**) and 2 μm in (**b**)

4.15 Narrowing the Length Distribution of Ge Nanowires

Another interesting non-linear growth effect of narrowing the length distribution of Au-catalyzed Ge nanowires with growth time was reported in [413]. This surprizing narrowing originates from different directions of the diffusion flux of sidewall adatoms in differently sized inclined nanowires. We first describe the experimental data and then explain the observed growth behavior within the diffusion growth model. The MBE growth procedure was the following. The Au droplets were formed directly by Au deposition onto a heated Si(111) surface in an ultra high vacuum chamber. The growth of Ge nanowires was then achieved by the sublimation of Ge at a deposition pressure of 10^{-9} mbar. In this MBE system, the Ge beam is perpendicular to the substrate. The equivalent deposition rate V in all growth runs was 14 ± 1 nm/min. The Si(111) substrate was heated up to $T = 350$ °C, which has been found to be the best growth temperature to obtain long and regular $< 110 >$-oriented Ge nanowires. The growth duration was varied from 15 to 70 min. The nanowire morphology was then investigated by SEM.

Figure 4.79 shows SEM images of three $< 110 >$-oriented Ge nanowires, obtained for the growth time of 15, 30 and 70 min. Such nanowires are known to have a hexagonal cross-section and the measurement of twice the apothem $2R$ yields a nanowire side of 109, 98 and 104 nm respectively. As the formation of Au islands

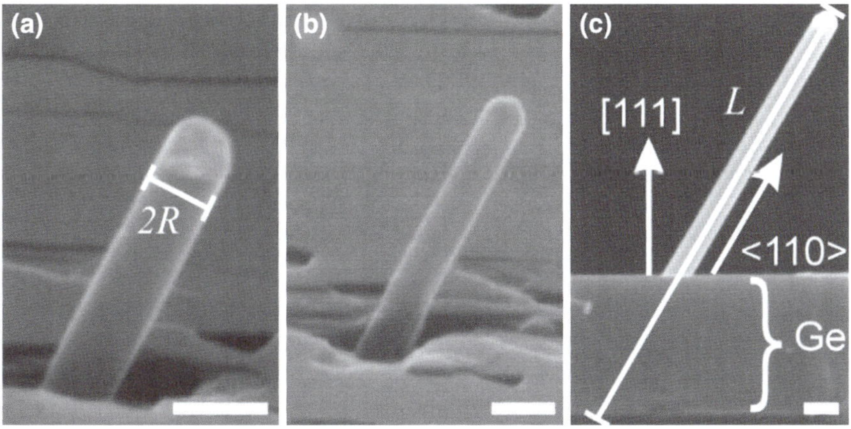

Fig. 4.79 SEM images of three < 110 >-oriented Ge nanowires grown at 350 °C for three different growth times of **a** 15 min, **b** 30 min, **c** 70 min. The substrate orientation and growth direction are indicated, as well as the thickness of the Ge layer. L and R yield for the length above the Ge layer and the apothem of the nanowire respectively. Scale bar is 150 nm

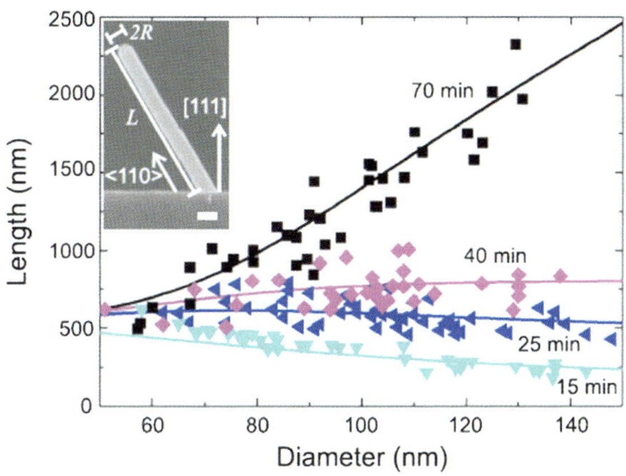

Fig. 4.80 Experimental (*symbols*) length-diameter dependences of < 110 > Ge nanowires obtained with different growth times. *Solid lines* are theoretical fits obtained from (4.176) with the parameters summarized in Table 4.8. Scale bar in the insert is 100 nm

leads to a significant variation in their size, a statistical analysis of distribution of the nanowire length [i.e., the length emerging above the overgrown Ge layer, as shown in Fig. 4.79c] as a function of R was performed for different growth times. Figure 4.80 reveals a decreasing $L(R)$ dependence after 15 and 25 min. Then the nanowire length becomes almost R-independent at 40 min and is eventually converted to an increasing $L(R)$ curve after 70 min of growth.

As discussed previously, a decreasing $L(R)$ dependence is typical for the diffusion-induced growth. Conversely, an increasing length-radius correlation is explained by the Gibbs-Thomson effect or the nucleation-limited regime. As demonstrated in Sect. 4.11, the $L(R)$ curves of Si, Ge and III-V nanowires may feature both types of dependences at different R. However, the conversion of decreasing into increasing

$L(R)$ dependence for nanowires of the same radii with the growth time was reported in [413] for the first time. We now show how this puzzling behavior can be explained by (4.108), adapted to the case of inclined nanowires.

The growth model is illustrated in Fig. 4.81, where V denotes the deposition flux with the incident angle α, while φ is the nanowire tilt angle. As above, θ_s, θ_f and θ_l are the activities of Ge on the main surface, nanowire sidewalls and in the Au-Ge alloy in the catalyst particle, respectively. In the case of inclined nanowires, the expression for θ_f should be modified to $\theta_f = (1/\pi)J\sin(\alpha+\varphi)\sigma_f\tau_f$ to account for the tilt φ. The λ_s is the adatom diffusion length on the main surface, which is assumed as being much larger than the wire radius R. We also assume that the adatom diffusion length on the sidewalls λ_f is much larger than the wire length L and therefore does not enter the growth equation. Finally, we consider the case where the adsorption-desorption term A (with the 2D growth rate subtracted) is negligible compared to the diffusion-induced terms B and C. This should be possible under our experimental growth conditions, because the direct impingement is almost exactly compensated by the surface growth, while the desorption from the droplet can be neglected at 350 °C. In these conditions, the nanowire length growth rate is given by (4.108), which can be re-written in the form

$$L = 2\Lambda_s\sqrt{\frac{\exp(2ag_fV(t-t_0)/R)-1}{2ag_f}}. \quad (4.176)$$

Here, $a = \sin(\alpha+\varphi)/(\pi\cos\alpha)$ is the geometrical factor of MBE growth for a tilted nanowire, $\Lambda_s = \lambda_s\sqrt{g_s/b}$ is the effective diffusion length which is of the order of λ_s and t_0 is the incubation time required to initiate the nanowire growth. The g_s and g_f factors are given by $g_f = 1-\theta_l/\theta_f$ and $g_s = 1-\theta_l/\theta_s$ and depend on R through $\theta_l(R)$. As discussed in Sect. 4.9, (4.176) yields infinite growth at $\theta_s > \theta_l$ and $\theta_f > \theta_l(g_s > 0, g_f > 0)$, while at $\theta_s > \theta_l, \theta_f < \theta_l(g_s > 0, g_f < 0)$ we observe limited growth to a finite critical length $L_* = \Lambda_s\sqrt{2/\left(a\left|g_f\right|\right)}$.

We now turn to the analysis of the experimental data using (4.176). In these growth experiments, the incident angle $\alpha = 0$, and the tilt angle of $< 110 >$-oriented nanowires $\varphi = 35.3°$, yielding $a = 0.184$. The incubation time t_0 amounts to 6 min on average. In addition to the dependences of L on $2R$ shown in Fig. 4.80, we plot out the time evolution of L for different R in Fig. 4.82. It is seen that the narrowest nanowires grow faster in the beginning but then stop growing. Conversely, the 130 nm-thick nanowires that have evolved slowly during the first 15 min overtake the length of the 60 nm-thick nanowires after 30 min and become more than 4 times longer than the narrowest nanowires after 60 min of growth. Equation (4.176) provides a good description of infinite and limited nanowire growth modes at Λ_s of about 100 nm, which is consistent with previous estimates for the diffusion lengths. The fitting parameters are summarized in Table 4.8 and yield the theoretical curves shown by solid lines in Figs. 4.80 and 4.82. The θ_f/θ_l ratios and corresponding values of $\Delta\mu_{fl} = k_BT\ln(\theta_f/\theta_l)$ are estimated from g_f. It is seen that theoretical curves reproduce quite well the experimental data. The initial stage with the square root law

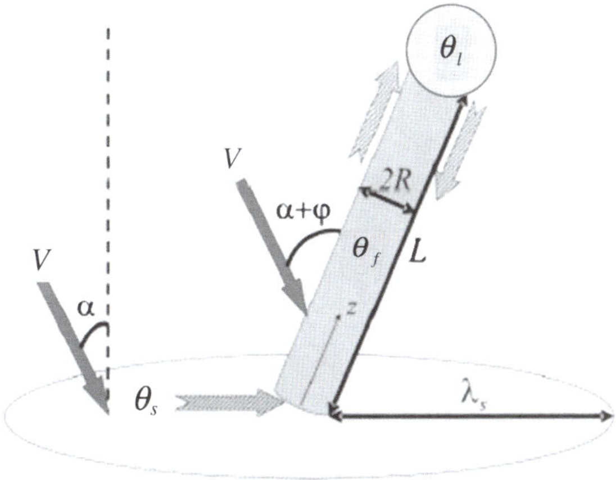

Fig. 4.81 Illustration of the growth model of a tilted nanowire. The cross-hatched *arrows* indicate the direction of the adatom fluxes that depend on the magnitude of the surface, sidewall, and droplet activities, θ_s, θ_f and θ_l

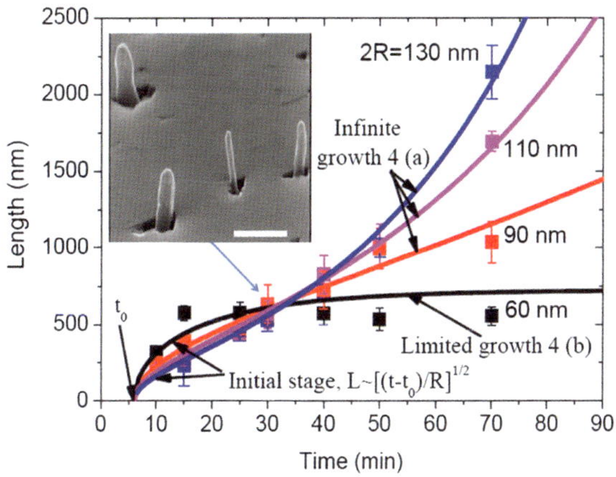

Fig. 4.82 Experimental length-time dependences of < 110 >-oriented Ge nanowires for four different diameters (*symbols*), and theoretical fits (*lines*) obtained from (4.176) with the parameters summarized in Table 4.8. *Inset:* SEM image of < 110 >-oriented Ge nanowires grown for a time of 30 min. The scale bar is 600 nm. Different growth regimes are indicated by *arrows*

$L \propto \sqrt{(t - t_0)/R}$, the regimes of infinite exponential growth and the limited growth to a finite length are indicated in Fig. 4.82.

Crucial to this focusing effect is the value of g_f, as it controls the magnitude and the direction of the corresponding diffusion flux. As seen from Table 4.8, g_f increases from −0.30 for 60 nm diameter nanowires (corresponding to a negative diffusion flux from the droplet) to 0.75 for 130 nm diameter nanowires. One obvious reason for the increase of g_f and the corresponding $\Delta\mu_{fl}$ is the Gibbs-Thomson effect. We note, however, that the increase of g_f obtained from the fits of < 110 >-oriented nanowires is faster than that given by the Arrhenius-like radius dependence of the liquid activity $\theta_l = \theta_l^\infty \exp(R_{GT}/R)$ at a constant θ_l^∞. Therefore, the effective supersaturation of sidewall adatoms with respect to the droplet is influenced by other physical effects, such as (i) variation of the Ge concentration in differently sized droplets, (ii) different atomic structure of sidewalls of differently sized nanowires, (iii) possible time variation of the Ge percentage caused by the Au migration, and (iv) scattering of

Table 4.8 Fitting parameters for differently sized < 110 >-oriented Ge nanowires

$2R$ (nm)	g_f	θ_f/θ_l	$\Delta\mu_{fl}$ (meV)
60	–0.30	0.77	–14
90	0.13	1.15	7.5
110	0.45	1.82	32
130	0.75	4.0	74

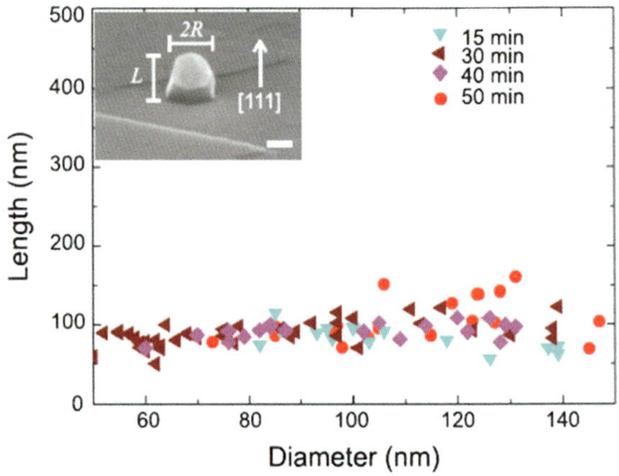

Fig. 4.83 Length-diameter dependences of < 111 >-oriented Ge nanowires after 15, 30, 40, and 50 min of growth. Scale bar is 50 nm

the Ge beam from the sidewalls. Regardless of physical mechanisms changing the activity of sidewall adatoms, a negative diffusion can be the only explanation for the timelimited nanowire growth. The slight decrease of length after thesaturation observed for the thinnest < 110 >-oriented nanowires in Fig. 4.82 can be explained by their negative growth via the diffusion from the top to the substrate.

Finally, we would like to point out that the narrowing effect is expected to be noticeable only when $\sin(\alpha + \phi)$ readily differs from zero. As shown in Fig. 4.83, < 111 >-oriented Ge nanowires also emerge on the Si(111) substrate during growth. Their length is much smaller than the length of the < 110 >-oriented nanowires and is found to be almost constant with R and t. When $\sin(\alpha + \phi)$ tends toward zero, the product $ag_f = -\theta_l/(J\sigma_f\tau_f)$ becomes negative regardless of the nanowire radius and a finite nanowire length, independent of R and t, is expected, as described above. Therefore, all the < 111 >-oriented nanowires should follow the scenario of limited growth. Such a behavior is consistent with almost invariant nanowire length seen in Fig. 4.83. We suspect this growth regime to occur quickly due to a small diffusion length on the sidewalls, as they are highly faceted. It is noteworthy that, whenever the liquid activity becomes larger than that of sidewall adatoms, the minimum of chemical potential is reached at the nanowire sidewalls. This should lead to a radial growth of nanowires after the length saturation.

To sum up the results of this section, the Au-catalyzed Ge nanowires obtained by MBE on Si(111) substrates exhibit a rather complex growth behavior. Narrow

$< 110 >$ Ge nanowires grow faster than wider ones at the beginning, but almost stop growing after a certain growth time. Consequently, the length-time dependences of differently sized nanowires cross at a certain time, after which the length-diameter dependences become reversed. Central to this effect is the change in the direction of the flow of adatoms on the nanowire sidewalls during the growth, due to the droplet activity dependence on the radius of the seed particle. It sets a fundamental limitation on the maximum length of MBE grown nanowires that should be considered as a rather general phenomenon, particularly in the setups where the direction of the beam is close to the substrate normal. However, such a limitation can be used to narrow the length distribution of nanowires at a certain time regardless of the initial distribution of growth seeds.

Chapter 5
Special Topics of Nanowire Growth and Morphology

In this chapter, we discuss some special questions of growth and morphology of different freestanding nanowires. Perhaps the most important feature of modern nanowires is their ability to dramatically enhance the critical dimension for plastic relaxation via misfit dislocations, which makes nanowires very promising for overcoming the lattice mismatch issue in monolithic integration of dissimilar semiconductor materials. We therefore consider in detail the critical dimensions for plastic relaxation in different nanowire geometries. In Sect. 5.1, we present the analysis of axial nanowire heterostructures at a given constant radius. It is shown that, for each mismatch, there always exists a certain critical nanowire radius below which an infinitely thick coherent layer can be grown on a cylindrical stem and thus the dislocation formation is completely suppressed. Section 5.2 presents similar analysis for different nanowire geometries (cylinder, full cone, truncated cone and reverse truncated cone) on lattice mismatched substrates. Section 5.3 treats radial nanowire heterostructures where one material surrounds the other. We show that the critical shell thickness for plastic relaxation is again considerably enhanced relative to the 2D limit due to the curvature of the core. In this case, the elastic relaxation mechanism is even more efficient than in the axial nanowire heterostructures.

Sections 5.4 and 5.5 are devoted to the radial growth and morphology of catalyzed nanowires. In Sect. 5.4, we discuss a model for the well-known cylindrical-to-conical shape modification of nanowires due to the sidewall nucleation at low temperatures, with some examples of the Au-catalyzed III-V nanowires. Section 5.5 treats the vertical elongation and radial growth simultaneously, with a simplified model for the radial growth rate which is only linearly dependent on the adatom concentration. This model is used to fit the data on the growth kinetics of CdTe nanowires by close-space sublimation on Mo foils.

Section 5.6 presents very interesting experimental data and a model of nucleation statistics in catalyzed nanowires. It is shown that, due to a small volume of a catalyst nanoparticle, the nucleation events producing 2D islands and finally the nanowire monolayers are not random but temporary anti-correlated. The analysis of Sect. 5.7 shows why and under which conditions the catalyzed nanowire growth proceeds in a self-regulatory mode with rather short nucleation pulses. These results

V. G. Dubrovskii, *Nucleation Theory and Growth of Nanostructures*, NanoScience and Technology, DOI: 10.1007/978-3-642-39660-1_5,

have far-reaching applications for understanding the nucleation-mediated nanowire growth, oscillatory behavior of nanowire morphology, their length uniformity and some important features of nucleation in confined volumes in general.

Section 5.8 presents experimental data and a model for a non-homogeneous radial distribution of B dopant in the Au-catalyzed Si nanowires. It is shown that the dopind level rapidly decreases toward the nanowire center, and different dopant incorporation pathways are discussed to explain this effect. In Sect. 5.9, we discuss Ga-catalyzed VLS growth of GaAs nanowires and in particular the steady state conditions that are required to maintain a constant droplet volume. When growth proceeds under Ga-rich conditions, the droplet volume and consequently the nanowire radius gradually increase. After termination of Ga flux, it is then possible to resume the VLS growth of a secondary nanowire at the top.

Catalyst-free, self-induced growth of GaN nanowires on silicon substrates has recently drawn much attention as an attractive way for monolithic integration of optical GaN-based heterostructures with the existing silicon electronic platform. Compared to the VLS approach, it is admittedly more challenging to maintain the necessary control over self-induced nanowires. In Sect. 5.10, we present relevant experimental data showing that GaN nanowires obtained by MBE usually start from the Volmer-Weber islands, both on amorphous Si_xN_y and lattice mismatched AlN interlayers on Si(111). Section 5.11 presents a scaling thermodynamic model for the self-induced nucleation of GaN nanowires. It is shown that the island-to-wire shape transformation is indeed preferred on surface energetic grounds for plausible parameters provided that the nanowire radius scales superlinearly with its radius. In Sect. 5.12 we give some kinetic arguments to justify the scaling length-radius dependence at the nanowire growth stage. We argue that the scaling growth kinetics might be due to a step flow radial growth at a low step density or the shadow effect in a dense nanowire array. Section 5.13 presents a model for the elongation rate of self-induced GaN nanowires that fits very well the data on the time, Ga flux and temperature dependences of the growth rate during MBE.

Finally, in Sect. 5.14 we consider some cooperative effects in nanowire growth. The first one is the shadow effect originating from the neighboring nanowires in directional deposition methods such as MBE or CBE. Shadowing blocks out the lower parts of the nanowires from the incoming flux. It is shown that the shadow effect should be responsible for the radius uniformity as well as particular shapes of dense arrays of InAs nanowires grown by MBE. The second cooperative effect is re-emission of material from the sidewalls of the neighboring nanowires, which may act as an effective secondary source of group V or even group III element in III-V nanowire growth. While the complete theoretical description of re-emission is a very complex problem in the general case, we present a simple model that explains why the growth rate of organized arrays of InSb nanowires grown by CBE is enhanced with increasing the nanowire density.

5.1 Axial Nanowire Heterostructures

As already discussed in Chap. 3 and in Sect. 4.1, reduced lateral size of freestanding nanowires largely improves their ability to release strain in two dimensions. This enables a radical increase of the critical thickness for plastic deformation via the formation of misfit dislocations compared to thin films and even quantum dots. For basic studies as well as applications, it is important to fabricate nanowires containing strained heterostructures of lattice mismatched materials. In this section, we consider axial nanowire heterostructures shown in Fig. 4.14a, where the materials are stacked along the wire axis. Glas [187] demonstrated theoretically the increase of the critical thickness in such heterostructures due to free lateral sidewalls. The model geometry considered consists of a cylindrical wire stem ($-\infty < z < 0$) and the top layer ($0 < z < H$), of the same radius $R = const$, as illustrated in Fig. 5.1a. Between these two parts, there exists a purely dilatational misfit ε_0, taken as positive for layers under compression. In cubic materials, ε_0 is the relative difference of lattice parameter between the two materials. As in Sect. 3.3, the linear isotropic elasticity theory gives the elastic energy stored in the region $-\infty < z < h$ in the form

$$W_e = \frac{E}{1-\nu}\pi R^2 H \varepsilon_0^2 z(\eta) \cong \frac{E}{1-\nu}\frac{\pi R^2 H \varepsilon_0^2}{1 + A\eta}. \tag{5.1}$$

Here, E is the Young's modulus and ν is the Poisson's ratio, taken identical for both materials, $\eta = H/(2R)$ is the aspect ratio of the top nanowire layer, and the elastic energy relaxationz $z(\eta)$ is taken in the simplest form (3.40). As discussed in Sect. 3.3, the value of A depends on the Poisson's ratio but neither on the Young's modulus nor mismatch. For this particular cylindrical geometry, A equals approximately 27.4 for cubic materials with $\nu = 1/3$. Similarly to the case of a cylinder seated on an infinite 2D substrate, W_e quickly saturates with increasing H, while the elastic energy of a 2D layer of the same volume would scale linearly with H.

Equation (5.1) was obtained in [187] not only from fitting the results of finite-element calculations at different η, but also based on the Barton solution for a misfitting layer of finite height in an infinite circular cylinder [482]. This solution satisfies the boundary conditions of our problem (see Fig. 5.1a) as regards elastic

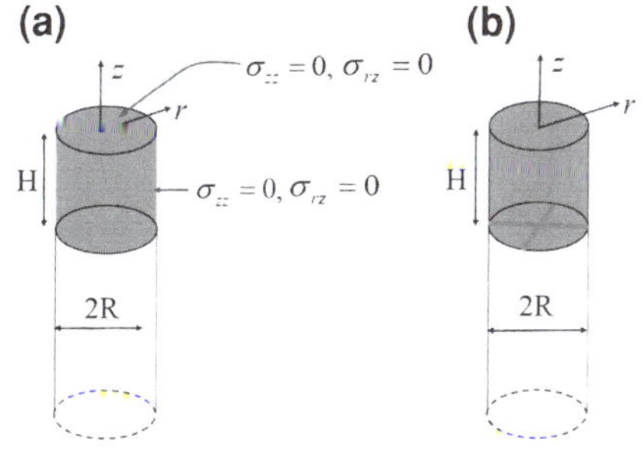

Fig. 5.1 **a** Model geometry of axial nanowire heterostructure showing the boundary conditions for the elastic stress σ; **b** a pair of orthogonal linear dislocations intersecting at the nanowire axis

stresses on the lateral free surface ($\sigma_{rr} = 0,\ \sigma_{rz} = 0$ for $r = R$) but not on the upper free surface ($\sigma_{zz} = 0,\ \sigma_{rz} = 0$ for $z = H$). However, one may insert additional misfitting layers into the wire above $z = H$ and optimize numerically their parameters so that the resultant stresses nearly cancel the stresses generated by the original layer on section $z = H$, which becomes nearly strain-free. The total elastic energy $W_e(R, H, \varepsilon_0)$ is then computed numerically from the strain and stress fields. This procedure yields the results that are very close to the finite element calculations.

To obtain the critical thickness for the formation of misfit dislocations as a function of R at a given lattice mismatch ε_0, we must compare the elastic energy of the dislocation-free heterointerface with the energy of the system at the onset of plastic relaxation. As in a 2D layer, interfacial dislocations reduce the mismatch between the substrate and layer to a smaller value ε_0, thereby reducing W_e. On the other hand, the own dislocation strain field and core energy increase the total energy of the system. To insure biaxial in-plane strain relaxation, we place a single pair of orthogonal dislocations in the heterointerface, intersecting at the wire axis as shown in Fig. 5.1b. We denote by $\vec{b}$ their Burgers vector and by b_{eff} the edge component of the latter in the interface. Following the procedure for the calculation of the critical thickness in 2D layers [124], we assume that the total energy is the sum of the energy W_d of the same dislocations in a homogeneous wire and the elastic energy $W_e(R, H, \varepsilon_0')$ calculated from (5.1) for the reduced mismatch ε_0'.

No exact solution could be found for the strain in a wire with such dislocations, so that we have to resort to some approximation. The elastic energy for dislocations parallel to an infinite planar free surface can be calculated as a function of the distance of the dislocations from the surface [483]. Following Ovid'ko [484], we use the same formulas after replacing the distance between dislocation and planar free surface by the effective distance $\overline{h}$ between the dislocation and the nearest free surface, defined as

$$\overline{h} = \begin{cases} H,\ H \le 2R/\pi \\ 2R/\pi,\ H > 2R/\pi \end{cases}. \tag{5.2}$$

This formula means that for a thin layer at the top of a wire, the relevant distance is the distance H to the top surface, whereas for a thicker layer it is the average distance to the lateral surface $2R/\pi$. We then substitute $\overline{h}$ by the distance between dislocation and planar free surface in the standard formula giving the energy in the 2D case [124] to obtain the total dislocation energy

$$W_d = \frac{Eb^2R}{\pi(1-\nu^2)}\left(1 - \nu\cos^2\theta\right)\left(\ln\frac{\overline{h}}{b} + 1\right). \tag{5.3}$$

Here, θ is the angle between the dislocation line and its Burgers vector, and $b = \left|\vec{b}\right|$ is as usual taken as the core cutoff radius for the calculation of the elastic energy [126, 483]. In (5.3), the terms of the sum in parentheses correspond, respectively, to the core and elastic energies of the dislocations.

We must now compute the elastic energy W_e for the reduced mismatch ε_0'. However, defining which part $\varepsilon_a = \varepsilon_0 - \varepsilon_0'$ of the mismatch is accommodated by the dislocation pair requires some care. For a 2D layer with a square grid of interfacial dislocations of period d, we simply have $\varepsilon_a = b_{eff}/d = b_{eff}l/2$, where l is the length of dislocation per unit surface [126]. Keeping the same formula in the wire case ensures in particular that we recover the standard 2D critical thickness when $R \to \infty$. Since, in the nanowire case, $l = 4/(\pi R)$, we are led to consider that the misfit accommodated is $\varepsilon_a = \alpha b_{eff}/(2r_0)$ with $\alpha = 4/\pi$. Under these assumptions, using (5.1) and (5.3), the excess energy of the state with the dislocation pair with respect to the state with a fully coherent interface is given by

$$\Delta W(R,H) = \frac{ER}{1-\nu}\left[z\left(\frac{H}{2R}\right)\pi RH\left(\frac{\alpha^2 b_{eff}^2}{4R^2} - \frac{\alpha b_{eff}\varepsilon_0}{R}\right) + C\left(\ln\frac{\overline{h}(R,H)}{b}+1\right)\right], \tag{5.4}$$

with $C = \left(1-\nu\cos^2\theta\right)b^2/\left[2\pi(1+\nu)\right]$. Equation (5.4) includes terms of second order in b_{eff}, which are essential for critical thickness calculations. Obviously, the coherent state is stable if $\Delta W > 0$ whereas if $\Delta W < 0$, it is favorable to introduce dislocations.

Hence, the equation

$$\Delta W(R,H) = 0 \tag{5.5}$$

defines the critical thickness $H_c(R)$ for the onset of dislocations. The heterointerface remains dislocation-free as long as $H < H_c(R)$, whereas plastic relaxation should occur above $H_c(R)$. This radius-dependent critical thickness can be calculated for pure edge ($\theta = \pi/2, b_{eff} = b$) and 60° dislocations ($\theta = \pi/3, b_{eff} = b/2$) [187]. Both types pertain to face-centered cubic crystals. For a given radius, the critical thickness is less for edge dislocations since they relieve strain more efficiently. These defects are seldom found in low-mismatch 2D films but may occur for higher misfits; the same should happen in nanowires.

The variations of the critical thickness with R for 60° dislocations are shown in Fig. 5.2 for several values of the mismatch ε_0, taking b $= 0.4$ nm, a value pertaining to GaAs and a good approximation for many semiconductors [187]. For each ε_0, the $H_c(R)$ curve separates the (R, H) plane into the coherency domain (below, left) and the domain of plastic relaxation (above, right). Symbols in Fig. 5.2 correspond to some experimental data on the axial nanowire heterostructures grown beyond their 2D critical thicknesses. In each case, the calculations correctly predict the state (coherent or with dislocations) experimentally observed.

Very importantly, the curves in Fig. 5.2 show that, as the lateral dimension is reduced, the critical thickness tends to infinity at a certain critical radius $R_c(\varepsilon_0)$. This gives a unique opportunity to grow infinitely thick dislocation-free layers if the nanowire radius is below $R_c(\varepsilon_0)$. Examination of (5.4) confirms that, since

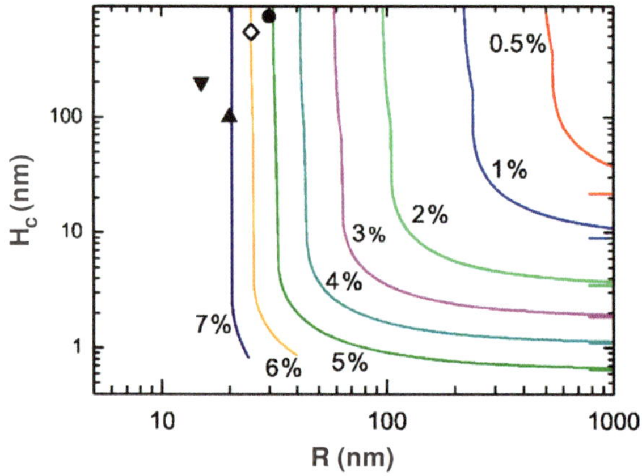

Fig. 5.2 Variations of the critical thickness of a mismatched layer growing on top of a nanowire as a function of the wire radius, for various values of the mismatch ε_0 (given in percent near each curve) and $\nu = 1/3$. In the plastically relaxed state, misfit is accommodated by 60° dislocations (b = 0.4 nm). Each segment at the right gives the asymptote of the curve immediately above for $R \to \infty$, which is the 2D critical thickness. Symbols give the dimensions of layers grown without (*full symbols*) or with (*open symbol*) dislocations, for $\varepsilon_0 = 0.9\,\%$ (*disk*) [485], 3.2 % (*up triangle*) [338], 3.7 % (*down triangle*) [486], and 7.1 % (*diamond*) [255] ([187])

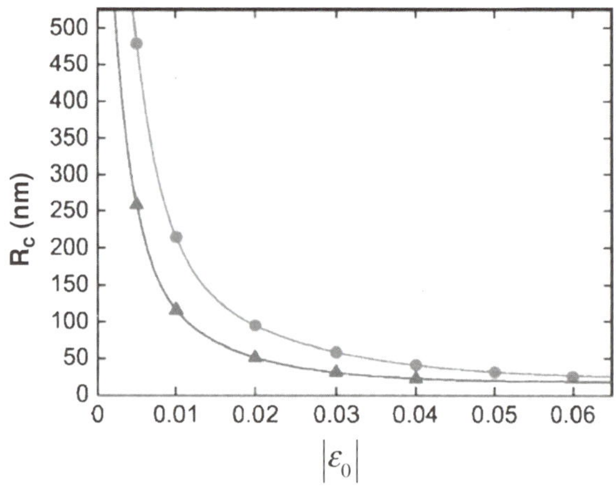

Fig. 5.3 Critical radius above which infinitely thick mismatching layers could grow coherently, as a function of mismatch. *Triangles* and *disks* correspond to plastic relaxation by, respectively, edge and 60° dislocations. b and ν are as in Fig. 5.2 [187]

$\eta z(\eta) \to 1/A$ when $\eta \to \infty$, a critical radius indeed exists (provided the mismatch is not too high), and is the solution of the following equation:

$$\frac{2\pi}{A}\left(\frac{\alpha^2 b_{eff}^2}{4} - \alpha b_{eff}\varepsilon_0 R_c\right) + C\left(\ln\frac{2R_c}{\pi b} + 1\right) = 0. \tag{5.6}$$

The variations of the critical radius with misfit calculated from (5.6) for edge and 60° dislocations are shown in Fig. 5.3. These values are of the order of the typical radii of current nanowires. Ertekin *et al.* [193] estimated a similar quantity via a different method: they considered a heterostructure made of two semi-infinite mismatched cylinders and posited an elastic displacement field depending on parameters that they found by minimizing the associated elastic energy. They considered only edge dislocations, the energy of which was calculated for an infinite matrix. The results of Glas given in Fig. 5.3 for the edge dislocations agree with that of [193] within $+6\,\%$. On the other hand, the critical radii for 60° dislocations are about twice larger.

Moreover, the very existence of a critical radius can be demonstrated independently of the particular expressions chosen for the energies involved, as follows. We saw earlier that for any mismatch, the elastic energy W_e scales with R as R^3 when $\eta \rightarrow \infty$. On the other hand, the total dislocation energy is larger than the energy of the dislocation cores, because of the added elastic energy, whatever its form. Since the core energy depends on the atomic structure of the dislocation, it is proportional to the dislocation length and hence to R (neglecting further relaxation at the intersections with the lateral surface, which involves a few atoms at most) and is independent of layer height; this remains true irrespective of the precise location of the defects in the interface and even of their number. From these different power dependencies upon R it follows that, for infinitely thick layers, the dislocation energy always dominates, provided that R is small enough. In other words, there exists a critical radius below which plastic relaxation is forbidden and coherent layers of arbitrary thickness should be obtainable. The general features of the $H_c(R)$ curves in Fig. 5.2, in particular the vertical and horizontal asymptotes, are thus independent of the expressions used in the calculations. These simple dimensional arguments explain why, for strained layers at the top of nanowires, the critical thickness for the introduction of misfit dislocations depends on the radius and becomes infinite below some critical value of the latter.

5.2 Nanowires on Lattice Mismatched Substrates

In this section, we will first discuss relevant experimental data on the critical dimensions for dislocation-free growth of III-V nanowires on lattice-mismatched substrates by different epitaxy techniques and then consider theoretically these results. An extensive study of the critical dimensions for the Au-catalyzed MOCVD growth of III-V nanowires on lattice-mismatched substrates was reported by Chuang *et al.* in [232]. Colloidal Au nanoparticles with sizes ranging from 10 to 160 nm were used as catalysts in a low-pressure MOCVD reactor under the VLS growth mode. Four lattice mismatched material combinations were used to study the growth as a function of lattice mismatch and catalyst size, for which the critical diameters are listed in the MOCVD row of Table 5.1. The substrates were Si(111) or (111)B in the case of III-V. The growth temperature (430–470 °C), pre-growth annealing temperature (610–660 °C), V/III ratio, growth rate, and growth pressure were the experimental parameters.

Table 5.1 Material-substrate combinations, lattice mismatches and critical diameters for coherent growth of III-V nanowires

NW/substrate	InAs/Si	InP/Si	InAs/GaAs	GaAs/Si	InP/GaAs	GaP/Si
Lattice mismatch (%)	11.6	8.1	7.1	4.0	3.9	0.4
Critical diameter MOCVD (nm)	26	36			96	–
Critical diameter MBE (nm)	24	39	44	106		

Figure 5.4a shows a typical SEM picture for InAs nanowires grown on a Si(111) substrate, with 11.6 % lattice mismatch. The widest epitaxial nanowire found has a 26 nm diameter. The Au nanoparticle size was determined by SEM in a separate, anneal-only run to be nominally 20 nm, with a wide size distribution from 10 to 40 nm. Hence it can be deduced that the maximum epitaxial nanowire diameter of 26 nm was due to the existence of a critical diameter. Figure 5.4b–d show SEM images of InP nanostructures grown on a Si(111) substrate using nominal 20 (b), 60 (c) and 120 (d) nm Au nanoparticles, respectively. With $\varepsilon_0 = 8.1\,\%$, only the 20 nm Au region supports epitaxial nanowire growth with vertically aligned nanowires. In the 60 and 120 nm Au nanoparticle regions, spiky star-like structures are created. For this case, the critical diameter was found to be 36 nm. Figure 5.4e shows a typical TEM image for a 17 nm InP nanowire on Si. No dislocations were observed along the 450 nm length of the nanowire.

Figure 5.5a shows the SEM picture of InP nanowires grown on a GaAs(111)B substrate with a mixture of nominally 20 and 60 nm Au nanoparticles. With 3.9 % lattice mismatch, well-aligned, epitaxial nanowires are obtained with both 20 and 60 nm nanoparticles. In the 120 nm Au nanoparticle region (shown in Fig. 5.5b), branching is again observed. The critical diameter in this case is 96 nm. For GaP nanowires on Si, epitaxial nanowires could grow even with the largest Au particles used, more than 100 nm in diameter. For the homoepitaxial InP/InP experiment, epitaxial nanowires as wide as 480 nm were obtained. For both experiments, it is confirmed that wide epitaxial nanowires can be synthesized, with no critical dimension constraint observed within the Au nanoparticle size limits used.

Figure 5.6 summarizes these growth results. The black dashed curve in this figure is the theoretical curve of the misfit-dislocation-free critical diameter taken from [193] for comparison. When the nanowire diameter is less than critical, the nanowire will be coherent everywhere solely via lateral relaxation. The epitaxial critical diameters experimentally obtained were fitted by the solid blue curve. It shows a similar trend to the black curve, but the magnitude is approximately two times larger for a given lattice mismatch. The discrepancy should mainly be due to the fact mentioned in the previous section: Ertekin *et al.* [193] considered only edge dislocations, while Fig. 5.3 shows that the critical dimensions for 60° dislocations are about twice larger. Also, the models discussed now apply to the case of two mismatching cylinders rather than a cylinder on a semi-infinite mismatching substrate. The latter case will be considered in detail shortly.

We now present similar data reported in [235] for the Au-catalyzed MBE growth of different III-V nanowires. Growth experiments were carried out in EP1203

Fig. 5.4 20°-tilt SEM pictures for nanostructures grown on Si(111) substrates. **a** InAs nanowires with nominal 20 nm Au nanoparticle catalysts. **b** InP nanowires grown with nominal 20 nm Au nanoparticles. **c** and **d** Star-like InP structures using nominal 60 (**c**) and 120 nm (**d**) Au nanoparticles, respectively. **e** TEM image of a 17 nm diameter InP wire showing clear planes perpendicular to the growth axis with 3.4 Å spacing. No dislocations were observed along the entire 450 nm nanowire length [232]

and Riber 32P MBE reactors equipped with the effusion Au cell, on the Si(111) and GaAs(111)B substrates. After the desorption of an oxide layer in the MBE growth chamber, a 100 nm thick GaAs buffer layer was grown on the GaAs(111)B substrate. In the case of Si(111) substrates, the buffer layer was not grown at all. To promote the nanowire formation by the growth catalyst, the deposition of ~0.5 to

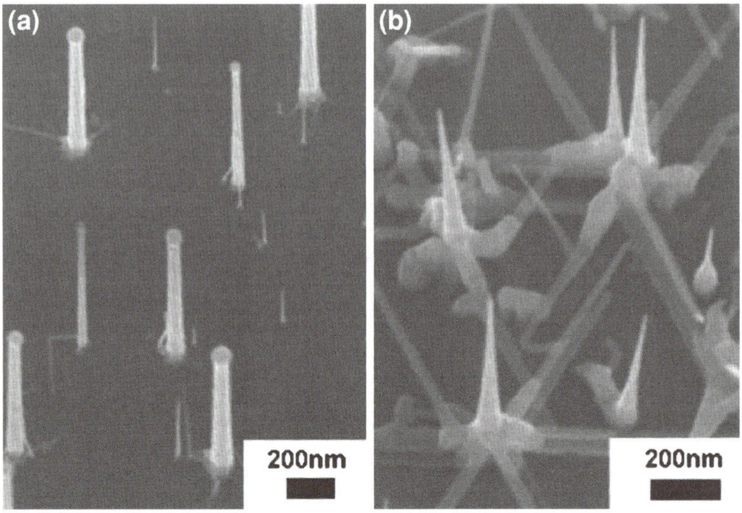

Fig. 5.5 20°-tilt SEM pictures for InP nanowires grown on a GaAs(111)B substrate with **a** a mixture of nominal 20 and 60 nm Au nanoparticle catalysts. **b** Region with nominal 120 nm Au catalysts. The branching of nanowires is clearly seen [232]

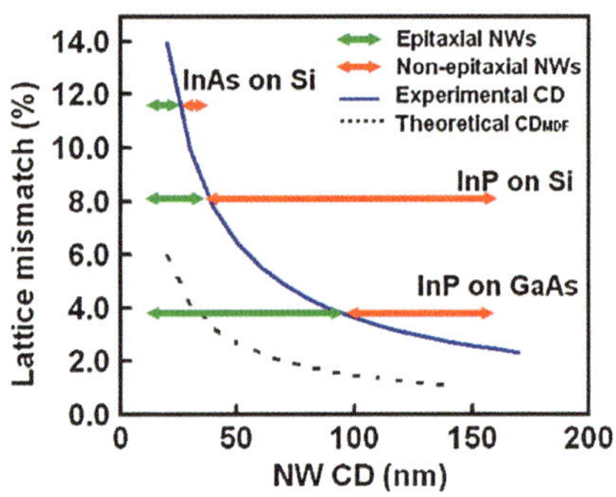

Fig. 5.6 Experimental nanowire critical diameters (solid line) and theoretical misfit-dislocation-free critical diameter from [193] (dashed line), as function of lattice mismatch. Green arrows denote that epitaxial nanowires within this size range were observed. *Red arrows* denote that Au nanoparticles within this size range were used as catalysts but no corresponding epitaxial nanowires could be found. Hence the boundary between a *green arrow* and a *red arrow* is the experimental critical diameter for that particular lattice mismatch

1 nm thick Au layer was performed. The samples were then annealed in order to form liquid drops of alloy of Au with the semiconductor material of the substrate. The MBE growth of different III-V nanowires was carried out by the conventional MBE. The substrate temperature, the V/III fluxes ratio and the deposition rate were varied depending on the material/substrate combination. After growth, the samples were studied by SEM.

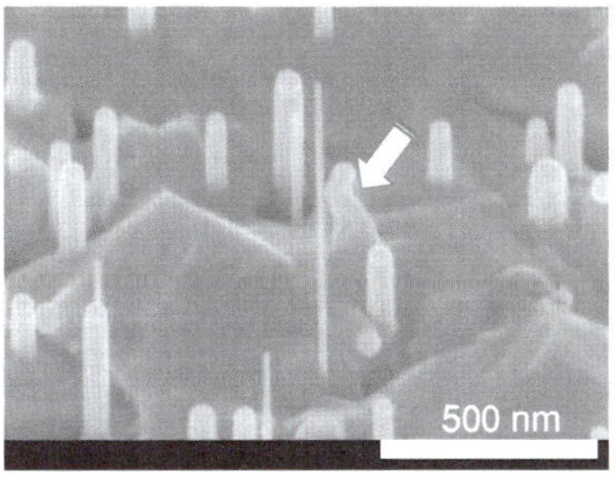

Fig. 5.7 SEM image of InP nanowires on Si(111) substrate. The growth temperature and InP deposition thickness amount to 360 °C and 100 nm, respectively

In the case of InAs/Si(111) system, the nanowires are very thin (typically 10–20 nm in diameter), but can be grown up to several micrometers in length. Maximum diameter for coherent epitaxial nanowires does not exceed 24 nm. If the diameter is larger, nanowires grow with dislocations or do not develop at all. Another important observation is a very narrow range of growth temperatures where these wires can be formed. This window has been found to extend only from 320 to 340° C. For InP nanowires grown on Si(111), the measured critical diameter equals 39 nm. Above this critical value, nanowires either do not grow at all or bend out of shape due to dislocations, as demonstrated by the highlighted wire in Fig. 5.7. The range of substrate temperatures for nanowire growth slightly widens compared to the previous case, typically extending from 330 to 360° C. Similar performance is observed for InAs wires on the GaAs(111)B substrates ($\varepsilon_0 = 7.1\,\%$). The critical diameter increases to 44 nm, while the corresponding temperature range extends to 370–420 °C. The largest critical diameter of 106 nm corresponds to the case of GaAs/Si(111) material system with $\varepsilon_0 = 4.0\,\%$. The temperature window for this system is much wider (380–540 °C). These results are summarized in the MBE row of Table 5.1.

The data in Table 5.1 and Fig. 5.6 clearly demonstrate a rapidly decreasing critical dimension with increasing ε_0, following the general tendency seen already from Fig. 5.3 for a cylindrical geometry. We note, however, that elastic relaxation in a nanowire on lattice mismatched substrate should be less efficient than in a confined axial heterostructure, because the latter has additional lateral surface to relieve strain. This feature follows from comparing the values of A in (5.40) for the elastic relaxation $z(\eta)$, which equals 27.4 for a cylindrical heterostructure and only 15.0 for a cylinder on rigid substrate (see Fig. 3.14). It is noteworthy that the experimental critical dimensions are close for nanowires obtained by two different epitaxy techniques. This fact supports the validity of our equilibrium approach and shows that the kinetic growth effects do not significantly influence the critical dimension for plastic deformation.

We now present theoretical analysis of critical dimensions for coherent growth of differently shaped nanostructures considered in Sect. 3.3: full cone, truncated cone, cylinder and reverse truncated cone [183]. With the simplest approximation of the elastic stress relaxation given by (3.40) (this choice was justified in Sect. 3.3), the elastic energy in these isotropic geometries can be put in the form

$$W_e = \frac{E}{(1-\nu)} z(\eta) V \varepsilon_0^2 \cong \frac{E}{(1-\nu)} \frac{\pi R^2 H \varepsilon_0^2}{(1+A\eta)} \frac{(a^2+a+1)}{3}. \tag{5.7}$$

Here, the total volume of a nanostructure is represented by $V = (a^2 + a + 1)\pi R^2 H/3$, with a equal to 1 and 0 for a cylinder and a full cone, respectively. For a more complex truncated geometry, a is just the ratio of the upper radius to the base radius, with $a < 1$ corresponding to a truncated cone and $a > 1$ to a reverse truncated cone.

Using (5.3) for the energy of the same system with two orthogonal dislocations at the base, the excess energy of the system with a dislocation pair with respect to a fully coherent state is obtained as

$$\Delta W(R,H) = \frac{ER}{1-\nu}\left[z\left(\frac{H}{2R}\right)\pi R H\left(\frac{\alpha^2 b_{eff}^2}{4R^2} - \frac{\alpha b_{eff}}{R}\varepsilon_0\right)\frac{(a^2+a+1)}{3} + C\left(\ln\frac{\bar{h}(R,H)}{b}+1\right)\right]. \tag{5.8}$$

Here, $\bar{h}$ can be again obtained from (5.2), while C and α are the same coefficients as in (5.4), which is now generalized to the case of nanostructures of different shapes seated on a lattice mismatched substrate. As in Sect. 5.1, a coherent state is stable at $\Delta W > 0$, whereas at $\Delta W < 0$ the dislocations are introduced. Equation $\Delta W(R,H) = 0$ therefore defines the critical thickness H_c for the onset of plastic deformation as a function of the base radius R at a given mismatch. From (3.40), $\eta Z(\eta) \to 1/A$ when $\eta \to \infty$ regardless of a particular nanostructure geometry. Using this asymptote in (5.8), we readily obtain the transcendent equation for the critical dimension of the form

$$\frac{2\pi(a^2+a+1)}{3A}\left(\frac{\alpha^2 b_{eff}^2}{4} - \alpha b_{eff}\varepsilon_0 R_c\right) + C\left(\ln\frac{2R_c}{\pi b}+1\right) = 0, \tag{5.9}$$

which generalizes (5.6).

The critical dimensions for plastic relaxation in nanostructures of different geometries, calculated as a function of lattice mismatch from (5.9) (in the case of 60° dislocations), are shown in Fig. 5.8. Squares and triangles correspond to the experimental critical diameters for cylindrical III-V nanowires from Table 5.1. It is seen that our calculations at $A = 15$ are in a good quantitative correlation with the experimental data. Fig. 5.9 shows the diameter dependence of the critical thickness for plastic deformation in a cylindrical nanowire on a mismatched rigid substrate, for different values of ε_0 at fixed $\nu = 1/3$ and $b = 2$ nm, in the case of 60° dislocations. As discussed, the critical thickness tends to infinity at the critical diameter, below which an infinitely long coherent nanowires can be grown. The lines in Fig. 5.9 separate

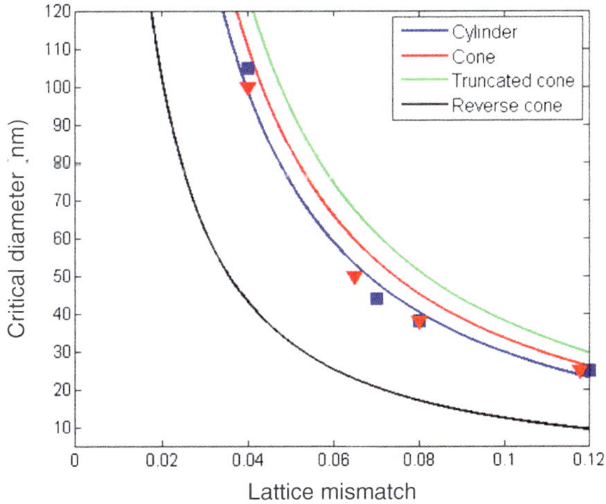

Fig. 5.8 Critical diameters for plastic deformation in different nanostructure geometries, plotted against the lattice mismatch, in the case of 60° dislocations with $b = 2$ nm. The dots represent experimental results of [232] (*triangles*) and [235] (*squares*)

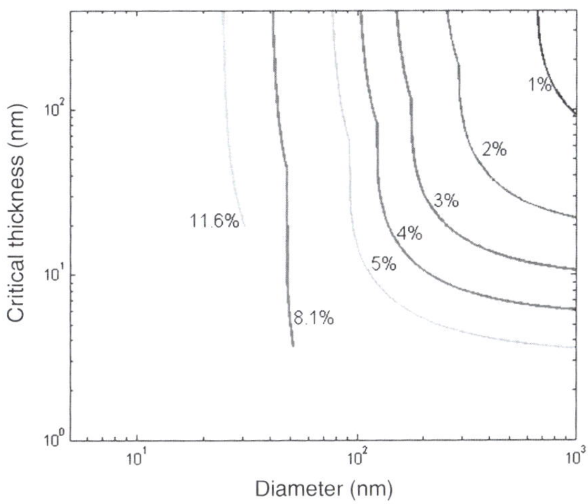

Fig. 5.9 Radius dependences of critical thickness in cylindrical nanowires for different lattice mismatches ε_0. Curves at $\varepsilon_0 = 4\,\%$, $8.1\,\%$ and $11.6\,\%$ relate to GaAs, InP and InAs nanowires on Si(111) substrates, respectively

the thickness-radius plane into two domains, the elastic domain (left and below) and the plastic domain (above and right). The values of lattice mismatch correspond to the case of GaAs nanowires on the Si(111) substrate ($\varepsilon_0 = 4\,\%$), InP nanowires on the Si(111) substrate ($\varepsilon_0 = 8.1\,\%$) and InAs nanowires on the Si(111) substrate ($\varepsilon_0 = 11.6\,\%$).

As for general tendencies regarding the critical base dimensions for plastic relaxation, we would like to point out the following. As discussed in Sect. 3.3, the strain relief is most efficient in a reverse tapered cone. Such geometry is experimentally observed in some Ga-catalyzed GaAs nanowires [48], most probably because of droplet inflation under excess Ga flux. This reverse tapering can be therefore considered as the major advantage: it does not extend the dimension of base heterointerface which would remain dislocation-free if its initial diameter is smaller than critical and, on the other hand, enables increasing the material volume towards the wire top which is crucial for the optical gain. Cylindrical wires exhibit less efficient strain relaxation. The critical base dimensions for their plastic relaxation are of the order of several tens of nanometers, which is within the range of diameters of III-V nanowires typically grown on Si. This explains the presence of some misfit dislocations at the bases [487]. Conical geometry is the worst regarding the elastic relaxation, i.e., the critical diameter for the formation of dislocations at their base is the smallest, and rapidly decreases as the taper angle increases. However, III-V nanoneedles grown on roughened Si and planar sapphire substrates, discussed in Sects. 3.14 and 3.15 emerge as tiny nanoislands with only few nanometers in diameter, which is why they should not contain dislocations initially. As shown in Sect. 3.4, the energetically preferred aspect ratio of the order of ten is acquired at the short scale nucleation stage and maintained throughout the subsequent growth stage by growing in a core-shell fashion. Sooner or later, such growth would develop dislocations at the base, as demonstrated by the experimental results discussed in Sect. 3.14.

5.3 Core-Shell Nanowire Heterostructures

We now consider radial, or core-shell, nanowire heterostructures depicted in Fig. 4.14b. High material quality core-shell nanowires with shell thickness exceeding 2D critical values for strain relaxation via dislocations were discussed in Sect. 4.1 and more recently reported also by Kavanagh *et al.* [488] in the case of InAs/GaAs. Another example is core-shell III-V tapered nanopillars considered in Sect. 3.14. Dayeh *et al.* [489] carried out a detailed in situ investigation of coherency limits and dislocation development in Si/Ge nanowires finding a factor of 3 increase over the 1 nm equilibrium critical thickness for planar Si/Ge heteroepitaxial growth roughening and grain formation with further shell growth.

Theoretically, Raychaudhuri and Yu studied critical dimensions in coherently strained coaxial nanowire heterostructures for two coaxial cylinders [490]. They considered two dislocation types with perpendicular Burgers vectors, one of which is tangential to the core-shell interface cross-section and the other is parallel to the cylinder axis. The existence of non-basal dislocation has been since proved experimentally [488, 489, 491, 492]. The results of [490] include a certain critical core radius and a critical shell thickness. When both values are exceeded, the dislocation formation is expected to onset. However, the model used assumes lattice constants invariable with position, which is almost certainly not the case. Trammell *et al.* [493] used a more complex model with variable lattice constants, in a coaxial geometry. It should be noted that pressure vessel theory as presented in [494, 495] does account for axial stress, unlike [493] seems to suggest. Experimental data is presented (and, in fact, the model by Liang *et al.* [494] used) by Goldthorpe, Marshall, and McIntyre [496], who carried out investigation on dislocation formation of Ge/Si nanowires. Interestingly, the authors found that loop dislocations appear more easily, which contradicts [490]. The effects may be partly due to periodic width modulation of the Ge core reported in [496], which would indeed favor loop dislocation formation, however our results predict that loop dislocations indeed appear easily even without width modulation. Unlike previously mentioned theoretical results, which used mostly analytical techniques, Søndergaard *et al* [497] applied the finite element method, which allowed them to study strain field of hexagonal core-shell nanowires in great detail. However, these authors did not study dislocation formation. The same can be said about the work by Grönqvist *et al.* [498] using even more rigorous atomistic valence force-field approach. The results obtained by valence force-field and continuum elastic model agree remarkable well. The authors also studied finite length wires. Along with experimental data on both WZ and ZB InAs/GaAs core-shell nanowires Kavanagh *et al.* [488] presented molecular statics calculations on relaxation in axial and radial directions. The model presented considers square and hexagonal nanowires of infinite length and different directions of dislocations, but does not deal with the critical shell thickness.

We now present our own theoretical analysis of elastic and plastic relaxation in the coaxial cylindrical geometry, which can be then generalized to include tapered geometry of core-shell nanopillars [499]. The simplest case of two long coaxial cylinders is shown in Fig. 5.10a. Our analysis is based on the known solution for a single hollow cylinder subjected to internal and external pressure [495]. Introducing the cylindrical coordinates (r, φ, z) with the axis z directed along the cylinder length, the governing equation of linear elasticity theory for the radial displacement field is reduced to

$$\frac{\partial^2 u_r}{\partial r^2} + \frac{1}{r}\frac{\partial u_r}{\partial r} - \frac{u_r}{r^2} = \frac{\partial}{\partial r}\left[\frac{1}{r}\frac{\partial}{\partial r}(r u_r)\right] = 0. \tag{5.10}$$

We assume that the core and the shell have identical Young's modulus and the Poisson's ratio ν and that both the base and the top of the cylinder are flat. The boundary conditions are given by

$$\begin{aligned} u_{sr}\,|_{r=R_0} - u_{cr}|_{r=R_0} &= \varepsilon_0 R_0; \\ u_{sz}\,|_{z=0} - u_{cz}|_{z=0} &= 0; \\ u_{sz}\,|_{z=L} - u_{cz}|_{z=L} &= \varepsilon_0 L, \end{aligned} \tag{5.11}$$

where the indexes "s" and "c" denote the shell and the core, respectively. The ε_0 is the lattice mismatch, and R_0, R, and L are the inner core radius, the outer shell radius and the cylinder length, respectively.

Using the results given in [495] for a single cylinder, we arrive at the generalized solution for the shell strain tensor ε and the corresponding stress tensor σ at a given pressure at the core-shell interface p:

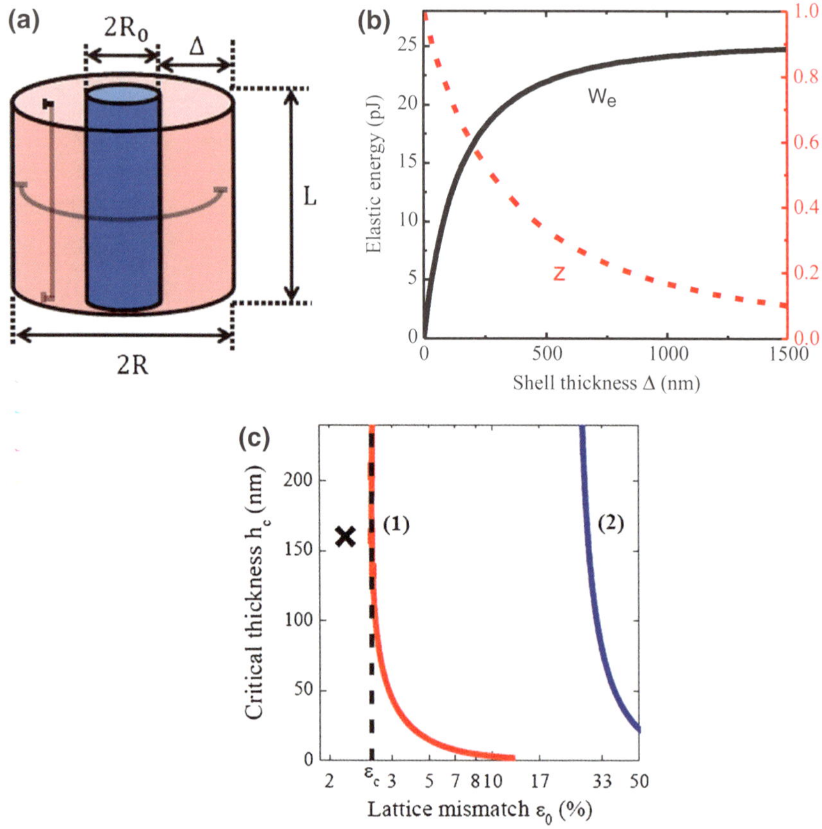

Fig. 5.10 **a** Coaxial cylinder model showing two possible types of dislocation, R_0 is the core radius, Δ is the shell thickness, L is the nanopillar length, and R is the shell outer radius. **b** Total elastic energy W_e (*left axis*) and relative elastic energy z (*right axis*) as a function of the shell thickness Δ in the cylindrical model with the material parameters corresponding to the $In_{0.2}Ga_{0.8}As/GaAs$ heterointerface (2 % mismatch in WZ structure). **c** The critical thickness h_c defined by the circular (*1*) and edge (*2*) dislocation formation versus the lattice mismatch ε_0. The cross symbol shows the experimentally achieved dislocation-free thickness at 2 % mismatch; the critical mismatch $\varepsilon_c \approx$ 2.7 % corresponds to infinite critical thickness

$$\varepsilon_{rr} = \frac{(1+\nu)\,C^2R^2}{E\left(R^2-R_0^2\right)}\left\{-\frac{p}{r^2}+(1-2\nu)\,\frac{p}{R^2}\right\} - \nu\varepsilon_0\frac{R_0^2}{R^2};$$
$$\varepsilon_{\varphi\varphi} = \frac{(1+\nu)\,R_0^2R^2}{E\left(R^2-R_0^2\right)}\left\{\frac{p}{r^2}+(1-2\nu)\,\frac{p}{R^2}\right\} - \nu\varepsilon_0\frac{R_0^2}{R^2};$$
$$\varepsilon_{zz} = \varepsilon_0\frac{R_0^2}{R^2}; \qquad (5.12)$$

$$\sigma_{rr} = \frac{pR_0^2}{R^2 - R_0^2}\left(1 - \frac{R^2}{r^2}\right);$$

$$\sigma_{\varphi\varphi} = \frac{pR_0^2}{R^2 - R_0^2}\left(1 + \frac{R^2}{r^2}\right);$$

$$\sigma_{zz} = \frac{2\nu pR_0^2}{R^2 - R_0^2} + \varepsilon_0 E\frac{R_0^2}{R^2}.$$

Both ε and σ are the diagonal matrices due to axial symmetry of the problem. The core stress and strain are calculated in a similar way. To obtain the final solution, we need to find p by using the boundary condition given by (5.11). The result is obtained in the form

$$p = E\varepsilon_0\frac{R^2 - R_0^2}{2\,(1-\nu)\,R^2}. \tag{5.13}$$

The total elastic energy W_e can now be computed as

$$W_e = \int_0^L dz \int_0^{2\pi} d\varphi \int_{R_0}^{R} \frac{\varepsilon_{rr}\sigma_{rr} + \varepsilon_{\varphi\varphi}\sigma_{\varphi\varphi} + \varepsilon_{zz}\sigma_{zz}}{2} rdr. \tag{5.14}$$

To compare the core-shell geometry with the thin film case, we consider the elastic energy in a uniformly strained 2D layer of the same volume, grown on a semi-infinite substrate with the same lattice mismatch, the Young's modulus and the Poisson's ratio:

$$W_{2D} = \frac{E}{1-\nu}\pi\left(R^2 - R_0^2\right)L\varepsilon_0^2. \tag{5.15}$$

The relative elastic energy, or the elastic stress relaxation, is defined as before, $z = W_e/W_{2D}$. Let us now consider z as a function of shell thickness $\Delta = R - R_0$. In the limit case where Δ tends to infinity, the elastic relaxation has the following asymptotic behavior:

$$z = \frac{\left(1-\nu^2\right)R_0^2}{\Delta^2} + O\left(\frac{R_0^3}{\Delta^3}\right) \quad \Delta \to \infty, \tag{5.16}$$

where $O(x)$ denotes a quantity of the order of x. This result for a very thick shell shows that the elastic stress relaxation scales as $1/\Delta^2$ due to core-shell geometry. Therefore, the total elastic energy saturates at large Δ:

$$W_e = \pi E\varepsilon_0^2\,(1+\nu)\,R_0^2 L + O\left(\frac{R_0}{\Delta}\right) \quad \Delta \to \infty, \tag{5.17}$$

while it would scale as Δ^2 in the 2D case. Equation (5.17) shows that the elastic stress in the shell is localized near the interface and is almost fully relaxed when the shell radius approaches the quantity of the order of core radius. While the internal part on the shell is stressed as it would be in the planar case, the strain in the outer regions is effectively released at free outer surface of the shell. This behavior is well illustrated by Fig. 5.10b showing the total and relative elastic energy as functions of the shell thickness. The material parameters used in calculations relate to the GaAs/$In_{0.2}Ga_{0.8}As$ heterointerface: E = 86 GPa, $\nu = 0.31$ with $\varepsilon_0 = 2\,\%$. The nanowire length L is fixed at 5 μm, and the core radius R_0 is 300 nm.

Such an effective elastic relaxation means that the dislocations might be effectively suppressed in the core-shell geometry. Since the elastic energy must scale linearly with the layer thickness in a planar geometry, it would reach the energy of a dislocated interface at a certain critical thickness. In contrast, the elastic energy saturates at a certain value in the core-shell geometry. This value might be greater or smaller than the energy of a dislocated interface. As in the previous sections, this feature naturally leads to the existence of a certain critical mismatch parameter for the given core radius below which the dislocation formation is entirely suppressed on energetic grounds.

To estimate the critical shell thickness in the core-shell geometry, we now consider two types of dislocations: the edge dislocation with the Burgers vector pointing along the tangential direction and the loop dislocation with the Burgers vector in the axial direction (see Fig. 5.10a). The dislocation formation would relax either the tangential strain $\varepsilon_{\theta\theta}$ or the edge strain ε_{zz}. With this choice of dislocation types, we compare the total energy of dislocated system with the elastic energy of the same system without dislocation. As in the previous sections, this is equivalent to comparing the pure dislocation energy of dislocation W_d with the elastic energy gain upon the dislocation formation due to the decrease of lattice mismatch ΔW_e. We use the results presented in [500] and [501] for the pure dislocation energy W_d for the two dislocation types considered. The ΔW is obtained from the results presented above by decreasing the lattice mismatch in horizontal and vertical planes due to nucleation of the edge and loop dislocations, respectively. The critical thickness is then found by setting $W_d = \Delta W_e$ separately for the edge and loop dislocations.

The results obtained are shown in Fig. 5.10c, where the two major effects are demonstrated. First, the values of critical thickness induced by the edge and loop dislocations feature a rather similar behavior, but the edge dislocations result in a much higher critical thickness. We can therefore conclude that the critical thickness is determined by nucleation of the loop dislocation. Second, the critical thickness becomes infinite at a certain critical lattice mismatch ε_0. Our estimates yield the critical mismatch of about 2.7 % for the $In_{0.2}Ga_{0.8}As$/GaAs system at a core radius of 300 nm, the material system considered in Sect. 3.14.

To develop a more realistic model of tapered nanopillars of Sect. 3.14, we consider coaxial cone geometry shown in Fig. 5.11a in the small slope approximation, where the first order perturbation theory in the taper angle is used. Now, the quantities R_0, p, ε and σ in (5.12) and (5.13) are treated as functions of z. While the analytical expressions are much more complicated in this case, the final results remain qualitatively the same as in the cylindrical case. The major effect regarding the efficient

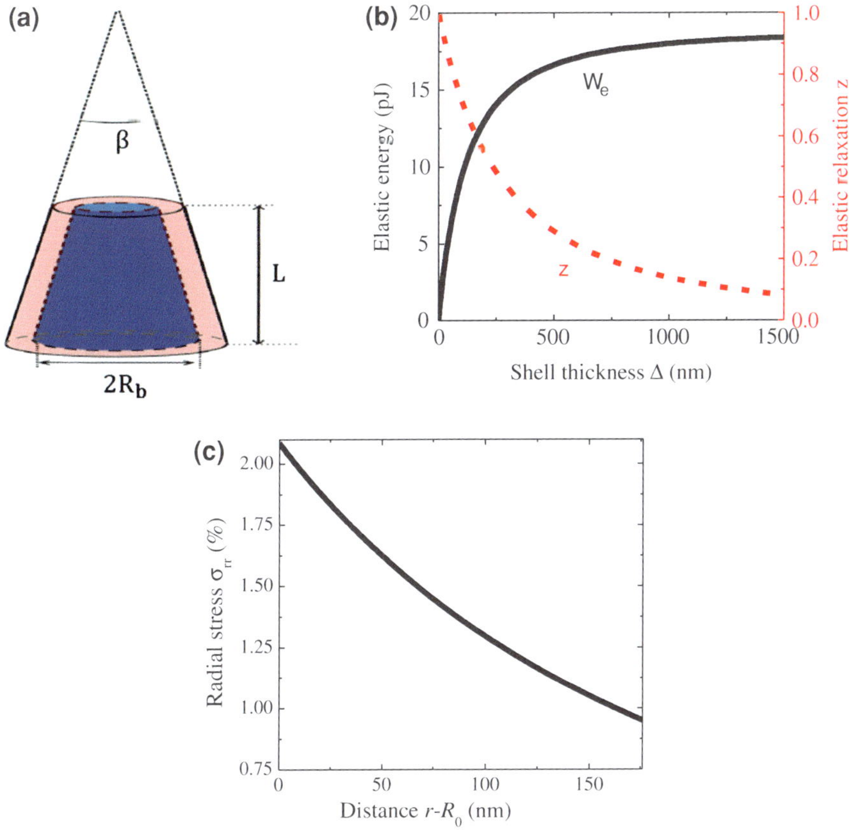

Fig. 5.11 **a** Truncated cone model with R_b as the core radius at the base, L as the nanopillar length, and β as the taper angle. **b** Relative elastic energy z as a function of the shell thickness R_0 with the material parameters corresponding to the $In_{0.2}Ga_{0.8}As/GaAs$ system. **c** Radial elastic stress in the shell σ_{rr} as a function of radial coordinate for the parameters described in text

stress relaxation is demonstrated in Fig. 5.11b and c. In calculations, we consider a nanopillar with a length of 2 μm and a taper angle of 5°, the core radius of 300 nm at the base, with all other parameters the same as before. Comparing Fig. 5.11b to Fig. 5.10b, it is seen that the elastic energy relaxation in the nanopillar is very close to the cylindrical case, so that the details of only slightly tapered nanopillar geometry do not matter too much. Fig. 5.11c presents the radial strain σ_{rr} depending on the distance from the core $r - R_0$. The core radius at the cross-section considered is 220 nm and the shell thickness is 160 nm. It is seen that the strain gradually decreases down to less than 1 % at the outer core interface, which correlates with the experimental observations described in Sect. 3.14.

Finally we note that the core-shell geometry is even more advantageous in terms of elastic energy relaxation than the laterally confined axial heterostructures in freestanding nanowires. Indeed, while nanowires do enable an infinite coherent growth below a certain critical diameter, their relative elastic energy asymptotically scales inversely proportional to the layer thickness, whereas in the core-shell geometry the relative elastic energy is inversely proportional to the squared layer thickness. This important property follows directly from our asymptotic analysis resulting in (5.16).

To conclude this section, the core-shell geometry enables a very efficient elastic stress relaxation at the interface, resulting in a radical increase of critical thickness as compared to the thin film case. Taking $In_{0.2}Ga_{0.8}As$/GaAs core-shell nanopillaes of Sect. 3.14 as an example, this stress relaxation enables growing a coherent 160 nm thick shell without any dislocations in spite of 2 % lattice mismatch that would result in the 10 nm critical thickness in 2D limit. The model presented here explains the observed effect and shows that the stress relief in the core-shell structures is even more efficient than in axial nanowire heterostructures. The obtained value of critical lattice mismatch at 300 nm core radius is 2.7 %, which is larger than in the material system considered. This would enable growing even thicker shells if required.

5.4 Cylinder-to-Cone Shape Modification

As follows from Fig. 4.64, the Au-catalyzed GaAs nanowires grown by MBE on GaAs(111)B substrates at low temperatures below 500 °C develop a specific conical shape so that their radius increases toward the base and is no longer defined by the size of catalyst droplet, while at higher temperatures their shape is cylindrical with a constant radius from base to top. Similar behavior is shown in Fig. 5.12 for MBE grown GaP nanowires [408] and in Fig. 5.13 for MOCVD grown InP nanowires [502] (the catalyst in both cases was Au). All these nanowires acquire conical shape at low temperatures, where the shape modification occurs due to the radial growth on the sidewalls. Radial growth is unwanted in most cases, because the control over lateral dimensions is lost. On the other hand, growth models described in Chap. 4 require modifications, since the assumption of $R = const$ is no longer valid.

While studying the diffusion of sidewall adatoms, one needs to introduce an additional adatom sink due to the radial (or lateral) wire growth, and to consider the corresponding kinetic equation for the radial growth at a given distance from the substrate. The sink terms in the diffusion equation for n_f have different forms depending on the mechanism of adatom incorporation. If the radial growth is due to a step flow, the appropriate term writes as $-n_f/\tau_{inc}$, i.e., is linear in concentration. When the radial growth is mediated by 2D nucleation on the sidewall facets, the sink can be described by $-D_f\alpha_f n_f^2$ (with D_f as the diffusion coefficient and α_f as the capture number) if the critical size is close to one (see Sect. 2.15). In a more general case, the nucleation-mediated sink is defined by the supersaturation-dependent Zeldovich nucleation rate $J(\eta)$, with $\eta = n_f/n_f^{eq} - 1$ as the adatom supersaturation.

We now discuss the model of [408] where the radial growth is assumed as being mediated by 2D nucleation. The wire considered is growing with an arbitrary

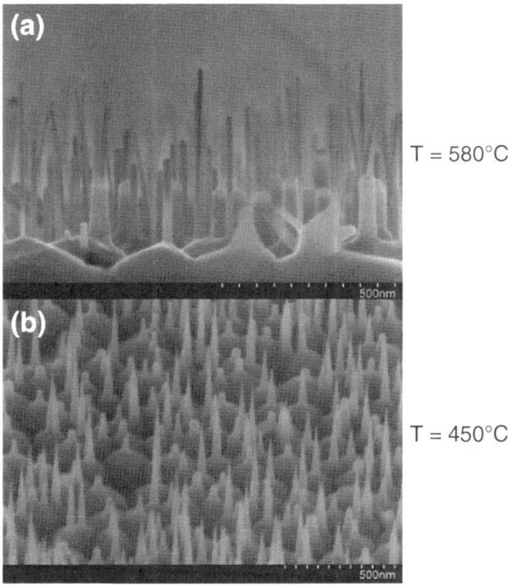

Fig. 5.12 SEM images of GaP wires grown by MBE at 580 °C (*cross view*) (**a**) and 450 °C (45° *tilted view*) (**b**) [408]

shape that can be characterized by radius $R(z, t)$ and length $L(t)$, where z is the distance from the surface layer ("height") and t is time (Fig. 5.14). Obviously, the point $z = 0$ corresponds to the wire base and $z = L(t)$ to the top. At $t = 0$, the droplet of base radius R_d is on the surface of a bare substrate and the wire length $L(t = 0) = 0$. We assume that the droplet base radius remains constant during growth, implying $R(z = L(t), t) = R_d$ at any time. The wire cross-section is assumed as being a circle or a regular polygon, in the latter case $R(z, t)$ is the radius of the circle inscribed in the polygon. Vertical and radial growths of the wire are described by the following equations:

$$\frac{dL}{dt} = V_L(L);\ L(t = 0) = 0; \tag{5.18}$$

$$\frac{dR(z, t)}{dt} = V_R(z, L);\ R(z = L(t), t) = R_d, \tag{5.19}$$

where V_L and V_R are, by definition, the vertical and radial growth rates, respectively. In the steady state, the radius R can be treated as function of z and L instead of z and t, by using the one-to one correspondence between L and t. In this case, (5.18) and (5.19) result in

$$R(z, L) = R_d + \int_z^L dL' \frac{V_R(z, L')}{V_L(L')}. \tag{5.20}$$

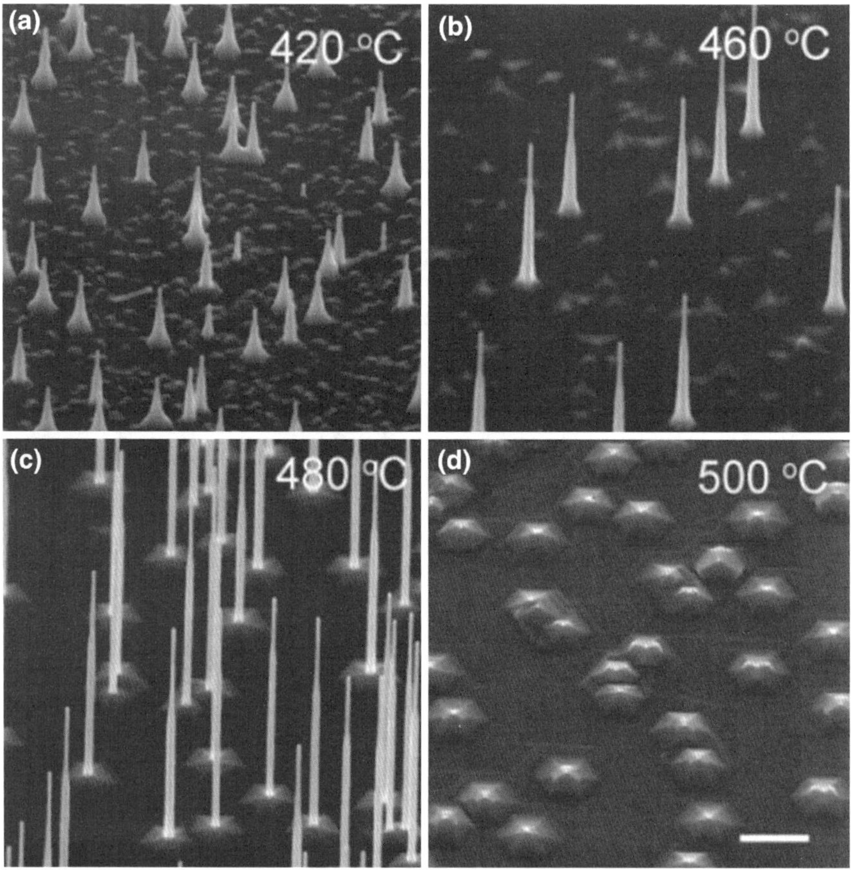

Fig. 5.13 SEM images of Au-assisted InAs nanowires grown by MOCVD at different temperatures: 420 °C (**a**), 460 °C (**b**), 480 °C (**c**), and 500 °C (**d**). The scale bar is 1 μm. For all samples the growth time was 8 min [502]

For the following analysis, we require some models for the vertical and radial growth rates. In the simplest approximation, the vertical growth rate in the diffusion-induced mode (see Sect. 4.8) depends on the equivalent deposition rate $V = I\Omega\cos\alpha_0$ (α_0 is the incident angle of the beam, see Fig. 5.14.), and the diffusion flux to the droplet as

$$V_L(L) = (\varepsilon - \xi_{des})V + \frac{\Omega}{\pi R_d^2} j_{diff}(L). \tag{5.21}$$

The parameter ξ_{des} is the rate of desorption from the drop in the units of V, Ω is the elementary volume in the solid phase and the parameter $\varepsilon = 1 - V_s/V$ accounts for the vertical growth of the surface layer at a rate V_s. The diffusion flux of adatoms to

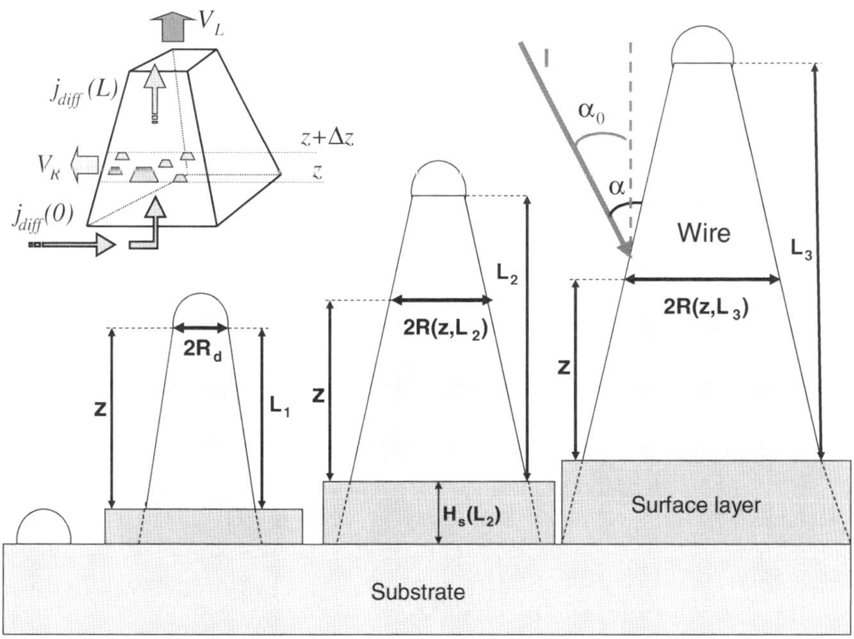

Fig. 5.14 Schematics of the growth model including the wire, the surface layer and the droplet; I is the impingement flux, α_0 is the incident angle of the beam to the flat surface and α is the local incident angle. The inset at the *top left* illustrates the polynuclear lateral growth of the side facet

the top, driven by the difference of chemical potentials on the surface layer and in the droplet, is defined by

$$j_{diff}(L) = -2\pi R_d D \left.\frac{dn}{dz}\right|_{z=L}. \tag{5.22}$$

Here, we omit the index "f" for the sidewall adatoms for brevity, i.e., D is the diffusion coefficient on the sidewalls and n is the local adatom concentration. As discussed in Sect. 4.8, the flux given by (5.22) generally consists of two fluxes, one resulting from the atoms directly impinging the sidewalls and the other formed by the atoms impinging the surface, then migrating to the wire base and finally to its top along the sidewalls. While considering the latter we assume that the flux to the wire base per unit length of the base $2\pi R_0 = 2\pi R(z = 0)$ remains constant during the growth. This is true if R_0 is much larger than the effective diffusion length of adatoms on the main surface. Our boundary condition should therefore hold for a sufficiently thick wire foot with $R_0 > 100$ nm, because the diffusion length on the surface is usually limited by the nucleation and is of the order of few tens of nanometers or even smaller (see the corresponding data of Sects. 4.11 and 4.13). Using the equation of continuity at the wire foot, we then obtain

$$-D\frac{dn}{dz}\bigg|_{z=0} = \frac{j_{diff}(0)}{2\pi R_0} = const. \tag{5.23}$$

The nucleation-mediated radial growth of the wire may proceed in the mononuclear or polynuclear mode, depending on the wire radius at a given height and the adatom supersaturation. For the sake of simplicity, also taking into consideration that the effect of sidewall nucleation is pronounced at high supersaturations, below we use the radius-independent expression for V_R in the polynuclear mode:

$$V_R(\eta) = Ch\Big[J(\eta)v^2(\eta)\Big]^{1/3}. \tag{5.24}$$

Here, h is the height of a monolayer, J is the nucleation rate, v is the lateral growth rate of a 2D island on the sidewalls, and C is a constant for which $C = 1$ is a good approximation. The functions J and v depend on the adatom supersaturation $\eta = n/n_{eq} - 1$. The equilibrium concentration of adatoms n_{eq} is determined by the temperature T, further assumed as being constant along the wire, which is always fulfilled for sufficiently short wires with the length of several μm [352]. As usual in classical nucleation theory, the dependence of V_R on η is mainly governed by a very steep exponential dependence of J on η:

$$J(\eta) \propto J_0 \exp\left[-\frac{A_*}{\ln(\eta+1)}\right]. \tag{5.25}$$

Here $A_* \equiv c\sigma^2 (\gamma_{SV}/k_B T)^2$ is a thermodynamic constant, σ is the area of adsorption site on the sidewalls, γ_{SV} is the surface energy of the newly formed solid-vapor interface of monolayer height around the island, and c is the island shape constant.

Steep exponential dependence of the nucleation rate on supersaturation allows one to simplify (5.24) by using the Γ- method described in Chap. 2:

$$V_R(U) \cong V_R(0)\exp(-U). \tag{5.26}$$

Here, the new unknown function is defined as $U = (\Gamma/3)[(\eta_0 - \eta)/\eta_0]$, while $V_R(0) = V_R(\eta_0)$ and $\eta_0 = \eta(z = 0)$ are the radial growth rate and the adatom supersaturation at the base, respectively. As usual, the parameter Γ has the same order of magnitude as the critical number of atoms in classical nucleation theory at η_0, and therefore is much larger than one. When the adatom diffusion flux is directed toward the wire top, η must decrease with the height. Therefore, we are looking for the solutions for U increasing with z.

Finally, the steady state kinetic equation for the adatom concentration $n(z)$ on the curved surface of local radius $R(z, L)$ writes down as

$$-\frac{d}{dz}\left[2\pi RD\frac{dn}{dz}\right] = 2\pi R\left[I_{eff} - \frac{n}{\tau_A} - \frac{V_R}{\Omega}\right]. \tag{5.27}$$

The left hand side describes the diffusion, the first term in the right hand side stands for the adsorption on the sidewalls, the second term describes the desorption from the sidewalls with a characteristic lifetime τ_A and the third term gives the adatom sink due to 2D nucleation. The effective impinging flux at height z in the case of directional deposition is determined as $I_{eff} = (V/\pi\Omega)\sin\alpha/\cos\alpha_0$, where $\alpha = \alpha_0 + \Delta\alpha$ is the local incident angle of the beam to the wire surface and $\Delta\alpha$ is the adjunct caused by the curvature of the surface as shown in Fig. 5.14. From geometrical considerations, the latter is given by $\tan\Delta\alpha = -dR/dz$.

Using (5.21) to (5.24), (5.26), (5.27) for n and (5.20) for $R(z, L)$ can be re-written in terms of the unknown U in the form:

$$\frac{d^2U}{dx^2} + \frac{1}{\rho}\frac{d\rho}{dx}\frac{dU}{dx} = -e^{-U} + \frac{V}{\pi V_R(0)}\frac{\sin(\alpha_0 + \Delta\alpha)}{\cos\alpha_0}; \tag{5.28}$$

$$U(x = 0) = 0; \left.\frac{dU}{dx}\right|_{x=0} = \sqrt{A}; \tag{5.29}$$

$$\rho(x, l) = 1 + \frac{e^{-U(x)}}{2}\int\limits_x^l \frac{dx'}{dU/dx' + \nu}. \tag{5.30}$$

Here, the normalized radius ρ is given by $\rho = R/R_d$, and the normalized vertical coordinates $x = z/L_*$; $l = L/L_*$ are expressed in the units of the effective diffusion length on the sidewalls

$$L_* = \left[\frac{3\Omega D n_{eq}\eta_0}{V_R(0)\Gamma}\right]^{1/2}, \tag{5.31}$$

limited by nucleation. Since, in the cases of interest, the value of L_* is always much smaller than the diffusion length of a single adatom on the sidewall surface, $\lambda = \sqrt{D\tau_A}$ (limited by desorption), we omit the $(L_*/\lambda)^2$ term in (5.28). The normalized diffusion flux to the wire base equals

$$\sqrt{A} = \frac{\Gamma}{3Dn_{eq}\eta_0}\frac{j_{diff}(0)}{2\pi R_0}L_*. \tag{5.32}$$

The local curvature of the surface in the right hand side of (5.28) is obtained from $\tan\Delta\alpha = -\frac{R_d}{L_*}\frac{d\rho}{dx}$. The constant $\nu = \frac{(\varepsilon-\xi_{des})VR_d}{2V_R(0)L_*}$ in (5.30) stands for the adsorption-desorption contribution to the vertical growth rate. In these non-dimensional coordinates, the wire shape is entirely determined by two constants, the diffusion flux to the base $\sqrt{A}$ and the coefficient ν. The solution to (5.28) for U with boundary conditions (5.29) is used in (5.30) for computing the wire shape. The non-linear equation for $\rho(x, l)$ in its general form can only be solved numerically.

Let us now consider the simplified version of (5.28) at a small curvature of the surface, putting the local incident angle α to α_0. The resulting two-parametric equation describes the diffusion with the constant rate of birth and the exponential rate of extinction at the fixed particle supplying flux at x = 0:

$$d^2U/dx^2 = -\exp(-U) + \beta$$
$$U(x = 0) = 0;\ U'(x = 0) = \sqrt{A}. \tag{5.33}$$

Here, the parameter $\beta = [V \tan \alpha_0]/[\pi V_R(0)]$ represents the direct impingement to the sidewalls, while $U'(0) = \sqrt{A}$, as above, stands for the adatom supply at the wire foot. Equation (5.33) has the first integral of the form

$$\frac{1}{2}\left(\frac{dU}{dx}\right)^2 + G(U) = \frac{A}{2} \tag{5.34}$$

with

$$G(U) = 1 - \beta U - \exp(-U). \tag{5.35}$$

We now show how the main properties of the system at different values of A and β can be analyzed by considering (5.34) and (5.35). It is noteworthy that (5.34) has the form of the energy conservation for a classical point "particle" with "coordinate" U, "time" x, "total energy" $A/2 = const$, "potential energy" $G(U)$ and a unit "mass". The total energy is defined by $U'(0)$ and the form of the potential depends on β. Initial conditions to (5.33) yield that our hypothetic particle has zero coordinate and the fixed velocity $U'(0)$ at zero time. As already mentioned, relevant solutions for $U(x)$ must increase with x, since the diffusion flux would change its direction otherwise. In terms of particle motion we are therefore searching for the cases when the particle, starting its motion at $U = 0$ with a given velocity, travels to point $U = \infty$ at $x \to \infty$. Such an unlimited motion takes place only if the total energy A/2 is larger than the potential barrier, which equals the maximum of $G(U)$ at $U = U_*$. The values of U_* and $G(U_*)$ are readily calculated by maximizing (5.35): $U_* = -\ln\beta$; $G(U_*) = 1-\beta+\beta\ln\beta$. This yields $U_* > 0$ at $\beta > 1$. The height of the potential barrier decreases with increasing β ; the barrier disappears at $\beta = 1$. Potentials $G(U)$ at different β are shown in Fig. 5.15. At a given β, the case of $A/2 < G(U_*)$ corresponds to the repulsion from the potential barrier at a turning point U_1 where $A/2 = G(U_1)$. After that, the particle will move to $U = -\infty$ Therefore, the condition $A_c/2 = G(U_*)$ determines the critical diffusion flux to the wire base $\sqrt{A_c}$, below which the wires cannot be grown by the surface diffusion. For fluxes lower than $\sqrt{A_c}$, the dominant fraction of adatoms cannot reach the wire top because of very efficient sidewall nucleation. The radial growth dominates over the vertical one and the crystals grow in the form of islands rather than wires. The resulting morphology is a corrugated laterally overgrown surface instead of expected ensemble of elongated structures. The curve $\sqrt{A_c} = \sqrt{2(1-\beta+\ln\beta)}$ in the $(\sqrt{A}, \beta)$ plane, shown in Fig. 5.16, separates the region of wire growth (above the curve) from the islanding region

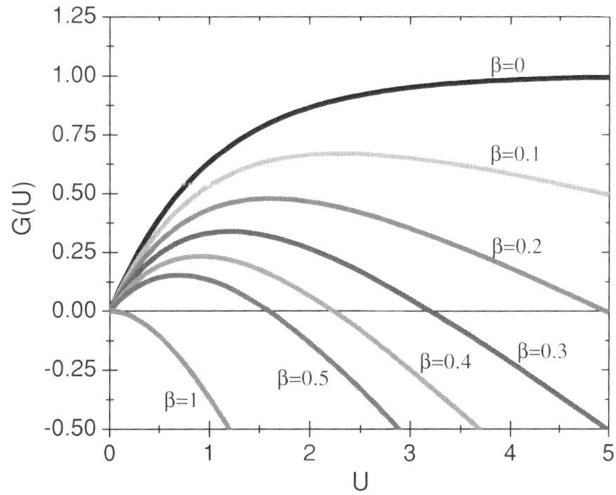

Fig. 5.15 Forms of "potential" $G(U)$ at different β

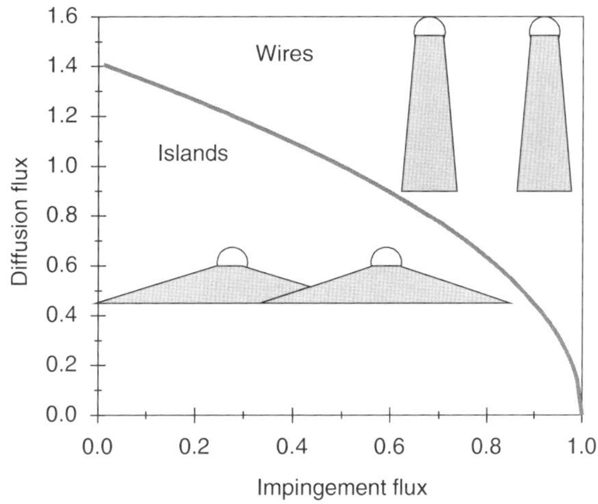

Fig. 5.16 Kinetic phase diagram illustrating the domains for the preferred island and wire growth

(below the curve). The decrease of the critical flux with β is well understood intuitively, because the direct impingement to the sidewalls at the wire top promotes the vertical growth.

The wire shape is obtained in a simple analytical form in the case $\beta = 0$, where the beam is strictly perpendicular to the surface or the direct impingement is much smaller than the surface diffusion flux. At $\beta = 0$, (5.33) can be integrated analytically:

$$U(x) = 2 \ln \left[\frac{\sinh(a(x + x_0))}{\sqrt{2}a} \right], \tag{5.36}$$

where the constants are given by

$$a = \frac{\sqrt{A-2}}{2};\ x_0 = -\frac{1}{2a}\ln\left(\frac{\sqrt{1/2+a^2}-a}{\sqrt{1/2+a^2}+a}\right). \tag{5.37}$$

Using (5.36) in (5.30) and integrating, we obtain the exact solution for the wire shape $\rho(x,l)$. Below we present the resulting expression only in the case of $\nu = 0$, where the diffusion induced contribution to the vertical growth rate is much larger than the flux directly impinging the wire top:

$$\rho(x,l) = 1 + \frac{1}{2\sinh^2[a(x+x_0)]}\ln\left[\frac{\cosh(a(l+x_0))}{\cosh(a(x+x_0))}\right]. \tag{5.38}$$

The wire shape in normalized coordinates now depends on the only parameter $\sqrt{A}$. The cylinder to cone shape transformation, induced by the reduction in the supplying flux to the wire base, is demonstrated in Fig. 5.17. The critical diffusion flux now equals $\sqrt{2}$. According to the second (5.37), $x_0 \to \sqrt{2}$ at $a \to 0$, and (5.38) is reduced to

$$\rho_c(x,l) = \frac{3}{4} + \frac{1}{4}\left(\frac{l+\sqrt{2}}{x+\sqrt{2}}\right)^2. \tag{5.39}$$

This function, shown as a dash-dotted curve in Fig. 5.17, does not depend on any of system parameters and is determined entirely by the form of model equations. The critical curve describes the limit case of the thickest possible wire of a given length l that can be grown by the surface diffusion. The radius of the wire base $\rho(0,l)$ scales with length as l^2 at large l, so that the lateral growth is much faster than the vertical one. The growth of the wire having the shape (5.39) would require infinite time, i.e., is physically unreachable.

The temperature behavior of the wire shape is dictated by the temperature dependence of the diffusion flux and the effective diffusion length on the sidewalls. Although it is not critical, we restrict the following analysis to the case $\beta \to 0$ when the wire radius is given by (5.38). The boundary condition at the wire foot of the form (5.23) yields $j_0/(2\pi R_0) = \varepsilon V/(\Omega P_W)$, where P_W is the total perimeter of the wire bases per unit surface area. At a small desorption from the surface, the parameter ε can be estimated as $\varepsilon = P_W/(P_W + P_I)$, where P_I is the appropriately averaged perimeter of islands emerging in the surface layer. From (4.154) and (4.155), the latter depends on the growth temperature T and the deposition flux V as $P_I \propto V\exp\left[\frac{3\Lambda_s/2+E_D^s}{k_BT}\right]$, with Λ_s being the specific condensation heat and E_D^s the diffusion barrier for the surface adatoms. To estimate the unknown supersaturation at the wire base, we use the model of an amorphous crystal growing uniformly in all directions. In this approximation the lateral growth rate at $z = 0$ equals the vertical growth rate of the surface layer itself: $V_R(0) \cong V_s = (1-\varepsilon)V$. Using this in (5.24), (5.25) at $\eta = \eta_0$, and also that $J_0 \propto 1/\tau$, with τ as the characteristic time of island growth, with logarithmic accuracy we get $\eta_0 \cong \exp\left[\frac{A_*}{3\ln(h/((1-\varepsilon)V\tau))}\right]$. This exponent contains the ratio of two very large values: the thermodynamic

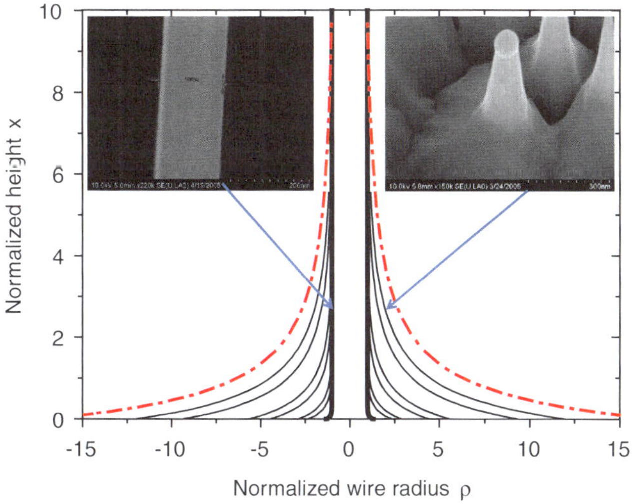

Fig. 5.17 The wire shape ρ *(x,l)* as given by (5.38) for fixed $l = 10$ and different $\sqrt{A} = 13$ (bold cylinder), 7.8, 4.5, 3.3, 2.4, 1.9, 1.5 and $\sqrt{2}$. The lateral growth rate increases as $\sqrt{A}$ decreases. The dash-dotted curve is the critical curve. The inserts show SEM images of the Au-catalyzed nanowires grown by MBE at $T = 550\,°\mathrm{C}$ (*left*) and 420 °C (*right*)

constant A_* to the logarithmic ratio of the macroscopic deposition time h/V and the microscopic time τ. The latter is defined as $\tau = t_D/\theta_{eq}$, where t_D is the diffusion time (such that the diffusion coefficient $D = \Omega/(ht_D)$, and θ_{eq} is the equilibrium adatom coverage of the surface at a given temperature.

Substitution of this expression for η_0 into (5.31) and (5.32), together with the conventional temperature dependences $t_D = \nu_D^{-1}\exp(E_D/k_BT)$ and $\theta_{eq} = \exp(-\Lambda/k_BT)$ leads to

$$L_* \cong \left[\frac{3\Omega\eta_0}{4(1-\varepsilon)\Gamma V\tau}\right]^{1/2} \propto \frac{1}{\sqrt{V}}\exp\left[-\frac{(E_D+\Lambda)}{k_BT}\right]; \tag{5.40}$$

$$\sqrt{A} \cong \frac{1}{L_* P_I} \propto \frac{1}{\sqrt{V}}\exp\left[-\frac{(E_D+2\Lambda)}{2k_BT}\right]. \tag{5.41}$$

When deriving these semi-quantitative expressions, we again assume that the characteristics of the surface layer and the sidewalls at the wire base are approximately the same. (5.40) and (5.41) demonstrate that both the diffusion length and the diffusion flux to the base decrease toward lower temperatures as the Arrhenius exponents, and scale with flux as $1/\sqrt{V}$.Therefore, the critical diffusion flux relates to a certain minimum temperature, below which the wires cannot be grown.

We now consider experimental data for the Au-assisted GaAs and GaP nanowires grown by MBE at different temperatures, shown in Figs. 4.64 and 5.12, respectively.

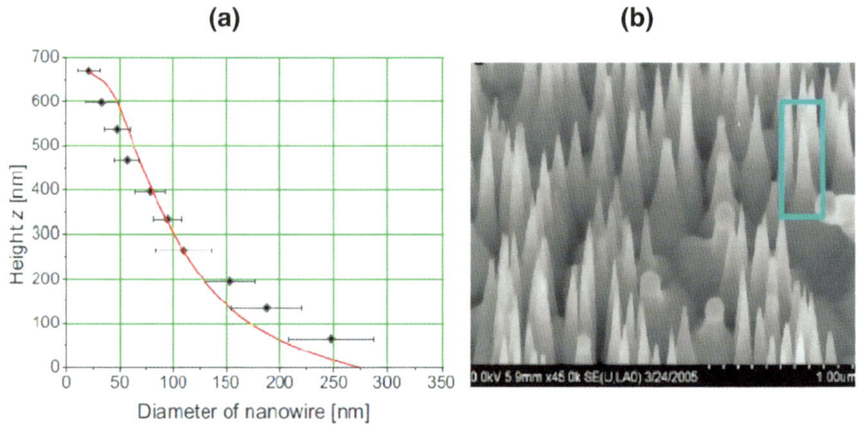

Fig. 5.18 **a** Experimental (*symbols*) and theoretical (*line*) wire shape $2R(z)$ for the GaAs wire shown in (**b**); $T = 420\,°\mathrm{C}$

As discussed, GaAs wires grow with approximately constant radius up to approximately 3 microns of their length. When T is below 500 °C, the wires adopt a specific conical shape, shown in Fig. 4.64a for $T = 420\,°\mathrm{C}$. The shape modification of GaP nanowires, demonstrated by Fig. 5.12, also occurs at T decreasing from 580 to 450 °C, although the droplet radius R_d is typically 3 times smaller. Figure 5.18a shows the typical example of experimental and theoretical wire shapes $R(z)$ at low T. The GaAs wire chosen for comparison with calculations is highlighted in the SEM image of Fig. 5.18b. The error bars indicate uncertainty in the measured $R(z)$ dependences of different wires with the same length, obtained from the statistical analysis of this SEM image. Theoretical curves are obtained from the numerical solution of (5.28) to (5.30). Normalized coordinates ρ, x are then transformed into R, z by fitting the value of L_* at the known R_d. The best fits to the experimental shapes are obtained at the values of parameters given in Table 5.2, where the MBE growth conditions are also presented. We find that $\nu \cong 0$ in (5.30) is a good approximation, i.e., the direct impingement to the drop can be neglected. In both cases, the best fit to the experimental data is obtained at low values of $\sqrt{A} = 1.13 - 1.2$, demonstrating that the growth parameters are quite close to the critical flux. This is consistent with our experimental findings, because regular epitaxial wires can hardly be observed at temperatures lower than 420 °C for GaAs and 450 °C for GaP. Finally, in all our simulations of different GaAs and GaP wires with a pronounced conical shape, the effective diffusion length L_* was not larger than 300 nm, i.e. at least ten times smaller than the diffusion length of Ga atoms on the sidewalls at 580 °C, limited by desorption.

Table 5.2 Growth conditions and fitting parameters of conical III-V nanowires

Material	Temperature (°C)	Top radius $R_d(nm)$	Length L (nm)	$\sqrt{A}$	β	Diffusion length L_* (nm)
GaAs	420	30	670	1.13	0.525	160
GaP	450	10	165	1.2	0.525	100

5.5 Growth Modeling of CdTe Nanowires

Here, we consider the Au-assisted growth of CdTe nanowires by close-space sublimation on Mo foils, the system interesting for photovoltaic applications. On the other hand, this VLS process involves a simultaneous vertical and radial growth at an almost uniform nanowire radius (with only a slight tapering toward the top), which requires a development of the growth models. It is well documented that the use of semiconductor nanowire arrays within photovoltaic devices may yield reduced manufacturing costs and superior performance. Indeed, improvements within each stage of solar cell operation—light absorption, charge generation, separation and collection—are viable [503]. For example, by using an array of core-shell radial p-n junctions, as opposed to a single planar junction, charge carriers can be separated more efficiently, thus electron hole recombination can be reduced and open-circuit voltages (V_{OC}) enhanced. CdTe cells are commonly limited by low values of V_{OC} (relative to the theoretical maximum), therefore nanowire structures are of particular interest for this material [504]. This explains the potential of CdTe nanowires on metal foils for photovoltaics. For such advantages to be fully realized however, it is critical that nanowire lengths, radii, spacing and shape are fully optimized. It is paramount therefore to have a full appreciation of the kinetic and thermodynamic processes involved during "bottom-up" nanowire growth and to understand how deposition conditions affect structure morphology.

The growth procedure of CdTe nanowires is the following [505]. Au films (5 nm thick) were deposited onto 0.1 mm thick Mo foils by thermal evaporation. Samples were then placed into the close space sublimation chamber, 10 mm above a powdered CdTe source. A static pressure of 25 Torr of N_2 was then introduced to the chamber and samples were annealed at 360 °C for 30 min to break the Au film up into an array of nanoparticles. For nanowire growth, the N_2 pressure was kept constant at 25 Torr, the CdTe source temperature was raised to $T_0 = 500$–600 °C and the substrate was held at temperatures T of 490–550 °C. Deposition took place for 10–60 min. Structural characterization was carried out using a Hitachi SU-70 FEG SEM.

Figure 5.19 shows a SEM image of a typical CdTe nanowire array grown on a Mo substrate. For this particular sample, deposition took place for 60 min in a N_2 pressure of 25 Torr, using source and substrate temperatures of 550 and 520 °C, respectively. The Au catalyst droplets appear brighter than the CdTe wires because the back-scattered electron detector was used which highlights elemental contrast. As the Au catalysts are located at the tips of the nanowires, growth was deemed to

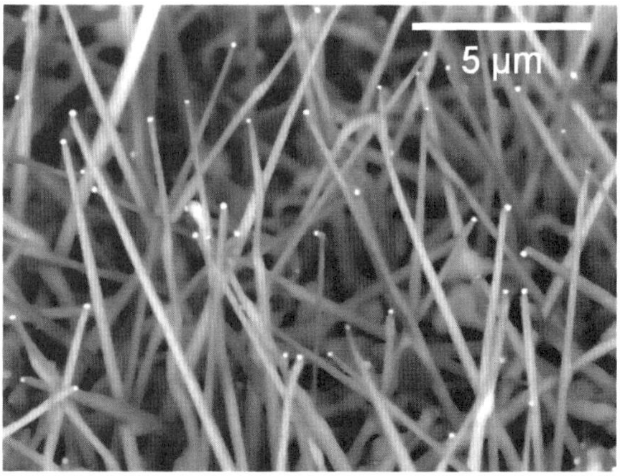

Fig. 5.19 Back-scattered electron image of a typical CdTe nanowire array grown on Mo foil

have proceeded via the VLS regime. For each sample, nanowire lengths (of up to 20 μm) and radii (in the range 100–400 nm) were measured using image measurement software on the SEM. Since some of our nanowires were slightly tapered, the diameters were averaged over their length. Average dimensions (calculated from the measurement of 100 nanowires from each sample) for samples grown at 550 and 520 °C are plotted as a function of growth time in Fig. 5.20.

In growth modeling of these CdTe nanowires, we use the conventional diffusion-induced model of VLS growth with one significant modification that takes into account the radial extension of cylindrical nanowires, as opposed to a conical shape considered in the previous section. As discussed in Chap. 4, the adatom concentration

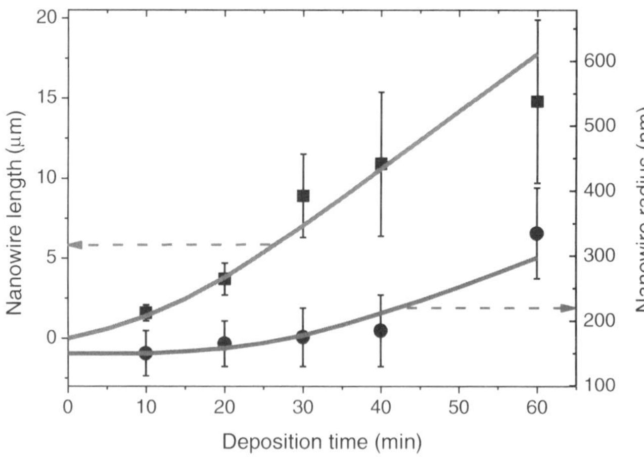

Fig. 5.20 Experimental (*dots*) and theoretical (*lines*) length-time and radius-time dependences of CdTe nanowires grown at the source temperature $T_0 = 550\,°\mathrm{C}$ and the substrate temperature $T = 520\,°\mathrm{C}$

on the sidewalls $n(z)$ (where z is the coordinate along the wire axis), is expected to obey the steady state diffusion equation of the form: $Dd^2n(z)/dz^2+\chi I-n(z)/\tau=0$. Here, χI denotes the adsorption from the vapor flux, I, with the adsorption coefficient χ, and $-n/\tau$ is the adatom sink. The latter term is limited either by desorption or incorporation. The effective lifetime is therefore given by $1/\tau = 1/\tau_{des} + 1/\tau_{inc}$, where τ_{des} and τ_{inc} are the lifetimes before desorption and incorporation, respectively. For the adatom incorporation rate, we thus use the linear approximation $-n/\tau_{inc}$ which is most relevant for a step flow radial growth. The boundary conditions are chosen in the simplified form: $n(z=0)=n_s$ and $n(z=L)=n_l \equiv \exp(\mu_l/k_BT)/\sigma$, where n_s is the adatom concentration on the substrate surface, μ_l is the chemical potential of CdTe dissolved in the droplet and σ is the elementary surface area on the sidewalls. With these boundary conditions, the solution for $n(z)$ is given by

$$n(z)=n_f+\frac{(n_l-n_f)\sinh(z/\lambda)+(n_s-n_f)\sinh\left[(L-z)/\lambda\right]}{\sinh(L/\lambda)}, \tag{5.42}$$

in which $n_f=\chi\tau I$ is the effective adatom activity on the sidewalls and $\lambda=\sqrt{D\tau}$ is the diffusion length.

The transport-limited elongation rate, dL/dt, of a hexahedral nanowire having the radius R at the top is obtained from the material balance equation

$$\frac{3\sqrt{3}R^2}{2\Omega}\frac{dL}{dt}=\frac{3\sqrt{3}R^2}{2}CI-6RD\left(\frac{dn}{dz}\right)_{z=L}. \tag{5.43}$$

Here, Ω is the volume of CdTe pair in the solid, and C is a coefficient accounting for: the droplet shape, the adsorption coefficient on the droplet surface, desorption from the droplet and growth of the surface layer. By substituting (5.42) into (5.43) we obtain

$$\frac{dL}{dt}=CV+\frac{4\chi V\lambda}{\sqrt{3}R}\frac{(1-n_l/n_f)[\cosh(L/\lambda)-1]+(n_s-n_l)/n_f}{\sinh(L/\lambda)}, \tag{5.44}$$

where $V=I\Omega$ is the equivalent deposition rate.

As for the radial growth, its rate in our model is given by

$$\frac{3\sqrt{3}RL}{\Omega_s}\frac{dR}{dt}=6R\int_0^L\frac{n(z)}{\tau_{inc}}dz, \tag{5.45}$$

in which we have again assumed a hexahedral nanowire shape. The radius R here is considered as being constant along the entire nanowire length including the top region. This assumption seems reasonable because most of our nanowires contain only slightly tapered sections at their tips. This tapering is most probably associated with a much lower fraction of adatoms trapped by the sidewalls in the top region.

Table 5.3 Parameters of CdTe nanowires used in fitting their growth behavior

n_s/n_f	n_l/n_f	R_0 (nm)	$\lambda(\mu m)$	CV(nm/min)	χV(nm/min)	τ/τ_{inc}
0	0	150	2	100	8.4	0.61

However, the integral effect on the averaged radius is almost negligible for long nanowires considered here. By integrating (5.42) and substituting the result into (5.45), we obtain

$$\frac{dR}{dt} = \frac{2\chi V}{\sqrt{3}} \frac{\tau}{\tau_{inc}} \left[1 + \frac{(\cosh(L/\lambda) - 1)}{(L/\lambda)\sinh(L/\lambda)} \left(\frac{n_s + n_l}{n_f} - 2 \right) \right]. \tag{5.46}$$

The quantities n_s/n_f and n_l/n_f entering (5.44) and (5.46) describe the contributions of the substrate adatoms and the reverse diffusion flux from the catalyst droplet onto the sidewalls, relative to the contribution of the sidewall adatoms into the vertical elongation and radial extension. We now show that these effects are almost negligible under our growth conditions. Indeed, at the beginning of growth where $L/\lambda << 1$, (5.46) is reduced to: $dR/dt \propto (n_s + n_l)/n_f$. Also, it is seen experimentally (Fig. 5.20), that while during the early stages of growth the nanowires grow in length, they do not initially increase in width, i.e. radially. Therefore, the quantities n_s/n_f and n_l/n_f can be set to zero with reasonable accuracy in modeling the experimental data shown in Fig. 5.20.

The lines in Fig. 5.20 are theoretical fits to the data obtained by numerically solving (5.44) and (5.46) with the initial conditions $L(t = 0) = 0$ and $R(t = 0) = R_0 = 150$ nm. The initial condition for R was taken from the measured average radii of the wires grown for 10 min. It is seen that our theoretical $R(t)$ curve represents fairly well the data: the radial growth is almost completely suppressed during the first 20 min of growth but leads to ~2 times increase of radius during the last 40 min. The fitting parameters are summarized in Table 5.3. As discussed, the adatom activity on the sidewalls is much larger than its boundary values, thus n_s/n_f and n_l/n_f can be set to zero. The effective diffusion length of sidewall adatoms of 2 μm is consistent with the values presented in Chap. 4 for III-V nanowires. The effective vapor flux onto the droplet, CV, far exceeds that to the sidewalls, χ V, showing that the droplet acts as an active chemical catalyst. The obtained value of $\tau/\tau_{inc} = 0.61$ signifies that the deposited CdTe incorporates to and desorbs from the sidewalls with probabilities of 61 and 39 % respectively, at $T = 520\,°C$.

5.6 Nucleation Statistics

The method of composition profiling of InPAs nanowires described in Sect. 4.13 permits the length of each composition oscillation to be determined precisely. As shown in [118], this allows one not only to investigate the growth chronology of

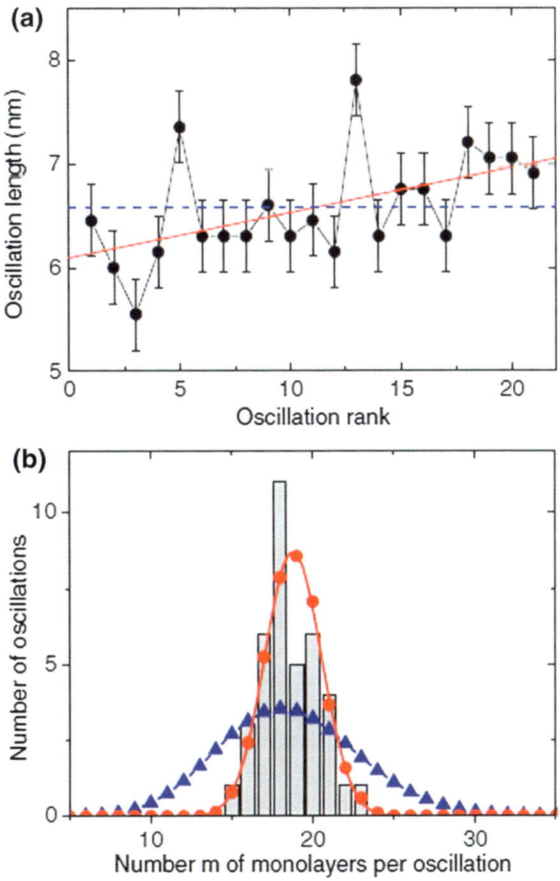

Fig. 5.21 Experimental determination of the nucleation statistics. **a** Length of successive oscillations (segment 3 of Table 5.4). The *dashed line* gives the mean length; the *solid line* is a linear regression ($dL/dt \propto t$). **b** Histogram of the numbers of nucleation events per oscillation (cumulated from segments 2 and 3; see Table 5.4). The two *curves* represent distributions calculated for a large number of nucleation events and normalized to the total number of experimental counts: Poissonian statistics (*triangles*), self-regulated statistics (*disks*) [118]

individual nanowire, but also to study the nucleation statistics in catalyst-assisted nanowire growth. Experimental data for the individual InAsP nanowire used for further analysis are shown in Fig. 5.21a and Table 5.4. Figure 5.21a presents the length of successive oscillations (segment 3 of Table 5.4). As discussed already in Sects. 4.3 and 4.6, the formation of nanowires of small enough radii requires one and only one nucleation event for forming each new monolayer, the growth thus proceeding in the mononuclear regime. In this regime, the growth rate is fixed by the rate of nucleation at the catalyst-solid interface, while the number of monolayers formed equals the number of nucleation events. The length of each oscillation (in monolayers) thus equals the number of nucleation events within the time period τ.

If these nucleation events occurred independently of each other, the probability of a given number m of nucleations taking place during τ should follow Poisson statistics. Figure 5.21b compares the distribution of the oscillation lengths with a Poisson distribution normalized to the same total number of oscillations. The nucleation statistics are markedly sub-Poissonian. In the three nanowire segments studied, the

Table 5.4 Data from 3 nanowire segments of radius R, containing S oscillations.

Segment	R(nm)	S	$\overline{m}$	σ	σ_i	$\overline{m}^{1/2}$
1	24.8	17	24.81	2.26	5.10	4.98
2	25.3	17	18.03	1.98	4.44	4.25
3	24.7	21	18.99	1.49	4.46	4.36
2 and 3	(25)	38	18.57	1.75	4.39	4.31

$\overline{m}$ is the mean number of monolayers per oscillation and σ the measured standard deviation σ_i and $\overline{m}^{1/2}$ are the inhomogeneous and homogeneous Poissonian standard deviations. Cumulated data are given for segments 2 and 3, which have similar R and $\overline{m}$

experimental standard deviations σ are less than half the Poisson standard deviations (Table 5.4). This proves that the nucleation events are not independent.

Sub-Poissonian nucleation statistics observed is the feature of nucleation in nano-volumes. The narrowness of measured distribution of nucleation events per oscillations can be qualitatively interpreted by the time dependence of supersaturation in the mononuclear growth regime shown in Fig. 4.22. As discussed already, the formation of one nanowire monolayer quickly empties growth constituencies from the droplet. The nucleation probability immediately after the nucleation event is therefore smaller than before. During a much longer refill stage which occurs due to a temporary lack of nucleation, the nucleation probability increases. Thus, the nucleation events become temporally anti-correlated. To quantify the analysis, Glas *et al.* [118] considered the particular case of the Au-assisted VLS growth of III-V nanowires, where the nucleation statistics should be controlled by group V element because of its much lower concentration in the droplet. For the sake of simplicity, the two group V species in the InPAs nanowire are not distinguished. We saw earlier that the Zeldovich nucleation probability $P \propto J$ is dominated by the factor $\exp(-\Delta G/k_BT)$, where $\Delta G = A/\Delta\mu$, and $A = C\Omega h\gamma^2$, with Ω as the volume of III-V pair, h the monolayer height, γ as the specific surface energy, and C as a shape constant. The $\Delta\mu(c_{\mathrm{III}}, c_V, T)$ is the difference of chemical potential per III-V pair between liquid with relative atomic concentrations $c_i (i = \mathrm{III, V})$ and solid [353], where now we need to consider both elements. To the first order,

$$\Delta\mu = \Delta\overline{\mu} + \alpha_{III}\frac{\delta c_{III}}{\overline{c}_{III}} + \alpha_V\frac{\delta c_V}{\overline{c}_V}, \tag{5.47}$$

where overbars indicate some average reference state, $\alpha_i = \overline{c}_i(\partial\mu/\partial c_i)_{\overline{c}_i}$, and $\delta c_i = c_i - \overline{c}_i$. Hence,

$$P \cong \overline{P}\exp\left[\frac{A}{\Delta\overline{\mu}^2}\left(\frac{\alpha_{III}}{k_BT}\frac{\delta c_{III}}{\overline{c}_{III}} + \frac{\alpha_V}{k_BT}\frac{\delta c_V}{\overline{c}_V}\right)\right]. \tag{5.48}$$

Let δN_i be the difference in numbers of group i atoms between current and reference liquid states, and $\overline{N}$ the total number of atoms in the reference liquid state. When one monolayer forms in a nanowire of radius R, then $\delta N_{III} = \delta N_V =$

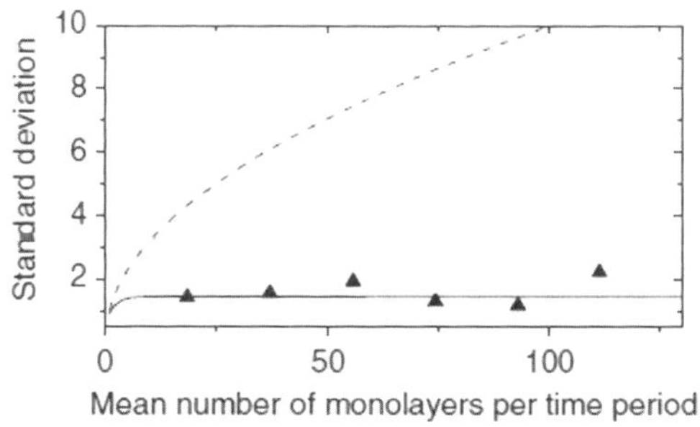

Fig. 5.22 Standard deviation of the number of monolayers per time period when this period is varied. *Triangles*: experimental deviations σ_* obtained by counting the number of nucleation events within sets of 1 to 6 consecutive oscillations. Simulations for Poissonian statistics (*dashed line*) and self-regulated statistics (*solid line*) of nucleation events [118]

$-(\pi R^2 h)/\Omega$. For simplicity we assume that droplet refills with both constituents at the same rate. Then, at any time, $\delta N_{III} = \delta N_V \equiv \delta N$. Since $c_i = N_i/(N_0 + N_i)$, where $N_0 = const$, we have $\delta c_i \cong (1 - \bar{c}_i)\delta N/\overline{N}$. Using this in (5.48), one obtains

$$P \cong \overline{P} \exp\left[\left(B_{III}\frac{1-\bar{c}_{III}}{\bar{c}_{III}} + B_V\frac{1-\bar{c}_V}{\bar{c}_V}\right)\frac{\delta N}{\overline{N}}\right], \tag{5.49}$$

where the dimensionless factors $B_i = (C\Omega h\gamma^2\alpha_i)/(k_B T\Delta\bar{\mu}^2)$ contain parameters which are not precisely known but can be estimated. For example, if $\Delta\mu$ contains terms $k_B T \ln c_i$, the coefficients $\alpha_i = k_B T$ for both group III and group V elements. According to thermodynamic calculations of [353], α_{III} and α_V are indeed of the same order between one and few $k_B T$ in the interesting range of concentrations. Thus, B_{III} and B_V are also close to each other. According to the data presented in Sect. 4.1, c_{III} is typically of the order of 0.5, while $c_V \ll 1$ due to a low solubility of group V elements in liquid catalyst. This yields $B_{III}(1 - \bar{c}_{III})/\bar{c}_{III} \ll B_V(1 - \bar{c}_V)/\bar{c}_V$, showing that the group V term in the exponent of (5.49) is dominating. As expected, the group V atoms influence the probability of nucleation much more than the group III atoms. Moreover, the thinner the nanowire, the larger the effect, since $\delta N/\overline{N}$ scales as R^{-1}.

The nucleation statistics was simulated from (5.49), considering only the effect of the variations of c_V on the nucleation probability. Mean concentration $\bar{c}_V$ was taken in the range of a few percent. As a first approximation, a constant refilling rate was assumed, matching the average growth rate. The simulated nucleation events are markedly anticorrelated in time. The distributions of oscillation lengths and their standard deviations are consistent with the experimental ones (Fig. 5.21b, Table 5.4), which can for instance be reproduced with $B_V = 2.5$, for $\bar{c}_V = 0.02$ and $B_V = 3.8$, for $\bar{c}_V = 0.03$. Such B_V values seem reasonable: for example, $B_V = 3.8$ may be realized with plausible values of $\gamma = 0.4\,\mathrm{J/m^2}$, $\Delta\bar{\mu} = 0.44\,\mathrm{eV}$, and $\alpha_V = 2k_B T$.

Hence, the sub-Poissonian character of the nucleation statistics is perfectly explained by the negative feedback of the variation of the group V concentration in the liquid on the nucleation rate. Different droplet refilling scenarios would lead to different quantitative estimates but would not alter this conclusion, provided that the refilling of the droplet is slower than its depletion upon the formation of one monolayer. This self-regulation mechanism is remarkably efficient against length fluctuations in long nanowires. To find out how the standard deviation varies for longer time periods, (or higher average numbers $\overline{m}$ of nucleations per period), the nucleation events were counted within sets of 1 to 6 consecutive oscillations. The deviation σ_* is calculated with respect to the average number of nucleation events in each multiple time period given by a linear fit to the full data set. The prediction of the model agrees with the experiments: The self-regulated growth quickly saturates to a $\overline{m}$-independent value of σ_*, whereas Poisson nucleation would lead to $\sigma_* = \overline{m}^{1/2}$, as shown in Fig. 5.22.

The observed difference between random statistics at $R \to \infty$ and sub-Poissonian statistics at small R can be further emphasized by considering kinetic equations for probabilities $p_m(t)$ that m monolayers are grown by the time moment t [506]. As above, we assume that an element that limits the nucleation statistics (group V element in the Au-assisted III-V nanowire growth or group IV material in the Au-assisted growth of elemental semiconductor nanowires) refills at a constant rate $F(1/s)$. We denote $\Delta N = (\pi R^2 h)/\Omega$ the number of atoms consumed from the droplet to form one monolayer of nanowire. Then the average probability of monolayer formation per unit time equals $\overline{P} = F/\Delta N$. The variation δN in numbers of atoms between current and reference liquid states is given by

$$\delta N = Ft - m\Delta N, \tag{5.50}$$

where Ft is the segment length in monolayers. Substitution of this into (5.49) with neglect of the second element gives the conditional probability for the formation of the $m + 1$-th monolayer at time t if m monolayers are formed:

$$Q_m(t) = \overline{P}\exp[\Psi(Ft - m\Delta N)], \tag{5.51}$$

with $\Psi = B(1 - \overline{c})/(\overline{c}\overline{N}) \propto 1/R^3$.

From balance considerations, $p_m(t)$ obey the following set of equations:

$$\begin{aligned} \frac{dp_0}{dt} &= -Q_0(t)p_0; \\ \frac{dp_m}{dt} &= Q_{m-1}(t)p_{m-1} - Q_m(t)p_m, m = 1, 2, 3 \ldots \end{aligned} \tag{5.52}$$

Initial conditions can be chosen in the form $p_m(t = 0) = \delta_{m0}$, i.e., no monolayers are present at $t = 0$. We know already from Sect. 1.9 that the solution at $Q_m(t) = \overline{P}$ for all m are given by Poisson distribution with the mean value $\overline{m} = \overline{P}t$ and standard deviation $\overline{m}^{1/2}$ scaling as $t^{1/2}$. From (5.51), this corresponds to the case of large droplet where the anti-correlation effect disappears ($\nu \equiv \Psi\Delta N \to 0$). In the general case, integration of (5.52) yields

$$p_0(\overline{m}) = \exp\left[-\frac{\exp(\nu\overline{m}) - 1}{\nu}\right];$$

$$p_m(\overline{m}) = \exp\left[-\frac{\exp(\nu(\overline{m} - m))}{\nu}\right]\int_0^{\overline{m}} dx p_{m-1}(x)$$

$$\times \exp\left[\frac{\exp \nu(x - m)}{\nu} + \nu(x - m) + \nu\right], \tag{5.53}$$

where the distribution parameter $\nu = \Psi\Delta N \propto \Delta N/N$ scales as R^{-1}. The decrease in dispersion of this distribution of nucleation probabilities for $\overline{m} = 6$ with increasing ν is shown in Fig. 5.23.

5.7 Self-Regulated Pulsed Nucleation in Catalyzed Nanowires

We now consider the results of the previous section in the framework of kinetic growth theory. Only one element is considered, which in the case of III-V nanowires should be group V element due to its much lower concentration in the droplet. Spherical cap droplet with contact angle β is seated on top of cylindrical nanowire with radius R and consists of N atoms of an element that limits the nucleation statistics and of $N_0 \cong const$ other atoms. The atomic concentration of the element of interest is $c = N/(N + N_0)$. As usual, we define supersaturation of the liquid alloy ζ by $\Delta\mu = k_B$ T ln $(\zeta + 1)$, where $\Delta\mu$ is the chemical potential of semiconductor material in the droplet with respect to the reference state in the solid state. Both $\Delta\mu$ and ζ can be obtained as functions of atomic concentration c for a given temperature [355]. Although it is not critical, below we use the assumption $N \ll N_0$ (and therefore $c \ll 1$), which definitely holds for As in Au-Ga-As melt. We also use the approximation

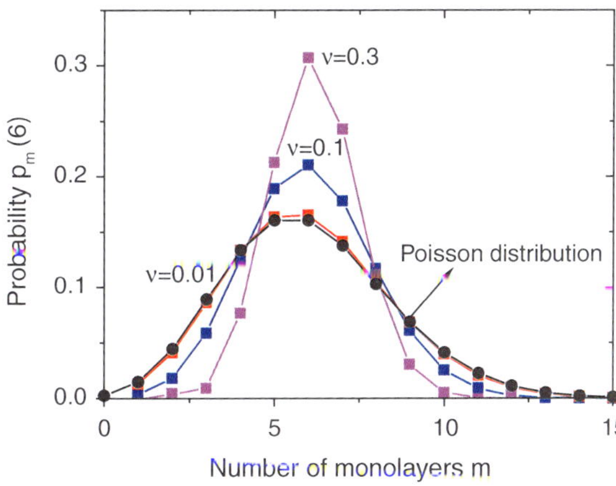

Fig. 5.23 Probabilities $p_m(\overline{m})$ at $\overline{m} = 6$ and different ν : the distribution at $\nu = 0.01$ almost exactly matches the Poissonian spectrum for independent nucleation events; the distribution rapidly narrows as ν increases

$\zeta = c/c_{eq} - 1$ corresponding to the perfect diluted alloy with equilibrium concentration c_{eq}. In this case, $c \cong N/N_0$, $\zeta \cong N/(N_0 c_{eq}) - 1$ and there is the linear scaling between the variations $\Delta\zeta$ and $\Delta N : \Delta\zeta \cong \Delta N/(N_0 c_{eq})$.

When one ML is removed from the droplet, the number of atoms decreases by $\Delta N = \pi R^2 h/\Omega$, with the corresponding decrease of supersaturation

$$\Delta\zeta = \frac{h}{c_{eq} R f(\beta)}, \tag{5.54}$$

where β is the droplet contact angle and $f(\beta)\pi R^3 = N_0\Omega = const$ is the droplet volume.

Let the droplet be fed by the vapour flux I and the diffusion flux from the sidewalls j_{diff} (per unit length per unit time), with negligible desorption. From

$$\frac{dN}{dt} = \frac{2\pi R^2}{1+\cos\beta} I + 2\pi R j_{diff}, \tag{5.55}$$

the time required to refill ΔN atoms equals

$$\Delta t_r = \frac{1}{2V[(1+\cos\beta)^{-1} + \lambda/R]}, \tag{5.56}$$

where $V = \Omega I/h$ is the impingement flux in ML/s and $\lambda = \Omega j_{diff}/(Vh)$ is the effective diffusion length. This refill time is limited by material transport, with $1/\Delta t_r$ being the average nanowire growth rate in ML/s. Equation (5.55) is written in the case of MOCVD, but can be easily modified for MBE by changing the geometrical factor in the adsorption term.

The probability density of nucleation at time t is given by $(h\pi R^2/\Omega)P(t)$, where $P \propto \exp[-A_*/\ln(\zeta+1)]$ and $A_* = \pi\Omega h(\gamma/k_B T)^2$. From material balance in absence of desorption, we can write

$$N(0) + \left[\frac{2\pi R^2 I}{1+\cos\beta} + 2\pi R j_{diff}\right] t = N(t) + \frac{h\pi R^2}{\Omega} \int_0^t dt' P(t') \left[\int_{t'}^t dt'' \frac{\zeta(t'')}{\tau}\right]^2. \tag{5.57}$$

Here, the integral term in the right hand side describes the statistically averaged number of atoms in a 2D island whose radius grows at a rate $r_0\zeta/\tau$, with r_0 as the 2D lattice spacing and τ the characteristic diffusion time in the liquid phase. We assume that $\zeta(0) = \Phi_* - \Delta\zeta$ at $t = 0$, where Φ_* is the ideal supersaturation (i.e., supersaturation in absence of nucleation given by the left hand side of (5.57) divided to $N_0 c_{eq}$ minus one) at the moment of time t_* corresponding to the maximum nucleation probability. At $t = 0$, the supersaturation is minimum and thus the nucleation probability can be put to zero. At $t = t_*$, the supersaturation is maximum, therefore, $t_* = \Delta t_r$ is the average duration of one growth cycle. Dividing (5.57) to $N_0 c_{eq}$, the material balance takes the form

$$\Phi_* + \frac{t - t_*}{t_\infty} = \zeta(t) + G(t), \tag{5.58}$$

with $t_\infty = \Delta t_r/\Delta\zeta$ as the macroscopic pumping time and

$$G(t) = \Delta\zeta \int_0^t dt' P(t') \left[\int_{t'}^t dt'' \frac{\zeta(t'')}{\tau} \right]^2 . \tag{5.59}$$

These equations are of the same type as studied in Sects. 2.2 and 2.3, and therefore can be solved by the Γ—method described there. We first present the nucleation probability as

$$P(\zeta(t)) = P(\Phi_*) \exp\left[-\frac{\Gamma}{\Phi_*}(\Phi_* - \zeta(t)) \right], \tag{5.60}$$

where

$$\Gamma = \frac{\Phi_*}{\Phi_* + 1} \frac{A_*}{\ln^2(\Phi_* + 1)} \tag{5.61}$$

is the large parameter. In (5.59), we use linear approximation $\zeta(t'') \cong \Phi_*$ near the maximum of the integrant. This yields the double exponential solution

$$\mathrm{P}(t) = P(\Phi_*) \exp\left[\frac{t - t_*}{\Delta t_n} - \exp\left(\frac{t - t_*}{\Delta t_n} \right) \right]. \tag{5.62}$$

The characteristic nucleation time is given by

$$\Delta t_n = \frac{\Phi_*}{\Gamma} t_*, \tag{5.63}$$

and should be much smaller than the refill time t_* in view of the strong inequality $\Gamma \gg 1$. The nucleation probability density given by (5.62) has the double exponential form which is indeed well known result of macroscopic nucleation theory for the nucleation rate, as discussed in Chap. 2. This function reaches its maximum at $t = t_*$ and is asymmetric around t_*, with the characteristic width $2\Delta t_n$. The latter determines the time interval where a 2D island can nucleate. The maximum supersaturation is obtained from the self-consistency condition as described in detail in Sect. 2.2. In the first approximation, the solution for Φ_* is given by

$$\Phi_* \cong \exp\left[\frac{A_*}{3 \ln(t_\infty/\tau)} \right] - 1. \tag{5.64}$$

Using (5.63), (5.56) and (5.54), we find

$$\frac{t_\infty}{\tau} = \frac{c_{eq} f(\beta)}{2Vh\tau} \frac{R^2}{R/(1+\cos\beta)+\lambda}; \tag{5.65}$$

$$\frac{\Delta t_n}{\Delta t_r} = \frac{\Phi_*}{\Gamma} \frac{c_{eq} f(\beta) R}{h}. \tag{5.66}$$

At $\zeta \cong \Phi_*$, the time Δt_g required to fill the complete monolayer equals $\tau R/(\Phi_* r_0)$, where r_0 is the lattice spacing. The ratio between Δt_g and Δt_g is then obtained as

$$\frac{\Delta t_g}{\Delta t_n} = \frac{\Gamma}{\Phi_*^2} \frac{\tau}{t_\infty} \frac{R}{r_0}. \tag{5.67}$$

With the known c_{eq}, β, V, λ, τ and A_* for a given material system grown at given deposition conditions (temperature and flux), one can now find the refill time by (5.56), the maximum supersaturation by (5.65) and (5.64), the parameter Γ by (5.61), and then the nucleation time and the island growth time by (5.66) and (5.67), as functions of the nanowire radius R. The normalized nucleation probability $Q(t)$ is obtained by integration of (5.62) at $P(\Phi_*) = 1/\Delta t_n$:

$$Q(t) = 1 - \exp\left[-\exp\left(\frac{t-t_*}{\Delta t_n}\right)\right]. \tag{5.68}$$

Regardless of the growth conditions, the ratio t_∞/τ must be much larger than one for small enough nanowire radius in view of $V\tau \ll 1$, and further decreases toward smaller R. The Φ_* value given by (5.64) thus increases as R decreases, i.e., smaller droplets require lower threshold supersaturation to nucleate the island. This decreases the Γ values at smaller R; the ratio $\Delta t_n/\Delta t_r$ given by (5.66) also becomes smaller. On the other hand, (5.67) shows that the ratio $\Delta t_g/\Delta t_n$ is much smaller than one in view of $\tau/t_\infty \ll 1$, provided that the radius is not too large. As a result, a certain critical nanowire radius always exists below which the following timescale hierarchy should be observed:

$$\Delta t_g \ll 2\Delta t_n \ll \Delta t_r. \tag{5.69}$$

The first inequality shows that the island growth is the fastest process in the entire growth cycle: once nucleated, the island fills the monolayer almost instantaneously. This requires, however, a sufficient number of semiconductor atoms to be present in the droplet to complete the monolayer without refill. The second inequality shows that the island nucleation can only take place during the time interval which is much shorter than the refill time. Whenever this property holds, any two consecutive nucleation events become temporary anti-correlated as suggested in [118], because the probability densities in different growth cycles do not overlap with each other.

This time scale hierarchy leads to the sawtooth time dependence of supersaturation discussed in Sects. 4.3 and 4.6, with analmost instantaneous decrease immediately after nucleation. Even more importantly, the nucleation events are not distributed randomly within the growth cycle Δt_r, but rather tend to occur very near mt_* for the m-th cycle. Such a self-regulated mechanism suppresses the Poissonian broadening of the nanowire length distribution and transforms it to a delta-like spectrum with theoretical standard deviation within 1 monolayer. Without this self-regulatory effect, all lithographical methods of droplet organization would never produce regular nanowire arrays, and studies of the growth rate dependence on the nanowire radius and growth conditions would be useless, because relevant deterministic correlations would be quickly degenerated by randomness. As discussed, the narrowing occurs only at small R, while at $R \to \infty$ the polynuclear growth from a supersaturated alloy would proceed under the condition $2\Delta t_n \ll \Delta t_g \ll t_\infty$, shown in Chaps. 1 and 3 for ensembles of nanoobjects. Here, the nucleation stage of an ensemble of 2D islands becomes the fastest, followed by a much longer size relaxation stage with characteristic time Δt_g, while the characteristic time of material influx is the longest.

We now consider a particular example of the Au-catalyzed growth of GaAs nanowires with the elementary volume per GaAs pair $\Omega = 0.045\,\text{nm}^3$ and $h = 0.326\,\text{nm}$. With a typical growth temperature between 450 and 620 °C used in different epitaxy techniques and $\gamma = 0.35\ \text{J/m}^2$, the A_* value ranges between 40 and 56. We therefore use a plausible value of $A_* = 45$ for our estimates. For the equilibrium As concentration in the droplet, we take $c_{eq} = 0.005$. We also use a hemispherical droplet geometry ($\beta = \pi/2$) and neglect the surface diffusion of As ($\lambda = 0$) [457]. With these parameters, the radius dependences of $2\,\Delta t_n/\Delta t_r$ and $\Delta t_g/2\Delta t_n$ are shown in Fig. 5.24 for two different $V\tau$ values. As expected, both curves increase with R. The island growth time indeed appears much smaller than the nucleation time for R between 10 and 150 nm, and is not very sensitive to $V\tau$. In contrast, the ratio between the nucleation and refill times increases significantly as $V\tau$ rises from 5×10^{-5} to 2×10^{-4}. In the latter case, the time scale hierarchy between the nucleation and refill stages is broken already at R ~ 50 nm. Therefore, the second inequality (5.69) is expected to be more sensitive to the growth conditions and requires low enough deposition rate and fast enough diffusion in the liquid state.

Figure 5.25 shows the statistically smoothed supersaturation profile (dotted line) and the nucleation probability density (black line) for 3 consecutive growth cycles, obtained with the parameters of Fig. 5.24 at $V\tau = 5 \times 10^{-5}$ and $R = 25\,\text{nm}$. The arsenic rate V is chosen at 5 ML/s, relating to a typical growth rate of GaAs nanowires of ~3.2 nm/s. In fact, the absolute V value does not influence the growth behaviour, which is affected only by the dimensionless parameter $V\tau$. As discussed, the nucleation probabilities in different growth cycles do not overlap, which is the best case regarding the length uniformity originating from truly anticorrelated nucleation statistics. The sawtooth red curve shows a particular realization of supersaturation kinetics, varying only a little within the time interval $2\Delta t_n$.

Figure 5.26a and b show how the situation changes for $R = 150\,\text{nm}$ and $V\tau = 2 \times 10^{-4}$. The corresponding probability densities in Fig. 5.26a largely overlap. Furthermore, the nucleation probability $Q(t)$ in Fig. 5.26b does not start from zero at the beginning of each growth cycle. In this case, our method cannot be rigorously justified, and the obtained solutions describe the system only qualitatively. However,

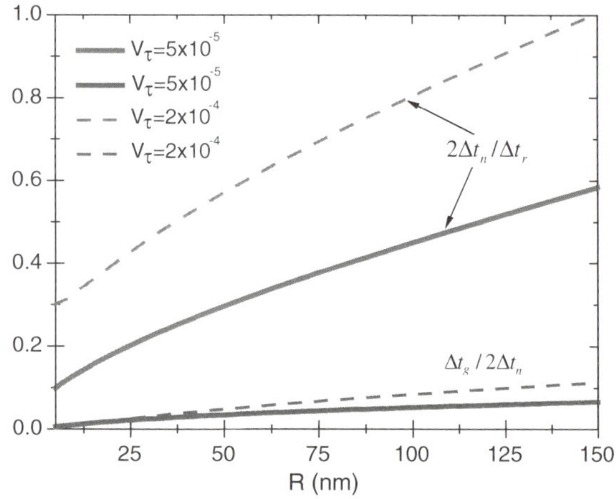

Fig. 5.24 Ratios of characteristic times versus R at $A_* = 45$ and two different values of $V\tau$

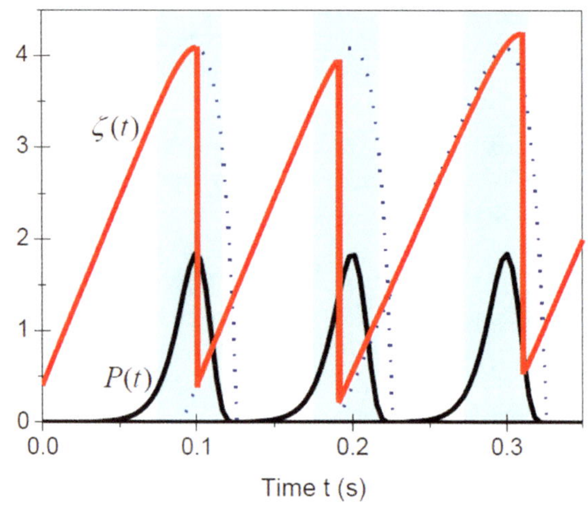

Fig. 5.25 Nucleation probability density $P(t)$ (*black curve*) and the corresponding supersaturation $\zeta(t)$ (*dotted curve*) for a 25 nm radius nanowire at $V = 5$ ML/s. The average duration of each growth cycle is 0.1 s. The maxima of $P(t)$ and $\zeta(t)$ relate to the time moments mt_*. The *shadowed areas* show the nucleation intervals of width $2\Delta t_n$ near mt_* within each growth cycle. The sawtooth curve corresponds to a particular realization of nucleation events

the overlapping nucleation probabilities clearly indicate a transition to a temporary independent statistics of nucleation events, where the Poissonian behaviour is most anticipated.

Our analysis referred to the case of VLS-grown nanowires. The liquid state of the droplet is however not critical for our conclusions, although a much lower diffusivity in the frozen nanoparticle on top of the vapor–solid–solid nanowires can narrow the window of parameters for non-overlapping nucleation probabilities . The assumption $c \ll 1$, which should always hold for As in Au but is not valid for Si or Ge in Au, is also not absolutely necessary. Indeed, the characteristic diffusion time τ is expected to be directly proportional to c_{eq}, showing that the ratio t_∞/τ given by our (5.64) and therefore the maximum supersaturation Φ_* are actually independent of c_{eq}. In this case, (5.66) yields: $\Delta t_n/\Delta t_r \propto \Phi_* c_{eq} = c_* - c_{eq}$, i.e., thecharacteristic time

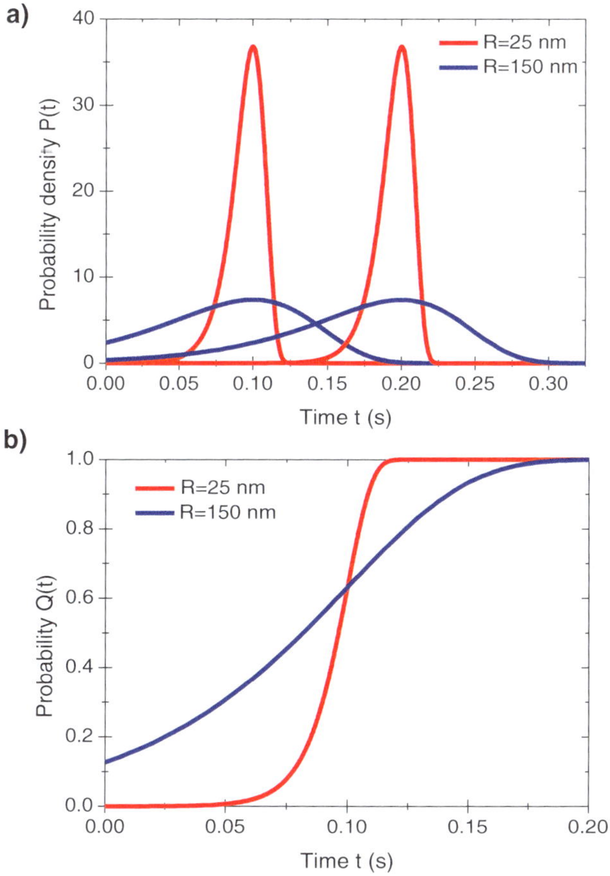

Fig. 5.26 **a** Non-overlapping and overlapping nucleation probability densities at $R = 25\,\text{nm}$, $V\tau = 5 \times 10^{-5}$ and $R = 150\,\text{nm}$, $V\tau = 2 \times 10^{-4}$, respectively. **b** Nucleation probabilities, obtained by integration of the *curves* shown in **a**

ratio is proportional to the difference between the maximum and the equilibrium concentration and does not contain the absolute concentration. Of course, the relative amplitude of supersaturation oscillations is smaller for higher c, but the nucleation probability should not be strongly affected by the absolute c value. Finally, neglect of desorption might only underestimate the refill time Δt_r but is not expected to considerably increase the nucleation time Δt_n so that the time scale hierarchy should be maintained.

5.8 Boron Distribution in Si Nanowire

The aim of this section is studying the inhomogeneous radial distribution of B dopant in Si nanowires, observed by laser-assisted atom probe tomography (APT) technique. Studies of dopant spatial distribution in individual nanowires is very important for understanding the doping mechanisms as well as the resulting electrical properties. Several methods have been used for characterization of doped nanowires, for example, the capacitance-voltage measurement of B-doped Si nanowires [507], the Kelvin probe force microscopy of P-doped Si nanowires [508], and APT of P-doped Ge nanowires [509]. An inhomogeneous dopant distribution in those nanowires has been found to be a common phenomenon. In particular, Koren and coauthors have demonstrated that the concentration of P atoms in straight Si wires can decrease by two orders of magnitude from the sidewall surface to the core [508]. The authors hypothesized that P atoms could diffuse directly from the vapor phase into the body of Si wires, with an enhanced diffusion coefficient compared to bulk.

In [510], a convenient and effective method was reported to perform the APT characterization of B-doped Si nanowires. Using this technique, quantitative information on the boron spatial distribution in the 30 nm wide core region was obtained. Si nanowires were grown by the Au-catalyzed CVD via the VLS mechanism, at 500 °C. To prepare Si nanowires for the APT investigation, welding a single Si nanowire on a supported tip was used, as shown in Fig. 5.27a. The welding process is realized by gas injection system in the electron beam-induced deposition mode, which is an effective *in situ* nanoscale deposition process (Fig. 5.27b). The APT analyzing conditions are set at 80 K with a chamber pressure of about 10^{-10} mbar. The laser is operating in the ultraviolet range (343 nm) with a power of 2.8 mW. This enables an excellent and controlled evaporation rate. By taking the advantage of APT, different elements are detected and viewed in 3D space.

The Si nanowires were first investigated by high resolution HRTEM to determine their structure. For the first 600–1000 nm below the Au droplet, nanowires are covered by Au-rich clusters, as already mentioned in Sect. 4.1. Such clusters are most probably caused by the further growth of the nanowires from residual silane gas after the interruption of the gas flow in the growth chamber and the subsequent diffusion of gold due to the high-temperature inertia of the chamber. Below this region, the sidewalls are not contaminated by Au clusters, and a thin shell appears as the wires are analyzed further away from its apex. The shell becomes thicker and thicker towards the cleaved end of the wires, as lateral growth occurs over a longer time at the bottom. In the region of the cleavage that is usually located 6–8 mm below the Au droplet, the typical shell thickness is between 5 and 10 nm as shown in Fig. 5.28. In contrast to the core, the shell appears polycrystalline (Fig. 5.28b). The high resolution and dark-field TEM images do not reveal any significant structural change of the shell structure along the wire growth direction, suggesting that the axial dopant concentration in the shell is constant.

In parallel to the TEM analysis, APT characterizations were performed in the region of the cleaved end of the nanowires that were also located between 6 and

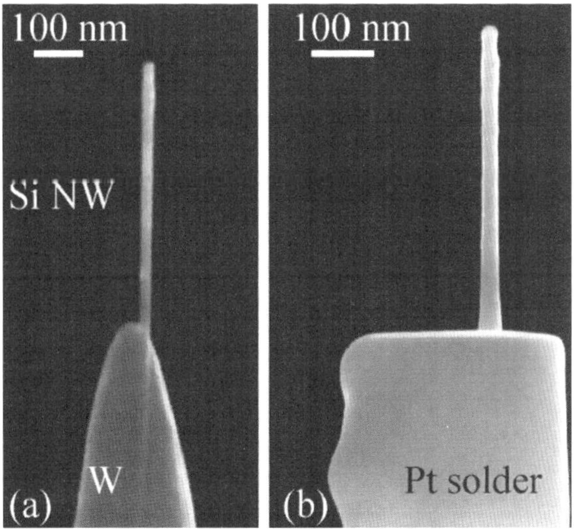

Fig. 5.27 SEM image of a Si nanowire welded on a W tip **a** before and **b** after Pt deposition by gas injection system. The bottom end of nanowire is at the top of the specimen

8 mm from the Au droplet. Figure 5.29a shows the SEM image of the cleaved portion of a Si wire that has been analyzed by APT. This nanowire has a diameter of 58 nm, but only the cylindrical core with a diameter of 30 nm was probed, due to the limitation explained above. Figure 5.29b shows a 3D reconstruction of the core that extends from the cleavage point over 116 nm along the growth axis. This reconstruction is representative of the chemical composition of the rest of the nanowire core, with

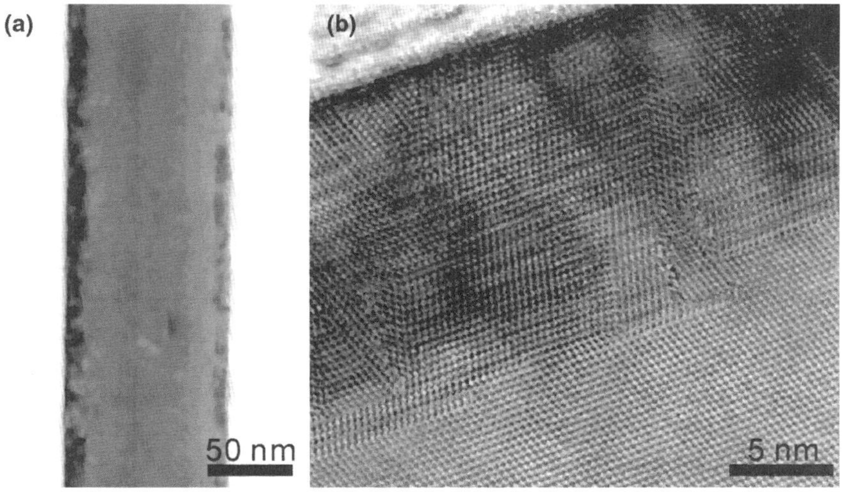

Fig. 5.28 **a** TEM image of a portion of B-doped Si nanowires obtained midway between the top and bottom of the wire. **b** HRTEM image of the shell structure

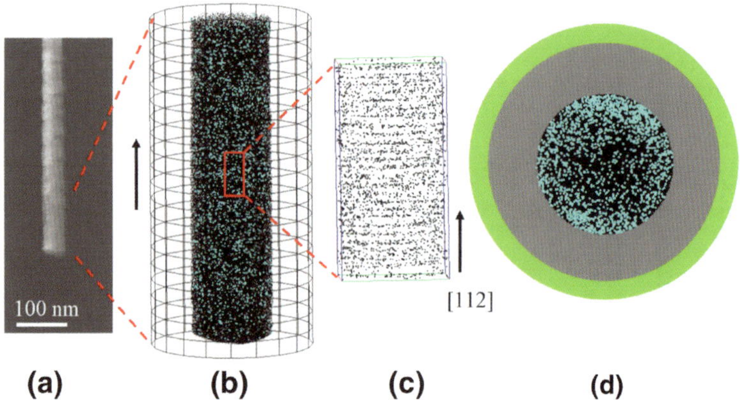

Fig. 5.29 APT investigation of the core region of an individual B-doped Si nanowire. **a** SEM image of the cleaved part of the wire that was analyzed. **b** 3D tomographic reconstruction of B atoms (*blue dots*) and Si atoms (*black matrix*) in the core of a 58 nm diameter Si nanowire; the analyzed volume is $30 \times 30 \times 116\,\text{nm}^3$. **c** Enlarged view of the 3D tomographic reconstruction showing the Si inter-plane distance of 0.22 ± 0.01 nm. The arrows point in the growth direction. **d** Cross-sectional view of radial B distribution and schematic illustration of the nanowire radial structure. The *grey* and *green* region represent the part of the Si nanowire that is not probed and the silica layer, respectively

the exception of the top and bottom part, where the growth conditions are usually different. The good crystallinity of the core is demonstrated from the resolution of the inter-plane distance of 0.22 ± 0.01 nm along the [112] growth direction (Fig. 5.29c), consistent with the HRTEM images of the core. In addition to the Si species, boron is also detected. The concentration of boron impurities establishes at $1 \times 10^{19}\,\text{cm}^{-3}$ in the center of the Si nanowire and gradually increases towards the edge of the core. The observed non-uniform spatial distribution is shown in Fig. 5.29d. As discussed in [510], it is not due to atomic surface migration under laser pulses during the APT measurements, but rather originates from the dopant incorporation mechanism during the VLS growth.

One can consider three different pathways for the dopant incorporation into a nanowire, as illustrated in Fig. 5.30. The first pathway (shown by arrow 1) corresponds to the direct incorporation from the vapor phase through the sidewalls. This is the vapor–solid (VS) incorporation by the solid diffusion, considered previously, for example, in [508]. The second pathway (arrow 2) is the incorporation through the catalyst droplet in the VLS fashion. As discussed in [509], the dopant profile inside the nanowire might be governed by the VS incorporation rather than by the VLS-like diffusion because of a higher dissociative chemisorption rate on the conformal deposition layer of tapered sidewalls with respect to the droplet surface. Another reason for the suppression of VLS incorporation might be a low solubility of dopant in molten Au. Also, the VLS mechanism could not satisfactory explain the observed spatial inhomogeneity of boron concentration that gradually increases towards the nanowire sidewalls. In Fig. 5.30, the third possible pathway is illustrated

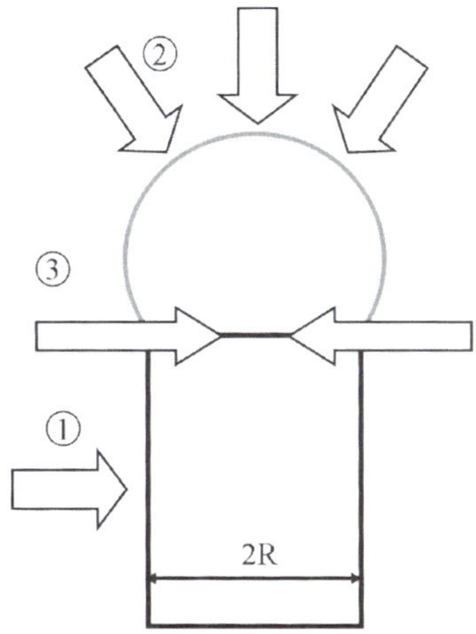

Fig. 5.30 Schematic illustration of three pathways of dopant incorporation into a VLS nanowire: (*1*) VS mechanism; (*2*) VLS mechanism; (*3*) ID mechanism

by arrow 3. This mechanism involves two-dimensional (2D) interfacial diffusion (ID) along the solid-liquid interface between the droplet and the wire.

Therefore, if we assume that the pathway "2" is negligible in the incorporation process of the dopant, the modeling of the diffusion profile of dopant in the nanowire is simply given by the resolution of the non-stationary diffusion equation. The non-stationary diffusion equation in the nanowire as a cylinder can be written as:

$$\frac{\partial C}{\partial t} = \frac{1}{r}\frac{\partial}{\partial r}\left(rD\frac{\partial C}{\partial r}\right). \tag{5.70}$$

Here, $C(r, t)$ is the radius dependent dopant concentration at time t, r is the distance from the wire centre and D is the dopant diffusion coefficient.

The initial and boundary conditions to (5.70) are generally given by

$$C(r, t = 0) = 0;\ C(r = 0, t) < \infty;\ C(r = R_0, t) = M(t). \tag{5.71}$$

The initial condition at $t = 0$ is obvious. The first boundary condition at $r = 0$ yields a finite boron concentration at the nanowire axis. The second boundary condition is usually put at the nanowire sidewalls ($R_0 = R$), with the function M increasing linearly with time due to material deposition: $M = kt$ [508]. The solution to (5.70) with the boundary conditions given by (5.71) is readily obtained in the form

$$C(r, t) = k\left(t - \frac{R^2 - r^2}{4D}\right) + \frac{2k}{RD}\sum_{n=1}^{\infty} \exp\left(-\alpha_n^2 Dt\right)\frac{J_0(\alpha_n r)}{\alpha_n^3 J_1(\alpha_n R)}. \tag{5.72}$$

Here, J_k are the Bessel functions of the first kind of the order k and α_n are the roots of the equation $J_0(\alpha_n R) = 0$. These roots, along with the corresponding values of $J_0(\alpha_n R)$ and $J_1(\alpha_n R)$ are given, for example, in [511]. Only the first ten of an infinite set of Bessel function roots can be taken into account. This ensures a high accuracy of approximation due to a strong decay of exponential terms of (5.72) with increasing α_n.

The boundary condition with $M = kt$ can be justified by considering the surface concentration of boron, c, on the wire sidewalls, driven by the material influx from the vapor phase, I: $dc/dt = I$, yielding $c = It$. However, the linear increase with time can be broken by boron desorption from the sidewalls that should occur at the typical growth temperatures around 500 °C. When desorption is included, the kinetic equation for c is changed to $dc/dt = I - c/\tau$, with τ being the effective boron lifetime on the sidewalls. In this case, the solution is given by $c = I\tau[1 - \exp(-t/\tau)]$, showing that, rather than growing infinitely, the surface concentration saturates at a constant value of $I\tau$ at $t \gg \tau$. Therefore, a more reasonable boundary condition at large enough t would be a constant concentration

$$C(r = R_0, t) = C_0 = const. \tag{5.73}$$

With this condition, the solution to (5.70) is given by

$$\frac{C(r,t)}{C_0} = 1 - \frac{2}{R_0} \sum_{n=1}^{\infty} \exp\left(-\alpha_n^2 Dt\right) \frac{J_0(\alpha_n r)}{\alpha_n J_1(\alpha_n R_0)}. \tag{5.74}$$

As already mentioned, our APT characterization is performed only within the nanowire core region of 15 nm radius. The TEM characterization of the nanowire shows the existence of three different regions: a core radius, a 5–10 nm thick shell with a high concentration of boron impurities, and a 3–4 nm thick SiO_2 layer, yielding the total nanowire radius of about 29 nm. Therefore, the outermost region in our APT reconstruction (15 nm) is located close to the interface between the core and the overgrown shell. Assuming, in the first approximation, that the boron concentration in the shell is constant, we put the boundary condition given by (5.73) at $R_0 = 15$ nm, with C_0 being equal to the outermost boron concentration measured by APT ($7.1 \times 10^{19}\,\mathrm{cm}^{-3}$). With this determination made, the fitting of boron concentration profile depends on only one parameter, i.e., the squared effective diffusion length of boron, Dt. Figure 5.31 shows the profile of the average radial boron distribution measured with APT and theoretical curves obtained from (5.74) with different Dt. The best fit is obtained at $Dt = 20.25\,\mathrm{nm}^2$.

If boron incorporates with the VS fashion, time t should be equal to the exposition time of the analyzed section to the B precursor. The exposure time can be estimated as $t = L/V$, where L is the distance between the cleavage point (the analyzed section) and the Au droplet and V is the nanowire growth rate. The length L is measured to be 8 μm and V equals 1 μm/min in our CVD experiments, yielding $t = 8$ min, which is supposed to be much longer than τ. As a consequence, the best fit given in Fig. 5.31 corresponds to $D = 4.2 \times 10^{-16}\,\mathrm{cm}^2/\mathrm{s}$. This value is much larger than the reported bulk diffusivity of B in Si at 500 °C ($\approx 10^{-18}\,\mathrm{cm}^2/\mathrm{s}$) [512]. With $D = 10^{-18}\,\mathrm{cm}^2/\mathrm{s}$, the penetration length of B atoms would be only 1 nm, and the VS mechanism could never work in this case. Therefore, the VS incorporation can explain our data only

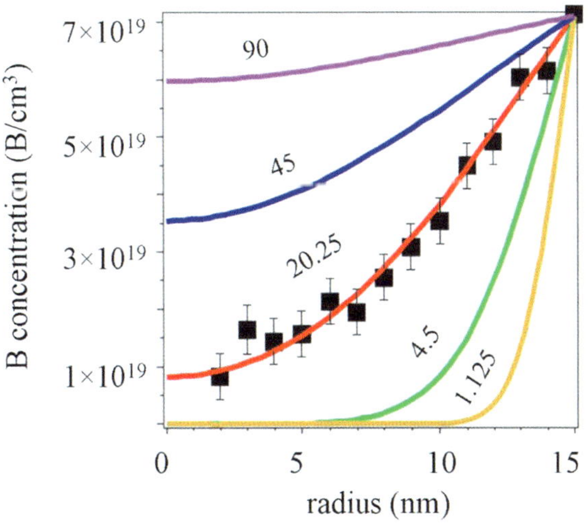

Fig. 5.31 Measured B concentration (*dots*) and theoretical concentration profiles (*lines*) obtained from (5.74) at fixed $C_0 = 7.1 \times 10^{19}\,\text{cm}^{-3}$, $R_0 = 15\,\text{nm}$, and different Dt (nm^2)

if the boron diffusivity is dramatically enhanced (by the factor of 400). It has been argued that P atoms can diffuse into Ge nanowires with an enhanced diffusivity as high as $\approx 10^{-15}\,\text{cm}^2/\text{s}$ at 460 °C [508]. As for B atoms in Si nanowires, the overgrown shell of the nanowires has been shown to exhibit defects (see Fig. 5.28) that could enhance the diffusion of the B impurities. For example, if a supersaturation of Si self-interstitial exists in the shell, it would favor the migration of B-Si interstitial defect complexes that would migrate towards the core and spontaneously break into an interstitial and a substitutional B [513–515]. The solid diffusivity of B could also be enhanced by the elastic stress, an example being the stress-driven diffusion during growth of Stranski-Krastanow quantum dots considered in Chap. 3. In the case of nanowiress, the stress might be induced by the Laplacian pressure caused by their small radius or by the excess of Boron impurities in the shell.

As discussed above, the formation of a complete monolayer slice during the VLS growth of nanowires is almost instantaneous. In this case, the rate of nucleation-mediated nanowire growth in the [112] direction, $V_{NW} = 1/t_*$ (ML/s), is determined by the waiting time t_* between two consecutive nucleation acts. While considering the ID pathway 3, we therefore assume that B atoms incorporate into the wire and diffuse along the 2D liquid–solid interface during time t_*. Such process is repeated upon the completion of each Si monolayer, leading to an axially homogeneous B distribution, since all monolayers are identical. Time t_* can be easily estimated from the growth rate and equals 0.033 s. Putting $t = t_*$ in (5.74), the best fit to the experimental profile corresponds to $D = 6.15 \times 10^{-12}\,\text{cm}^2/\text{s}$. This value is much larger than the solid diffusion coefficient, because the ID diffusion is two dimensional. On the other hand, it is much smaller that the diffusivity at the solid Si surface in contact with vapor (typically of the order of $10^{-8}\,\text{cm}^2/\text{s}$ [516]). This also seems reasonable, because the diffusion coefficient at the solid–liquid interface should be much smaller than that on the solid-vapor interface. However, the boundary condition with a constant concentration is less easily justified for a short-scale monolayer growth. Overall, the ID mechanism has to be considered to account for the dopant distribution in the Si nanowire core, although we cannot make a definite commitment regarding its stronger contribution with respect to the VS mechanism.

5.9 Ga-catalyzed GaAs Nanowires

Triggered by the well-known incompatibility of gold with Si technology, great effort has been devoted to the development of Au-free growth protocols for nanowire synthesis. In Sect. 4.1, we mentioned the self-catalyzed III-V nanowires whose VLS growth is promoted by group III metal rather than by a foreign metal catalyst. Effective results have been obtained, in case of GaAs, by self-assisted methods as reported, e.g., in [293] and [517]. By a suitable choice of either growth conditions or substrates, the formation of Ga droplets is promoted, and these droplets induce the epitaxial growth of GaAs nanowires either on GaAs [517, 518] or on Si [48, 49, 293, 323]. In both cases, the substrate surface is covered with a thin layer of Si oxide. Concerning the growth process, an important difference between Au- and Ga-catalyzed growths is that the size of the metal droplet, which in case of Au does not change appreciably during the growth under the optimized conditions, is not necessarily constant in the case of Ga, and is highly sensitive to the V/III flux ratio. When a balanced As/Ga ratio is used, the radius of Ga droplet stays constant [457]. In Ga-rich conditions, the droplet size is found to increase linearly with time, together with the nanowire radius, which is uniform along the nanowire axis [294, 519]. The contact angle β that the Ga droplet forms with the top facet is maintained constant and larger than 110° [519]. In Ga-poor conditions, the droplet volume gradually shrinks, and eventually the droplet gets completely consumed [518].

We first study the conditions that should be satisfied to maintain a constant droplet volume during the self-catalyzed growth, referring to the case of the Ga-catalyzed GaAs nanowires obtained by MBE. Let us consider the As and Ga fluxes reaching the catalyst droplet. The net flux of As atoms into the droplet equals

$$\frac{dN_{As}}{dt} = I_{As,eff}\pi R^2. \tag{5.75}$$

Here, $I_{As,eff}$ summarizes adsorption, desorption and re-emission of arsenic for a given geometry and neglects its surface diffusion, as suggested by the data of [457]. For Ga atoms, in contrast, lateral sidewall diffusion is taken into account, while diffusion from the substrate is neglected because of a much greater nanowire length as compared to the Ga diffusion length on the sidewalls λ at the high growth temperatures typically used (T ~620–640 °C). The net influx of Ga atoms is therefore given by

$$\frac{dN_{Ga}}{dt} = I_{Ga,eff}\pi R^2 + I_{Ga}\sin\alpha_{Ga}2R\lambda\left(1 - \frac{\theta_l}{\theta_f}\right), \tag{5.76}$$

with $I_{Ga,eff}$ summarizing adsorption, desorption and possible re-emission of gallium. The diffusion-induced term contains the impingement rate of Ga onto the sidewalls, $I_{Ga}\sin\alpha_{Ga}$ (with I_{Ga} as the direct Ga flux and α_{Ga} the incidence angle of Ga beam), onto the equivalent surface area $2R\lambda$ seen by the beam. The ratio of gallium activities in the liquid phase (θ_l) and on the sidewalls (θ_f) accounts for the reverse diffusion flux from the droplet, as discussed in Chap. 4. Obviously, the droplet volume remains constant during growth when the net As and Ga fluxes equal each other:

$$\frac{dN_{Ga}}{dt} = \frac{dN_{As}}{dt}. \tag{5.77}$$

Under the assumption $\beta > \alpha + \pi/2$ for both As and Ga fluxes, (4.36) gives $I_{eff} = I/\sin^2\beta$ for both direct fluxes. Taking into account the As re-emission flux (J^r_{As}) and neglecting it for Ga, we have: $I_{As,eff} = I_{As}/\sin^2\beta + I^r_{As}$ and $I_{Ga,eff} = I_{Ga}/\sin^2\beta$ (with neglect of desorption), where I_{As} denotes the primarily As atomic flux. Using these expressions in (5.75) and (5.76), the steady state condition (5.77) yields

$$\sin^2\beta = \frac{x-1}{(2\Lambda\sin\alpha_{Ga}/\pi R) - x_r}. \tag{5.78}$$

Here, $\Lambda \equiv \lambda(1-\theta_l/\theta_f)$ is the effective diffusion length at given growth conditions, which might also depend upon R due to nanoscale size effects, $x = I_{As}/I_{Ga}$ is the actual As/Ga flux ratio (which can be deduced from the beam equivalent pressures for a given beam geometry as described in [457]), and $x_r = I^r_{As}/I_{Ga}$ is the additional contribution from As re-emission. Equation (5.78) gives the condition for a steady state self-catalyzed growth with a constant radius R. For example, it shows that, at $\Lambda = 0$ (negligible diffusion of Ga) but with some As re-emission ($x_r > 0$), the steady state growth can proceed only in excess of Ga ($x < 1$). On the other hand, As-rich conditions in the direct fluxes ($x > 1$) can be compensated by a Ga diffusion flux, which should be large enough to ensure a positive denominator in (5.78). This flux can be self-regulated by adjusting either Λ or R at the initial, non-stationary growth step to reach the stationary state as will be discussed shortly.

We now discuss a very interesting data of [520] allowing one to quantify some important parameters of the Ga-catalyzed MBE growth of GaAs nanowires under Ga-rich conditions. Nanowires were grown by standard solid-source MBE on Si-treated [518] GaAs(111)B substrates. Self-catalyzed nanowire growth was carried out at 640 °C with a V/III beam equivalent pressure (BEP) ratio of 5. The Ga flux was set to a BEP of 3.6×10^{-7} Torr, as measured by an ion gauge placed next to the substrate holder. This flux corresponds to a two-dimensional growth rate of 16.8 nm/min on a GaAs(100) surface. The As_4 flux was therefore set to a BEP value of 1.8×10^{-6} Torr. Growth was initiated by simultaneous exposure to Ga and As_4 fluxes. After 60 min, the Ga shutter was closed and the sample was kept for 10 min under As flux to obtain the complete consumption of the nanoparticle. After this step, the Ga shutter was opened again for different durations. Growth was in all cases stopped by shutting off both beams and by letting the sample to cool down outside the growth chamber. The nanowire morphology was then studied by SEM, using a Zeiss SUPRA 40, operated at 10 keV, with a nominal point resolution of 1.5 nm. Nanowires were investigated both by direct imaging of the as-grown samples and after being mechanically transferred on an n-doped Si(111) wafer.

Figure 5.32 shows the evolution of the nanowire tip after different exposure times to the Ga and As beams. Wire A represents the starting point: after the complete consumption of the Ga droplet, the tip appears flat and it is possible to notice a slight diameter reduction. This morphology is characteristic of the Ga droplet consumption by exposure to As [521]. After 30 s of exposure to Ga and As beams (wires B, C and D) the morphology of the tip changes. A spherical cap droplet, with diameter smaller than that of the wire, becomes visible. This droplet is now sitting on top of a pyramid-like structure, which morphology resembles that of wires terminating with (110) facets [293, 518]. Wires B–D belong to the same sample, and the differences between them can be ascribed to local fluctuations of the effective beam fluxes. After 60 s (wire E) a new wire portion has grown and the droplet has already reached its equilibrium geometry, with a contact angle exceeding 110°, in agreement to what is expected for the regular self-catalyzed growth. For increased deposition times (wire F after 150 s and G after 300 s), the wire continues to elongate. Nanowires B–F show two diameter discontinuities along their bodies: the first related to the consumption of the droplet, as in A, and the second due to the growth resumption with smaller diameter. By increasing the deposition time, the separation between the two discontinuities tends to decrease. After 420 s, (wire H), only one discontinuity is observed, with an irregular shape.

In Fig. 5.33, the average length and diameter of the wire portion due to the regrowth are plotted together with the length and diameter of Ga-assisted wires grown under the same conditions (the regular growth). The values are to be intended as the average values across the sample. Fits of the experimental data obtained according to the model described below are also plotted. Assuming that the wires length depends linearly on the growth time, one can estimate a regrowth rate of 130 ± 5 nm/min, that agrees well with the corresponding value obtained for the regular Ga-catalyzed growth (133 ± 1 nm/min).

The radial growth behaves differently, as shown in Fig. 5.34. In regular growth of self-catalyzed nanowires the growth rate is constant, and is estimated to be 1.5 ± 0.1 nm/min starting from a value at nucleation of about 20 nm. By contrast, the radial growth rate in the regrowth process is not constant. At the early stage it equals approximately 9 nm/min, and slows down for increased regrowth times. From Fig. 5.34, we observe that a regrowth of 10 min is sufficient to recover the initial diameter. After 15 min, the diameter of the secondary wire is indistinguishable from that one would obtain after 75 min of uninterrupted Ga-catalyzed growth (135 nm).

In our modeling, we address the following points:

(i) Description of the regular growth stage and in particular the observed linear length-time [$L(t)$] and radius-time [$R(t)$] dependences shown in Figs. 5.33 and 5.34;
(ii) Explanation of why it is on the wire top where the Ga nanodroplet re-appears;

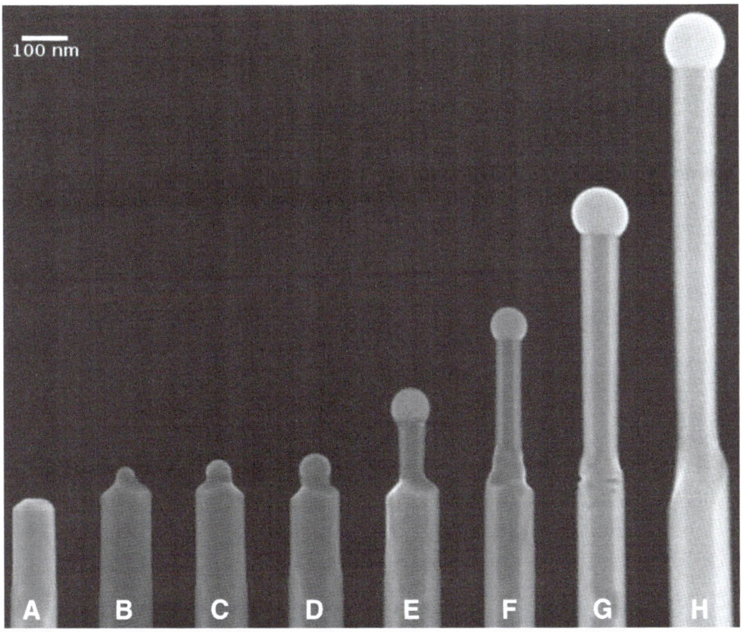

Fig. 5.32 Collage of SEM images of the wires coming from different samples after the consumption of the Ga nanoparticle (*A*) and after exposure to Ga and As beams for 30 s (*B–D*), 60 s (*E*), 150 s (*F*), 300 s (*G*) and 420 s (*H*), respectively

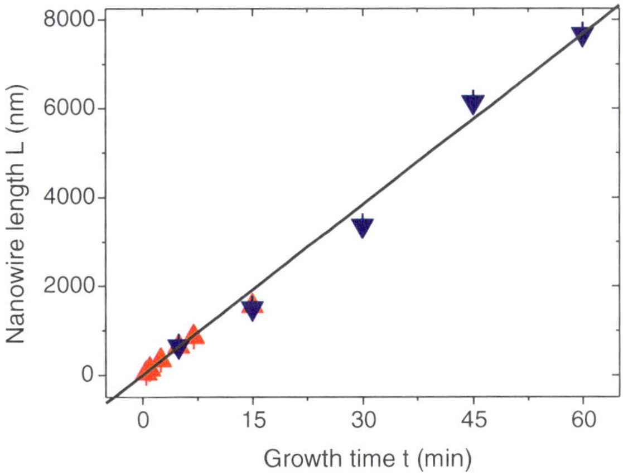

Fig. 5.33 Regrowth length as measured from the interface (*up triangles*) and length of the regular Ga-catalyzed nanowires grown in the same conditions (*down triangles*) as a function of time. *Solid line* is the fit according to the model described in the text

(iii) Modeling of the re-growth process and explanation of the observed non-linear radius-time $[r(t)]$ dependence shown in Fig. 5.34, while the length-time correlation remains almost exactly identical to that at the regular growth stage (Fig. 5.33).

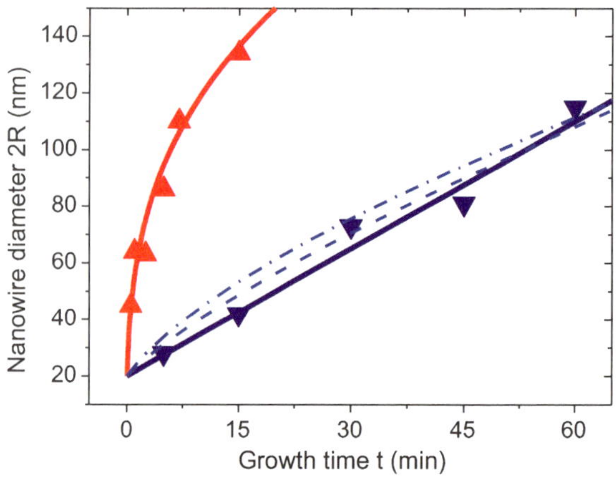

Fig. 5.34 Diameters of the regrown nanowires (*up triangles*) and diameters of the Ga-catalyzed nanowires at the regular growth stage (*down triangles*) in the same conditions, as a function of time. *Lines* are theoretical fits according to the model described in the text

The model of the regular growth is schematized in Fig. 5.35a. The wire cross-section is approximated by a cylinder of a constant radius R, which equals the base radius of a spherical cap droplet with a fixed contact angle β. The regular growth stage is described by (5.75) and (5.76). However, in these experiments the diameter of the Ga droplets increases with time. We therefore consider the case of excessive Ga influx where

$$\frac{dN_{Ga}}{dt} > \frac{dN_{As}}{dt}. \tag{5.79}$$

The nanowire elongation is arsenic-limited and is simply proportional to the effective atomic As influx, therefore $\pi R^2 dL/dt = \Omega_{GaAs} dN_{As}/dt$, where Ω_{GaAs} is the elementary volume of a GaAs pair in the solid phase. Using (5.75) and integrating, we obtain the observed linear length-time dependence

$$L = \Omega_{GaAs} I_{As,eff} t \equiv Ct. \tag{5.80}$$

This is used to fit the experimental data, as shown in Fig. 5.33, with C = 130 nm/min.

The excessive Ga influx leads to the increase of the liquid volume V according to $dV/dt = \Omega_{Ga}\,(dN_{Ga}/dt - dN_{As}/dt)$, where Ω_{Ga} is the elementary volume of Ga in the liquid phase. Using (5.75) and (5.76), and the condition $\beta = const$, we arrive at the differential equation

$$\frac{dR}{dt} = A + \frac{B}{R}. \tag{5.81}$$

The coefficients A and B are given by

$$A = \frac{\Omega_{Ga}}{f(\beta)}(I_{Ga,eff} - I_{As,eff});\; B = \frac{\Omega_{Ga}}{f(\beta)} \frac{2\lambda I_{Ga} \sin\alpha_{Ga}}{\pi}\left(1 - \frac{\theta_l}{\theta_f}\right), \tag{5.82}$$

with $f(\beta) = [(1-\cos\beta)(2+\cos\beta)]/[(1+\cos\beta)\sin\beta]$. The parameter A is proportional to the Ga/As effective flux imbalance, while B is proportional to the diffusion flux going into the droplet, or out of it. Its sign, corresponding to the direction of the flux, is determined by the relative activities of the Ga adatoms in the liquid phase and on the sidewalls. If the activity of the adatoms on the facets is larger than that of the liquid, the diffusion flux is directed towards the droplet, and B is positive.

Let us now assume that B is positive. The A value in (5.81) can be of either signs: $A < 0$ (excess $J_{As,eff}$, or $A > 0$ (excess $J_{Ga,eff}$). In the former case, large droplets with $R > -B/A$ shrink and small droplets with $R < -B/A$ inflate, until the steady state condition is reached at $R_s = -B/A$. This state can only be achieved by Ga diffusion assistance, which compensates the excessive As droplet flux. In the latter case, all droplets inflate regardless of their initial size. However, the radius-time relationship depends drastically on the values of A and B. Indeed, integration of (5.81) with the initial condition $R(t=0) = R_0$ readily yields

$$At = R - R_0 + \frac{B}{A}\ln\left(\frac{R_0 + B/A}{R + B/A}\right). \tag{5.83}$$

Whenever the ratio $B/(AR)$ is large compared to unity, the radius scales with time as $R^2 = R_0^2 + 2Bt$ (actually regardless of the sign of A). In the opposite case (negligible diffusion, $B/(AR) \ll 1$), the radius-time correlation is linear: $R = R_0 + At$.

Our experimental data demonstrate a linear $R(t)$ correlation during the regular growth, shown in Fig. 5.34. Blue curves in the figure are theoretical curves obtained from (5.83) at $A = 0.75\,\text{nm/min}$, $B = 0$ (bold solid line), $A = 0.43\,\text{nm/min}$, $B = 8.75\,\text{nm}^2/\text{min}$ (dashed line), and $A = 0.275\,\text{nm/min}$, $B = 15\,\text{nm}^2/\text{min}$ (dash-dotted line). It is seen how increasing B leads to non-linearity of the curves and worsens the correspondence. We therefore conclude that (i) our regular growth proceeds in the regime where the Ga droplet flux exceeds that of As, and (ii) sidewall diffusion of Ga is ineffective. Both conclusions look quite surprising. At a high As_4 to Ga BEP ratio of 5, one could anticipate an excessive As influx to the droplet (probably with a significant contribution of re-emitted arsenic), which is compensated by the sidewall diffusion of Ga. However, this is not the case: our results reveal an excessive Ga influx that increases the droplet volume at a negligible diffusion. Rather small actual ratio of $I_{As,eff}/I_{Ga,eff}$ is most probably due to a high desorption rate of As caused by its low solubility in liquid Ga. On the other hand, a high resulting elongation rate does not exclude an efficient secondary source of Ga, also caused by re-emission. As regards an effective absence of Ga diffusion, it cannot be explained solely by Ga re-evaporation, because the diffusion length λ should be more than at least 100 nm even at 640 °C. We note, however, that the diffusion length can be considerably reduced by the sidewall incorporation. Indeed, a cylindrical nanowire shape can only be preserved by the successive nucleation and growth of shells around the nanowire, a process occurring simultaneously with the droplet inflation. Smallness of the B value may also be due to the factor $1 - \theta_l/\theta_f$ which can be very close to zero or even negative if the chemical potentials of Ga atoms in the liquid phase and on the sidewalls are close to each other. In any case, an excessive Ga droplet flux seems to be the only reasonable explanation of the linear $R(t)$ dependence observed. For $\beta = 130^0$, we estimate the relative Ga excess from the obtained $A/C = 0.00577$ at $(I_{Ga,eff} - I_{As,\,eff})/I_{As,\,eff} \cong 0.09$, a substantial value.

Next, we consider the nucleation of Ga nanoparticle on the nanowire top, a process shown in Fig. 5.35b. From similar considerations as above, the net fluxes of As and Ga to the top facet are given by

$$\frac{dN_{As}}{dt} = \left(\frac{I_{As}}{n}\cos\alpha_{As} - I_{As}^{des}\right)\pi R^2;$$
$$\frac{dN_{Ga}}{dt} = (I_{Ga}\cos\alpha_{Ga} - I_{Ga}^{des})\pi R^2 + I_{Ga}\sin\alpha_{Ga}2R\lambda\left(1 - \frac{\theta_{top}}{\theta_f}\right). \quad (5.84)$$

Here, I_{As} is the direct impingement flux of arsenic, α_{As} is the corresponding incidence angle, and I_{des} describe desorption from the top facet. The diffusion term is similar to that in (5.76), where the Ga activity in the liquid is changed to that on the top facet (θ_{top}). In contrast to the previous case, the quantity $n > 1$ accounts for the re-evaporation of a certain portion of impinging As species from a solid surface. It is commonly accepted that $n = 2$ for As_4, i.e., every second dimer re-evaporates after dissociation of arsenic tetramers at the solid surface [522]. From (5.84), it can be seen that the difference $dN_{Ga}/dt - dN_{As}dt$ of Ga and As atoms precipitating to the top facet without droplet is always larger than that to the droplet. There are four main reasons for that: (i) the $1/n \cong 1/2$ factor excluding one half of arriving As atoms from the very beginning, (ii) absence of As re-emission flux which is blocked by the nanowire corners; (iii) purely geometrical factor of a much larger inclination of As beam to the substrate normal ($\alpha_{Ga} = 14°$ and $\alpha_{As} = 45°$ in this particular MBE chamber), and (iv) possibly larger contribution of Ga sidewall diffusion in this case (we saw earlier that it is almost negligible with the droplet). As a result, if the regular growth proceeds under Ga rich conditions, the Ga excess should always be larger at the flat facet, which explains well a rapid nucleation of Ga droplet on top of the nanowire.

Considering the regrowth process, after a short incubation time required to reach the stationary droplet configuration (which is actually reached after only 60 s of deposition), the conditions for the As droplet flux are equivalent to the regular growth. Therefore we can conclude that nanowire length depends linearly on the growth time ($l = Ct$) with the same time constant, in agreement with the experimental data in Fig. 5.33. However, if we consider the nanowire diameter, the experimental $r(t)$ correlation is drastically different from $R(t)$ for short growth times. The only difference between the two configurations shown in Fig. 5.35a, c is the additional diffusion flux coming from Ga impinging onto the top facet, which definitely dominates at small enough l, shorter than Ga diffusion length λ. Indeed, Ga atoms arriving there can subsequently reach the droplet and contribute to the droplet inflation. We will therefore use the equation

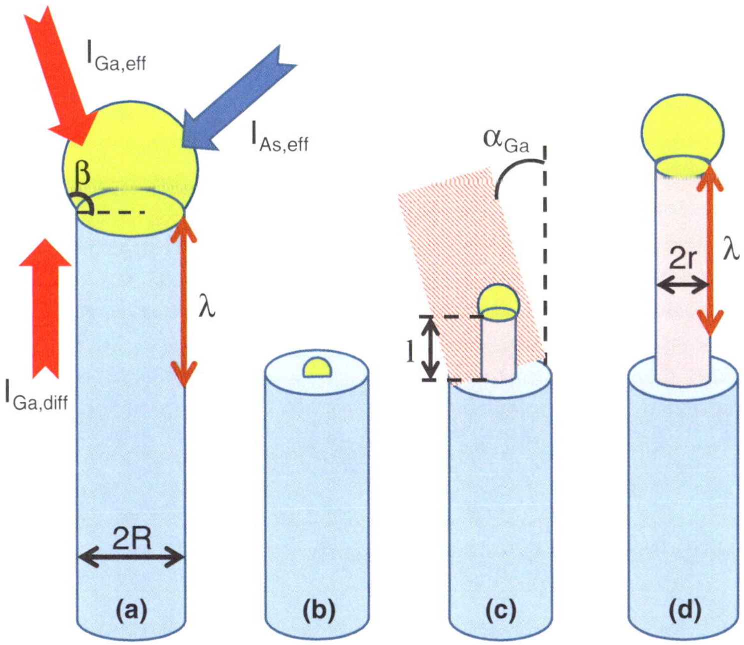

Fig. 5.35 Schematic of the nanowire growth process as described in the text

$$\frac{dv}{dt} \cong \Omega_{Ga}\pi r^2(I_{Ga,eff} - I_{As,eff}) + \Omega_{Ga}\pi(R^2 - r^2)I_{Ga}\cos\alpha_{Ga}\left(1 - \frac{\theta_l}{\theta_s}\right), r < R \tag{5.85}$$

for the increase of the droplet volume v on top of the secondary nanowire. The first term here is the same as at the regular growth stage. The second term contains the factor $1 - \theta_l/\theta_s$, accounting for the supersaturation of the surface Ga adatoms with respect to the droplet. The term $\pi(R^2 - r^2)$ accounts for the free portion of the top facet and vanishes at $r = R$, i.e., when the radius of the top section reaches that of the stem. Diffusion of atoms impinging on the sidewall is ignored according to the results obtained for the regular growth process. Using the relationship $dv/dt = f(\beta)\pi r^2 dr/dt$ at $\beta = const$, we arrive at

$$\frac{dr}{dt} = (A - b) + b\frac{R^2}{r^2}, r < R;$$

$$\frac{dr}{dt} = A, r \geq R, \tag{5.86}$$

with

$$b = \frac{\Omega_{Ga}}{f(\beta)} I_{Ga}\cos\alpha_{Ga}\left(1 - \frac{\theta_l}{\theta_s}\right).$$

The solution to the first (5.86) contains two distinctly different regimes, namely the linear regime $r = r_0 + (A - b)t$ and the non-linear regime (at $R \cong const$)

$$r^3 = r_0^3 + 3bR^2t. \tag{5.87}$$

The resulting $r(t)$ correlation depends on which of the two terms in the right hand side of the first (5.86) dominates. Since our experimental $r(t)$ curve is essentially non-linear, the b value should be large enough compared to A. The red line in Fig. 5.34 shows the fit obtained from (5.87) with $b = 2.34$ nm/min, which is noticeably larger than A. Finally, when r reaches the R value, since there is no contribution from Ga atoms impinging on the remaining area of the top facet, the conditions are exactly identical to the regular growth stage, where the kinetic equation for dr/dt is given by the second (5.86). This explains the tendency for saturation observed in the experimental $r(t)$ dependence of Fig. 5.34.

5.10 Self-Induced GaN Nanowires

Catalyst-free, self-induced GaN nanowires grown on Si substrates [312, 419–437] present another interesting opportunity for heterogeneous integration of optical nanosemiconductors with Si electronic platform. To understand the origin of the self-induced formation of GaN nanowires, several studies have been devoted to the initial nucleation growth step. In particular, very important data have been reported by Riechert and coauthors [431, 432] and Daudin and coauthors [428, 429], showing that GaN nanowires grown by MBE usually start with the Volmer-Weber 3D islands. In this section, we present relevant experimental data that will then be used to develop theoretical models of the self-induced formation of GaN nanowires.

In [432], GaN nanowires were obtained by radio-frequency plasma-assisted MBE of Si(111) substrates covered with a thin amorphous Si_xN_y interlayer. To form the latter, the Si(111) substrate was exposed for 5 min to an active nitrogen flux to induce a nitridation process. GaN wires were subsequently grown at a high substrate temperature of 780 °C under highly nitrogen-rich conditions with a V/III ratio of 6.2. *In situ* RHEED measurements were carried out during the entire growth process. One RHEED pattern was recorded every 12 s: The RHEED intensity was determined by monitoring in situ the diffraction intensity in a fixed rectangular area at a given position that involves several GaN 3D spots, as depicted by the dashed white rectangle in the inset of Fig. 5.36. The evolution of the GaN spot intensity was then assessed by integrating and summing the overall area over each GaN 3D spot. In order to combine the RHEED measurements with a microstructural analysis, cross-sectional TEM specimens were prepared from dedicated samples grown at different durations of 7, 11, 16, and 24 min. HRTEM experiments were performed with a JEOL 3010 microscope operating at 300 kV.

The evolution of the RHEED intensity is shown in Fig. 5.36 and reveals in correlation with the HRTEM images that the growth process consists of four consecutive

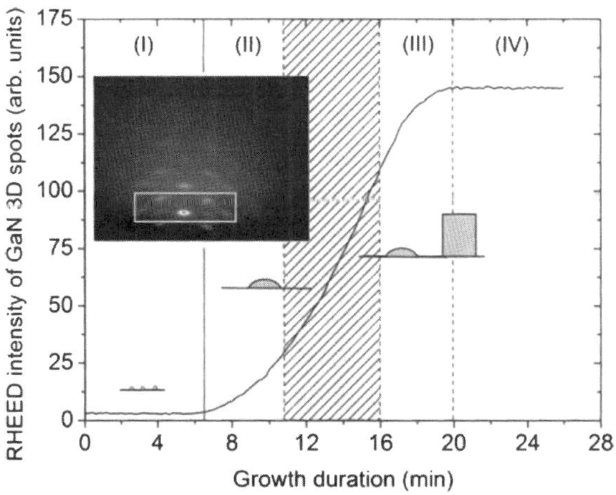

Fig. 5.36 Evolution of the RHEED intensity, revealing four distinct consecutive regions. The first insets represent simple sketches of GaN island shape, as observed by HRTEM images in Fig. 5.37 recorded on dedicated samples. The *dashed zone* points out that the first spherical cap undergoing a shape transition toward the nanowire morphology takes place between 11 and 16 min according to HRTEM imaging. The second inset represents a typical RHEED pattern of the GaN nanowires (40 min). The *rectangular area* depicted by *dashed white lines* yields the RHEED intensity from the GaN 3D spots [432]

stages. Typically, the RHEED intensity is initially close to zero and then increases and eventually saturates. At the beginning, the 7×7 surface reconstruction of the Si(111) substrate is observed and rapidly vanishes once the nitridation process begins. Stage (I) starts with the subsequent opening of the gallium shutter (i.e., at $t = 0$). In this stage, no RHEED signal is detected, revealing the occurrence of an incubation period. A delay in the nucleation process owing to a massive gallium desorption at the high growth temperature of 780 °C could be involved during this period.

The start of stage (II) takes place after an incubation time of $\cong$ 6.5 min and is marked by the occurrence of the first GaN spots along rings on the RHEED patterns as shown in the inset of Fig. 5.36. This stage is associated with the formation of the first stable 3D GaN islands. A typical HRTEM image of a GaN island is shown in Fig. 5.37: spherical cap-shaped islands are nucleated on a continuous SixNy amorphous interlayer with a thickness of 1.5–2 nm. The start of stage (III) corresponds to the formation of the very first GaN nanowires after a nucleation time of roughly 16 min and is deduced from HRTEM images collected on dedicated samples. An example of all the collected HRTEM images is presented in Fig. 5.37. Nanowire-shaped islands having $\{0001\}$ top facet and $\{1\bar{1}00\}$ vertical sidewalls (i.e., *m*-planes) are observed. The GaN nanowires are expected to be gallium-polar as reported in [435]. GaN islands thus undergo a shape transition from spherical caps to nanowires during stage (III). Eventually, stage (IV) starts with the constancy of the RHEED intensity after 20 min. It should be noted that the presence of spherical caps is still

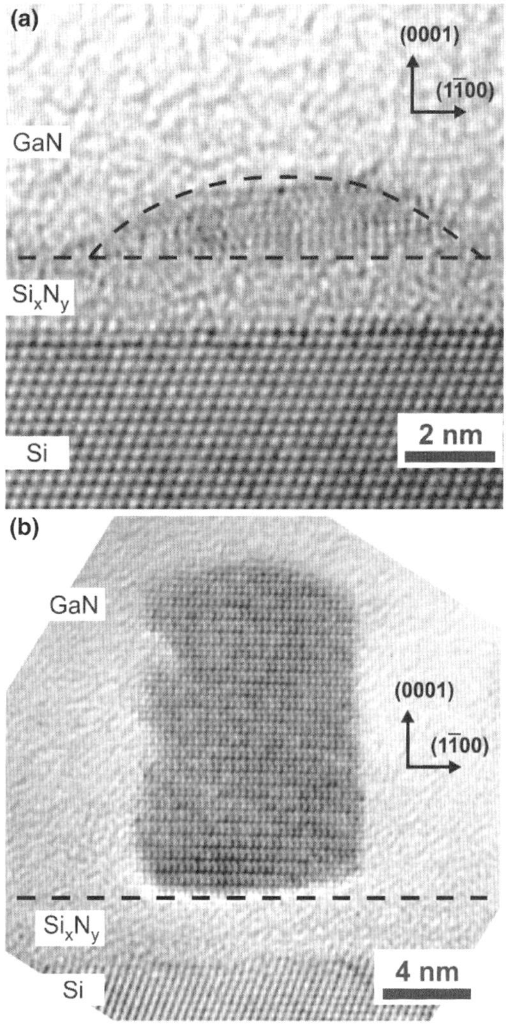

Fig. 5.37 HRTEM images collected on dedicated samples grown for 7 and 16 min, revealing the following respective GaN island shapes at the onset of the nucleation process: spherical cap-shaped island (**a**) and nanowires (**b**). The nanowires are hexahedral. The shape of spherical caps is *outlined* for the sake of clarity [432]

detected in this stage. However, the nanowire density continuously increases with respect to the spherical cap density for longer growth durations.

The height h and radius r_0 of spherical caps and nanowires are systematically determined from HRTEM images, as shown in Fig. 5.38. There exist two separate ranges of specific sizes for spherical caps and nanowires. The spherical caps have a typical radius and height systematically smaller than 5 and 2 nm, respectively. The wetting angle θ equals $42 \pm 7°$ in the entire range of their sizes. The shape transition from spherical caps toward the nanowire morphology is therefore discontinuous and of first-order type. The formation of the very first GaN nanowires abruptly occurs at an experimental critical radius of 5 nm. From geometrical considerations, a critical

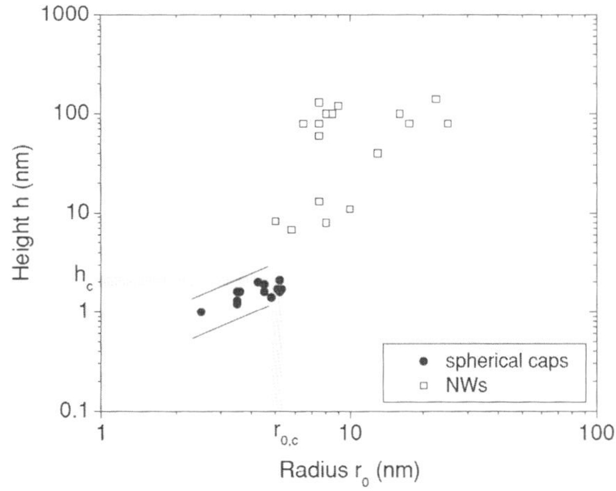

Fig. 5.38 Height h as a function of radius r_0 obtained from HRTEM images for the two different shapes of GaN islands as growth proceeds: the spherical caps and nanowires are observed on dedicated samples grown for 7, 11, 16, 24, 16 and 24 min, respectively. The gray color regions correspond to the experimental critical height and radius [432]

island height of about 2 nm is also determined. After that, the nanowire aspect ratio gradually increases with time.

In [431], similar studies were reported for GaN nanowires grown on a lattice mismatching AlN interlayers on Si(111). As opposed to the previous case (where the expitaxial constraint should be very week), the GaN island formation on AlN should be essentially stress-driven as discussed in Chap. 3. MBE growth experiments were as follows. First, a 5-nm-thick AlN buffer layer was deposited on the Si(111) substrate in order to ensure an epitaxial relation with the GaN nanowires: the aluminum atoms were accumulated on the substrate surface at 800 °C and then exposed to the active atomic nitrogen flux for 5 min to induce a nitridation process. The resulting AlN buffer layer is completely relaxed, as revealed by x-ray diffraction measurements. GaN were then were grown at a high substrate temperature of 800 °C under highly nitrogen-rich conditions, with a V/III ratio of 10.

The evolution of the two distinct RHEED intensities is presented in Fig. 5.39. The insets represent simple sketches of the GaN island morphology, as obtained from HRTEM images Fig. 5.40. From the evolution of the two distinct RHEED intensities, the nucleation process can now be divided into six stages. During the stage (I), the AlN intensity decreases corresponding to a gradually increasing AlN surface coverage by Ga adatoms and 2D GaN nuclei. The start of stage (II) after an incubation time of about 3.5 min is marked by an increase in the overall intensity in Fig. 5.39. This time corresponds to the occurrence of the first 3D transmission spots belonging to GaN on the RHEED patterns. A typical HRTEM image of a GaN 3D island, shown in Fig. 5.40a, reveals the spherical cap morphology similar to that on an amorphous interlayer, with a radius of 2 nm and a height of about 1 nm. No sign for the presence of a complete continuous 2D wetting layer is observed by HRTEM images either during the stage (I) or during the stage (II), showing that the islands are formed in the Volmer-Weber mode.

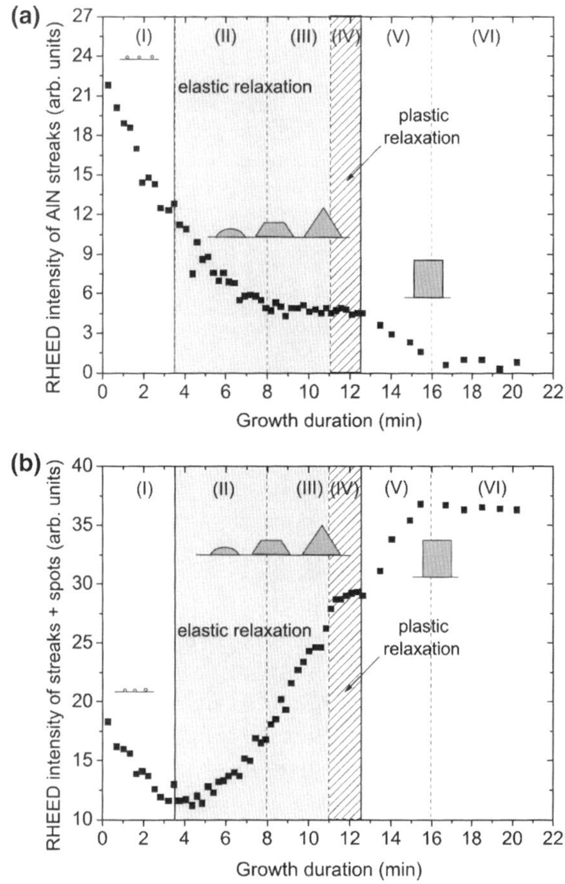

Fig. 5.39 Evolution of the RHEED intensity (**a**) of AlN streaks and (**b**) of its sum with the GaN spot intensity as growth proceeds. The insets represent simple sketches of GaN island shape, as observed by HRTEM images in Fig. 5.40 collected on dedicated samples. Six distinct stages are distinguished within such an analysis. The two consecutive regions involving elastic and plastic relaxation processes are shown [431]

Figure 5.40b reveals the presence of truncated pyramid shaped islands. GaN islands thus undergo a shape transition from spherical caps to truncated pyramids during stage (II). Pyramidal islands were not seen on an amorphous Si_xN_y interlayer and should be due to the strain induced by lattice mismatch. As discussed in Chap. 3, the formation of flat top facets and of the lateral facets should aim at elastically relieving this strain: the total free energy per unit volume of each island is reduced by decreasing the stored elastic strain energy more than increasing the surface energy through the newly formed surfaces. During stages (I) and (II), the spherical cap and pyramidal islands remain dislocation-free, as shown in Fig. 5.39.

Stage (III) starts after a growth time of 8 min and is characterized by an abrupt change in the evolution of the AlN streak intensity. The intensity stops decreasing and reaches a plateau around this growth time. The saturation of the AlN streak intensity reveals that the island density also saturates and remains constant for longer growth durations. Figure 5.40c reveals the presence of full pyramid-shaped islands with

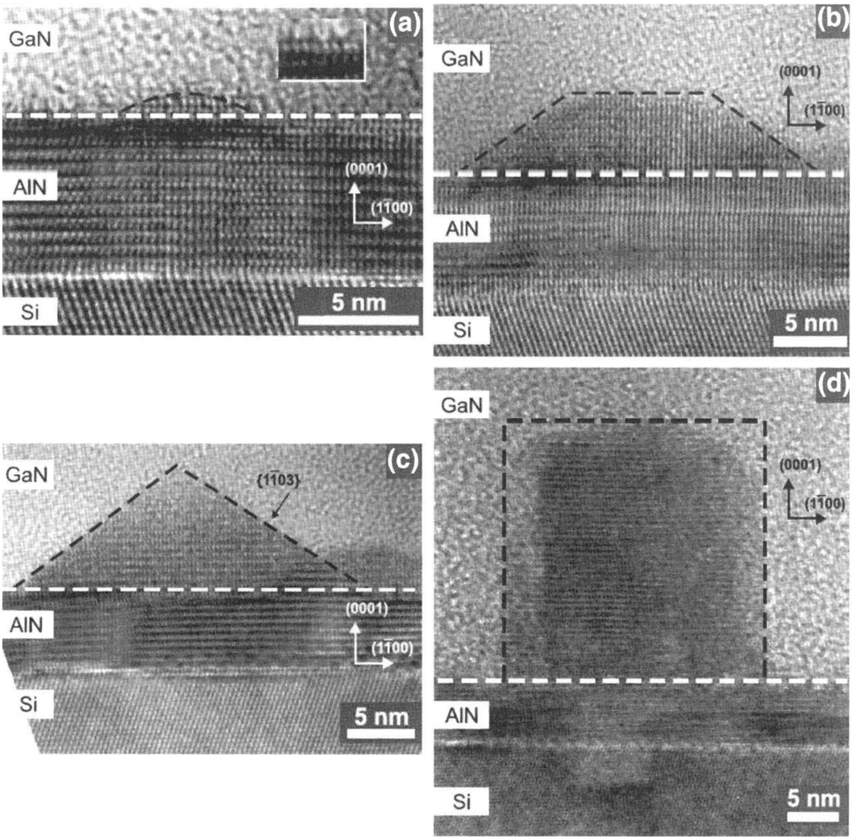

Fig. 5.40 HRTEM images collected on dedicated samples grown during 4.5, 6, 9, 10, and 15 min, revealing the following respective GaN island shapes: **a** spherical-cap-shaped island with an inset representing a high magnification of the first AlN monolayers at the interface, **b** truncated pyramid-shaped island, **c** full-pyramid-shaped island, and **d** nanowire. The pyramid-shaped island and nanowire are hexahedral: the full pyramid- shaped island presents a facet angle of 32° with respect to the (0001) plane, suggesting that the $\{1\bar{1}03\}$ side facets are involved. All the island shapes are outlined for clarity [431]

larger dimensions compared to the sizes of truncated pyramids. GaN islands thus undergo a new shape transition from truncated to full pyramids during this stage. Subsequently, stage (IV) starts with the plastic relaxation process occurring between 11.5 and 12.5 min. The dislocation nucleation becomes predominant and induces an abrupt strain relaxation toward a completely strain-free state. A typical example of full pyramid island with dislocation at the heterointerface is shown in Fig. 5.41. Interestingly, the plastic relaxation occurs at the end of the plateau for the AlN streak intensity, as depicted by the hatched region in Fig. 5.39a. The overall intensity also slightly saturates upon the plastic relaxation process as shown in Fig. 5.39b, revealing that such a process directly affects the growth mechanisms. An example among all

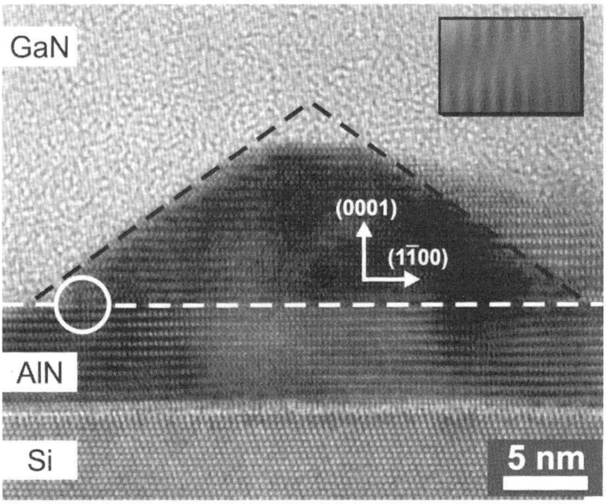

Fig. 5.41 HRTEM image of a dislocated full-pyramid-shaped island collected on a sample grown during 12 min. The misfit dislocation is located within the white circle, which is close to the island edges. The inset reveals a Fourier-filtered enhancement showing only the $\{1\,\bar{1}\,00\}$ lattice planes. The dislocation is mixed with a Burgers vector of $(1/3) < 1\,\bar{1}\,20 >$ and retains a dominant edge component. Note that the full-pyramid-shaped island exhibits one top facet due to its cut during the TEM sample preparation

the collected HRTEM images is shown in Fig. 5.40d and reveals the presence of nanowire-shaped hexahedral islands.

The start of stage (V) is revealed by the rapid decrease in the AlN streak intensity after a growth time of 12.5 min, especially once the plastic relaxation process has been completed. The single GaN spot intensity increases again at the same time. The nanowire height rapidly increases due to the preferential axial growth along the [0001] direction compared to the radial growth along the [1$\bar{1}$00] direction, as in the case of growth on an amorphous interlayer. This is expected to result in more intense 3D transmission spots belonging to GaN. Simultaneously, shadowing processes prevent the surface of the AlN buffer layer from being probed by the electron beam, resulting in the vanishing of the AlN streak intensity. Finally, stage (VI) is defined by the constancy of the main final features of the RHEED patterns after a growth time of 16 min, as nanowire growth proceeds. The electron beam probes only the nanowire top region, whose crystallographic structure does not change with time.

Based on these results and also considering the data presented in Sect. 4.2, we can formulate the most important points regarding the self-induced growth of GaN nanowires on Si(111) by MBE. First, no Ga droplet is detected at the nanowire top by electron microscopy imaging, so the self-catalytic effect cannot explain their formation according to the VLS mode. Second, the self-induced approach commonly employs specific MBE growth conditions: a highly nitrogen-rich vapor phase is required and often combined with high substrate temperature. Decreasing the nitrogen to gallium fluxes ratio often leads to a more pronounced radial growth [423].

Surface diffusion plays a crucial role in the nanowire elongation [312, 422]. Third, GaN never nucleates in the nanowire morphology [419, 428, 431, 432]. Rather, the nanostructures emerge as 3D nanoislands having a fixed shape and a complex free surface composed of high-index atomic planes, which can be interpolated by a curved spherical cap surface. On a lattice mismatched AlN layer, 3D nanoislands undergo a series of shape transformations, whereby the misfit dislocations are developed at the interface prior to the nanowire formation. It can therefore be concluded that GaN nanowires are relaxed from the very beginning. During growth on an amorphous interlayer, the spherical cap islands transform directly to nanowires. This growth transformation occurs at a critical radius of about 5 nm. Fourth, GaN nanowires are hexahedral, restricted by six equivalent m-plane vertical sidewalls that are the low energy planes. Fifth, self-induced GaN nanowires usually grow in both vertical and radial directions, but the vertical growth rate is much faster [422, 431, 432]. These features will be used for modeling the self-induced formation of GaN nanowires, discussed in the next three sections.

5.11 Nucleation of GaN Nanowires

The initial nucleation stage of self-induced GaN nanowires from 3D islands should be driven by the energetic factors. Here, we use a thermodynamic approach [436] where the preferred growth configuration corresponds to the minimum formation energy for different geometries. Based on the observations given in the previous section, 3D GaN nanoislands (e.g., the spherical cap islands on Si_xN_y) grow at a fixed aspect ratio. In contrast, the aspect ratio of GaN nanowires should increase as they grow, otherwise they would never become nanowires. We will therefore distinguish between the two major modes of self-induced growth:

(i) Isotropic growth with a fixed shape, the case of the Stranski-Krastanow islands, catalyst-free nanoneedles (see Chap. 3), and GaN 3D islands at the initial nucleation step;
(ii) Anisotropic growth with the aspect ratio increasing with time, the case of GaN nanowires. Illustrations of the two growth modes, along with the data on the spherical cap-to-nanowire shape transformation on an amorphous interlayer, are presented in Fig. 5.42.

In the foregoing analysis, we ignore the strain induced by the lattice mismatch, which is directly applicable on an amorphous layer. While the energetically preferred island shape can drastically depend on the elastic relaxation and plastic deformation as described above, our approach would also apply to the growth on a lattice mismatched layer, because the strain is released prior to the nanowire formation and the elastically relaxed islands have a fixed pyramidal shape and aspect ratio. We consider the shape transformations due to the anisotropy of surface energies [431], which must be the dominant driving force for any self-induced growth in absence of lattice mismatch. We then couple this with the mentioned nanowire growth anisotropy

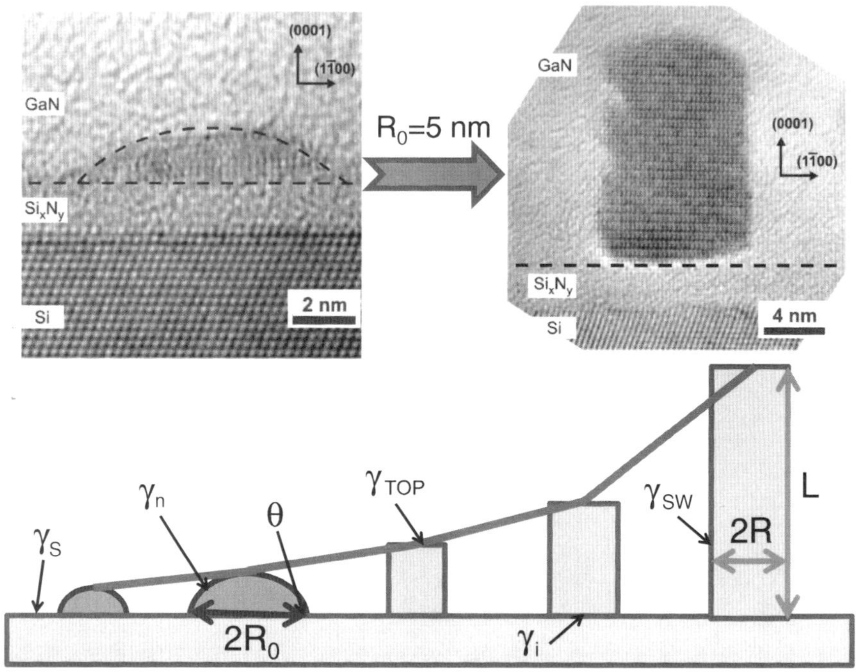

Fig. 5.42 Cross-sectional TEM images of GaN spherical cap island and nanowire on a Si_xN_y amorphous layer from Fig. 5.37, demonstrating the shape transformation at the spherical cap base radius of 5 nm (top); model schematics illustrating the parameters used and the nanowire growth anisotropy (bottom). The shape transformation occurs instantaneously at a given volume so that the nanowire radius R decreases with respect to R_0

in order to access different growth scenarios. Thermodynamic approach is based on the comparison of the formation energies of a 3D island and a nanowire having the same volume under identical growth conditions. The driving force for the island-to-wire shape transformation is defined by the difference of the corresponding formation energies. The kinetic volume term $\Delta\mu V$ (where $\Delta\mu$ is the difference of chemical potentials in the metastable phase and in the solid state and $V = const$ is the volume) should be identical in strain-free islands and nanowires and cancels in the driving force. It is therefore sufficient to consider only the surface and edge contributions (the surface energy for brevity).

We now assume the scaling dependence of the nanowire length (or height) on its radius:

$$L = \nu R^{\delta}, \tag{5.88}$$

where ν is a constant and δ is the growth index. The case of $\delta = 1$ relates to an isotropic growth; the islands elongate with time at $\delta > 1$ (the nanowire case where the vertical growth is faster than the radial extension) and flatten at $\delta < 1$. Equation (5.88) expresses the evolution of a given nanowire and does not contradict

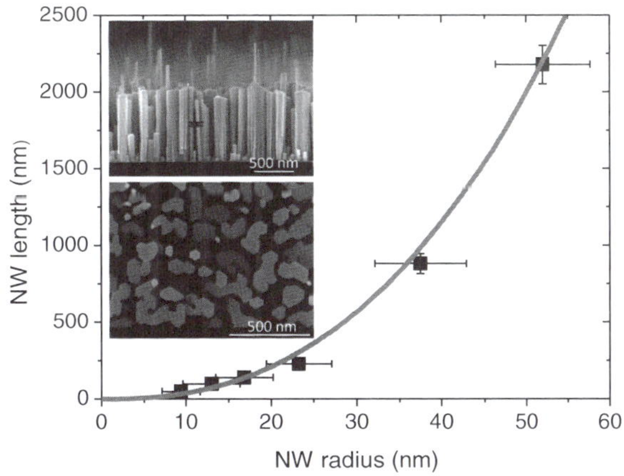

Fig. 5.43 Scaling length-radius dependence of self-induced GaN nanowiress: the experimental data (*symbols*) at $T = 780\,°C$, N rate = 0.28 nm/s, Ga rate = 0.045 nm/s, fitted by the power law $L = 0.132R^{2.46}$; the inserts showing typical plan and top view SEM images of samples grown for 3 h

the well-known relation $L \sim 1/R$ characteristic for the diffusion-induced growth as discussed in Sect. 4.2, which describes an ensemble of nanowires at a given moment in time. By analyzing TEM and SEM images of GaN nanowires grown at typical MBE conditions (T = 780 °C, N rate = 0.28 nm/s, Ga rate = 0.045 nm/s) with different growth durations, the length—radius dependence can be thoroughly deduced [436]. The nanowire length and radius were assessed among a population of more than 50 wires, taking into account both the incubation and the transition time that are needed before the nanowire growth starts, and the coalescence effects at a later growth stage. As seen from Fig. 5.43, the averaged $L(R)$ curve remarkably follows the scaling (5.88) with $\delta = 2.46$ and $\nu = 0.132$ under these particular MBE growth conditions.

The volume of a 3D island growing in an isotropic mode is given by $V_{ISL} = k_V R_0^3$, where k_V is a shape constant and R_0 is the base radius. In particular, $k_V = [\pi f(\theta)]/3$ with $f(\theta) = [(1 - \cos\theta)(2 + \cos\theta)]/[(1 + \cos\theta)\sin\theta]$ in the case of spherical cap islands with the contact angle θ (see Fig. 5.42). The volume of a hexahedral nanowire with side R and height L writes as $V_{NW} = (3\sqrt{3}/2)R^2 L$. From the condition $V_{ISL} = V_{NW}$ and (5.88), one obtains

$$R_0 = \left(\frac{3\sqrt{3}\nu}{2k_V}\right)^{1/3} R^{\frac{2+\delta}{3}}. \tag{5.89}$$

The surface energy generated upon the formation of a 3D island can generally be put in the form

$$G_{ISL} = \left(\sum_n k_n \gamma_n + k_i(\gamma_i - \gamma_S)\right) R_0^2 + k_\varepsilon \varepsilon_{ISL} R_0. \tag{5.90}$$

Here, γ_n are the surface energies of planes terminating the island surface, γ_i is the interfacial energy, γ_S is the surface energy of the substrate, and ε_{ISL} is the specific edge energy of the island, with k being the shape constants. For example, $\sum_n k_n \gamma_n = 2\pi\gamma_{SC}/(1+\cos\theta)$, $k_i = \pi$ and $k_\varepsilon = 2\pi$ for the spherical cap islands, in which γ_{SC} is the corresponding surface energy. The surface energy of a hexahedral nanowire is given by

$$G_{NW} = 6\gamma_{SW} RL + \frac{3\sqrt{3}}{2}(\gamma_{TOP} + \gamma_i - \gamma_S)R^2 + 6\varepsilon_{NW} R, \tag{5.91}$$

with γ_{SW} as the sidewall surface energy, γ_{TOP} as the surface energy of the top facet, and ε_{NW} as the specific edge energy at the nanowire base.

Expressing R_0 through R in (5.90) by means of (5.89) and re-arranging different terms, we arrive at

$$g_\delta(R) = bR^{\frac{\delta-1}{3}} + cR^{-\frac{2(\delta-1)}{3}} + dR^{-\frac{(2\delta+1)}{3}} - eR^{-\frac{(\delta+2)}{3}} - 1. \tag{5.92}$$

Here, $g_\delta(R) = (G_{NW} - G_{ISL})/\left(AR^{2(2+\delta)/3}\right)$ is the normalized difference of surface energies. The superlinear nanowire growth is preferred at $g_\delta(R) < 0$ and suppressed at $g_\delta(R) > 0$. The coefficients are defined as follows:

$$b = \frac{B}{A};\, c = \frac{C}{A};\, d = \frac{D}{A};\, e = \frac{E}{A}, \tag{5.93}$$

where

$$A = \left(\frac{3\sqrt{3}\upsilon}{2k_V}\right)^{2/3}\left(\sum_n k_n\gamma_n + k_i(\gamma_i - \gamma_S)\right).$$

$$B = 6\upsilon\gamma_{SW};\, C = \frac{3\sqrt{3}}{2}(\gamma_{TOP} + \gamma_i - \gamma_S);$$

$$D = 6\varepsilon_{NW};\, E = \left(\frac{3\sqrt{3}\upsilon}{2k_V}\right)^{1/3} k_\varepsilon \varepsilon_{ISL}. \tag{5.94}$$

The analysis of the driving force given by (5.92) simplifies with neglect of the edge terms. Since the latter are short-range, such an analysis remains qualitatively correct in the general case. At $d = e = 0$, the function $g_\delta(R)$ is negative between the two critical radii $R_{1,2} = x_{1,2}^{3/(\delta-1)}$, where $x_{1,2}$ are the positive roots of the cubic equation

$$bx^3 - x^2 + c = 0. \tag{5.95}$$

The two roots exist provided that

$$b^2 c < \frac{4}{27}. \tag{5.96}$$

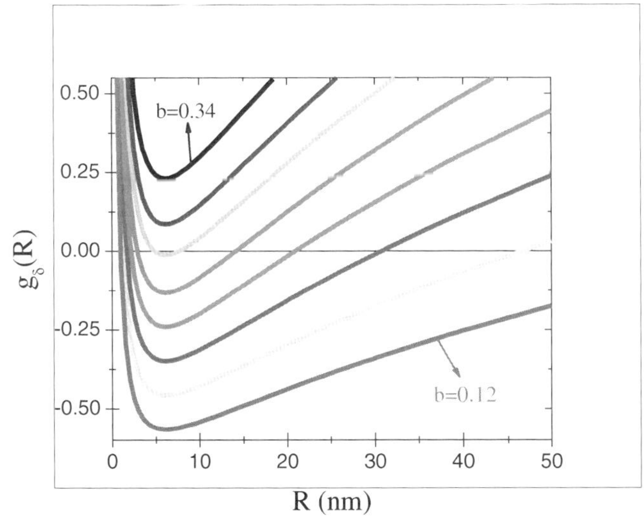

Fig. 5.44 Graphs of g_δ (R) obtained from (5.92) with different b decreasing from 0.34 to 0.12 at fixed $\delta = 2.46$, $\nu = 0.132$, $c/b = 7$ and $d = e = 0$

This inequality should be treated as the necessary condition for any self-induced anisotropic growth driven by the surface energetics and starting from 3D islands. It is noteworthy that (5.96) is much more sensitive to the sidewall energy b than to the in-plane energy c.

Graphs of the driving force for a hypothetical model system at $\delta = 2.46$, $\nu = 0.132$, fixed $c/b = 7$, zero edge terms and different b are presented in Fig. 5.44. The curves demonstrate a very strong dependence of the system performance on b. With the sidewall surface energy coefficient decreasing from 0.34 to 0.12, the energetically preferred nanowire region in R extends from zero to a wide range between ~1 and ~75 nm. The critical radius R_1 of the island-to-wire shape transformation decreases from 5 to 1 nm as the b decreases from 0.273 to 0.12. It should be noted that the existence of the maximum dimension for the energetically preferred anisotropic growth is inevitable in our quasi-equilibrium model, since the b term in (5.92) increases infinitely at larger for any $\delta > 1$. However, such transformations are most probably kinetically forbidden since they require a re-arrangement of too many atoms. Also, sufficiently thick nanowire may coalesce due to their high surface density, which makes the second critical radius R_2 physically unreachable (at $R_2 \gg R_1$).

As discussed, the driving force given by (5.92) applies whenever the volume contribution into the formation enthalpy is identical for the islands and nanowires, while the nanowire height-radius dependence is given by the scaling (5.88). The scaling index δ as well as the surface energies defining the coefficients in (5.92) may depend on the growth kinetic effects. In this sense, the model treats the driving force for the energetically preferred nanowire formation in a particular growth experiment. A study of the nanowire growth kinetics enabling the determination of δ at the

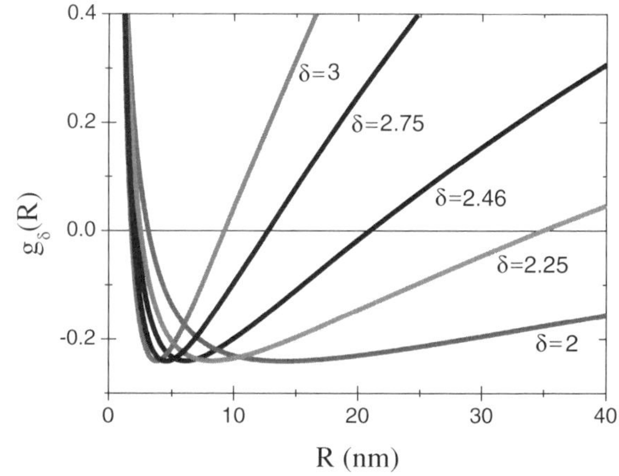

Fig. 5.45 Driving force for the island-to-wire shape transformation at different δ

given growth conditions is presented in the next section. We note, however, that the energetic tendency for the island-to-wire shape transformation will be preserved in a certain domain of radii for any $\delta > 1$ provided that the inequality (5.96) is satisfied, i.e., at a low enough surface energy of the nanowire sidewalls. Indeed, the condition for the preferred anisotropic nanowire growth given by (5.96) does not depend on δ at all, while its dependence on ν cancels in view of (5.94). This feature is illustrated in Fig. 5.45, where the driving force is plotted against the nanowire radius at fixed $b = 0.21$, $c = 1.47$, zero edge contributions and different δ. It is seen that the first critical radius does not change significantly, while the energetically preferred nanowire region extends drastically as the scaling index decreases from 3 to 2.

Let us now consider the parameters of GaN nanowires on an amorphous Si_xN_y layer, corresponding to well-defined experimental growth conditions [431]. As shown in 5.43, the island-to-wire transformation in this case typically occurs at $R_0 = 5$ nm, relating to the first critical radius of nanowire $R_1 = 3.4$ nm by (5.89) at $\theta = 42°$. As for the surface energies, only the values of γ_{SW} for the sidewall m-planes of 118 meV/A^2 (with neglect of surface reconstruction) [523] and $\gamma_S = 137$ meV/Å^2 [524] are known with reasonable accuracy. We therefore study the driving force given by (5.92)–(5.94) within a plausible range of surface energies, sticking to the experimental scaling indices deduced from the data of Fig. 5.43.

We first use the values given in [431]: $\gamma_i = 40$ meV/Å^2 by the analogy with the Si/SiO_2 interface, and $\gamma_{SC} = 130 - 176$ meV/Å^2 from the Young's equation. The surface energy of the nanowire top facet must be larger than γ_{SW}, because the nucleation-mediated vertical growth (involving the formation of m-plane facets of a monolayer height) is faster than the radial one (involving the formation of c-planes). We therefore use the value of γ_{TOP} between 120 and 130 meV/Å^2. With the experimental contact angle of the spherical cap islands $\theta = 42^o$, and the scaling parameters $\delta = 2.46$ and $\nu = 0.132$, this yields the following average values of coefficients: $A = 228$, $B = 99$, $C = 86$, $b = 0.43$ and $c = 0.38$. The corresponding driving force with zero edge terms is shown in Fig. 5.46 (curve 1). It is seen that the energetically preferred nanowire region is quite narrow, extending from 1 to 5 nm only. Curve 2 in Fig. 5.46 corresponds to the same surface energies with the edge contributions at $d = 12$ and $e = 6$. Inclusion of the edge terms corrects the critical radius R_1 of the shape transformation to the experimental value of 3.4 nm, but the nanowire region remains narrow. However, we definitely observe a thermodynamic tendency for the nanowire formation in the correct range of critical radii.

Further, the value of interfacial energy γ_i at 40 meV/Å^2 might be largely underestimated, because the GaN/Si_xN_y interface should be much more energetic than the Si/SiO_2. This argument should hold since GaN and Si_xN_y materials form a heterogeneous interface, which is distinctly different from the Si/SiO_2 homogeneous interface. We therefore use the value of $\gamma_i = 150$ meV/Å^2 as the upmost estimate. Since the dangling bonds of m-plane sidewalls can be passivated by N under nitrogen-rich conditions, we take the value of $\gamma_{SW} = 90$ meV/Å^2 as the lowest estimate. The spherical cap island surface is composed of high-index atomic planes of high surface energy, and many edges separating the planes, which may result in a much higher surface energy γ_{SC} than that given in [431]. We thus use the value of $\gamma_{SC} = 200$ meV/Å^2 as the upmost estimate. With these parameter set, the coefficients entering (5.92) and (5.94) are changed to: $A = 526$, $B = 76$, $C = 372$, $b = 0.144$ and $c = 0.71$, the case relating to the widest range of the energetically preferred GaN nanowire growth on an amorphous layer. The corresponding driving force at $d = e = 0$ is given by curve 3 in Fig. 5.46. As in the previous case, the curve without the edge terms predicts a lower critical radius R_1 ($\cong$1 nm), while the curve with the edge terms corrects it to the experimental value. This corresponds to curve 4 in Fig. 5.46 obtained with $d = 14$ and $e = 6$. While nothing can be said about the edge energy, it usually increases with the inclination angle of island facets [148]. In view of a small ν in (5.94) for E, the inequality $d > e$ (obtained for both limit cases considered) seems reasonable. The edge terms are thus shown to be important at the beginning of growth, in particular, for the determination of the first critical radius. Otherwise, the two curves match at large R, showing that the second critical radius R_2 for the reverse transformation to an isotropic mode is determined entirely by the surface energies.

As mentioned already, the surface energies of GaN nanowire sidewall planes are sensitive to the vapor environment in a particular growth experiment, while the scaling index δ may depend on the kinetic factors. The results shown in Figs. 5.45 and 5.46 demonstrate that the value of R_1 is not considerably affected by varying these values. This yields the island-to-wire shape transformation within a plausible range of parameters. We note, however, that the scaling growth law given by (5.88) with $\delta > 1$ is absolutely necessary to explain why GaN tends to nucleate in the form of Volmer Weber nanoislands rather than nanowires at the beginning of growth, and undergoes the shape transformation only when the base dimension exceeds a certain critical value. Indeed, at $\delta = 1$ (5.92) with zero edge terms becomes radius-independent: $g_1 = b + c - 1$ showing that, while one of isotropic geometries might be preferred to another, no critical radius for the shape transformation exists in the isotropic case. Hence, the anisotropy of surface energies must be coupled with the scaling growth anisotropy to observe the self-induced nucleation of nanowires from

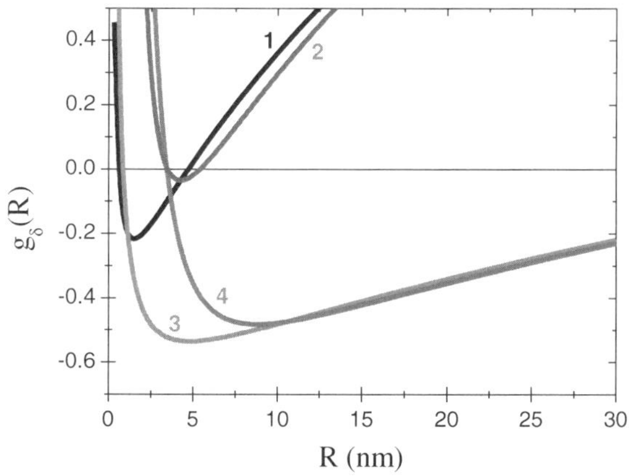

Fig. 5.46 Driving forces for the superlinear nanowire growth at $\delta = 2.46$ and $\nu = 0.132$ for different parameters of GaN on an amorhous layer

pre-existing islands, which is the main message of this section. On the other hand, changing the growth conditions strongly affect the R_2 value, however, the reverse transformation to an isotropic growth should be suppressed on kinetic grounds. It can therefore be said that the initial stage of the nanowire nucleation is driven by the surface energetics in a wide range of deposition conditions, while the follow-up nanowire growth itself proceeds in the kinetically controlled mode.

To sum up the results of this section, whenever the condition for the anisotropic growth given by (5.96) is fulfilled, the model predicts two distinct stages of system evolution for $\alpha > 1$: (i) *Nucleation and isotropic growth of islands at* $R < R_1$. Since $g_\delta(R) > 0$ at small R, the nanostructures cannot nucleate as nanowires. Rather, they emerge as nanoislands and then grow according to an isotropic mode. At a certain time, the base dimension hits the first critical radius at which g_δ becomes zero. This eventually leads to the development of straight nanowire sidewalls and onsets the growth anisotropy. Since the shape transformation occurs almost instantaneously, the base dimension discontinuously decreases and the height discontinuously increases at the transition point. This can result in an overlap between the island and wire sizes for the critical radius as suggested by the experimental data shown in Fig. 5.38. (ii) *Anisotropic nanowire growth* is preferred between R_1 and R_2, because the growth anisotropy decreases the surface energy at a given volume. The nanowire length increases faster than the radius as given by (5.88). Anisotropic elongation again becomes more energetically costly than isotropic growth at $R = R_2$, since, at this time, the increase of sidewall surface area outweighs their low energy in the overall energy balance. However, kinetic limitations should prevent this second transformation that has never been observed experimentally. More generally, the model presented might be useful for understanding the self-induced growth of other elongated structures (III-V and II-VI nanowires, metal whiskers etc.), as it reveals general conditions necessary to observe such a growth in a given material system.

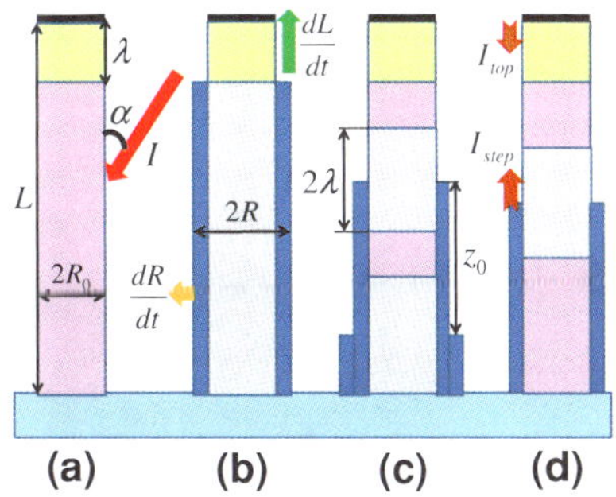

Fig. 5.47 Schematics of different growth scenarios for a single nanowire (low nanowire density): *yellow part*—nanowire surface contributing to the elongation, *magenta parts*—desorption areas, *grey parts*—radial growth, *blue layers* show the growing shells; (*a*)—no radial growth, (*b*)—radial growth by the sidewall nucleation, (*c*)—radial growth by the step flow with many steps, (*d*)—radial growth by the step flow with just one step anytime

5.12 Scaling Growth Kinetics

Experimental data presented in Sects. 5.11 and 5.12 clearly demonstrate that self-induced GaN nanowires have a small aspect ratio of the order of one right after the shape transformation, but then become highly anisotropic because their length increases with time much faster than the radius. This feature was used in thermodynamic modeling of the island-to-wire shape transformation, however, no explanation of the superlinear length-radius correlation given by (5.88) can be proposed based on this analysis. Here, we describe the kinetic model of [437] that shows why the scaling growth laws may occur.

The model of self-induced, Ga-limited nanowire growth is illustrated in Figs. 5.47 and 5.48. In Fig. 5.47, we consider a single nanowire, while Fig. 5.48 shows the shadow effect on a given nanowire originating from the neighboring wires. Cylindrical nanowire having the length L and radius R at time t elongates due to the direct impingement from the beam intercepting the top facet and the surface diffusion. The diffusion length of Ga adatoms on the wire sidewalls, denoted as λ, is limited by either desorption or incorporation and amounts to only few tens of nanometers at the typical MBE temperatures [312, 422, 430,] This is the main difference between the growth of GaN and other III-V nanowires, where the sidewall diffusion length is usually very large, of the order of several micrometers as discussed in Sect. 4.13. As long as L is smaller than λ, all Ga atoms will reach the top and contribute into the nanowire elongation. Radial extension at $L < \lambda$ is unlikely.

After this short stage (say, after time t_0 from the beginning of deposition), the nanowire length will exceed the length $L_0 \sim \lambda$, and the growth will acquire the steady state character. Gallium atoms collected by the topmost nanowire part of heightλ migrate to the top facet. The elongation rate of a nanowire with $L \geq L_0$ is therefore given by the conventional equation discussed in detail in Chap. 4:

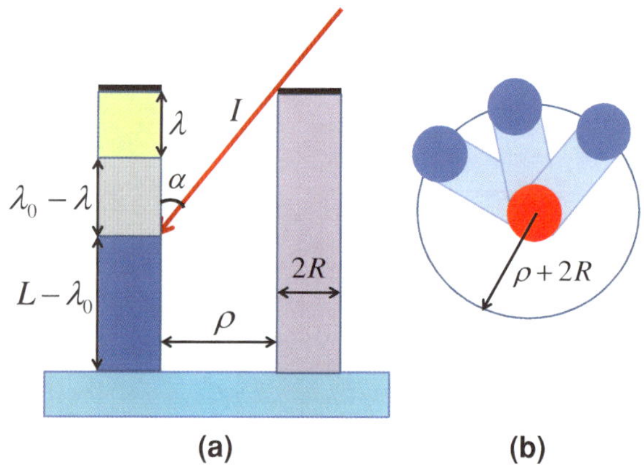

Fig. 5.48 Shadow effect on a given nanowire (high nanowire density): **a** shadowing decreases the sidewall collection height from $L - \lambda$ to $\lambda_0 - \lambda$, **b** shadowed zones showing that in a dense nanowire forest the influx onto the bottom part of the height $L - \lambda_0$ in (**a**) gets almost completely blocked

$$\frac{\pi R^2}{\Omega}\frac{dL}{dt} = \left(\frac{\chi_f I \sin\alpha}{\pi} - I_{top}\right) 2\pi R\lambda + (\chi_{top} I \cos\alpha - I_{des})\pi R^2. \tag{5.97}$$

The left hand side describes the net flux of Ga adatoms into the wire contributing to its elongation, with Ω as the elementary volume of a GaN pair in the solid. The first term in the right hand side stands for the resulting diffusion flux collected from the area $2\pi R\lambda$ of the topmost nanowire part at the Ga flux I making the angle α to the substrate normal. The χ_f is the accommodation coefficient at the sidewalls accounting for possible scattering of Ga. The parameter I_{top} represents the reverse flux of gallium from the top facet to the sidewalls, which equals the arrival rate at the equilibrium conditions. All gallium atoms diffusing from the top will evaporate. The second term accounts for the direct impingement onto the top facet, with χ_{top} as the accommodation coefficient at the top, and desorption from the top facet (I_{des}).

Gallium atoms collected by the remaining surface area of nanowire sidewalls, $2\pi R(L-\lambda) \cong 2\pi RL$ at $L > L_0$, may contribute to the radial growth. Denoting the percentage of Ga adatoms that can reach the incorporation sites as ψ, we can write quite generally:

$$\frac{2\pi RL}{\Omega}\frac{dR}{dt} = \left(\frac{\chi_f I \sin\alpha}{\pi} - I_{SW}\right) 2\pi RL\psi. \tag{5.98}$$

Here, I_{SW} is the desorption flux of gallium from the sidewalls, which equals the arrival rate when no lateral growth occurs. Introducing the equivalent deposition rate, $V = \Omega I \cos\alpha$, (5.97) and (5.98) can be put in the form

$$\frac{1}{V}\frac{dL}{dt} = \frac{a}{R} + c; \; \frac{1}{V}\frac{dR}{dt} = B\psi, \tag{5.99}$$

where $a = (2\chi_f g_f \lambda \tan\alpha)/\pi$, $c = \chi_{top} g_{top}$, $B = (\chi_f g_{SW} \tan\alpha)/\pi$, $g_f = 1 - (\pi I_{top})/(\chi_f I \sin\varphi)$, $g_{top} = 1 - I_{des}/(\chi_{top} I \cos\alpha)$ and $g_{SW} = 1 - (\pi I_{SW})/$

($\chi_f I \sin\alpha$). The constant growth rate c in the first (5.99) can often be neglected, because the sidewall collection is more efficient than that of the top facet for sufficiently small radii. With a time-independent R, this yields the typical diffusion-like length-radius dependence of the form $L \propto 1/R$ at a given moment of time.

We first consider the incorporation probability ψ for a single nanowire. Four possible scenarios of sidewall incorporation are illustrated in Fig. 5.47. If the diffusion length is limited by desorption ($\psi = 0$, Fig. 5.47a), all adatoms collected by the lower part of the nanowire would re-evaporate and the growth would proceed at a constant radius determined at the nucleation stage (where the island-to-wire growth transformation occurs). Growth with a constant radius is not typical for self-induced GaN nanowires, as demonstrated by Fig. 5.43. Under the conditions shown in Fig. 5.47b, all adatoms collected by the sidewalls would be incorporated into the nanowire, yielding $\psi = 1$. Integrating the second (5.99) with the initial condition $R(t = t_0) = R_0$ (where R_0 is the radius acquired at the nucleation stage) readily gives $R(t) = R_0 + BV(t - t_0)$. Inserting this into the first (5.99) and integrating, we obtain a combination of linear and logarithmic growth laws, $t - t_0$ and $\ln(t - t_0)$, showing that the length cannot increase super-linearly with R. Since the case of $\psi = 1$ corresponds to the diffusion length limited by the sidewall nucleation, which normally happens at low temperatures, our considerations well explain why self-induced GaN nanowires can be obtained only at relatively high growth temperatures.

The scenarios shown in Fig. 5.47c and d relate to the core-shell radial growth by the step flow. The shells most probably start from the base, since the nanowire-substrate rectangular interface provides a perfect site to create the step [430]. According to the results of Sect. 1.6, the incorporation probability is given by $\psi = (2\lambda)/z_0$, where z_0 is the distance between the steps (Fig. 5.47c). If z_0 is smaller than L, the nanowire would become tapered, with the inclination angle of the sidewalls to the substrate normal given by $\tan\beta = h/z_0$ (where h is the height of the step). Tapering is not typical for self-induced GaN nanowires. The only possibility to preserve the cylindrical geometry in the step flow model thus yields $z_0 = L$ and $\psi = (2\lambda)/L$, that is, just one surface step on the sidewalls anytime (Fig. 5.47d).

Let us now see how the shadow effect influences the nanowire growth by MBE. As shown in Fig. 5.48, the bottom part of a given nanowire is shadowed by the neighboring nanowires so that only the top part of length $\lambda_0 = \rho \cot an\varphi$ (where $\rho \cong 1/\sqrt{\pi N}$ is the average spacing between the wires and N is their density) is exposed to the Ga flux. The lower part is blocked out when the Ga beam is intercepted with one of the neighboring wires (Fig. 5.48b). In a dense nanowire forest, it can be said that almost no atoms impinge the sidewalls below the height $L - \lambda_0$. If the surface nucleation is enabled everywhere on the sidewall surface exposed to the flux, the incorporation probability in a dense nanowire array is given by $\psi = (\lambda_0 - \lambda)/L$.

With neglect of c in the first (5.99), the kinetic equations in the growth modes presented in Figs. 5.47d and 5.48 (more generally, whenever ψ is inversely proportional to L) take the form

$$\frac{1}{V}\frac{dL}{dt} = \frac{a}{R};\ \frac{1}{V}\frac{dR}{dt} = \frac{b}{L}. \tag{5.100}$$

Here, $b = (2\chi_f g_{SW}\lambda \tan\alpha)/\pi$ for the step flow growth and $b = [\chi_f g_{SW}(\lambda_0 - \lambda)\tan\alpha/\pi]$ for the nucleation-mediated growth of a shadowed nanowire. Integrating this system with the initial conditions $L(t = t_0) = L_0$ and $R(t = t_0) = R_0$ leads to the scaling growth laws of the form

$$L = L_0\left[1 + \frac{(\delta+1)}{\delta}\frac{aV(t-t_0)}{L_0R_0}\right]^{\delta/(\delta+1)};$$
$$R = R_0\left[1 + \frac{(\delta+1)}{\delta}\frac{aV(t-t_0)}{L_0R_0}\right]^{1/(\delta+1)}; \tag{5.101}$$

$$L = L_0\left(\frac{R}{R_0}\right)^{\delta}. \tag{5.102}$$

The power index δ is given by

$$\delta = \frac{a}{b} = \frac{\chi_f I\sin\alpha - \pi I_{top}}{\chi_f I\sin\alpha - \pi I_{step}};\ \delta = \frac{2[\chi_f I\sin\alpha - \pi I_{top}]}{(\lambda_0/\lambda - 1)[\chi_f I\sin\alpha - \pi I_{SW}]} \tag{5.103}$$

for the step-flow growth and for the shadowed nanowires, respectively. Therefore, the scaling index depends not only on the material parameters (that determine the equilibrium fluxes), but also on the growth conditions. The quantity I_{step} in (5.103) equals I_{SW} for the step flow growth. This term originates from the reverse diffusion flux from the step (see Fig. 5.47d). Obviously, (5.102) is equivalent to (5.88), where ν is related to the initial condition, i.e., the average nanowire length and radius right after nucleation.

The general condition for the super-linear growth is given by the inequality $\delta > 1$. For the step flow, this is reduced to $I_{step} > I_{top}$, the condition which seems reasonable. Indeed, the reverse diffusion fluxes from the top and the step should be proportional to the equilibrium adatom concentrations on the top facet and on the sidewalls, respectively. The nanowire elongation must proceed by successive nucleation of nanowire monolayers, which requires sufficient adatom supersaturation on the top facet. This supersaturation is achieved by the gallium deposition from the beam. Same deposition process does not provide high enough supersaturation on the sidewalls to onset the nucleation there, so that the radial growth proceeds by the step flow alone. This strongly suggests that the equilibrium adatom concentration on the sidewalls is noticeably larger than on the top facet, yielding the inequality $I_{step} > I_{top}$ during the entire growth process. As for the shadow effect, typical values of $\lambda_0 = 147$ nm at $N = 10^{10}\,\text{cm}^{-2}$, $\alpha = 21°$ and $\lambda = 45$ nm ([312]) yield $2/(\lambda_0/\lambda - 1) \cong 1$, so that the inequality $\alpha > 1$ can be easily ensured at $I_{SW} > I_{top}$.

The scaling growth behavior is very important for deducing useful information regarding the kinetic parameters. Indeed, measuring the length-radius, length-time

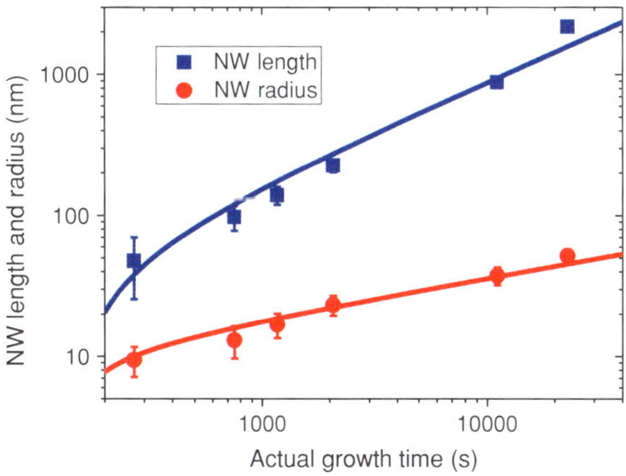

Fig. 5.49 Experimental length-time (**a**) and radius-time (**b**) dependences (symbols) fitted by (5.101) at $\delta = 2.46$, $a = 51$ nm, $L_0 = 38$ nm, $R_0 = 10$ nm, $t_0 = 270$ s, and $V = 0.045$ nm/s

and radius-time dependences and fitting them by (5.101) and (5.102) enables the determination of the scaling index δ and the effective diffusion length a. Self-induced GaN nanowires described in the previous section follow the scaling growth law with $\delta = 2.46$. This yields the values of $\delta/(\delta + 1) = 0.711$ and $1/(\delta + 1) = 0.289$ for the scaling indices of the time dependences of the length and radius, respectively. Symbols in Fig. 5.49 are the measured $L(t)$ and $R(t)$ dependences. Lines in Fig. 5.49 demonstrate the fits by (5.101), obtained at $a = 51$ nm, $L_0 = 38$ nm, $R_0 = 10$ nm, and $t_0 = 270$ s, with $V = 0.045$ nm/s. It is seen that the scaling (5.101) describe quite well the time evolutions of the averaged nanowire length and radius which are indeed sub-linear in the entire time domain. Furthermore, these parameters yield exactly the scaling $L(R)$ dependence shown by line in Fig. 5.43. The discrepancy observed for the longest growth time of 22700 s in Fig. 5.49 is most probably explained by the influence of direct impingement onto the nanowire top facet (the omitted c term in (5.99)), whose contribution increases for larger R.

Thus, we have demonstrated the scaling length-time, radius-time and consequently length-radius growth laws of self-induced GaN nanowires obtained by MBE. Our kinetic model shows that the scaling laws can originate from the core-shell radial growth by the step flow for low nanowire density (i.e., short growth duration) and from the shadow effect for high nanowire density (i.e., long growth duration). A combination of both mechanisms is expected and the transition between the two predominant growth modes could be observed by changing the nanowire density. The latter is determined by the density of the Volmer-Weber islands emerging at the initial growth step, and can be controlled by the surface temperature and deposition flux as described in Chap. 3. The scaling growth properties of self-induced GaN nanowires should be considered fundamental; however, the scaling indices can be tuned by deposition conditions.

5.13 Vertical Growth Rate of GaN Nanowires

In [312], the vertical growth rate of self-induced GaN nanowires grown by MBE under various conditions was determined with high precision from SEM and TEM data by taking into account *in situ* measurements of the initial incubation time, which is needed before the nanowire growth starts. GaN nanowires were grown on Si(111) substrates covered with a continuous Si_xN_y amorphous interlayer with a thickness of about 2 nm. The nominal gallium and nitrogen rates correspond to the growth rate of GaN planar layers deposited on SiC(111) substrates under N- and Ga-rich conditions, respectively. The growth temperature and gallium rate were varied in the ranges of 770–815 °C and 0.025–0.07 nm/s, respectively. The nitrogen rate was equal to 0.28 nm/s, yielding a V/III ratio in the range of 4 to 11.2. All the growth steps were monitored in situ by RHEED. As discussed above, the total growth time for the self-induced approach is the sum of the incubation time necessary for the nucleation of spherical-cap shaped islands on Si_xN_y, of the transition time required for the shape transition to nanowires, and of the actual (i.e. elongation) growth time corresponding to the growth phase. The total growth time was therefore adjusted for each experiment such that the sum of the transition and actual growth times (denoted as the pseudo-actual growth time) was systematically 3 h: the gallium shutter was closed 3 h after the end of the incubation stage as indicated by the appearance of GaN three-dimensional spots in RHEED. In other words, the transition time of typically several seconds to several minutes and its dependence on the growth conditions is considered as negligible with respect to the actual growth time. The nanowire mean elongation rate was experimentally measured as the ratio between the nanowire length L and the pseudo-actual growth time $t_{pseudo-actual}$, as follows: $dL/dt = L/t_{pseudo-actual}$. The nanowire length L was assessed among a population of more than 50 nanowires from cross-sectional or slightly tilted field-emission SEM images, such as shown in Fig. 5.50.

Another series of samples was grown with different durations at a substrate temperature and gallium rate of 780 °C and 0.045 nm/s, respectively. For these growth conditions, the incubation time was found to be 6.5 min as shown by the appearance of GaN three-dimensional spots in RHEED. Additionally, the transition time was determined by performing TEM imaging on all the samples of the growth time series. While only spherical cap-shaped islands are initially formed, it was revealed that the very first nanowires are nucleated after about 5 min following the appearance of GaN three-dimensional spots in RHEED. This strongly indicates that the transition time was about 5 min under such growth conditions. As a result, the actual growth time between 270 and 22770 s is considered here for this series instead of the pseudoactual growth time for the gallium rate and growth temperature series. The nanowire mean growth rate was further experimentally deduced as the ratio between the nanowire length L and the actual growth time t_{actual}, : $dL/dt = L/t_{actual}$.

In the growth modeling, one can use (5.43) for dL/dt, re-arranged for the case of MBE and flat top facet of a hexahedral nanowire with apothem R (Fig. 5.51):

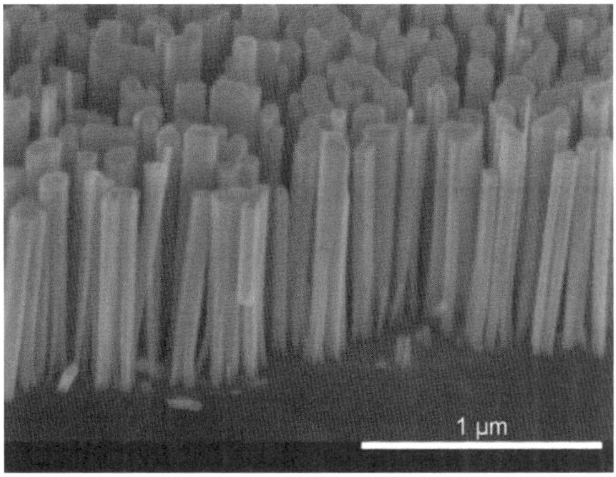

Fig. 5.50 20° tilted SEM image of GaN nanowires grown for 3 h at a substrate temperature and gallium rate of 780 °C and 0.045 nm/s, respectively

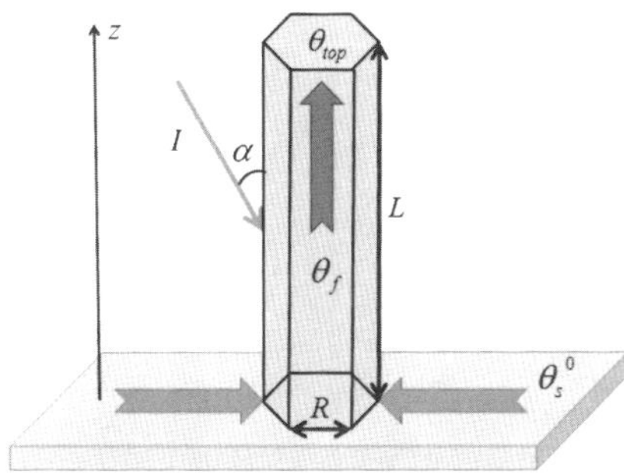

Fig. 5.51 Schematics of the model for the diffusion-controlled nanowire growth in the self-induced approach. The parameters given in the schematic are defined in the text

$$\frac{3\sqrt{3}R^2}{2\Omega_s}\frac{dL}{dt} = \frac{3\sqrt{3}R^2}{2}(I\cos\alpha - I_{des}) - 6RD_f\left(\frac{dn_f}{dz}\right)_{z=L}. \tag{5.104}$$

Here, the radius of the nanowire cross-section is assumed as being constant in the first approximation, I is the Ga flux, α is the Ga incidence angle, Ω_s is the elementary volume in the nanowire, I_{des} is the desorption rate from the nanowire top facet, D_f and n_f are the diffusion coefficient and the concentration of gallium adatoms on the nanowire vertical sidewalls composed of m-planes. For the self-induced growth, the top facet at a lower chemical potential relative to that of surface Ga adatoms effectively replaces the catalyst droplet, as depicted in Fig. 5.51. It is therefore expected that $n_f(z)$ obeys the standard stationary diffusion equation: $D_f d^2n_f/dz^2 + (I/3)\sin\alpha - n_f/\tau_f = 0$, in which the diffusion flux is balanced by the impingement flux onto the sidewalls and the gallium sink with the characteristic lifetime τ_f. The chemical potential of gallium adatoms on the sidewalls equals $k_BT\ln\theta_f$, where $\theta_f = \sigma_f n_f << 1$ is the surface filling factor, σ_f is the elementary area, T is the surface temperature and k_B is the Boltzmann constant.

We use the following boundary conditions for the diffusion equation as seen in Fig. 5.51: (i) The equality of the activity of gallium adatoms at $z = L$ to its activity on the top facet results in $\theta_f(z = L) = \theta_{top}$; (ii) The equality of the activity of gallium adatoms at the nanowire base to its activity on the substrate surface leads to $\theta_f(z = 0) = I\tau_s\sigma_s \cos\alpha \equiv \theta_s^0$. As discussed previously, such a condition at $z = 0$ is valid when the diffusion length of gallium adatoms on the planar substrate surface is much larger than the nanowire radius. By solving the diffusion equation for $n_f(z)$ and by calculating the corresponding diffusion flux at $z = L$, (5.104) can be reduced to

$$\frac{dL}{dt} = V\left\{1 - v_{des} + \frac{4}{\sqrt{3}R}\left[\frac{(\lambda_s^2/b\lambda_f)g_s + \lambda_f g_f \tan\alpha\left(\cosh(L/\lambda_f) - 1\right)}{\sinh(L/\lambda_f)}\right]\right\}. \tag{5.105}$$

Here, $V = I\Omega_s \cos\alpha$ is the equivalent gallium growth rate and $v_{des} = V_{des}/V$ is the relative desorption rate. The quantities $\lambda_s = \sqrt{D_s\tau_s}$ and $\lambda_f = \sqrt{D_f\tau_f}$ are the effective diffusion lengths on the substrate surface and on the nanowire sidewalls, respectively. The coefficient $b = (D_s\sigma_f/D_f\sigma_s)$ is of the order of one. The parameters $g_s = 1 - \theta_{top}/(I\tau_s\sigma_s \cos\alpha)$ and $g_f = 1 - 3\theta_{top}/(I\tau_f\sigma_f \sin\alpha)$ represent the driving forces for the diffusion of gallium adatoms to the wire top.

At small and large nanowire lengths L, (5.105) has the following asymptotes:

$$\frac{dL}{dt} = V\left\{1 - v_{des} + \frac{4}{\sqrt{3}R}\left[\frac{\lambda_s^2 g_s}{bL} + \frac{g_f \tan\alpha}{2}L\right]\right\}, L << \lambda_f; \tag{5.106}$$

$$\frac{dL}{dt} = V\left\{1 - v_{des} + \frac{4\lambda_f}{\sqrt{3}R}g_f \tan\alpha\right\}, L >> \lambda_f, \tag{5.107}$$

which are very similar to those discussed in Sect. 4.9 for VLS nanowires. From (5.106), it follows that the elongation rate at $L << \lambda_f$ is independent of λ_f and comprises three distinct contributions: a constant adsorption-desorption growth rate, the $1/L$ diffusion term originating from the gallium adatoms on the substrate surface and the L term coming from the gallium adatoms on the sidewalls. From (5.107), it turns out that the elongation rate at $L >> \lambda_f$ does not contain the contribution from the substrate surface since all the gallium adatoms are either desorbed or trapped by the islands or steps on the sidewalls before they can reach the nanowire top. The diffusion term is hence proportional to λ_f/R and does not depend on L.

The dependence of the nanowire growth rate on the gallium rate is linear in the first approximation, where the bracket term in (5.105) is independent of V. We also note that the diffusion length of gallium adatoms on the sidewalls at typical GaN growth temperatures is small, of the order of several tens of nanometers [422, 430]. Therefore, for sufficiently long nanowires, the nanowire growth rate is roughly given by (5.107). By assuming that the temperature dependence of g_f is weak, the dependence of the growth rate on temperature is driven by λ_f and v_{des}. For the former, we use (4.165) discussed in Sect. 4.12, with $N_l(T_0, V_0)$, $\lambda_f^0(T_0)$, $F_f' \equiv F$ and $G_f \equiv G$ as fitting parameters corresponding to the reference temperature T_0 and gallium flux V_0. For the latter, the standard Arrhenius-like approximation is used such that $v_{des} \propto \theta_{top} \propto \exp(-E_{top}/k_B T)$ with the corresponding energy barrier E_{top}. This can be represented in the form

$$v_{des} = v_{des}(T_0) \exp\left[-C\left(\frac{T_0}{T} - 1\right)\right], \tag{5.108}$$

where $C = E_{top}/k_B T_0$.

The present theoretical modeling focuses in an ideal case on the growth rate of an isolated nanowire within the self-induced approach. Still, in most practical cases, this approach typically results in the formation of highly dense nanowire ensembles, which can lead to shadow effects between nanowires. The shadowing is dependent upon the geometrical parameters such as both the nanowire density and radius (i.e., nanowire spacing) and the gallium beam angle, as discussed in the previous section. However, the obtained value of $\lambda_0 \cong 147$ nm shows that gallium beam with an angle of 21°, for instance, impinges on a nanowire height which is much larger than the effective diffusion length of about 40–45 nm as reported in [422, 439] and deduced hereafter. As a consequence, although shadowing effects can play a significant role, the present theoretical modeling can account for the nanowire elongation rate in most practical cases, because the effective diffusion length of gallium adatoms on the nanowire sidewalls is fairly short.

The evolution of the mean nanowire growth rate with the nanowire length, gallium rate and growth temperature is presented in Figs. 5.52, 5.53 and 5.54. Interestingly, the mean growth rate is systematically larger than the gallium rate but smaller than the nitrogen rate, showing that the surface diffusion of gallium adatoms on the substrate and then along the nanowire vertical sidewalls to their top does play a significant role as discussed in our theoretical model. This also confirms the Ga-limited growth regime as required in our model.

The evolution of the mean growth rate as a function of the nanowire length at fixed $T = 780\,°\text{C}$ and $V = 0.045$ nm/s is presented in Fig. 5.52. The nanowire mean growth rate for the length of about 50 nm reveals that the nucleation events spread over a certain time as shown by the large error bars. As the nanowire length increases from about 50 to 880 nm (corresponding to the increase in the actual growth time from 270 to 11070 s), the mean growth rate initially decreases drastically. Subsequently, for the length larger than 880 nm (corresponding to the actual growth time longer than 11070 s), the mean growth rate slightly increases from 0.079 to 0.095 nm/s. It should be noted that the opposite trend has been reported in [525, 526], indicating the deleterious effect of not considering the incubation and transition times. The experimental dependence of the mean growth rate on the nanowire length follows a $1/L$-type law for short nanowires and is overall well fitted by (5.105). The fit shown in Fig. 5.52 is obtained with the following parameters: $\lambda_f = 45$ nm, $g_f = 0.794$, $v_{des} = 0.1$, $\sqrt{g_s/b}\lambda_s = 70$ nm, $R = 30$ nm, at the gallium incident angle $\alpha = 21°$. The value of λ_f of 45 nm at 780 °C is consistent with the reported value of 40 nm according to [422, 439]. Thevalue of $\lambda_s \sim 100$ nm justifies the assumption

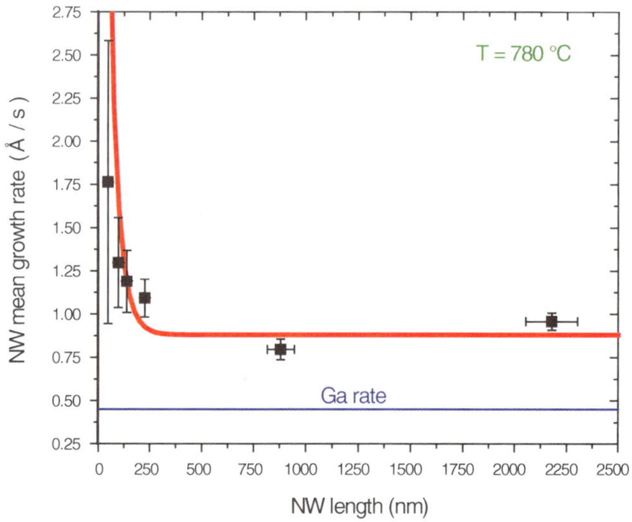

Fig. 5.52 Evolution of the nanowire mean growth rate as a function of the nanowire length: *dots*—experimental data, *red line*—theoretical fit obtained from (5.105) with $\lambda_f = 45\,\text{nm}$, $g_f = 0.794$, $v_{des} = 0.1$, $\sqrt{g_s/b}\lambda_s = 70\,\text{nm}$ and $R = 30\,\text{nm}$. The blue line indicates the value of the gallium rate

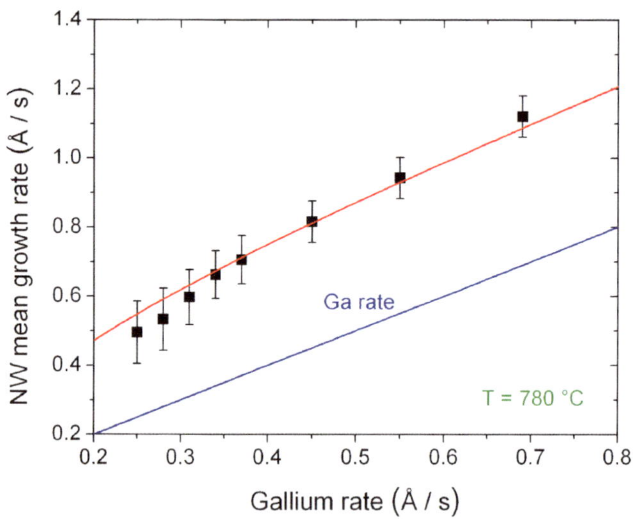

Fig. 5.53 Evolution of the nanowire mean growth rate as a function of the gallium rate: *dots*—experimental data, *red line*—theoretical fit obtained from (5.107) and (4.165) with $\lambda_f = 45\,\text{nm}$, $g_f = 0.794$, $v_{des} = 0.1$, $R = 30\,\text{nm}$, $\lambda_f^0(T_0, V_0) = 118\,\text{nm}$, and $N_{isl}(T_0, V_0) = 6 \times 10^{-5}\,\text{nm}^{-2}$ at the reference values $T_0 = 780\,°\text{C}$ and $V_0 = 0.045\,\text{nm/s}$

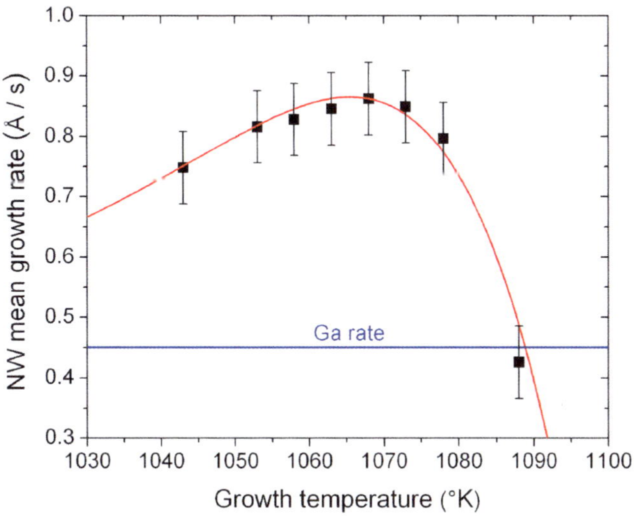

Fig. 5.54 Evolution of the nanowire mean growth rate as a function of the substrate temperature: *dots*—experimental data, *red line*—theoretical fit obtained from (5.107), (4.165) and (5.108) with $\lambda_f = 45\,\text{nm}$, $g_f = 0.794$, $R = 30\,\text{nm}$, $\lambda_f^0(T_0, V_0) = 118\,\text{nm}$ and $N_{isl}(T_0, V_0) = 6 \times 10^{-5}\,\text{nm}^{-2}$ at the reference values $T_0 = 780\,°\text{C}$ and $V_0 = 0.045\,\text{nm/s}$, $v_{des}\,(T_0) = 0.1$, $G = 28$, $F = 60$, and $C = 80$

made in the derivation of the growth equation: the diffusion length of gallium adatoms on the substrate surface is indeed much larger that the mean nanowire radius of 30 nm.

In the following, the pseudo-actual growth time of 3 h is systematically selected and, since λ_f is small, the mean growth rate can be fitted by (5.107) at $L \gg \lambda_f$. The evolution of the nanowire mean growth rate is presented as a function of the gallium rate in Fig. 5.53. As the gallium rate is increased between 0.025 and 0.07 nm/s, the nanowire mean growth rate continuously rises from 0.042 to 0.112 nm/s. Its experimental dependence is again in good agreement with the theoretical model and well fitted by (5.107) and (4.165). The fit shown in Fig. 5.53 is obtained with the following parameters: $\lambda_f = 45\,\text{nm}$, $g_f = 0.794$, $v_{des} = 0.1$, $R = 30\,\text{nm}$, $\lambda_f^0(T_0, V_0) = 118\,\text{nm}$, and $N_{isl}(T_0, V_0) = 6 \times 10^{-5}\,\text{nm}^{-2}$ at the reference values $T_0 = 780\,°\text{C}$ and $V_0 = 0.045\,\text{nm/s}$. It is expected that a larger amount of gallium adatoms is involved in the adsorption and diffusion contribution to the growth rate as the gallium rate is increased. Therefore, the growth rate is higher at a larger gallium rate. The dependence presented in Fig. 5.53 is however not completely linear in contrast to [525, 527, 528]. A small deviation from the linearity towards higher gallium rates is accounted for by a faster capture of adatoms at the nanowire sidewalls. These results strongly suggest that the GaN nanowire growth should be performed at a very high gallium rate. Still, increasing the gallium rate is not conceivable since the nanowire nuclei would not be formed. Indeed, these nuclei are formed following the shape transitions from pre existing 3D islands asdescribed above, which are energetically favorable due to an energy balance in favor of the nanowire morphology only in a certain range of gallium rate (i.e. V/III ratio). Also, even during the growth stage, the gallium rate cannot be increased deliberately since this would eventually lead to an enhancement of radial growth, and the nanowire morphology would be lost.

Most important is the dependence of the nanowire mean growth rate on substrate temperature, as depicted in Fig. 5.54. The evolution of the growth rate follows two consecutive steps: it initially increases to 0.086 nm/s as the growth temperature is raised from 770 to 795 °C and decreases rapidly for higher growth temperatures. The occurrence of a maximum growth rate for intermediate growth temperatures is of high interest for technological applications in which the highest nanowire growth rate is required. Interestingly, a maximum nanowire density has also been reported at about 785 °C [529] which points out that such a growth temperature is optimal for the nanowire morphology. The experimental temperature dependence of the nanowire mean growth rate is again very well fitted by the model. The fit shown in Fig. 5.54 is obtained with the same values of parameters as for the growth time and gallium rate dependences at $v_{des}(T_0) = 0.1$, $G = 28$, $F = 60$ and $C = 80$.

The obtained value of $G \equiv \frac{E_A^f - E_D^f}{2k_B T_0}$ leads to $E_A^f - E_D^f = 2.54\,\text{eV}$, which is in accordance with the available data on the activation energies. In GaN, the diffusion barrier for gallium adatoms on the nanowire vertical sidewalls composed of m-planes is 0.21 eV [530], which reveals that $E_{Df}^f = 0.21\,\text{eV}$ and hence that $E_A^f = 2.75\,\text{eV}$. No value for the activation energy of desorption of gallium adatoms on the GaN m-planes has so far been reported but it should be noted that the corresponding activation energy lies in the typical range of about 2–5.1 eV on the GaN c-plane [531–533]. Consequently, the value of 2.75 eV for E_A^f seems quite reasonable. The values of F and C result in$E_{nucl}^f = 5.44\,\text{eV}$ and $E_{top} = 7.26\text{eV}$. The former is noticeably larger, showing that the nucleation barrier on the nanowire vertical sidewalls is twice higher than the activation energy of desorption of gallium adatoms. The higher value of E_{top} is most likely due to the fact that it simultaneously accounts for the surface nucleation and desorption. Interestingly, desorption activation energy on the GaN c-planes is expected to be larger than 2 eV [531–533]. As a result, the nucleation barrier is lower on the top facet than on the vertical sidewalls. It is further worth noticing that the surface diffusion of gallium adatoms along the sidewalls is enhanced by raising the growth temperature, whereby a larger amount of gallium adatoms reaches the nanowire top. However, the gallium desorption from the top facet and vertical sidewalls may be prevalent for growth temperatures higher than 790 °C, accounting for the eventual decrease in the mean growth rate.

5.14 Cooperative Effects in Nanowire Growth

We will now briefly discuss some important cooperative effects in nanowire growth that originate from indirect or direct interactions between nanowires. In fact, some of them have already been mentioned above: the gold diffusion from the smaller

Au-Si droplets to the larger ones resulting in the nanowire radius re-distribution [356] (Sect. 4.1, Fig. 4.10); overlapping surface adatom collection areas leading to a reduced nanowire growth rate [404, 447, 534] (Sect. 4.9, Fig. 4.43); the shadow effect in directional deposition methods [414, 437] (Sect 5.12, Fig. 5.48); re-emission of material as a secondary source of group V [457] or even group III [536] element for VLS III V nanowires (Sect. 5.9), direct coalescence of dense arrays of self-induced GaN nanowires [312] (Sect. 5.13). In all these cases, neighboring nanowires influence each other's growth and morphology. Cooperative effects may lead to rather counterintuitive growth behavior such as described, e.g., in [397] where the growth rate of VLS GaP nanowires by MOCVD is enhanced with decreasing the interwire spacing. Theoretical analyses of cooperative growth phenomena should deal either with an ensemble of nanowires or certain "mean-field" terms in individual growth equations that describe the influence of neighboring nanowires.

5.14.1 Shadow Effect

Here, we present the model of [414] that accounts for the shadow effect and allows one to describe unusual, barrel-like shape observed in dense arrays of InAs nanowires grown by MBE. We start the analysis of the shadow effect with standard kinetic equations for the vertical and radial growth rates of an individual nanowire:

$$\frac{1}{\Omega_s}\frac{dL}{dt} = I\frac{1+\cos\alpha}{2} - 2I_{des} - 2D_f\frac{1}{R_d}\left.\frac{\partial n_f}{\partial z}\right|_{z=L}; \tag{5.109}$$

$$\frac{1}{\Omega_s}\frac{dR(z,L)}{dt} = J\left(n_f(z,L)\right). \tag{5.110}$$

The vertical growth rate dL/dt is equivalent to (5.43), re-arranged for a hemispherical droplet and directional flux I with the incident angle α. The radial growth is assumed to be driven by nucleation on the sidewalls similarly to Sect. 5.4, and described by a local function of the adatom concentration on the sidewalls $n_f(z, L)$. The latter also depends on time via $L(t)$. Other parameters are the same as usual, and the droplet radius R_d is assumed as being constant during growth. As in Sect. 5.4, Zeldovich nucleation rate in the right hand side of (5.110) can be approximated by

$$J\left(n_f\right) = J_0\exp\left(\Gamma\frac{n_f - n_{\max}}{n_{\max}}\right), \tag{5.111}$$

where Γ is the large parameter of the order of number of atoms (or pairs) in the critical nucleus and n_{max} is the maximum adatom concentration on the sidewalls. Unlike in sect. 5.4, we now do not impose that the maximum concentration is reached at the base (i.e., at $z = 0$).

These standard growth equations do not contain any shadowing. The shadow effect is introduced in the diffusion equations for the adatom concentrations on the substrate (n_s) and sidewall surfaces, written in the form

$$D_s \Delta n_s + I \cos\alpha \left[1 - s(L)\right] - \frac{n_s}{\tau_s} = 0;$$
$$D_f \frac{\partial^2 n_f}{\partial z^2} + I \sin\alpha \left[1 - s\left(L - z\right)\right] - \frac{n_f}{\tau_f} - J\left(n_f\right) = 0. \tag{5.112}$$

Compared to the conventional linear model given by (4.64) and (4.65), these equations contain the effective shadowing coefficient $s(y)$ and the nucleation-mediated sink of the sidewall adatoms $-J(n_f)$, while τ_f stands for the desorption-limited lifetime. The $s(y)$ function gives the percentage of atoms from the overall flux captured by the neighboring nanowires as shown in Fig. 5.55, with $y = L - z$. The boundary conditions to (5.112) have been considered in detail in Sect. 4.8 and, for a particular case of InAs nanowires discussed below, are chosen in the form (4.69), (4.70), (4.71) and (4.72).

To calculate the shadowing coefficient $s(y)$, we assume that the nanowires are randomly distributed on the substrate. The value of $s(y)$ can be found by integration of shadows from neighboring nanowires located within a circle of the radius $\rho = y \tan\alpha$ (see Fig. 5.55). The expected number of nanowires intercepting the incident flux equals the product of the area where one nanowire can shadow another one, $2\int_0^\rho R\left(z + r\mathrm{cotan}\alpha\right)dr$, and the nanowire surface density N_W. To exclude the nanowires that are masked by other ones standing closer, we introduce the statistical probability that no other nanowires are present between two given nanowires. For a random nanowires distribution, this probability follows the Poisson law and can be put as $\exp\left(-2N_W \int_0^\rho R\left(z + r\mathrm{cotan}\alpha\right)dr\right) \approx \exp\left[-(2N_W R \tan\alpha)y\right]$. Therefore, the shadowing coefficient can be approximated by

$$S(y) = 1 - \exp(-y/y_*), \tag{5.113}$$

with $y_* = 1/(2N_W R \tan\alpha)$ as the characteristic nanowire height exposed to the flux. Taking for estimates the values $N_W = 3 \times 10^9\,\mathrm{cm}^{-2}$, $R = 40\,\mathrm{nm}$, and $\alpha = 30°$, the effective height y_* amounts to 720 nm. When y_* is larger than the diffusion length of adatoms on the nanowire sidewalls, the diffusion flux from the latter is determined by shadowing.

We now apply the general model described above to the specific case of InAs nanowires grown by Au-catalyzed MBE under As-rich conditions. The parameters are chosen as follows. The lattice constant of InAs is 0.606 nm, yielding $\Omega_s = 0.056\,\mathrm{nm}^3$, $\sigma_f = 0.26\,\mathrm{nm}^2$, and $\sigma_s = 0.16\,\mathrm{nm}^2$. We neglect desorption of In from the droplet ($I_{des} = 0$) and from the sidewalls ($\tau_f \to \infty$), which should be reasonable at a typical MBE growth temperature of ~400 °C. The parameter Γ is set to 50. The In flux I corresponds to the experimental deposition rate of 0.4 nm/s from the In beam with the incident angle $\alpha = 16.7°$. We assume that the incorporation-

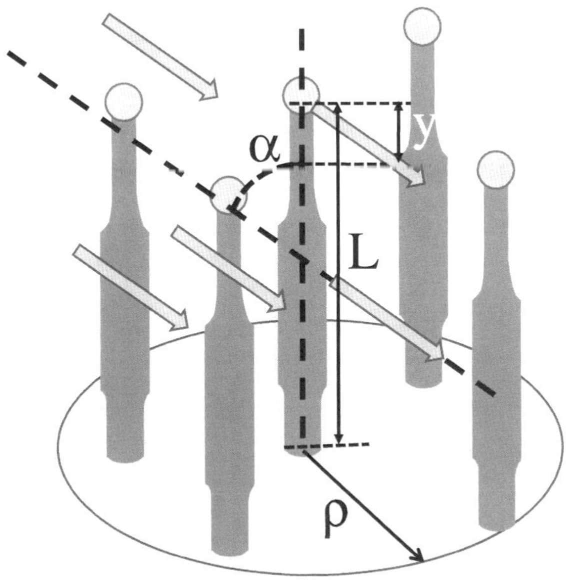

Fig. 5.55 Schematics of barrel-like nanowire ensemble illustrating the shadowing effect with the parameters described in the text

limited diffusion length of In λ_s equals 10 nm on the substrate surface. The diffusion coefficient on the nanowire sidewalls is fixed at $D_f = 10^{-8}\,\text{cm}^2/\text{s}$.

Equations (5.112) with the boundary conditions (4.69) to (4.72) and the shadowing coefficient given by (5.113) are solved numerically by the fourth-order Runge–Kutta method. The obtained dependences of the vertical growth rate (given by (5.109)) on the nanowire density are presented in Fig. 5.56 for different incident angles of the impinging flux. These dependences show two distinct growth modes in limit cases. The first one characterizes a dilute nanowire ensemble, where the shadowing is negligible and the growth rate is limited by the diffusion length. The vertical growth rate in a dilute case can be obtained from (5.109)–(5.112) at $s \equiv 0$ under the assumption that no lateral growth occurs until the concentration reaches n_{max}:

$$\frac{1}{\Omega_s}\frac{dL}{dt} \cong I\frac{1+\cos\alpha}{2} + 2\frac{\sqrt{2n_{\max}D_f I\sin\alpha}}{R_d}. \tag{5.114}$$

This expression contains the conventional $1/R_d$ diffusion like dependence on the droplet radius and is independent of the nanowire density.

For a dense nanowire array, the shadow effect gains a tremendous significance. In particular, almost no lateral growth is observed, since a vast majority of deposited In impinges directly the droplet surface and a small sidewall region just beneath the nanowire top, from where all In adatoms can migrate to the droplet. The vertical growth rate in this case can also be found from (5.109) to (5.112) by setting the nucleation rate to zero and assuming that the nanowire length is much larger than y_*:

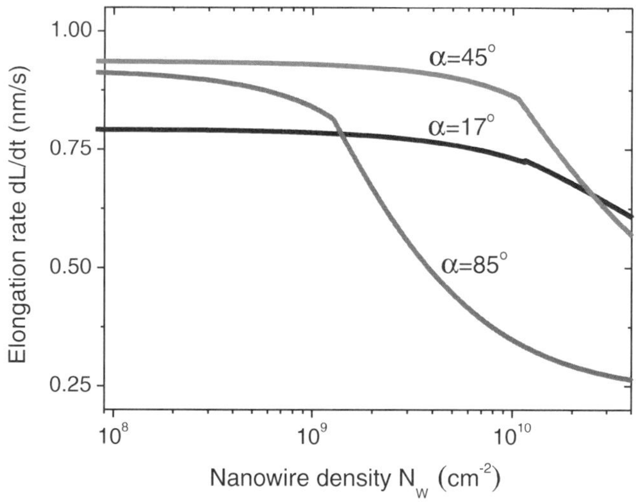

Fig. 5.56 Dependence of the elongation rate on the nanowire density at different incident angles α

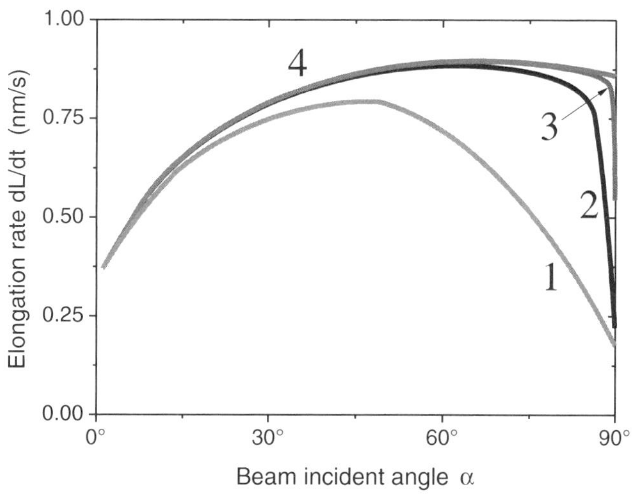

Fig. 5.57 Dependence of the axial growth rate on the incident angle at different nanowire densities: *1*—10^{10} cm^{-2}; *2*—10^{9} cm^{-2}; *3*—10^{8} cm^{-2}; *4*—10^{6} cm^{-2}

$$\frac{1}{\Omega_s}\frac{dL}{dt} \cong I\frac{1+\cos\alpha}{2} + \frac{I\cos\alpha}{R_d^2 N_W}. \tag{5.115}$$

The adsorption term here is the same as in (5.114), while the diffusion term is inversely proportional to the nanowire density and decreases with increasing the incidence angle.

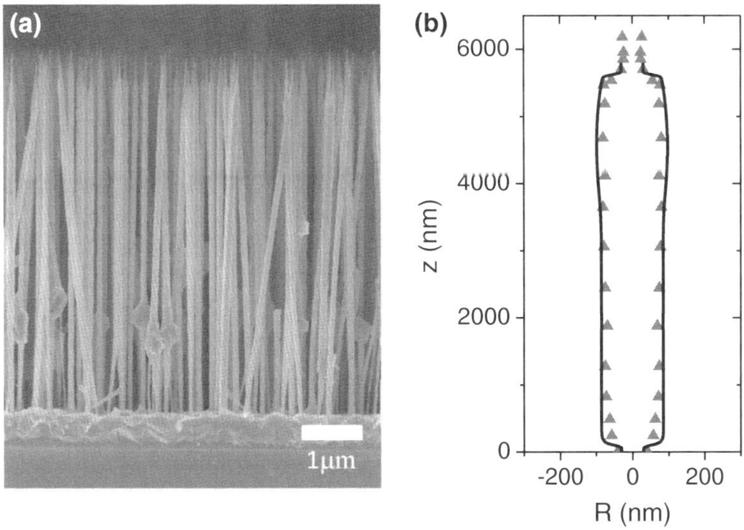

Fig. 5.58 SEM cross-section view image (**a**) and barrel-like shape (**b**) of individual InAs nanowire. *Triangles* in (**b**) represent experimental shape, and *solid line* is theoretical fit

Figure 5.57 shows theoretical dependences of the elongation rate on the incident angle for different nanowire densities. All the curves have a maximum which increases with decreasing the density. Qualitative interpretation of such dependences is rather simple. Indeed, consider two limit cases with near glancing and near normal material fluxes. When the incident angle approaches glancing incidence, the growth species cannot impinge the sidewalls due to the shadowing effect. In the opposite case where the flux is near perpendicular to the substrate, its projection to the sidewalls tends to zero. In both cases, almost no atoms are collected by the sidewalls, thus disabling the diffusion-induced contribution into the overall nanowire growth rate. The maximum of the elongation rate is therefore reached at an intermediate value of the incident angle determined by the nanowire surface density. It should be noted that there is a slope change in the $dL/dt(N_W)$ and $dL/dt(\alpha)$ curves caused by the onset of the radial growth. The transition between the two limit cases becomes more abrupt with increasing the value of Γ.

A typical cross-sectional SEM image of InAs nanowires grown by the Au-assisted MBE for 80 min at a temperature of 410 °C, deposition rate of 0.4 nm/s and V/III flux ratio of 3 is shown in Fig. 5.58a. It is seen that the majority of these nanowires exhibit a barrel-like shape. This shape is characterized by the following features: at the bottom part, the nanowire presents a slow widening, followed by a long segment where the diameter is almost uniform, and finally the nanowire is tapered toward the catalyst particle. By analyzing SEM images, the experimental $R(z)$ dependence can be extracted for individual nanowires. An example of such a dependence for the nanowire having a typical shape is shown in Fig. 5.58b. Theoretical shape $R(z)$, aslo shown in Fig. 5.58b, is computed by means of (5.109) to (5.113) with the same parameters as before, for the experimental nanowire density $N_W = 1.3 \times 10^9\,\mathrm{cm}^{-2}$. It is seen that the model reproduces quite well the main features of the barrel-like nanowire shape.

The explanation of the observed shape is attributed entirely to the shadow effect. In MBE growth of dense InAs nanowires, the maximum concentration of sidewall adatoms is reached at some point remote from the surface, where the radial growth rate is larger than at the base. This leads to the widening of the bottom part of the nanowire. After that, each point at the wire sidewall begins to extend radially at the rate determined by the maximum concentration. Without the shadow effect, the nanowires would acquire a conical shape as described in Sect 5.4, because, at roughly the same radial growth rate, the lower parts would grow for a longer time. However, the conical shape is suppressed by shadowing: the growth time at each height is now determined by the period when the given point is exposed to the beam before it gets shadowed by the neighboring nanowires. As a result, the conditions for the radial growth become identical everywhere at the sidewalls, yielding an almost cylindrical shape of the middle nanowire part. This confirms the importance of the shadow effect in the avoidance of usually unwanted radial growth. As for the upper spire, it is always exposed to the beam, but most atoms collected by the top part tend to diffuse to the droplet rather than be trapped by the sidewalls. The upward diffusion flux decreases the local concentration and consequently the radial growth rate at the top, leading to the narrowing of the nanowire tip.

Thus, for dense nanowire arrays and at large enough incident angles of the beam in the directional deposition methods, the elongation and radial growth rates are limited not only by the effective diffusion length but also by the shadowed height. This leads to the decrease of the wire length compared to the case of dilute nanowires. More importantly, the shadow effect is very efficient in suppressing the radial growth and thus helps to preserve the cylindrical nanowire shape by blocking out the material influx onto the shadowed areas of the sidewalls. Without the shadow effect in the systems where the sidewall nucleation is otherwise enabled everywhere, the diameter uniformity would be unattainable.

5.14.2 Re-emission Assisted Growth of InSb-InAs Nanowires

In [536], the growth rate of InSb sections of hetertostructured InSb-InAs nanowires grown by the Au-catalyzed CBE was found to increase with increasing nanowire density. This is opposite to the commonly observed behavior, in which the nanowires compete for the incoming material and therefore their growth rate is typically reduced for high nanowire densities. The InSb-InAs nanowires were grown as described in Sect. 4.11 but employing electron beam lithography defined patterns on InAs (111)B substrates with highly controlled nanowire diameter and density. For all samples, the InAs stems were grown for 45 min at 415 °C with precursor line pressures of 0.3 and 1 Torr for TMIn and TBAs, respectively. After the growth of the InAs stems, the sample was cooled down to 400 °C under TBAs flow and the growth of InSb was started. The precursor line pressures were 0.45 and 0.75 Torr for TMIn and TDMASb, respectively, and the growth time of InSb was 60 min. In order to study the growth of InSb nanowires, patterns with different diameter and density were fabricated on the same sample. The hexagonal arrays of Au discs have center to center distance (pitch) a of 240, 340, 500, and 1060 nm, corresponding to nanowire densities of ~20, 10, 4.6, and 1 μm^{-2}. Figure 5.59 shows the SEM images of one as-grown pattern with InAs diameter of 77 ± 3 and 240 nm pitch. A 45°-tilt SEM image of the whole pattern with area of 15 × 15 μm^2 is shown in Figure 5.59a, while Fig. 5.59b shows a high magnification image of the central part of the pattern.

In Fig. 5.60a the average length of InSb segments versus InSb diameter is plotted at different pitches. There are clearly two different trends visible: for a fixed pitch (i.e., fixed nanowire density), the length of the InSb segments increases with increasing InSb diameter, before it saturates to a constant value for large diameters, following the behavior modeled in Sect. 4.11. On the other hand, the length of the InSb segments increases when the pitch is decreasing (i.e., increasing the nanowire density). This is also clearly evident from the SEM images in Fig. 5.60b and c, which were obtained under a tilt angle of 20° and correspond to the arrays with pitch of (b) 240 nm and (c) 1060 nm. Green (blue) color in the insets indicates the InAs (InSb) segments. Clearly, for the smaller pitch, the InSb segment is longer than for the larger pitch.

To describe a rather unusual increase of the InSb length with the nanowire density, we consider the growth model schematized in Fig. 5.61. Besides the usual kinetic processes for a single nanowire (direct impingement onto and desorption from the droplet surface, adatopm diffusion from the sidewalls and possibly from the substrate), we include an additional material transport pathway: the adsorbed adatoms re-emitting from nanowire sidewalls and landing on the neighboring nanowires, as schematically depicted in Fig. 5.61e. The corresponding growth equation can be written in the form

$$\frac{dL}{dt} = \left(\frac{dL}{dt}\right)_{ind} + V_{re}\omega, \tag{5.116}$$

where the first term gives the growth rate for an individual nanowire and the second is the re-emission induced contribution. In the latter, V_{re} is a certain dimensional coefficient (nm/s) and ω is the re-emission probability, defined as the ratio of the solid angle Ω_i (comprising all the directions in which re-emitted adatoms would be intercepted by other catalyst droplets) and the solid angle Ω_t of all possible re-emission directions. We assume here that re-emission direction is fully random and re-emitted species travel ballistically, while any (positive) contribution from the substrate can be ignored.

The solid angle Ω_i projected onto the catalyst droplet of a neighboring nanowire roughly equals the ratio of the droplet cross-section area (πR^2) and the squared distance $y^2 + a^2$ between the droplet and a point P on the nanowire sidewall at a distance y from the top. The re-evaporation solid angle Ω_t can be roughly set to π, since the solid angle of all possible desorption directions is 2π, and a half of this sector is directed down to the substrate. Thus, in the first order approximation, the re-emission probability from a given point P is given by $\Omega_i/\Omega_t \cong R^2/(y^2 + a^2)$.

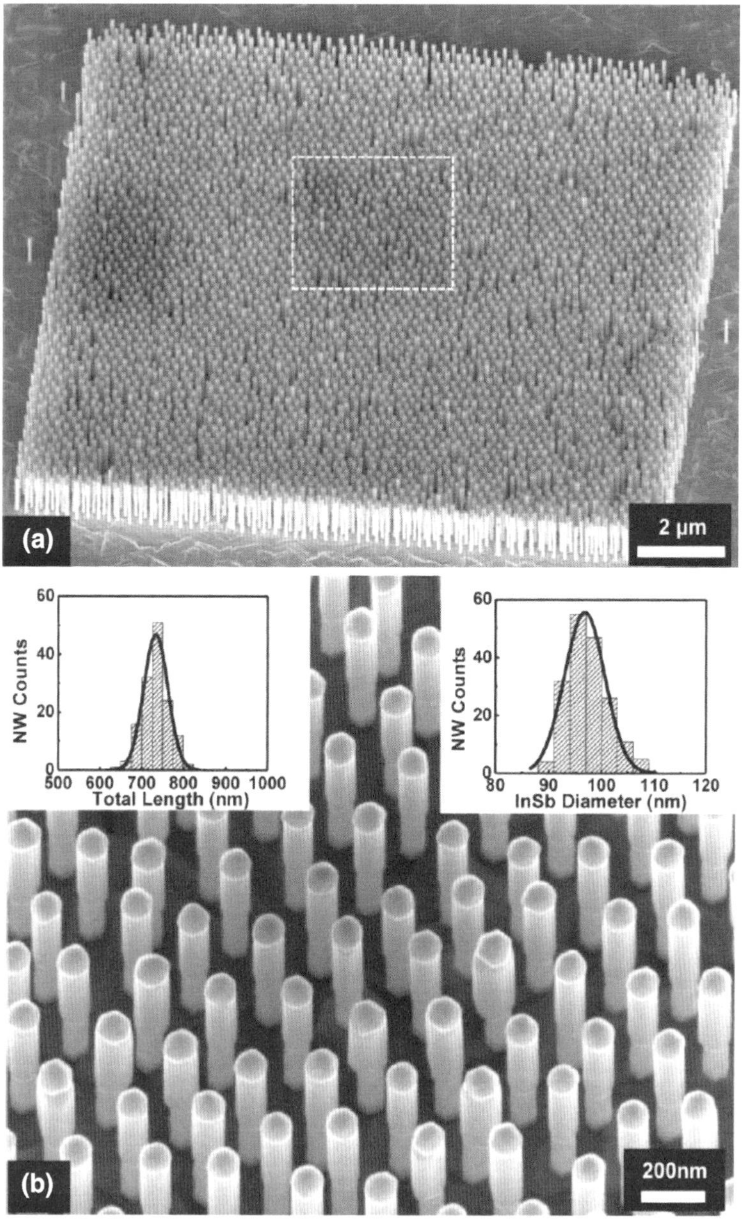

Fig. 5.59 SEM images of a highly uniform heterostructured InAs/InSb array with area of 15 × 15 μm^2 and InAs diameter of 77 ± 3 nm. **a** 45°-tilt SEM image of the nanowire array; **b** 20°-tilt high magnification SEM image of the array. The *histograms* of total nanowire lengths and InSb diameters are shown in the insets of (**b**)

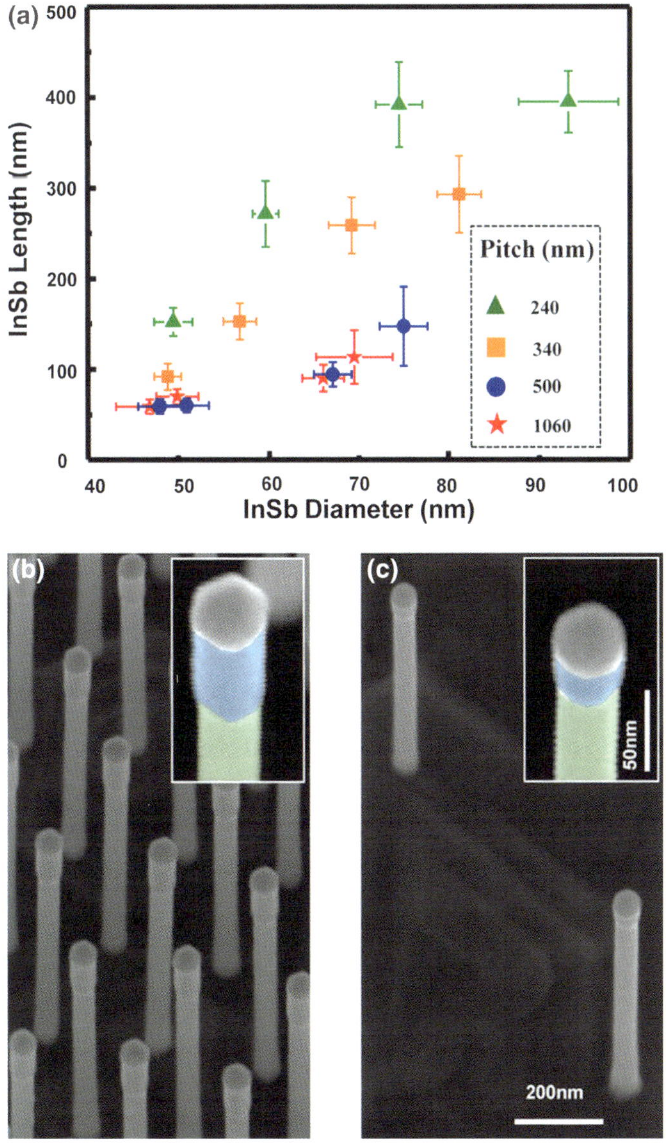

Fig. 5.60 Pitch and diameter dependence of the length of InSb segments. **a** Length of InSb segments as a function of InS diameter for different pitches. 20°-tilt SEM image of InAs/InSb nanowire arrays with fixed InSb diameter of 48±3 nm and pitch of **b** 240 nm and **c** 1060 nm. In the high magnification insets, *green color* indicates the InAs stems while *blue color* indicates the InSb segments

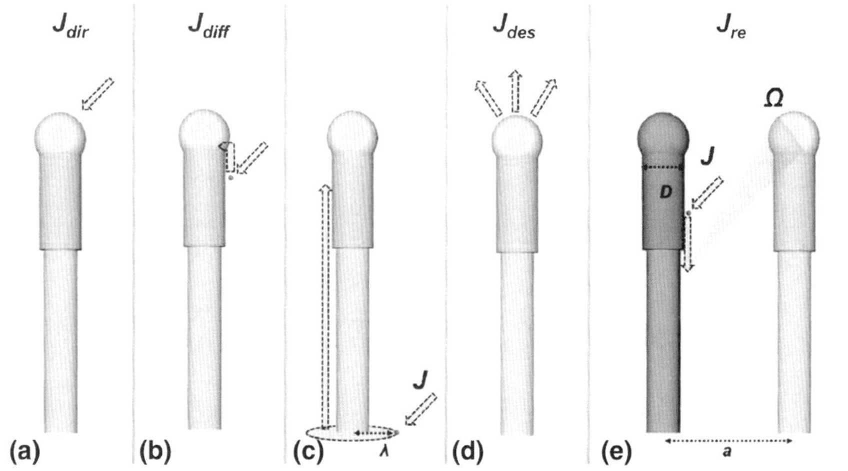

Fig. 5.61 The main processes involved in nanowire growth: **a** direct impingement, **b** diffusion from the sidewalls, **c** diffusion from the substrate, **d** evaporation from the droplet, and **e** readsorption from the nanowire sidewalls

The mean value of the readsorption probability along the sidewalls is

$$\omega \cong \frac{R^2}{L} \int_0^L \frac{dy}{y^2 + a^2} = \frac{R^2}{La} \arctan\left(\frac{L}{a}\right). \tag{5.117}$$

Since the length of the InAs stems is much longer than the length of the InSb segments in these growth experiments, we can approximate that the total nanowire length does not change during the InSb growth. We also neglect the radius dependence of all other fluxes in (5.116) in the first approximation, ending up with a growth rate that is linearly dependent on readsorption probability ω.

Figure 5.62 shows the experimental points of Fig. 5.60 rescaled using (5.117). It can be seen that larger ω corresponds to a smaller pitch and longer wires. Figure 5.62 also shows the theoretical curve obtained from (5.116) and (5.117). A reasonable agreement between our model and experiment is observed. The increase of InSb length with readsorption probability is striking, considering that the plot of Fig. 5.62 includes all of the data of Fig. 5.60, i.e., any diameter and density. This result indicates that re-emission from neighboring nanowires plays a fundamental role as a secondary material supply source. Similar results have been reported, e.g., by Dalacu *et al.* [537] who considered re-emission of the source materials from the SiO_2 mask on the substrate and their subsequent capture by InP nanowires grown by CBE. Another example is given by Ramdani *et al.* in [457], showing that re-emission of As species from the substrate and from nanowire sidewalls can act as an efficient secondary arsenic source during the Ga-cayalyzed growth of GaAs nanowires by MBE. Re-distribution of material between the neighboring nanowires due to re-emission

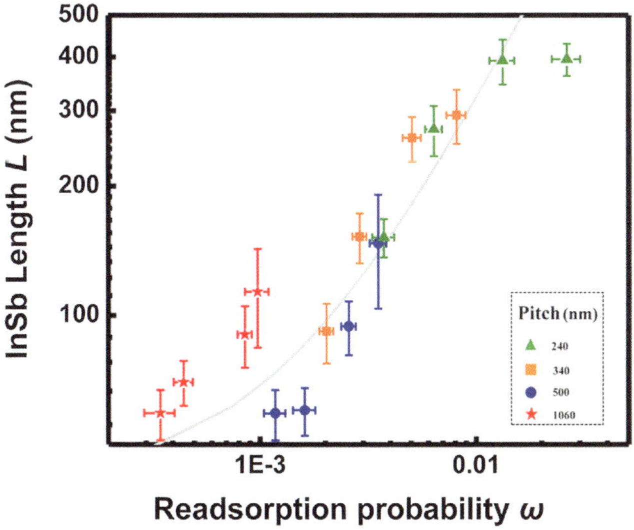

Fig. 5.62 Dependence of InSb segment length L on readsorption probability ω. The *gray line* is the theoretical *curve* obtained from (5.117), and the *color dots* are the same experimental data as in Fig. 5.60a

may lead to many interesting effects including narrowing the length distribution [538]. Indeed, longer nanowires emit more material than receive from the sidewalls of shorter nanowires and thus the growth rate of the latter is enhanced. More precise models of re-emission should include the inhomogeneous length distribution of nanowires, the droplet configuration, beam incident angles, the shadow effect and the contribution from adatoms scattered from the substrate surface.

5.15 Chemical Potentials for Au-assisted VLS Growth of III–V Nanowires

In our analysis of the VLS nanowire growth, the liquid supersaturation ζ is generally defined as $\Delta\mu_{LS} \equiv \Delta\mu = k_B T \ln(\zeta + 1)$, where $\Delta\mu$ is the difference of chemical potential between a liquid alloy formed by the growth species dissolved in the catalyst and the solid semiconductor material in the nanowire. In the case of the Au-catalyzed VLS growth of elementary semiconductors, $\Delta\mu$ and ζ are the functions of the atomic concentration c of the semiconductor material dissolved in gold and the growth temperature T. The simplest approximations $\zeta = c/c_{eq} - 1$ and $\Delta\mu = k_B T \ln(c/c_{eq})$, where $c_{eq}(T)$ is the temperature-dependent equilibrium concentration of a given semiconductor material in the melt, pertain for the case of a diluted alloy with $c \ll 1$. The situation is more complex for the Au-assisted VLS growth of III-V nanowires, where $\Delta\mu$ and ζ become the functions of two atomic concentrations: $c_{III} \equiv c_A, c_V \equiv c_B$ in the $A - B - Au$ liquid, and the temperature of the melt. Glas [354] has provided analytical expressions for $\Delta\mu\ (c_A, c_B, T)$ and calculated explicitly the liquid chemical potentials for seven binary arsenide and phosphide compounds by using the following scheme.

Let us consider the formation of a stoichiometric binary solid compound $A_{III}B_V$ from an $A-B-Au$ melt at temperature T. We denote μ_X^{ϕ} the chemical potential per atom of element X (A, B or Au) in phase ϕ. The phases considered are the ternary liquid $A-B-Au$ with atomic concentrations c_A, c_B and $c_{Au} = 1-c_A-c_B (\phi = L)$, solid AB with ZB structure ($\phi = AB$), pure liquid X ($\phi = pL$), or pure solid X ($\phi = pS$). The difference of chemical potential per III-V pair between the liquid alloy and bulk solid AB is given by

$$\Delta\mu = \mu_A^L + \mu_B^L - 2g_{A_{0.5}B_{0.5}}^{AB}, \tag{5.118}$$

where $g_{A_{0.5}B_{0.5}}^{AB}$ is half the Gibbs energy of an *AB* pair in the solid at temperature T. The chemical potentials of species X in the liquid alloy can be expressed through their chemical potentials in the pure liquid and their activities α_X^L as

$$\mu_X^L = \mu_X^{pL} + k_B T \ln \alpha_X^L. \tag{5.119}$$

Introducing the atomic enthalpies $h_{X,0}^{pS}$ of element X in the solid at the temperature $T_0 = 295.16$ K, which are taken as standard energy reference values in the literature ([539, 540]), and substituting (5.119) into (5.118), one obtains

$$\begin{aligned}\Delta\mu = k_B T \ln \alpha_A^L + k_B T \ln \alpha_B^L + (\mu_A^{pL} - h_{A,0}^{pS}) + (\mu_B^{pL} - h_{B,0}^{pS})\\ - 2\left(g_{A_{0.5}B_{0.5}}^{AB} - 0.5h_{A,0}^{pS} - 0.5h_{B,0}^{pS}\right).\end{aligned} \tag{5.120}$$

Here, all terms excluding the enthalpies $h_{X,0}^{pS}$ refer to temperature T. Moreover, all terms apart from the activities are easily computed. Indeed, the differences $\mu_X^{pL} - h_{A,0}^{pS}$ and $g_{A_{0.5}B_{0.5}}^{AB} - 0.5h_{A,0}^{pS} - 0.5h_{B,0}^{pS}$, which refer to single-element or stoichiometric phases, only depend on temperature and the corresponding molar expressions are tabulated in the compilations of Dinsdale [539] and Ansara *et al.* [540], respectively. Thus, the calculation of $\Delta\mu$ is reduced to the calculation of the activities of A and B in the ternary liquid.

This calculation is based on the regular solution model

$$k_B T \ln \alpha_X^L = k_B T \ln c_X + c_Y^2 \omega_{XY} + c_Z^2 \omega_{XZ} + c_Y c_Z (\omega_{XY} + \omega_{XZ} - \omega_{YZ}), \tag{5.121}$$

where Y and Z denote the two other species in the liquid phase and $\omega_{\alpha\beta}$ the interaction parameter per atom for couple α, β of elements. According to Stringfellow [541], this parameter is given by

Table 5.5 Parameters for use in Stringfellow's calculations

Element	ΔH_α^{at} (kJ/mol) ([543])	χ
Al	326.4	1.61
Ga	277.0	1.81
In	243.3	1.78
P	314.6	2.19
As	302.5	2.18
Sb	262.3	2.05
Au	366.1	2.54

$$\omega_{\alpha\beta} = \frac{1}{N_A} \frac{V_\alpha V_\beta}{c_\alpha V_\alpha + c_\beta V_\beta} \left[\left(\delta_\alpha - \delta_\beta\right)^2 - \frac{C(\chi_\alpha - \chi_\beta)^2}{(V_\alpha V_\beta)^{1/2}} \right]. \tag{5.122}$$

Here, V_α, χ_α and δ_α denote, respectively, the molar volume in the liquid state, the Pauling electronegativity, and the Hildebrand solubility parameter [542] of species α, N_A is Avogadro's number and $C = 1.256 \times 10^5$ if all quantities are in SI units.

As a refinement of Stringfellow's scheme, Glas [354] considered temperature-dependent molar volumes and solubility parameters. The molar volume was calculated as a function of temperature by using expressions for this quantity or for the density available in the literature for different group III and V elements. For P, no simple atomic liquid exists, and the temperature-independent value is used proposed by Stringfellow [541]. The solubility δ_α is taken as $\delta_\alpha = \sqrt{(\Delta H_\alpha^{at} - RT)/V_\alpha}$, where ΔH_α^{at} is the molar heat of atomization of α. Table 5.5 lists the relevant parameters used in calculations.

As discussed above, typical values of concentrations of the group III elements in gold during the VLS growth of III–V nanowires amount to several tens of percent. On the contrary, for arsenide and posphide nanowires, the energy dispersive XRD analyses reveal no significant amount of As and P in the droplets after growth. This sets an upper limit of the order of 1 % to the post-growth group V concentration. Hence, the concentration of group V atoms should be of a few percents at most. Therefore, Figs. 5.63, 5.64 and 5.65 shows the calculated liquid chemical potentials for different III-V materials as functions of atomic concentrations of group V elements between 0.1 and 6.1 %, for three different temperatures within the typical growth domains of these nanowires and at three different group V concentrations of 20, 35 and 70 %. Figure 5.66 shows the liquid chemical potential per GaAs pair in the Ga-As-Au alloy depending on c_{Ga}, for three different $c_{As} = 0.5$, 2 and 5% and at three different temperatures.

The curves shown in Figs. 5.63 to 5.66 as well as the results of [354] for other Au-catalyzed III–V nanowires can be well fitted by the expressions that present $\Delta\mu$ as functions of the atomic concentrations c_{III}, c_V and the absolute temperature T. In these expressions, the $\Delta\mu$ is split into the material-independent terms depending logarithmically on the concentrations, and the complementary polynomial of c_{III}, c_V and T:

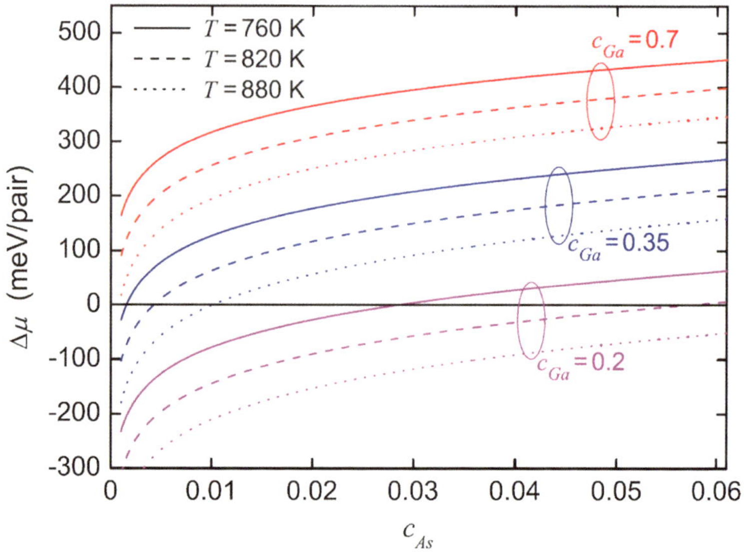

Fig. 5.63 Variation with the concentration of group V element (As) in the liquid of the difference of chemical potential per III-V pair between a ternary Ga–As–Au liquid and solid ZB GaAs, calculated for a set of temperatures and group III element (Ga) concentrations in the liquid

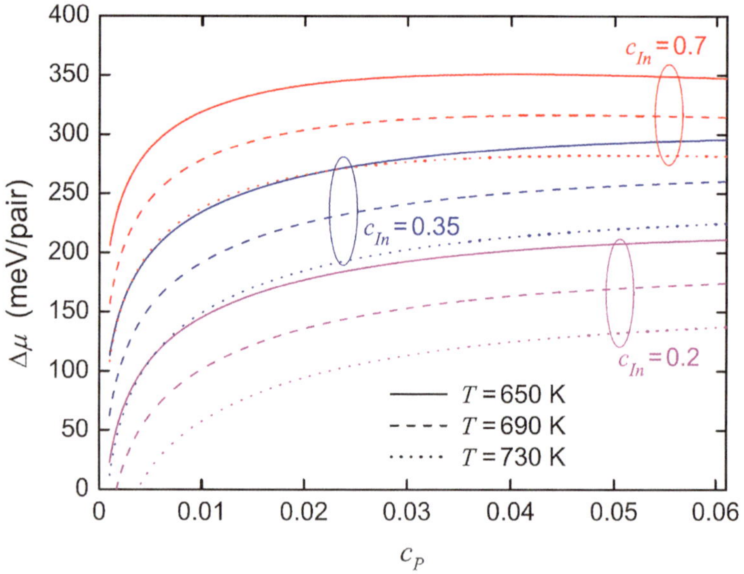

Fig. 5.64 Same as Fig. 5.63 for InP

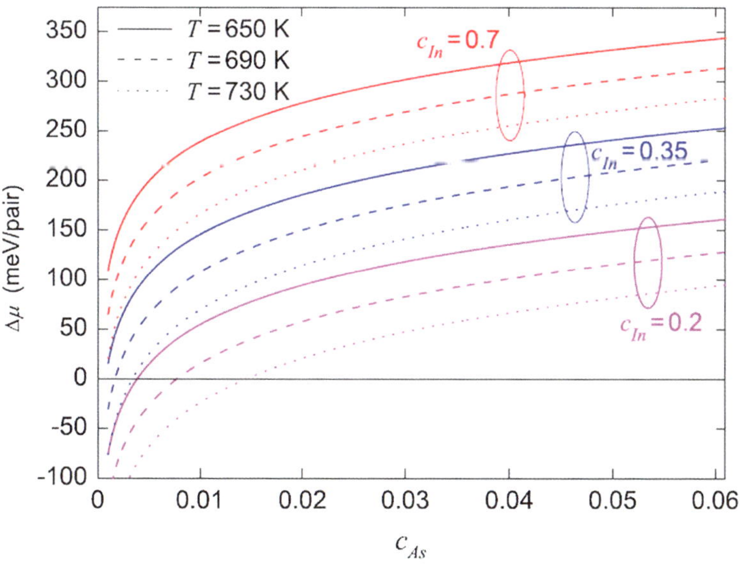

Fig. 5.65 Same as Fig. 5.63 for InAs

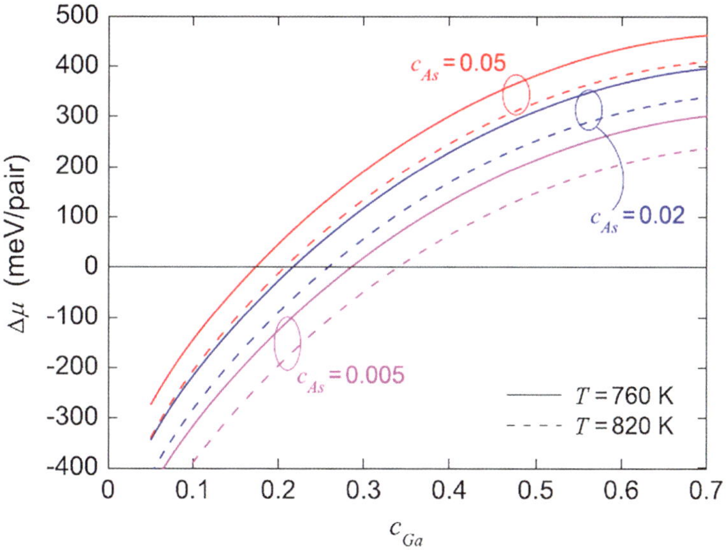

Fig. 5.66 Variation with the concentration of Ga in the liquid of the difference of chemical potential per III-V pair between a ternary Ga-As-Au liquid and solid ZB GaAs, calculated for a set of temperatures and As concentrations in the liquid

$$\Delta\mu(c_{III}, c_V, T) = k_B T \ln c_{III} + k_B T \ln c_V + \sum_{i=0}^{2} p_i(c_{III}, T) c_V^i. \tag{5.123}$$

Here,

$$p_i(c_{III}, T) = \sum_{j=0}^{m_i} A_{ij}(T) c_{III}^j. \tag{5.124}$$

with $m_0 = 3, m_1 = m_2 = 4$, and

$$A_{ij}(T) = \sum_{k=0}^{2} B_{ijk} T^k. \tag{5.125}$$

The dependence of $\Delta\mu$ on three independent variables thus obliges one to introduce a sizable number of 42 parameters B_{ijk} for each binary, which can be found in [354].

To conclude the results of this section, we point out the following important features. First, even if the Au-catalyzed VLS growth of a given III-V binary is limited by the material transport of group III element, the value of $\Delta\mu$ at a given temperature depends on both c_V and c_{III}, and can be considered as a function of a single concentration c_{III} only under the assumption of $c_V = const$. Second, $\Delta\mu$ generally increases with group III and group V concentrations. In general, $\Delta\mu$ is most sensitive to the group V concentration when the latter is less than about 1 % (Figs. 5.63, 5.64 and 5.65), whereas the variations with group III concentrations are more uniform in the ranges considered (Fig. 5.66). Third, at fixed concentrations, $\Delta\mu$ always decreases with increasing temperature.

Chapter 6
Crystal Structure of III–V Nanowires

Perhaps the most surprising feature of freestanding nanowires of cubic zincblende (ZB) III-V materials is their ability to form in hexagonal wurtzite (WZ) crystal phase. For GaAs, the bulk energy difference between WZ and ZB equals approximately 24 meV per III-V pair, a substantial gap that can be overpassed by applying a huge pressure of the order of several tens of GPa. However, the hexagonal WZ structure has been reported for most <111>-oriented III-V nanowires obtained by different epitaxy techniques and on a variety of substrates, both in Au-catalyzed and selective area approaches. Such an effect never occurs in 2D layers or quantum dots and is specific to the nanowire geometry and the growth method. Surprisingly, the kinetically controlled polytypism of VLS nanowires appears to be closely related to the character of 2D nucleation from a supersaturated catalyst particle. This leads to many interesting possibilities for tuning the crystal structure by intelligent manipulations with the growth conditions and the catalyst size. The present chapter gives a systematic account of theoretical approaches describing the polytypism of III-V nanowires, along with relevant experimental data that demonstrate the achieved level of crystal phase engineering. Overall, this topic is where the crystallography, surface energetics, nanowire growth method and physical chemistry of nucleation in a confined catalyst volume meet together and shed new light on a rather complex physical phenomenon.

Section 6.1 describes the bulk crystal structures to be investigated. We consider the stacking sequences and structural properties, the calculated bandgaps and bulk energy differences of ZB and WZ phases of different III-V materials. Mixed polytypic structures are also discussed. Section 6.2 presents some relevant experimental data on the nanowire polytypism. It is shown that the observed crystal structure of Au-catalyzed GaAs nanowires can change during growth and therefore is not entirely determined by the equilibrium energetic balance. This feature is of paramount importance for further understanding the polytypism of III-V nanowires and for constructing relevant theoretical models.

Since the bulk cohesive energy is larger for WZ than for ZB phase, the only physical reason for the WZ phase formation in nanowires could be a lower surface

V. G. Dubrovskii, *Nucleation Theory and Growth of Nanostructures*, NanoScience and Technology, DOI: 10.1007/978-3-642-39660-1_6,

energy of the particular WZ sidewalls. Unfortunately, these surface energies cannot be measured directly, while different theoretical calculations give discrepant results. Consequently, in Sect. 6.3 we present a simplified method for calculating the surface energies of relevant side facets, based simply on counting the number of dangling bonds in the second-nearest neighbor approximation. Section 6.4 presents a thermodynamic analysis of the critical radius for the structural transition from ZB to WZ and also to different polytypes. However, these simple considerations within a fully formed nanowire appear insufficient, yielding the critical radius of the structural transition which is by the order of magnitude smaller than the observed experimental values.

Section 6.5 describes a kinetic model of the WZ phase formation in VLS nanowires. Rather than considering the free energies of fully formed nanowires with different structures, kinetic theory is based on comparing the nucleation barriers for forming ZB or WZ islands of the critical size. The orientation positioning of 2D critical nucleus will then dictate the stacking sequence of the whole nanowire in the mononuclear growth regime. The key effect responsible for the WZ phase formation is the triple phase line nucleation, where a fraction of the nucleus perimeter is in contact with vapor. Such an approach has far-reaching implications in the field and in is capable of describing the supersaturation dependence of the crystal structure. In Sect. 6.6, we further develop the kinetic approach, in particular, by introducing the radius dependence through the available nucleation areas for nucleation at or away from the triple phase line. In Sect. 6.7, by pursuing a self-consistent model for the VLS nanowire growth and crystal phase, we calculate the kinetic radii of the ZB-WZ structural transition (relating to 50 % phase mixing) for different Au-catalyzed III-V nanowires. Section 6.8 presents some impressive experimental results of the kinetically controlled crystal phase engineering in III-V nanowires, most of which are designed with the help of theoretical modeling. It is shown that crystal phase can be changed abruptly from WZ to ZB and back even within a single nanowire, thus enabling the production of well-controlled ZB/WZ superlattices. In Sect. 6.9 we consider somewhat unexpected benefits of high nanowire growth rates (that are achievable in MOCVD and HVPE) on the defect-free crystal capability. It is discussed why and how the structure of Au-catalyzed III-V nanowires can be retained to pure ZB when the elongation rate is increased.

Self-catalyzed III-V nanowires are grown from the group III metal droplets instead of gold. Changing the catalyst metal has the major effect on the crystal structure. In Sect. 6.10, we show that the polytypism of Ga-catalyzed GaAs nanowires is suppressed on surface energetic grounds. On the other hand, the droplet stability at the nanowire top becomes questionable when the catalyst surface energy is decreased. Consequently, in Sect. 6.11 we describe a new, wetting mode of the VLS nanowire growth where the droplet wets the sidewalls and surrounds the nanowire tip. In a model cylindrical nanowire geometry with a planar growth interface, this yields an interesting droplet shape which is elongated in the growth direction. Finally, Sect. 5.12 presents a model which explains the single crystalline WZ structure of self-induced III-V nanoneedles.

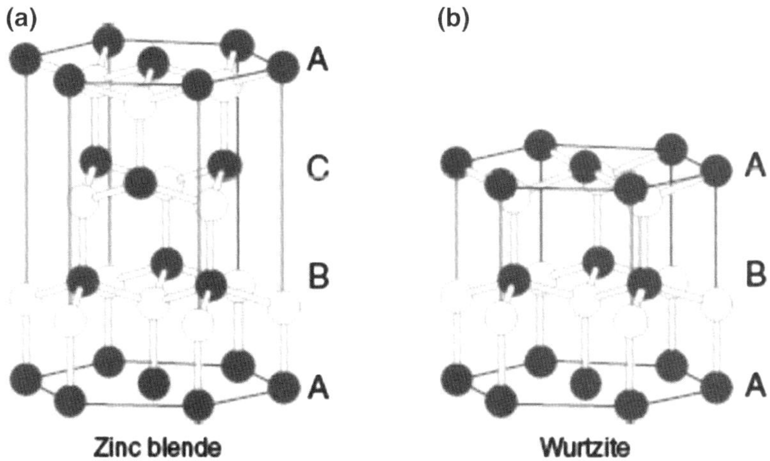

Fig. 6.1 ZB (**a**) and WZ (**b**) crystal structures showing their ABCABC. . . and ABAB. . . stacking sequence along the nanowire growth direction

6.1 Crystal Structures of III–V Materials

It is well known that, under bulk form, all III-V materials except for nitrides exhibit a cubic ZB crystal structure. In contrast, III-N materials exhibit a hexagonal WZ structure. As discussed preliminary in Sect. 3.14 and will be studied in detail in this chapter, different III-V nanowires [45, 184, 228–234] growing in <111> ZB (which is closely related to <0001> WZ) crystallographic direction usually exhibit predominant WZ structure. In fact, high purity Au-catalyzed ZB nanowires are the exception rather than the rule [234]. In particular, dominant WZ nanowires have been reported for GaAs [45, 229–233, 365] and InP [234, 449] materials. Long WZ nanowire segments have been found in GaP [83]. The WZ structure in InAs nanowires is far more common than the ZB one, and virtually all InAs nanowires form in the WZ phase [292, 316, 543]. The only exception seems to be InSb and other antimonite nanowires [234, 412] which adopt the ZB structure regularly.

Let us now describe the crystal structures to be investigated. Cubic ZB (or sphalerite) and hexagonal WZ lattices differ only in their stacking sequence along the close-packed <111> or <0001> directions, as shown in Fig. 6.1. The ZB structure (or *3C*) has *ABCABC*... stacking sequence, where each letter corresponds to one of the three possible positions of the pairs of group III and group V atomic layers aligned in <111> direction. In the WZ (or *2H*) structure, only two of the three possible positions are occupied, giving *ABAB*... stacking sequence aligned in <0001> direction. *ABAB*... stacking can be obtained from *ABCABC*... by rotating all the C layers by 60° around the growth direction. In a more precise notation, each atomic layer can be denoted separately by using lower case and upper case letters for the two different groups of atoms, giving *AaBbCcAaBbCc*... and *AaBbAaBb*... for ZB and

WZ stacking sequences, respectively (Fig. 6.2). In situations where the structures are intermixed and/or different faults in their stacking sequence arise, this more detailed notation can be necessary to fully describe and distinguish between different defects and faults [234].

Atoms in both ZB and WZ crystal structures are held together by tetrahedral bonds. The energetic difference between the two structures originates from the difference in the third-nearest-neighbor bond length which is shorter for the WZ than for the ZB structure (see the zoomed-in views in Fig. 6.2). The unit-cell vectors for the ZB structure are: $\vec{a} = (0, \frac{1}{2}, \frac{1}{2})a$, $\vec{b} = (\frac{1}{2}, 0, \frac{1}{2})a$, $\vec{c} = (\frac{1}{2}, \frac{1}{2}, 0)a$, where a is the cubic lattice constant. There are two atoms per unit cell, A at $(0, 0, 0)a$ and B at $(\frac{1}{4}, \frac{1}{4}, \frac{1}{4})a$. For the WZ structure, the lattice vectors are: $\vec{a} = (\frac{1}{2}, \frac{\sqrt{3}}{2}, 0)a$, $\vec{b} = (\frac{1}{2}, -\frac{\sqrt{3}}{2}, 0)a$, $\vec{c} = (0, 0, \frac{c}{a})a$, where c/a is the axial ratio. There are four atoms per hexagonal unit cell. Their positions in the units of $\vec{a}$, $\vec{b}$ and $\vec{c}$ are A at $(0, 0, 0)$ and $(\frac{2}{3}, \frac{1}{3}, \frac{1}{2})$ and B at $(0, 0, u)$ and $(\frac{2}{3}, \frac{1}{3}, u + \frac{1}{2})$, where u denotes the cell-internal structural parameter. For an ideal structure, $c/a = \sqrt{8/3}$ and $u = 3/8$. The lattice parameters of the ZB and WZ structures can be optimized to reach a minimum in the total energies by *ab-initio* methods [226]. For the ZB structure, only the value of the lattice constant

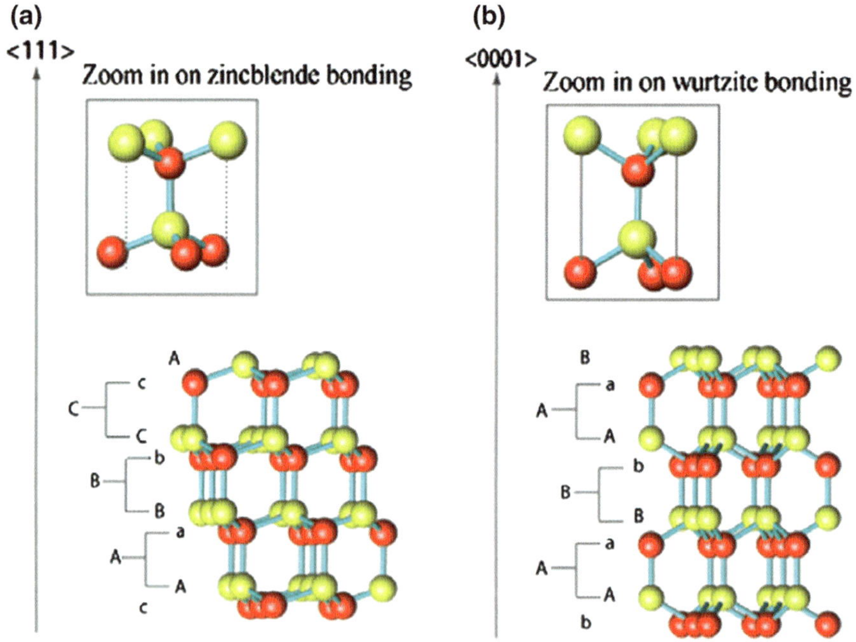

Fig. 6.2 Stacking sequences in ZB (**a**) and WZ (**b**) crystal structures in the <110> viewing direction. Different atom colors denote groups III and V atoms, while upper and lower case letters indicate layers of these two atoms types. The insets show zoomed-in views of the bonding configuration in the ZB and WZ structures. It is clear that the third-nearest-neighbor bond length is shorter in the WZ structure [234]

Table 6.1 Structural parameters and difference in bulk cohesive energies $\psi = E_{WZ} - E_{ZB}$ for different III-Vs

Material	ZB a (nm)	ZB Ω_S per pair (nm^3)	WZ a (nm)	WZ c/a	WZ Ω_S per pair (nm^3)	ψ_{WZ} (meV/pair)
AlN	0.4365 [226]	0.0208	**0.311**	**1.600**	**0.0208**	−36.8 [226]
GaN	**0.453**	**0.0233**	**0.319**	**1.628**	**0.0229**	−19.8 [226]
InN	**0.498**	**0.0309**	**0.354**	**1.609**	**0.0309**	−22.9 [226]
GaAs	**0.565**	**0.0452**	0.391 [226]	1.647 [226]	0.0426	24.0 [226]
			0.396 [546]	1.646 [546]	0.0442	23.1 [546]
InAs	**0.606**	**0.0556**	–	–	–	10.6 [544]
			0.426 [546]	1.642 [546]	0.0549	19.1 [546]
GaP	**0.545**	**0.0405**	–	–	0.0400	12.4 [544]
InP	**0.587**	**0.0506**	–	–	–	6.8 [571]
			0.411 [546]	1.641 [546]	0.0493	10.6 [546]
GaSb	**0.610**	**0.0567**	–	–	–	19.8 [544]
InSb	**0.648**	**0.0680**	–	–	–	16.4 [544]
			0.455 [546]	1.642 [546]	0.0669	20.7 [546]

a needs to be optimized. The independent structural parameters which need to be determined for the WZ structure are a, c/a and u.

In this way, the difference in bulk cohesive energies of WZ and ZB structures at zero pressure, $\psi_{WZ} \equiv E_{WZ} - E_{ZB}$, can be calculated for different materials [226, 544]. Usually, the optimized c/a and u are close to their values in an ideal WZ structure, and the piece of ψ_{WZ} due to $c/a \neq \sqrt{8/3}$ and $u \neq 3/8$ relaxation is relatively small (e.g., it equals −0.7 meV for GaAs with ψ = 24 meV/pair as reported by Yeh *et al.* [226]). Table 6.1 lists the values of ψ_{WZ} per III-V pair in III-nitrides and in several ZB III-V compounds of interest. We also present relevant structural parameters, including elementary volumes Ω_S per III-V pair which equals $a_{ZB}^3/4$ in ZB and $(\sqrt{3}/4)a_{WZ}^2c$ in WZ. Bold numbers are experimental data [545], the others are theoretical calculations of [226, 544, 546] (where available). It is seen that WZ is the favored structure in small, ionic nitride materials for which ψ_{WZ} is negative, while ZB is favored in other III-V materials.

The ZB-WZ stacking disorder should be distinguished from another kind of disorder associated with different twin orientations. For example, the ZB stacking sequence in nanowires can suddenly change from $ABCABC\ldots$ to $ABC\overline{A}CBACB\ldots$. The stacking sequence is still of ZB type stacking after $\overline{A}$, but the sequences before and after this layer are mirrored in $\overline{A}$. The "mirror" plane $\overline{A}$ is called a twin plane and the segments above and below $\overline{A}$ have different twin orientations, as shown in Fig. 6.3 [234]. However, the small part of the sequence surrounding the twin plane could also be viewed as following a WZ-type stacking sequence. Therefore, it was suggested [234] that the term "twin" or "twin plane" should be used only for this type of defects, while the term WZ is used only if the WZ stacking sequence occurs over at least four bilayers.

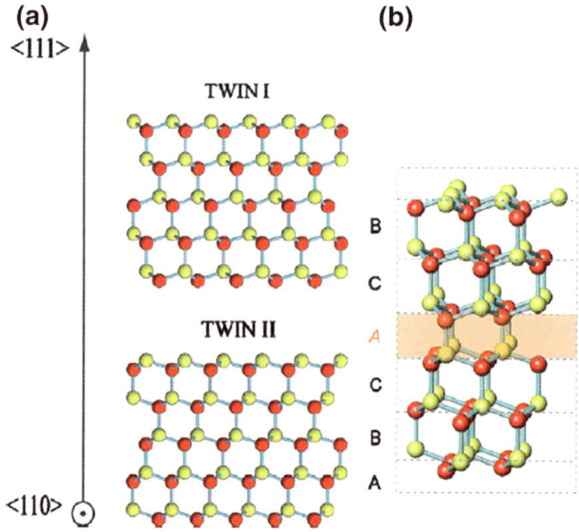

Fig. 6.3 **a** Two different orientations of the ZB stacking sequence, denoted as "I" and "II". **b** When the two orientations of ZB are combined, the common plane $\overline{A}$ is a twin plane or mirror plane

Table 6.2 lists the known polytypes between *3C* ZB and *2H* WZ structures with the periods up to 18 monolayers [547, 548]. Crystal structures of 4H and 6H polytypes, which have also been found in III-V nanowires [351, 549–553], are shown in Fig. 6.4. In column 1, we present the standard notations for different polytypes, where the number gives the lattice periodicity and the letter denotes the symmetry group. Column 2 describes the standard stacking sequence in <111> direction in terms of atom positions. Column 3 presents the same sequence in terms of hexagonal WZ (h) and cubic ZB (c) positions (orientations) of particular layers. The Zhdanov symbols [554] given in column 4 are readily obtained from the layer sequence as the number of layers preceding the WZ layer in the sequence, e.g., *(22)* = *chch* for *4H* polytype. The hexagonality of the polytype α, given in column 5, is simply the fraction of WZ layers in the total number of layers, e.g., $\alpha = 2/4 = 0.5$ for *4H* polytype.

In [546], Panse *et al.* presented *ab initio* study of the energetic, structural, and elastic properties of the most important hexagonal polytypes *2H*, *4H*, and *6H* in comparison to the well-known unstrained ZB 3C structure of GaAs, InAs, InP, and InSb. In particular, the results for the cohesive energies have been derived, shown in Fig. 6.5. It is seen that the difference in bulk cohesive energies between WZ and ZB follows a monotonous increase along the row *3C*, *6H*, *4H*, and *2H*. The variation of the relative energy per pair however varies nonlinearly with hexagonality.

From the data of Table 3.1 and Fig. 6.5, GaAs has the lowest tendency to crystallize in hexagonal polytypes in bulk. The calculated energy difference per pair between WZ and ZB given by different authors is almost identical and lies between 23.1 and 24.0 meV (the value of $\psi_{WZ} = 23.4$ meV was also reported by Murayama and Nakayama [555]). For other materials, the discrepancy between the calculated

Table 6.2 Polytypes between the ZB and WZ structures

1	2	3	4	5
PT	Stacking sequence	Layer sequence	Zhdanov symbol	α
2H	AB	hh	(1 1)	1.000
3C	ABC	ccc	∞	0.000
4H	ABCB	chch	(2 2)	0.500
6H	ABCACB	cchcch	(3 3)	0.333
8H	ABCACBAC	ccchccch	(4 4)	0.250
8L	ABCABCAC	cccccchh	(7 1)	0.250
10L	ABCABCABAC	cccccccchch	(8 2)	0.200
10L	ABCBACABAC	cchcchchch	(3 3 2 2)	0.400
10H	ABCABACBAC	cccchcccch	(5 5)	0.200
12R	ABCBCABABCAC	cchhcchhcchhcchh	(3 1)∗3	0.500
12H	ABCABCBACBAC	ccccchccccch	(6 6)	0.167
12L	ABCABCABCBAC	ccccccccchcch	(9 3)	0.167
12L	ABCACBACABAC	ccchccchchch	(4 4 2 2)	0.333
14H	ABCABCACBACBAC	cccccchcccccch	(7 7)	0.143
14L	ABCABCABCABCAC	cccccccccccchh	(13 1)	0.143
14L	ABCABACBABCBAC	cccchccchchcch	(5 4 2 3)	0.286
14L	ABCABACBACACAC	cccchccccchhhhh	(5 5 1 1 1 1)	0.429
14L	ABCABACBCBCBAC	cccchcchhhhcch	(5 3 1 1 1 3)	0.429
14L	ABCABACBACABAC	cccchccccchchch	(5 5 2 2)	0.286
14L	ABCACBACABCBAC	ccchccchcchcch	(4 4 3 3)	0.286
14L	ABCBACABCBACAC	cchcchcchcchhh	(3 3 3 3 1 1)	0.429
16H	ABCABCABCBACBACB	cccccccchccccccch	(8 8)	0.125
16L	ABCABCABCABCABAC	cccccccccccccchch	(14 2)	0.125
16L	ABCABACBACABCBAC	cccchccccchcchcch	(5 5 3 3)	0.250
16L	ABCBACABCBACABAC	cchcchcchcchchch	(3 3 3 3 2 2)	0.375
18R	ABCACBCABCBABCABAC	ccchchccchchccchch	(4 2)∗3	0.333

ψ_{WZ} is much larger. Whereas the energy deviations from the 3C value are close for InAs, InSb and GaSb, the energy differences for phosphides are smaller. The smallest ψ_{WZ} between 6.8 and 10.6 meV is predicted for InP, showing that stacking fluctuations under bulk should be easier for InP in comparison with other III-V compounds. However, these tendencies under equilibrium conditions and in bulk form are different from the experimentally observed structural trends in nanowires. As mentioned at the beginning, both InAs and InP nanowires tend to adopt WZ structure more easily than GaAs and especially GaP, while antimonite nanowires appear to be almost purely ZB. This is due to a non-trivial interplay between the volume and surface energies of ZB and WZ nanowires coupled with position-dependent nucleation probabilities, as will be discussed later on.

We now only briefly discuss the electronic band structure of III-V polytypes, which has been a subject of extensive theoretical [554–558] and experimental [232, 233, 558–568] investigations. Low temperature energy gaps of unstrained ZB and WZ III-V compounds are quite close to each other, as demonstrated by the data

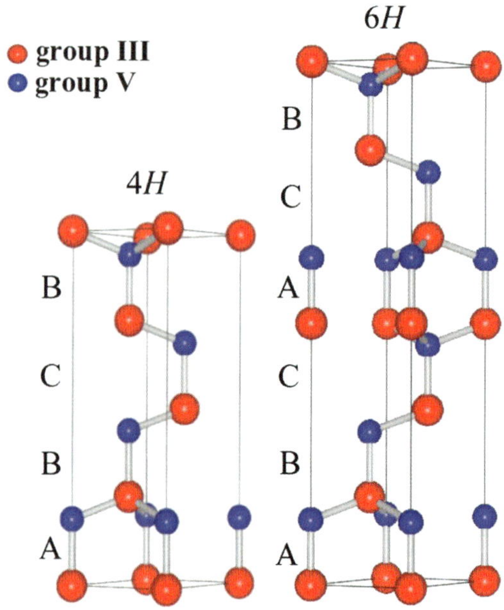

Fig. 6.4 Crystal structures of *4H* and *6H* polytypes. Cations: red spheres, anions: blue spheres. The stacking sequence of the cation-anion bilayers are indicated by the symbols *A*, *B* and *C* [554]

Fig. 6.5 Cohesive energy per III-V pair relative to 3C value. Lines are given as guide to the eye only [546]

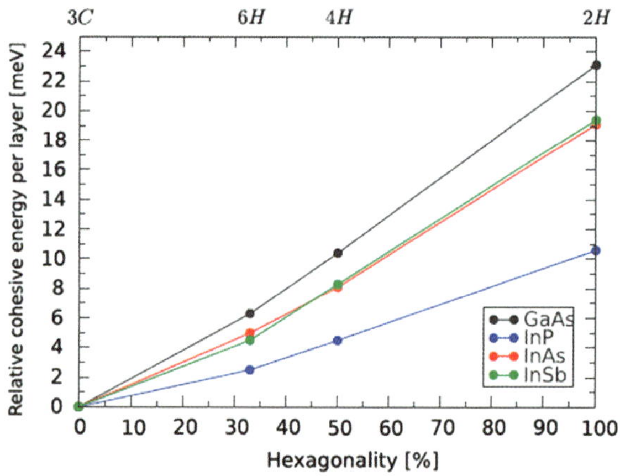

of Table 6.3 and Fig. 6.6. In column 2 of Table 6.3, theoretical values for the WZ band gap E_g^{WZ} are summarized for three materials [557]. These values have been obtained by different methods, such as *ab-initio* techniques based on density-functional theory in the local density approximation (LDA) [554] and methods based on empirical pseudopotentials including spin-orbit coupling [557], and show a large discrepancy. Column 3 summarizes experimental data, all obtained on nanowires and some of them extrapolated from the photocurrent spectroscopy measurements on $InAs_{1-x}P_x$ (for $0.14 < x < 0.48$) [565] in x to obtain the band gaps of InAs and InP. This gives a

Table 6.3 Summary of theoretical and experimental data on energy gaps in GaAs, InP and InAs

1	2	3	4	5	6	7
Material	E_g^{WZ}, theory (eV)	E_g^{WZ}, exp. (eV)	E_g^{ZB}, theory (eV) [554]	E_g^{ZB}, exp. (eV)	$E_g^{WZ}-E_g^{ZB}$, theory (eV)	$E_g^{WZ}-E_g^{ZB}$, exp. (eV)
GaAs	1.381–1.623	1.476–1.553	1.421	1.52	−0.139–0.132	−0.044–0.033
InP	1.474–1.576	1.440–1.645	1.475	1.42	0–0.156	0.020–0.225
InAs	0.481–0.642	0.54	0.411	0.43	0.051–0.212	0.110

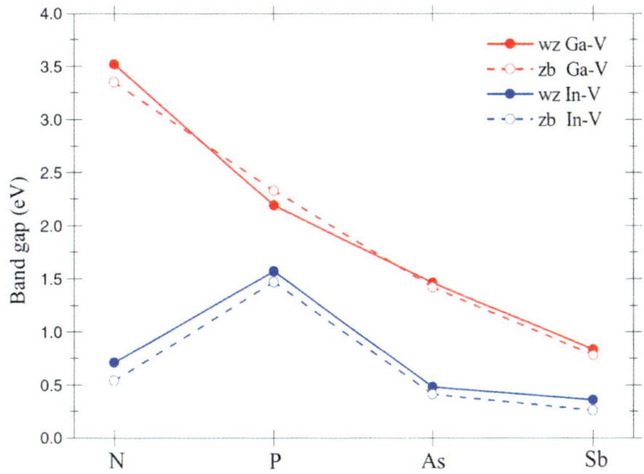

Fig. 6.6 The gaps E_g^{WZ} and E_g^{ZB} versus the group V anion as calculated in [554] by LDA

band gap of 1.645 eV for InP, which is higher than the values from low-temperature photoluminescence measurements and the calculated results. Also, measurements on nanowires may give slightly higher values than E_g^{WZ} due to confinement. In column 4, we give theoretical values for the ZB energy gap E_g^{ZB} obtained in [554] by LDA including spin-orbit coupling. Column 5 gives the reference experimental values for E_g^{ZB} at $T = 0$. In columns 6 and 7, we present theoretical and experimental energy differences between the two structures with maximum discrepancy.

Since the LDA usually underestimates band gaps (as seen from two of the three underestimated theoretical E_g^{ZB} given in column 4), calculations of WZ and ZB are typically compared to obtain the differences between the two structures. However, even the relative values $E_g^{WZ} - E_g^{ZB}$ calculated by different methods have different signs for GaAs. For example, De and Pryor predicts 17 meV redshift from empirical pseudopotential [557], while [554] predicts 32 meV blueshift for WZ GaAs based on LDA for WZ GaAs. Experimental data for GaAs are also qualitatively different: while low temperature PL measurements of [561] give 41 meV redshift for WZ GaAs, it is blueshifted by 33 meV from the PL data of [559]. Of course,

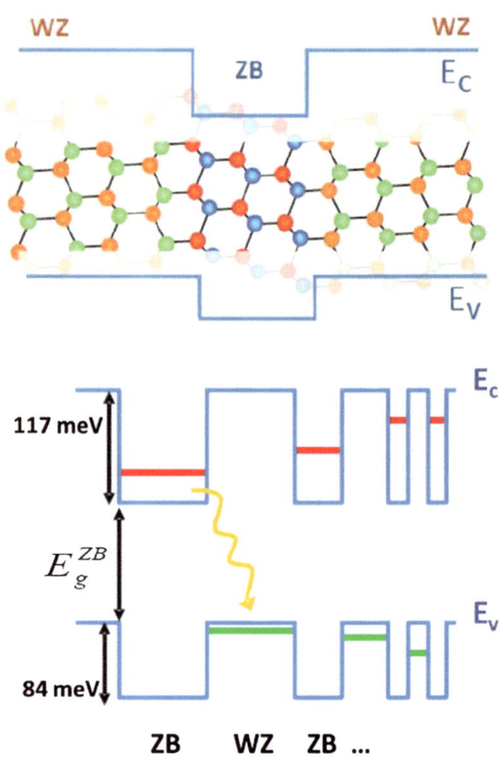

Fig. 6.7 **a** An atomistic model of a WZ/ZB/WZ double heterostructure along with schematics of the band diagram. The Ga and As atoms are marked in *orange* and *green*, respectively, for the WZ domains, and in *red* and *blue*, respectively, for the ZB quantum well. **b** Theoretical band alignment of WZ and ZB GaAs, along with the values of conduction-band and valence-band discontinuities are shown. Some selected examples of quantum heterostructures with different widths are presented, along with the first quantized level [558]

both calculated and measured $E_g^{WZ} - E_g^{ZB}$ within each work are not as large as given in columns 6 and 7 which quote the maximum error bar from the data of different authors. Nevertheless, these diverse data clearly show the major trend: WZ structure should have larger band gap for InP and InAs, which is confirmed by most experiments [233, 560, 565, 568], while the relationship between WZ and ZB energy gaps in GaAs remains an open question. There are several reasons that complicate the analysis of the optical data obtained on nanowires, such as the influence of strain on the optical properties, size quantization, surface states, and spatially indirect recombination in nanowires with mixed phases, as will be discussed shortly.

Figure 6.6 presents the LDA energy gaps of different III-V materials depending on the group V anion [554]. Clearly, the WZ structure has a larger gap for both nitrides, while in phosphide materials the difference between the WZ and ZB gaps is predicted positive for InP (with $E_g^{WZ} - E_g^{ZB} = 101$ meV) but negative for GaP. In arsenides and antimonites, the WZ gaps are predicted larger than ZB, with the difference $E_g^{WZ} - E_g^{ZB}$ of 32 meV for GaAs, 70 meV for InAs, 52 meV for GaSb and 34 meV for InSb.

As for the band alignment, it is generally believed that the valence band is higher in WZ GaAs, with theoretically predicted conduction and valence-band discontinuities of 117 and 84 meV, respectively [555]. In this case, WZ/ZB crystal phase heterostructures will exhibit a type II band alignment as shown schematically in Fig. 6.7, which is rather unusual for GaAs. It is then expected that an electron-hole pair confined to the WZ/ZB interface will give rise to a spatially indirect recombination, as shown in Fig. 6.7b. According to the data of [555], the recombination energy amounts to 1.431 eV for a high WZ percentage in nanowire, and increases toward 1.515 eV for thinner WZ quantum wells due to confinement. Therefore, the controlled production of WZ/ZB layers may open new possibilities for structural band gap engineering of III-V nanowires and produce the so-called crystal phase quantum dots where the heterostructure is based on the crystallographic polytypism rather than chemical composition.

6.2 Experimental Observations of Polytypism

Indeed the twins, stacking faults and polytypes associated with WZ-ZB phase instability are better known for the case of II-VI semiconductors, for which the difference in bulk formation energies is generally lower than for III-Vs [569]. As mentioned already in Sect. 3.14, formation of the WZ crystal structure in III-Vs has been observed almost exclusively on <111> oriented nanowires and nanoneedles with hexagonal cross-section. The WZ phase has been found in the Au-assisted VLS nanowires [45, 228–232], catalyst-free nanowires grown by selective area method [233] and self-induced nanoneedles [185, 186], obtained by different epitaxy techniques and on a variety of substrates.

During MBE, the crystal structure can be monitored *in situ* by RHEED. Typical examples of RHEED patterns from an ensemble of GaAs nanowires are shown in Fig. 6.8 [552]. It should be noted that the diffraction patterns in $<1\bar{1}2>$ zone axis should be identical for ZB, WZ and most polytypes. The differences appear in $<1\bar{1}0>$ zone axis. The spotty RHEED pattern in Fig. 6.8a is observed at the initial growth step and contains "diagonal" ZB (hki) spots with integer indices. As the growth proceeds, the RHEED pattern transforms into the one shown in Fig. 6.8b, corresponding to a superposition of diffractions from ZB and WZ structures (the latter is characterized by $(000l)$ spots shown in the figure). The $(hkil+1/2)$ spots, also present in this pattern, can originate only from *4H* and some other polytypes.

TEM studies of MBE-grown GaAs nanowires [552] allowed us to identify at least three types of crystal structures: ZB, WZ and faulted sections with crystal phase intermixing, as shown in Fig. 6.9a, b and c, respectively. The structure in (c) is close, however, to 4H polytype, because it can be presented as *chch*... stacking sequence. The FFT image from this structure contains the $(hkil + 1/2)$ spots, which also confirms the presence of 4H lattice. Estimates of interlayer distances from the HRTEM images of pure WZ sections give the values of 0.326 and 0.341 nm for longitudinal and transverse directions, respectively, which corresponds to the monolayer heights of the WZ GaAs along [0002] and $[1\bar{1}00]$.

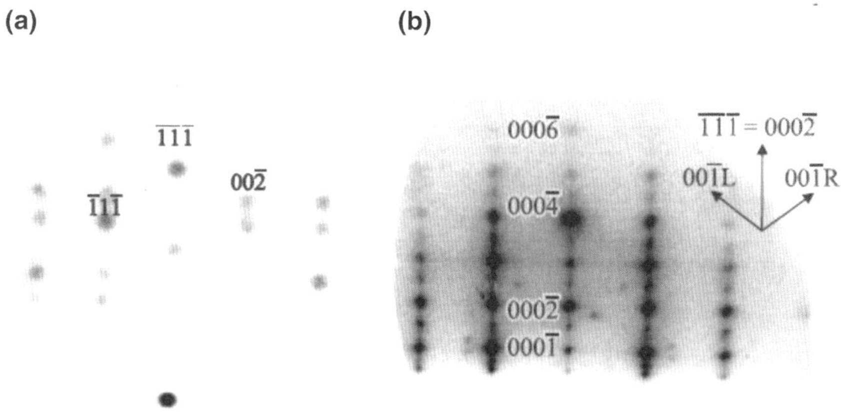

Fig. 6.8 RHEED patterns for <110> zone axis, demonstrating pure ZB crystal structure at the beginning of growth (**a**) and a mixed ZB, WZ and 4H polytype structures at a later growth stages (**b**) for the ensembles of GaAs nanowires obtained by the Au-assisted MBE on GaAs(111)B substrates

Fig. 6.9 HRTEM images of GaAs nanowire segments with ZB (**a**), WZ (**b**) and 4H polytype (**c**) crystal structures, with the inserts showing the corresponding electron diffraction images

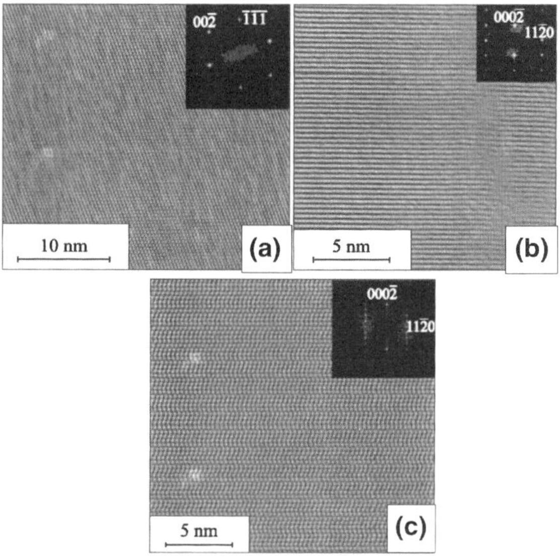

Figure 6.10a displays a HRTEM image of an InAs nanowire grown by the Au-assisted MBE on InAs(111)B substrate at 410 °C for 20 min, deposited on a carbon membrane [292]. The inset shows the corresponding electron diffraction for the $[1\overline{2}10]$ zone axis. The crystalline structure is WZ. The only defects present in the nanowire are stacking faults perpendicular to the growth axis. However, the density of these defects remains low in comparison to that typically observed in GaAs wires:

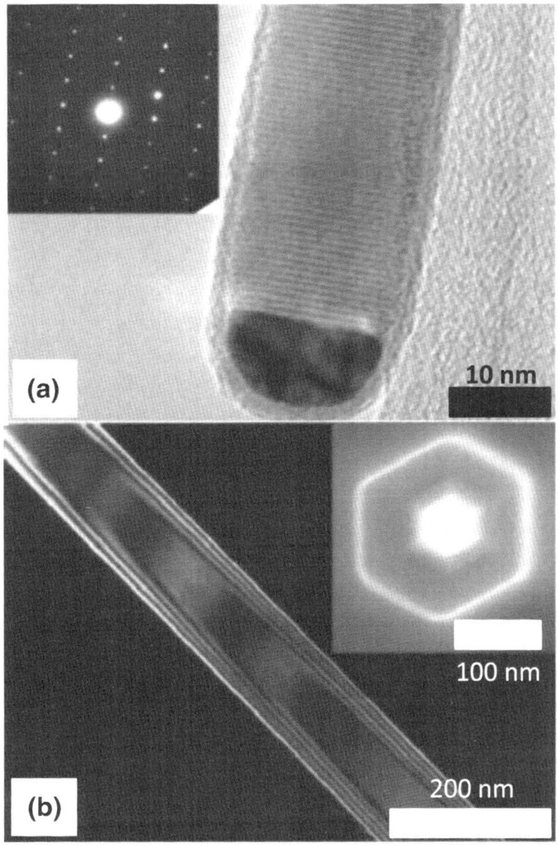

Fig. 6.10 **a** HRTEM image for [$1\bar{2}10$] zone axis of an InAs nanowire grown by the Au-assisted MBE at 410 °C for 20 min. Inset: corresponding electron diffraction. **b** Dark-field TEM image for [$1\bar{1}00$] zone axis of an InAs nanowire grown at 410 °C for 75 min. Inset: top-view SEM image of a nanowire from the same sample

pure WZ segments longer than 0.5 μm could be found. Dark-field views were used to determine the WZ facet orientation. Figure 6.10b shows the image of InAs nanowire grown at 410 °C for 75 min, taken with a [$1\bar{1}00$] zone axis using a ($1\bar{2}10$) spot. In this view, the inner part of the wire has a constant contrast, while the equal thickness fringes parallel to the edges can be seen near the nanowire borders. This means that the nanowire facet is parallel to the image plane. From top-view SEM images [shown in the inset to Fig. 6.10b], it was found that the nanowires have a hexagonal cross section defined by six facets. Since these InAs nanowires are highly tapered in the top region, top-view images show the hexagonal cross-section of the regular nanowire part. Combining TEM and SEM observations, it can be concluded that the six lateral facets are defined by ($1\bar{1}00$) planes in this case.

Figure 6.11a shows pure WZ crystal structure of a 20 nm diameter InP nanowire grown by the Au-catalyzed MOCVD on Si(111), confirmed by the FFT of TEM image displayed in Fig. 6.11b [232]. The interlayer distance in the [0001] growth direction *c/2* is measured to be 0.339 nm, which is close to its theoretical value of 0.337 nm from Table 6.1. Low temperature micro-PL measurements of individual NW (Fig. 6.11c) demonstrate a narrow linewidth of only 1.4 meV and a strong blueshift of 190 meV

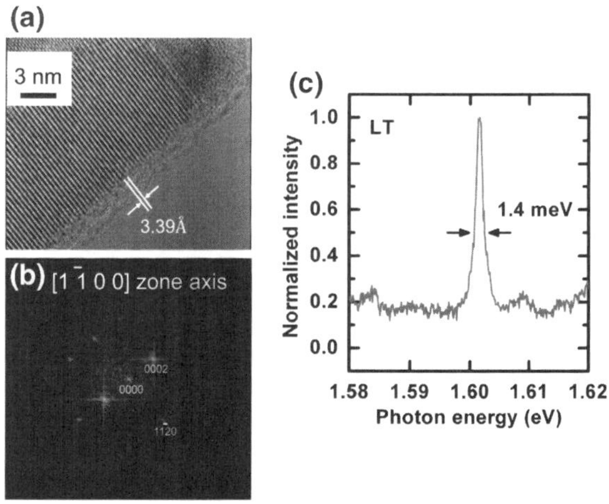

Fig. 6.11 (**a**) Single-crystalline WZ structure of a 20 nm diameter InP nanowire grown by Au-catalyzed MOCVD on Si(111), confirmed by the FFT of the TEM image shown in (**b**). (**c**) Low temperature micro-PL revealing a narrow linewidth and a 190 eV blueshift relative to bulk ZB structure [232]

relative to bulk ZB InP. This correlates with the corresponding data given in Table 6.3. Only a part of this blueshift can be attributed to quantum confinement, while the major difference should be due to a larger E_g^{WZ}.

Figure 6.12 from [233] clearly shows that the formation of WZ phase in III-V nanowires is not at all restricted to the case of Au catalyst. These InP nanowires were grown by catalyst-free selective area MOCVD on partially masked InP(111)A substrates following the procedure described in Sect. 4.1, with 70 nm diameter. HRTEM analysis and electron diffraction measurements indicated that the nanowires were single-crystalline WZ (Fig. 6.12a, b). The nanowires were oriented along the c axis of the WZ structure and the lattice constant c was estimated to be about 0.66 nm. Typical 4 K PL spectra of rod-like InP nanowire arrays are shown in Fig. 6.12c. The InP nanowire arrays exhibited a single intense peak at 1.49 eV, which is 62 meV blueshifted with respect to bulk ZB InP according to this study.

As already follows from the data shown in Figs. 6.8 and 6.9, the structure of GaAs nanowires grown by the Au-assisted MBE is predominantly ZB at the very beginning of GaAs deposition, and then changes to a mixed WZ-ZB phase observed throughout the steady state growth stage; this happens in the same growth experiment. At low enough surface temperatures the steady state structure is usually predominantly WZ. Close examination of the initial MBE growth step was performed in [45, 570], with the results displayed in Fig. 6.13a and b. By growing GaAs nanowires for only a few seconds, the foot of each wire can be observed before its burying by the growth of the non-activated surface. The samples were thinned to obtain cross-sectional views, with the nanowires still epitaxially attached to their (111B) substrate. The images identify unambiguously ZB as the sole phase formed in this initial stage (Fig. 6.13a). When the nanowire height reaches about 30 nm, growth switches abruptly to WZ stacking (Fig. 6.13b).

The second instance where ZB structure forms systematically is the final growth stage, after the MBE growth is stopped by switching off the Ga flux while maintaining

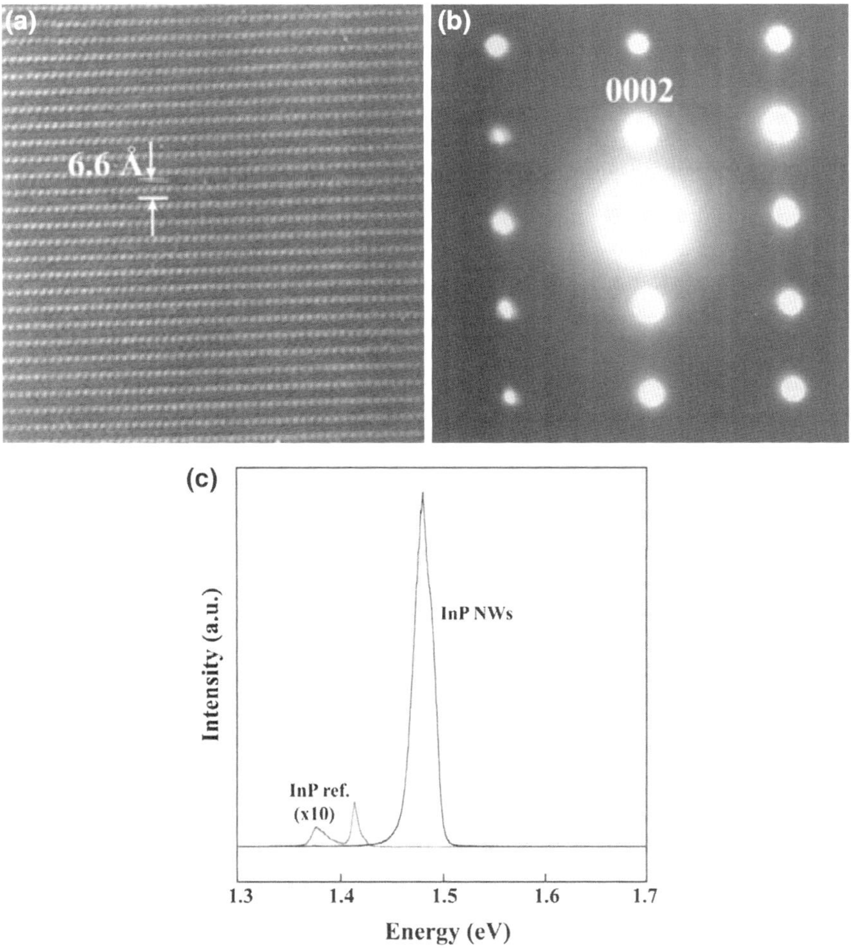

Fig. 6.12 **a** HRTEM image and **b** electron diffraction pattern of selective area MOCVD-grown InP nanowire. **c** 4 K PL spectra of InP nanowire arrays and reference InP planar sample [233]

an As flux during cooling. Over a few tens of nanometers, the topmost section of the nanowire features ZB structure (Fig. 6.13c). The formation of short terminal ZB sections of GaAs nanowires was also reported by Persson *et al.* in the case of CBE [229], and Soshnikov *et al.* [230] in the case of MBE. The latter data are shown in Fig. 6.14 and are very similar to Fig. 6.13c. In all cases, growth continues in the absence of a fresh gallium supply by consuming the Ga atoms from the droplet, whereas As is provided by the beam. Since the volume and the base area of the droplet decrease as the Ga is consumed, the final sections of the nanowires shown in Figs. 6.13c and 6.14a are tapered.

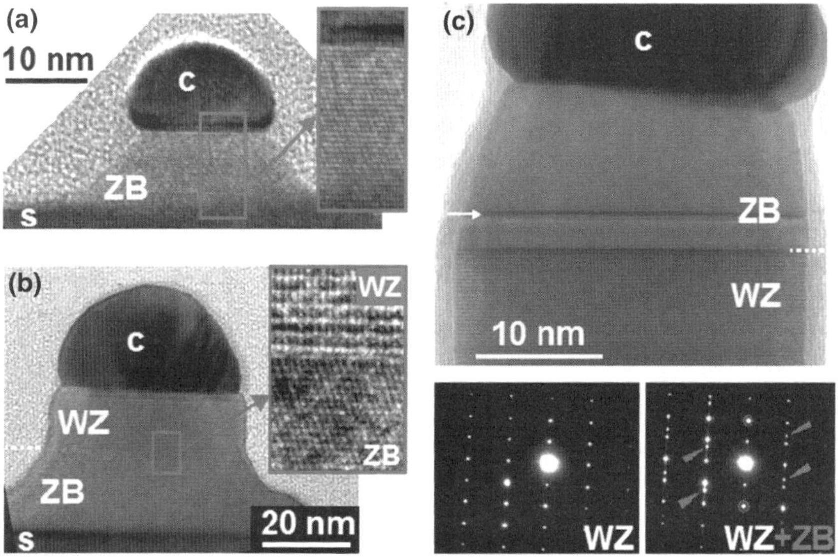

Fig. 6.13 **a** TEM image of a very short GaAs nanowire with HREM close-up showing the pure ZB structure (s: substrate, c: catalyst). **b** A short GaAs nanowire soon after the ZB-WZ transition, with HREM image of the transition zone. **c** When growth is terminated under As only, the final tapered section of the GaAs nanowire (*above dashed line*) is pure ZB (with a stacking fault, pointed by *arrow*): the TEM diffraction pattern from the main body of the nanowire (*left*) shows only WZ whereas the pattern from an area including also the terminal section (*right*) shows a superposition of WZ and ZB (*arrowed spots*) [570]

The TEM data of Figs. 6.13 and 6.14 clearly reveal that ZB GaAs structure forms systematically at the beginning and at the end of MBE growth, while the middle part of the nanowires is WZ. This shows that crystal structure of GaAs and other III-V nanowires is not entirely determined by the nanowire radius and thermodynamic parameters such as temperature, but is influenced by the growth kinetics. More precisely, the two transient growth steps that favor ZB phase relate to a low Ga concentration in the liquid droplet (corresponding to a low liquid supersaturation) compared to the steady state. Indeed, immediately after the beginning of GaAs deposition, the Ga concentration in the droplet must increase from its equilibrium value to a stationary value which is then maintained in the steady state. When the Ga flux is terminated at the end of growth, the VLS process continues by consuming the residual Ga from the droplet, and the liquid supersaturations decreases. We can therefore speculate that ZB structure forms at low liquid supersaturations, while WZ structure is preferred at high supersaturations. This conclusion has been paramount in explaining the kinetically controlled ZB-WZ polytypism [45] and different methods for the crystal phase control and engineering as will be discussed in detail in the foregoing sections.

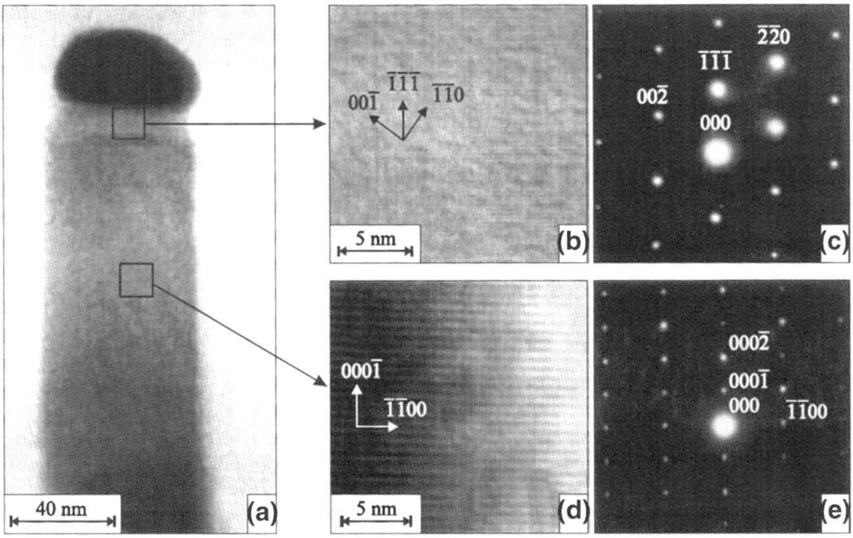

Fig. 6.14 HRTEM images of upper (**b**) and middle (**d**) parts of a GaAs nanowire shown in (**a**) with the corresponding TEM diffraction patterns [(**c**) and (**e**)], demonstrating the WZ-to-ZB crystal phase transition at the end of MBE growth

6.3 Surface Energies of Nanowire Sidewalls

Since the difference in the bulk formation energies ψ is positive for the high-pressure WZ phase as well as for all *pH* polytypes, it is reasonable to assume that polytypism of ZB III-V nanowires is physically explained by a lower effective surface energy of WZ nanowires. In [571, 572] and then in [407] it has been shown that unpassivated WZ nanowires indeed feature a lower number of dangling bonds on relevant sidewall facets or the edges separating the facets. Later on, different authors calculated the radius-dependent equilibrium formation energies of WZ and ZB III-V nanowires including the contributions from the sidewall facets and the edges (see, for example, [573–575] and references therein). Here, we concentrate on modeling the surface energies of relevant sidewall planes which are absolutely necessary for understanding and quantitative description of WZ-ZB polytypism in III-V nanowires.

As discussed in Sect. 4.1, regular hexahedral nanowires growing in <111>/[0001] direction are restricted by either (110) or (211) vertical ZB planes and their $(11\bar{2}0)$ or $(1\bar{1}00)$ WZ counterparts. In Fig. 6.15 taken from the work by Rosini and Magri [574], WZ nanowires with sidewalls belonging to the $(1\bar{1}00)$ family and ZB nanowires with sidewalls belonging to the (211) family are labeled type I. WZ nanowires with sidewalls belonging to the $(11\bar{2}0)$ family and ZB nanowires with sidewalls belonging to the (110) family are labeled type II. The corners of the hexagonal nanowire cross-sections consist of a III-V pair, each atom having either one dangling bond (same as on the facets), or two dangling bonds. Therefore, letters (a), (b) and (c) in Fig. 6.15

distinguish the composition of the edges. For example, WZ nanowires type I(a) have only one dangling bond per surface atom and WZ nanowires type I(b) have surface atoms with two dangling bonds at the edges. The ZB type II structure having (110) facets has only one representative with a III-V pair at the edges. For ZB type I nanowires restricted by (211) planes, type I(a) structure has a III-V pair at the edge position, type I(b) structure contains a minifacet of three pairs at the edges, and type I(c) structure has two pairs at the edges. Among the nanowires having the same stacking sequence and same sidewall orientation, the nanowires with smallest formation energy are those whose surface atoms have the smallest number of dangling bonds at the edges.

Generally, the equilibrium nanowire formation energy relative to bulk ZB structure can be defined as [574]

$$E_f = \frac{E_{NW}}{N} - \mu_b^{ZB}, \tag{6.1}$$

where E_{NW} is the total energy of the nanowire relating to the minimum energy configuration (obtained, e.g., by DFT), N is the number of III-V pairs, and μ_b^{ZB} is the total energy of the bulk ZB primitive cell. Obviously, N can be related to the nanowire radius and thus the R—dependent formation energies can be calculated for a given radius, crystal structure, facet type and edge composition. The asymptote at $N \to \infty$ should correspond to the difference in bulk formation energies for all WZ structures. In order to separate contributions from bulk, sidewall facets and edges, the formation energy per pair can be put in the form

$$E_f^{ij} = \psi_i + \frac{2}{R}\gamma^{ij}(R)\Omega_S + \frac{1}{R^2}A_{ij}(R)\Omega_S. \tag{6.2}$$

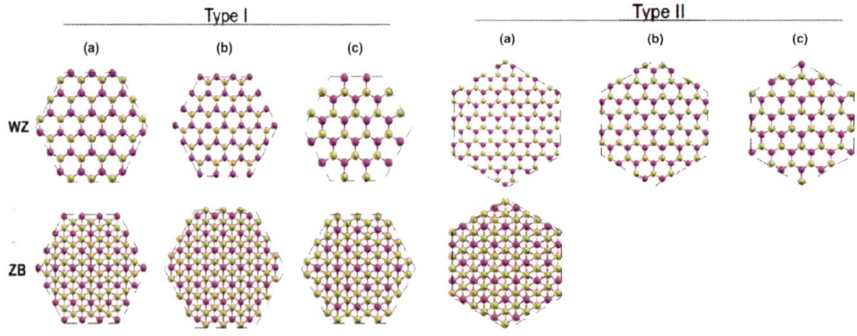

Fig. 6.15 Cross sections of the structural models of the bare III-V nanowires. The WZ structures are shown at the *top line* and the zinc blende structures at the *bottom line*. Type I structures have $(1\bar{1}00)$ WZ or (211) ZB facets, while type II structures have $(11\bar{2}0)$ WZ or (110) ZB facets. Dashed hexagons emphasize the nanowire shape. *Yellow* and *purple* atoms represent cations and anions, respectively. Letters (**a**), (**b**) and (**c**) distinguish the edge composition as described in the text [574]

Here, i = WZ, ZB and j denotes the facet orientation (type I or II), ψ_i is the total energy difference per pair in bulk (i.e., $\psi_{ZB} = 0$), Ω_S is the elementary volume per III-V pair in bulk (similar for ZB and WZ from the data of Table 6.1), γ_{ij} is the surface energy per unit area and A_{ij} is the edge energy per unit length, both allowed to depend on R for small wires.

Considering E_f^{ij} at large enough R enables one to deduce the bulk energy difference and the surface energies of the four sidewall facets. In particular, the calculations of [574] yield the value of 22.8 meV for GaAs, which is very close to the previously obtained ψ_{ZB} given in Table 6.1. The calculated GaAs surface energies of [574] are the following: $\gamma_{(110)} = 0.825\ \text{J/m}^2$, $\gamma_{(211)} = 1.07\ \text{J/m}^2$, $\gamma_{(11\bar{2}0)} = 0.730\ \text{J/m}^2$ and $\gamma_{(1\bar{1}00)} = 0.693\ \text{J/m}^2$. These values are very close to the results of [369]: $\gamma_{(110)} = 0.798\ \text{J/m}^2$, $\gamma_{(211)} = 1.07\ \text{J/m}^2$, $\gamma_{(11\bar{2}0)} = 0.706\ \text{J/m}^2$ and $\gamma_{(1\bar{1}00)} = 0.700\ \text{J/m}^2$. It is seen that the surface energies for both WZ facets are lower than for the ZB ones, with the maximum $\tau = \gamma_{WZ}/\gamma_{ZB}$ between 0.648 and 0.654 for the direct (211) $\rightarrow$ ($1\bar{1}00$) transition without 30° angular rotation. *Ab initio* calculations include full surface atomic relaxation; however, the obtained surface energies seem largely underestimated: the earlier theoretical value for GaAs $\gamma_{(110)}$ amounts to 1.5 J/m^2 [576] instead of 0.798–0.825 J/m^2. The ZBγ are also much lower than the measured [577, 578] and calculated [579–581] values for (110) and (211) surface energies of cubic Si ($\gamma_{(110)} = 1.43$–2.04 J/m^2 for different surface reconstructions, $\gamma_{(211)} = 1.31\ \text{J/m}^2$) and Ge ($\gamma_{(110)} = 1.17$–$1.51$ J/m^2 for different surface reconstructions).

In the following, we will use the surface energies obtained by the following approach [453]. Consider a crystal surface created by the dissection of crystal by a plane which is parallel to one of the crystallographic planes with Miller's indices (hkl). For an ideal crystal without surface reconstruction, the surface energy per unit area can be defined as the ratio between the energy spent on such a dissection φ_{hkl} and the created surface area:

$$\gamma_{hkl} = \frac{\varphi_{hkl}}{S_{hkl}}. \tag{6.3}$$

We then assume that the energy spent on the atom detachment equals the energy of the corresponding dangling bond on the newly formed surface. Only the first and the second nearest neighbors are taken into account, belonging to the first and the second coordination sphere of the crystal as shown in Fig. 6.16. With these approximations, the energy of dangling bonds on a given crystallographic plane is presented as

$$\varphi_{hkl} = n_1\varphi_1 + n_2\varphi_2. \tag{6.4}$$

Here, n_1 and n_2 are the numbers of the first and the second nearest neighbors for a given atom, respectively, and φ_1, φ_2 are the corresponding interaction energies. To calculate the surface energies from (6.3) and (6.4), we thus need to know the density of dangling bonds per unit surface area of a given plane and the interaction energies.

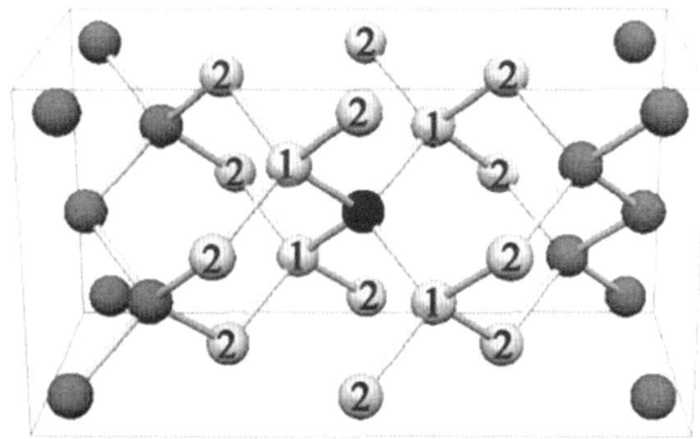

Fig. 6.16 FCC cubic lattice: the first (*1*) and the second (*2*) nearest neighbors for the *black* atom in the *center*, other atoms are *grey*

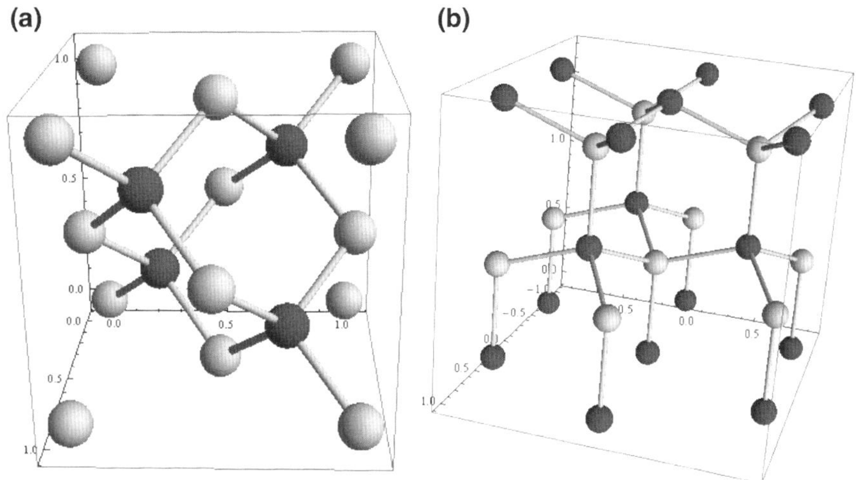

Fig. 6.17 Cubic ZB (**a**) and hexagonal WZ (**b**) crystal lattices subjected to the dissection by relevant planes

Counting the number of the dangling bonds on a given plane is performed as follows. All atoms belonging to the plane are obtained by translation of the primitive cell. As is seen from Fig. 6.17, the surface atoms can hold different positions in the cell. The number of dangling bonds is identical for the atoms holding the same positions relative to the cell. It is therefore possible to dissect the lattice into two parts, one of which describes the atoms remaining in the crystal and the other corresponds to the atoms cut off from the crystal upon the formation of (hkl) or $(hkli)$ plane.

The dissection of cubic crystal by the (211) plane is illustrated in Fig. 6.18. The newly formed (211) plane contains atoms that have different positions in the primitive cell (Fig. 6.19a). The plane is therefore divided into the disjoint sets of atoms holding the same positions in the primitive cell. The example of such a set is shown in Fig. 6.19b. Counting the number of dangling bonds is performed separately for each set. The total number of dangling bonds is their sum over the sets. The surface area

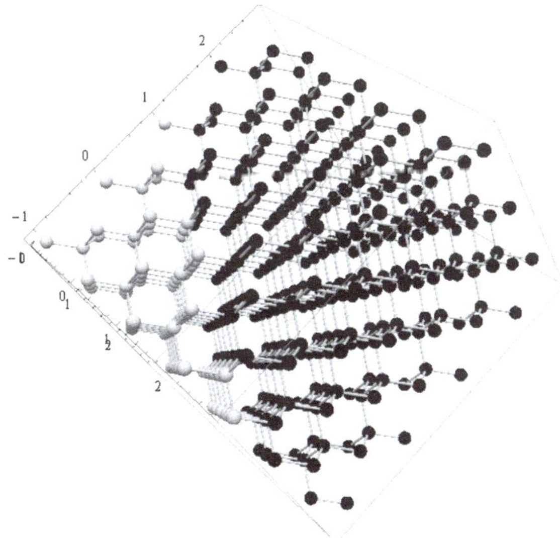

Fig. 6.18 Dissected cubic crystal, (211) section. Gray atoms are removed from crystal by the dissection

Table 6.4 Calculated surface densities of dangling bonds per a^2 for different facets

Crystal structure	Surface	Number of dangling bonds per a^2	
		First nearest neighbors	Second nearest neighbors
ZB	(1 0 0)	0.75	1
	(1 1 0)	0.53	3.18
	(1 1 1)	0.43	2.60
	(2 1 1)	0.61	3.06
	(2 2 1)	0.50	3.00
	(3 0 1)	0.71	3.32
	(3 1 1)	0.68	3.17
	(5 0 1)	0.74	3.24
WZ	$(11\bar{2}0)$	0.53	3.18
	$(1\bar{1}00)$	0.46	2.76

per one surface atom is obtained by constructing the Wigner-Seitz cell for a given set of identical atoms on the facet. When counting the number of dangling bonds, we also included the atoms belonging to all the neighboring planes with non-compensated bonds. The number of planes contributing to the surface energy depends on Miller's indices (hkl) and rapidly increases with increasing h, k and l. Therefore, counting the dangling bonds tangles for high-index surfaces, where many planes contribute into the resulting surface energy. In Table 6.4, we summarize the obtained surface densities of dangling bonds per a^2(a is the lattice constant of the cubic crystal) for different crystallographic planes of ZB and WZ structures.

To obtain the interaction energies φ_1 and φ_2 between the first and the second neighbors, we assume that φ depends only on the interatomic distance r and use

Table 6.5 Surface energies of ZB and WZ facets of Si, Ge and III-V materials

Material	Structure	Surface	Surface energy γ (J/m^2)	
			Previous results [576–581]	Our calculations [453]
Si	CUB	(1 0 0)	**1.41**	1.58
	CUB	(1 1 0)	**1.50**	1.48
	CUB	(1 1 1)	**1.36**	1.21
	CUB	(2 1 1)	1.31	1.50
	CUB	(301)	**1.62**	1.66
	CUB	(311)	**1.38**	1.58
	CUB	(501)	1.56	1.65
	HEX	$(11\bar{2}0)$		1.48
	HEX	$(1\bar{1}00)$		0.96
Ge	CUB	(1 0 0)	**1.00**	1.28
	CUB	(1 1 0)	**1.20**	1.22
	CUB	(1 1 1)	**1.01**	1.00
	CUB	(2 1 1)		1.22
	CUB	(301)	**1.14**	1.35
	CUB	(311)	0.90	1.29
	CUB	(501)	1.01	1.34
	HEX	$(11\bar{2}0)$		1.22
	HEX	$(1\bar{1}00)$		0.78
GaAs	ZB	(1 0 0)	2.2	2.11
	ZB	(1 1 0)	1.5	1.543
	ZB	(1 1 1)	1.3	1.27
	ZB	(2 1 1)		1.79
	WZ	$(11\bar{2}0)$		1.543
	WZ	$(1\bar{1}00)$		1.30
InAs	ZB	(1 0 0)	1.4	1.42
	ZB	(1 1 0)	1.0	1.06
	ZB	(1 1 1)	0.84	0.86
	ZB	(2 1 1)		1.19
	WZ	$(11\bar{2}0)$		1.06
	WZ	$(1\bar{1}00)$		0.91
AlAs	ZB	(1 0 0)	2.6	2.55
	ZB	(1 1 0)	1.8	1.86
	ZB	(1 1 1)	1.5	1.51
	ZB	(2 1 1)		2.11
	WZ	$(11\bar{2}0)$		1.86
	WZ	$(1\bar{1}00)$		1.6

(continued)

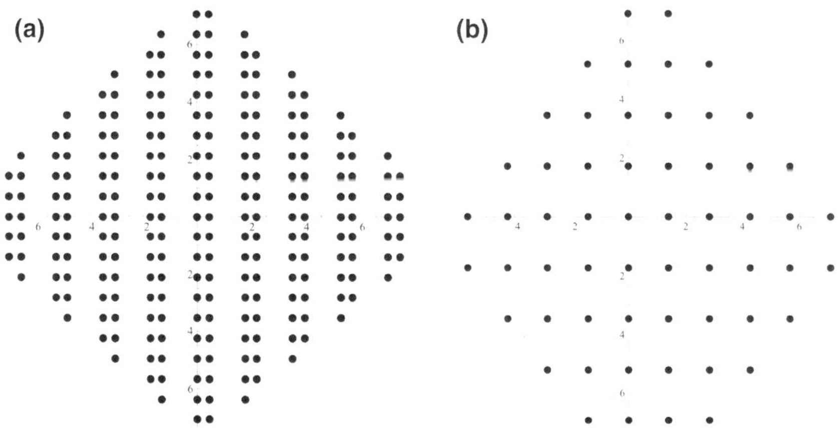

Fig. 6.19 **a** (211) section of a cubic crystal. **b** (211) plane with atoms having the same positions in the primitive cell

the linear approximation for $\varphi(r)$ based on the known values for the (100), (110) and (111) surface energies of the cubic crystals [576–581]. Since the first and the second nearest neighbor distances are identical for ZB and WZ, the obtained φ_1 and φ_2 are used for both structures. We start by calculating the surface energies for cubic Si and Ge, for which the existing experimental data and theoretical values for non-reconstructed surfaces are averaged and a reasonable correlation with present calculations is found. We then apply the same method for arsenides, phosphides and antimonites. The results are summarized in Table 6.5, along with the previously obtained values for the particular ZB surfaces (experimental data for Si and Ge are shown in bold). Overall, these results show a good agreement with all the available data on the ZB surface energies, suggesting that the WZ values could also serve as a reasonable first approximation.

Thus, the surface energy of WZ sidewalls of hexahedral nanowires appears to be lower than of their ZB counterparts. We note that this property holds not only for III-Vs but also for Si and Ge nanowires. Therefore, the formation of hexagonal Si nanowires reported, e.g., in [295] can be explained within the same approaches as for III-Vs. Calculations show that the surface energy of (211) ZB side facets is higher than that of (110) facets. However, GaAs and other III-V nanowires are often terminated by (211) facets [49,311], which is thermodynamic non-sense and should be due to kinetic factors. For the WZ structure, $(1\bar{1}00)$ facets have lower energy than $(11\bar{2}0)$, and most WZ nanowires have $(1\bar{1}00)$ sidewalls as shown in Fig. 6.10. The surface energy gain in the ZB-WZ transitions can be characterized by the surface energy ratio $\tau = \gamma_{WZ}/\gamma_{ZB}$ for a given pair of sidewall facets. Theoretical values of τ for the four possible transitions according to different authors are given in Table 6.6 and qualitatively correlate with each other (although the absolute energy values are very different). According to our estimates, the maximum surface energy gain with $\tau \simeq 0.73$ relates to the direct $(211) \rightarrow (1\bar{1}00)$ transition without rotation [407].

Table 6.5 (continued)

Material	Structure	Surface	Surface energy γ (J/m^2)	
			Previous results [576–581]	Our calculations [453]
GaP	ZB	(1 0 0)	2.9	2.85
	ZB	(1 1 0)	2.0	2.06
	ZB	(1 1 1)	1.7	1.69
	ZB	(2 1 1)		2.36
	WZ	$(11\bar{2}0)$		2.06
	WZ	$(1\bar{1}00)$		1.79
InP	ZB	(1 0 0)	1.9	1.87
	ZB	(1 1 0)	1.3	1.38
	ZB	(1 1 1)	1.1	1.12
	ZB	(2 1 1)		1.56
	WZ	$(11\bar{2}0)$		1.38
	WZ	$(1\bar{1}00)$		1.19
AlP	ZB	(1 0 0)	3.4	3.37
	ZB	(1 1 0)	2.4	2.44
	ZB	(1 1 1)	2.0	1.99
	ZB	(2 1 1)		2.79
	WZ	$(11\bar{2}0)$		2.44
	WZ	$(1\bar{1}00)$		2.11
AlSb	ZB	(1 0 0)	1.90	1.87
	ZB	(1 1 0)	1.30	1.38
	ZB	(1 1 1)	1.10	1.38
	ZB	(2 1 1)		1.56
	WZ	$(11\bar{2}0)$		1.38
	WZ	$(1\bar{1}00)$		1.19
GaSb	ZB	(1 0 0)	1.60	1.56
	ZB	(1 1 0)	1.10	1.12
	ZB	(1 1 1)	0.91	0.91
	ZB	(2 1 1)		1.29
	WZ	$(11\bar{2}0)$		1.12
	WZ	$(1\bar{1}00)$		0.97
InSb	ZB	(1 0 0)	1.10	1.12
	ZB	(1 1 0)	0.75	0.85
	ZB	(1 1 1)	0.60	0.69
	ZB	(2 1 1)		0.95
	WZ	$(11\bar{2}0)$		0.85
	WZ	$(1\bar{1}00)$		0.73

The indirect $(211) \rightarrow (11\bar{2}0)$ and $(110) \rightarrow (1\bar{1}00)$ transitions, associated with the rotation by an angle of $30\,^\circ$, give a lower gain at $\tau \cong 0.85$. The direct $(110) \rightarrow (11\bar{2}0)$ transition does not gain any energy.

Table 6.6 Surface energy ratios τ related to the ZB to WZ transitions in GaAs nanowires

Transition	τ from [574]	τ, our calculations
Direct $(211) \to (1\bar{1}00)$	0.648	0.726
$30°(211) \to (11\bar{2}0)$	0.682	0.862
$30°(110) \to (1\bar{1}00)$	0.840	0.843
Direct $(110) \to (11\bar{2}0)$	0.884	1

6.4 Equilibrium Radius of Zincblende-Wurtzite Transition

Many authors [407, 558, 571, 574, 575] considered the equilibrium radius of ZB-WZ transition below which the WZ of III-V nanowires is energetically preferred. These considerations rely simply upon the comparison of free energies of fully formed nanowires having the same size and shape. For a ZB hexahedral nanowire of length L and radius R, the free energy of its formation from vapor is given by

$$\Delta G_{WIRE}^{ZB} = -\frac{3\sqrt{3}}{2}\frac{R^2 L}{\Omega_S}\Delta\mu_{VS} + 6RL\gamma_{ZB}. \tag{6.5}$$

Here, $\Delta\mu_{VS} = \mu_V - \mu_S$ is the chemical potential difference between the vapor and the ZB solid phase, and γ_{ZB} is the sidewall surface energy in the ZB phase (i.e., $\gamma_{ZB} = \gamma_{(100)}$ or $\gamma_{(211)}$). The vapor-solid difference of chemical potentials is smaller for WZ, because $\mu_S^{WZ} = \mu_S + \psi_{WZ}$. Therefore, the free energy of a WZ nanowire equals

$$\Delta G_{WIRE}^{WZ} = -\frac{3\sqrt{3}}{2}\frac{R^2 L}{\Omega_S}(\Delta\mu_{VS} - \psi_{WZ}) + 6RL\gamma_{WZ}, \tag{6.6}$$

where $\gamma_{WZ} = \gamma_{(11\bar{2}0)}$ or $\gamma_{WZ} = \gamma_{(1\bar{1}00)}$ is the sidewall energy in the WZ phase. Since the volume and surface terms scale differently with R, this naturally leads to the equilibrium critical radius R_{eq} at which $\Delta G_{WIRE}^{ZB}(R_{eq}) = \Delta G_{WIRE}^{WZ}(R_{eq})$:

$$R_{eq} = \frac{4}{\sqrt{3}}\frac{\Omega_S(\gamma_{ZB} - \gamma_{WZ})}{\psi_{WZ}}. \tag{6.7}$$

The WZ structure is favored below R_{eq}, while ZB should form above R_{eq}. For a cylindrical nanowire, the coefficient $4/\sqrt{3}$ in this formula is changed to 2. Of course, the critical radius exists only if $\gamma_{ZB} > \gamma_{WZ}$; the WZ structure can never form without this inequality. When $\gamma_{ZB} > \gamma_{WZ}$ but $\psi_{WZ} < 0$ as in nitrides, free energy is always lower for WZ and thus polytypism is not expected.

Similarly, the equilibrium critical radius can be introduced for polytypes [407], with ψ_{WZ} changed to ψ_{PT} and γ_{WZ} to γ_{PT} in (6.6) The bulk energy difference ψ_{PT} increases monotonously with the hexagonality α [554]. In the first approximation, we assume the linear scaling of ψ_{PT} with the fraction of WZ layers: $\psi_{PT} \cong \psi_{WZ}\alpha$. The surface energy ratio can also be defined for polotypes as $\tau = \gamma_{PT}/\gamma_{ZB}$, and estimated by counting the number of dangling bonds on the sidewalls of polytype nanowires. With these definitions, R_{eq} can be written as

$$R_{eq} = \frac{2(1-\tau)}{\alpha} R_0; \quad R_0 \equiv c \frac{\Omega_S \gamma_{ZB}}{\psi_{WZ}} \tag{6.8}$$

uniformly for pure WZ and all the polytype phases (with $c = 2/\sqrt{3}$ for hexahedral and 1 for cylindrical geometry). The characteristic radius R_0 is material-related (for a given type of ZB facets) and determines the characteristic scale of the ZB to WZ phase transition in different materials. Materials with larger R_0 should adopt WZ phase more often than the others.

Table 6.7 lists the characteristic radii R_0 for different III-Vs, obtained for the lowest energy ZB (110) facet. The values of $\gamma_{(110)}$ (the same as in the last column of Table 6.5) and ψ_{WZ} used in calculation are also displayed. The values of ψ_{WZ} for Al-V materials are taken from [571]. The fifth column gives the equilibrium critical radii for the (110) $\rightarrow$ ($1\bar{1}00$) transition which corresponds to the lowest energy facets in both phases. In this case, $\alpha = 1$ and $\tau = 0.84$. It is seen that the predicted critical radius is the smallest for GaAs and antimonites (lower than 8.1 nm) and the largest for InP. This tendency is supported by the results of Akiyama *et al.* [571], plotted in Fig. 6.20 in the normalized variable $2R_{eq}/a$ versus the ionicity of semiconductors. It is noteworthy that our estimate for the equilibrium ZB-WZ transition in GaAs nanowires at 5.6 nm is close to the values obtained by different methods: $R_{eq} = 5.6$ nm based on an empiric potential [571], and 3.2–6.3 nm by DFT calculations [574, 575]. For the direct (211) $\rightarrow$ ($1\bar{1}00$) transition with $\tau \cong 0.73$, the equilibrium critical radii listed in the last column are 1.72 times larger.

As for polytypes, Fig. 6.21 illustrates the calculation of the dangling bond densities for (211) ZB, WZ and 4H polytype facets of the same type [407]. This simplified method yields $\tau = 0.75$ for the (211) $\rightarrow$ ($1\bar{1}00$) transition. Same procedure can be applied for all the polytypes listed in Table 6.2. Figure 6.22 presents graphically the results. From (6.8), the maximum equilibrium radius for the ZB to polytype phase

Table 6.7 Characteristic radii R_0 and R_{eq} in different semiconductor nanowires

Material	$\gamma_{(110)}$ (J/m^2)	ψ_{WZ} (meV/pair)	$R_0 = \frac{\Omega_s \gamma_{(100)}}{\psi_{WZ}}$ (nm)	R_{eq} for (110) $\rightarrow$ ($1\bar{1}00$) (nm)	R_{eq} for (211) $\rightarrow$ ($1\bar{1}00$) (nm)
AlSb	1.38	19.0	25	8.1	13.9
GaSb	1.12	19.8	20	6.4	11.0
InSb	0.85	16.4	22	7.1	12.2
AlAs	1.86	16.4	32	10.2	17.5
GaAs	1.54	24.0	17	5.6	9.6
InAs	1.06	10.6	34	10.9	18.7
AlP	2.44	11.4	52.5	16.8	28.9
GaP	2.06	12.4	41	13.4	23.0
InP	1.38	6.8	64	20.4	35.0

Fig. 6.20 Normalized critical diameter d_c/a (a is the ZB lattice constant) between ZB and WZ structures as a function of ionicity f_i

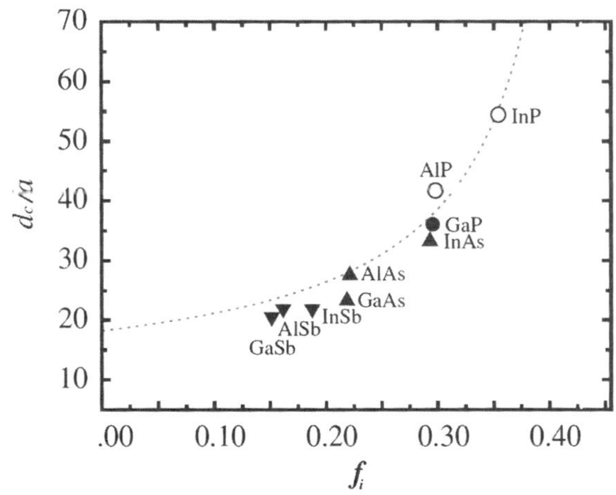

transition is reached at the maximum value of $2(1 - \tau)/\alpha$. For all the polytypes considered, this expression is not larger than 1 and equals 1 for 4 structural types: 4H, 10H (10L), 14L and 16H (16L) (coefficients α and τ are identical for pairs 10H, 10L and 16H, 16L). The corresponding (α, τ) coordinates are placed on the straight line shown in Fig. 6.22. For these polytypes, $R_{eq} \cong R_0$, while for all other polytypes $R_{eq} < R_0$. To find the most energetically favorable crystallographic phase at $R < R_{eq}$, we compare the difference of the formation energies given by (6.5) and (6.6) for all polytypes having the maximum R_{eq}. The minimum formation energy is reached at the maximum α, i.e., in the case of 4H polytype. The differences of formation energies (relative to ZB) for all polytypes with the maximum R_{eq} versus the wire radius are displayed in Fig. 6.23, where the corresponding curve for the WZ structure is also shown for reference. It is seen that the curves for all polytypes cross at $R = R_0$. In the preferred polytype range below the critical radius, 4H polytype has the lowest energy, which is also lower than that of pure WZ.

To conclude this section, the equilibrium crossing radius between ZB and WZ can be easily determined at the known surface energies and the bulk energy difference for a given material. Similar considerations can be applied for polytypes, and predict 4H structure as being most energetically favored. The dependence of formation energies on the growth conditions through $\Delta\mu_{VS}$ cancels in the energetic balance, while a weak temperature dependence can be only due to that of the surface energies and ψ_{WZ}. Different estimates for GaAs yield the structural transition below 10 nm. These conclusions contradict with what is observed experimentally: polytypism of GaAs nanowires occurs at least up to 75 nm radius and is known to be extremely sensitive to temperature and other growth parameters [184]. The crystal structure also changes during growth depending on the Ga concentration in the droplet as demonstrated in Sect. 6.2. Therefore, the adequate description of ZB-WZ polytypism in III-V nanowires requires a different perspective, which should be based on the kinetic analysis rather than thermodynamic arguments. On the other hand, kinetic

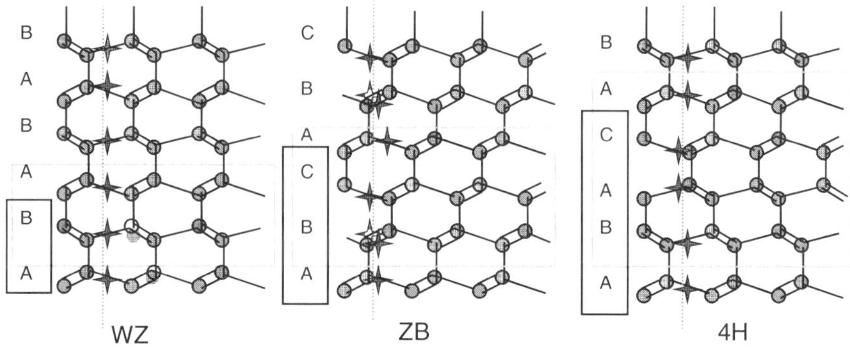

Fig. 6.21 Side view of WZ, ZB and 4H polytype structures. Star symbols indicate the dangling bonds on the lateral (211) or $(1\bar{1}00)$ facets. The number of dangling bonds in the WZ phase equals 1 per one layer in the WZ and 4H phases and 4/3 in the ZB phase

Fig. 6.22 (α, τ)—coordinates of different polytypes between WZ and ZB. Points on the straight line $\tau = 1-\alpha/2$ relate to the maximum critical radius for the structural transition

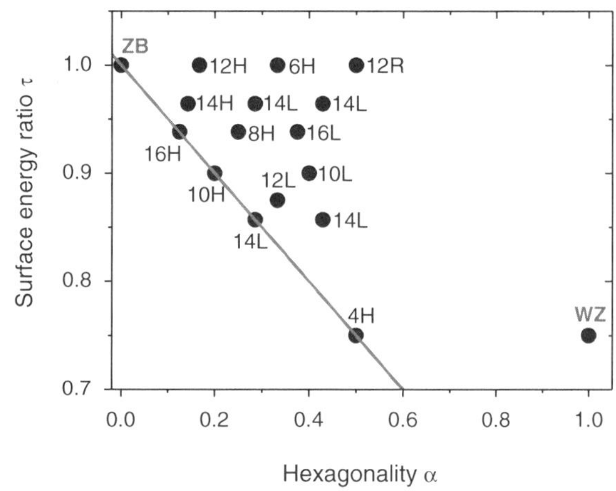

studies rely on theoretical values of the material constants, which is why further studies are required to decrease the existing uncertainties in bulk and surface energies.

6.5 Kinetic Theory of Wurtzite Phase Formation

A pioneering step in understanding ZB-WZ polytypism in III-V nanowires was taken by Glas *et al.* in [45], where a kinetic approach was established based on the Zeldovich nucleation theory. As discussed in Sect. 6.2, experimental results show that ZB phase in GaAs nanowires is systematically observed during transient growth steps at the beginning and at the end of deposition during which the Ga concentration in the liquid is less than in the steady nanowire growth (where the crystal

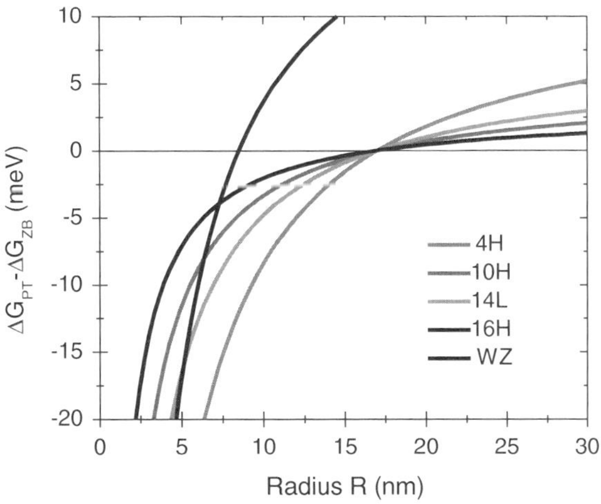

Fig. 6.23 Free energies of 4H, 10H, 14L, 16H polytypes and WZ relative to the ZB structure at $\psi_{WZ} = 24\,\text{meV}$, $\Delta\mu_{VS} = 290\,\text{meV}$ and $R_0 = 17\,\text{nm}$. The value of $\Delta\mu_{VS}$ approximately corresponds to MBE growth of GaAs on GaAs(111)B substrate at $T = 580\,°\text{C}$ with a nominal growth rate of 0.2 nm/s

structure is predominantly WZ). Higher Ga concentration relates to a larger liquid supersaturation with respect to the solid state. We can therefore conclude that, while the WZ structure is prevalent for the Au-catalyzed GaAs grown by MBE at typical growth conditions, the ZB structure is systematically observed at low liquid supersaturations. Based on this important observation, it has been proposed that rather than free energy of fully formed nanowire, a more realistic critical parameter is the nucleation barrier for the formation of WZ or ZB critical nuclei. The nucleation barrier depends on the surface energetic but also includes supersaturation. This view has rapidly gained much support. The kinetic theory of polytypism has been developed by several authors [48, 49, 82, 83, 234, 300, 316, 323, 351, 387, 410, 582] and is now widely used for tuning the crystal structure of III-V nanowires [49, 83, 234, 297, 300, 311, 316, 323, 351, 410, 582–585]. Furthermore, this kinetic approach remains the only relevant theoretical explanation of the kinetically controlled polytypism to our knowledge.

In Chaps. 4 and 5, several arguments have been presented supporting the mononuclear character of the VLS nanowire growth. In mononuclear regime, the crystal structure of each monolayer will be determined by the orientation positioning of a single 2D nucleus with respect to the previous layer. This positioning determines if the stack of 2 monolayers formed by adding the nucleus is of the type found in ZB or WZ crystals (Fig. 6.24). Using (4.25) and assuming for the moment equal elementary volumes in the liquid and solid phases ($\Omega_S = \Omega_L$), we write the formation energy of a 2D island of linear size r, perimeter $c_1 r$ and surface area $c_2 r^2$ uniformly for both phases:

$$\Delta G = -\left[\Delta\mu - \psi'\right]\frac{c_2 r^2 h}{\Omega_S} + \gamma_{eff} c_1 r h. \tag{6.9}$$

Here, $\Delta\mu = \Delta\mu_{LS}^{\infty} \equiv k_B T\,\ln(\zeta_\infty + 1)$ is the liquid chemical potential with respect to the bulk ZB solid, ζ_∞ is the liquid supersaturation (without the Gibbs-Thomson

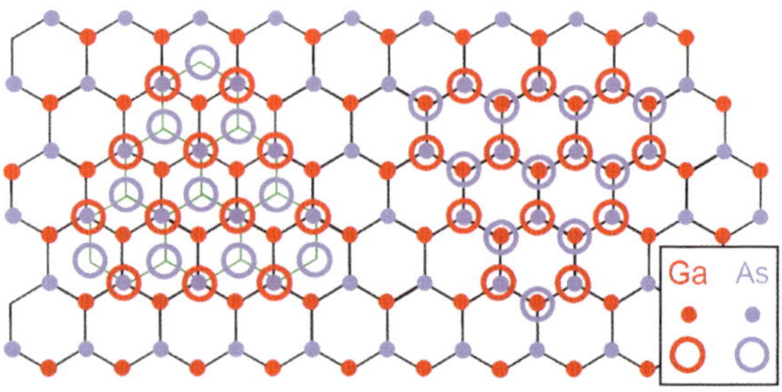

Fig. 6.24 A given nucleus (open symbols) in ZB (*left*) and WZ (*right*) positions on top of a (111)B monolayer (disks) [45]

modification), and h is the height of a monolayer. The quantity ψ' equals zero for ZB and ψ'_{WZ} for WZ. The latter value gives the excess energy of WZ phase for monolayers, which might be lower than the bulk energy difference ψ_{WZ} and is sometimes taken as the stacking fault energy $\psi'_{WZ} = \psi_{SF} \cong \psi/2$ for the very first 2D WZ nucleus on a ZB substrate [45]. In what follows, we will use the value $\psi' \cong \psi$, because it corresponds to the correct limit case for a fully formed nanowire at $r = R$ and $h = L$.

The effective surface energy γ_{eff} depends on the nucleus position and is given by (4.26). This can be put in the form which is more convenient for the foregoing analysis:

$$\gamma_{eff} = (\tilde{\gamma}_j - \gamma_{LV} \sin\beta)x + \gamma^l_{SL}(1 - x); \tag{6.10}$$

$$\tilde{\gamma}_j = \frac{\gamma_j}{\cos\theta_j} - (\gamma_{SL} + \gamma_{LV}\cos\beta)\tan\theta_j. \tag{6.11}$$

Here, $\gamma_j \equiv \gamma_{WV}$ is the solid–vapor surface energy of the nanowire sidewall facet j that depends on the crystal phase, θ_j is the facet angle with the vertical in the case of tilted ZB facets, γ_{LV} is the phase-independent liquid–vapor surface energy of the droplet, β is the droplet contact angle, and γ^l_{SL}, γ_{SL} are the solid–liquid surface energies of the lateral and planar nucleus-liquid boundaries, respectively. The x gives the fraction of the island perimeter at the triple phase line as discussed in Sect. 4.5 (e.g., $x = 1/3$ for regular triangle and 1/6 for regular hexagon island). When comparing the nucleation barriers for 2D islands with WZ and ZB orientations, j relates to $(1\bar{1}00)$ or $(11\bar{2}0)$ vertical WZ sidewalls, (110) or (211) vertical ZB sidewalls, or (111)A, (111)B for tilted ZB sidewall facets as originally considered by Glas *et al.* [45] (Fig. 6.25).

If nucleation occurs away from the triple phase line, $x = 0$ in (6.10) and the island surface energy simply equals γ^l_{SL}. This value can be considered phase independent, i.e.

$$\gamma^l_{SL}(ZB) = \gamma^l_{SL}(WZ), \tag{6.12}$$

due to close atomic environments of ZB and WZ islands surrounded by liquid, actually independent on the metal catalyst. Phase independence of γ^l_{SL} is supported by the absence of polytypism in III-V materials grown by the conventional liquid phase epitaxy, with all islands being entirely surrounded by the liquid phase. When the

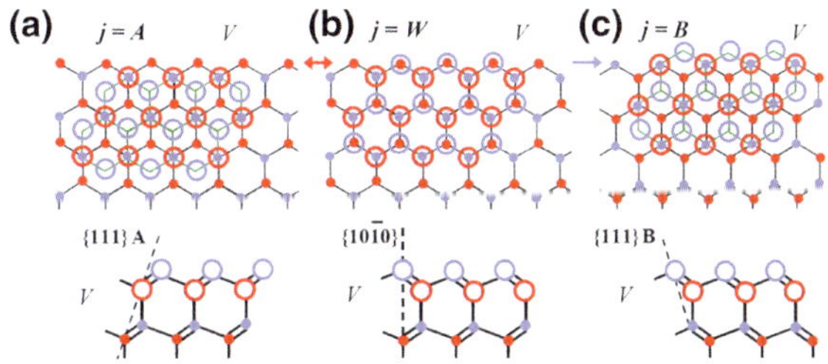

Fig. 6.25 Top (*top row*) and side (*bottom row*) views of a given nucleus in ZB (**a**) and WZ (**b**) positions at a Ga edge and in ZB position at an As edge (**c**). Edges are indicated by arrows (*top*), the vapor side by *V* and non-vertical dangling bonds by segments (*bottom*). Key is the same as in Fig. 6.25 [45]

solid–liquid surface energies are the same for ZB and WZ nuclei, (6.9) at $\psi' > 0$ for WZ nuclei yields a higher nucleation barrier for WZ, i.e., ZB nucleation is always favored. Therefore, polytypism of III-V nanowires necessarily requires the triple phase line nucleation where a fraction of the island perimeter is in contact with vapor, leading to a surface energy gain for WZ structure.

Whatever is the facet type, the analysis of Sect. 4.5 gives the following general condition for the preferred nucleation at the triple phase line

$$\tilde{\gamma}_j < \gamma_{LV} \sin\beta + \gamma_{SL}^l \tag{6.13}$$

for any lateral facet j. The same inequality should hold for any nanowire which grows in the stable VLS configuration with a planar growth interface and the droplet seated on the top (see Sects. 4.4 and 4.5). Polytypism of III-V nanowires can be thus observed only when (6.13) is satisfied for at least one of the two straight WZ facets. According to the data of Tables 6.4 and 6.6, inequality (6.13) should be more easily satisfied for $(1\bar{1}00)$ sidewalls with a lower surface energy. Surface energy of the two vertical ZB facets is higher than that of $(1\bar{1}00)$ WZ, and is also expected to be higher for the alternating (111)A and (111)B sidewalls.

Let us now consider the particular case of the Au-assisted GaAs nanowires grown at a temperature of 550 °C. Taking a constant value of As concentration in an Au-Ga-As droplet at 0.02 for further estimates, the equilibrium Ga concentration $c_{eq} = 0.25$ from Fig. 5.66. Neglecting the influence of arsenic on the droplet surface energy, we use a linear interpolation

$$\gamma_{LV} = c_{Ga}\gamma_{Ga} + (1 - c_{Ga})\gamma_{Au} \tag{6.14}$$

for the liquid surface energy between the values of $\gamma_{Ga} = 0.675\ \text{J/m}^2$ and $\gamma_{Au} = 1.248\ \text{J/m}^2$ at 550 °C. The graphs of

$$\Delta\gamma = \gamma_j - \gamma_{LV}\sin\beta - \gamma_{SL}^l \tag{6.15}$$

for the four vertical facets j with the solid–vapor surface energies from Table 6.4 are shown in Fig. 6.26, within a plausible range of the unknown γ_{SL}^l between 0.5 and

0.6 J/m^2, as estimated in [82] from Young's equation. The value of β is chosen at 110°, because the Au-Ga droplets measured after growth usually have contact angles between 90° and 130° [45].

With these model parameters, the triple phase line nucleation is preferred in the entire growth region ($0.25 < c_{Ga} < 0.7 - 0.89$ depending on the value of γ_{SL}^{l}) for (1$\bar{1}$00) WZ facets, and in the region $0.25 < c_{Ga} < 0.43$ for (11$\bar{2}$0) WZ and (110) ZB facets at $\gamma_{SL}^{l} = 0.6\text{J/m}^2$. For (211) ZB facets having the highest surface energy, the triple phase line nucleation is energetically suppressed everywhere in the growth region. Similar considerations at typical VLS growth temperatures with the surface energies listed in Table 6.4 show that the triple phase line nucleation should be even more favorable for the Au-catalyzed InAs, InSb and InP nanowires with less energetic sidewall facets than those of GaAs. On the other hand, it is questionable for GaP nanowires whose facets have a higher surface energy in comparison with GaAs.

Whenever WZ islands nucleate at the triple phase line, their effective surface energy can be lower than for ZB islands (which can be formed either in TPL or C positions depending on the sign of their $\Delta\gamma$). In this case, the nucleation barrier for WZ may become lower than for ZB due to the surface energy gain in the WZ structure. Writing (6.9) at $\psi' = \psi$ separately for ZB and WZ nuclei and maximizing it in r, we obtain the nucleation barriers in the form

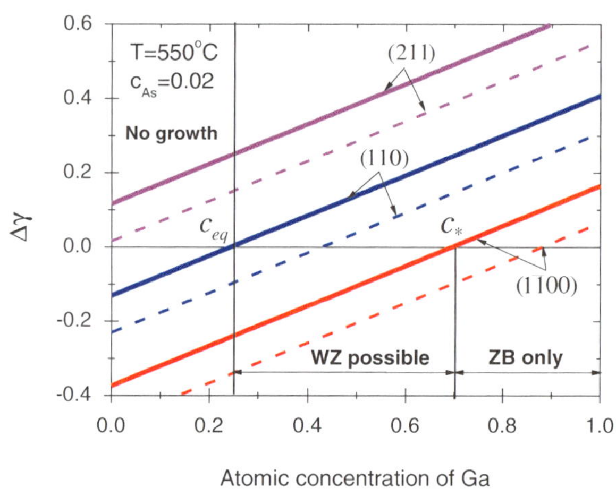

Fig. 6.26 $\Delta\gamma$ as a function of Ga atomic concentration c_{Ga} for the four vertical lateral facets at $\gamma_{(1\bar{1}00)} = 1.30$ J/m^2, $\gamma_{(100)} = \gamma_{(11\bar{2}0)} = 1.543$ J/m^2 and $\gamma_{(211)} = 1.79$ J/m^2, with γ_{SL}^{l} between 0.5 J/m^2 (*solid lines*) and 0.6 J/m^2 (*dashed lines*). Lines for (110) ZB and (11$\bar{2}$0) WZ facets are the same. The equilibrium Ga concentration c_{eq} equals 0.25 at $T = 550$ °C and $c_{As} = 0.02$, which is why the VLS growth is possible only at $c_{Ga} > 0.25$. The triple phase line nucleation is possible at negative $\Delta\gamma$, yielding the possible WZ region in c_{Ga} between $c_{eq} = 0.25$ and $c_* = 0.7$ at $\gamma_{SL}^{l} = 0.5$ J/m^2 or 0.89 at $\gamma_{SL}^{l} = 0.6$ J/m^2

$$\Delta G^*_{ZB} = \frac{c_1^2}{4c_2}\Omega_S h \frac{\gamma^2_{eff,\,ZB}}{\Delta\mu}; \tag{6.16}$$

$$\Delta G^*_{WZ} = \frac{c_1^2}{4c_2}\Omega_S h \frac{\gamma^2_{eff,\,WZ}}{\Delta\mu - \psi_{WZ}}. \tag{6.17}$$

With neglect of pre-exponential factors in the Zeldovich nucleation rate on the nucleation probabilities, the WZ structure dominates when $\Delta G^*_{WZ} < \Delta G^*_{ZB}$, while ZB is favored at $\Delta G^*_{WZ} > \Delta G^*_{ZB}$. The critical liquid chemical potential $\Delta\mu_c$ corresponding to $\Delta G^*_{WZ} = \Delta G^*_{ZB}$ is readily obtained from (6.16) and (6.17). The WZ structure is thus kinetically favored when the liquid chemical potential is larger than critical:

$$\Delta\mu > \Delta\mu_c = \frac{\psi_{WZ}}{1 - (\gamma_{eff,\,WZ}/\gamma_{eff,\,ZB})^2}. \tag{6.18}$$

This inequality, obtained in [45], sets the fundamental kinetic criterion for the WZ phase formation in nanowires. Equation (6.18) actually requires two conditions. The first condition, $\gamma_{eff,\,WZ} < \gamma_{eff,\,ZB}$, is material-related, and can be satisfied only when nucleation occurs at the triple phase line (at least for WZ islands), i.e., when the inequality (6.13) is met for relevant lateral facets. It can therefore be called the necessary condition for polytypism, because the unstable crystal phase can only be stabilized by its lower surface energy. The second, sufficient condition for the formation of WZ structure requires liquid chemical potential to be high enough to overcome the bulk energy difference ψ_{WZ}.

Estimates of [45] give the critical chemical potential $\Delta\mu_c$ of the order of several hundreds meV, which is in the range of typical $\Delta\mu$ for the Au-catalyzed GaAs nanowires at MBE growth temperatures (see Sect. 5.15). For our numerical example, assuming $c_{Ga} = 0.4$ during steady state growth and $\gamma^l_{SL} = 0.60$ J/m^2, (6.10) yields $\gamma_{eff,\,ZB} = 0.595$ J/m^2 and $\gamma_{eff,\,WZ} = 0.514$ J/m^2 for straight (110) and ($1\bar{1}00$) sidewall facets, respectively. From (6.18) with $\psi_{WZ} = 24$ meV/pair, $\Delta\mu_c$ approximately equals 100 meV/pair. This value is much lower than the chemical potential of GaAs in the Au-Ga-As droplet at $T = 550°$C and $c_{As} = 0.02$, which is higher than 200 meV/pair from Fig. 5.66. Therefore, the steady state growth proceeds at $\Delta\mu > \Delta\mu_c$ and the structure of these GaAs nanowires is expected to be predominantly WZ.

Despite its simplicity, criterion given by (6.18) explains very well why the Au-assisted GaAs nanowires adopt the ZB phase at the beginning and at the end of MBE deposition, while their structure is predominantly WZ during the steady growth as described above. The supersatutration-dependent polytypism gives a powerful tool for controlling the crystal structure of III-V nanowires by the growth conditions, because liquid supersaturation could be altered by the growth temperature, fluxes and the substrate properties (see Sect. 6.7). On the other hand, the radius dependence is lost in (6.19), while it is known that the structure of sufficiently wide nanowires should retain to the bulk-stable ZB. Therefore, some modifications of the theory are required, which are described in the next section.

6.6 Further Development of Kinetic Approach

The kinetic criterion given by (6.18) determines the critical chemical potential during growth above which the WZ structure is expected to be prevalent; however, it does not provide explicitly the percentages of WZ and ZB phases in nanowires. Also, it does not show any radius dependence. The simplest modification which enables both generalizations is the following. We consider the particular case where WZ islands emerge at the triple phase line while ZB islands form at the liquid–solid interface. This is relevant when the condition for the TPL nucleation is satisfied for WZ but broken for ZB sidewalls due to a higher surface energy of the latter. Looking at Fig. 6.26, such a situation is most likely to occur for the model parameters of GaAs considered. Since the nucleation at the liquid–solid interface can occur anywhere on the top facet excluding the ring of the critical radius and the triple phase line nucleus can emerge only inside such a ring, the entropic factor discussed in Sect. 4.5 yields the ratio of WZ and ZB nucleation-mediated growth rates in the form

$$\frac{P_{WZ}}{P_{ZB}} = \frac{2r_c^{WZ}}{R} \exp\left(\frac{\Delta G^*_{ZB} - \Delta G^*_{WZ}}{k_B T}\right). \tag{6.19}$$

Here, R is the nanowire radius and r_c^{WZ} is the critical linear dimension obtained from (4.44):

$$r_c^{WZ} = \frac{c_1}{2c_2} \frac{\Omega_s \gamma_{eff,\,WZ}}{\Delta\mu - \psi_{WZ}}. \tag{6.20}$$

Equation (6.19) has the correct asymptotic behavior $P_{WZ}/P_{ZB} \to 0$ at $R \to \infty$. The percentage of WZ structure can be deduced from (6.19) as

$$p_{WZ} = \frac{P_{WZ}}{P_{ZB} + P_{WZ}} = \frac{P_{WZ}/P_{ZB}}{1 + P_{WZ}/P_{ZB}}. \tag{6.21}$$

While considering the structural phase transitions, it is convenient to introduce the dimensionless liquid chemical potential measured in the units of ψ_{WZ}, i.e. $f \equiv \Delta\mu/\psi_{WZ}$. Using (6.16) to (6.20), the critical dimensionless chemical potential $f_c = \Delta\mu_c/\psi_{WZ}$ and the P_{WZ}/P_{ZB} ratio can be put as

$$f_c = \frac{1}{1 - (\gamma_{eff,\,WZ}/\gamma^l_{SL})^2}; \tag{6.22}$$

$$\frac{P_{WZ}}{P_{ZB}} = \frac{2R_s}{R} \frac{\sqrt{1 - 1/f_c}}{f - 1} \exp\left[U_* \left(\frac{1/f_c - 1/f}{f - 1}\right)\right]. \tag{6.23}$$

The parameters entering these expressions are defined as follows

$$R_s = \frac{c_1}{2c_2} \frac{\Omega_S \gamma^l_{SL}}{\Psi_{WZ}}; \quad U_* = \frac{c_1^2}{4c_2} \frac{\Omega_S h}{k_B T} \frac{(\gamma^l_{SL})^2}{\psi_{WZ}} = \frac{k_B T}{\psi_{WZ}} A_*, \tag{6.24}$$

Table 6.8 Model parameters of Au-catalyzed GaAs nanowires used in calculations

T (°C)	c_{As}	ψ_{WZ} (meV)	$\gamma_{(1\bar{1}00)}$ (J/m^2)	γ_{SL}^l (J/m^2)	Island shape	A_*	U_*	R_s(nm)
550	0.02	24	1.3	0.5	Triangle	147	435	20.3

where A_* is the energetic constant used in Chaps. 4 and 5 at $\gamma = \gamma_{SL}^l$. In (6.22) and (6.24), $\gamma_{eff,ZB}$ is changed to γ_{SL}^l because ZB islands are entirely surrounded by the liquid phase. Equation (6.23) together with (6.21) presents the WZ percentage as a function of dimensionless chemical potential f and radius R, with a parametric dependence on f_c. Since $k_BT/\psi_{WZ} > 1$ for all III-V materials at their typical growth temperatures and A_* is the large parameter of the order of several tens at least, the U_* value is very large and can reach several hundreds.

Figure 6.27 shows the critical chemical potential f_c versus Ga atomic concentration, obtained from (6.22), (6.10) and (6.14) for straight $(1\bar{1}00)$ WZ and (110) ZB nanowire sidewalls with the same model parameters as above, at $\gamma_{SL}^l = 0.5$ J/m^2. It is seen that f_c increases drastically as c_{Ga} approaches its boundary value $c_* = 0.7$. Above this concentration, WZ phase formation becomes completely suppressed on surface energetic grounds. Figure 6.28 shows the graphs of the WZ percentage obtained from (6.23), (6.24) and (6.21) with the parameters of the Au-catalyzed GaAs nanowires listed in Table 6.8, for 20 nm radius and different f_c. The value of $A_* = 147$ in this case appears noticeably larger than in Sect. 5.7, where A_* was equal 45. This is due to a combined effect of triangle island geometry instead of circular, a much larger surface energy of 0.5 J/m^2 instead of 0.35 J/m^2 and a slightly lower temperature, yielding together a multiplying factor of 3.3. Very importantly, the obtained dependences $p_{WZ}(f)$ demonstrate a very strong parametric dependence on f_c due to a huge value of $U_* = 435$ in the exponent of (6.23). The curve at low $f_c = 4$ rapidly increases with f, shows the 100 % WZ region for intermediate chemical potentials f between 4 and 15, and then decreases toward larger f. As f_c increases, the regions with the maxima of the WZ percentage gradually become less pronounced and are shifted toward larger f. Finally, at $f_c = 15$ the WZ percentage remains smaller than 10 % everywhere. For each f_c, the WZ percentage tends to zero for large enough f. This new effect, which does not follow from (6.18), is seen already from (6.23), where the exponent tends to one and the pre-exponential factor (defined by the radius of the critical nucleus) tends to zero at $f \to \infty$.

Figure 6.29 shows the graphs $p_{WZ}(f)$ for different nanowire radii at fixed $f_c = 6$. As expected, the radius dependence is given by the *1/R* factor in (6.19), while other terms are radius-independent when $\Omega_S = \Omega_L$. Quite naturally, WZ is most favorable for 10 nm radius nanowires and almost disappears in 300 nm radius nanowires. This radius dependence is however weaker compared to what is observed in most experiments [234, 315]. A steeper dependence can be obtained by introducing the Gibbs-Thomson correction under the exponent as will be discussed shortly. However, simple (6.22) and (6.23) clearly show major structural trends:

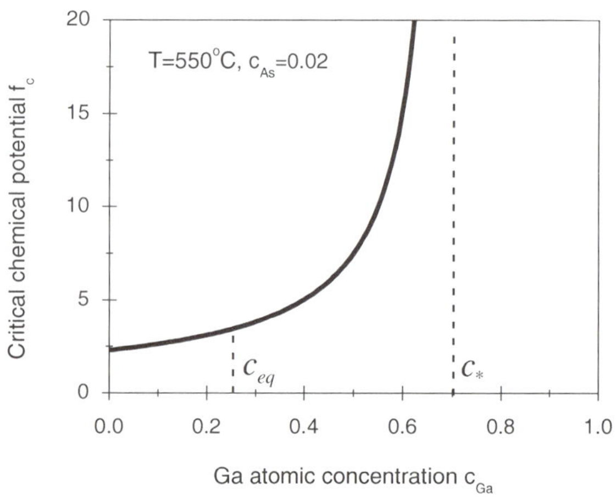

Fig. 6.27 Dimensionless critical chemical potential of GaAs in the Au-Ga-As droplet versus Ga atomic concentration at 550 °C and fixed $c_{As} = 0.02$; f_c tends to infinity as c_{Ga} approaches the boundary value of 0.7 corresponding to $\Delta\gamma = 0$

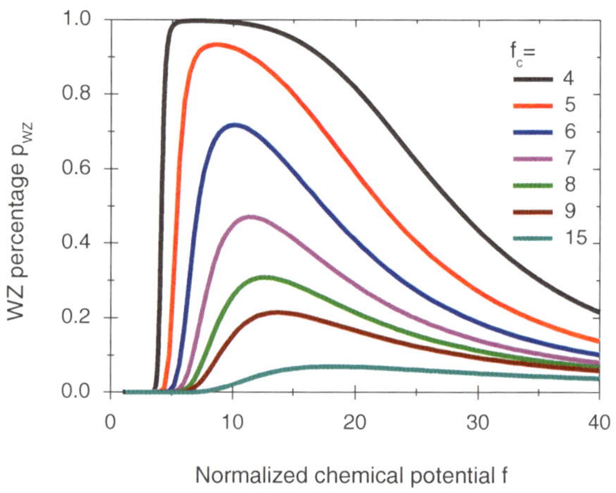

Fig. 6.28 WZ percentage versus dimensionless chemical potential in a 20 nm radius GaAs nanowire at different f_c. The model parameters are summarized in Table 6.8.

(1) WZ phase formation is suppressed kinetically at higher group III concentrations in the droplet due to increasing f_c. WZ phase disappears completely above a certain threshold concentration of group III element due to a low surface energy of the droplet which no longer favors the triple phase line nucleation.
(2) As a particular case, Ga-catalyzed GaAs nanowires and other group III-catalyzed III-V nanowires are expected to be almost purely ZB.
(3) Supersaturation dependence of the WZ percentage is generally a function with a maximum: p_{WZ} is close to zero at low supersaturations, then rapidly increases to its maximum value and tends to zero at very high supersaturations.
(4) In MBE case, liquid supersaturation cannot be very high (see below). ZB is thus favored at low supersaturations and WZ dominates at the highest possible supersaturations accessible in MBE, where nanowires can actually grow.

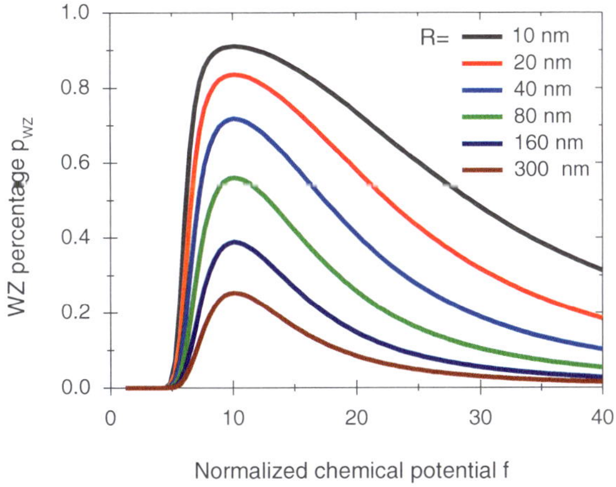

Fig. 6.29 WZ percentage versus dimensionless chemical potential for differently sized Au-catalyzed GaAs nanowires of radius R, at fixed $f_c = 6$

(5) As a consequence of non-monotonous dependence of p_{WZ} on f, III-V nanowires grown with exceptionally high elongation rates (relating to a very high liquid supersaturation) are expected to be single crystalline ZB.

A more detailed kinetic model of polytypism [82] is based on the nucleation barriers given by (4.45):

$$\Delta G^*_{ZB} = \frac{c_1^2}{4c_2} h\Omega_S \frac{\gamma^2_{eff,\,ZB}}{(\Delta\mu - 2\gamma_{LV}(\Omega_S - \Omega_L)\sin\beta/R)}; \tag{6.25}$$

$$\Delta G^*_{ZB} = \frac{c_1^2}{4c_2} h\Omega_S \frac{\gamma^2_{eff,\,WZ}}{(\Delta\mu - \psi_{WZ} - 2\gamma_{LV}(\Omega_S - \Omega_L)\sin\beta/R)}, \tag{6.26}$$

and includes four possible nucleation scenarios for WZ and ZB islands emerging either at the liquid–solid interface or at the triple phase line. Unlike (6.16) and (6.17), these formulas contain the Gibbs-Thomson correction which is proportional to $\Omega_S - \Omega_L$ and cancels only when the elementary volumes in the solid and liquid phases equal each other. Since Ω_S are almost exactly identical for ZB and WZ, the Gibbs-Thomson correction is the same for both phases. Comparing (6.25) and (6.26), the critical dimensionless chemical potential in the liquid phase is now given by

$$f_c = \frac{1}{1-\eta^2} + \frac{(\omega - 1)\delta}{\rho}. \tag{6.27}$$

Here, we introduce dimensionless values $\eta \equiv \gamma_{eff,\,WZ}/\gamma_{eff,\,ZB}$, $\delta \equiv 2\gamma_{LV}\Omega_L/\gamma_{ZB}\Omega_S$, $\omega = \Omega_S/\Omega_L$ and measure the nanowire radii ρ in the units of the characteristic radius $R_0 \equiv \Omega_s\gamma_{ZB}/\psi_{WZ}$ (where γ_{ZB} denotes the surface energy for a given type of ZB lateral facets). The parameter η is related to $\tau = \gamma_{WZ}/\gamma_{ZB}$ (with γ_{WZ} as the surface energy for a given type of WZ facets) through (6.10).

The following considerations will apply to the case of MBE. We saw in Chap. 4 that MBE growth of III-V nanowires is mainly induced by surface diffusion of adatoms. Diffusion-induced growth is possible only when liquid chemical potential modified by the Gibbs-Thomson effect, $\Delta\mu_{LS} = \Delta\mu - (2\gamma_{LV}\Omega_L \sin\beta)/R$, is smaller than chemical potential of adatoms on the nanowire sidewalls, $\Delta\mu_{AS}$. This yields the upper limit for the liquid chemical potential below which the nanowires can still be grown by surface diffusion

$$\Delta\mu < \Delta\mu_{\max} = \Delta\mu_{AS} - \frac{2\gamma_{LV}\Omega_L \sin\beta}{R}. \tag{6.28}$$

This maximum chemical potential can be put in dimensionless form as

$$f_{\max} = a - \frac{\delta}{\rho}, \tag{6.29}$$

with $a = \Delta\mu_{AS}/\psi_{WZ}$ as the measure of the driving force for surface diffusion.

The dependences of f_c and $f_{\max}$ on the dimensionless nanowire radius ρ shown in Fig. 6.30 are obtained for the following parameters of the Au-assisted MBE of GaAs nanowires: $\psi_{ZB} = 24$ meV/pair, $\gamma_{ZB} = 1.54$ J/m^2, $\tau = 0.875$, $\Omega_S = 0.0452$ nm^3, $\Omega_L = 0.038$ nm^3 (leading to $\delta = 1.12$, $\omega = 1.19$ and $R_0 = 17$ nm), regular triangle island ($x = 1/3, c_1 = 3, c_2 = \sqrt{3}/4$), $\gamma_{SL}^l = 0.594$ J/m^2 and $\beta = 100°$. The value of a is set to 8, corresponding to $\Delta\mu_{AS} = 192$ meV/pair. The minimum chemical potentials $f_{\min}$ for different crystal phases are obtained by equating the nucleation-mediated nanowire growth rates to the 2D growth rate on GaAs(111)B substrate between the nanowires [82].

Generally, ZB phase is prevalent between the curves $f_{\min}^{ZB}(\rho)$ and $f_c(\rho)$, while WZ phase dominates between the curves $f_c(\rho)$ and $f_{\max}(\rho)$. This again confirms the role of supersaturation in MBE: at a given nanowire radius, ZB phase is formed at low and WZ phase at high liquid supersaturations. The existence of $f_{\max}(\rho)$ sets the upper limit for supersaturation above which the nanowire length will be either saturated or diminished by negative diffusion as explained in Sects. 4.14 and 4.15. Looking at the $p_{WZ}(f)$ curve in Fig. 6.28 at $f_c = 5$, MBE growth region in Fig. 6.30 corresponds to the increasing part of the curve. Therefore, under MBE growth conditions, the WZ percentage is expected to increase monotonously with supersaturation.

The ZB and WZ percentages in the general case can be defined as

$$p_{ZB} = p_{ZB,C} + p_{ZB,TPL} \quad ; \quad p_{WZ} = p_{WZ,C} + p_{WZ,TPL}, \tag{6.30}$$

with $p_k = P_k/\sum_{k=1}^{4} P_k$ being the normalized probabilities of the "nucleation scenarios" $k =$ (ZB,C), (ZB,TPL), (WZ,C) and (WZ, TPL). The nucleation-mediated growth rates P_k in the four possible configurations are given by

$$P_C^{(j)} = \rho^2 \exp\left[-\frac{\Delta G_C^{(j)}}{k_B T}\right]; \quad P_{TPL} = 2\rho\rho_c^{(j)} \exp\left[-\frac{\Delta G_{TPL}^{(j)}}{k_B T}\right], \tag{6.31}$$

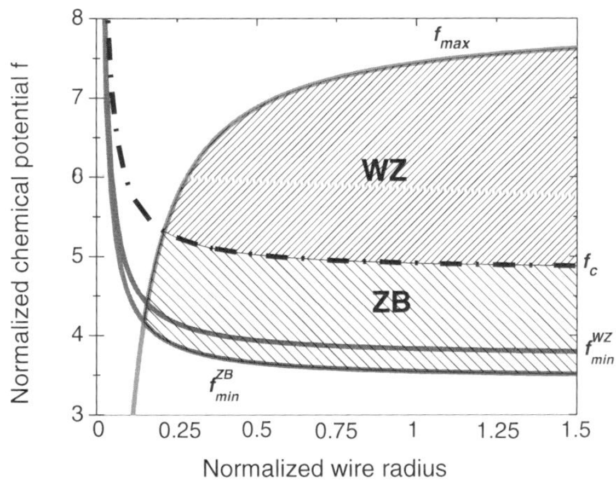

Fig. 6.30 Structural phase diagrams in the case of the triple phase line nucleation of both phases at $a = 8$ and $\tau = 0.875$. Shadowed domains, separated by the critical curve $f_c(\rho)$, correspond to the prevalence of ZB or WZ crystal structure

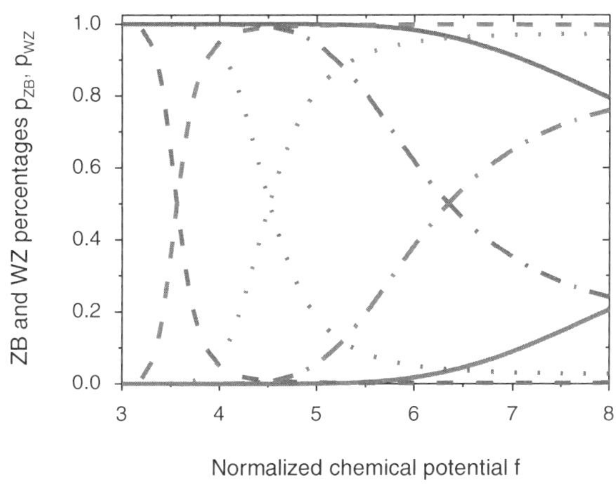

Fig. 6.31 ZB (decreasing *blue curves*) and WZ (increasing *red curves*) percentages as functions of f at fixed radius $\rho = 1.14$: $\tau = 0.95$—*solid lines*, $\tau = 0.91$—*dash-dotted lines*, $\tau = 0.875$—*dotted lines* and $\tau = 0.83$—*dashed lines*

where j is for the crystal phase. Figure 6.31 shows the ZB and WZ percentages obtained for the same parameters as above. The curves are plotted against f at a fixed $\rho = 1.14$ (relating to 20 nm radius in the case of GaAs) and different $\tau = \gamma_{WZ}/\gamma_{ZB}$ between 0.83 and 0.95, corresponding to different facet types.

Calculations show that nanowires adopt the ZB structure in the entire growth domain with nearly 100 % probability at $\tau > 0.97$. The curves at $\tau = 0.95$ show the onset of WZ phase formation only at the highest possible f with ~20 % probability. The curves at $\tau = 0.91$ and $\tau = 0.875$ in Fig. 6.31 demonstrate phase mixing for intermediate values of chemical potential, with 50 % mixing reached at $f = 6.4$ for $\tau = 0.91$ and $f = 4.5$ at $\tau = 0.875$. In these regions, spontaneous switching between the two structures and stacking faults are most anticipated. Lower chemical potential values correspond to the pure ZB phase, and higher values to the pure WZ phase. At $\tau = 0.83$, the ZB structure is observed only for very low chemical potentials ($f < 3.5$), while WZ is prevalent in the rest of the growth domain. As discussed above, the WZ percentage appears monotonously increasing with liquid chemical potential in the entire growth region.

6.7 Kinetic Radius of Structural Transition

Critical radius for ZB-WZ structural transformation in III-V nanowires within the self-consistent kinetic approach can be estimated from the following model [311, 453]. Consider the VLS nanowire of a constant radius R growing by the direct impingement from the vapor flux with supersaturation Φ. Using the results of Sect. 4.7, axial nanowire growth rate is given by

$$\frac{\mathrm{d}L}{\mathrm{d}t} = \frac{\pi}{c}\frac{2}{1+\cos\beta} I_{\mathrm{eq}}\Omega_{\mathrm{S}}\left[\Phi + 1 - (\zeta+1)\exp\left(\frac{R_{\mathrm{GT}}}{R}\right)\right]. \tag{6.32}$$

Here, c is the geometrical shape constant ($c = \pi$ for cylinder and $c = 2\sqrt{3}$ for regular hexagon), I_{eq} is the desorption flux from the droplet, $\zeta \equiv \zeta_\infty$ is the liquid supersaturation at $R \to \infty$ and R_{GT} is the Gibbs-Thomson radius describing the curvature-induced modification of the liquid chemical potential. If the adatom diffusion is important, the right hand side of (6.32) should contain the corresponding diffusion fluxes obtained from the diffusion equations for surface and sidewall adatoms [586]. On the other hand, the nucleation-mediated growth rate under the assumption of a constant ζ_∞ in each nanowire writes

$$\frac{\mathrm{d}L}{\mathrm{d}t} = \frac{h}{\tau}\frac{V_{\mathrm{mono}}}{(1+\Theta V_{\mathrm{mono}}/V_{\mathrm{poly}})}, \tag{6.33}$$

with τ as the microscopic island growth time, V_{mono} as the appropriately normalized growth rate in the mononuclear mode, where only one island succeeds in nucleation in one layer, and V_{poly} as the vertical growth rate in the polynuclear mode, where many islands arise in one layer. The Θ factor accounts for the suppression of polynucleation by nucleation antibunching as discussed in Chap. 4.

From (6.32) and (6.33), the unknown liquid supersaturation ζ can be obtained as a function of radius and growth conditions from numerically solving the following equation:

$$\frac{V_{\mathrm{mono}}(\zeta)}{1+\Theta V_{\mathrm{mono}}(\zeta)/V_{\mathrm{poly}}(\zeta)} = \varepsilon\left[\Phi + 1 - (\zeta+1)\exp\left(\frac{R_{\mathrm{GT}}}{R}\right)\right]. \tag{6.34}$$

Non-dimensional parameter ε at $c = 2\sqrt{3}$ is defined as $\varepsilon = (\pi/\sqrt{3})(1 + \cos\beta)\tau[J_{\mathrm{eq}}\Omega_{\mathrm{S}}/h]$. Since $(J_{\mathrm{eq}}\Omega_{\mathrm{S}})/h$ is the equilibrium deposition rate in monolayers per second (a quantity less than 1 s^{-1}) and τ is the microscopic time of island growth (usually 10^{-9}–10^{-5} s), the value of ε must be very small. The growth rates V_{mono} are given by the same Zeldovich exponents and geometrical pre-factors as in the previous section, written for hexahedral nanowires:

$$V_{\text{mono}} = \frac{h}{\Omega_S} Z\left[(3\sqrt{3}/2)R^2 \exp\left(-F_C\right) + 6Rr_c \exp\left(-F_{\text{TPL}}\right)\right]. \tag{6.35}$$

Here, $Z(\zeta) = [(\zeta + 1)\ln^{1/2}(\zeta + 1)]/\sqrt{\pi}$ as the non-equilibrium Zeldovich factor, $r_c(\zeta) = (2\sqrt{3}\gamma_{eff}\Omega_S)/[k_B T\Delta\bar{\mu}(\varsigma)]$ as the critical size of regular triangle island, and

$$\Delta\bar{\mu}(\zeta) = \ln(\zeta + 1) - \frac{\psi}{k_B T} - (\Omega_S/\Omega_L - 1)\frac{R_{GT}}{R}. \tag{6.36}$$

The dimensionless nucleation barriers $F \equiv \Delta G/k_B T$ are given by (6.25), (6.26) for different crystal phases and island positions:

$$F(\zeta) = \frac{3\sqrt{3}\Omega_S h\gamma_{eff}^2}{(k_B T)^2\Delta\bar{\mu}(\zeta)}; \quad \gamma_{eff} = (1 - x)\gamma_{SL}^l + x(\gamma_{WV} - \gamma_{LV}\sin\beta). \tag{6.37}$$

with $x = 0$ or 1/3. For polynuclear growth, the corresponding expression writes as

$$V_{\text{poly}} = \left(\frac{Z\zeta^2}{4\sqrt{3}}\right)^{1/3} \exp\left(-F_C/3\right). \tag{6.38}$$

At a given set of material constants for a particular system, numerical analysis of (6.34) enables one to find ζ as function of R and Φ. Substitution of this ζ into (6.32) or (6.33) [yielding identical results in view of (6.34)] gives the nanowire growth rate as a function of transparent parameters R and Φ. To investigate the crystal phase of different III-V nanowires grown with Au catalysts, this procedure was performed separately for ZB and WZ nanowires at different Φ. Since the kinetic constant ε is generally unknown, it was varied in the range from 10^{-9} to 10^{-5}. In varying Φ and ε we stick, however, to the experimental values of the growth rates for different Au-catalyzed III-V nanowires (typically in the range 1–10 nm/s) [453]. Supersaturations and the corresponding growth rates of ZB and WZ nanowires [$P_{ZB} \equiv (dL/dt)_{ZB}$ and $P_{WZ} \equiv (dL/dt)_{WZ}$, respectively] were calculated for the parameters summarized in Table 6.9, for the typical growth temperatures of Au-catalyzed III-V nanowires. The values of γ_{Sl}^l were estimated from Young's equation for equilibrium Au-III droplets on different semiconductor surfaces when available [453] and then the probabilities of ZB and WZ phase formation were obtained as $p_{ZB} = P_{ZB}/(P_{ZB} + P_{WZ})$; $p_{WZ} = P_{WZ}/(P_{ZB} + P_{WZ})$. These probabilities are the functions of two variables, R and Φ. The critical curve $R_c(\Phi)$, relating to 50 % phase mixing (at $P_{ZB} = P_{WZ}$), separates the regions of prevalence of ZB and WZ phase in the (R, Φ) plane.

The typical graph for $R_c(\Phi)$, obtained at $\varepsilon = 10^{-6}$ with the parameters of Au-catalyzed GaAs nanowires at 465 °C (which was the MOCVD growth temperature in [311]), is given in Figure 6.32, where the critical supersaturation curve $\zeta_c(\Phi)$ is also shown. The curve $R_c(\Phi)$ exhibits the non-monotonous behavior as we saw

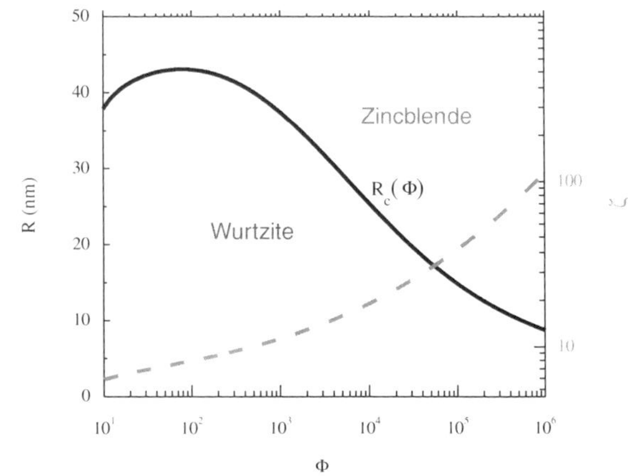

Fig. 6.32 Critical radius $R_c(\Phi)$ (*solid line*), separating ZB (*above the curve*) and WZ (*below the curve*) regions in the (R, Φ) plane. *Dashed line* is the critical liquid supersaturation

earlier with the simplified model described in the previous section (see Fig. 6.28). The critical radius increases with Φ for the modest vapor supersaturations, thus extending

Table 6.9 Parameters of Au-catalyzed III-V nanowires and kinetic radii for the structural transition at typical growth temperatures

Material	γ_{SL}^{l}(J/m^2)	γ_{LV}(J/m^2)	γ_{ZB}(J/m^2)	γ_{WZ}(J/m^2)	ψ_{WZ} (meV)	T (°C)	R_c (nm)
GaAs	0.5	1	1.79	1.3	24	450	45.5
						500	38.8
						550	33.7
						600	29.7
InAs	0.4	1	1.19	0.91	10.6	380	86.5
						400	73.2
						420	62.6
GaP	0.7	1	2.36	1.79	18.3	450	8.1
						500	6.9
						550	6.8
						600	6.7
InP	0.28	1	1.56	1.19	7	380	73.4
						400	69.4
						420	65.8
						470	50.3
GaSb	0.3	1	1.12	0.97	19.8	450	8.6
						500	8.2
						550	7.9
						600	7.7
InSb	0.3	1	0.95	0.73	16.4	380	5.2
						400	5.2
						420	5.2

Fig. 6.33 TEM image of a GaAs nanowire grown with two temperature steps, revealing structural phase transitions along the nanowire length. From left to right: pure WZ phase corresponding to 530 °C growth (*1*), WZ with twins (*2*) and highly faulted WZ/ZB mix up (*3*) corresponding to temperature ramping, pure ZB phase corresponding to 630 °C (*4*), significant reduction of the diameter followed by phase mixing (*5*). Scale bar represents 40 nm

the WZ region below the curve. The maximum of $R_c(\Phi)$ is reached at $\Phi \cong 100$, which corresponds to 300 meV at $T = 465\,°C$. The maximum critical radius amounts to approximately 43 nm, which also looks feasible. This kinetic value is much larger than the equilibrium critical radius of 5.6 nm. At larger $\Phi > 100$, the situation is reversed and the critical radius rapidly decreases with Φ, reaching only 9 nm at $\Phi \cong 10^6$ and $\varsigma_c \cong 100$, relating to $\Delta\mu_{LS}^{\infty} \cong 300$ meV (these values are however too high to be reached in experimental growth conditions). Numerical analysis shows that the maximum of $R_c(\Phi)$ is shifted to smaller values at larger ε and disappears completely at $\varepsilon \sim 10^{-5}$, so that the critical curve converts to monotonously decreasing. Within the range of $\varepsilon = 10^{-9} - 10^{-6}$, the maximum of critical radius is shifted to larger values of Φ for smaller ε.

Numerical results for the kinetic radii for ZB-WZ structural transitions in Au-catalyzed III-V nanowires given in Table 6.9 relate to their maximum value in the entire range of Φ (e.g., 43 nm in Fig. 6.33). These data show the following major trends. First, the critical radius in a given system usually decreases with increasing the growth temperature, i.e., it is expected that higher temperatures suppress the WZ phase formation. Second, antimonite nanowires (InSb and GaSb) have the lowest radius of transition (<9 nm) and should form in pure ZB phase in most cases. Third, theoretical values of R_c are lower for GaAs (30–46 nm) than InAs (63–87 nm) at their typical growth temperatures, therefore, InAs nanowires are expected to be more WZ than GaAs. Fourth, for phosphide nanowires, the influence of group III element is the largest: GaP nanowires have a low critical radius smaller than 9 nm, while for InP nanowires the values of R_c range between 50 and 73 nm, i.e., close to that of InAs. These theoretical conclusions will be compared to experimental data in the next section. Finally we note that the constant value of $\gamma_{LV} = 1\ J/m^2$ used in these calculations does not account for the effect of liquid supersaturation on the energetic balance, which would finally lead to complete suppression of WZ at high enough ζ.

6.8 Crystal Phase Engineering

In this section, we present some experimental data showing how the achieved theoretical understanding of kinetically controlled polytypism helps to control the crystal structure of different III-V nanowires by tuning the growth conditions. We start with the Au-assisted GaAs nanowires grown by MBE using the two-step growth procedure with temperature ramping. The growth procedure is described and explained in Sect. 4.14. An interesting feature is the impact of the high temperature growth step on the crystal structure of GaAs nanowires. As discussed above, crystal structure of III-V nanowires is strongly affected by mononuclear nucleation at the triple phase line: growth at high supersaturation during MBE (small nuclei with prevalence of their edge energy) should onset the formation of WZ phase while the low supersaturation (larger nuclei with prevalence of their cohesive energy) conditions should lead to the prevalence of ZB structure. Since in MBE case supersaturation generally decreases with increasing the surface temperature, a high temperature growth via the continuing growth scenario (see Sect. 4.14) could provide a method for the production of stacking fault free ZB nanowires.

The experimental data of [410] fully confirm these theoretical considerations. High resolution TEM images of a GaAs nanowire of 37 nm diameter in the bottom part, detached from sample C of Fig. 4.76, are shown in Fig. 6.33. Figure 6.34 presents the enlarged TEM images of four sections of the nanowire shown in Fig. 6.33, and the corresponding Fourier transforms of the real space images. The bottom part of the wire (Section 1 in Figs. 6.33 and 6.34a) corresponds to the low temperature growth at 530 °C and it has pure WZ structure, as seen from the TEM image of Section 1 and the corresponding Fourier transform in Fig. 6.34a. Then, the WZ structure presents an increasing number of stacking faults in a region which corresponds to the temperature ramping, where supersaturation starts to decrease as the temperature increases from 530 to 630 °C (Section 2, 3 in Figs. 6.33 and 6.34b). At the end of this region, the crystal structure switches to ZB structure with a few twins. Finally, a 200 nm segment of pure ZB is found corresponding to the high temperature growth at 630 °C [Section 4 in Figs. 6.33 and 6.34a], as demonstrated by the TEM image and the corresponding Fourier transform in Fig. 6.34a. In this ZB segment, the nanowire diameter first increases slightly and then reduces significantly. These changes of diameter follow the changes of catalyst droplet size. Below 25 nm diameter, stacking faults and phase mixing reappear (Fig. 6.33), consistent with WZ phase being more probable for smaller nanowire diameter.

Next, we consider the InSb/InAs axial nanowire heterostructures grown by the Au-catalyzed CBE, shown in Figs. 4.60 and 4.61 of Sect. 4.11 [412]. While InAs stems feature WZ structure with <0001> direction along the growth axis, the InSb segments have pure ZB structure with <111> axis. The two segments are rotated with respect to each other by 30°. The InAs nanowire section is terminated by six equivalent $(1\bar{1}00)$ WZ facets. After the WZ to ZB crystal phase transition and 30° angular rotation, the InSb section is defined by six (110) ZB facets, as illustrated in Fig. 4.61. The 30° angular rotation of ZB InSb section with respect to WZ InAs stem is clearly explained by the data on the surface energies given in the last column of Table 6.5. Indeed, the WZ InAs nanowire with a regular hexagonal cross-section tends to acquire the lowest surface energy and, therefore, must be terminated by six equivalent $(1\bar{1}00)$ WZ facets, whose energy is 14 % lower than that of $(2\bar{1}\bar{1}0)$ facets (0.91 against 1.06 J/m^2). After the crystal phase is switched to ZB, the lower surface

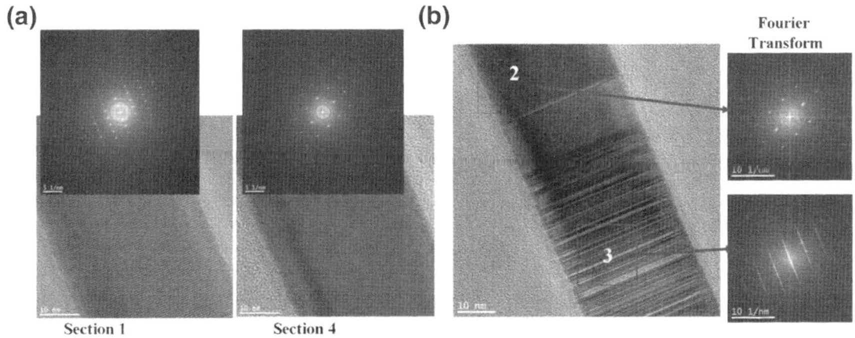

Fig. 6.34 **a** HRTEM close-ups of pure WZ (Section 1) and ZB (Section 4) of the GaAs nanowire shown in Fig. 6.33; **b** enlarged TEM images of WZ with stacking faults (Section 2) and highly faulted WZ/ZB mix up (Section 3). Inserts show the corresponding Fourier transforms

energy of ZB (110) facets (0.85 J/m^2) compared to that of (211) planes (0.95 J/m^2) drives the 30° angular rotation of the InSb part.

As for the preferred crystal structure, switching from WZ phase in InAs to ZB phase in InSb NW part is well explained by very different critical radii of WZ to ZB phase transition: 63 nm in InAs compared to only 5.2 nm in InSb at 420–435 °C (see Table 6.9). Since most of InAs nanowire stems are about 30 nm in radius, they are well below the transition point at 63 nm and must be almost pure WZ. The critical dimension to observe the WZ phase (R_c = 5.2 nm) in InSb segments appears to be even smaller than the GT minimum dimension (R_{min} ∼ 10 nm) to grow an InSb wire at these conditions. Therefore, all InSb nanowires should be pure ZB. A large difference in the critical radii for InAs and InSb is explained by the two factors. First, the difference of bulk cohesive energies between WZ and ZB structure in InSb is much higher: 16.4 meV against 10.6 meV (see Table 6.9). Second, while the relative gain in surface energy of $(1\bar{1}00)$ WZ facets with respect to (110) ZB facets is 13–14 % for both materials, the absolute values are smaller for InSb. This yields only 0.12 J/m^2 absolute gain in InSb compared to 0.15 J/m^2 in InAs. Since the nucleation barriers are extremely sensitive to the surface energies and effective supersaturations, the critical radius of WZ to ZB transition is more than 12 times smaller in InSb. We point out again that, according to theoretical analysis of the previous section, InSb has the smallest radius of WZ to ZB transition among all III-V nanowires and, therefore, is most stable in ZB phase.

Shtrikman *et al.* [297] used the idea of supersaturation-dependent polytypism to obtain stacking fault-free ZB GaAs nanowires by applying the Au-assisted MBE on highly terraced (311)B surfaces. This helps to achieve low values of supersaturation during growth. The (311)B surface was formed by tilting the (100) surface by 25° toward the (111)B surface; hence it is a highly terraced vicinal surface with typically ∼6 Å wide terraces. Atoms impinging on such small terraces are quickly trapped by the step ledges as well as at the kinks along the terraces. When the Au layer is annealed and GaAs is deposited, long GaAs nanowires are obtained with a tilt angle of ∼70°

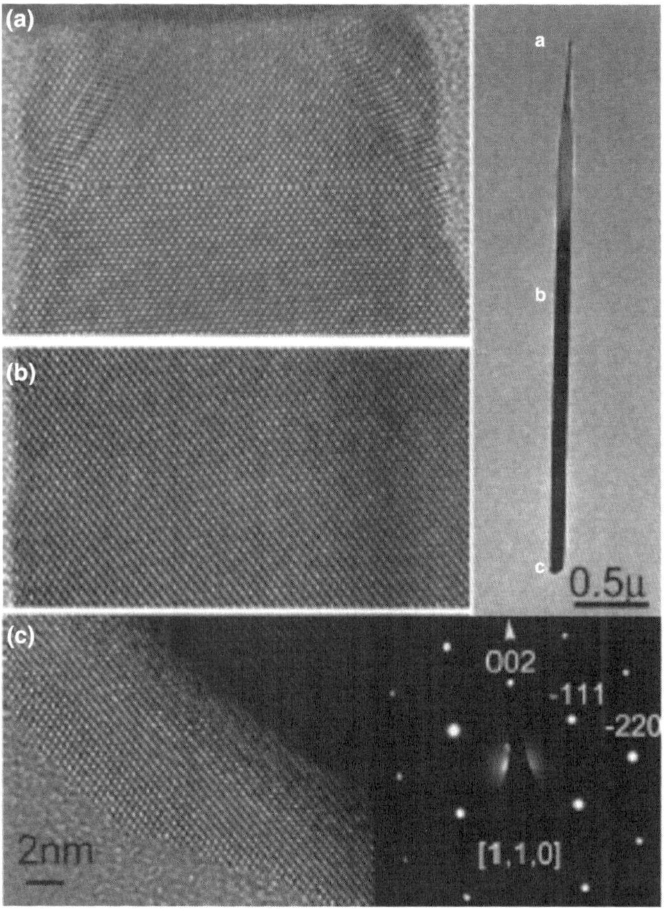

Fig. 6.35 HRTEM images taken from the top, center, and bottom of a ZB GaAs nanowire grown on a (311)B surface as marked in the low-magnification TEM of the wire (*top right*). *Bottom right* is the electron diffraction from the center of the wire; the *arrow* indicates the growth direction of the wire [297]

to the substrate. This angle is similar to the angle between the <311> and the <100> directions which is 72.4°. Long GaAs growing in the <111> direction gradually develop four identical (110) side facets, which grow at the same rate, thus forming a square cross-section. A combined effect of surface terraces and a pronounced 2D growth of the sidewall facets, both adsorbing the impinging adatoms, decreases the Ga diffusion flux to the catalyst droplets and thus reduces the effective liquid supersaturation. Indeed, TEM micrographs proved that the crystal structure of the wires is a pure, completely free of stacking faults, ZB crystal lattice, as shown in Fig. 6.35.

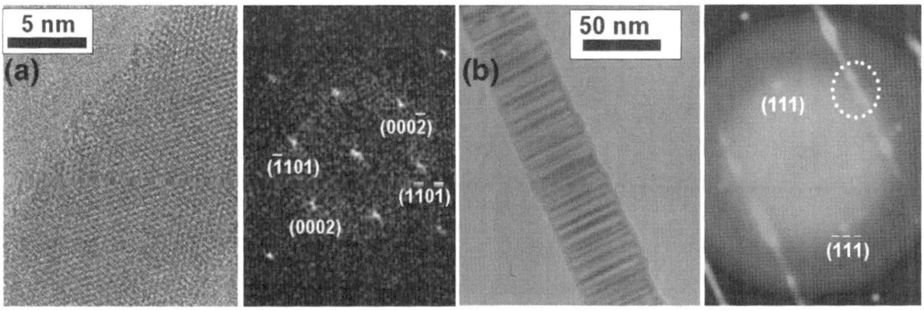

Fig. 6.36 TEM images of InP nanowires: **a** shows a nanowire with 6 nm radius and a pure WZ phase as indicated by the fast Fourier transform to the *right*, **b** shows a larger 23 nm radius nanowire. Twinning defects are clearly seen in this image. The electron diffraction pattern also shows the hybrid of typical ZB and WZ patterns

We now consider the radius dependence of ZB-WZ polytypism in the particular case of InP nanowires grown by MOCVD with differently sized Au catalyst nanoparticles on GaAs(111)B substrates at 470 °C [449]. Figure 6.36a shows a typical InP nanowire with a radius of 6 nm imaged on the <11–20> zone axis. For smaller nanowires, only the WZ phase is typically observed, without twinning defects. This is shown in the FFT on the right of the image, which reveals a clear WZ pattern. For nanowires larger than ~20 nm radius, the twinning between WZ and ZB phases becomes quite frequent, which can clearly be seen in the image. The diffraction pattern also shows a spread between diffraction peaks seen in both WZ and ZB phases, as indicated by the circled section of the pattern. However, WZ phase remains prevalent, consistent with the critical radius of 50 nm relating to 50 % phase mixing in InP at 470 °C (see Table 4.9).

Johansson *et al.* [315] published a detailed analysis of the crystal structure of Au-catalyzed InAs nanowires depending on nanowire radius and growth temperature. The nanowires were grown in a MOCVD reactor at 10 kPa on InAs(111)B substrates. Samples were grown at temperatures of 420, 440, 460, and 480 °C without special annealing steps, using gold particles with nominal diameters ranging from 10 to 100 nm. Gold particles with different sizes within this interval were deposited on the same substrates, and nanowires were grown in a single growth run for each temperature, to allow for direct comparison between the resulting nanowires. The molar fractions of the precursors were 3×10^{-6} for TMIn and 4×10^{-4} for arsine, yielding a V/III ratio of 133.

Figure 6.37 shows InAs nanowires with different radii, grown at 420 °C. It is seen that the fraction of WZ in the nanowires decreases with increasing radius, in favor of ZB. The thinnest wire measured (with a radius of 7 nm) is pure WZ. The planar defects in the wires increase in frequency as the nanowire radius increases. The arrows in the wires with diameters 75 and 92 nm indicate stacking faults. The thickest wire that is illustrated (diameter 125 nm) is dominated by ZB and has only 15 % WZ. In this case, most of the planar defects can be characterized as twin planes. In Fig. 6.38, InAs nanowires grown at 460 °C are displayed. At this higher growth temperature,

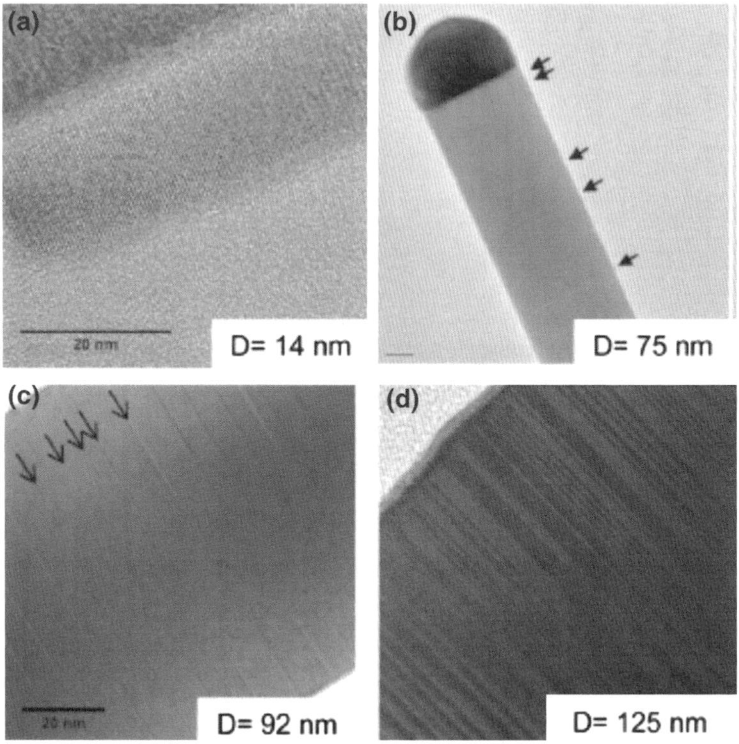

Fig. 6.37 TEM images viewed along the ⟨110⟩ zone axis of InAs nanowires with different diameters D, grown at 420 °C. The fraction of WZ decreases as the wire diameter increases: **a** 100 % WZ, **b** 97 % WZ, **c** 86 % WZ, and **d** 15 % WZ. The arrows in panels (**b**) and (**c**) indicate single stacking faults in the WZ matrix. In panel (**d**), where there is only 15 % WZ, the structure is closer to ZB and the faulty stacked layers are predominantly twin planes [315]

the general trend is the same; i.e., thin wires have WZ structure and thick wires are ZB dominated. However, at higher growth temperatures, the ZB phase starts to dominate at smaller nanowire diameters. The pure WZ nanowires exhibit $(1\bar{1}00)$ side facets. The nanowires containing both WZ and ZB are terminated by (211)-oriented sidewalls, which consist of alternating (111)A and (111) B microfacets for twinned ZB segments and $(1\bar{1}00)$ facets for WZ segments. The zig-zag pattern appearing in the thickest nanowires grown at higher temperature (see Fig. 6.38d), is an additional effect arising in wires with pure ZB structure. The zigzag morphology, arising from alternating (111)A and (111)B facets, is a direct consequence of the periodic twinning in these nanowires, as discussed in Sect. 4.1.

Experimental data for the diameter dependences of WZ percentage in nanowires grown at four different temperatures, 420, 440, 460 and 480 °C are indicated by data points in Fig. 6.39. It is seen that the trend is qualitatively the same as follows from the kinetic model of Sect. 6.6 (see the increasing parts of the curves in Figs. 6.28

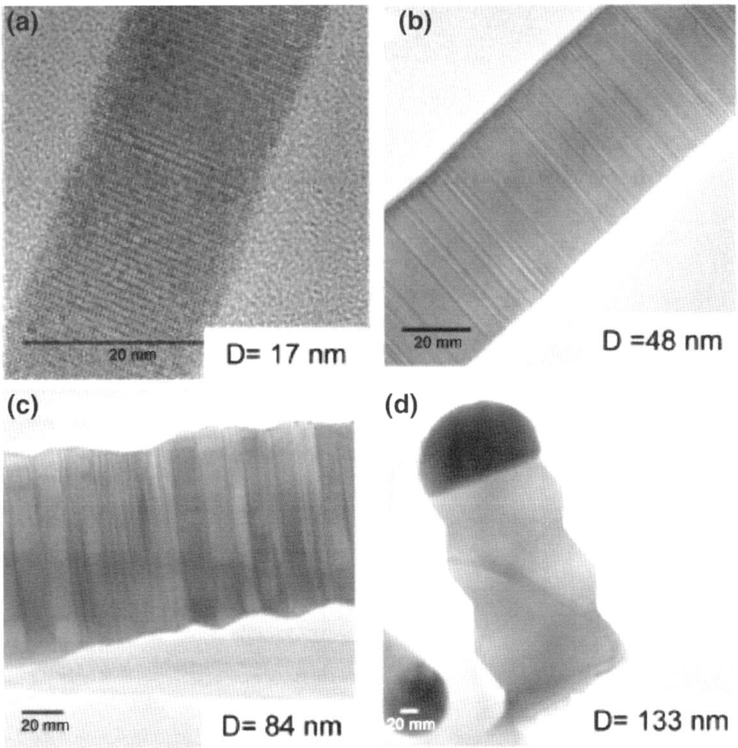

Fig. 6.38 TEM images of InAs nanowires corresponding to the ones in Fig. 6.37, but the wires were grown at 460 °C. The crystal structure ranges from pure WZ (**a**) to pure ZB (**d**), via the mixed phases exemplified in (**b**) with 85 % WZ and in (**c**) with 25 % WZ, as the diameter is increased [315]

and 6.29). Since the effective supersaturation in the catalyst particle is expected to increase toward lower temperatures also in the case of MOCVD (due to a reduced desorption rate), the WZ percentage is higher at lower temperatures for a given radius. The WZ fraction decreases with increasing radius at a fixed temperature. The curve at 420 °C shows the 50 % ZB-WZ phase mixing at approximately 120 nm diameter, which is in excellent agreement with the 63 nm critical radius at this temperature (see Table 6.9).

We note, however, that our radius-dependent curves are much smoother than the experimental curves shown in Fig. 6.39 for the entire temperature range, where the WZ percentage decreases from almost 100 % to zero within ~50 nm diameter range. Such an abrupt behavior is very well fitted by theoretical curves obtained from the nucleation model of [315]. This model is similar to what is presented in Sect. 6.6, but has one important difference. Johansson *et al.* [315] considered the liquid chemical potential $\Delta\bar{\mu}$ (in $k_B T$ units) in the Zeldovich exponent $\exp[-A_*/\Delta\bar{\mu}]$ in the form

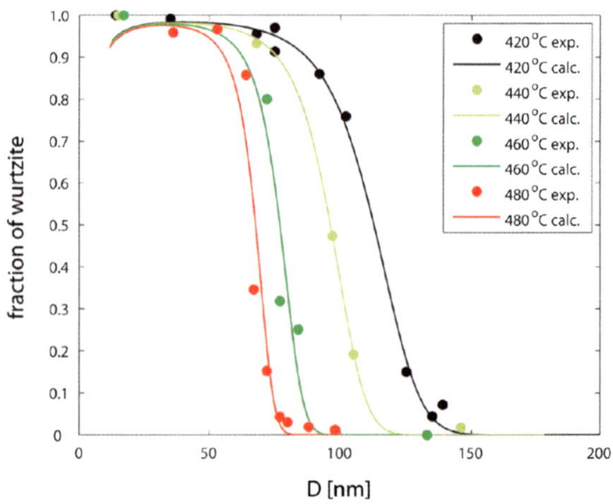

Fig. 6.39 The fraction of WZ structure as a function of InAs nanowire diameter as observed in experiment (*data points*) and estimated from the nucleation model of [315] (*lines*)

$$\Delta\bar{\mu} = \ln(\zeta_\infty + 1) + \frac{2\gamma_{LV}\Omega_L \sin\beta}{k_B TR}, \tag{6.39}$$

thus accounting for the Gibbs-Thomson effect in the droplet but neglecting it in the wire. Our expressions (6.25) and (6.26) contain the Gibbs-Thomson terms that are proportional to $\Omega_L - \Omega_S$ instead. This difference has been already discussed in Sect. 4.5. We are therefore not absolutely convinced that the Gibbs-Thomson effect alone is responsible for such an abrupt change in the crystal structure of InAs nanowires with radius and believe that this interesting feature needs to be investigated in more detail.

The comprehensive analysis of [234] presents the variation of the crystal structure of gold-seeded III-V nanowires with MOCVD growth parameters (temperature, fluxes ratio and absolute pressure of the growth system) and Au nanoparticle radius (5–40 nm) in six materials: GaAs, InAs, GaP, InP, GaSb and InSb. In brief, the results can be summarized as follows:

1. **GaSb and InSb**

   GaSb and InSb nanowires, grown on top of GaAs, InAs and InP nanowire templates, have a hexagonal cross-section with (110) side facets and pure ZB crystal structure free of stacking faults or twin defects. This fully agrees with theoretical predictions of Sects. 6.3 and 6.7 giving the lowest ZB-WZ transition radius for antimonites and the most stable (110) facets in ZB nanowires.
2. **GaAs**
   The crystal structure of GaAs nanowires has been investigated over the temperature range of 350–550 °C, using molar fractions of 1.3×10^{-5} for TMGa and 5.5×10^{-4} for AsH_3. Nanowires grown at low temperature exhibit a ZB crystal structure with low-density twin planes. As the growth temperature is increased, twin plane density increases, with short WZ segments first appearing around

450 °C. The side facets of GaAs nanowires with the highly twinned ZB structure consist of (111)A and (111)B microfacets, such that nanowire segments take the form of octahedral slices. Extended ZB segments in GaAs nanowires exhibit (211)-oriented facets. At sufficiently high temperature (around 550 °C), the WZ structure begins to dominate the nanowires. WZ GaAs nanowires are terminated by $(1\bar{1}00)$ facets, as all other nanowires studied in this work. GaAs nanowires with the predominantly WZ structure generally exhibit the ZB structure in the neck region directly under the particle, as the MBE-grown GaAs nanowires in Figs. 6.13c and 6.14.

3. **GaP**
   Similar temperature effect is observed for GaP nanowires, with twin plane and stacking fault density increasing with temperature. However, the predominant crystal structure remains ZB, consistent with the much smaller transition radius for GaP than for GaAs in Table 4.9. The effect of supersaturation has also been investigated, by "pulsing" the Ga precursor source such that the concentration of supply atoms at the interface fluctuated. It was observed that lowering the Ga supersaturation decreases the twin plane density substantially, consistent with ZB phase being more stable at lower supersaturations.
4. **InAs**
   For this material, the structural trend is the same as shown in Fig. 6.39 for the temperature range between 400 and 480 °C: InAs shows a shift from the predominantly WZ structure towards the predominantly ZB structure with increasing temperature. By varying the radius it was confirmed that the structure variation with temperature extends over the entire range (WZ to ZB). However, at lower temperatures, a completely opposite behavior is observed. From 370 to 390 °C, the nanowires exhibit a predominantly ZB structure, with twin plane and stacking fault densities increasing with temperature. There is a sharp discontinuity in the temperature dependence between 390 and 400 °C, when the structure switches from very predominantly ZB to nearly pure WZ. One very interesting effect is that the neck region of ZB InAs nanowires contains first a short WZ segment, then a short ZB segment; this structure was observed for all nanowires investigated at 390 °C or below. On the other hand, the WZ nanowires grown at 400 °C have a short ZB neck region.
5. **InP**
   It was observed that the predominantly crystal structure of InP nanowires changes from ZB at 380 °C to WZ at 430 °C, for molar fractions of 4.6×10^{-6} for TMIn and and 3.8×10^{-5} for AsH_3. However, interpretation of this effect must be made with care. It was also observed, for InP nanowires grown at 430 °C and V/III ratio of 207, that the total molar fraction affects the crystal structure. Nanowires grown at the highest molar fractions exhibit the mixed WZ and ZB structure, while nanowires grown at the lowest molar fractions exhibit the WZ structure with low-density stacking faults. This effect was not observed for the lower temperature. Additionally, it was observed that nanowires grown at 430 °C with a reduced V/III ratio were entirely ZB with dense twin planes.

These results indicate that the effects of growth parameters on the crystal structure are complicated and must be interpreted carefully. For example, the growth temperature dependences of the structure of GaAs and InAs nanowires in these MOCVD experiments show different behaviors. The WZ percentage in GaAs nanowires increases with temperature in the entire temperature range 350–550 °C. For InAs nanowires, the temperature correlation is non-monotonic, with the twin plane and stacking fault density increasing with temperature between 370 to 390 °C and the WZ percentage decreasing with temperature between 400 and 480 °C. We note that the temperature behavior of effective supersaturation in catalyst particles during MOCVD can be affected by many factors, such as the temperature-dependent pyrolysis efficiencies of different precursors at different surfaces and, vey importantly, the state of catalyst particles (liquid or solid). However, these trends can be qualitatively understood from the non-monotonic WZ percentage versus supersaturation curves shown in Fig. 6.28. We assume that the temperature dependence of $\Delta\mu$ and therefore $f = \Delta\mu/\psi_{WZ}$ is decreasing for a given material, i.e., is determined by a higher desorption rate from catalyst particles at higher T. For GaAs nanowires, the experimental deposition conditions then most probably relate to a larger chemical potential region with a higher group III concentration in catalyst particles, where the WZ fraction decreases with f and thus increases with T. The same trend for InAs nanowires within 370–390° should also correspond to larger f and therefore to the decreasing parts of $p_{WZ}(f)$ curves in Fig. 6.28. Further increasing T leads to the conversion of $p_{WZ}(f)$ curves to a more conventional increasing dependence in Fig. 6.28 and thus the $p_{WZ}(T)$ dependence converts to a decreasing one. An abrupt change in the temperature dependence between 390 and 400 °C could be caused, for example, by solidification of Au-In-As alloy. This would result in a corresponding discontinuous change in $\Delta\mu$, surface energies and therefore $\Delta\mu_c$ below and above the threshold temperature.

In Sect. 4.1, it was demonstrated that Au-In particles are liquid above 425 °C and solid below 360 °C, while the transient state between solid and liquid corresponds to the intermediate region between 360 and 425 °C. We therefore assume that the solid-liquid phase transition occurs at 400 °C. For the liquid state, the triple phase line nucleation is preferred for both $(1\bar{1}00)$ WZ and (110) ZB side facets with $\gamma_{(1\bar{1}00)} = 0.91\ \mathrm{J/m^2}$, $\gamma_{(110)} = 1.06\ \mathrm{J/m^2}$, $\gamma_{LV} = 1.0\ \mathrm{J/m^2}$ and $\gamma_{SL}^{l} = 0.4\ \mathrm{J/m^2}$ from Tables 6.5 and 6.9, for any reasonable contact angle β, for which we will use the value of 110° for further estimates. When both types of islands nucleate at the triple phase line, (6.22), (6.23) and (6.24) should be re-written as

$$f_c = \frac{1}{1 - (\gamma_{eff,\,WZ}/\gamma_{eff,\,ZB})^2}; \tag{6.40}$$

$$U_* = \frac{c_1^2}{4c_2}\frac{\Omega_S h}{k_B T}\frac{(\gamma_{eff,\,ZB})^2}{\psi_{WZ}}; \tag{6.41}$$

$$\frac{P_{WZ}}{P_{ZB}} = \exp\left[U_*\left(\frac{1/f_c - 1/f}{f - 1}\right)\right], \tag{6.42}$$

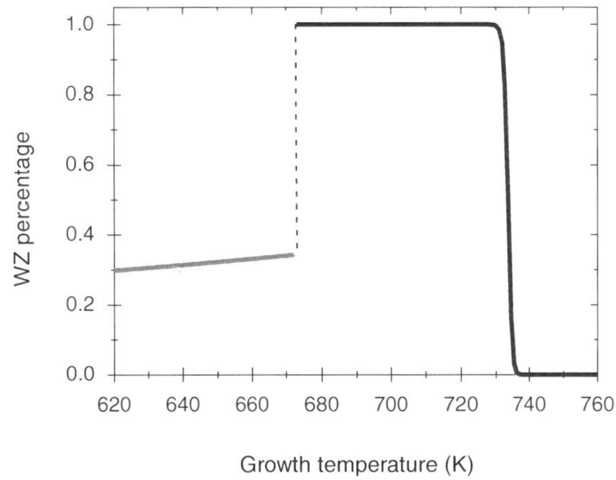

Fig. 6.40 WZ percentage in Au-catalyzed InAs nanowires with a discontinuous change from predominantly ZB to pure WZ at 400 °C and the reverse transition from pure WZ to pure ZB at ~460 °C

with $f = \Delta\mu/\psi_{WZ}$. For the liquid-solid chemical potential, we use the simplest approximation $\Delta\mu = k_B T \ln(c/c_{eq})$, yielding a linearly decreasing temperature dependence

$$\Delta\mu = E_L - k_B T \ln(c_0/c), \tag{6.43}$$

assuming that the equilibrium concentration is given by $c_{eq} = c_0 \exp(-E_L/k_B T)$ with a pre-exponent $c_0 >> c$ and a certain activation energy E_L. For a regular triangle island, $\Omega_S = 0.0556\,\text{nm}^3$, $h = 0.35$ nm and $\psi_{WZ} = 10.6$ meV/pair, the parameter values of InAs from (6.10), (6.40) and (6.41) are: $\gamma_{eff,\,WZ} = 0.26$ J/m^2, $\gamma_{eff,\,ZB} = 0.30$ J/m^2, yielding $f_c \cong 4$ and $U_* \cong 600$. For $\Delta\mu$, we use $E_L = 1.0$ eV and $\ln(c_0/c) = 11$, yielding a reasonable $\Delta\mu \cong 100$ meV/pair at 400 °C.

The black curve in Fig. 6.40 shows the temperature dependence of the WZ percentage above 400 °C for liquid particles, obtained from (6.21), (6.42) with these parameters of InAs. This theoretical dependence exhibits a sharp transition from 100 % WZ to 100 % ZB at ~455–460 °C, which is observed experimentally. It is reasonable to assume that the island nucleation from a frozen particle still occurs at the triple phase line and that the desorption rate from the solid phase is lower than from liquid. The red curve below 400 °C is obtained from the same expressions for frozen particles, with $f_c = 11$ and $E_L = 1.0$ eV changed to $E_S = 1.27$ eV in (6.43) and shows a slight increase of WZ percentage in predominantly ZB structure toward higher T. This simple example qualitatively reproduces the two abrupt structural transitions with a discontinuous change from predominantly ZB to pure WZ at the melting temperature.

Such abrupt changes in the crystal structure of III-V nanowires at varying temperatures and fluxes can be used for the controlled fabrication of ZB/WZ heterostructures. One of the impressive examples was demonstrated by Dick *et al.* in [582], in the case of Au-catalyzed MOCVD of InAs nanowires. As mentioned above, the neck region of InAs nanowires grown at these low temperatures invariably has the opposite crystal structure to the remainder of the nanowire. Within a crystal phase stabilized

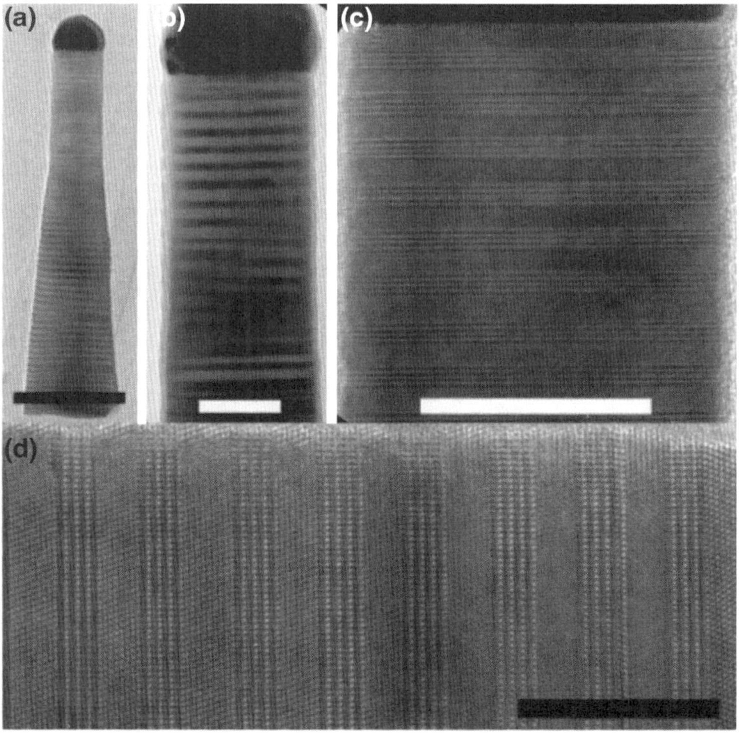

Fig. 6.41 TEM images of InAs nanowire superlattices, defined by 60 periods of alternating ZB and WZ structure (each eight bilayers thick), viewed along the <110> zone axis. Nanowires are grown at 380 °C with WZ segments being formed during growth interrupts under group V precursor flow. **a** Full-length nanowire with alternating segments showing up as striped contrast. **b** Higher-magnification image showing the regular length of the striped segments; the scale bar is 20 nm. **c** High-resolution image, with different crystal structure visible in different regions; the scale bar is 20 nm. **d** High-resolution image with different positions of atom columns in ZB and WZ structure apparent; the scale bar is 10 nm [582]

by optimized temperature, the indium concentration within the catalyst particle can be varied by introducing time-controlled growth interrupts under group V flow. At 380 °C, the dominant crystal structure in InAs nanowires is ZB, and therefore the neck region is WZ. WZ layers would then act as barriers in InAs, suitable for electron transport experiments. Figure 6.41 demonstrates a polytypic superlattice consisting of alternating ZB and WZ segments of precisely eight atomic bilayers each. This structure is repeated over 60 periods in a single InAs wire with misplaced bilayers accounting for only 2 % of the total.

6.9 Benefits of High Growth Rate

Joyce *et al.* [583], Ramdani *et al.* [450], Huang *et al.* [299] and Ren *et al.* [311] reported on the unexpected benefits of a high growth rate on the phase purity of III-V nanowires. All these works used chemical epitaxy techniques (MOCVD or HVPE) with Au catalyst particles. The typical growth rates of GaAs nanowires were about 10 nm/s or even higher, while the crystal structure was pure ZB without any stacking faults along the entire length and actually regardless of the nanowire radius [450, 299, 311]. In particular, TEM and HRTEM images of the Au-catalyzed, MOCVD grown GaAs nanowires from [311] clearly show pure ZB structures of the three wires with 84, 34 and even 12 nm radius (Fig. 6.42).

Initially, the formation of pure ZB phase at high growth rates (corresponding to high liquid supersaturations during growth) was considered contradictory to the Glas theory [45], which predicts the preferred WZ structure at $\Delta\mu > \Delta\mu_c$ according to (6.18). More precise examination shows, however, that the simple criterion given by (6.18) applies within a low chemical potential zone, where a certain threshold value of the liquid chemical potential should be exceeded to overcome the bulk energy difference. Such a situation is more common for MBE technique with moderate values of the VLS growth rates, low Ga concentrations and growth temperatures. Further increase of supersaturation leads to suppression of the WZ phase formation for two reasons. The first reason is the decrease of the critical radius of WZ nucleus and the corresponding reduction of the available nucleation area as given by (6.26) and illustrated in Figs. 6.28 and 6.32. This suppression mechanism has the kinetic origin and is expected to pertain at an intermediate supersaturation range. Finally, very high supersaturation values should relate to high Ga content in the catalyst droplet. Here, the triple phase line nucleation condition given by (6.13) is broken simply due to a decreased droplet surface energy, as shown in Fig. 6.26. This second reason for the WZ phase suppression is purely energetic and should hold for nanowires of any radius.

To quantify our analysis, we consider a particular case of the VLS-HVPE growth of GaAs nanowires with Au catalyst, developed by Gil's group [450, 587, 588]. This deposition method involves chloride precursors for Ga, unusually high nanowire growth temperature near 715 °C and yields the constant rod-like shape with the record growth rate $>100\,\mu$m/h (>28 nm/s) in <111> direction. Experimental results shown in Fig. 6.43 clearly reveal the defect-free ZB structure of the GaAs nanowires with 22.5 and 10 nm diameters, as attested by Raman spectroscopy and TEM.

As usual, we define the vapor supersaturation with respect to the equilibrium liquid phase in the droplet according to: $\Phi = V/V_{eq} - 1 = \chi V_{GaCl}/V_{eq} - 1$, where V is the effective arrival rate of Ga into the droplet, V_{eq} is the Ga desorption rate from the droplet with the equilibrium Ga concentration at a given temperature, V_{GaCl} is the impingement rate of GaCl, and χ is the temperature-dependent pyrolysis efficiency at the liquid surface. This Φ depends on the growth temperature, V/III ratio and also on the pressure of the carrier gas (H_2). Assuming that the VLS growth is induced only by the direct impingement from vapor and proceeds in the mononuclear mode,

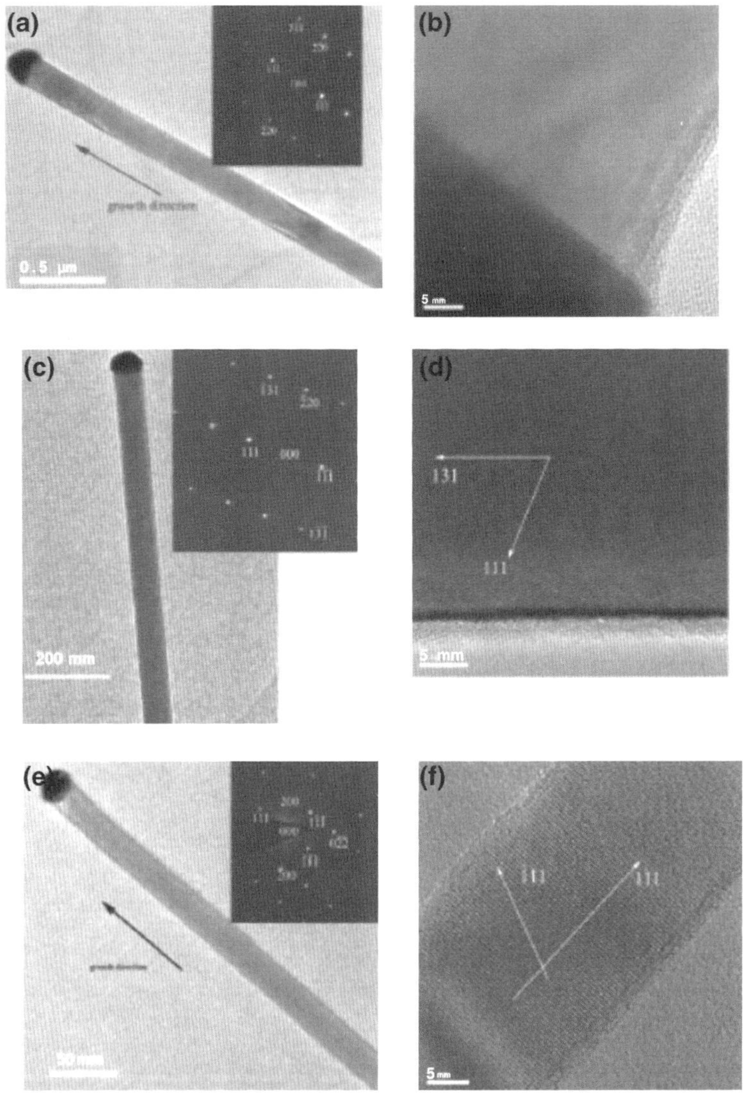

Fig. 6.42 TEM and HRTEM images of GaAs nanowires with 84 nm [(**a**) and (**b**)], 34 nm [(**c**) and (**d**)] and 12 nm [(**e**) and (**f**)] radius. These wires were grown by the Au-catalyzed MOCVD at 464 °C with an elongation rate of 10.3 nm/s. The inserts in (**a**) (**c**) and (**e**) show the selected area electron diffraction patterns. The zone axis is [11–2] for (**a**) and (**c**) and [01–1] for (**e**)

the nanowire elongation rate is given by

$$\frac{dL}{dt} = \frac{2V_{eq}(\Phi - \zeta)}{1 + \cos\beta} = h\pi R^2 J(\zeta). \qquad (6.44)$$

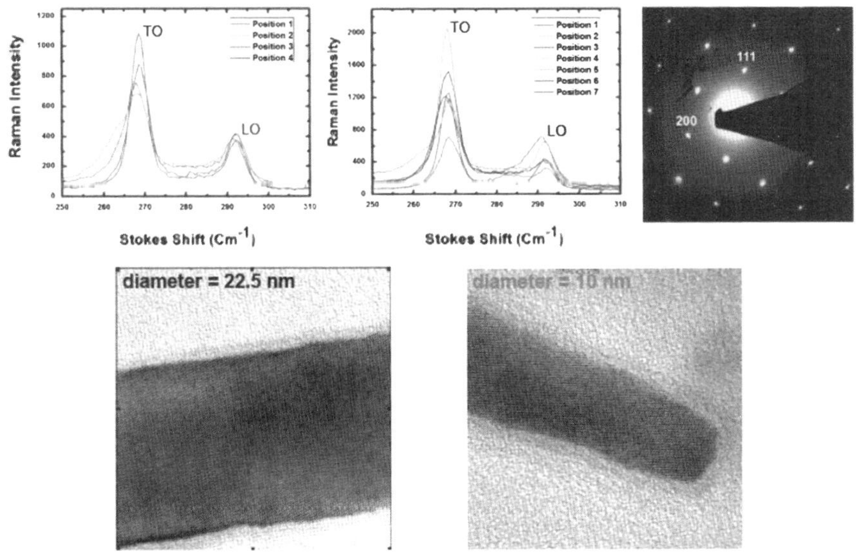

Fig. 6.43 Raman spectra, TEM images and TEM diffraction pattern demonstrating pure ZB crystal structure of 22.5 and 10 nm GaAs nanowires grown by the Au-catalyzed HVPE at 715 °C and III/V ratio of 5 [587]

Here, ζ is the Ga-controlled liquid supersaturation (under the assumption of a constant As concentration in the droplet), related to the liquid chemical potential as $\Delta\mu = k_B T \ln(\zeta + 1)$, β is the droplet contact angle, h is the monolayer height, and R is the nanowire radius. The ζ-dependent nucleation rate is given by

$$J(\zeta) = \frac{h}{\sqrt{\pi}\Omega\tau}(\zeta + 1)\ln^{1/2}(\zeta + 1)\exp\left[-\frac{A_*}{\ln(\varsigma + 1)}\right], \tag{6.45}$$

where τ is the characteristic time of 2D island growth, and the energetic constant $A_* = c\Omega h\left(\gamma_{SL}^l/k_B T\right)^2$ with c as the shape factor.

Within a plausible range of the unknown solid-liquid surface energy γ_{SL}^l between 0.3 and 0.5 J/m^2 and at $c = \pi$, the parameters of ZB GaAs nanowires ($h = 0.326$ nm, $\Omega = 0.0452$ nm^3) yield the values of A_* between 45 and 113 at $T = 715$°C. Using the value of $D = 3 \times 10^{-9}$m^2/s for the Ga diffusion coefficient in the liquid Au-Ga alloy [589] at 715 °C and estimating the characteristic time τ from $\tau^{-1} \cong c_{eq}D/\sqrt{\Omega h}$ with the equilibrium atomic Ga concentration $c_{eq} = 0.4$, the second (6.44) and (6.45) yield the nanowire growth rates shown in Fig. 6.44 for two different $A_* = 45$ and 60. At a fixed liquid supersaturation ζ, the nanowire growth rate increases with the radius as R^2, and increases exponentially with ζ for a given radius. Of course, the same growth rate requires larger ζ for larger A_*. Since the experimental growth rates for differently sized nanowires are typically larger than 100 μm/h (marked by the horizontal line in Fig. 6.44), the liquid supersaturation in the droplets during growth

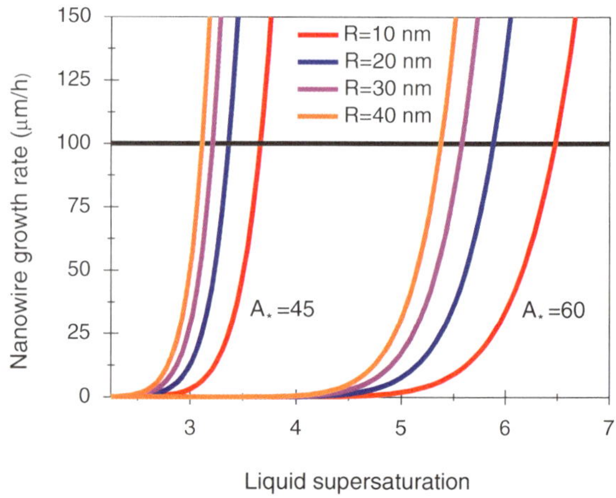

Fig. 6.44 GaAs nanowire growth rates *dL/dt* for different nanowire radii *R* as given by the second (6.44) and (6.45), at $A_* = 45$ and 60

should be systematically larger than 3, relating to the minimum value of $A_* = 45$. We will therefore use the value of $\zeta = 3$, corresponding to $\Delta\mu_{LS} = 125$ meV/pair, as the lowest estimate for the liquid supersaturation during growth.

On the other hand, the material transport growth rate given by the first (6.44) at $\beta = 110°$, $\zeta = 3$ and $V_{eq} = 0.597$μm/h (estimated from the data shown in Fig. 4.69 as an equilibrium Ga desorption flux from GaAs(111)B surface) yields the value of the vapor supersaturation $\Phi = 58$ for $dL/dt = 100$ μm/h. The nanowire growth thus proceeds at $\Phi >> \zeta$, i.e., with almost negligible influence of desorption. This is due to an extremely high vapor supersaturation in HVPE, because the desorption rate itself is quite large at a high growth temperature of 715 °C.

The necessary condition for the WZ phase formation is nucleation at the triple phase line, which is energetically preferred when $\Delta\gamma = \gamma_j - \gamma_{LV}\sin\beta - \gamma_{SL}^l < 0$. Using again the linear interpolation in c_{Ga} given by (6.14) for γ_{LV}, we re-plot Fig. 6.26 at $T = 715$ °C for the four relevant sidewall facets with the surface energies $\gamma_{(211)} = 1.79$ J/m^2, $\gamma_{(100)} = \gamma_{(2\bar{1}\bar{1}0)} = 1.543$ J/m^2, and $\gamma_{(1\bar{1}00)} = 1.30$ J/m^2. From Fig. 6.45 (corresponding to $\gamma_{SL}^l = 0.5$ J/m^2), it is seen that the triple phase line nucleation cannot occur for ZB wires terminated by (211) facets, which are observed in HVPE-grown GaAs nanowires. The triple phase line nucleation is also suppressed for (110) ZB and $(2\bar{1}\bar{1}0)$ WZ sidewalls (having the same surface energy) in the growth region at $c_{Ga} > c_{eq} = 0.47$ for $c_{As} = 0.01$. The only possibility for forming the WZ structure is thus $(1\bar{1}00)$ WZ sidewalls with the lowest surface energy. In this case, the Ga atomic concentration during growth should be smaller than $c_* = 0.67$.

Figure 6.46 shows the atomic concentrations of Ga corresponding to $\Delta\mu = 0$ (the equilibrium Ga concentration) and $\Delta\mu = 125$ meV/pair (which is the minimum difference between the liquid and solid chemical potentials of GaAs that yields a growth rate of 100 μm/h) at 715 °C, calculated from the model described in Sect. 5.15 at different c_{As} within a plausible range between 0.01 and 0.03. These data show that the equilibrium Ga concentration is systematically higher than 0.35 and increases to 0.47

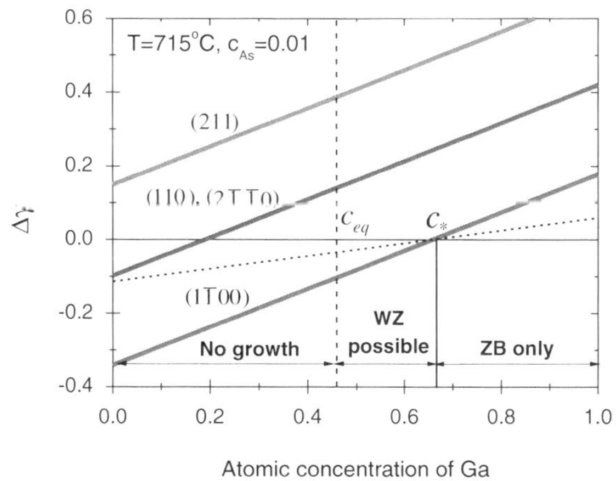

Fig. 6.45 $\Delta\gamma$ versus c_{Ga} for the four different side facets of ZB and WZ GaAs nanowires. $c_{eq} = 0.47$ relates to the equilibrium Ga concentration in the Au-Ga-As liquid alloy at $c_{As} = 0.01$; the VLS growth is possible only at $c_{Ga} > c_{eq}$. The $\Delta\gamma$ is positive in the growth region for both ZB and $(2\bar{1}\bar{1}0)$ WZ sidewalls, where the TPL nucleation is suppressed on surface energetic grounds. The critical Ga concentration $c_* = 0.67$ relates to $\Delta\gamma = 0$ for $(1\bar{1}00)$ WZ sidewalls having the lowest surface energy. WZ structure in this case is possible only for the low enough Ga concentrations between 0.47 and 0.67. The *dotted line* corresponds to the surface energy difference $\gamma_{WZ}^{eff} - \gamma_{SL}^{l}$, with γ_{WZ}^{eff} given by (6.46)

for smaller c_{As}. The minimum growth concentration of Ga is systematically higher than 0.5, and also increases toward lower c_{As}, so that it is higher than $c_* = 0.67$ at $c_{As} = 0.01$ but becomes lower than c_* for $c_{As} > 0.012$. Higher arsenic concentrations therefore yield lower concentrations of Ga in the droplet during growth, where the WZ phase formation becomes possible because the triple phase line nucleation is enabled (see Fig. 6.45).

We now show that the WZ phase formation is kinetically forbidden even for higher As concentrations in the droplet where c_{Ga} is lower than c_*. Indeed, whenever the condition for the triple phase line nucleation given by (6.15) is fulfilled, WZ structure could form only when the liquid-solid chemical potential is larger than a certain critical value which is required to create a stacking fault. This critical value is given by (6.18), which in our case takes the form

$$\Delta\mu_c = \frac{\psi}{1 - (\gamma_{WZ}^{eff}/\gamma_{SL}^{l})^2}; \quad \gamma_{WZ}^{eff} = (1-x)\gamma_{SL}^{l} + x(\gamma_{WV} - \gamma_{LV}\sin\beta). \qquad (6.46)$$

Figure 6.47 shows the $\Delta\mu_c$ obtained from (6.46) for a triangle island ($x = 1/3$) and $(1\bar{1}00)$ WZ sidewalls with the same parameters as before. It is seen that the formation of WZ structure is kinetically forbidden in the entire growth region with Ga concentration higher than 0.5.

Fig. 6.46 c_{eq} and c_{Ga} at a 125 meV/pair chemical potential difference at $T = 715°C$ for different c_{As}

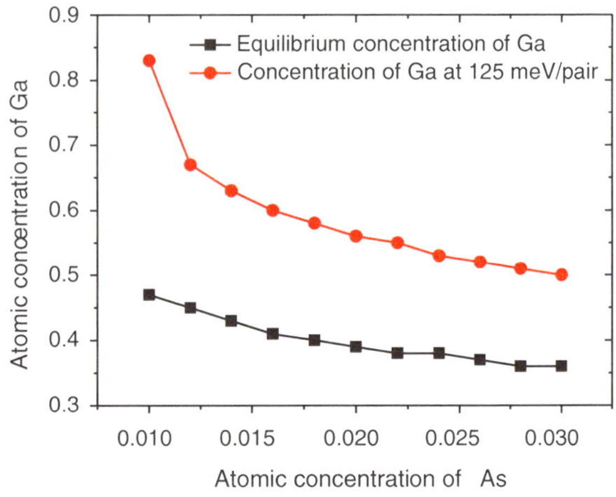

Fig. 6.47 Critical liquid-solid chemical potential $\Delta\mu_c$ such that WZ structure is favored above $\Delta\mu_c$ and suppressed below $\Delta\mu_c$, compared to the chemical potential during growth ($\Delta\mu = 125$meV, *horizontal line*). WZ structure is forbidden in the entire growth region above $c_{Ga} = 0.5$

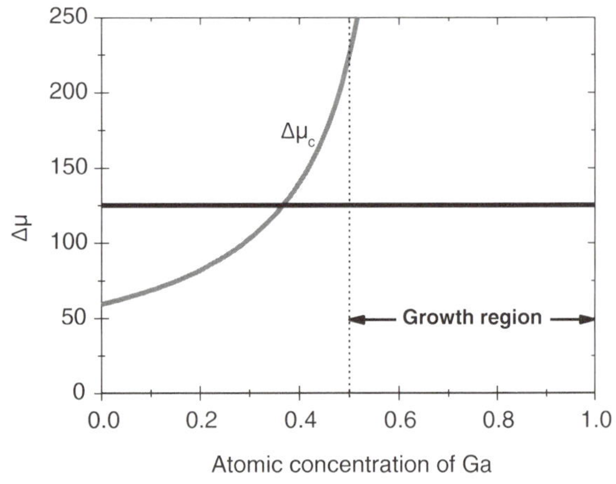

Thus, the Au-catalyzed, high temperature HVPE growth of GaAs nanowires proceeds at a high effective vapor supersaturation of 58, and a high liquid supersaturation above 3. These high supersaturation values yield a high Ga content in the Au-Ga-As droplets during growth, where the Ga concentration is higher than 0.5 for $c_{As} \leq 0.03$. High Ga concentration in the droplets substantially decreases their surface energy. Formation of WZ structure is then suppressed either on surface energetic grounds (at a low As content) or kinetically (at a higher As content), regardless of the nanowire radius. This well explains why the VLS growth at a high Ga concentration in the Au-Ga-As alloy yields the exceptional elongation rate and the defect-free crystal capability simultaneously.

6.10 Zincblende Structure of Ga-catalyzed GaAs Nanowires

Self-catalyzed, or Ga-catalyzed, VLS growth of GaAs nanowires has been considered in Sects. 4.1 and 5.9. Besides the obvious advantage of safely avoiding the unwanted Au contamination during the VLS procedure, the self-catalyzed approach also helps to suppress the uncontrolled ZB-WZ polytypism. This important feature readily follows from the condition for the triple phase line nucleation given by (6.13) and is illustrated in Fig. 6.26: with neglect of a trace amount of As, the catalyst droplet during the Ga-assisted growth should consist of almost pure Ga. Since the surface energy of Ga is much lower than that of Au, the inequality (6.13) cannot hold for any reasonable γ_{SL}^{l} and thus the WZ phase formation is suppressed on surface energetic grounds. Therefore, Ga-catalyzed GaAs nanowires and other self-catalyzed III-V nanowires are expected to be almost purely ZB under optimized deposition conditions.

Self-catalyzed MBE growth of GaAs nanowires is usually performed at slightly elevated temperatures, typically between 580 and 650 °C. In particular, the GaAs nanowires discussed in [49] were obtained at 590 °C. For this temperature, $\gamma_{LV} = 1.24\ \mathrm{J/m^2}$ for pure Au and $\gamma_{LV} = 0.67\ \mathrm{J/m^2}$ for pure Ga. The value of $\gamma_{LV} = 1.0\ \mathrm{J/m^2}$ corresponds to approximately 40 % Ga concentration in the Au-Ga alloy during the growth. We use the values $\gamma_{SL}^{l} = 0.59\ \mathrm{J/m^2}$ and $\gamma_j = \gamma_{(1\bar{1}00)} = 1.30\ \mathrm{J/m^2}$ for the lowest energy WZ side facet of a hexahedral nanowire, and the contact angles β between 110 and 125° according to the post-growth measurements. With these parameters, the $\Delta\gamma$ given by (6.15) takes the negative values between −0.23 and −0.11 J/m^2 for the Au-assisted growth and the positive values between 0.08 to 0.16 J/m^2 for the Ga-catalyzed growth. As discussed above, changing the catalyst to a lower surface energy liquid makes the substitution of its surface to a more energetically costly wire-vapor interface unfavorable, which should retain the crystal structure of Ga-catalyzed GaAs nanowires to ZB. Basically, this is the extreme case of the WZ suppression mechanism described in the previous section at $x_{Ga} \to 1$, where no Au is present in the catalyst particle.

Experimental findings of different groups confirm this theoretical conclusion: Ga-catalyzed GaAs nanowires indeed adopt the ZB phase much more often than the Au-catalyzed wires [48, 49, 293, 323, 324, 558, 590, 591]. While the WZ percentage may depend on the growth conditions such as the As flux during MBE [558], or be present in a narrow region just beneath the droplet [49], 100 % pure ZB structure in GaAs NWs can never be obtained under otherwise identical conditions when catalyzed by Au.

In particular, Fig. 6.48 presents typical TEM images of the Ga-catalyzed GaAs nanowires grown on Si(111) substrates at 590°C [49]. It was found that the majority of the length of the nanowires adopted the ZB phase, but that the region immediately beneath the Ga droplet had the WZ crystal structure (Fig. 6.49a, region B), the material having undergone a transition (region C). Close examination of region C in Fig. 6.49 shows how the purely ZB part transforms via the introduction of single stacking faults, then narrow twin lamellae, followed by random stacking, the WZ structure with somestacking faults and ultimately the pure WZ phase. The phase

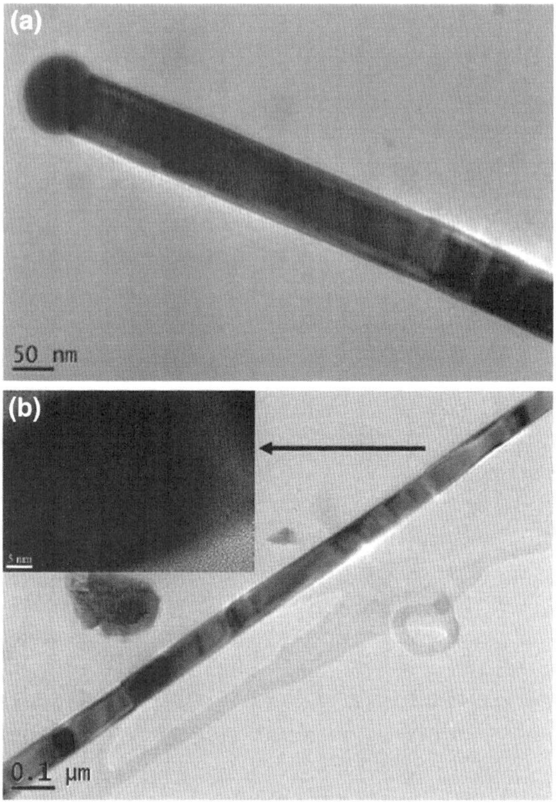

Fig. 6.48 Multi-beam bright-field diffraction contrast TEM images of the upper part of a nanowire terminated by a Ga droplet preceded by a region characterized by planar faults on the (111) growth planes (**a**) and the main part of a nanowire showing the general absence of stacking faults and mixed-phase regions. The dark bands are diffraction contrast features (bend contours) that are not associated with stacking disorder (**b**). High resolution TEM close-up, shown in the insert, demonstrates pure ZB phase in the wire

mixing (stacking transition) happens only when the Ga-supplying flux is stopped, and is therefore associated with an abrupt change in growth conditions.

The explanation of ZB to WZ transition at the end of the growth (opposite to the well-known WZ to ZB transition in the Au-catalyzed case described in Sects. 6.2 and 6.5) is the following. After the Ga flux is turned off, there is still enough arsenic in the residual atmosphere to form stoichiometric solid GaAs at the expense of liquid Ga emptied from the droplet. This necessarily decreases the droplet volume, the contact angle of the droplet and may even lead to its breakdown. The remaining upper part of the droplet at the nanowire top may then enable the TPL nucleation producing the WZ structure. In later experiments [48], we took special care to preserve at most the stationary droplet configuration at the end of the growth. The key growth modification involves periodical turn on the Ga source before or while the substrate is being cooled down.

Figure 6.50 shows typical TEM images of the GaAs nanowires grown by the modified MBE procedure [48]. The TEM image of the main part of the nanowire in Fig. 6.50a, with the corresponding selected area diffraction pattern (acquired from a region far removed from the droplet) demonstrates the general absence of stacking

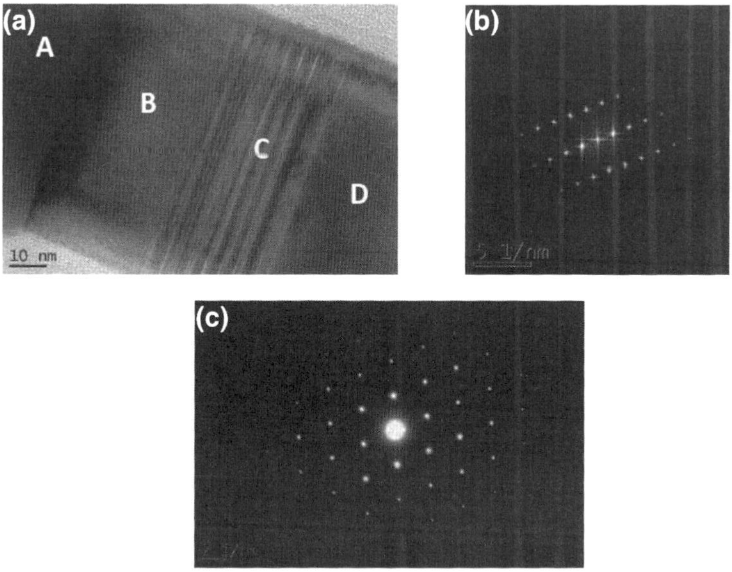

Fig. 6.49 High resolution TEM image of the upper part of the nanowire (**a**) and corresponding diffraction patterns. Region *A*—the Ga catalyst, region *B*—the WZ phase region (the Fourier transform in figure (**b**) shows the $(11\bar{2}0)$ WZ zone axis). In the transition region *C*, the structure contains twins, and region *D* has the pure ZB phase (the selected area diffraction pattern in figure (**c**) shows the (110) ZB zone axis and was acquired from a region far removed from the Ga droplet). The scales in (**b**) and (**c**) are 5 and 2 nm^{-1} respectively

faults and mixed-phase regions. Figure 6.50b shows the nanowire region just beneath the droplet, which, in contrast to the earlier results shown in Fig. 6.49, also appears to be pure ZB, although featuring few rotational twins. Thus, the use of turn on the Ga source at the end of growth enables to avoid the polytypism of the nanowire tip.

Figure 6.51 shows typical SEM images of Ga-catalyzed GaAs nanowires. Plan views in Fig. 6.51a, b clearly demonstrate the elongated shape of most droplets. Since such a shape is impossible if the droplet rests on the top facet, one can assume that the droplet surrounds the nanowire, as described by the theoretical model of Sect. 1.4 and will be explored further in the next section (although the buried tips of the nanowires cannot be seen). The short nanowire highlighted in Fig. 6.51b should have lost part of Ga droplet catalyzing its growth. This effect is observed in many samples and is probably explained by a rather small Young's equilibrium contact angle (110° according to our estimates), so some droplets may diffuse to the sidewalls as in the complete wetting case. Figures 6.50a and 6.51c demonstrate a lateral extension of ~ 2000 nm long nanowires toward their tops, which should be caused by the droplet inflation under Ga-rich conditions.

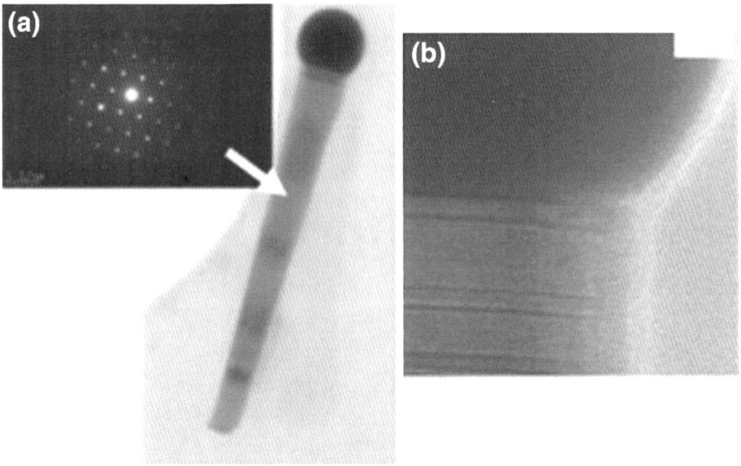

Fig. 6.50 TEM image of the main part of a nanowire terminated by a Ga droplet (**a**). The selected area diffraction pattern, acquired from a region far removed from the droplet, shows the (110) ZB zone axis and demonstrates pure ZB structure. The scale in the insert is 5 nm^{-1}. Figure (**b**) shows pure ZB phase of the nanowire section just beneath the droplet, with few rotational twins. The scale bar in (**b**) represents 10 nm

Fig. 6.51 Plan view SEM images of GaAs nanowires grown after 15 min of growth (**a**) and (**b**), showing an elongated droplet shape; scale bar represents 500 nm. Arrow in figure (**b**) points at the nanowire having lost its droplet. Figure (**c**) shows the nanowire extending towards the top

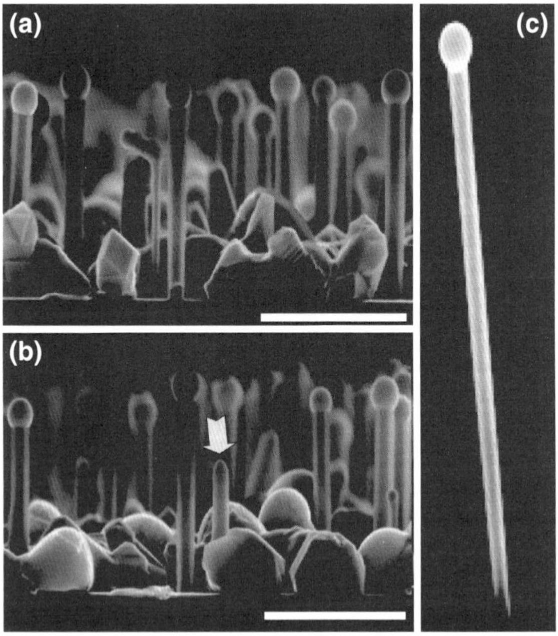

6.11 Wetting Mode of VLS Growth

In Sect. 1.4 it was shown that, whenever the Nebol'sin-Shetinin-Glas condition given by (6.13) is broken for straight sidewalls ($\tilde{\gamma}_j = \gamma_j \equiv \gamma_{WV}$) of a cylindrical nanowire, i.e., when

$$\gamma_{WV} > \gamma_{LV} \sin \beta_0 + \gamma_{SL}^l, \tag{6.47}$$

and the top nanowire facet is planar, the non-wetting droplet configuration at the nanowire top becomes unstable. The droplet will therefore slide down until the nanowire cylinder hits the droplet surface ($y = y_{max}$, $\beta = \beta_{max}$), as shown in Fig. 1.11b. This VLS configuration relates to the minimum surface energy of a spherical droplet under the condition of a fixed liquid volume (given by (1.52)). The droplet cannot be pinned for the same energetic reasons that drive its sliding down. However, as seen from Fig. 1.11c, the force F_1 acting upon the bottom TPL on the wire sidewall remains uncompensated and tends to decrease the contact angle β_{max}. Such force gives rise to further evolution of system morphology, with the droplet losing its spherical geometry. The new, stationary droplet shape should be obtained from the minimization of surface energy functional, which is not a simple problem in the general case. However, this new, non-wetting VLS configuration is of major importance for understanding the growth of nanowires catalyzed by low surface energy metals (such as Ga with $\gamma_{LV} = 0.71$ J/m^2, In with $\gamma_{LV} = 0.57$ J/m^2 and Al with $\gamma_{LV} = 0.87$ J/m^2 at their respective melting temperatures) and, more generally, by group III-rich alloys with Au when γ_{LV} gets low enough to ensure the inequality given by (6.47).

We start with a simplified model of the droplet shape proposed in [48]. The non-spherical shape can be computed under the following reasonable assumptions, illustrated in Fig. 6.52a: (i) a constant Laplacian pressure along the droplet surface, yielding the constant curvature $1/R_1 + 1/R_2 = const$, with R_1 and R_2 as the main radii of surface curvature; (ii) droplet surface given by a figure of rotation, $r^2 = f^2(z)$; (iii) the Young's condition for the equilibrium contact angle $\gamma_{WV} = \gamma_{LV} \sin \beta_* + \gamma_{SL}^l$; (iv) small enough β_* such that$-\cot an\beta_* << 1$ and (v) $[f(z) - R]/R << 1$. It can be shown that, after the nanowire cylinder hits the droplet surface, the Young's condition becomes corresponding to the minimum surface energy at a fixed liquid volume, as in the standard 2D case. The parabolic equation describing the surface of a droplet resting on the sidewalls can be written down as

$$f(z) = \frac{z^2}{4}\left(\frac{1}{R} - C\right) - z \cot an\beta_* + R, \tag{6.48}$$

where C is the integration constant. The part of a droplet resting on the nanowire top facet should be spherical, of the radius $R_d = 2/C$, yielding $\sin \beta' = R/R_d = RC/2$ for the top contact angle β'. Together with the condition for the surface smoothness at the top corners, $\beta' = \pi - \beta_*$ (see Fig. 6.52a), this gives $C = (2 \sin \beta_*)/R$, completing the unique determination of the curve given by (6.48). As seen from Fig. 6.52, the

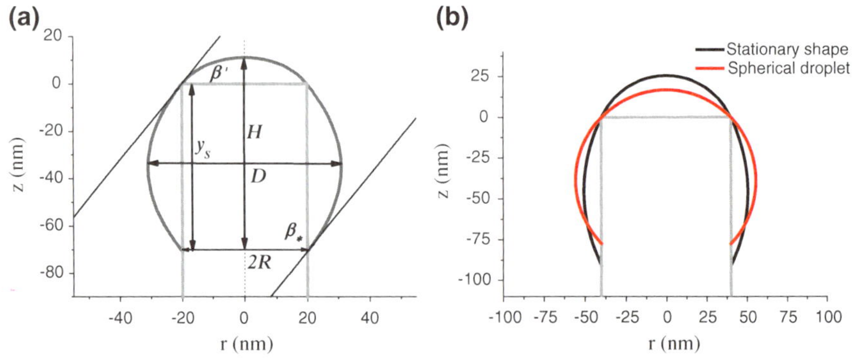

Fig. 6.52 Geometry of elongated droplet with $H > D$ (**a**); shape transformation of Ga droplet for the parameters described in the text (**b**)

stationary wetted part of the nanowire y_s is larger than $y_{\max}$. The height of such a parabolic droplet H is larger than its width D, which can never happen to a spherical droplet.

We now consider the parameters approximately corresponding to the Au-catalyzed and Ga-catalyzed MBE growths of ZB, <111> oriented GaAs nanowires: $\gamma_{WV} = \gamma_{(100)} = 1.5$ J/m^2, $\gamma_{SL}^{l} = 0.7$ J/m^2, $\gamma_{LV} = 1.0$ J/m^2 for an Au-Ga liquid droplet and $\gamma_{LV} = 0.8$ J/m^2 for a Ga-As liquid droplet, with a typical radius of 40 nm. Spherical to parabolic shape transformation in the Ga-catalyzed case is shown in Fig. 6.52b. Figure 6.53 presents the curves of the β-dependent surface energy difference $\Delta G(\beta) = G(\beta) - G(\beta_0)$, obtained from (1.51) and plotted against $\Delta\beta = \beta - \beta_0$ for the Au-catalyzed nanowires (for three different values of the initial contact angle β_0 in the non-wetting VLS configuration) and for the Ga-catalyzed nanowires at $\beta_0 = 115°$. As is seen from the figure, these typical parameters of Au-catalyzed growth relate to the standard VLS case with a positive derivative of the surface energy at $\beta = \beta_0$. However, the increase of β_0 transforms the increasing $\Delta G(\Delta\beta)$ dependence to the curves having the maximum at $\beta = \beta_*$. When catalyst metal is changed to Ga, the surface energy decreases in the entire range of contact angles, with the minimum energy of spherical droplet reached at $\beta_{\max}$. The circular dot in Fig 6.53 represents the elongated droplet with the minimum energy at a constant Laplacian pressure, calculated for the surface geometry given by (6.48). The arrow represents the energetically favorable shape transformation shown in Fig. 6.52b.

It should be noted that the stationary state with the elongated droplet is not always possible. Indeed, at a very low γ_{LV} such that $\gamma_{WV} > \gamma_{LV} + \gamma_{SL}^{l}$, the Young's equation on the sidewalls can never be satisfied. This situation corresponds to a complete wetting, with Ga atoms diffusing from the droplet and finally forming a thin film on the sidewalls. Since the droplet disappears completely, the VLS growth can no longer proceed. As mentioned above, this might be the explanation of a minor fraction of GaAs nanowires having lost their catalyst particle, the effect seen in Fig. 6.51.

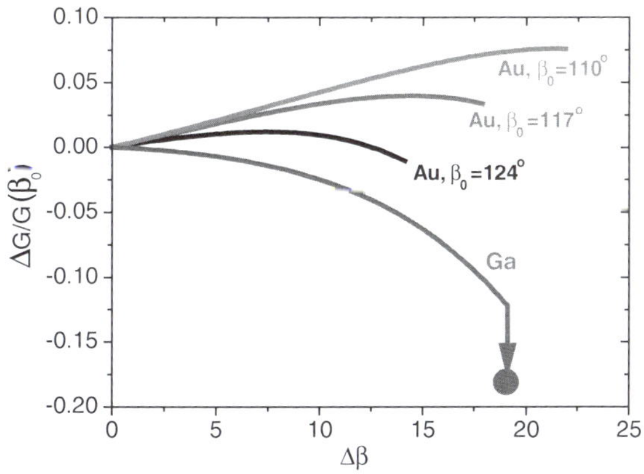

Fig. 6.53 Dependences $\Delta G(\Delta\beta)$ for Au-catalyzed GaAs nanowires at $\beta_0 = 124, 117$ and $110°$ and Ga-catalyzed GaAs nanowires at $\beta_0 = 115°$, obtained from (1.51) and (1.52). The circular dot represents the elongated droplet with minimum energy at Young's contact angle $\beta_* = 110°$

In a more precise approach [592], the stationary droplet configuration in the wetting case can be obtained from either balancing the surface tensions or by minimizing the surface energy functional. We consider again a cylindrical nanowire of radius R, growing perpendicular to a substrate from a spherical cap droplet with the contact angle β, under the constraint of a fixed droplet volume (1.52) in the wetting case where the inequality given by (6.47) is satisfied at $y = 0$. As above, the top nanowire facet is assumed planar. When the droplet slides down and rests at the corners of the nanowire cylinder (see Fig. 6.54), the balance of vertical surface forces yields the Young's condition $\gamma_{WV} = \gamma_{LV} \sin\beta_* + \gamma^l_{SL}$, or $\gamma_{WV} = \gamma_{LV}\cos\alpha + \gamma^l_{SL}$, where $\alpha = \beta_* - \pi/2$ is the equilibrium contact angle on the nanowire sidewalls (see Fig. 6.54a). Pascal's law requires pressure to be constant at all points of the droplet surface. Due to a small size of the droplet, its pressure equals the Laplacian pressure $P = \gamma_{LV} K$, where the mean curvature of the droplet surface is determined through its principal radii of curvature as $2K = 1/R_1 + 1/R_2$ [38]. In view of axial symmetry of our model geometry, the droplet curvature is given by

$$2K = \nabla\left(\frac{\nabla S}{\sqrt{1 + (\nabla S)^2}}\right) = -\frac{1}{r(z)\sqrt{1 + r'(z)^2}} + \frac{r''(z)}{\sqrt{1 + r'(z)^2}^3}. \qquad (6.49)$$

Here, the z axis is directed along the nanowire axis, and $r(z)$ is the distance from the nanowire axis to the droplet surface. There are different smooth surfaces of a constant curvature such as sphere, cylinder, nodoid, catenoid and onduloid (also called an unduloid) [593]. It can be shown that, in our geometry, the condition of a constant Laplacian pressure yields the droplet consisting of two parts, a spherical cap on the nanowire top (as in Fig. 6.52) and an onduloid surrounding the nanowire cylinder. In geometry, an onduloid is a surface with a constant nonzero mean curvature obtained as a surface of revolution of elliptic catenary: that is, by rolling an ellipse along a fixed line, tracing the focus and revolving the resulting curve around the line. Mathematical representation of $r(z)$ for an onduloid will be given below.

On the other hand, the droplet shape should relate to the minimum surface energy of the system at a constant volume. The surface energy functional can be written down as

$$G\left(r(z), y, h\right) = \gamma_{LV} \int_{-y}^{h} 2\pi r\left(z\right) \sqrt{1 + r'\left(z\right)^2} dz - \gamma_{WV}\left(2\pi Ry + \pi R^2\right) + \gamma_{SL}^{l}\left(2\pi Ry + \pi R^2\right), \quad (6.50)$$

where y is the height of the wetted part of the nanowire and h is the droplet height above the top nanowire facet. The optimal droplet shape corresponds to the minimum of this G under the condition of a constant droplet volume V given by (1.52). To find the minimum, we use the Lagrange multiplier method. Variation with respect to the lower droplet boundary y gives exactly the Young's equilibrium condition:

$$\frac{\delta\left(G + \lambda V\right)}{\delta y} = -2\pi R\left[\gamma_{LV}\frac{1}{\sqrt{1 + r'\left(-y\right)^2}} + \gamma_{SL}^{l} - \gamma_{WV}\right] = 0. \quad (6.51)$$

Variation with respect to the droplet surface $r(z)$ results in the condition of a constant pressure along the surface, while the corresponding Lagrange multiplier equals exactly the Laplacian pressure in the droplet:

$$\frac{\delta\left(G + \lambda V\right)}{\delta r} = -2\pi\gamma_{LV}\left[\frac{1}{\sqrt{1 + r'(z)^2}} - \frac{r''(z)r(z)}{\sqrt{1 + r'(z)^2}^3}\right] + \lambda 2\pi r(z) = 0 \quad (6.52)$$

In the general case, the solution to this equation can be expressed through the elliptic integrals. For our geometry, the upper part of the droplet will have the shape of a spherical cap, while the side surface will have the onduloid shape described about the nanowire axis. The latter is given by the following parametric representation for $r(z)$ [593]:

$$\begin{aligned} z\left(u\right) &= aF\left(\frac{\mu u}{2} - \frac{\pi}{4}, k\right) + cE\left(\frac{\mu u}{2} - \frac{\pi}{4}, k\right) - aF\left(\frac{\mu u_0}{2} - \frac{\pi}{4}, k\right) \\ &\quad + cE\left(\frac{\mu u_0}{2} - \frac{\pi}{4}, k\right) \\ r\left(u\right) &= \sqrt{m \sin \mu u + n} \\ \mu &= \frac{a + c}{2}; \quad k^2 = \frac{c^2 - a^2}{c^2}; \quad m = \frac{c^2 - a^2}{2}; \quad n = \frac{c^2 + a^2}{2}. \end{aligned} \quad (6.53)$$

Here, $F(X, k)$ and $E(X, k)$ are the elliptic integrals of the first and the second type, respectively [101], a is the maximum distance from the nanowire axis to the droplet surface, while the parameters c and u_0 are defined by the Young's angle and the

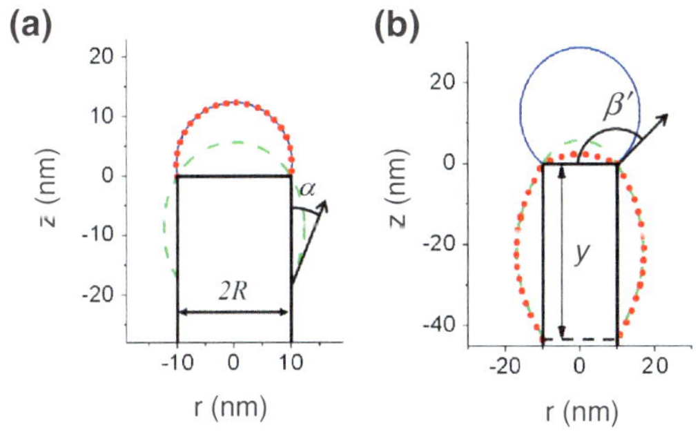

Fig. 6.54 Droplet shapes in different VLS configurations: non-wetting mode with the droplet on top (*solid lines*), the smoothly linked wetting droplet (*dashed lines*), and the droplet with the minimum surface energy (*dotted lines*). The calculations were performed with the following model parameters: $\gamma_{SL}^{l} = 0.7$ J/m^2, $\gamma_{LV} = 0.7$ J/m^2, $\gamma_{WV} = 1.3$ J/m^2, $R = 10$ nm, for two different droplet volumes: $V = 0.7(4\pi/3)R^3$ (**a**) and $V = 4(4\pi/3)R^3$ (**b**)

nanowire radius as follows:

$$z'(u_0) = \cos\alpha; \quad r(u_0) = R. \tag{6.54}$$

The only free parameter remaining in the obtained solutions for the droplet shape in the wetting VLS mode is the contact angle β' of the spherical cap on the wire top. Since the droplet should have a convex surface, the angle β' must be smaller than or equal $\pi/2 - \alpha$. The onduloid and the spherical cap surfaces will be linked smoothly at $\beta' = \pi/2 - \alpha$. Direct calculations show that the minimum surface energy of the entire droplet is reached when the Laplacian pressures at the spherical cap surface and at the side onduloid surface equal each other, consistent with the force balance considerations.

Figure 6.54 shows the calculated droplet shapes in the standard and wetting VLS configurations, for the parameters given in the insert and two different droplet volumes V. Figure 6.55 presents the comparison of the corresponding surface energies. It is seen that the energetically preferred droplet configurations depend drastically on V. The standard non-wetting VLS configuration with the droplet seated on the top is expected for small enough V, while for larger V the system tends to acquire the wetting configuration where the droplet surrounds the nanowire top. Since the wetting configuration always favors the ZB structure and the non-wetting droplets are expected to produce WZ islands, this well explains the ZB-to-WZ phase transition at the end of growth as the droplet size shrinks (see Fig. 6.49).

The described wetting VLS configurations have been observed experimentally in Ga-catalyzed GaAs nanowires [48, 594, 595]. In particular, Fig. 6.56 from [595] by García Núñez *et al.* presents very important results, demonstrating both the direct experimental evidence of the wetted growth front and the suppression of ZB-WZ stacking defects beneath the droplet in the wetting VLS configuration. These GaAs nanowires were grown by Ga-assisted CBE on Si(111) substrates covered by a thin

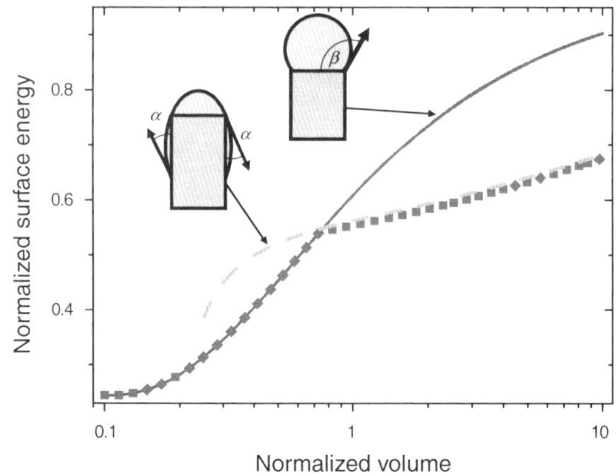

Fig. 6.55 Dependences of the normalized surface energy $G/[(4\pi)^{1/3}(3V)^{2/3}\gamma_{LV}]$ on the normalized droplet volume $3V/(4\pi R^3)$, in different VLS configurations: non-wetting mode with the droplet on top (*solid line*), the smoothly linked wetting droplet (*dashed line*), and the droplet with the minimum surface energy (*dotted line*)

oxide layer using different substrate temperatures and growth times. Oliveira *et al.* [596] observed that thin InP nanowires generated via the VLS mechanism can display periodic oscillations of the droplet shapes between the non-wetting and wetting configurations, associated with the nanowire diameter oscillations.

We would finally like to point out that our model for the elongated droplet shape applies only to the case of the planar growth interface. Obviously, equilibrium shapes do not account for any kinetic effects. In Sect. 4.6, we discussed the experimental evidence and the model of the non-planar (and non-stationary) growth interfaces of different VLS nanowires including the Au-catalyzed GaP nanowires, where a stable truncated facet is present [360]. However, regarding the avoidance of polytypism, the particular geometry of the wire top is not crucial, as schematized in Fig. 6.57. In both cases shown in the figure, one with a planar top facet and the droplet surrounding the nanowire top, and the other with a truncated top and the droplet surrounding only the truncated part of the wire, the triple phase line nucleation cannot occur and therefore the WZ phase formation is not expected. This might explain why all the nanowires studied in [360, 391], including GaP, had the ZB crystal structure.

6.12 Wurtzite Structure of III–V Nanoneedles

Self-induced GaAs and other III-V nanoneedles and nanopillars considered in Sects. 3.14, 3.15 and 5.3, have the defect-free, single crystalline WZ structure shown in Figs. 3.53 and 3.55–3.57. As discussed in Sects. 3.4 and 3.15, these nanoneedles emerge due to strain induced by the lattice mismatch. They start as tiny Volmer-

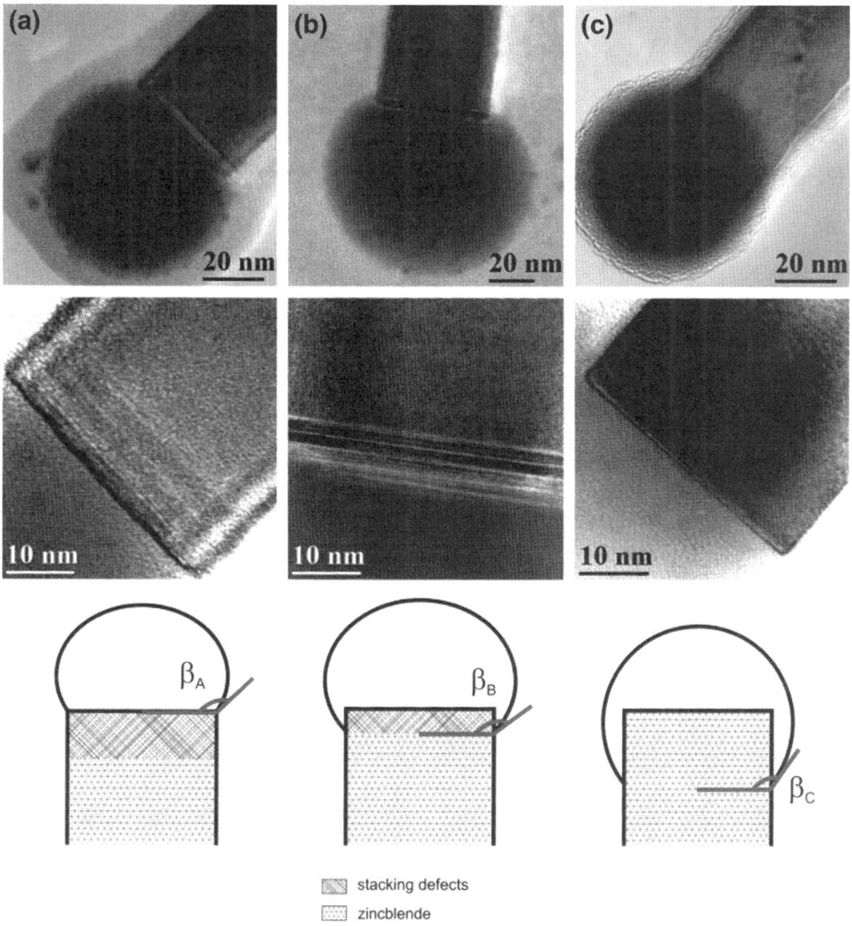

Fig. 6.56 TEM images of the Ga-catalyzed GaAs nanowires grown at $T = 600\,^\circ\mathrm{C}$ [(**a**) and (**c**)] and $T = 580\,^\circ\mathrm{C}$ (**b**). The sketch shows how the wetting droplet shape suppresses stacking disorder [595]

Weber islands with a high aspect ratio of the order of ten and a sharp tip, and then continue growing in a core-shell mode with a fixed shape. Their WZ crystal structure is maintained up to micrometer sizes, which is difficult to achieve in the VLS-grown nanowires. It is clear that the formation of WZ structure in self-induced nanostructures should have the same physical origin as in III-V nanowires, that is, a lower surface energy in WZ phase relative to ZB. However, since the growth procedure does not involve any catalyst and the nanoneedle geometry is tapered, the mechanism of WZ phase formation should be different from the one considered above.

The HRTEM image shown in Fig. 3.52 clearly demonstrates that the sidewalls of the GaAs nanoneedles are composed of atomic steps, which results in tapering.

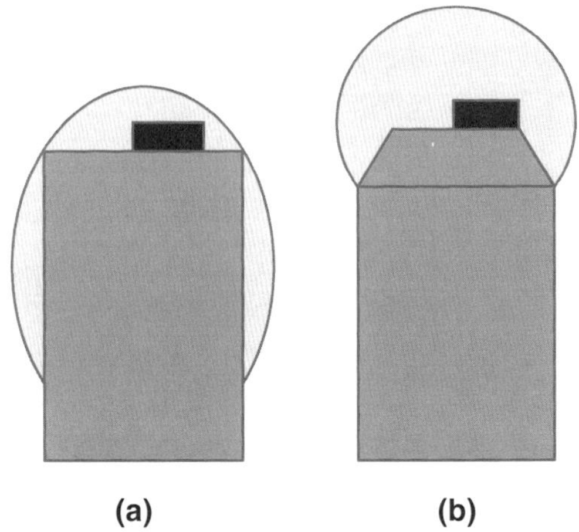

Fig. 6.57 **a** Wetting VLS configuration with a planar growth interface and a non-spherical elongated droplet; **b** truncated growth interface where the catalyst droplet wets only the truncated top. These configurations have a very similar effect on the crystal structure: in both cases, the emerging 2D islands are surrounded by the liquid phase and thus the WZ phase is not expected

The tapered sidewall consists of $(1\bar{1}00)$ vertical planes separated by (0001) terraces [186]. From the results given in Table 6.6, we already know that the gain in surface energy $\tau = \gamma_{WZ}/\gamma_{ZB}$, originating from the lower number of gangling bonds on $(1\bar{1}00)$ WZ sidewalls, is 0.843 and 0.726 for their respective (110) and (211) ZB counterparts. We therefore use this condition for the surface energy of the vertical planes γ_F in (3.65): $\gamma_F^{WZ} = \tau\gamma_F^{ZB}$ with $\tau < 1$. The analysis of Sect. 3.4 shows that, for tall enough nanoneedles with the aspect ratio η of the order of ten, the nucleation barrier $\Delta G_*^{3D}(\eta)$ reaches its minimum in η at $\eta_* \cong 3B/A$. Here, the coefficient B is defined as $B = w_{2D}/\Delta\mu_0$, $\Delta\mu_0$ is the vapor chemical potential with respect to the reference state in ZB solid phase, and the elastic energy relaxation $z(\eta) = w(\eta)/w_{2D}$ equals $1/(1 + A\eta) \cong 1/A\eta$. The minimum nucleation barrier, corresponding to the energetically preferred aspect ratio, is now reduced to

$$\Delta G_{\min} = \Delta G_0 \frac{1}{A\chi^2}\frac{27B}{4} = \frac{const}{A}\left(\frac{\gamma_F}{\Delta\mu_0}\right)^3 w_{2D}, \tag{6.55}$$

with a phase independent constant.

The chemical potential difference for a WZ nanoneedle decreases to$\Delta\mu_0 - \psi$, with ψ as the difference in bulk cohesive energies of WZ and ZB GaAs. The condition for the prevalence of WZ phase obviously writes as $\Delta G_{\min}^{WZ} < \Delta G_{\min}^{ZB}$. From (6.55), the latter condition is equivalent to the inequality [123]

$$\Delta\mu_0 > \Delta\mu_c = \psi/(1 - \omega\tau), \tag{6.56}$$

where $\omega = [(A_{ZB}/A_{WZ})(w_{2D,WZ}/w_{2D,ZB})]^{1/3}$. As in the case of VLS nanowires considered in Sect. 6.5, III-V nanoneedles tend to adopt WZ phase at large enough chemical potential above a certain threshold value $\Delta\mu_c$. However, the form of condition for $\Delta\mu_0$ is different from the VLS case. Taking typical values of τ between

0.726 and 0.843 and $\psi = 24\,\text{meV}$ per GaAs pair at $\omega \cong 1$, we arrive at the estimate for $\Delta\mu_c$ between 88 and 153 meV per pair. This critical chemical potential is thus even less than in the VLS case [typically few hundreds of meV obtained from (6.18)].

Very importantly, the value of $\Delta\mu_0$ in the case of self-induced nanoneedles relates to a vapor phase, because the nanoneedles are fed either directly from vapor or from the substrate surface where the chemical potential should be close to that in vapor. We saw earlier that, in the VLS case, the $\Delta\mu$ in (6.18) relates to the liquid phase, where the chemical potential is generally much lower than in vapor.

It is therefore much easier to exceed the threshold $\Delta\mu_c$ in vapor, which is why the MOCVD-grown III-V nanoneedles should nucleate in WZ phase even more systematically than the VLS nanowires. Once nucleated, nothing would prevent them to continue in WZ phase at the follow-up growth stages. Same considerations may apply to other self-induced nanostructures in supersaturated surroundings, provided that their surface energy is lower in WZ phase.

References

1. R. Brout, *Phase Transitions* (University of Brussels, New York, 1965)
2. M.E. Fisher, *The Nature of Critical Points* (University of Colorado Press, Boulder, 1965)
3. S.A. Kukushkin, A.V. Osipov, Phys. Usp. **168**, 1083 (1998)
4. A.Z. Patashinsky, V.L. Pokrovsky, *Fluctuation Theory of Phase Transitions* (Pergamon Press, Oxford, 1979)
5. L.P. Kadanoff, Physics **2**, 263 (1966)
6. K.G. Wilson, Phys. Rev. B **4**, 3174 (1971)
7. D. Kashchiev, *Nucleation: Basic Theory with Applications* (Butterworth-Heinemann, Oxford, 2000)
8. J.W.P. Schmelzer, *Nucleation: Theory and Applications* (Wiley, New York, 2005)
9. F.F. Abraham, *Homogeneous Nucleation Theory* (Academic Press, New York, 1974)
10. F.M. Kuni, A.K. Shchekin, A.P. Grinin, Phys. Usp. **171**, 345 (2001)
11. N.N. Tunitskii, Zh. Fiz. Khim. **15**, 1061 (1941)
12. J. Zeldovich, Sov. Phys. JETP (Eng. Transl) **12**, 525 (1942)
13. O.M. Todes, Zh. Fiz. Khim. **20**, 629 (1946)
14. V.A. Shneidman, J. Chem. Phys. **115**, 8141 (2001)
15. V.A. Shneidman, Phys. Rev. Lett. **101**, 205702 (2008)
16. S.A. Kukushkin, A.V. Osipov, J. Chem. Phys. **107**, 3247 (1997)
17. F.M. Kuni, Colloid J. **46**, 791 (1984)
18. V.V. Slezov, YaJ Tkatch, J. Schmelzer, J. Mater. Sci. **32**, 3739 (1997)
19. V.V. Slezov, J. Schmelzer, Phys. Rev. E **65**, 031506 (2002)
20. D.B. Dadyburjor, E. Ruckenstein, J. Cryst. Growth **40**, 279 (1977)
21. V. Holten, M.E.H. van Dongen, J. Chem. Phys. **130**, 014102 (2009)
22. V.K. Shen, P.G. Debenedetti, J. Chem. Phys. **118**, 768 (2003)
23. D.T. Wu, J. Chem. Phys. **97**, 1922 (1992)
24. A.V. Osipov, S.A. Kukushkin, F. Schmitt, P. Hess, Phys. Rev. B **64**, 205421 (2001)
25. A.V. Osipov, F. Schmitt, S.A. Kukushkin, P. Hess, Appl. Surf. Sci. **188**, 156 (2002)
26. V.G. Dubrovskii, G.E. Cirlin, V.M. Ustinov, Phys. Rev. B **68**, 075409 (2003)
27. T. Lummen, T. Kraska, Aerosol Sci. **36**, 1409 (2005)
28. D. Kashchiev, J. Chem. Phys. **129**, 164701 (2008)
29. J. Wolk, R. Strey, C.H. Heath, B.E. Wyslouzi, J. Chem. Phys. **117**, 4954 (2002)
30. K. Binder, Phys. Rev. B **15**, 4425 (1977)
31. U. Gasser, E.R. Weeks, A. Schonfield, P.N. Pusey, D.A. Weitz, Science **292**, 258 (2001)
32. V.G. Dubrovskii, D.A. Bauman, V.V. Kozachek, V.V. Mareev, G.E. Cirlin, Phys. A **260**, 349 (1998)

V. G. Dubrovskii, *Nucleation Theory and Growth of Nanostructures*,
NanoScience and Technology, DOI: 10.1007/978-3-642-39660-1,

33. J.W. Cahn, Acta Metal. **9**, 795 (1961)
34. J.S. Langer, Ann. Phys. **65**, 53 (1971)
35. P. Scott, Phys. Rev. A **42**, 7447 (1990)
36. C. Unger, W. Klein, Phys. Rev. B **29**, 2698 (1984)
37. K. Binder, D. Stauffer, Adv. Phys. **25**, 343 (1976)
38. A.V. Osipov, J. Phys. D: Appl. Phys. **28**, 1670 (1995)
39. L.G. Landau, E.M. Lifshitz, *Statistical Physics (Vol. 5 of Course of Theoretical Physics)* (Pergamon, Oxford, 1980)
40. J.J. Jasper, J. Phys. Chem. Ref. Data **I**(4), 841 (1972)
41. D. Bimberg, M. Grundmann, N.N. Ledentsov, *Quantum Dot Heterostructures* (Wiley, New York, 1999)
42. V.M. Ustinov, A.E. Zhukov, A.Yu. Egorov, N.A. Maleev, *Quantum Dot Lasers* (Oxford University Press, New York, 2003)
43. V.A. Shchukin, N.N. Ledentsov, D. Bimberg, *Epitaxy of Nanostructures* (Springer, New York, 2003)
44. V.A. Nebol'sin, A.A. Shchetinin, Inorg. Mater. **39**, 899 (2003)
45. F.M. Ross, J. Tersoff, M.C. Reuter, Phys. Rev. Lett. **95**, 146104 (2005)
46. F. Glas, J.C. Harmand, J. Patriarche, Phys. Rev. Lett. **99**, 146101 (2007)
47. R.S. Wagner, W.C. Ellis, Appl. Phys. Lett. **4**, 89 (1964)
48. E.I. Givargizov, *Highly Anisotropic Crystals* (Springer, Berlin, 1987)
49. V.G. Dubrovskii, G.E. Cirlin, N.V. Sibirev, F. Jabeen, J.C. Harmand, P. Werner, Nano Lett. **11**, 1247 (2011)
50. G.E. Cirlin, V.G. Dubrovskii, Yu.B. Samsonenko, A.D. Bouravleuv, K. Durose, Y.Y. Proskuryakov, B. Mendis, L. Bowen, M.A. Kaliteevski, R.A. Abram, D. Zeze, Phys. Rev. B **82**, 035302 (2010)
51. V.G. Dubrovskii, J. Chem. Phys. **131**, 164514 (2009)
52. V.G. Dubrovskii, M.V. Nazarenko, J. Chem. Phys. **132**, 114507 (2010)
53. V.G. Dubrovskii, M.V. Nazarenko, J. Chem. Phys. **132**, 114508 (2010)
54. V.G. Dubrovskii, J. Phys.: Condens. Matter **16**, 6929 (2004)
55. Z. Kozisek, P. Demo, Aerosol Sci. **40**, 44 (2009)
56. J. Lothe, M. Pound, J. Chem. Phys. **36**, 2080 (1962)
57. K. Reichelt, Vacuum **38**, 1083 (1988)
58. I. Markov, *Crystal Growth for Beginners*, 2nd edn. (World Scientific, New Jersey, 2003)
59. E.A. Flood (ed.), *The Solid-Gas Interface* (Marcel Dekker, New York, 1967)
60. D. Kashchiev, Surf. Sci. **220**, 428 (1989)
61. D. Kashchiev, Surf. Sci. **225**(1), 107 (1990)
62. M.J. Ludowise, J. Appl. Phys. **58**, R31 (1985)
63. C.W. Tu, J.S. Harris (eds.), *Molecular Beam Epitaxy* (North Holland, Amsterdam, 1991)
64. E.H.C. Parker, *The Technology and Physics of Molecular Beam Epitaxy* (Plenum Press, New-York, 1985)
65. W.K. Burton, N. Cabrera, F.C. Frank, Philos. Trans. R. Soc. London **243**, 299 (1951)
66. V.G. Dubrovskii, Phys. Stat. Sol. (b) **171**, 345 (1992)
67. S.A. Kukushkin, A.V. Osipov, Prog. Surf. Sci. **51**, 1 (1996)
68. S. Brunauer, P.H. Emmett, E. Teller, J. Am. Chem. Soc. **60**, 309 (1938)
69. H.J. Kreuzer, Phys. Rev. B **44**, 1232 (1991)
70. A. Wierzbicki, H.J. Kreuzer, Surf. Sci. **257**, 417 (1991)
71. J.-S. McEwen, *Kinetic Lattice Gas Models for Adsorption Desorption and Diffusion* (VDM Verlag, Saarbrücken, 2009)
72. S. Kotake, I.I. Glass, Prog. Aerospace Sci. **19**, 129 (1988)
73. V.G. Dubrovskii, N.V. Sibirev, I.E. Eliseev, S.Yu. Vyazmin, V.M. Boitsov, Yu.V. Natochin, M.V.Dubina, J. Chem. Phys., submitted (2013)
74. M.V. Dubina, S.Yu. Vyazmin, V.M. Boitsov, E.N. Nikolaev, I.A. Popov, A.S. Kononikhin, I.E. Eliseev, Yu.V. Natochin, Orig. Life Evol. Biosph. doi:10.1007/s11084-013-9326-5 (2013)
75. V.G. Dubrovskii, Theor. Math. Phys. **108**, 1110 (1996)

76. E.W. Montroll, K.E. Shuler, J. Chem. Phys. **26**, 454 (1957)
77. C.C. Rankin, J.C. Light, J. Chem. Phys. **46**, 1305 (1967)
78. R. Glauber, J. Math. Phys. **4**, 294 (1963)
79. F.M. Kuni, A.P. Grinin, Colloid J. **46**, 412 (1984)
80. V.A. Shneidman, Phys. Rev. A **44**, 2609 (1991)
81. V.G. Dubrovskii, N.V. Sibirev, Phys. Rev. E **70**, 031604 (2004)
82. D. Kashchiev, Cryst. Growth Des. **6**, 1154 (2006)
83. V.G. Dubrovskii, N.V. Sibirev, J.C. Harmand, F. Glas, Phys. Rev. B **78**, 235301 (2008)
84. J. Johansson, L.S. Karlsson, K.A. Dick, J. Bolinsson, B.A. Wacaser, K. Deppert, L. Samuelson, Cryst. Growth Des. **9**, 766 (2009)
85. V.G. Dubrovskii, N.V. Sibirev, J. Cryst. Growth **304**, 504 (2007)
86. V.N. Detsik, S.A. Kukushkin, A.V. Osipov, Phys. Sol. State **39**, 104 (1997)
87. I.M. Lifshitz, V.V. Slezov, Sov. Phys. JETP **8**, 331 (1959)
88. I.M. Lifshitz, V.V. Slezov, J. Phys. Chem. Solids **19**, 35 (1961)
89. A.M. Streets, S.R. Quake, Phys. Rev. Lett. **104**, 178102 (2010)
90. I.L. Maximov, M. Sanada, K. Nishioka, J. Chem. Phys. **113**, 3323 (2000)
91. B. Giron, B. Meerson, P.V. Sasorov, Phys. Rev. E **58**, 4213 (1998)
92. V.P. Morozov, I.L. Maximov, Inorg. Mat. **35**, 869 (1990)
93. B. Meerson, Phys. Rev. E **60**, 3072 (1999)
94. A.N. Kolmogorov, Bull. Acad. Sci. URSS (Cl. Sci. Math. Nat.) **3**, 355 (1937)
95. W.A. Johnson, R.F. Mehl, Trans. Am. Inst. Min. Metall. Pet. Eng. **135**, 416 (1939)
96. M. Avrami, J. Chem. Phys. **7**, 1103 (1939)
97. M. Avrami, J. Chem. Phys. **8**, 212 (1940)
98. D. Kashchiev, J. Cryst. Growth **40**, 29 (1977)
99. B. Lorentz, Cryst. Res. Technol. **22**, 869 (1987)
100. M. Fanfoni, M. Tomellini, M. Volpe, Phys. Rev. B **65**, 172301 (2002)
101. L. Gránásy, T. Börzönui, T. Pusztai, Phys. Rev. Lett. **88**, 206105 (2002)
102. M. Abramowitz, I.A. Stegun (ed.) *Handbook of mathematical functions*, *National Bureau of Standards, Applied Mathematics Series 55* (1964)
103. V.G. Dubrovskii, G.E. Cirlin, Semiconductors **39**, 1267 (2005)
104. F.M. Kuni, A.P. Grinin, Colloid J. **46**, 460 (1984)
105. M.A. Kazanskii, M.V. Nazarenko, V.G. Dubrovskii, Tech. Phys. Lett. **37**(3), 282 (2011)
106. J. Crank, P. Nicolson, Proc. Camb. Phil. Soc. **43**, 50 (1947)
107. V.G. Dubrovskii, M.A. Kazansky, M.V. Nazarenko, L.T. Adzhemyan, J. Chem. Phys. **134**, 094507 (2011)
108. R. Powell, in *Fundamental Formulas of Physics*, ed. by D.H. Menzel (Prentice-Hall Inc., New York, 1955)
109. L.V. Asryan, R.A. Suris, Semicond. Sci. Technol. **11**, 554 (1996)
110. W.B. Russel, D.A. Saville, W.R. Schowalter, *Colloidal Dispersions* (Cambridge University Press, Cambridge, 1989)
111. D.E. Wolf, J. Villian, Europhys. Lett. **13**, 389 (1990)
112. M. Siegert, M. Plischke, Phys. Rev. Lett. **68**, 2035 (1992)
113. D.A. Kessler, H. Levine, L.M. Sandler, Phys. Rev. Lett. **69**, 100 (1992)
114. P. Smilauer, M. Kotrla, Europhys. Lett. **27**, 261 (1994)
115. M. Kardar, G. Parisi, Y-Ch. Zhang, Phys. Rev. Lett. **56**, 889 (1986)
116. V.G. Dubrovskii, Phys. A **307**, 228 (2002)
117. V.I. Trofimov, H.S. Park, Appl. Surf. Sci. **219**, 93 (2003)
118. V.G. Dubrovskii, N.V. Sibirev, R.A. Suris, G.E. Cirlin, J.C. Harmand, V.M. Ustinov, Surf. Sci. **601**, 4395 (2007)
119. F. Glas, J.C. Harmand, G. Patriarche, Phys. Rev. Lett. **104**, 135501 (2010)
120. ZhI Alferov, Semiconductors **32**, 1 (1998)
121. X. Ren, Opt. Photonics J., doi:10.4236/opj.2013
122. R. Dingle, C.H. Henry, *Quantum Effects in Semiconductor Lasers*, US Patent 3982207 (1976)
123. Y. Arakawa, H. Sakaki, Appl. Phys. Lett. **40**, 939 (1982)

124. V.G. Dubrovskii, N.V. Sibirev, X. Zhang, R.A. Suris, Cryst. Growth Des. **10**, 3949 (2010)
125. J.W. Matthews, A.E. Blakeslee, J. Cryst. Growth **27**, 118 (1974)
126. S.F. Fang, K. Adomi, S. Iyer, H. Morkoc, H. Zabel, C. Choi, N. Otsuka, J. Appl. Phys. **68**, R31 (1990)
127. E.A. Fitzgerald, Mater. Sci. Rep. **7**, 87 (1991)
128. L. Goldstein, F. Glas, J.Y. Marzin, M.N. Charasse, G. Le Roux, Appl. Phys. Lett. **47**, 1099 (1985)
129. R. Apetz, L. Vescan, A. Hartman, C. Deicker, H. Luth, Appl. Phys. Lett. **66**, 445 (1995)
130. G.E. Cirlin, V.G. Dubrovskii, V.N. Petrov, N.K. Polyakov, N.P. Korneeva, V.N. Demidov, A.O. Golubok, S.A. Masalov, D.V. Kurochkin, O.M. Gorbenko, N.I. Komyak, V.M. Ustinov, A.Yu. Egorov, A.R. Kovsh, M.V. Maximov, A.F. Tsatusul'nikov, B.V. Volovik, A.E. Zhukov, P.S. Kop'ev, Zh.I. Alferov, N.N. Ledentsov, M. Grundmann, D. Bimberg, Semicond. Sci. Technol. **13**, 1262 (1998)
131. D. Leonard, K. Pond, P.M. Petroff, Phys. Rev. B **50**, 11687 (1994)
132. A.E. Zhukov, A.Yu. Egorov, A.R. Kovsh, V.M. Ustinov, N.N. Ledentsov, M.V. Maksimov, A.F. Tsatsulnikov, S.V. Zaitsev, N.Y. Gordeev, P.S. Kopev, Zh.I. Alferov, D. Bimberg, Semiconductors **31**, 411 (1997)
133. R. Leon, S. Fafard, D. Leonard, J.L. Merz, P.M. Petroff, Appl. Phys. Lett. **67**, 521 (1995)
134. A.F. Tsatsul'nikov, A.Yu. Egorov, P.S. Kop'ev, A.R. Kovsh, N.N. Ledentsov, M.V. Maximov, V.M. Ustinov, B.V. Volovik, A.E. Zhukov, M. Grundmann, D. Bimberg, ZhI Alferov, Appl. Surf. Sci. **123–124**, 381 (1998)
135. V.M. Ustinov, N.A. Maleev, A.E. Zhukov, A.R. Kovsh, A.Yu. Egorov, A.V. Lunev, B.V. Volovik, I.L. Krestnikov, Yu.G. Musikhin, N.A. Bert, P.S. Kop'ev, Zh.I. Alferov, N.N. Ledentsov, D. Bimberg, Appl. Phys. Lett. **74**, 2815 (1999)
136. N. Carlsson, W. Siefert, A. Petersson, P. Castrillo, M.E. Pistol, L. Samuelson, Appl. Phys. Lett. **65**, 3093 (1994)
137. M.K. Zundel, P. Specht, K. Eberl, N.Y. Jin-Philipp, F. Phillipp, Appl. Phys. Lett. **71**, 2972 (1997)
138. A. Ponchet, A. Le Corre, H. L'Haridon, B. Lambert, S. Salaün, Appl. Phys. Lett. **67**, 1850 (1995)
139. S. Fafard, Z. Wasilewski, J. McCarey, S. Raymond, S. Charbonneau, Appl. Phys. Lett. **68**, 991 (1996)
140. V.M. Ustinov, E.R. Weber, S. Ruvimov, Z. Liliental-Weber, A.E. Zhukov, A.Yu. Egorov, A.R. Kovsh, A.F. Tsatsul'nikov, P.S. Kop'ev, Appl. Phys. Lett. **72**, 362 (1998)
141. J.-W. Lee, A.T. Schremer, D. Fekete, J.M. Ballantyne, Appl. Phys. Lett. **69**, 4236 (1997)
142. B. Junno, T. Junno, M.S. Miller, L. Samuelson, Appl. Phys. Lett. **72**, 954 (1998)
143. V.G. Dubrovskii, Phys. Stat. Sol (b) **238**, R1 (2003)
144. V.G. Dubrovskii, G.E. Cirlin, V.M. Ustinov, Phys. Stat. Sol. (b) **241**, R42 (2004)
145. V.G. Dubrovskii, G.E. Cirlin, Yu.G. Musikhin, Yu.B. Samsonenko, A.A. Tonkikh, N.K. Polyakov, V.A. Egorov, A.F. Tsatsul'nikov, N.A. Krizhanovskaya, V.M. Ustinov, P. Werner, J. Cryst. Growth **267**, 47 (2004)
146. O.P. Pchelyakov, YuB Bolkhovityanov, A.V. Dvurechenskii, A.I. Nikiforov, A.I. Yakimov, B. Voigtländer, Thin Solid Films **367**, 75 (2000)
147. J.M. Moison, F. Houzay, F. Batthe, L. Leprince, E. Andr, O. Vatel, Appl. Phys. Lett. **64**, 196 (1994)
148. P. Müller, R. Kern, Appl. Surf. Sci. **102**, 6 (1996)
149. V.A. Shchukin, N.N. Ledentsov, P.S. Kop'ev, D. Bimberg, Phys. Rev. Lett. **75**, 2968 (1995)
150. I. Daruka, A.-L. Barabasi, Phys. Rev. Lett. **79**, 3708 (1997)
151. V.A. Shchukin, D. Bimberg, Rev. Mod. Phys. **71**, 1125 (1999)
152. J. Tersoff, R.M. Tromp, Phys. Rev. Lett. **70**, 2782 (1993)
153. J. Tersoff, F.K. LeGoues, Phys. Rev. Lett. **72**, 3570 (1994)
154. J. Tersoff, C. Teichert, M.G. Lagally, Phys. Rev. Lett. **76**, 1675 (1996)
155. G. Medeiros-Riberio, A.M. Bratkovski, T.I. Kamins, D.A.A. Ohlberg, R.S. Williams, Science **279**, 353 (1998)

156. R.E. Rudd, G.A. Briggs, A.P. Sutton, G. Medeiros-Riberio, R.S. Williams, Phys. Rev. Lett. **90**, 146101 (2002)
157. Q. Xie, A. Madhukar, P. Chen, N.P. Kobayashi, Phys. Rev. Lett. **75**, 2542 (1995)
158. R.S. Williams, G. Medeiros-Riberio, T.I. Kamins, D.A.A. Ohlberg, Annu. Rev. Phys. Chem **51**, 527 (2000)
159. Y. Chen, J. Washburn, Phys. Rev. Lett. **77**, 4046 (1996)
160. A.-L. Barabasi, Appl. Phys. Lett. **70**, 2562 (1997)
161. I. Daruka, A.-L. Barabasi, Appl. Phys. Lett. **72**, 2102 (1998)
162. D.E. Jesson, G. Chen, K.M. Chen, S.J. Pennycook, Phys. Rev. Lett. **80**, 5156 (1998)
163. M. Meixner, E. Schöll, V.A. Shchukin, D. Bimberg, Phys. Rev. Lett. **87**, 236101 (2001)
164. V.A. Shchukin, D. Bimberg, T.P. Munt, D.E. Jesson, Phys. Rev. Lett. **90**, 076102 (2003)
165. V.A. Shchukin, A.I. Borovkov, N.N. Ledentsov, D. Bimberg, Phys. Rev. B **51**, 10104 (1995)
166. V.A. Shchukin, A.I. Borovkov, N.N. Ledentsov, P.S. Kop'ev, Phys. Rev. B **51**, 17767 (1995)
167. S. Kiravittaya, Y. Nakamura, O.G. Schmidt, Phys. E **13**, 224 (2002)
168. R. Songmuang, S. Kiravittaya, M. Sawadsaringkam, S. Panyakeow, O.G. Schmidt, J. Cryst. Growth **251**, 166 (2003)
169. P.B. Joyce, T.J. Krzyzewski, G.R. Bell, T.S. Jones, S. Malik, D. Childs, R. Murray, Phys. Rev. B **62**, 10891 (2000)
170. G. Abstreiter, P. Schittenhelm, C. Engel, E. Silveria, A. Zrenner, D. Meertens, W. Jäger, Semicond. Sci. Technol. **11**, 1521 (1996)
171. V.P. Evtikhiev, V.E. Torkanov, A.K. Kryganovskii, A.M. Boiko, R.A. Suris, A.N. Titkov, J. Cryst. Growth **201–202**, 1154 (1999)
172. V.P. Evtikhiev, A.M. Boiko, I.V. Kudryashov, R.A. Suris, A.N. Titkov, V.E. Torkanov, Semicond. Sci. Technol. **17**, 545 (2002)
173. G.E. Cirlin, V.N. Petrov, A.O. Golubok, SYa. Tipissev, V.G. Dubrovskii, G.M. Guryanov, N.N. Ledentsov, D. Bimberg, Surf. Sci. **377–379**, 895 (1997)
174. F. Arciprete, E. Placidi, V. Sessi, M. Fanfoni, F. Patella, A. Balzarotti, Appl. Phys. Lett. **89**, 041904 (2006)
175. A.A. Tonkikh, V.G. Dubrovskii, G.E. Cirlin, V.A. Egorov, V.M. Ustinov, P. Werner, Phys. Stat. Sol. (b) **236**, R1 (2003)
176. O.P. Pchelyakov, Y.B. Bolkhovityanov, A.V. Dvurechenskii, L.V. Sokolov, A.I. Nikiforov, A.I. Yakimov, B. Voigtlander, Semiconductors **34**, 1229 (2000)
177. O.P. Pchelyakov, A.I. Nikiforov, B.Z. Shanetsky, S.A. Teys, A.I. Yakimov, S.I. Chikichev, J. Phys. Chem. **69**, 669 (2008)
178. A. Polimeni, A. Patane, M. Cappizi, F. Martelli, L. Nazi, G. Salvani, Phys. Rev. B **53**, R4213 (1996)
179. A. Patane, M.G. Alessi, F. Intonti, A. Polimeni, M. Capizzi, F. Martelli, L. Nasi, L. Lazzarini, G. Salviati, A. Bosacchi, S. Franchi, J. Appl. Phys. **83**, 5529 (1998)
180. U. Perinetti, N. Akopian, Yu.B. Samsonenko, A.D. Bouravleuv, G.E. Cirlin, V. Zwiller, Appl. Phys. Lett. **94**, 163114 (2009)
181. M. Kästner, B. Voigtländer, Phys. Rev. Lett. **82**, 2745 (1999)
182. T.J. Kamins, G. Medeiros-Riberio, D.A.A. Ohlberg, R.S. Williams, J. Appl. Phys. **85**, 1159 (1999)
183. V.G. Dubrovskii, V.A. Egorov, G.E. Cirlin, N.K. Polyakov, YuB Samsonenko, N.V. Kryzhanovskaya, A.F. Tsatsul'nikov, V.M. Ustinov, Semiconductors **37**, 855 (2003)
184. X. Zhang, V.G. Dubrovskii, N.V. Sibirev, X. Ren, Cryst. Growth Des. **11**, 5441 (2011)
185. V.G. Dubrovskii, G.E. Cirlin, V.M. Ustinov, Semiconductors **43**, 1539 (2009)
186. M. Moewe, L.C. Chuang, S. Crankshaw, C. Chase, C. Chang-Hasnain, Appl. Phys. Lett. **93**, 023116 (2008)
187. L.C. Chuang, M. Moewe, W.N. Kar, T.T.D. Tran, S. Crankshaw, R.G. Chen, W.S. Ko, C. Chang-Hasnain, Appl. Phys. Lett. **98**, 123101 (2011)
188. F. Glas, Phys. Rev. B **74**, 121302 (2006)
189. C. Ratsch, A. Zangwill, Surf. Sci. **293**, 123 (1993)
190. S.P.A. Gill, A.C.F. Cocks, Proc. R. Soc. A **462**, 3523 (2006)

191. F. Glas, Private communication.
192. S.P. Timoshenko, J.N. Goodier, *Theory of Elasticity*, 3rd edn. (McGraw-Hill, New York, 1970)
193. L.D. Landau, E.M. Lifshitz, *Theory of Elasticity* (Pergamon, London, 1970)
194. E. Ertekin, P.A. Greaney, D.C. Chrzan, J. Appl. Phys. **97**, 114325 (2005)
195. N.A. Bert, A.B. Freidin, A.L. Kolesnikova, I.K. Korolev, A.E. Romanov, Phys. Stat. Sol. A **207**, 2323 (2010)
196. E. Penev, Ph.D. Thesis, Technischen Universität, Berlin, 2002
197. V.A. Shchukin, D. Bimberg, T.P. Munt, D. Jesson, Phys. Rev. B **70**, 085416 (2004)
198. S.P.A. Gill, Appl. Phys. Lett. **89**, 203115 (2006)
199. A.P. Grinin, F.M. Kuni, Theor. Math. Phys. **80**, 968 (1989)
200. V.G. Dubrovskii, M.V. Nazarenko, L.C. Chuang, W.S. Ko, K.W. Ng, C. Chang-Hasnain, Appl. Phys. Lett. **98**, 153113 (2011)
201. H. Mariette, C.R. Phys. **6**, 23 (2005)
202. G. Constantini, A. Ratelli, C. Manzano, P. Acosta-Diaz, G. Katsaros, R. Songmuang, O.G. Schmidt, H.V. Kanel, K. Kern, J. Cryst. Growth **278**, 38 (2005)
203. G.E. Cirlin, N.K. Polyakov, V.N. Petrov, V.A. Egorov, Yu.B. Samsonenko, D.V. Denisov, V.M. Busov, B.V. Volovik, V.M. Ustinov, Zh.I. Alferov, N.N. Ledentsov, D. Bimberg, N.D. Zakharov, P. Werner, Czech. J. Phys. **49**, 1547 (1999)
204. J.S. Wang, C.S. Yang, P.I. Chen, C.F. Su, W.J. Chen, K.C. Chiu, W.C. Chou, Appl. Phys. A **97**, 553 (2009)
205. H. Landolt, R. Bormstein, *Numerical Data and Functional Relationships in Science and Technology, New Series*, ed. by W. Martienssen, Group III: Condensed Matter vol. 41a1a and vol. 42a2
206. M. Grundmann, O. Stier, D. Bimberg, Phys. Rev. B **52**, 11969 (1995)
207. M. Grundmann, N.N. Ledentsov, O. Stier, J. Bohrer, D. Bimberg, V.M. Ustinov, P.S. Kop'ev, Zh.I. Alferov, Phys. Rev. B **53**, R10509 (1996)
208. O. Stier, M. Grundmann, D. Bimberg, Phys. Rev. B **59**, 5688 (1999)
209. A.I. Yakimov, A.V. Dvurechenskii, YuYu. Proskuryakov, A.I. Nikiforov, O.P. Pchelyakov, S.A. Teys, A.K. Gutakovskii, Appl. Phys. Lett. **75**, 1413 (1999)
210. V.G. Dubrovskii, N.V. Sibirev, Tech. Phys. Lett. **31**(2), 161 (2005)
211. V.G. Dubrovskii, X. Zhang, Tech. Phys. Lett. **37**(9), 797 (2011)
212. B.L. Liang, Zh.M. Wang, J.H. Lee, K. Sablon, Yu.I. Mazur, G.J. Salamo, Appl. Phys. Lett. **89**, 043113 (2006)
213. A. Huggenberger, S. Heckelmann, C. Schneider, S. Höfling, S. Reitzenstein, L. Worschech, M. Kamp, A. Forchel, Appl. Phys. Lett. **98**, 131104 (2011)
214. E. Stock, T. Warming, I. Ostapenko, S. Rodt, A. Schliwa, J.A. Töfflinger, A. Lochmann, A.I. Toropov, S.A. Moshchenko, D.V. Dmitriev, V.A. Haisler, D. Bimberg, Appl. Phys. Lett. **100**, 093111 (2012)
215. F. Zhan, S.S. Huang, Z.C. Niu, H.Q. Ni, Y.H. Xiong, Z.D. Fang, H.Y. Zhou, Y. Luo, J Nanosci. Nanotechnol. **9**, 844 (2009)
216. Y.G. Musikhin, G.E. Cirlin, V.G. Dubrovskii, Yu.B. Samsonenko, A.A. Tonkikh, N.A. Bert, V.M. Ustinov, Semiconductors **39**, 820 (2005)
217. I. Mukhametzhanov, Z. Wei, R. Heitz, A. Madhukar, Appl. Phys. Lett. **75**, 85 (1999)
218. V.G. Dubrovskii, N.V. Kryzhanovskaya, V.M. Ustinov, A.A. Tonkikh, V.A. Egorov, N.K. Polyakov, Yu.B. Samsonenko, G.E. Cirlin, Tech. Phys. Lett. **30**(4), 272 (2004)
219. V.G. Dubrovskii, G.E. Cirlin, P.A. Brunkov, U. Perimetti, N. Akopyan, Semiconductors, submitted (2013)
220. G.E. Cirlin, V.G. Dubrovskii, A.A. Tonkikh, N.V. Sibirev, V.M. Ustinov, P. Werner, Semiconductors **39**, 547 (2005)
221. A.A. Tonkikh, G.E. Cirlin, V.G. Dubrovskii, V.M. Ustinov, P. Werner, Semiconductors **38**, 1202 (2004)
222. K.W. Ng, W.S. Ko, T.T.D. Tran, R. Chen, M.V. Nazarenko, F. Lu, V.G. Dubrovskii, M. Kamp, A. Forchel, C.J. Chang-Hasnain, ACS Nano **7**, 100 (2013)

223. M. Moewe, L.C. Chuang, S. Crankshaw, K.W. Ng, C.J. Chang-Hasnain, Opt. Express **17**, 7831 (2009)
224. L.C. Chuang, F.G. Sedgwick, R. Chen, W.S. Ko, M. Moewe, K.W. Ng, T.T.D. Tran, C.J. Chang-Hasnain, Nano Lett. **11**, 385 (2011)
225. R. Chen, T.T.D. Tran, K.W. Ng, W.S. Ko, L.C. Chuang, F.G. Sedgwick, C.J. Chang-Hasnain, Nat. Photonics **5**, 170 (2011)
226. C.-Y. Yeh, Z.W. Lu, S. Froyen, A. Zunger, Phys. Rev. B **46**, 10086 (1992)
227. M.I. McMahon, R.J. Nelmes, Phys. Rev. Lett. **95**, 215505 (2005)
228. M. Koguchi, H. Kakibayashi, M. Yasawa, K. Hiruma, T. Katsuyama, Jpn. J. Appl Phys. **31**, 2061 (1992)
229. A.I. Persson, M.W. Larsson, S. Stenstrom, B.J. Ohlsson, L. Samuelson, L.R. Wallenberg, Nat. Mater. **3**, 678 (2004)
230. I.P. Soshnikov, G.E. Cirlin, A.A. Tonkikh, Y.B. Samsonenko, V.G. Dubrovskii, V.M. Ustinov, O.M. Gorbenko, D. Litvinov, D. Gerthsen, Phys. Sol. State **47**, 2213 (2005)
231. J.C. Harmand, G. Patriarche, N. Péré-Laperne, M.-N. Mérat-Combes, L. Travers, F. Glas, Appl. Phys. Lett. **87**, 203101 (2005)
232. L.C. Chuang, M. Moewe, S. Crankshaw, C. Chase, N.P. Kobayashi, C. Chang-Hasnain, Appl. Phys. Lett. **90**, 043115 (2007)
233. P. Mohan, J. Motohisa, T. Fukui, Nanotechnology **16**, 2903 (2005)
234. K.A. Dick, P. Caroff, J. Bolinsson, M.E. Messing, J. Johansson, K. Deppert, L.R. Wallenberg, L. Samuelson, Semicond. Sci. Technol. **25**, 024009 (2010)
235. G.E. Cirlin, V.G. Dubrovskii, I.P. Soshnikov, N.V. Sibirev, YuB Samsonenko, A.D. Bouravleuv, J.C. Harmand, F. Glas, Phys. Stat. Sol. RRL **3**, 112 (2009)
236. K. Hiramatsu, K. Nishiyama, A. Motogaito, H. Miyake, Y. Iyechika, T. Maeda, Phys. Stat. Sol. (a) **176**, 535 (1999)
237. J. Zou, D.J.H. Cockayne, B.F. Usher, J. Appl. Phys. **73**, 619 (1993)
238. M.T. Hill, Y.S. Oei, B. Smalbrugge, Y. Zhu, T. Vries, P.J. Veldhoven, F.W. Otten, T.J. Eijkemans, J.P. Turkiewicz, H. Waardt, E.J. Geluk, S.H. Kwon, Y.H. Lee, R.; Nötzel, M.K. Smit, Nat. Photon. **1**, 589 (2007)
239. C.Y. Lu, S.W. Chang, S.L. Chuang, T.D. Germann, D. Bimberg, App. Phys. Lett. **96**, 251101 (2010)
240. J.Y. Lee, M. Khajavikhan, A. Simic, Q. Gu, O. Bondarenko, B. Slutsky, M.P. Nezhad, Y. Fainman, Opt. Express **19**, 21524 (2011)
241. J. Venables, G.D.T. Spiller, M. Hanbucken, Rep. Prog. Phys. **47**, 399 (1984)
242. J. Robins, Appl. Surf. Sci. **33–34**, 379 (1988)
243. J. Zhou, Y.C. Kang, D.A. Chen, Surf. Sci. **537**, L429 (2003)
244. K. Høјrup-Hansen, S. Ferrero, C.R. Henry, Appl. Surf. Sci. **226**, 167 (2004)
245. C. Mocuta, A. Barbier, G. Renaud, B. Dieny, Thin Solid Films **336**, 160 (1988)
246. F. Silly, M.R. Castell, Appl. Phys. Lett. **87**, 053106 (2005)
247. K.R. Heim, S.T. Coyle, G.G. Hembree, J. Venables, M.R. Scheinfein, J. Appl. Phys. **80**, 1161 (1996)
248. N. Mattoso, D.H. Mosca, W.H. Schreiner, I. Mazzaro, S.R. Teixeira, Thin Solid Films **272**, 83 (1996)
249. N. Yakovlev, A. Kaveev, N. Sokolov, B. Krichevtsov, A. Huan, Curr. Appl. Phys. **6**, 575 (2006)
250. Y. Shiratsuchi, M. Yamamoto, S. Bader, Prog. Surf. Sci. **82**, 121 (2007)
251. A.D. Gates, J.L. Robins, Surf. Sci. **116**, (1982)
252. B.W. Lee, R. Alsenz, A. Ignatiev, M.A. Van Hove, Phys. Rev. B. **17**, 1510 (1978)
253. P.M. Petroff, A.C. Gossard, W. Wiegmann, Appl. Phys. Lett. **45**, 620 (1984)
254. R. Bhat, E. Kapon, S. Simhony, E. Colas, D.M. Hwang, N.G. Stoffel, M.A. Koza, J. Cryst. Growth **107**, 716 (1991)
255. K. Hiruma, M. Yazawa, T. Katsuyama, K. Ogawa, K. Haraguchi, M. Koguchi, J. Appl. Phys. **77**, 447 (1995)

256. E.I. Givargizov, A.A. Chernov, Kristallografiya **18**, 147 (1973)
257. E.I. Givargizov, J. Cryst. Growth **31**, 20 (1975)
258. Y. Cui, C.M. Lieber, Science **91**, 851 (2001)
259. G. Zheng, W. Lu, S. Jin, C.M. Lieber, Adv. Mater. **16**, 1890 (2004)
260. A.B. Greytak, L.J. Lauhon, M.S. Gudiksen, C.M. Lieber, Appl. Phys. Lett. **84**, 4176 (2004)
261. Y. Huang, C.M. Lieber, Pure Appl. Chem. **76**, 2051 (2004)
262. E. Patolsky, G. Zheng, O. Hayden, M. Lakadamyali, X. Zhuang, C.M. Lieber, Proc. Nat. Acad. Sci. USA **101**, 14017 (2004)
263. R.S. Friedman, M.C. McAlpine, D.S. Ricketts, D. Ham, C.M. Lieber, Nature **434**, 1085 (2005)
264. Y. Li, J. Xiang, F. Qian, S. Gradecak, Y. Wu, H. Yan, H. Yan, D.A. Blom, C.M. Lieber, Nano Lett. **6**, 1468 (2006)
265. S. Gradecak, F. Qian, Y. Li, H.G. Park, C.M. Lieber, Appl. Phys. Lett. **87**, 173111 (2005)
266. O. Hayden, G.F. Zheng, P. Agarwal, C.M. Lieber, Small **3**, 2048 (2007)
267. J. Xiang, W. Lu, Y.J. Hu, Y. Wu, H. Yan, C.M. Lieber, Nature **441**, 489 (2006)
268. G.C. Liang, J. Xiang, N. Kharche, G. Klimeck, C.M. Lieber, M. Lundstrom, Nano Lett. **7**, 642 (2007)
269. W. Lu, C.M. Lieber, Nat. Mater. **6**, 841 (2007)
270. B.J. Ohlsson, M.T. Björk, M.H. Magnusson, K. Deppert, L. Samuelson, Appl. Phys. Lett. **79**, 3335 (2001)
271. M.T. Bjork, B.J. Ohlsson, T. Sass, A.I. Persson, C. Thelander, M.H. Magnusson, K. Deppert, L.R. Wallenberg, L. Samuelson, Appl. Phys. Lett. **80**, 1058 (2002)
272. M.T. Bjork, B.J. Ohlsson, C. Thelander, A.I. Persson, K. Deppert, L.R. Wallenberg, L. Samuelson, Appl. Phys. Lett. **81**, 4458 (2002)
273. B.J. Ohlsson, M.T. Bjork, A.I. Persson, C. Thelander, L.R. Wallenberg, M.H. Magnusson, K. Deppert, L. Samuelson, Phys. E **13**, 1126 (2002)
274. K.A. Dick, K. Deppert, M.W. Larsson, T. Martensson, W. Seifert, L.R. Wallenberg, L. Samuelson, Nat. Mater. **3**, 380 (2004)
275. J. Johansson, L.S. Karlsson, C.P.T. Svensson, T. Martensson, B.A. Wacaser, K. Deppert, L. Samuelson, W. Seifert, Nat. Mater. **5**, 574 (2006)
276. T. Bryllert, L.E. Wernersson, T. Lowgren, L. Samuelson, Nanotechnology **17**, S227 (2006)
277. T. Martensson, P. Carlberg, M. Borgstrom, L. Montelius, W. Seifert, L. Samuelson, Nano Lett. **4**, 699 (2004)
278. C. Thelander, H.A. Nilsson, L.E. Jensen, L. Samuelson, Nano Lett. **5**, 635 (2005)
279. D. Li, Y. Wu, P. Kim, P. Yang, A. Majumdar, Appl. Phys. Lett. **83**, 3186 (2003)
280. M. Law, L.E. Greene, A. Radenovic, T. Kuykendall, J. Liphardt, P. Yang, J. Phys. Chem. B **110**, 22652 (2006)
281. P.J. Pauzauskie, P. Yang, Mater. Today **9**, 36 (2006)
282. R. Fan, P. Yang, Nano. Lett. **6**, 973 (2006)
283. P. Pauzauskie, D. Sirbuly, P. Yang, Phys. Rev. Lett. **96**, 143903 (2006)
284. Y. Nakayama, P.J. Pauzauskie, A. Radenovic, R.M. Onorato, R.J. Saykally, J. Liphardt, P. Yang, Nature **447**, 1098 (2007)
285. P. Yang, R. Yan, M. Fardy, Nano Lett. **10**, 1529 (2010)
286. L. Schubert, P. Werner, N.D. Zakharov, G. Gerth, F.M. Kolb, L. Long, U. Gösele, T.Y. Tan, Appl. Phys. Lett. **84**, 4968 (2004)
287. B. Fuhrmann, H.S. Leipner, H.-R. Hoche, L. Shubert, P. Werner, U. Gösele, Nano Lett. **5**, 2524 (2005)
288. J.B. Baxter, A.M. Walker, K. van Ommering, E.S. Aydil, Nanotechnology **17**, S304 (2006)
289. J.C. Harmand, F. Glas, G. Patriarche, Phys. Rev. B **81**, 235436 (2010)
290. V.G. Dubrovskii, G.E. Cirlin, I.P. Soshnikov, A.A. Tonkikh, N.V. Sibirev, Yu.B. Samsonenko, V.M. Ustinov, Phys. Rev. B **71**, 205325 (2005)
291. M. Tchernycheva, L. Travers, G. Patriarche, J.C. Harmand, G.E. Cirlin, V.G. Dubrovskii, J. Appl. Phys. **102**, 094313 (2007)
292. F. Jabeen, V. Grillo, S. Rubini, F. Martelli, Nanotechnology **19**, 275711 (2008)

293. C. Colombo, D. Spirkoska, M. Frimmer, G. Abstreiter, A. Fontcuberta i Morral, Phys. Rev. B **77**, 155326 (2008)
294. A. Fontcuberta i Morral, J. Arbiol, J. D. Prades, A. Cirera, J.R. Morante, Adv. Mater. **19**, 1347 (2007)
295. M.C. Plante, R.R. LaPierre, J. Cryst. Growth **286**, 394 (2006)
296. H. Shtrikman, R. Popovitz-Biro, A. Kretinin, M. Heiblum, Nano Lett. **9**, 215 (2009)
297. T. Xu, J.-P. Nys, B. Grandidier, D. Stievenard, Y. Coffinier, R. Boukherroub, R. Larde, E. Cadel, P. Pareige, J. Vac. Sci. Technol. B **26**, 1960 (2008)
298. H. Huang, X. Ren, X. Ye, J. Guo, Q. Wang, Y. Yang, S. Cai, Y. Huang, Nano Lett. **10**, 64 (2010)
299. H.J. Joyce, J. Wong-Leung, O. Gao, H. Hoe Tan, C. Jagadish, Nano Lett. **10**, 908 (2010)
300. S.K. Chan, Y. Cai, N. Wang, I.K. Sou, J. Cryst. Growth **301–302**, 866 (2007)
301. M.N. Jung, J.E. Koo, S.J. Oh, B.W. Lee, W.J. Lee, S.H. Ha, Y.R. Cho, J.H. Chang, Appl. Phys. Lett. **94**, 041906 (2009)
302. Z.M. Wang (ed.), *One Dimensional Nanostructures, Lecture Notes in Nanoscale Science and Technology 3* (Springer, New York, 2008)
303. K.A. Dick, Prog. Cryst. Growth Charact. Mater. **54**, 138–173 (2008)
304. F.M. Ross, Rep. Prog. Phys. **73**, 114501 (2010)
305. T.B. Massalski (ed.) *Binary Alloy Phase Diagrams*, 2nd edn, vol. 1 (ASM International, Metals Park, 1990), p. 369
306. W. Dittmar, K. Neumann, in *Growth and Perfection of Crystals*, eds. by R.H. Doremus, B.W. Roberts, D. Turnball (Wiley, New York, 1958), p. 121
307. J.M. Blakely, K.A. Jackson, J. Chem. Phys. **37**, 428 (1962)
308. V. Ruth, J.R. Hirth, J. Chem. Phys. **41**, 31 (1964)
309. A.A. Tonkikh, G.E. Cirlin, Yu.B. Samsonenko, I.P. Soshnikov, V.M. Ustinov, Semiconductors **38**, 1217 (2004)
310. X. Ren, H. Huang, V.G. Dubrovskii, N.V. Sibirev, M.V. Nazarenko, A.D. Bolshakov, X. Ye, Q. Wang, Y. Huang, X. Zhang, J. Guo, X. Liu, Semicond. Sci. Technol. **26**, 014034 (2011)
311. V. Consonni, V.G. Dubrovskii, A. Trampert, L. Geelhaar, H. Riechert, Phys. Rev. B **85**, 155313 (2012)
312. J. Arbiol, A. Fontcuberta i Morral, S. Estradé, F. Peiró, B. Kalache, P.R.I Cabarrocas, J.R. Morante, J. Appl. Phys. **104**, 064312 (2008)
313. R.E. Algra, M.A. Verheijen, M.T. Borgstrom, L. Feiner, G. Immink, W.J.P. van Enckevort, E. Vlieg, E.P.A.M. Bakkers, Nature **456**, 369 (2008)
314. J. Johansson, K.A. Dick, P. Caroff, M.E. Messing, J. Bolinsson, K. Deppert, L. Samuelson, J. Phys. Chem. C **114**, 3837 (2010)
315. P. Carrof, K.A. Dick, J. Johansson, M.E. Messing, K. Deppert, L. Samuelson, Nat. Nanotechnol. **4**, 50 (2009)
316. P. Caroff, J.B. Wagner, K.A. Dick, H.A. Nilsson, M. Jeppsson, K. Deppert, L. Samuelson, L.R. Wallenberg, L.-E. Wernersson, Small **4**, 878 (2008)
317. P. Krogstrup, S. Curiotto, E. Johnson, M. Aagesen, J. Nygard, D. Chatain, Phys. Rev. Lett. **106**, 125505 (2011)
318. K. Brakke, Exp. Math. **1**, 141 (1992)
319. V.G. Dubrovskii, N.V. Sibirev, G.E. Cirlin, J.C. Harmand, V.M. Ustinov, Phys. Rev. E **73**, 021603 (2006)
320. J.C. Harmand, M. Tchernycheva, G. Patriarche, L. Travers, F. Glas, G.E. Cirlin, J. Cryst. Growth **301–302**, 853 (2007)
321. M. Tchernycheva, J.C. Harmand, G. Patriarche, L. Travers, G.E. Cirlin, Nanotechnology **17**, 4025 (2006)
322. P. Krogstrup, R. Popovitz-Biro, E. Johnson, M.H. Madsen, J. Nygard, H. Shtrikman, Nano Lett. **10**, 4475 (2010)
323. A. Fontcuberta i Morral, IEEE J. Sel. Top. Quantum Electron. **17**, 819 (2011)

324. B. Mandl, J. Stangl, E. Hilner, A.A. Zakharov, K. Hillerich, A.W. Dey, L. Samuelson, G. Bauer, K. Deppert, A. Mikkelsen, Nano Lett. **10**, 4443 (2010)
325. B. Mandl, A.W. Dey, J. Stangl, M. Cantoro, L.E. Wernersson, G. Bauer, L. Samuelson, K. Deppert, C. Thelander, J. Cryst. Growth **334**, 51 (2011)
326. M. Roussel, W.H. Chen, E. Talbot, R. Larde, E. Cadel, F. Gourbilleau, B. Grandidier, D. Stievenard, P. Pareige, Nanoscale Res. Lett. **6**, 271 (2011)
327. T. Xu, J. Sulerzycki, J.P. Nys, G. Patriarche, B. Grandidier, D. Stievenard, Nanoscale Res. Lett. **6**, 113 (2011)
328. P. Kratzer, S. Sakong, V. Pankoke, Nano Lett. **12**, 943 (2012)
329. V. Pankoke, S. Sakong, P. Kratzer, Phys. Rev. B **86**, 085425 (2012)
330. V.G. Dubrovskii, R.R. LaPierre et al., to be published.
331. T. Bryllert, L.-E. Wernersson, L.E. Froberg, L. Samuelson, IEEE Electron Dev. Lett. **27**, 323 (2006)
332. H. Wang, M. Upmanyu, C.V. Ciobanu, Nano Lett. **8**, 4305 (2008)
333. N. Skolld, L.S. Karlsson, M.W. Larsson, M.-E. Pistol, We Seifert, J. Tragardh, L. Samuelson, Nano Lett. **5**, 1943 (2005)
334. E. Uccelli, J. Arbiol, J.R. Morante, A. Fontcuberta i Morral, ACS Nano **4**, 5985 (2010)
335. R. Popovitz-Biro, A. Kretinin, P. Von Huth, H. Shtrikman, Cryst. Growth Des. **11**, 3858 (2011)
336. S.G. Ghalamestani, M. Heurlin, L.-E. Wernersson, S. Lehmann, K.A. Dick, Nanotechnology **23**, 285601 (2012)
337. M.T. Björk, B.J. Ohlsson, T. Saas, A.I. Persson, C. Thelander, M.H. Magnusson, K. Deppert, L.R. Wallenberg, L. Samuelson, Nano Lett. **2**, 87 (2002)
338. M. Tchernycheva, G.E. Cirlin, G. Patriarche, L. Travers, V. Zwiller, U. Perinetti, J.C. Harmand, Nano Lett. **7**, 1500 (2007)
339. X. Yan, X. Zhang, X. Ren, X. Lu, J. Li, Q. Wang, S. Cai, Y. Huang, Nano Lett. **12**, 1851 (2012)
340. R.S. Wagner, W.C. Ellis, Trans. Metall. Soc. AIME **233**, 1053 (1965)
341. R.S. Wagner, in *Whisker Technology*, ed. by E.P. Levitt (Wiley-Interscience, New York, 1970), pp. 47–119
342. E.I. Givargizov, A.A. Chernov, J. Cryst. Growth **31**, 20 (1975)
343. G.A. Bootsma, H.J. Gassen, J. Cryst. Growth **10**, 223 (1971)
344. W. Seifert, M. Borgstrom, K. Deppert, K.A. Dick, J. Johansson, M.W. Larsson, T. Martensson, N. Skold, C.P.T. Svensson, B.A. Wacaser, L.R. Wallenberg, L. Samuelson, J. Cryst. Growth **272**, 211 (2004)
345. M. Borgstrom, K. Deppert, L. Samuelson, W. Seifert, J. Cryst. Growth **260**, 18 (2004)
346. J. Johansson, C.P.T. Svensson, T. Mårtensson, L. Samuelson, W. Seifert, J. Phys. Chem. B **109**, 13567 (2005)
347. L.E. Fröberg, W. Seifert, J. Johansson, Phys. Rev. B **76**, 153401 (2007)
348. J. Johansson, B.A. Wacaser, K.A. Dick, W. Seifert, Nanotechnology **17**, S355 (2006)
349. A.I. Persson, L.E. Fröberg, S. Jeppesen, M.T. Björk, L. Samuelson, J. Appl. Phys. **101**, 034313 (2007)
350. J. Johansson, J. Bolinsson, M. Ek, P. Caroff, K.A. Dick, ACS Nano **6**, 6142 (2012)
351. F. Glas, J.C. Harmand, Phys. Rev. B **73**, 155320 (2006)
352. F. Glas, J. Appl. Phys. **104**, 093520 (2008)
353. F. Glas, Phys. Stat. Solidi B **247**, 254 (2010)
354. F. Glas, J. Appl. Phys. **108**, 073506 (2010)
355. J.B. Hannon, S. Kodambaka, F.M. Ross, R.M. Tromp, Nature **440**, 69 (2006)
356. S. Kodambaka, J. Tersoff, M.C. Reuter, F.M. Ross, Phys. Rev. Lett. **96**, 096105 (2006)
357. K.W. Schwarz, J. Tersoff, Nano Lett. **12**, 1329 (2012)
358. C.-Y. Wen, J. Tersoff, M.C. Reuter, E.A. Stach, F.M. Ross, Phys. Rev. Lett. **105**, 195502 (2010)
359. C.-Y. Wen, J. Tersoff, K. Hillerich, M.C. Reuter, J.H. Park, S. Kodambaka, E.A. Stach, F.M. Ross, Phys. Rev. Lett. **107**, 025503 (2011)

360. F.M. Kolb, H. Hofmeister, R. Scholz, M. Zacharias, U. Gosele, D.D. Ma, S.-T. Lee, J. Electrochem. Soc. **151**, G472 (2004)
361. D.S. Kim, U. Gosele, M. Zacharias, J. Cryst. Growth **311**, 3216 (2009)
362. V. Schmidt, J.V. Wittemann, U. Gosele, Chem. Rev. **110**, 361 (2010)
363. M.C. Plante, R.R. LaPierre, J. Cryst. Growth **310**, 356 (2008)
364. M.C. Plante, R.R. LaPierre, Nanotechnology **19**, 495603 (2008)
365. M.C. Plante, R.R. LaPierre, J. Appl. Phys. **105**, 114304 (2009)
366. E. De Jong, R.R. LaPierre, J.Z. Wen, Nanotechnology **21**, 045602 (2010)
367. P. Kratzer, M. Scheffler, Phys. Rev. Lett. **88**, 036102 (2002)
368. V. Pankoke, P. Kratzer, S. Sakong, Phys. Rev. B **84**, 075455 (2011)
369. M.H. Madsena, P. Krogstrup, E. Johnson, S. Venkatesan, E. Mühlbauer, C. Scheu, C.B. Sørensen, J. Nygård, J. Cryst. Growth **364**, 16 (2013)
370. R.E. Algra, M.A. Verheijen, L.-F. Feiner, G.G.W. Immink, W.J.P. van Enckevort, E. Vlieg, E.P.A.M. Bakkers, Nano Lett. **11**, 1259 (2011)
371. R.E. Algra, M.A. Verheijen, L.-F. Feiner, G.G.W. Immink, R. Theissmann, W.J.P. van Enckevort, E. Vlieg, E.P.A.M. Bakkers, Nano Lett. **10**, 2349 (2010)
372. E. Russo-Averchi, M. Heiss, L. Michelet, P. Krogstrup, J. Nygard, C. Magen, J.R. Morante, E. Uccelli, J. Arbiol, A. Fontcuberta i Morral, Nanoscale **4**, 1486 (2012)
373. S.A. Dayeh, E.T. Yu, D. Wang, Nano Lett. **7**, 2486 (2007)
374. S.A. Dayeh, E.T. Yu, D. Wang, Nano Lett. **9**, 1967 (2009)
375. S.A. Dayeh, S.T. Picraux, Nano Lett. **10**, 4032 (2010)
376. I. Avramov, Nanoscale Res. Lett. **2**, 235 (2007)
377. W. Miller, S. Succi, Phys. Rev. E **76**, 031601 (2007)
378. A.A. Golovin, S.H. Davis, P.W. Voorhees, J. Appl. Phys. **104**, 074301 (2008)
379. W.H. Chen, R. Lardé, E. Cadel, T. Xu, B. Grandidier, J.P. Nys, D. Stiévenard, P. Pareige, J. Appl. Phys. **107**, 084902 (2010)
380. H. Huang, X. Ren, X. Ye, J. Guo, Q. Wang, X. Zhang, S. Cai, Y. Huang, Nanotechnology **21**, 475602 (2010)
381. S. Noor Mohammad, J. Chem. Phys. **131**, 204703 (2009)
382. S. Noor Mohammad, J. Appl. Phys. **107**, 114304 (2010)
383. S. Noor Mohammad, J. Appl. Phys. **106**, 104311 (2009)
384. S. Noor Mohammad, J. Chem. Phys. **131**, 224702 (2009)
385. S. Noor Mohammad, Nano Lett. **8**, 1532 (2008)
386. S. Breuer, PhD Thesis, Humboldt-University Berlin, 2011
387. K.W. Schwarz, J. Tersoff, Nano Lett. **11**, 316 (2011)
388. K.W. Schwarz, J. Tersoff, Phys. Rev. Lett. **102**, 206101 (2009)
389. A.G. Nastovjak, I.G. Neizvestny, N.L. Shwartz, Pure Appl. Chem. **82**, 2017 (2010)
390. A.D. Gamalski, C. Ducati, S. Hofmann, J. Phys. Chem. C **115**, 4413 (2011)
391. M Galicka, M Bukała, R Buczko, P. Kacman, J. Phys.: Condens. Matter **20**, 454226 (2008)
392. D. Shakthivel, S. Raghavan, J. Appl. Phys. **112**, 024317 (2012)
393. K.K. Lew, J.M. Redwing, J. Cryst. Growth **254**, 14 (2003)
394. T.Y. Tan, N. Li, U. Gosele, Appl. Phys. A **78**, 519 (2004)
395. C.X. Wang, B. Wang, Y.H. Yang, G.W. Yang, J. Phys. Chem. **109**, 9966 (2005)
396. M.T. Borgström, G. Immink, B. Ketelaars, R. Algra, E.P.A.M. Bakkers, Nat. Nanotechnol. **2**, 541 (2007)
397. M. Suzuki, Y. Hidaka, T. Yanagida, A. Klamchuen, M. Kanai, T. Kawai, S. Kai, Phys. Rev. E **83**, 061606 (2011)
398. H. Mehdipour, K. Ostrikov, A.E. Rider, S.A. Furman, Nanotechnology **22**, 315707 (2011)
399. E.J. Schwalbach, S.H. Davis, P.W. Voorhees, J.A. Warren, D. Wheeler, J. Appl. Phys. **111**, 024302 (2012)
400. Y.Y. Lu, H. Cui, G.W. Yang, C.X. Wang, Nano Lett. **12**, 4032 (2012)
401. H. Ye, P.F. Lu, Z.Y. Yu, Y.X. Song, D.L. Wang, S.M. Wang, Nano Lett. **9**, 1921 (2009)
402. H. Ye, Z.Y. Yu, S. Kodambaka, V.B. Shenoy, Appl. Phys. Lett. **100**, 263103 (2012)

403. V.G. Dubrovskii, N.V. Sibirev, R.A. Suris, G.E. Cirlin, V.M. Ustinov, M. Tchernysheva, J.C. Harmand, Semiconductors **40**, 1075 (2006)
404. V.G. Dubrovskii, I.P. Soshnikov, G.E. Cirlin, A.A. Tonkikh, Yu.B. Samsonenko, N.V. Sibirev, V.M. Ustinov, Phys. Stat. Sol. B **241**, R30 (2004)
405. V.G. Dubrovskii, I.P. Soshnikov, N.V. Sibirev, G.E. Cirlin, V.M. Ustinov, J. Cryst. Growth **289**, 31 (2006)
406. V.G. Dubrovskii, N.V. Sibirev, Phys. Rev. B **77**, 035414 (2008)
407. V.G. Dubrovskii, N.V. Sibirev, G.E. Cirlin, M. Tchernycheva, J.C. Harmand, V.M. Ustinov, Phys. Rev. E **77**, 031606 (2008)
408. V.G. Dubrovskii, N.V. Sibirev, G.E. Cirlin, I.P. Soshnikov, W.H. Chen, R. Larde, E. Cadel, P. Pareige, T. Xu, B. Grandidier, J.-P. Nys, D. Stievenard, M. Moewe, L.C. Chuang, C. Chang-Hasnain, Phys. Rev. B **79**, 205316 (2009)
409. V.G. Dubrovskii, N.V. Sibirev, G.E. Cirlin, A.D. Bouravleuv, Yu.B. Samsonenko, D.L. Dheeraj, H.L. Zhou, C. Sartel, J.C. Harmand, G. Patriarche, F. Glas, Phys. Rev. B **80**, 205305 (2009)
410. X. Zhang, V.G. Dubrovskii, N.V. Sibirev, G.E. Cirlin, C. Sartel, M. Tchernycheva, J.C. Harmand, F. Glas, Nanoscale Res. Lett. **5**, 1692 (2010)
411. L. Lugani, D. Ercolani, L. Sorba, N.V. Sibirev, M.A. Timofeeva, V.G. Dubrovskii, Nanotechnology **23**, 095602 (2012)
412. V.G. Dubrovskii, T. Xu, Y. Lambert, J.-P. Nys, B. Grandidier, D. Stievenard, W. Chen, P. Pareige, Phys. Rev. Lett. **108**, 105501 (2012)
413. N.V. Sibirev, M. Tchernycheva, M.A. Timofeeva, J.C. Harmand, G.E. Cirlin, V.G. Dubrovskii, J. Appl. Phys. **111**, 104317 (2012)
414. K.A. Dick, K. Deppert, T. Martensson, S. Mandl, L. Samuelson, W. Seifert, Nano Lett. **5**, 761 (2005)
415. T.I. Kamins, R.S. Williams, D.P. Basile, T. Hesjedal, J.S. Harris, J. Appl. Phys. **89**, 1008 (2001)
416. Q. Xu, I.D. Sharp, C.W. Yuan, D.O. Yi, C.Y. Liao, A.M. Glaeser, A.M. Minor, J.W. Beeman, M.C. Ridgway, P. Kluth, J.W. Ager, D.C. Chrzan, E.E. Haller, Phys. Rev. Lett. **97**, 155701 (2006)
417. S. Kodambaka, J. Tersoff, M.C. Reuter, F.M. Ross, Science **316**, 729 (2007)
418. M. Yoshizawa, A. Kikuchi, M. Mori, N. Fujita, K. Kishino, Jpn. J. Appl. Phys. **36**, L459 (1997)
419. M.A. Sanchez-Garcia, E. Calleja, E. Monroy, F.J. Sanchez, F. Calle, E. Muñoz, R. Beresford, J. Cryst. Growth **183**, 23 (1998)
420. M. Wolz, V.M. Kaganer, O. Brandt, L. Geelhaar, H. Riechert, Appl. Phys. Lett. **98**, 261907 (2011)
421. R.K. Debnath, R. Meijers, T. Richter, T. Stoica, R. Calarco, H. Lüth, Appl. Phys. Lett. **90**, 123117 (2007)
422. M. Tchernycheva, C. Sartel, G. Cirlin, L. Travers, G. Patriarche, J.C. Harmand, Le Si Dang, J. Renard, B. Gayral, L. Nevou, F. Julien, Nanotechnology **18**, 385306 (2007)
423. R. Calarco, R. Meijers, R.K. Debnath, T. Stoica, E. Sutter, H. Lüth, Nano Lett. **7**, 2248 (2007)
424. A. Bertness, A. Roshko, L.M. Mansfield, T.E. Harvey, N.A. Sanford, J. Cryst. Growth **300**, 94 (2007)
425. T. Stoica, E. Sutter, R.J. Meijers, R.K. Debnath, R. Calarco, H. Lüth, D. Grützmacher, Small **4**, 751 (2008)
426. H.W. Seo, Q.Y. Chen, L.W. Tu, C.L. Hsiao, M.N. Iliev, W.K. Chu, Phys. Rev. B **71**, 235314 (2005)
427. O. Landre, C. Bougerol, H. Renevier, B. Daudin, Nanotechnology **20**, 415602 (2009)
428. R. Songmuang, T. Ben, B. Daudin, D. Gonzalez, E. Monroy, Nanotechnology **21**, 295605 (2010)
429. E. Galopin, L. Largeau, G. Patriarche, L. Travers, F. Glas, J.C. Harmand, Nanotechnology **22**, 245606 (2011)

430. V. Consonni, M. Knelangen, L. Geelhaar, A. Trampert, H. Riechert, Phys. Rev. B **81**, 085310 (2010)
431. V. Consonni, M. Hanke, M. Knelangen, L. Geelhaar, A. Trampert, H. Riechert, Phys. Rev. B **83**, 035310 (2011)
432. C. Chèze, L. Geelhaar, A. Trampert, H. Riechert, Appl. Phys. Lett. **97**, 043101 (2010)
433. V. Consonni, A. Trampert, L. Geelhaar, H. Riechert, Appl. Phys. Lett. **99**, 033102 (2011)
434. C. Chèze, L. Geelhaar, O. Brandt, W. Weber, H. Riechert, S. Munch, R. Rothemund, S. Reitzenstein, A. Forchel, T. Kehagias, P. Komninou, G.P. Dimitrakopulos, T. Karakostas, Nano Res. **3**, 528 (2010)
435. V.G. Dubrovskii, V. Consonni, A. Trampert, L. Geelhaar, H. Riechert, Phys. Rev. B **85**, 165317 (2012)
436. V.G. Dubrovskii, V. Consonni, L. Geelhaar, A. Trampert, H. Riechert, Appl. Phys. Lett. **100**, 153101 (2012)
437. M. Akabori, J. Takeda, J. Motohisa, T. Fukui, Nanotechnology **14**, 1071 (2003)
438. M. Inari, J. Takeda, J. Motohisa, T. Fukui, Phys. E **21**, 620 (2004)
439. J. Noborisaka, J. Motohisa, T. Fukui, Appl. Phys. Lett. **86**, 213102 (2005)
440. J. Noborisaka, J. Motohisa, S. Hara, T. Fukui, Appl. Phys. Lett. **87**, 093109 (2005)
441. T. Hamano, H. Hirayama, Y. Aoyagi, Jpn. J. Appl. Phys., Part 2, **36**, L286 (1997)
442. P.J. Poole, J. Lefebvre, J. Fraser, Appl. Phys. Lett. **83**, 2055 (2003)
443. M.T. Bjork, H. Schmid, C.M. Breslin, L. Gignac, H. Riel, J. Cryst. Growth **344**, 31 (2012)
444. K. Choi, M. Arita, Y. Arakawa, J. Cryst. Growth **357**, 58 (2012)
445. Y. Calahorra, Y. Greenberg, S. Cohen, D. Ritter, Nanotechnology **23**, 245603 (2012)
446. T. Gotschke, T. Schumann, F. Limbach, T. Stoica, R. Calarco, Appl. Phys. Lett. **98**, 103102 (2011)
447. V.V. Mamutin, Tech. Phys. Lett. **25**(9), 741 (1999)
448. M. Moewe, L.C. Chuang, V.G. Dubrovskii, C. Chang-Hasnain, J. Appl. Phys. **104**, 044313 (2008)
449. M.R. Ramdani, E. Gil, Ch. Leroux, Y. André, A. Trassoudaine, D. Castelluci, L. Bideux, G. Monier, C. Robert-Goumet, R. Kupka, Nano Lett. **10**, 1836 (2010)
450. V.G. Dubrovskii, Tech. Phys. Lett. **37**(1), 53 (2011)
451. V.G. Dubrovskii, M.V. Nazarenko, Tech. Phys. Lett. **37**(5), 427 (2011)
452. N.V. Sibirev, M.A. Timofeeva, A.D. Bol'shakov, M.V. Nazarenko, V.G. Dubrovskii, Phys. Sol. State **52**, 1531 (2010)
453. V.G. Dubrovskii, A.D. Bol'shakov, Tech. Phys. Lett. **38**(4), 311 (2012)
454. S.H. Oh, M.F. Chisholm, Y. Kauffmann, W.D. Kaplan, W. Luo, M. Rühle, C. Scheu, Science **330**, 489 (2010)
455. W.C. Carter et al., Acta Metall. Mater. **43**, 4309 (1995)
456. M.R. Ramdani, J.C. Harmand, F. Glas, G. Patriarche, L. Travers, Cryst. Growth. Des. **13**, 91 (2013)
457. T. Rieger, S. Heiderich, S. Lenk, M.I. Lepsa, D. Grützmacher, J. Cryst. Growth **353**, 39–46 (2012)
458. V.G. Dubrovskii, Tech. Phys. Lett. **39**(2), 157 (2013)
459. G.J. Janz, Natl. Bur. Standards, NSRDS-NBS 28, Washington (1969)
460. K. Ostrikov, H. Mehdipour, J. Amer. Chem. Soc. **134**, 4303 (2012)
461. K. Ostrikov, H. Mehdipour, ACS Nano **5**, 8372 (2011)
462. Yu.Yu. Hervieu, private communication
463. B. Voigtlander, Surf. Sci. Rep. **43**, 127 (2001)
464. Yu.Yu. Hervieu, V.G. Dubrovskii, to be published.
465. R.J. Yee, S.J. Gibson, V.G. Dubrovskii, R.R. LaPierre, Appl. Phys. Lett. **101**, 263106 (2012)
466. S.A. Fortuna, X. Li, Semicond. Sci. Technol. **25**, 024005 (2010)
467. M. Hilse, M. Ramsteiner, S. Breuer, L. Geelhaar, H. Riechert, Appl. Phys. Lett. **96**, 193104 (2010)
468. J.A. Czaban, D.A. Thompson, R.R. LaPierre, Nano Lett. **9**, 148 (2009)

469. S. Wirths, K. Weis, A. Winden, K. Sladek, C. Volk, S. Alagha, T.E. Weirich, M. von der Ahe, H. Hardtdegen, H. Luth, N. Demarina, D. Grutzmacher, Th Schapers, J. Appl. Phys. **110**, 053709 (2011)
470. L. Rigutti, A. De Luna Bugallo, M. Tchernycheva, G. Jacopin, F.H. Julien, G. Cirlin, G. Patriarche, D. Lucot, L. Travers, J.C. Harmand, J. Nanomater. **1**, (2009)
471. K. Sinniah, M.G. Sherman, L.B. Lewis, W.H. Weinberg, J.T. Yates, K.C. Janda, J. Chem. Phys. **92**, 5700 (1990)
472. T.R. Bramblett, Q. Lu, M.A. Hasan, S.K. Jo, J.E. Greene, J. Appl. Phys. **76**, 1884 (1994)
473. Y. Wu, R. Fan, P. Yang, Nano Lett. **2**, 83 (2002)
474. L. Lugani, D. Ercolani, F. Rossi, G. Salviati, F. Beltram, L. Sorba, Cryst. Growth Des. **10**, 4038 (2010)
475. C.S. Peng, Q. Huang, W.Q. Cheng, J.M. Zhou, Y.H. Zhang, T.T. Sheng, C.H. Tung, Appl. Phys. Lett. **72**, 2541 (1998)
476. C. Chen, M.C. Plante, C. Fradin, R.R. LaPierre, J. Mater. Res. **21**, 2801 (2006)
477. D.L. Dheeraj, G. Patriarche, H. Zhou, J.C. Harmand, H. Weman, B.O. Fimland, J. Cryst. Growth **311**, 1847 (2009)
478. C. Chatillon, F. Hodaj, A. Pisch, J. Cryst. Growth **311**, 3598 (2009)
479. M. Lopez, Y. Nomura, J. Cryst. Growth **150**, 68 (1995)
480. V.G. Dubrovskii, N.V. Sibirev, M.A. Timofeeva, Semiconductors **43**, 1267 (2009)
481. M.V. Barton, J. Appl. Mech. **8**, A97 (1941)
482. J.P. Hirth, J. Lothe, *Theory of Dislocations* (Wiley, New York, 1982)
483. I.A. Ovid'ko, Phys. Rev. Lett. **88**, 046103 (2002)
484. C.P.T. Svensson, W. Seifert, M.W. Larsson, L.R. Wallenberg, J. Stangl, G. Bauer, L. Samuelson, Nanotechnology **16**, 936 (2005)
485. M.A. Verheijen, G. Immink, T. de Smet, M.T. Borgström, E.P.A.M. Bakkers, J. Am. Chem. Soc. **128**, 1353 (2006)
486. L. Gao, R.L. Woo, B. Liang, M. Pozuelo, S. Prikhodko, M. Jackson, N. Goel, M.K. Hudait, D.L. Huffaker, M.S. Goorsky, S. Kodambaka, R.F. Hicks, Nano Lett. **9**, 2223 (2009)
487. K.L. Kavanagh, J. Salfi, I. Savelyev, M. Blumin, H.E. Ruda, Appl. Phys. Lett. **98**, 152103 (2011)
488. ShA Dayeh, W. Tang, F. Boioli, K.L. Kavanagh, H. Zheng, J. Wang, N.H. Mack, G. Swadener, J.Y. Huang, L. Miglio, K.-N. Tu, S.T. Picraux, Nano Lett. **13**, 1869 (2013)
489. S. Raychaudhuri, E.T. Yu, J. Appl. Phys. **99**, 114308 (2006)
490. F. Wu, E.C. Young, I. Koslow, M.T. Hardy, P.S. Hsu, A.E. Romanov, S. Nakamura, S.P. Den Baars, J.S. Speck, Appl. Phys. Lett. **99**, 251909 (2011)
491. G. Perillat-Merceroz, R. Thierry, P.-H. Jouneau, P. Ferret, G. Feuillet, Appl. Phys. Lett. **100**, 173102 (2012)
492. T.E. Trammell, X. Zhang, Y. Li, L.-Q. Chen, E.C. Dickey, J. Cryst. Growth **310**, 3084 (2008)
493. Y. Liang, W.D. Nix, P.B. Griffin, J.D. Plummer, J. Appl. Phys. **97**, 043519 (2005)
494. A.F. Bower, *Applied Mechanics of Solids* (CRC Press, New York, 2009)
495. I.A. Goldthorpe, A.F. Marshall, P.C. McIntyre, Nano Lett. **8**, 4081 (2008)
496. N. Søndergaard, Y. He, Ch. Fan, R. Han, Th Guhr, H.Q. Xub, J. Vac. Sci. Technol. B **27**, 827 (2009)
497. J. Grönqvist, N. Søndergaard, F. Boxberg, Th Guhr, S. Åberg, H.Q. Xu, J. Appl. Phys. **106**, 053508 (2009)
498. M.V. Nazarenko, N.V. Sibirev, K.W. Ng, F. Ren, W.S. Ko, V.G. Dubrovskii, C. Chang-Hasnain, J. Appl. Phys. **113**, 104311 (2013)
499. J. Dundurs, G.P. Sendeckyj, J. Mech. Phys. Solids **13**, 141 (1965)
500. D. Hull, D.J. Bacon, *Introduction to Dislocations* (Butterworth-Heinemann, Oxford, 2001)
501. K.A. Dick, K. Deppert, T. Mårtensson, B. Mandl, L. Samuelson, W. Seifert, Nano Lett. **5**, 761 (2005)
502. N. Tajik, Z. Peng, P. Kuyanov, R.R. LaPierre, Nanotechnology **22**, 225402 (2011)
503. R. Kapadia, Z. Fang, A. Javey, Appl. Phys. Lett. **96**, 103116 (2010)

504. V.G. Dubrovskii, A.D. Bolshakov, B.L. Williams, K. Durose, Nanotechnology **23**, 485607 (2012)
505. N.V. Sibirev, to be submitted. The author thanks F. Glas for discussing with him the nucleation statistics in nanowires.
506. E.C. Garnett, Y.-C. Tseng, D.R. Khanal, J. Wu, J. Bokor, P. Yang, Nat. Nanotechnol. **4**, 311 (2009)
507. E. Koren, N. Berkovitch, Y. Rosenwaks, Nano Lett. **10**, 1163 (2010)
508. D.E. Perea, E.R. Hemesath, E.J. Schwalbach, J.L. Lensch-Falk, P.W. Voorhees, L.J. Lauhon, Nat. Nanotechnol. **4**, 315 (2009)
509. W. Chen, V.G. Dubrovskii, X. Liu, T. Xu, R. Larde, J.P. Nys, B. Grandidier, D. Stievenard, G. Patriarche, P. Pareige, J. Appl. Phys. **111**, 094909 (2012)
510. E.T. Goodwin, J. Staton, Quart. J. Mech. Appl. Math. **1**, 220 (1948)
511. G.E. McGuire, *Semiconductor Materials and Process Technology Handbook* (Publication, Noyes, 1988), p. 501
512. N.E.B. Cowern, G.F.A. van de Walle, D.J. Gravesteijn, C.J. Vriezema, Phys. Rev. Lett. **67**, 212 (1991)
513. M. Uematsu, J. Appl. Phys. **82**, 2228 (1997)
514. D. De Salvador, E. Napolitani, S. Mirabella, G. Bisognin, G. Impellizzeri, A. Carnera, F. Priolo, Phys. Rev. Lett. **97**, 255902 (2006)
515. D. Srivastava, B.J. Garrison, Phys. Rev. B **46**, 1472 (1992)
516. A. Fontcuberta i Morral, C. Colombo, G. Abstreiter, J. Arbiol, J.R. Morante, Appl. Phys. Lett. **92**, 063112 (2008)
517. S. Ambrosini, M. Fanetti, V. Grillo, A. Franciosi, S. Rubini, AIP Adv. **1**, 042142 (2011)
518. T. Rieger, S. Heiderich, S. Lenk, M.I. Lepsa, D. Grützmacher, J. Cryst. Growth **353**, 39 (2012)
519. G. Priante, S. Ambrosini, V.G. Dubrovskii, A. Franciosi, S. Rubini, Cryst. Growth Des., submitted (2013)
520. Y.H. Kim, D.W. Woo Park, S.J. Lee, Appl. Phys. Lett. **100**, 033117 (2012)
521. C.T. Foxon, B.A. Joyce, Surf. Sci. **50**, 434 (1975)
522. J.E. Northrup, J. Neugebauer, Phys. Rev. B **53**, R10477 (1996)
523. J.C. Idrobo, H. Iddir, S. Ögüt, A. Ziegler, N.D. Browning, R.O. Ritchie, Phys. Rev. B **72**, 241301(R) (2005)
524. R. Songmuang, O. Landre, B. Daudin, Appl. Phys. Lett. **91**, 251902 (2007)
525. S.D. Carnevale, J. Yang, P.J. Phillips, M.J. Mills, R.C. Myers, Nano Lett. **11**, 866 (2011)
526. Y.S. Park, S.-H. Lee, J.-E. Oh, C.-M. Park, T.-W. Kang, J. Cryst. Growth **282**, 313 (2005)
527. S. Fernandez-Garrido, J. Grandal, E. Calleja, M.A. Sanchez-Garcıa, D. Lopez-Romero, J. Appl. Phys. **106**, 126102 (2009)
528. V. Consonni, M. Knelangen, A. Trampert, L. Geelhaar, H. Riechert, Appl. Phys. Lett. **98**, 071913 (2011)
529. P. Dogan, O. Brandt, C. Pfuller, A.-K. Bluhm, L. Geelhaar, H. Riechert, J. Cryst. Growth **323**, 418 (2011)
530. L. Lymperakis, J. Neugebauer, Phys. Rev. B **79**, 241308(R) (2009)
531. P. Hacke, G. Feuillet, H. Okumura, S. Yoshida, Appl. Phys. Lett. **69**, 2507 (1996)
532. S. Guha, N.A. Bojarczuk, D.W. Kisker, Appl. Phys. Lett. **69**, 2879 (1996)
533. L.E. Jensen, M.T. Björk, S. Jeppesen, A.I. Persson, B.J. Ohlsson, L. Samuelson, Nano Lett. **4**, 1961 (2004)
534. A. Li, N.V. Sibirev, D. Ercolani, V.G. Dubrovskii, L. Sorba, Cryst. Growth Des. **13**, 878 (2013)
535. D. Dalacu, A. Kam, D.G. Austing, X. Wu, J. Lapointe, G.C. Aers, P.J. Poole, Nanotechnology **20**, 395602 (2009)
536. V. Kaganer, Private communication.
537. T. Dinsdale, CALPHAD **15**, 317 (1991)
538. I. Ansara, C. Chatillon, H.L. Lukas, T. Nishizawa, H. Ohtani, K. Ishida, M. Hillert, B. Sundman, B.B. Argent, A. Watson, T.G. Chart, T. Anderson, CALPHAD **18**, 177 (1994)
539. G.B. Stringfellow, J. Phys. Chem. Solids **33**, 665 (1972)

540. J.H. Hildebrand, R.L. Scott, *Solubility of Nonelectrolytes*, 3rd edn. (Reinhold, New York, 1949)
541. A. James, M. Lord, *VNR Index of Chemical and Physical Data* (Van Nostrand Reinhold, New York, 1992)
542. H. Shtrikman, R. Popovitz-Biro, A. Kretinin, L. Houben, M. Heiblum, M. Bukała, M. Galicka, R. Buczko, P. Kacman, Nano Lett. **9**, 1506 (2009)
543. C.-Y. Yeh, Z.W. Lu, S. Froyen, A. Zunger, Phys. Rev. B **45**, R12130 (1992)
544. D.R. Lide (ed.) *CRC Handbook of Chemistry and Physics*, 80th edn. (CRC Press, Florida, 1999)
545. C. Panse, D. Kriegner, F. Bechstedt, Phys. Rev. B **84**, 075217 (2011)
546. I.T. Steinberger, Prog. Cryst. Growth Charact. **7**, 7 (1983)
547. Group IV Elements, IV-IV and III-V Compounds, ed. by U. Rössler, Landolt-Börnstein, New Series, Group III, vol. 41, Part A (Springer, Berlin, 2006)
548. D. Kriegner, C. Panse, B. Mandl, K.A. Dick, M. Keplinger, J.M. Persson, P. Caroff, D. Ercolani, L. Sorba, F. Bechstedt, J. Stangl, G. Bauer, Nano Lett. **11**, 1483 (2011)
549. B. Mandl, K.A. Dick, D. Kriegner, M. Keplinger, G. Bauer, J. Stangl, K. Deppert, Nanotechnology **22**, 145603 (2011)
550. D.L. Dheeraj, G. Patriarche, H. Zhou, T.B. Hoang, A.F. Moses, S. Gronsberg, A.T.J. van Helvoort, B.-O. Fimland, H. Weman, Nano Lett. **8**, 4459 (2008)
551. I.P. Soshnikov, G.E. Cirlin, N.V. Sibirev, V.G. Dubrovskii, Y.B. Samsonenko, D. Litvinov, D. Gerthsen, Tech. Phys. Lett. **34**, 538 (2008)
552. S.O. Mariager, C.B. Sørensen, M. Aagesen, J. Nygard, R. Feidenhans, P.R. Willmott, Appl. Phys. Lett. **91**, 083106 (2007)
553. A. Belabbes, C. Panse, J. Furthmuller, F. Bechstedt, Phys. Rev. B **86**, 075208 (2012)
554. M. Murayama, T. Nakayama, Phys. Rev. B **49**, 4710 (1994)
555. Z. Zanolli, F. Fuchs, J. Furthmüller, U. von Barth, F. Bechstedt, Phys. Rev. B **75**, 245121 (2007)
556. A. De, C.E. Pryor, Phys. Rev. B **81**, 155210 (2010)
557. T. Akiyama, K. Nakamura, T. Ito, Phys. Rev. B **73**, 235308 (2006)
558. D. Spirkoska, J. Arbiol, A. Gustafsson, S. Conesa-Boj, F. Glas, I. Zardo, M. Heigoldt, M.H. Gass, A.L. Bleloch, S. Estrade, M. Kaniber, J. Rossler, F. Peiro, J.R. Morante, G. Abstreiter, L. Samuelson, A. Fontcuberta i Morral, Phys. Rev. B **80**, 245325 (2009)
559. T.B. Hoang, A.F. Moses, H.L. Zhou, D.L. Dheeraj, B.O. Fimland, H. Weman, Appl. Phys. Lett. **94**, 133105 (2009)
560. N. Akopian, G. Patriarche, L. Liu, J.-C. Harmand, V. Zwiller, Nano Lett. **10**, 1198 (2010)
561. B.V. Novikov, S.Yu. Serov, N.G. Filosofov, I.V. Shtrom, V.G. Talalaev, O.F. Vyvenko, E.V. Ubyivovk, Yu.B. Samsonenko, A.D. Bouravleuv, I.P. Soshnikov, N.V. Sibirev, G.E. Cirlin, V.G. Dubrovskii, Phys. Stat. Sol. RRL **4**, 175 (2010)
562. M.H. Sun, E.S.P. Leong, A.H. Chin, C.Z. Ning, G.E. Cirlin, Yu.B. Samsonenko, V.G. Dubrovskii, L. Chuang, C. Chang-Hasnain, Nanotechnology **21**, 335705 (2010)
563. I.P. Soshnikov, G.E. Cirlin, V.G. Dubrovskii, A. Veretekha, A. Gladyshev, V.M. Ustinov, Phys. Solid State **48**, 786 (2006)
564. I. Regolin, V. Khorenko, W. Prost, F.J. Tegude, D. Sudfeld, J. Kästner, G. Dumpich, K. Hitzbleck, H. Wiggers, J. Appl. Phys. **101**, 054318 (2007)
565. J. Trägårdh, A.I. Persson, J.B. Wagner, D. Hessman, L. Samuelson, J. Appl. Phys. **101**, 123701 (2007)
566. M. Mattila, T. Hakkarainen, H. Lipsanen, H. Jiang, E.I. Kauppinen, Appl. Phys. Lett. **90**, 033101 (2007)
567. S. Reitzenstein, S. Münch, C. Hofmann, A. Forchel, S. Crankshaw, L.C. Chuang, M. Moewe, C. Chang-Hasnain, Appl. Phys. Lett. **91**, 091103 (2007)
568. Z. Zanolli, M.-E. Pistol, L.E. Froberg, L. Samuelson, J. Phys.: Condens. Matter **19**, 295219 (2007)
569. K. Durose, in *Narrow Gap II-VI Compounds for Optoelectronic and Electromagnetic Applications*, ed. by P. Capper (Chapman and Hall, London, 2007), pp. 239–250

570. F. Glas, G. Patriarche, J.C. Harmand, J. Phys: Conf. Ser. **209**, 012002 (2010)
571. T. Akiyama, K. Sano, K. Nakamura, T. Ito, Jpn. J. Appl. Phys., Part 2, **45**, L275 (2006)
572. R. Leitsmann, F. Bechstedt, J. Appl. Phys. **102**, 063528 (2007)
573. M. Galicka, M. Bukała, R. Buczko, P. Kacman, J. Phys.: Condens. Matter **20**, 454226 (2008)
574. M. Rosini, R. Magri, Nano Lett. **10**, 6021 (2010)
575. J.W. Cahn, R.E. Hanneman, Surf. Sci. **1**, 387 (1964)
576. R.J. Jaccodine, J. Electrochem. Soc. **110**, 524 (1963)
577. D.J. Eaglesham, A.E. White, L.C. Feldman, N. Moriya, D.C. Jacobson, Phys. Rev. Lett. **70**, 1643 (1993)
578. A.A. Stekolnikov, J. Furthmuller, F. Bechstedt, Phys. Rev. B **65**, 115318 (2002)
579. A.A. Stekolnikov, J. Furthmuller, F. Bechstedt, Phys. Rev. B **70**, 045305 (2004)
580. A.A. Stekolnikov, F. Bechstedt, Phys. Rev. B **72**, 125326 (2005)
581. K.A. Dick, C. Thelander, L. Samuelson, P. Caroff, Nano Lett. **10**, 3494 (2010)
582. H.J. Joyce, Q. Gao, H.H. Tan, C. Jagadish, Y. Kim, M.A. Fickenscher, S. Perera, T.B. Hoang, L.M. Smith, H.E. Jackson, J.M. Yarrison-Rice, X. Zhang, J. Zou, Nano Lett. **9**, 695 (2009)
583. S. Lehmann, D. Jacobsson, K. Deppert, K. Dick, Nano Res. **5**, 470 (2012)
584. D.L. Dheeraj, A.M. Munshi, M. Scheffler, A.T.J. van Helvoort, H. Weman, B.O. Fimland, Nanotechnology **24**, 015601 (2013)
585. M.V. Nazarenko, N.V. Sibirev, V.G. Dubrovskii, Tech. Phys. **56**, 311 (2011)
586. Y. Andre, E. Gill, A. Trassoudaine, G. Avit, K. Lekhal, Ch. Leroux, C. Bougerol, C. Varenne, G. Monier, C. Robert-Goumet, D. Castelluci, L. Bideux, Abstr. 6th Nanowire Growth Workshop, St. Petersburg, Russia, p. 32 (2012)
587. E. Gil, V.G. Dubrovskii, F. Glas et al., to be published.
588. P. Hoggan, private communication.
589. A.M. Munshia, D.L. Dheeraja, J. Todorovicb, A.T.J. van Helvoortb, H. Wemana, B.-O. Fimland, J. Cryst. Growth **372**, 163 (2013)
590. N.V. Sibirev, M.V. Nazarenko, V.G. Dubrovskii, Tech. Phys. Lett. **38**(3), 221 (2012)
591. M. Hadzhilazova, I.M. Mladenov et al., Arch. Math. (BRNO) **43**, 417 (2007)
592. X. Yu, H. Wang, J. Lu, J. Zhao, J. Misuraca, P. Xiong, S. von Molnár, Nano Lett. **12**, 5436 (2012)
593. C. García Núñeza, A.F. Brañaa, J.L. Paua, D. Ghitaa, B.J. Garcíaa, G. Shenb, D.S. Wilbertb, S.M. Kimb, P. Kung, J. Cryst. Growth **372**, 205 (2013)
594. D.S. Oliveira, L.H.G. Tizei, D. Ugarte, M.A. Cotta, Nano Lett. **13**, 9 (2013)

Index

V. G. Dubrovskii, *Nucleation Theory and Growth of Nanostructures*, NanoScience and Technology, DOI: 10.1007/978-3-642-39660-1,

Printed by Printforce, the Netherlands